Coherence

David Ramírez • Ignacio Santamaría •
Louis Scharf

Coherence

In Signal Processing and Machine Learning

Springer

David Ramírez
Universidad Carlos III de Madrid
Madrid, Spain

Ignacio Santamaría
Universidad de Cantabria
Santander, Spain

Louis Scharf
Colorado State University
Fort Collins, CO, USA

ISBN 978-3-031-13330-5 ISBN 978-3-031-13331-2 (eBook)
https://doi.org/10.1007/978-3-031-13331-2

This Springer imprint is published by the registered company Springer Nature Switzerland AG
The registered company address is: Gewerbestrasse 11, 6330 Cham, Switzerland

To Ana Belén, Carmen, and Merche
David Ramírez

To Isabel, Cristina, Inés, and Ignacio
Ignacio Santamaría

To the next generation
Louis Scharf

Preface

This book is directed to graduate students, practitioners, and researchers in signal processing and machine learning. A basic understanding of linear algebra and probability is assumed. This background is complemented in the book with appendices on matrix algebra, (complex) multivariate normal theory, and related distributions.

The book begins in Chap. 1 with a breezy account of coherence as it has commonly appeared in science, engineering, signal processing, and machine learning. Chapter 2 is a comprehensive account of classical least squares theory, with a few original variations on cross-validation and model-order determination. Compression of ambient space dimension is analyzed by comparing multidimensional scaling and a randomized search algorithm inspired by the Johnson-Lindenstrauss lemma. But the central aim of the book is to analyze coherence in its many guises, beginning with the correlation coefficient and its multivariate extensions in Chap. 3. Chapters 4–8 contain a wealth of results on maximum likelihood theory in the complex multivariate normal model for estimating parameters and detecting signals in first- and second-order statistical models. Chapters 5 and 6 are addressed to matched and adaptive subspace detectors. Particular attention is paid to the geometries, invariances, and null distributions of these subspace detectors. Chapters 7 and 8 extend these results to detection of signals that are common to two or more channels, and to detection of spatial correlation and cyclostationarity. Coherence plays a central role in these chapters.

Chapter 9 addresses subspace averaging, an emerging topic of interest in signal processing and machine learning. The motivation is to identify subspace models (or centroids) for measurements so that images may be classified or noncoherent communication signals may be decoded. The dimension of the average or central subspace can also be estimated efficiently and applied to source enumeration in array processing. In Chap. 10, classical quadratic performance bounds on the accuracy of parameter estimators are complemented with an account of information geometry. The motivation is to harmonize performance bounds for parameter estimators with the corresponding geometry of the underlying manifold of log-likelihood random variables. Chapter 11 concludes the book with an account of other problems and methods in signal processing and machine learning where coherence is an organizing principle.

This book is more research monograph than textbook. However, many of its chapters would serve as complementary resource materials in a graduate-level course on signal processing, machine learning, or statistics. The appendices contain comprehensive accounts of matrix algebra and distribution theory, topics that join optimization theory to form the mathematical foundations of signal processing and machine learning. Chapters 2–4 would complement textbooks on multivariate analysis by covering least squares, linear minimum mean-squared error estimation, and hypothesis testing of covariance structure in the *complex* multivariate normal model. Chapters 5–8 contain an account of matched and adaptive subspace detectors that would complement a course on detection and estimation, multisensor array processing, and related topics. Chapters 9–11 would serve as resource materials in a course on advanced topics in signal processing and machine learning.

It would be futile to attempt an acknowledgment to all of our students, friends, and colleagues who have influenced our thinking and guided our educations. But several merit mention for their contribution of important ideas to this book: Yuri Abramovich, Pia Abbaddo, Javier Álvarez-Vizoso, Antonio Artés-Rodríguez, Mahmood Azimi-Sadjadi, Carlos Beltrán, Olivier Besson, Pascal Bondon, Ron Butler, Margaret Cheney, Yuejie Chi, Edwin Chong, Doug Cochran, Henry Cox, Víctor Elvira, Yossi Francos, Ben Friedlander, Antonio García Marqués, Scott Goldstein, Claude Gueguen, Alfred Hanssen, Stephen Howard, Jesús Ibáñez, Steven Kay, Michael Kirby, Nick Klausner, Shawn Kraut, Ramdas Kumaresan, Roberto López-Valcarce, Mike McCloud, Todd McWhorter, Bill Moran, Tom Mullis, Danilo Orlando, Pooria Pakrooh, Daniel P. Palomar, Jesús Pérez-Arriaga, Chris Peterson, Ali Pezeshki, Bernard Picinbono, José Príncipe, Giuseppe Ricci, Christ Richmond, Peter Schreier, Santiago Segarra, Songsri Sirianunpiboon, Steven Smith, John Thomas, Rick Vaccaro, Steven Van Vaerenbergh, Gonzalo Vázquez-Vilar, Javier Vía, Haonan Wang, and Yuan Wang. Javier Álvarez-Vizoso, Barry Van Veen, Ron Butler, and Stephen Howard were kind enough to review chapters and offer helpful suggestions.

David Ramírez wishes to acknowledge the generous support received from the Agencia Española de investigación (AEI), the Deutsche Forschungsgemeinschaft (DFG), the Comunidad de Madrid (CAM), and the Office of Naval Research (ONR) Global.

Ignacio Santamaría gratefully acknowledges the various agencies that have funded his research over the years, in particular, the Agencia Española de investigación (AEI) and the funding received in different projects of the Plan Nacional de I+D+I of the Spanish government.

Louis Scharf acknowledges, with gratitude, generous research support from the US National Science Foundation (NSF), Office of Naval Research (ONR), Air Force Office of Scientific Research (AFOSR), and Defense Advanced Research Projects Agency (DARPA).

Madrid, Spain — David Ramírez
Santander, Spain — Ignacio Santamaría
Fort Collins, CO, USA — Louis Scharf
December 2021

Contents

Acronyms

Applications of signal processing and machine learning are so wide ranging that acronyms, descriptive of methodology or application, continue to proliferate. The following is an exhausting, but not exhaustive, list of acronyms that are germane to this book.

ACE	Adaptive coherence estimator
ASD	Adaptive subspace detector
BIC	Bayesian information criterion
BLUE	Best linear unbiased estimator
CBF	Conventional (or Capon) beamformer
CCA	Canonical correlation analysis
cdf	Cumulative density function
CFAR	Constant false alarm rate
CG	Conjugate gradient
chf	Characteristic function
CRB	Cramér-Rao bound
CS	Cyclostationary
DFT	Discrete Fourier transform
DOA	Direction of arrival
EP	Estimate and plug
ETF	Equi-angular tight frame
EVD	Eigenvalue decomposition
FA	Factor analysis
FIM	Fisher information matrix
GLR	Generalized likelihood ratio
GLRT	Generalized likelihood ratio test
GSC	Generalized sidelobe canceler
HBT	Hanbury-Brown-Twiss
i.i.d.	Independent and identically distributed
ISI	Intersymbol interference
JL	Johnson-Lindenstrauss (lemma)
KAF	Kernel adaptive filtering
KCCA	Kernel canonical correlation analysis

KL	Kullback-Leibler (divergence)
KLMS	Kernel least mean square
LASSO	Least absolute shrinkage and selection operator
LDU	Lower-diagonal-upper (decomposition)
LHS	Left hand side
LIGO	Laser Interferometer Gravitational-Wave Observatory
LMMSE	Linear minimum mean square error
LMPIT	Locally most powerful invariant test
LMS	Least mean square
LS	Least squares
LTI	Linear time-invariant
MAXVAR	Maximum variance
MCCA	Multiset canonical correlation analysis
MDD	Matched direction detector
MDL	Minimum description length
MDS	Multidimensional scaling
mgf	Moment generating function
MIMO	Multiple-input multiple-output
ML	Maximum likelihood
MMSE	Minimum mean square error
MP	Matching pursuit
MSC	Magnitude squared coherence
MSD	Matched subspace detector
MSE	Mean square error
MSWF	Multistage Wiener filter
MVDR	Minimum variance distortionless response
MVN	Multivariate normal
MVUB	Minimum variance unbiased (estimator)
OBLS	Oblique least squares
OMP	Orthogonal matching pursuit
PAM	Pulse amplitude modulation
PCA	Principal component analysis
pdf	Probability density function
PDR	Pulse Doppler radar
PMT	Photomultiplier tube
PSD	Power spectral density
PSF	Point spread function
RHS	Right hand side
RIP	Restricted isometry property
RKHS	Reproducing kernel Hilbert space
RP	Random projection
rv	Random variable
s.t.	Subject to
SAR	Synthetic aperture radar
SAS	Synthetic aperture sonar

SIMO	Single-input multiple-output
SNR	Signal-to-noise ratio
SUMCOR	Sum-of-correlations
SVD	Singular value decomposition
SVM	Support vector machine
TLS	Total least squares
ULA	Uniform linear array
UMP	Uniformly most powerful
UMPI	Uniformly most powerful invariant
wlog	Without loss of generality
WSS	Wide-sense stationary

NB: We have adhered to the convention in the statistical sciences that cdf, chf, i.i.d., mgf, pdf, and rv are lowercase acronyms.

Introduction

1

Coherence has a storied history in mathematics and statistics, where it has been attached to the concepts of correlation, principal angles between subspaces, canonical coordinates, and so on. In electrical engineering, frequency selectivity in filters and wavenumber selectivity in antenna arrays are really a story of constructive coherence between lagged or differentiated frequency components in the passband and destructive coherence between them in the stopband. Coherence is perhaps better appreciated in physics, where it is used to describe phase alignment in time and/or space, as in coherent light. In fact, the general understanding of coherence in physics and engineering is that it describes a state of propagation wherein propagating waves maintain phase coherence. In radar, sonar, and communication, this coherence is used to steer transmit and receive beams. Coherence in its many guises appears in work on radar, sonar, communication, microphone array processing, machine monitoring, sensor networks, astronomy, remote sensing, and so on. If you read Richard Feynman's delightful book, *QED: The Strange Theory of Light and Matter* [117], you might draw the conclusion that coherence describes a great many other phenomena in classical and modern physics.

But coherence is not limited to the physical and engineering sciences. It arises in many guises in a great many problems of inference where a model is to be fit to measurements for the purpose of estimation, tracking, detection, or classification. As we shall see, the study of coherence is really a study of invariances. This suggests that geometry plays a fundamental role in the problems we address. In many cases, results derived from statistical arguments may be derived from geometrical arguments, and vice versa.

In this opening chapter, we review several topics in communication, signal processing, and machine learning where coherence illuminates an important effect. We begin our story with the early days of wireless communication.

D. Ramírez et al., *Coherence*, https://doi.org/10.1007/978-3-031-13331-2_1

1.1 The Coherer of Hertz, Branly, and Lodge

A standard dictionary definition of coherence would speak to the quality of being logical and consistent, or the quality of forming a unified whole. Perhaps this is why Guglielmo Marconi and his contemporaries named their detector the *coherer*.

The *coherer* was a primitive form of signal detector used in the first radio receivers for wireless telegraphy at the beginning of the twentieth century.[1] Its use in radio was based on the 1890 findings of French physicist Édouard Branly. The device consisted of a tube or capsule containing two electrodes spaced a small distance apart with loose metal filings in the space between. When a radio frequency signal was applied to the device, the metal particles would cling together or *cohere*, reducing the initial high resistance of the device, thereby allowing direct current to flow through it. In a receiver, the current would activate a bell or a Morse paper tape recorder to record the received signal. The metal filings in the coherer remained conductive after the signal ended so that the coherer had to be *decohered* by tapping it with a clapper, such as a doorbell ringer, thereby restoring the coherer to its original state. Coherers remained in widespread use until about 1907, when they were replaced by more sensitive electrolytic and crystal detectors.

The story of the coherer actually began on June 1, 1894, a few months after the death of Heinrich Hertz. Oliver Lodge delivered a memorial lecture on Hertz where he demonstrated the properties of *Hertzian waves* (radio waves). He transmitted them over a short distance and used an improved version of Branly's filings tube, which Lodge had named the "coherer," as a detector. In May 1895, after reading about Lodge's demonstrations, the Russian physicist Alexander Popov built a Hertzian wave-based lightning detector using a coherer. That same year, Marconi demonstrated a wireless telegraphy system using radio waves, based on a coherer.

It is clear that the coherer was designed to be in a state of coherence or incoherence, not in an intermediate state of partial coherence. In fact, as the following account of the Van Cittert-Zernike work shows, the concept of partial coherence was not quantified until about 1938.

1.2 Interference, Coherence, and the Van Cittert-Zernike Story

Coherence arises naturally in the study of interference phenomena, where the problem is to characterize the sums $Z = A_1 e^{j\theta_1} e^{j\omega t} + A_2 e^{j\theta_2} e^{j\omega t}$ and $S = \sqrt{2}\mathrm{Re}\{Z\} = (1/\sqrt{2})(Z + Z^*) = \sqrt{2}A_1 \cos(\omega t + \theta_1) + \sqrt{2}A_2 \cos(\omega t + \theta_2)$. S is the real sum represented by the complex sum Z.

The magnitude squared of the complex sum Z is

$$\begin{aligned} |Z|^2 = ZZ^* &= A_1^2 + A_2^2 + 2\mathrm{Re}\left\{A_1 A_2 e^{j(\theta_1 - \theta_2)}\right\} \\ &= A_1^2 + A_2^2 + 2A_1 A_2 \cos(\theta_1 - \theta_2). \end{aligned}$$

[1] Excerpted from a Wikipedia posting under the title Coherer.

This is the law of cosines. The third term is the interference term, which we may write as $2A_1A_2\rho$, where $\rho = \cos(\theta_1 - \theta_2)$ is coherence. But what about the square of the real sum S? This may be written as

$$S^2 = \frac{1}{2}(Z + Z^*)(Z + Z^*) = ZZ^* + \mathrm{Re}\{ZZ\}$$
$$= A_1^2 + A_2^2 + 2A_1A_2\cos(\theta_1 - \theta_2) + \mathrm{Re}\left\{(A_1e^{j\theta_1} + A_2e^{j\theta_2})^2 e^{j2\omega t}\right\}.$$

The last term is a term in twice the frequency ω, so it will not be seen in a detector matched to a narrow band around frequency ω. Even when this double-frequency term cannot be ignored, a statistical analysis that averages the terms in S^2 will average the last term to zero, as many electromagnetic radiations (such as light) are commonly understood to be proper, meaning $E[ZZ] = 0$ for *proper* complex random variables Z (see Appendix E). The upshot is that, although the square of the real sum S has an extra *complementary* term ZZ, this is a double-frequency term that averages to zero. Consequently, coherence ρ for the complex sum describes coherence for the real sum it represents.

Our modern understanding of coherence is that it is a number or statistic, bounded between minus one and plus one, that indicates the *degree of coherence* between two or more variables. But in the opening to his 1938 paper [391], Zernike says: "In the usual treatment of interference and diffraction phenomena, there is nothing intermediate between coherence and incoherence. Indeed the first term is understood to mean complete dependence of phases, and the second complete independence. ...It would be an improvement in many respects if intermediate states of partial coherence could be treated as well."

The Van Cittert-Zernike story begins with a distributed source of light, each point of the light propagating as the wave $\frac{a_m}{r_m}e^{-jkr_m}e^{j\omega t}$ from a point m located at distance r_m from a fixed point P_1. The wave received at P_1 is $x_1 = \left(\sum_m \frac{a_m}{r_m}e^{-jkr_m}\right)e^{j\omega t}$. The wave received at the nearby point P_2 is the sum $x_2 = \left(\sum_m \frac{a_m}{s_m}e^{-jks_m}\right)e^{j\omega t}$, where s_m is the distance of point m from point P_2 (see Fig. 1.1). The complex coefficients a_m are uncorrelated, zero-mean, proper complex random amplitudes, $E[a_m a_n^*] = \sigma_m^2\delta_{mn}$ and $E[a_m a_n] = 0$, as in the case of incoherent light.

The variance of the sum $z = x_1 + x_2$ is $\mathrm{E}[zz^*] = \mathrm{E}[|x_1|^2] + \mathrm{E}[|x_2|^2] + 2\mathrm{Re}\{\mathrm{E}[x_1x_2^*]\}$, which, under the assumption that the r_m and s_m are approximately equal, may be approximated as

$$\mathrm{E}[|z|^2] = 2V(1 + \rho),$$

where V is variance and ρ is coherence:

$$V = \sum_m \frac{\sigma_m^2}{r_m s_m} \quad \text{and} \quad \rho = \sum_m \rho_m.$$

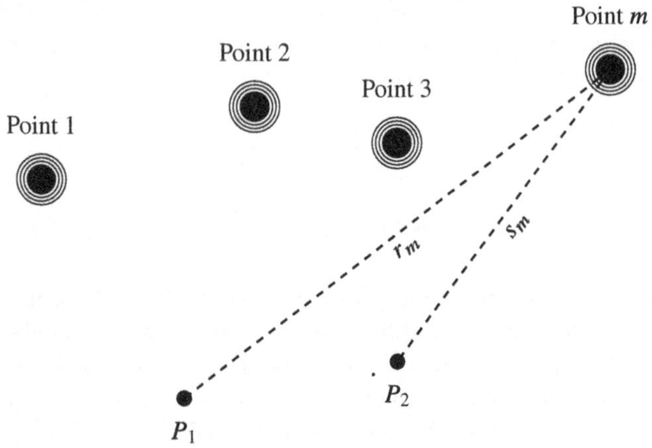

Fig. 1.1 Schematic view of the Van Cittert-Zernike story

The individual coherence terms are

$$\rho_m = \frac{\sigma_m^2/(r_m s_m)}{V} \cos[k(r_m - s_m)].$$

That is, coherence is the coherence ρ_m of each point pair, summed over the points. This coherence is invariant to common scaling of all points of light and individual phasing of each. We would say the coherence is invariant to complex scaling of each point of light. Moreover, it is bounded between minus one and plus one, with incoherence describing a case where the sum of the individual phasors $\frac{\sigma_m^2}{\sum_m \sigma_m^2} e^{-jk(r_m - s_m)}$ tends to the origin and full coherence when the phases align at $k(r_m - s_m) = 0$ or π (mod 2π).

To the modern eye, there is nothing very surprising in this style of analysis. But the importance of the Van Cittert-Zernike work was to establish a link between coherence, interference, and diffraction. With this connection, Zernike was able to interpret Michelson's method of measuring small angular diameters of imaged objects by analyzing the first zero in a diffraction pattern, corresponding to the first zero in the coherence.

1.3 Hanbury Brown-Twiss Effect

The Van Cittert-Zernike result seems to suggest that interference effects must, per force, arise as phasing effects between oscillatory signals. But the Hanbury Brown-Twiss (HBT) effect shows that interference may sometimes be observed between two non-negative intensities.

For our purposes, the HBT story begins in 1956, when Robert Hanbury Brown and Richard Q. Twiss published a test of a new type of stellar interferometer in which two photomultiplier tubes (PMTs), separated by about 6 meters, were aimed at the star Sirius. Light was collected into the PMTs using mirrors from searchlights. An interference effect was observed between the two intensities, revealing a positive correlation between the two signals, despite the fact that no apparent phase information was collected. Hanbury Brown and Twiss used the interference signal to determine the angular size of Sirius, claiming excellent resolution. The result has been used to give non-classical interpretations to quantum mechanical effects.[2] Here is the story.

Let $E_1(t)$ and $E_2(t)$ be signals entering the PMTs:

$$E_1(t) = E(t)\sin(\omega t) \text{ and } E_2(t) = E_1(t-\tau) = E(t-\tau)\sin(\omega(t-\tau)).$$

The narrowband signal $E(t)$ modulates the amplitude of the carrier signal $\sin(\omega t)$, and the two signals are time-delayed versions of each other. The measurements recorded at the PMTs are the *intensities*

$$i_1(t) = \text{low-pass filtering of } E_1^2(t) = \frac{1}{2}E^2(t) \geq 0$$

and

$$i_2(t) = \text{low-pass filtering of } E_2^2(t) = \frac{1}{2}E^2(t-\tau) \geq 0.$$

Assume $E(t) = A_0 + A_1 \sin(\Omega t)$, with $A_1 \ll A_0$, in which case we would say the signal $E(t)$ is an amplitude modulating signal with small modulation index. These low-pass filterings may be written as

$$\begin{aligned} i_1(t) &= \frac{1}{2}(A_0 + A_1\sin(\Omega t))^2 = \frac{1}{2}\left(A_0^2 + 2A_0A_1\sin(\Omega t) + A_1^2\sin^2(\Omega t)\right), \\ i_2(t) &= \frac{1}{2}(A_0 + A_1\sin(\Omega(t-\tau))^2 \\ &= \frac{1}{2}\left(A_0^2 + 2A_0A_1\sin(\Omega t - \phi) + A_1^2\sin^2(\Omega t - \phi)\right), \end{aligned}$$

where $\phi = \Omega\tau$. These may be approximated by ignoring the quadratic term in A_1^2 since $A_1^2 \ll A_0A_1 \ll A_0^2$. Then, by subtracting the constant terms in $i_1(t)$ and $i_2(t)$, we are left with the intensities

$$I_1(t) = A_0A_1\sin(\Omega t) \quad \text{and} \quad I_2(t) = A_0A_1\sin(\Omega t - \phi).$$

[2] This excerpt is extracted from Wikipedia.

Define the time-varying, zero-lag, correlation

$$\begin{aligned} r_{12}(t) &= I_1(t)I_2(t) = A_0^2 A_1^2 \sin(\Omega t)\sin(\Omega t - \phi) \\ &= \frac{A_0^2 A_1^2}{2}\left[\cos(\phi) - \cos(2\Omega t - \phi)\right]. \end{aligned}$$

The time average of this time-varying correlation produces the constant $R_{12} = \frac{A_0^2 A_1^2}{2}\cos(\phi)$. Recall $\phi = \Omega\tau$, so this constant encodes for the delay τ, determined entirely from the correlation between non-negative intensities. The coherence between the time-averaged intensities is

$$\rho_{12} = \frac{R_{12}}{\sqrt{R_{11}R_{22}}} = \cos(\phi) = \cos(\Omega\tau),$$

where $R_{ii} = \frac{A_i^4}{2}$. The phase $\phi = \Omega\tau$ is the *phase mismatch of the modulating signals* entering the PMTs, and this phase mismatch is determined from apparently phase-less intensity measurements. The trick, of course, is that the phase ϕ is still carried in the modulating signal $E(t-\tau)$ and in its intensity, $i_2(t)$. Why could this phase not be read directly out of $i_2(t)$ or $I_2(t)$? The answer is that the model for the modulating signal $E(t)$ is really $E(t) = A_0 + A_1 \sin(\Omega t - \theta)$, where the phase θ is unknown. The term $\sin(\Omega t - \phi)$ is really $\sin(\Omega t - \phi - \theta)$, and there is no way to determine the phase ϕ from $\phi + \theta$, except by computing correlations.

1.4 Tone Wobble and Coherence for Tuning

Perhaps you have heard two musicians in an orchestra tuning their instruments or a piano tuner "matching a key to a tuning fork." Near to the desired tuning, you have heard two slightly mis-tuned pure tones (an idealization) whose frequencies are close but not equal. The effect is one of a *beating phenomenon* wherein a pure tone seems to wax and wane in intensity. This waxing and waning tone is, in fact, a pure tone whose frequency is the average of the two frequencies, amplitude modulated by a low-frequency tone whose frequency is half the difference between the two mismatched frequencies. This is easily demonstrated for equal amplitude tones:

$$\begin{aligned} Ae^{j((\omega+\nu)t+\phi+\psi)} + Ae^{j((\omega-\nu)t+\phi-\psi)} &= Ae^{j(\omega t+\phi)}\left(e^{j(\nu t+\psi)} + e^{-j(\nu t+\psi)}\right) \\ &= 2Ae^{j(\omega t+\phi)}\cos(\nu t + \psi). \end{aligned}$$

The average frequency ω is half the sum of the two frequencies, and the difference frequency ν is half the difference between the two frequencies. The corresponding real signal is $2A\cos(\omega t+\phi)\cos(\nu t+\psi)$, a signal that waxes and wanes with a period

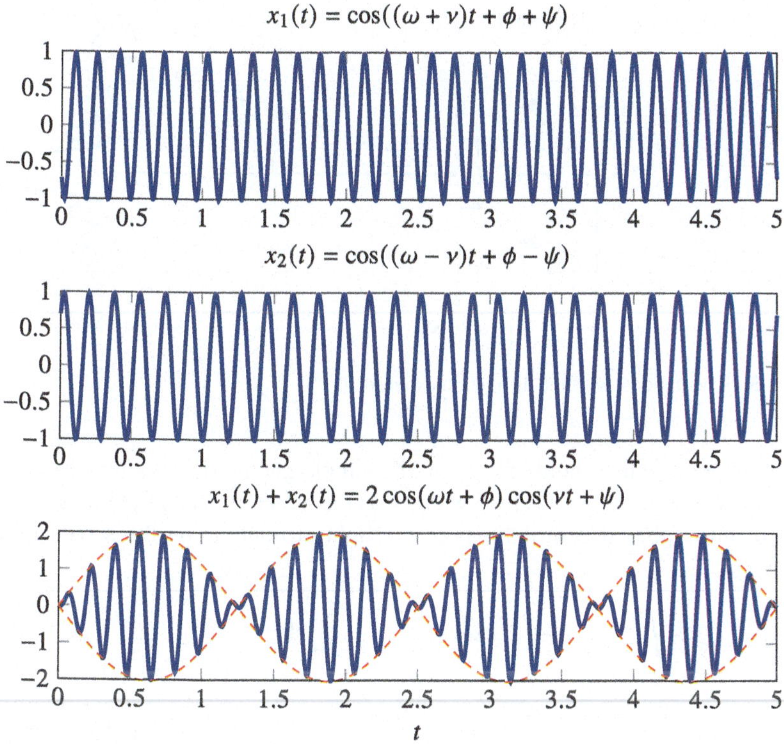

Fig. 1.2 Beating phenomenon when two tones of frequencies $\omega + \nu$ and $\omega - \nu$ are added. The resulting signal, shown in the bottom plot, waxes and wanes with period $2\pi/\nu$

of $2\pi/\nu$ (see Fig. 1.2). Constructive interference (coherence) produces a wax, and destructive interference produces a wane.

1.5 Beampatterns and Diffraction of Electromagnetic Radiation by a Slit

Assume that a single-frequency plane wave, such as laser light, passes through a slit in an opaque sheet. At any point $-L/2 < x \leq L/2$ within the slit, the complex representation of the real wave passing through the slit is $Ae^{j\phi}e^{j\omega t}$. The baseband, or phasor, representation of this wave is $Ae^{j\phi}$, independent of x. Then, according to Fig. 1.3, this wave arrives at point P on a screen as $Ae^{j\phi}e^{j\omega(t-\tau)}$, with phasor representation $Ae^{j\phi}e^{-j\omega\tau}$, where $\tau = r(x, P)/c$ is the travel time from the point x within the slit to an identified point P on the screen, $r(x, P)$ is the travel distance, and c is the speed of propagation. This may be written as $Ae^{j\phi}e^{-j(2\pi/\lambda)r(x,P)}$, where $\lambda = 2\pi c/\omega$ is the wavelength of the propagating wave. Under the so-called far-field

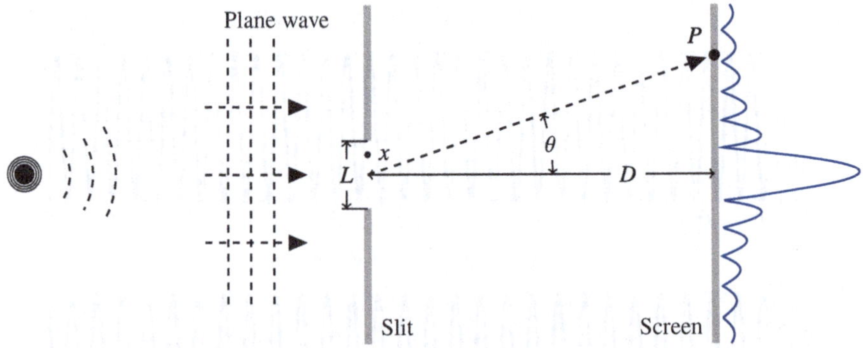

Fig. 1.3 Schematic view of the diffraction experiment

approximation, which holds for $L \ll D$, simple geometric reasoning establishes that $r(x, P) \approx r(0, P) - x \sin\theta$, where θ is the angle illustrated in the figure. The sum of these phased contributions to the light at point P on the screen may be written as

$$E(\theta) = Ae^{j\phi} \int_{-L/2}^{L/2} e^{-j\frac{2\pi}{\lambda}(r(0,P)-x\sin\theta)} dx = Ae^{j\left(\phi - \frac{2\pi}{\lambda} r(0,P)\right)} \int_{-L/2}^{L/2} e^{j\frac{2\pi}{\lambda} x \sin\theta} dx$$

$$= ALe^{j\left(\phi - \frac{2\pi}{\lambda} r(0,P)\right)} \frac{\sin\left(\frac{L\pi}{\lambda}\sin\theta\right)}{\frac{L\pi}{\lambda}\sin\theta} = ALe^{j\left(\phi - \frac{2\pi}{\lambda} r(0,P)\right)} \operatorname{sinc}\left(\frac{L\pi}{\lambda}\sin\theta\right),$$

for $-\pi/2 < \theta \leq \pi/2$. The squared intensity is

$$I^2(\theta) = |E(\theta)|^2 = A^2L^2 \operatorname{sinc}^2\left(\frac{L\pi}{\lambda}\sin\theta\right).$$

The squared intensity is zero for $\sin\theta = l\lambda/L$, $l = \pm 1, \pm 2, \ldots$ These are dark spots on the screen, where the phasor terms cancel each other. The bright spots are close to where $\sin\theta = (2l+1)\lambda/2L$, $l = 0, \pm 1, \ldots$, in which case the phasor terms tend to align. Tend to cohere. Only for $\theta = 0$ do they align perfectly, but for other values, they tend to cohere for values of $\sin\theta$ between the dark spots. This is also illustrated in the figure, where $10\log_{10} I^2(\theta)$ is plotted versus $D\tan\theta$, which is the lateral distance from the origin of the screen, parameterized by θ.

Beampatterns of a Linear Antenna. This style of analysis goes through for the analysis of antennas. For example, if $Ae^{j\phi}e^{j\omega t}$ is the uniform current distribution on an antenna, then the radiated electric field, ignoring square law spreading, is $E(\theta)$. The intensity $|E(\theta)|$, $-\pi < \theta \leq \pi$, is commonly called the *transmit beampattern* of the antenna. If the point P is now treated as a point source of radiation, and the slit is treated as a wire onto which an electric field produces a current, then the same phasing arguments hold, and $|E(\theta)|$ describes the magnitude of a received signal from direction θ. The intensity $|E(\theta)|, -\pi < \theta \leq \pi$, is called the *receive beampattern*. In both cases, the function $|E(\theta)|$ describes the coherence between components of the received signal, a coherence that depends on angle. The first zero of the sinc function, $\text{sinc}\left(\frac{L\pi}{\lambda}\sin\theta\right)$, is determined as $\frac{L\pi}{\lambda}\sin\theta = \pi$, which is to say $\sin\theta = \lambda/L$. This is the Rayleigh limit to resolution of *electrical angle* $\sin\theta$. For small λ/L, this is approximately the Rayleigh limit to resolution of physical angle θ. So for a fixed aperture L, short wavelengths are preferred over long wavelengths. Or, at short wavelengths (high frequencies), even small apertures have high resolution.

Beampattern of Linear Antenna Array. If the antenna is replaced by an antenna array, consisting of $2N - 1$ discrete dipoles, separated by uniform distance d, then the radiated field as a function of θ is

$$E(\theta) = Ae^{j\phi}\sum_{n=-(N-1)}^{N-1} e^{-j\frac{2\pi}{\lambda}(r(0,P)-nd\sin\theta)} = Ae^{j\left(\phi-\frac{2\pi}{\lambda}r(0,P)\right)}\sum_{n=-(N-1)}^{N-1} e^{j\frac{2\pi}{\lambda}nd\sin\theta}$$

$$= Ae^{j\left(\phi-\frac{2\pi}{\lambda}r(0,P)\right)}\frac{\sin\left((2N-1)\frac{\pi d}{\lambda}\sin\theta\right)}{\sin\left(\frac{\pi d}{\lambda}\sin\theta\right)},$$

for $-\pi/2 < \theta \leq \pi/2$. The magnitude $|E(\theta)|$ is the transmit beampattern. If the point P is now treated as a point source of radiation, and the slit is replaced by this set of discrete dipoles, then the same phasing arguments hold. If the responses of the dipoles are summed, then $|E(\theta)|$ is the receive beampattern. It measures coherence between components of the signal received at the individual dipoles, a coherence that depends on the angle to the radiator. The first zero of $|E(\theta)|$ is determined as $\frac{(2N-1)d\pi}{\lambda}\sin\theta = \pi$, which is to say $\sin\theta = \lambda/((2N-1)d)$. This is the Rayleigh limit to resolution of *electrical angle* $\sin\theta$. For small λ/L, this is approximately the Rayleigh limit to resolution of physical angle θ. So for a fixed aperture $(2N-1)d$, short wavelengths are preferred over long wavelengths. Or, at short wavelengths (high frequencies), even small apertures have high resolution.

The reader may wish to speculate that the wave nature of electrons may be exploited in exactly this way to arrive at electron microscopy. The *Rayleigh limit* is the classical limit to resolution of antennas, antenna arrays, and electron microscopes.

1.6 LIGO and the Detection of Einstein's Gravitational Waves

The recent detection of *gravitational waves*[3] is a story in *coherence*. The storyline is this. 1.3 billion years ago, two black holes combined into one black hole, launching a gravitational wave. 1.3 billion years later, this wave stretched and

[3] The Economist, February 13, 2016.

squeezed space-time in one leg of a laser interferometer at the Laser Interferometer Gravitational-Wave Observatory (LIGO) in Hanford, WA, USA, Earth, modulating the phase of the laser beam in that leg. This modulation changed the coherence between the laser beams in the two legs of the interferometer, and thus was a gravitational wave detected. The same effect was observed at the LIGO in Livingston, LA, USA, Earth, and a cross-validation of the detections made at these two facilities established that the observed effects were coherent effects not attributable to locally generated changes in the lengths of the interferometer legs. Einstein's prediction of 100 years ago was validated.

Let us elaborate on the idea behind the LIGO detector. Assume signals from two channels (the perpendicular legs of the laser channels) are calibrated so that in the quiescent state, $x_1(t) = x_2(t)$. The difference signal is $x_1(t) - x_2(t) = 0$. The square of this difference is zero, until a gravitational wave modulates space-time differently in the two channels. So $\int [x_1(t) - x_2(t)]^2 dt$ is a statistic to be monitored for its deviation from zero. Or more reasonably, this statistic should be normalized by the RMS values of the signals $x_1(t)$ and $x_2(t)$. Then, assuming the modulation to be lossless so $x_1(t)$ and $x_2(t)$ have the same energy, the normalized statistic is

$$\beta = \frac{\int |x_1(t) - x_2(t)|^2 \, dt}{\sqrt{\int |x_1(t)|^2 \, dt \int |x_2(t)|^2 \, dt}} = 2\,(1 - \mathrm{Re}(\rho)),$$

where ρ is the following coherence statistic:

$$\rho = \frac{\int x_1(t) x_2^*(t) dt}{\sqrt{\int |x_1(t)|^2 \, dt \int |x_2(t)|^2 \, dt}}.$$

Of course, this notional narrative does no justice to the scientific and engineering effort required to design and construct an instrument with the sensitivity and precision to measure loss in coherence due to the stretching of space-time.

1.7 Coherence and the Heisenberg Uncertainty Relations

In this subsection, we show that there is a coherence bound on the commutator formed from two Hermitian, but non-commuting, operators. When these operators are the time-scale operator $(Tf)(t) = tf(t)$ and the time-derivative operator $(\Omega f)(t) = \left(-j\frac{d}{dt}f\right)(t)$, then this result specializes to the so-called Heisenberg uncertainty principle, which in this case is a bound on time-frequency concentration.

In broad outline, the argument is this. Begin with a function $f(t) \in L_2(\mathbb{R})$ and define operators $O : L_2(\mathbb{R}) \longrightarrow L_2(\mathbb{R})$:

$$T : f(t) \longrightarrow (Tf)(t) = tf(t),$$

$$\Omega : f(t) \longrightarrow (\Omega f)(t) = \left(-j\frac{d}{dt}f\right)(t),$$

$$\mathcal{F} : f(t) \longrightarrow (\mathcal{F}f)(t) = \int_{-\infty}^{\infty} f(t)e^{-j\omega t}\frac{dt}{\sqrt{2\pi}}.$$

These are called, respectively, the time-scale, time-derivative, and Fourier transform operators. Two familiar identities from Fourier analysis are $\mathcal{F}T = -\Omega\mathcal{F}$ and $\mathcal{F}\Omega = T\mathcal{F}$.

Let $L_2(\mathbb{R})$ to be an inner-product space with inner product[4]

$$\langle f, g\rangle_{L_2(\mathbb{R})} = \int_{-\infty}^{\infty} f(t)g^*(t)dt.$$

The adjoint $\tilde{O}$ of an operator O is then defined as follows:

$$\langle Of, g\rangle = \langle f, \tilde{O}g\rangle.$$

If $\langle Of, g\rangle = \langle f, Og\rangle$, then O is said to be self-adjoint, written as $O = \tilde{O}$. Note $\langle Of, Of\rangle = \langle f, \tilde{O}Of\rangle$. If this equals $\langle f, f\rangle$, then O is said to be unitary, written $\tilde{O}O = \text{Id}$, where Id is the identity operator. The operators T and Ω are self-adjoint, but not unitary. The operator $\mathcal{F}$ is unitary, but not self-adjoint. Its adjoint $\tilde{\mathcal{F}}$ is the inverse Fourier transform operator. Two operators may be composed, in which case $\langle ABf, g\rangle = \langle Bf, \tilde{A}g\rangle = \langle f, \tilde{B}\tilde{A}g\rangle$. If A is self-adjoint, then $\langle ABf, g\rangle = \langle Bf, Ag\rangle = \langle Ag, Bf\rangle^*$.

The Commutator and a Coherence Bound. Define the commutator $[A, B]$ for two self-adjoint operators A and B as

$$[A, B] = AB - BA.$$

Then, $\langle f, [A, B]f\rangle$ is imaginary:

$$\langle f, [A, B]f\rangle = \langle f, ABf\rangle - \langle f, BAf\rangle = \langle Af, Bf\rangle - \langle Af, Bf\rangle^*.$$

That is, $\langle f, [A, B]f\rangle/2$ is the imaginary part of $\langle Af, Bf\rangle$. The Pythagorean theorem, together with the Cauchy-Schwarz inequality, says

$$\frac{1}{4}|\langle f, [A, B]f\rangle|^2 \le |\langle Af, Bf\rangle|^2 \le \langle Af, Af\rangle\langle Bf, Bf\rangle.$$

[4] If it is clear from the context, we will drop the subindex $L_2(\mathbb{R})$ in the inner products.

If each term in this chain of inequalities is normalized by $\langle f, f\rangle^2$, then this may be written as a bound on the coherence between $[A, B]f$ and f:

$$\frac{|\langle f, [A, B]f\rangle|^2}{\langle f, f\rangle^2} \leq 2\frac{\langle Af, Af\rangle}{\langle f, f\rangle} \cdot 2\frac{\langle Bf, Bf\rangle}{\langle f, f\rangle}.$$

Time-Frequency Concentration. Let $A = T$ and $B = \Omega$. Then $[T, \Omega] = \Omega T - T\Omega = -j\mathrm{Id}$. That is, $-\langle f, f\rangle/2$ is the imaginary part of $\langle Tf, \Omega f\rangle$, which is to say

$$\frac{1}{4} \mid \langle f, f\rangle \mid^2 \leq |\langle Tf, \Omega f\rangle|^2 \leq \langle Tf, Tf\rangle\langle \Omega f, \Omega f\rangle.$$

Now, use the unitarity of the operator $\mathcal{F}$, and the identity $\mathcal{F}\Omega = T\mathcal{F}$, to write

$$\frac{1}{4}|\langle f, f\rangle|^2 \leq |\langle Tf, \Omega f\rangle|^2 \leq \langle Tf, Tf\rangle\langle \mathcal{F}\Omega f, \mathcal{F}\Omega f\rangle = \langle Tf, Tf\rangle\langle T\mathcal{F}f, T\mathcal{F}f\rangle.$$

This may be written as

$$\frac{1}{4} \leq \frac{\int_{-\infty}^{\infty} t^2|f(t)|^2 dt}{\int_{-\infty}^{\infty} |f(t)|^2 dt} \cdot \frac{\int_{-\infty}^{\infty} \omega^2|F(\omega)|^2 d\omega}{\int_{-\infty}^{\infty} |F(\omega)|^2 d\omega},$$

where $F = \mathcal{F}f$ is the Fourier transform of f. That is, the product of time concentration and frequency concentration for a function $f \longleftrightarrow F$ is not smaller than $1/4$.

The Gaussian Pulse. Consider the Gaussian pulse

$$f(t) = \frac{1}{\sqrt{2\pi\sigma^2}} e^{-t^2/2\sigma^2} \longleftrightarrow e^{-\omega^2/(2/\sigma^2)} = F(\omega).$$

It is not hard to show that

$$\frac{\int_{-\infty}^{\infty} t^2|f(t)|^2 dt}{\int_{-\infty}^{\infty} |f(t)|^2 dt} = \frac{\sigma^2}{2} \quad \text{and} \quad \frac{\int_{-\infty}^{\infty} \omega^2|F(\omega)|^2 d\omega}{\int_{-\infty}^{\infty} |F(\omega)|^2 d\omega} = \frac{1}{2\sigma^2},$$

for a product of $1/4$. It is straightforward to show that $Tf = \beta\Omega f$ for $\beta = j\sigma^2$.

1.8 Coherence, Ambiguity, and the Moyal Identities

Ambiguity functions arise in radar, sonar, and optical imaging, where they are used to study the resolution of matched filters that are scanned in delay and Doppler shift or in spatial coordinates. The ambiguity function is essentially a point spread function that determines the so-called Rayleigh or diffraction limit to resolution. The Moyal identities arise in the study of inner products between ambiguity functions. The important use of the Moyal identities is to establish that the total volume of the ambiguity function is constrained and that the consequent aim of signal design or imaging design must be to concentrate this volume or otherwise shape it for purposes such as resolution or interference rejection.

Begin with four signals, f, g, x, y, each an element of $L_2(\mathbb{R})$. Define the cross-ambiguity function Γ_{fg} to be

$$\Gamma_{fg}(\nu, \tau) = \int_{-\infty}^{\infty} f(t-\tau)g^*(t)e^{-j\nu t}dt,$$

with $\nu \in \mathbb{R}$ and $\tau \in \mathbb{R}$. This definition holds for all pairs like (f, y), (g, x), etc. The ambiguity function may be interpreted as a correlation between the signals $f(t-\tau)$ and $g(t)e^{j\nu t}$. When normalized by the norms of f and g, this is a complex coherence.

The Moyal identity is

$$\langle \Gamma_{fg}, \Gamma_{yx} \rangle = (\Gamma_{fy} \cdot \Gamma^*_{gx})(0, 0) = \langle f, y \rangle \langle g, x \rangle^*.$$

This has consequences for range-Doppler imaging in radar and sonar. Let $f = y$ and $g = x$. Then, the Moyal identity shows that the volume of the ambiguity function Γ_{xy} is fixed by the energy in the signals (x, y):

$$\langle \Gamma_{yx}, \Gamma_{yx} \rangle = (\Gamma_{yy} \cdot \Gamma^*_{xx})(0, 0) = \langle y, y \rangle \langle x, x \rangle.$$

That is, $\langle \Gamma_{yx}, \Gamma_{yx} \rangle / (\langle y, y \rangle \langle x, x \rangle) = 1$. The special case $\langle \Gamma_{xx}, \Gamma_{xx} \rangle / (\langle x, x \rangle \langle x, x \rangle) = 1$ shows the volume of the normalized ambiguity function to be invariant to the choice of x. But, of course, the distribution of this volume over delay and Doppler depends on the signal x, and the point of signal design is to distribute this volume according to an imaging objective.

1.9 Coherence, Correlation, and Matched Filtering

The correlator and the matched filter are important building blocks in the study of coherence. And, in fact, there is no essential difference between them. There is a connection to ambiguity and to inner product.

Coherence and Cross-Correlation. The cross-correlation function $r_{fg}(\tau)$ is defined to be

$$r_{fg}(\tau) = \int_{-\infty}^{\infty} f^*(t-\tau)g(t)dt = \int_{-\infty}^{\infty} g(t+\tau)f^*(t)dt = r_{gf}^*(-\tau).$$

At $\tau = 0$, $r_{fg}(0)$ is the inner product $\int f^*(t)g(t)dt$, denoted $\langle g, f\rangle$. A special case is $r_{ff}(0) = r_{ff}^*(0)$, which shows $r_{ff}(0) = \langle f, f\rangle$ to be real.

The frequency-domain representation of the correlation function is

$$r_{fg}(\tau) = \int_{-\infty}^{\infty} F^*(\omega)G(\omega)e^{j\omega\tau}\frac{d\omega}{2\pi},$$

with special case $r_{ff}(\tau) = \int |F(\omega)|^2 e^{j\omega\tau}\frac{d\omega}{2\pi}$. In other words, $r_{ff}(\tau) \longleftrightarrow |F(\omega)|^2$, a Fourier transform pair commonly called a Wiener-Khinchin identity. The function $|F(\omega)|^2$ is called an energy spectrum because it shows how the energy of $f(t)$ is distributed over frequency: $r_{ff}(0) = \int |f(t)|^2 dt = \int |F(\omega)|^2 d\omega/2\pi$. Of course, this is a Parseval identity.

Define the *squared coherence*

$$\rho_{fg}^2(\tau) = \frac{|r_{fg}(\tau)|^2}{r_{ff}(0)r_{gg}(0)}.$$

It is a simple application of the Schwarz inequality that for fixed f and free g, this squared coherence is maximized when $f(t-\tau) = g(t)$. This suggests that $f(t)$ might be chosen to be $g(t)$, and then cross-correlation or coherence is scanned through delays τ to search for a maximum. The virtue of squared coherence over cross-correlation is that squared coherence is invariant to the scale of f and g, so there is no question of what is large and what is small. One is large and zero is small. This is just one version of coherence that will emerge throughout the book.

The Correlator for Maximizing Signal-to-Noise Ratio. A measurement $g(t)$ is thought to consist of a signal $as(t)$ plus noise $w(t)$, with the signal $s(t)$ known. The problem is to extract the unknown coefficient a. An idea suggested by the previous paragraph is to correlate this measurement with another signal $f(t)$:

$$r_{fg}(0) = \int_{-\infty}^{\infty} f^*(t)g(t)dt.$$

This evaluates to

$$r_{fg}(0) = a\int_{-\infty}^{\infty} f^*(t)s(t)dt + \int_{-\infty}^{\infty} f^*(t)w(t)dt.$$

If the noise is zero mean and white, which is to say that for all t, $\mathrm{E}[w(t)] = 0$ and $\mathrm{E}[w(t+\tau)w^*(t)] = \sigma^2\delta(\tau)$, with $\delta(\tau)$ the Dirac delta, then the mean and variance of $r_{fg}(0)$ are

$$\mathrm{E}[r_{fg}(0)] = ar_{fs}(0)$$

and

$$\mathrm{E}[(r_{fg}(0) - ar_{fs}(0))(r_{fg}(0) - ar_{fs}(0))^*] = \sigma^2 r_{ff}(0).$$

The output signal-to-noise ratio is commonly defined to be the magnitude squared of the mean, divided by the variance

$$\mathrm{SNR} = \frac{|a|^2|r_{fs}(0)|^2}{\sigma^2 r_{ff}(0)}.$$

We have no control over the input signal-to-noise ratio, $\mathrm{snr} = |a|^2/\sigma^2$, but we do have control over the choice of f. The SNR is invariant to the scale of f, so without loss of generality, we may assume $r_{ff}(0) = 1$. Then, by the Schwarz inequality, the output SNR is bounded above by $(|a|^2/\sigma^2)r_{ss}(0)$ with equality iff $f(t) = \kappa s(t)$. So the correlator that maximizes the output SNR at $\mathrm{SNR} = (|a|^2/\sigma^2)r_{ss}(0)$ is

$$r_{sg}(0) = \int_{-\infty}^{\infty} s^*(t)g(t)dt.$$

Of course, the resulting SNR does depend on the scale of as, namely, $|a|^2 r_{ss}(0)$.

Matched Filter. Define the matched filter to be a linear time-invariant filter whose impulse response is $\tilde{f}(t)$, where $\tilde{f}(t) = f^*(-t)$. This is called the complex conjugate of the time reversal of $f(t)$. If the input to this matched filter is $g(t)$, the output is the convolution

$$(\tilde{f} * g)(t) = \int_{-\infty}^{\infty} \tilde{f}(t-t')g(t')dt' = \int_{-\infty}^{\infty} f^*(t'-t)g(t')dt' = r_{fg}(t).$$

That is, the correlation between f and g at delay t may be computed as the output of filter $\tilde{f}$. Moreover, the correlation $r_{ff}(t)$ is $(\tilde{f} * f)(t)$. When g is the Dirac delta, then $(\tilde{f} * g)(t) = \tilde{f}(t)$, which explains why $\tilde{f}$ is called an impulse response.

The filter $\tilde{f}$ is sometimes called the adjoint of the filter f, because when treated as an operator, it satisfies the equality

$$\langle f * z, g\rangle = \langle z, \tilde{f} * g\rangle.$$

That is, to correlate g and $f * z$ is to correlate z and $\tilde{f} * g$.

These results hold at the sampled data times $t = kt_0$, in which case

$$(\tilde{f} * g)(kt_0) = r_{fg}(kt_0).$$

These fundamentally important ideas may be exploited to analyze and implement algorithms in signal processing and machine learning. For example, samplings of convolutions in a convolutional neural network may be viewed as correlations at lags equal to these sampling times.

Sampled-Data: Discrete-Time Correlation and Convolution. When the continuous-time measurement $g(t), t \in \mathbb{R}$, is replaced by a discrete-time measurement $g[n] = g(n), n \in \mathbb{Z}$, then the discrete-time measurement is said to be a *sampled-data* version of the continuous-time measurement. The discrete-time model of $g(t) = as(t) + w(t)$ is $g[n] = as[n] + w[n]$. A filter with unit-pulse response $\tilde{f}[n] = f^*[-n]$ is a discrete-time filter. Discrete-time convolution is

$$(\tilde{f} * g)[n] = \sum_k \tilde{f}[n-k]g[k] = \sum_k f^*[k-n]g[k] = r_{fg}[n]$$

That is, filtering of g by $\tilde{f}$ is correlation of f and g. It is easy to argue that $f = s$ maximizes SNR in the signal-plus-noise model $g[n] = as[n] + w[n]$, when $w[n]$ is discrete-time white noise, which is to say $\mathrm{E}[w[n+k]w^*[n]] = \sigma^2\delta[k]$.

Nyquist Pulses. Some continuous-time signals $f(t)$ are special: either they sample as Kronecker delta sequences, or their lagged correlations $r_{ff}(\tau)$ sample as Kronecker delta sequences. That is, $f(kt_0) = f(0)\delta[k]$ and $(\tilde{f} * f)(kt_0) = (\tilde{f} * f)(0)\delta[k]$. In the first case, the pulse is said to be Nyquist-I, and in the second, it is said to be Nyquist-II. The trivial examples are compactly supported pulses, which are zero outside the interval $0 < t \leq T$, with $T < t_0/2$. They and their corresponding lagged correlations are zero outside the interval $0 < \tau \leq 2T$, but these are by no means the only pulses with this property. For instance, in communications, Nyquist-I pulses are typically impulse responses of raised-cosine filters, whereas Nyquist-II pulses are impulse responses of root-raised-cosine filters. These two pulses are shown in Fig. 1.4. The Nyquist-I and Nyquist-II properties are exploited in many imaging systems like synthetic aperture radar (SAR), synthetic aperture sonar (SAS), and pulsed Doppler radar (PDR).

Consider the following continuous-time Fourier transform pairs:

$$\int_{-\infty}^{\infty} F(\omega)e^{j\omega t}\frac{d\omega}{2\pi} = f(t) \longleftrightarrow F(\omega) = \int_{-\infty}^{\infty} f(t)e^{-j\omega t}dt,$$

$$\int_{-\infty}^{\infty} |F(\omega)|^2 e^{j\omega t}\frac{d\omega}{2\pi} = (\tilde{f} * f)(t) \longleftrightarrow |F(\omega)|^2 = \int_{-\infty}^{\infty} (\tilde{f} * f)(t)e^{-j\omega t}dt.$$

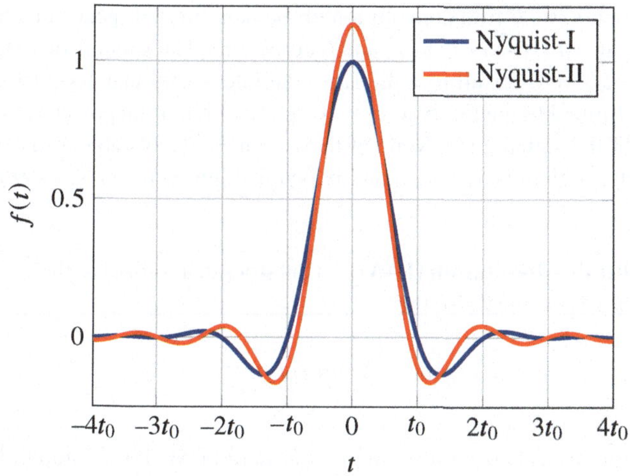

Fig. 1.4 Examples of Nyquist-I and Nyquist-II pulses

It follows from the Poisson sum formulas that the discrete-time Fourier transform pairs for the sampled-data sequences $f(kt_0)$ and $(\tilde{f} * f)(kt_0)$ are

$$\int_0^{2\pi/t_0} t_0 \sum_r F(\omega + r2\pi/t_0) e^{j\omega n t_0} \frac{d\omega}{2\pi} = f(nt_0)$$

$$\longleftrightarrow t_0 \sum_r F(\omega + r2\pi/t_0) = \sum_n f(nt_0) e^{-jn\omega t_0},$$

and

$$\int_0^{2\pi/t_0} t_0 \sum_r |F(\omega + r2\pi/t_0)|^2 e^{j\omega n t_0} \frac{d\omega}{2\pi} = (\tilde{f} * f)(nt_0)$$

$$\longleftrightarrow t_0 \sum_r |F(\omega + r2\pi/t_0)|^2 = \sum_n (\tilde{f} * f)(nt_0) e^{-jn\omega t_0}.$$

For the sampled-data sequences to be Kronecker delta sequences, the *aliased spectrum* $t_0 \sum_r F(\omega+r2\pi/t_0)$ or $t_0 \sum_r |F(\omega+r2\pi/t_0)|^2$ must be constant at unity on the *Nyquist band* $0 < \omega \leq 2\pi$. That is, the original Fourier transforms (spectral densities) $F(\omega)$ or $|F(\omega)|^2$ must *alias white*. There is a library of such spectra, and their corresponding signals $f(t)$, for which this holds. Perhaps the best known are the cardinal or generalized Shannon pulses for which

$$\left[(\pi/t_0)\frac{\sin(\pi t/t_0)}{\pi t/t_0}\right]^k = f(t) \longleftrightarrow F(\omega) = \chi(\omega)(*)^k \chi(\omega),$$

where $(*)^k$ denotes a k-fold convolution of the bandlimited spectrum $\chi(\omega)$ that is 1 on the interval $-\pi/t_0 < \omega \leq \pi/t_0$ and 0 elsewhere. The support of $f(t)$ is the real line, but $f(nt_0) = 0$ for all $n \neq 0$. This example shows that Nyquist pulses need not be time-limited to the interval $0 < t \leq t_0$. The higher the power k, the larger the bandwidth of the signal $f(t)$. None of these signals is realizable as a causal signal, so the design problem is to design nearly Nyquist pulses under a constraint on their bandwidth.

Pulse Amplitude Modulation (PAM). This story generalizes to the case where the measurement $g(t)$ is the pulse train

$$g(t) = \sum_n a[n] f(t - nt_0).$$

This is commonly called a pulse amplitude-modulated (PAM) signal, because the sequence of complex-valued symbols $a[n]$, $n = 0, \pm 1, \ldots,$ modulates the common pulse $f(t)$. If the pulse $f(t)$ is Nyquist-I, then the measurement $g(t)$ samples as $g(kt_0) = a[k]$. But more importantly, if the pulse is Nyquist-II, the matched filter output evaluates to

$$(\tilde{f} * g)(kt_0) = \sum_n a[n](\tilde{f} * f)((k - n)t_0) = a[k].$$

This is called PAM with no intersymbol interference (ISI). The key is to design pulses that are nearly Nyquist-II, under bandwidth constraints, and to synchronize the sampling times with the modulation times, which is nontrivial.

1.10 Coherence and Matched Subspace Detectors

Begin with the previously defined matched filtering of a signal $g(t)$ by a filter $\tilde{f}(t)$:

$$(\tilde{f} * g)(t) = \int_{-\infty}^{\infty} \tilde{f}(t - \tau) g(\tau) d\tau = \int_{-\infty}^{\infty} f^*(\tau - t) g(\tau) d\tau,$$

where $\tilde{f}(t) = f^*(-t)$ is the complex conjugate time reversal of $f(t)$. As discussed in Sect. 1.9, it is convention to call the LHS of the above expression the *output of a matched filter at time t* and the RHS the *output of a correlator at delay t*. The RHS is an inner product $\langle g, D_t f \rangle$, where D_t is a delay operator with action $(D_t f)(t') = f(t' - t)$. The matched filter $\tilde{f}(t)$ is non-causal when the signal f is causal, suggesting unrealizability. This complication is easily accommodated when f has compact support, by introducing a fixed delay into the convolution. This fixed delay is imperceptible in applications of matched filtering in radar, sonar, and communication. The aim of this filter is to detect the presence of a delayed version of f in the signal g, and typically this is done by comparing the output of the matched filter, or the squared magnitude of this output, to a threshold.

Coherence. Let's normalize the output of the matched filter as follows:

$$\rho(t) = \frac{\langle g, D_t f\rangle}{\sqrt{\langle f, f\rangle\langle g, g\rangle}}.$$

We shall call this *complex coherence* and define *squared coherence* to be

$$\rho^2(t) = \frac{|\langle g, D_t f\rangle|^2}{\langle f, f\rangle\langle g, g\rangle}.$$

Recall that $\mathcal{F}$ is the unitary Fourier transform operator. The inner product $\langle g, D_t f\rangle$ may be written as $\langle \mathcal{F}g, \mathcal{F}D_t f\rangle$. Let $F = \mathcal{F}f$ denote the Fourier transform of f. Then, the Fourier transform of the signal $D_t f$ is the complex Fourier transform $e^{-j\omega t}F(\omega)$. The coherence-squared may be written as

$$\rho^2(t) = \frac{|\langle G, e^{-j\omega t}F\rangle|^2}{\langle F, F\rangle\langle G, G\rangle},$$

which is a frequency-domain implementation of squared coherence. Throughout this book, time-domain formulas may be replaced by frequency-domain formulas. When the signal space is $L_2(\mathbb{R})$, then the Fourier transform is the continuous-time Fourier transform; when the signal space is $\ell_2(\mathbb{Z})$, then the Fourier transform is the discrete-time Fourier transform; when the signal space is the circle S_1, then the Fourier transform is the Fourier series; and when the signal space is the cyclic group $\mathbb{Z}/N$, then the Fourier transform is the discrete Fourier transform (DFT).

We gain more insight by writing $\rho^2(t)$ as

$$\rho^2(t) = \frac{\langle g, P_{D_t f} g\rangle}{\langle g, g\rangle},$$

where $P_{D_t f}$ is an idempotent projection operator onto the subspace $\langle D_t f\rangle$, with action

$$(P_{D_t f} g)(t') = (D_t f)(t')\frac{1}{\langle f, f\rangle}\langle g, D_t f\rangle.$$

So squared coherence measures the cosine-squared of the angle that the measurement g makes with the subspace $\langle D_t f\rangle$. By the Cauchy-Schwarz inequality, $0 \leq \rho^2(t) \leq 1$.

Matched Subspace Detector. When the signal to be detected is known only to consist of a linear combination of modes, $h_1(t), \ldots, h_p(t)$, with unknown mode weights, then the subspace $\langle D_t f \rangle$ is replaced by the subspace spanned by these modes. Then, squared coherence is

$$\rho^2(t) = \frac{\langle g, D_t\mathbf{h}\rangle^H \mathbf{M}^{-1} \langle g, D_t\mathbf{h}\rangle}{\langle g, g\rangle},$$

where the vectors and matrices in this formula are defined as follows:

$$D_t\mathbf{h} = [D_t h_1 \ \cdots \ D_t h_p]^T : \text{ a column vector of time-delayed modes,}$$

$$\langle g, D_t\mathbf{h}\rangle = [\langle g, D_t h_1\rangle \ \cdots \ \langle g, D_t h_p\rangle]^T : \text{ a column vector of inner products,}$$

$$(\mathbf{M})_{ij} = \langle h_j, h_i\rangle : \ (i, j)\text{th element in a matrix of inner products between modes.}$$

Proceeding as before, we may write squared coherence as

$$\rho^2 = \frac{\langle g, P_{D_t\mathbf{h}} g\rangle}{\langle g, g\rangle},$$

where $P_{D_t\mathbf{h}}$ is an idempotent operator onto the subspace $\langle D_t\mathbf{h}\rangle$, with action

$$(P_{D_t\mathbf{h}} g)(t') = D_t\mathbf{h}^T \mathbf{M}^{-1} \langle g, D_t\mathbf{h}\rangle.$$

The squared coherence ρ^2 is called a scale-invariant matched subspace detector statistic, as it is invariant to scaling of g.

From Continuous-Time Inner Products to Euclidean Inner Products. In the chapters to follow, it will be a commonplace to replace these continuous-time inner products by Euclidean inner products for windowings and samplings of signals f and g. These produce finite-dimensional vectors $\mathbf{f}$ and $\mathbf{g}$. Then, an operator-theoretic formula for $\rho^2(t)$ is mapped to quadratic form in a projection matrix,

$$\rho^2 = \frac{\mathbf{g}^H \mathbf{P_H} \mathbf{g}}{\mathbf{g}^H \mathbf{g}},$$

where $\mathbf{P_H}$ is an idempotent projection matrix: $\mathbf{P_H} = \mathbf{H}(\mathbf{H}^H\mathbf{H})^{-1}\mathbf{H}^H$. The matrix $\mathbf{H} = [\mathbf{h}_1 \cdots \mathbf{h}_p]$ is an $n \times p$ matrix of n-dimensional modes $\mathbf{h}_i \in \mathbb{C}^n$, with $n \geq p$. All interpretations remain unchanged. In some cases, a problem comes as a finite-dimensional problem in Euclidean space, and there are no underlying continuous-time measurements to be windowed and sampled. In other cases, the burden of mapping a continuous-time problem to a finite-dimensional Euclidean problem is left to the engineer, mathematician, or scientist whose aim is to estimate parameters, detect signals, or classify effects. This is where the burden belongs, as modifications of a general result are always required for a specific application, and these modifications require application-specific expertise.

1.11 What Qualifies as a Coherence?

In this book, we are flexible in our use of the term *coherence* to describe a statistic. Certainly, normalized inner products in a vector space, such as Euclidean, Hilbert, etc., qualify. But so also do all of the correlations of multivariate statistical analysis such as standard correlation, partial correlation, the correlation of multivariate regression analysis, and canonical correlation. In many cases, these correlations *do* have an inner product interpretation. But in our lexicon, coherence also includes functions such as the ratio of geometric mean to arithmetic mean, the Hadamard ratio of product of eigenvalues to product of diagonal elements of a matrix, and various functions of these. Even when these statistics do not have an inner product interpretation, they have support [0, 1], and they are typically invariant to transformation groups of scale and rotation. In many cases, their null distribution is a beta distribution or a product of beta distributions. In fact, our point of view will be that any statistic that is supported on the interval [0, 1] and distributed as a product of independent beta random variables is a fortiori a coherence statistic.

So, as we shall see in this book, a great number of detector statistics may be interpreted as coherence statistics. In some cases, their null distributions are distributed as beta random variables or as the product of independent beta random variables. To each of these measures of coherence, we attach an invariance. For example, the squared coherence $\frac{|\mathrm{E}[uv^*]|^2}{\mathrm{E}[uu^*]\,\mathrm{E}[vv^*]}$ is invariant to non-zero complex scaling of u and v and to common unitary transformation of them. The ratio of geometric mean of eigenvalues of a matrix $\mathbf{S}$ to its arithmetic mean of eigenvalues is invariant to scale and unitary transformation, with group action $\beta\mathbf{QSQ}^H\beta^*$, and so on. For each coherence we examine, we shall endeavor to establish invariances and explain their significance to signal processing and machine learning.

By and large, the detectors and estimators of this book are derived for signals that are processed as they are measured and therefore treated as elements of Euclidean or Hilbert space. But it is certainly true that measurements may be first mapped to a reproducing kernel Hilbert space (RKHS), where inner products are computed through a positive definite kernel function. This is the fundamental idea behind kernel methods in machine learning. In this way, it might be said that many of the methods and results of the book may serve as signal processing or machine learning algorithms applied to nonlinearly mapped measurements.

1.12 Why Complex?

To begin, there are a great number of applications in signal processing and machine learning where there is no need for complex variables and complex signals. In these cases, every complex variable, vector, matrix, or signal of this book may be taken to be real. This means a complex conjugate x^* may be read as x; an Hermitian transpose $\mathbf{x}^H$ or $\mathbf{X}^H$ may be read as transpose $\mathbf{x}^T$ or $\mathbf{X}^T$; and $\{x^*[n]\}$ may be read as $\{x[n]\}$. When a random variable $z = x + jy$ is said to be a proper complex Gaussian random variable of unit variance, then its real and imaginary parts are independent with respective variances of $1/2$. If the imaginary part is zero, then the random variable is a real Gaussian random variable with variance $1/2$. This kind of reasoning leads to the conclusion that distribution statements for real random variables, and for real functions of real random variables, may be determined from distribution statements for complex variables or real functions of complex variables. Typically, the parameter value in a distribution for a complex variable is divided by two to obtain the distribution for a real variable. So many readers of this book may wish to simply ignore complex conjugates and read Hermitian transposes as if they were transposes. No offense is done to formulas, and only a bit of caution is required in the adjustment of distribution statements.

But, as signal processing, machine learning, and data science find their way into communication, imaging, autonomous vehicle control, radar and sonar, remote sensing, and related technologies, they will encounter signals that are low-frequency modulations of high-frequency carriers. Consequently, they will encounter complex signals as *one-channel complex representations of two real signals*. That is, $z(t) = x(t) + jy(t)$. In this representation, the real channel $x(t)$ is called the *real channel* (duh?), and the real channel $y(t)$ is called the *imaginary channel* (huh?). The situation is entirely analogous to the construction of complex numbers as $z = x + jy$, where x is the *real part* and y is the *imaginary part* of the complex number z. So let's get real: complex signals are *representations* of real signals. This complexification of two real signals into one complex signal may be inverted for the original real signals: $x(t) = \frac{1}{2}(z(t) + z^*(t))$ and $y(t) = \frac{1}{2j}(z(t) - z^*(t))$. In this way, correlations, matched filterings, and inner products between complex signals may be, and actually are, computed from these same operations on real signals. Correspondingly, correlations, matched filterings, and inner products of real signals may be represented as these same operations on complex signals.

There is no avoiding these artificially constructed complex signals, as they dramatically simplify the algebra of signal processing and machine learning. So throughout this book, nearly every result is developed in the context of complex signals or complex data. Complex noise is assumed to be *proper*, which assumes a special structure on the covariance and cross-covariance of the two real noise channels that compose the complex noise. For many problems, this assumption is justified, but it should not be assumed without thought. The reader is directed to the appendices on the multivariate normal distribution (Appendix D) and the complex multivariate normal distribution (Appendix E) for a clarification of these assumptions.

When the signals $x(t)$ and $y(t)$ are replaced by vectors $\mathbf{x}, \mathbf{y} \in \mathbb{R}^n$, then the vector $\mathbf{z} = \mathbf{x} + j\mathbf{y}$ is a vector in the space $\mathbb{R}^n \oplus \mathbb{R}^n$, denoted $\mathbb{C}^n$. This makes $\mathbb{C}^n$ homeomorphic to $\mathbb{R}^{2n}$. This is a fancy way of saying the complex vector $\mathbf{x} + j\mathbf{y}$ may be encoded as the vector $[\mathbf{x}^T \ \mathbf{y}^T]^T$. The complex vector $\mathbf{z} = \mathbf{x} + j\mathbf{y}$ may be expanded as $\mathbf{z} = \sum_{k=1}^{n} x_k \mathbf{e}_k + \sum_{k=1}^{n} y_k j \mathbf{e}_k$. That is, the Euclidean basis for $\mathbb{C}^n$ is $\{\mathbf{e}_1, \ldots, \mathbf{e}_n, j\mathbf{e}_1, \ldots, j\mathbf{e}_n\}$, where the $\mathbf{e}_k$ are the standard basis vectors in $\mathbb{R}^n$.

Every complex scalar, vector, or matrix is composed of real and imaginary parts: $z = x + jy$, $\mathbf{z} = \mathbf{x} + j\mathbf{y}$, $\mathbf{Z} = \mathbf{X} + j\mathbf{Y}$, where $x, y, \mathbf{x}, \mathbf{y}, \mathbf{X}$, and $\mathbf{Y}$ are real. Sometimes, they are constrained. If $|z|^2 = 1$, then $x^2 + y^2 = 1$; if $\mathbf{z}^H\mathbf{z} = 1$, then $\mathbf{x}^T\mathbf{x} + \mathbf{y}^T\mathbf{y} = 1$; if $\mathbf{Z}$ is square, and $\mathbf{Z}^H\mathbf{Z} = \mathbf{I}$, then $\mathbf{X}^T\mathbf{X} + \mathbf{Y}^T\mathbf{Y} = \mathbf{I}$; z is said to be unimodular, $\mathbf{z}$ is said to be unit-norm, and $\mathbf{Z}$ is said to be unitary. If $z^* = z$, then $y = 0$; if $\mathbf{z} = \mathbf{z}^*$, then $\mathbf{y} = \mathbf{0}$; if $\mathbf{Z}^H = \mathbf{Z}$, then $\mathbf{X}^T = \mathbf{X}$ and $\mathbf{Y}^T = -\mathbf{Y}$; $\mathbf{X}$ is said to be symmetric and $\mathbf{Y}$ is said to be skew-symmetric. A matrix $\mathbf{W} = \mathbf{Z}\mathbf{Z}^H$ is Hermitian, and it may be written as $\mathbf{W} = (\mathbf{X}\mathbf{X}^T + \mathbf{Y}\mathbf{Y}^T) + j(\mathbf{Y}\mathbf{X}^T - \mathbf{X}\mathbf{Y}^T)$. The real part is symmetric and the imaginary part is skew-symmetric. Linear transformations of the form $\mathbf{Mz}$ may be written as $\mathbf{Mz} = (\mathbf{A} + j\mathbf{B})(\mathbf{x} + j\mathbf{y}) = (\mathbf{Ax} - \mathbf{By}) + j(\mathbf{Bx} + \mathbf{Ay})$. The corresponding transformation in real variables is

$$\begin{bmatrix} \mathbf{A} & -\mathbf{B} \\ \mathbf{B} & \mathbf{A} \end{bmatrix} \begin{bmatrix} \mathbf{x} \\ \mathbf{y} \end{bmatrix}.$$

This is not the most general linear transformation in $\mathbb{R}^{2n}$ as the elements in the transforming matrix are constrained. The linear transformations in real coordinates and in complex coordinates are said to be strictly linear.

A quadratic form in a Hermitian matrix $\mathbf{H}$ is real: $\mathbf{z}^H\mathbf{H}\mathbf{z} = (\mathbf{x}^T - j\mathbf{y}^T)(\mathbf{A} + j\mathbf{B})(\mathbf{x} + j\mathbf{y}) = \mathbf{x}^T\mathbf{A}\mathbf{x} + \mathbf{y}^T\mathbf{A}\mathbf{y} + 2\mathbf{y}^T\mathbf{B}\mathbf{x}$.

Among the special Hermitian matrices are the complex projection matrices, denoted $\mathbf{P_V} = \mathbf{V}(\mathbf{V}^H\mathbf{V})^{-1}\mathbf{V}^H$, where $\mathbf{V}$ is a complex $n \times r$ matrix of rank $r < n$. That is, the Gramian $\mathbf{V}^H\mathbf{V}$ is a nonsingular $r \times r$ Hermitian matrix. The matrix $\mathbf{P_V}$ is idempotent, which is to say $\mathbf{P_V} = \mathbf{P_V}\mathbf{P_V}$. Write complex $\mathbf{P_V}$ as $\mathbf{P_V} = \mathbf{A} + j\mathbf{B}$, where $\mathbf{A}^T = \mathbf{A}$ and $\mathbf{B}^T = -\mathbf{B}$. Then, for the projection matrix to be idempotent, it follows that $\mathbf{A}\mathbf{A}^T + \mathbf{B}\mathbf{B}^T = \mathbf{A}$ and $\mathbf{A}\mathbf{B} - \mathbf{B}^T\mathbf{A}^T = \mathbf{B}$.

Second-Order Random Variables. These same ideas extend to second-order random variables. Each of the real random variables x, y, may be viewed as a vector in the Hilbert space of second-order random variables. The inner product between them is $E[xy]$. Write $z = x + jy$, and consider the Hermitian inner product $E[zz^*] = E[xx] + E[yy]$ and the *complementary* inner product $E[zz] = E[xx] - E[yy] + j2E[xy]$. In this way, the real second moments $E[xx], E[yy]$, and $E[xy]$ may be extracted from complex Hermitian and complementary inner products.

For real random vectors $\mathbf{x}$ and $\mathbf{y}$, inner products are organized into correlation matrices $E[\mathbf{x}\mathbf{x}^T]$, $E[\mathbf{y}\mathbf{y}^T]$, and $E[\mathbf{x}\mathbf{y}^T]$. Then, the Hermitian and complementary correlation matrices for the complex random vector $\mathbf{z} = \mathbf{x} + j\mathbf{y}$ are $E[\mathbf{z}\mathbf{z}^H] = E[\mathbf{x}\mathbf{x}^T] + E[\mathbf{y}\mathbf{y}^T] + j(E[\mathbf{x}\mathbf{y}^T] - E[\mathbf{y}\mathbf{x}^T])$ and $E[\mathbf{z}\mathbf{z}] = E[\mathbf{x}\mathbf{x}^T] - E[\mathbf{y}\mathbf{y}^T] + j(E[\mathbf{x}\mathbf{y}^T] + E[\mathbf{y}\mathbf{x}^T])$. Real correlations may be extracted from these two complex correlations. The complex random vector $\mathbf{z}$ is said to be *proper* if the complementary correlation is $\mathbf{0}$, which is to say $E[\mathbf{x}\mathbf{x}^T] = E[\mathbf{y}\mathbf{y}^T]$ and $E[\mathbf{x}\mathbf{y}^T] = -E[\mathbf{y}\mathbf{x}^T]$. Then, the Hermitian correlation is $E[\mathbf{z}\mathbf{z}^H] = 2E[\mathbf{x}\mathbf{x}^T] + j2E[\mathbf{x}\mathbf{y}^T]$. When the real vectors $\mathbf{x}$ and $\mathbf{y}$ are uncorrelated, then the Hermitian correlation is $E[\mathbf{z}\mathbf{z}^H] = 2E[\mathbf{x}\mathbf{x}^T]$. To say a complex random vector is proper and *white* is to say its complementary correlation is zero and its Hermitian correlation is $E[\mathbf{z}\mathbf{z}^H] = \mathbf{I}_n$. That is, $\mathbf{z} = \mathbf{x} + j\mathbf{y}$, where $E[\mathbf{x}\mathbf{x}^T] = (1/2)\mathbf{I}_n$, $E[\mathbf{y}\mathbf{y}^T] = (1/2)\mathbf{I}_n$, and $E[\mathbf{x}\mathbf{y}^T] = \mathbf{0}$.

1.13 What is the Role of Geometry?

There are four geometries encountered in this book: Euclidean geometry, the Hilbert space geometry of second-order random variables, and the Riemannian geometries of the Stiefel and Grassmann manifolds . Are these geometries real, which is to say fundamental to signal processing and machine learning? Or are they only constructs for remembering equations, organizing computations, and building insight? And, if only the latter, then aren't they real? This is a paraphrase of Richard Price in his introduction to relativity in the book, *The Future of Spacetime* [265]: "If the geometry (of relativity) is not real, then it is so useful that its very usefulness makes it real." He goes on to observe that Albert Einstein in his original development of special relativity presented the Lorentz transformation as the only reality, with no mention of a geometry. It was Hermann Minkowski who showed Einstein that the Lorentz transformation could be viewed as a feature of what is now called Minkowski geometry. In this geometry, Minkowski distance is an invariant to the Lorentz transformation. Price continues: "At first the Minkowski geometry seemed like an interesting construct, but quickly this construct became so useful that the idea that it was only a construct faded. Today, Einsteinian relativity is universally viewed as a description of a spacetime of events with the Minkowski spacetime geometry, and the Lorentz transformation is a sort of rotation in that spacetime geometry."

Is it reasonable to suggest that the geometries of Euclid, Hilbert, and Riemann are so useful in signal processing and machine learning that their very usefulness makes them real? We think so.

1.14 Motivating Problems

Generally, we are motivated in this book by problems in the analysis of time series, space series, or space-time series, natural or manufactured. Such problems arise when measurements are made:

- In sensor arrays for radar, sonar, or geophysical detection and localization;
- In multi-input-multi-output communication systems and networks;
- In acoustics arrays for detection and localization;
- In multi-spectral sensors for hyperspectral imaging;
- In accelerometer arrays for intrusion detection;
- In arrays of strain gauges for structural health monitoring;
- In multi-mode sensor arrays for human health monitoring;

- In networks of phasor measurement units (PMUs) in the smart grid;
- In networks of servers and terminals on the internet;
- In hidden layers of a deep neural network;
- As collections of neuro signals where the aim is to establish causal influence of one or more time series on another;
- As multidimensional measurements to be compressed or classified;
- As kernelized functions of raw measurements;
- As financial time series of economic indicators.

In some cases, these measurements produce multiple time series, each associated with a sensor such as an antenna element, hydrophone, accelerometer, etc. But, in some cases, these measurements arise when a single time series is decomposed into polyphase components, as in the analysis of periodically correlated time series. In all such cases, a finite record of measurements is produced and organized into a *space-time data matrix*. This language is natural to engineers and physical scientists. In the statistical sciences, the analysis of random phenomena from multiple realizations of an experiment may also be framed in this language. Space may be associated with a set of random variables, and time may be associated with the sequence of realizations for each of these random variables. The resulting data matrix may be interpreted as a space-time data matrix. Of course, this framework describes longitudinal analysis of treatment effects in people and crop science.

Figure 1.5 illustrates the kinds of problems that motivate our interest in coherence. On the LHS of the figure, each sensor produces a time series, and in aggregate, they produce a space-time data matrix. If each sensor is interpreted as a generator of realizations of a random variable, then in aggregate, these generators produce a space-time data matrix for an experiment in which each time series is a surrogate for its corresponding random variable.

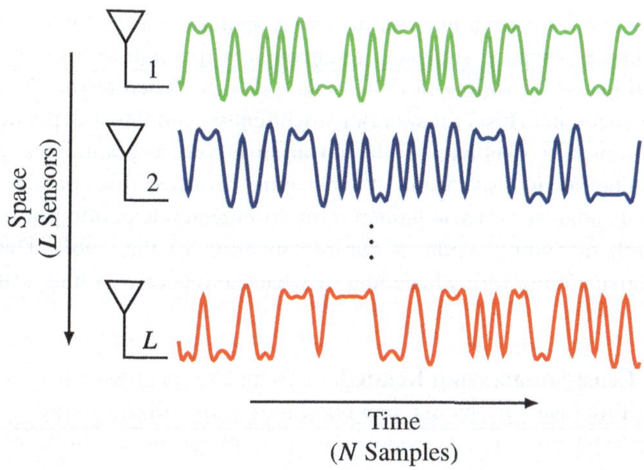

Fig. 1.5 Multi-sensor array of time series

The problem in all such applications is to answer questions about correlation or coherence between various components of the space-time series, using experimental realizations that are organized into a space-time data matrix. Sometimes, these questions are questions of estimation, and sometimes they are questions of detection or classification. But, in many cases we consider in this book, the estimator or detector may be interpreted as a function of a coherence statistic whose invariances reveal something fundamental about the application. Many of these statistics are invariant to linear transformations or unitary transformations in space and in time. In these cases, an $L \times N$ space-time matrix $\mathbf{X}$ may be replaced by the frequency-wavenumber matrix $\mathbf{F}_L\mathbf{X}\mathbf{F}_N$, where the matrix $\mathbf{F}_N \in \mathbb{C}^{N\times N}$ is the DFT matrix $(\mathbf{F}_N)_{kl} = e^{-j2\pi(k-1)(l-1)/N}$. Then $\mathbf{X}\mathbf{X}^H$ would be replaced by $\mathbf{F}_L\mathbf{X}\mathbf{F}_N\mathbf{F}_N^H\mathbf{X}^H\mathbf{F}_L^H = \mathbf{Z}\mathbf{Z}^H$ where $\mathbf{Z} = \mathbf{F}_L\mathbf{X}\mathbf{F}_N$ is a two-dimensional DFT of the matrix $\mathbf{X}$.

1.15 A Preview of the Book

From our point of view, one might say signal processing and machine learning begin with the fitting of a model according to a metric such as squared error, weighted squared error, absolute differences, likelihood, etc. Regularization may be used to constrain solutions for sparsity or some other favored structure for a feasible solution. This story may be refined by assigning means and covariances to model parameters and model errors, and this leads to methods of inference based on first- and second-order moments. A further refinement is achieved by assigning probability distributions to models and model errors. In the case of multivariate normal distributions, and related compound and elliptical distributions, to assign a probability distribution is to assign means and covariances to model parameters and to noises. Once first- and second-order moments are assigned, there arises the question of estimating these parameters from measurements, or detecting the presence of a signal, so modeled. This leads to a menagerie of results in multivariate statistical theory for estimating and detecting signals. For example, when testing for deterministic subspace signals, one encounters quadratic forms in projections that are used to form a variety of coherence statistics. When testing for covariance pattern, one encounters tests for sphericity, whiteness, and linear dependence. When testing for subspace structure in the covariance, one encounters a great many variations on factor analysis. Many of the original results in this book are extensions of multivariate analysis to two-channel or multi-channel detection problems.

This rough taxonomy explains our organization of the book. The following paragraphs give a more refined account of what the reader will find in the chapters to follow.

Chapter 2: Least Squares and Related. This chapter begins with a review of least squares and Procrustes problems, and continues with a discussion of least squares in the linear separable model, model order determination, and total least squares. A section on oblique projections addresses the problem of resolving a few modes in the presence of many. Sections on multidimensional scaling and the Johnson-Lindenstrauss lemma introduce two topics in ambient dimension reduction that are loosely related to least squares. There is an important distinction between model order reduction and ambient dimension reduction. In model order reduction, the dimension of the ambient measurement space is left unchanged, but the complexity of the model is reduced. In ambient dimension reduction, the dimension of the measurement space is reduced, under a constraint that distances or dissimilarities between high-dimensional measurements are preserved or approximated in a measurement space of lower dimension.

Chapter 3: Classical Correlations and Coherence. This chapter proceeds from a discussion of scalar-valued coherence to multiple coherence to matrix-valued coherence. Connections are established with principal angles and with canonical correlations. The study of factorizations of two-channel covariance matrices leads to filtering formulas for MMSE filters and their error covariances. When covariance matrices are estimated from measurements, then the filter and error covariance are random matrices. Their distribution is given. The multistage Wiener filter (MSWF), a conjugate gradient algorithm, is reviewed as a way to recursively update the order of the MMSE filter. Beamforming is offered as an illustration of these ideas. Half- and full-canonical coordinates are shown to be the correct bases for model order reduction. When three channels are admitted into the discussion, then the theory of partial coherence arises as a way to quantify the efficacy of a third channel when solving regression problems.

Chapter 4: Coherence and Classical Tests in the Multivariate Normal Model. In this chapter, we establish many basic results concerning inference and hypothesis testing in the proper, complex, multivariate normal distribution.[5] We consider in particular second-order measurement models in which the unknown covariance matrix belongs to a cone. This is often the case in signal processing and machine learning. Two important results concerning maximum likelihood (ML) estimators and likelihood ratios computed from ML estimators are reviewed. We then proceed to examine several classical hypothesis tests about the covariance matrix of measurements in multivariate normal (MVN) models. These are the *sphericity test* that tests whether or not the covariance matrix is a scaled identity matrix with unknown scale parameter; the *Hadamard test* that tests whether or not the variables in a MVN model are independent, thus having a diagonal covariance matrix with unknown diagonal elements; and the *homogeneity test* that tests whether or not the covariance matrices of independent vector-valued MVN models are equal. The chapter concludes with a discussion of the expected likelihood principle for cross-validating a covariance model.

[5] If measurements are real, read this as ... inference and hypothesis testing in the multivariate normal model.

Chapter 5: Matched Subspace Detectors. This chapter is devoted to the detection of signals that are constrained to lie in a subspace. The subspace may be known or known only by its dimension. The probability distribution for the measurements may carry the signal in a parameterization of the mean or in a parameterization of the covariance matrix. Likelihood ratio detectors are derived, their invariances are revealed, and their null distributions are derived where tractable. The result is a comprehensive account of matched subspace detectors in the complex multivariate normal model.

Chapter 6: Adaptive Subspace Detectors. This chapter opens with the estimate and plug (EP) adaptations of the detectors in Chap. 5. These solutions adapt matched subspace detectors to unknown noise covariance matrices by constructing covariance estimates from a secondary channel of signal-free measurements. Then the Kelly and ACE detectors, and their generalizations, are derived as generalized likelihood ratio detectors. These detectors use maximum likelihood estimates of the unknown noise covariance matrix, computed by fusing measurements from a primary channel and a secondary channel.

Chapter 7: Two-Channel Matched Subspace Detectors. This chapter considers the detection of a common subspace signal in two multi-sensor channels. This problem is usually referred to as passive detection. We study second-order detectors, where the unknown transmitted signal is modeled as a zero-mean Gaussian and averaged out or marginalized, and first-order detectors, where the unknown transmitted signal appears in the mean of the observations with no prior distribution assigned to it. The signal subspaces at the two sensor arrays may be known or unknown but with known dimension. In the first case, the resulting detectors are termed *matched subspace detectors*; in the second case, they are *matched direction detectors*. We study different noise models ranging from spatially white noises with identical variances to arbitrarily correlated Gaussian noises. For each noise and signal model, the invariances of the hypothesis testing problem and its GLR are established. Maximum likelihood estimation of unknown signal and noise parameters leads to a variety of coherence statistics.

Chapter 8: Detection of Spatially Correlated Time Series. This chapter extends the problem of null hypothesis testing for linear independence between random variables to the problem of testing for linear independence between times series. When the time series are approximated with finite-dimensional random vectors, then this is a problem of null hypothesis testing for block-structured covariance matrices. The test statistic is a coherence statistic. Its null distribution is the distribution of a double product of independent beta-distributed random variables. In the asymptotic case of wide-sense stationary time series, the coherence statistic may be written as a broadband coherence, with a new definition for broadband coherence. Additionally, this chapter addresses the problem of testing for block-structured covariance, when the block structure is patterned to model cyclostationarity. Spectral formulas establish the connection with the cyclic spectrum of a cyclostationary time series.

Chapter 9: Subspace Averaging and Its Applications. All distances between subspaces are functions of the principal angles between them and thus can ultimately be interpreted as measures of coherence between pairs of subspaces. In this chapter, we first review the geometry of the Grassmann and Stiefel manifolds, in which q-dimensional subspaces and q-dimensional frames live, respectively. Then, we assign probability distributions to these manifolds. We pay particular attention to the problem of subspace averaging using the projection (a.k.a. chordal) distance. Using this metric, the average of orthogonal projection matrices turns out to be the central quantity that determines, through its eigendecomposition, both the central subspace and its dimension. The dimension is determined by thresholding the eigenvalues of the average projection matrix, while the corresponding eigenvectors form a basis for the central subspace. We discuss applications of subspace averaging to subspace clustering and to source enumeration in array processing.

Chapter 10: Performance Bounds and Uncertainty Quantification. This chapter begins with the Hilbert space geometry of quadratic performance bounds and then specializes these results to the Euclidean geometry of the Cramér-Rao bound for parameters that are carried in the mean value or the covariance matrix of a MVN model. Coherence arises naturally. A concluding section on information geometry ties the Cramér-Rao bound on error covariance to the resolvability of the underlying probability distribution from which measurements are drawn.

Chapter 11: Variations on Coherence. In this chapter, we illustrate the use of coherence and its generalizations to other application domains, namely, compressed sensing, multiset CCA, kernel methods, and time-frequency modeling. The concept of coherence in compressed sensing and matrix completion is made clear by the restricted isometry property and the concept of coherence index, which are discussed in the chapter. We also consider in this chapter multi-view learning, in which the aim is to extract a low-dimensional latent subspace from a series of views of common information sources. The basic tool for fusing data from different sources is multiset canonical correlation analysis (MCCA). Coherence is a measure that can be extended to any reproducing kernel Hilbert space (RKHS). We present in the chapter two kernel methods in which coherence between pairs of nonlinearly transformed vectors plays a prominent role: the kernelized versions of CCA (KCCA) and the LMS adaptive filtering algorithm (KLMS). The chapter concludes with a discussion of a complex time-frequency distribution based on the coherence between a time series and its Fourier transform.

Chapter 12: Epilogue. Many of the results in this book have been derived from maximum likelihood reasoning in the multivariate normal model. This is not as constraining as it might appear, for likelihood in the MVN model actually leads to the optimization of functions that depend on sums and products of eigenvalues, which are themselves data dependent. Moreover, it is often the case that there is an illuminating Euclidean or Hilbert space geometry. Perhaps it is the geometry that is fundamental and not the distribution theory that produced it. This suggests that geometric reasoning, detached from distribution theory, may provide a way to address vexing problems in signal processing and machine learning, especially when there is no theoretical basis for assigning a distribution to data. This suggestion is developed in more detail in the concluding epilogue to the book.

Appendices. In the appendices, important results in matrix theory and multivariate normal theory are collected. These results support the body of the book and provide the reader with an organized account of two of the topics (the other being optimization theory) that form the mathematical and statistical foundations of signal processing and machine learning. The appendix on matrix theory contains several important results for optimizing functions of a matrix under constraints.

1.16 Chapter Notes

1. Sir Francis Galton first defined the correlation coefficient in a lecture to the Royal Institution in 1877. Generalizations of the correlation coefficient lie at the heart of multivariate statistics, and they figure prominently in linear regression. In signal processing and machine learning, linear regression includes more specialized topics such as normalized matched filtering, inversion, least squares and minimum mean-squared error filtering, multi-channel coherence analysis, and so on. Even in detection theory, linear regression plays a prominent role when detectors are to be adapted to unknown parameters.
2. Important applications of coherence began to appear in the signal processing literature in the 1970s and 1980s with the work of Carter and Nuttall on coherence for time delay estimation [64, 65, 249] and the work of Trueblood and Alspach on multi-channel coherence for passive sonar [345]. An interesting review of classical (two-channel) coherence may be found in Gardner's tutorial [127]. In recent years, the theory of multi-channel coherence has been significantly advanced by the work of Cochran, Gish, and Sinno [76, 77, 133] and Leshem and van der Veen [216]. The authors' own interests in coherence began to develop with their work on matched and adaptive subspace detectors [204, 205, 302, 303] and their work on multi-channel coherence [201,268,273,274]. This work, and its extensions, will figure prominently in Chaps. 5–8, where coherence is applied to problems of detection and estimation in time series, space series, and space-time series.
3. In the study of linear models and subspaces, the appropriate geometries are the geometries of the Stiefel and Grassmann manifolds. So the question of model identification or subspace identification becomes a question of finding distinguished points on these manifolds. Representative recent developments may be found in [8–10, 108, 114, 229].
4. Throughout this book, detectors and estimators are written as if measurements are recorded in time, space, or space-time. This is natural. But it is just as natural, and in some cases more intuitive, to replace these measurements by their Fourier transforms. One device for doing so is to define the N-point DFT matrix $\mathbf{F}_N$

and DFT a linear transformation $\mathbf{y} = \mathbf{A}\mathbf{x}$ as $\mathbf{F}_N\mathbf{y} = N^{-1}\mathbf{F}_N\mathbf{A}\mathbf{F}_N^H\mathbf{F}_N\mathbf{x}$. Then $\mathbf{F}_N\mathbf{x}$ and $\mathbf{F}_N\mathbf{y}$ are frequency-domain versions of $\mathbf{x}$ and $\mathbf{y}$, and $\mathbf{F}_N\mathbf{A}\mathbf{F}_N^H$ is a two-dimensional DFT of $\mathbf{A}$. A Hermitian quadratic form like $\mathbf{x}^H\mathbf{Q}\mathbf{x}$ may be written as $N^{-2}(\mathbf{F}_N\mathbf{x})^H\mathbf{F}_N\mathbf{Q}\mathbf{F}_N^H(\mathbf{F}_N\mathbf{x})$. Then if $\mathbf{Q}$ is Hermitian and circulant, $\mathbf{F}_N\mathbf{Q}\mathbf{F}_N^H$ is diagonal, and the quadratic form is a weighted sum of magnitude-squared Fourier coefficients.

5. The results in this chapter on interference effects, imaging, and filtering are classical. Our account aims to illuminate common ideas. As these classical results are readily accessible in textbooks, we have not given references. In many cases, effects are known by the names of those who discovered them. The subsection on matched subspace detectors (MSD) serves as a prelude to several chapters of the book where signals to be estimated or detected are modeled as elements of a subspace, known or known only by its dimension.

2 Least Squares and Related

This chapter is an introduction to many of the important methods for fitting a linear model to measurements. The standard problem of *inversion* in the linear model is the problem of estimating the signal or parameter $\mathbf{x}$ in the linear measurement model $\mathbf{y} = \mathbf{Hx} + \mathbf{n}$. The game is to manage the fitting error $\mathbf{n} = \mathbf{y} - \mathbf{Hx}$ by estimating $\mathbf{x}$, possibly under constraints on the estimate. In this model, the measurement $\mathbf{y} \in \mathbb{C}^L$ may be interpreted as complex measurements recorded at L sensor elements in a receive array, and $\mathbf{x} \in \mathbb{C}^p$ may be interpreted as complex transmissions from p sources or from p sensor elements in a transmit array. These may be called source symbols. The matrix $\mathbf{H} \in \mathbb{C}^{L\times p}$ may be interpreted as a channel matrix that conveys elements of $\mathbf{x}$ to elements of $\mathbf{y}$. An equivalently evocative narrative is that the model $\mathbf{Hx}$ for the signal component of the measurement may be interpreted as a *forward model* for the mapping of a source field $\mathbf{x}$ into a measured field $\mathbf{y}$.

But there are many other interpretations. The elements of $\mathbf{x}$ are *predictors*, and the elements of $\mathbf{y}$ are *response variables*; the vector $\mathbf{x}$ is an *input* to a multiple-input-multiple-output (MIMO) system whose *filter* is $\mathbf{H}$ and whose *output* is $\mathbf{y}$; the columns of $\mathbf{H}$ are *modes or dictionary elements* that are excited by the *initial conditions or mode parameters* $\mathbf{x}$ to produce the response $\mathbf{y}$; and so on.

To *estimate* $\mathbf{x}$ from $\mathbf{y}$ is to *regress* $\mathbf{x}$ onto $\mathbf{y}$ in the *linear model* $\mathbf{y} = \mathbf{Hx}$. To validate this model is to make a statement about how well $\mathbf{Hx}$ approximates $\mathbf{y}$ when $\mathbf{x}$ is given its regressed value. In some cases, the regressed value minimizes squared error; but, in other cases, it minimizes another measure of error, perhaps under constraints.

In one class of problems, the channel matrix is known, and the problem is to estimate the source $\mathbf{x}$ from the measurements $\mathbf{y}$. A typical objective is to minimize $(\mathbf{y}-\mathbf{Hx})^H(\mathbf{y}-\mathbf{Hx})$, which is the norm-squared of the residual, $\mathbf{n} \in \mathbb{C}^L$. This problem generalizes in a straightforward way to the problem of estimating a source matrix $\mathbf{X} \in \mathbb{C}^{p\times N}$ from measurements $\mathbf{Y} \in \mathbb{C}^{L\times N}$, when $\mathbf{H} \in \mathbb{C}^{L\times p}$ remains known and the measurement model is $\mathbf{Y} = \mathbf{HX} + \mathbf{N}$. Then a typical objective is to minimize $\mathrm{tr}[(\mathbf{Y} - \mathbf{HX})(\mathbf{Y} - \mathbf{HX})^H]$. The interpretation is that the matrix $\mathbf{X}$ is a matrix of N temporal transmissions, with the transmission at time n in the nth column of $\mathbf{X}$. The nth column of $\mathbf{Y}$ is then the measurement at time n.

D. Ramírez et al., *Coherence*, https://doi.org/10.1007/978-3-031-13331-2_2

When the dimension of the source $\mathbf{x}$ exceeds the dimension of the measurement $\mathbf{y}$, that is, $p > L$, then the problem is said to be under-determined, and there is an infinity of solutions that reproduce the measurements $\mathbf{y}$. Preferred solutions may only be extracted by constraining $\mathbf{x}$. Among the constrained solutions are those that minimize the energy of $\mathbf{x}$, or its entropy, or its spectrum, or its ℓ_0-norm. Methods based on (reweighted) ℓ_1 minimization promote *sparse* solutions that approximate minimum ℓ_0 solutions. Probabilistic constraints may be enforced by assigning a prior probability distribution to $\mathbf{x}$.

Another large class of linear fitting problems addresses the estimation of the unknown channel matrix $\mathbf{H}$. When the channel matrix is parametrized or constrained by a q-dimensional parameter $\boldsymbol{\theta}$, then the model $\mathbf{H}(\boldsymbol{\theta})\mathbf{x}$ is termed a *separable linear model*, and the problem is to estimate $\mathbf{x}$ and $\boldsymbol{\theta}$. This is commonly called a problem of modal analysis, as the columns of $\mathbf{H}(\boldsymbol{\theta})$ may be interpreted as *modes*. For example, the kth column of $\mathbf{H}$ might be a Vandermonde mode of the form $[1\ z_k^1\ \cdots\ z_k^{L-1}]^T$, with each of the complex mode parameters $z_k = \rho_k e^{j\theta_k}$ unknown and to be estimated. In a variation on this problem, it may be the case that there is no parametric model for $\mathbf{H}$. Then, the problem is to identify a channel that would synchronize simultaneous measurement of $\mathbf{Y}$ and $\mathbf{X}$ in the linear model $\mathbf{Y} = \mathbf{HX}+\mathbf{N}$, when $\mathbf{Y}$ is an $L \times N$ measurement matrix consisting of N measurements $\mathbf{y}_n$ and $\mathbf{X}$ is a $p \times N$ source matrix consisting of N source transmissions $\mathbf{x}_n$, measured simultaneously with $\mathbf{Y}$. This is a coherence idea. When there is an orthogonality constraint on $\mathbf{H}$, then this is a Procrustes problem.

All of these problems may be termed *inverse problems*, in the sense that measurements are inverted for underlying parameters that might have given rise to them. However, in common parlance, only the under-determined problem is called an inverse problem, to emphasize the difficulty of inverting a small set of measurements for a non-unique source that meets physical constraints or adheres to mathematical constructs.

This chapter addresses least squares estimation in a linear model. Over-determined and under-determined cases are considered. In the sections on over-determined least squares, we study weighted and constrained least squares, total least squares, dimension reduction, and cross-validation. A section on oblique projections addresses the problem of resolving a few modes in the presence of many and compares an estimator termed oblique least squares (OBLS) with ordinary least squares (LS) and with the best linear unbiased estimator (BLUE). In the sections on under-determined linear models, we study minimum-norm, maximum entropy, and sparsity-constrained solutions. The latter solutions are approximated by ℓ_1-regularized solutions that go by the name LASSO (for Least Absolute Shrinkage and Selection Operator) and by other solutions that use approximations to sparsity (or ℓ_0) constraints.

Sections on multidimensional scaling and the Johnson-Lindenstrauss lemma introduce two topics in ambient dimension reduction that are loosely related to least squares. There is an important distinction between *model order reduction* and *ambient dimension reduction*. In model order reduction, the dimension of the ambient measurement space is left unchanged, but the complexity of the model is reduced. In ambient dimension reduction, the dimension of the measurement space is reduced, under a constraint that distances or dissimilarities between high-dimensional measurements are preserved or approximated in a measurement space of lower dimension.

In the least squares story of this chapter, no probability distribution or moments are specified for the parameter $\mathbf{x}$. No probability distribution or moments are specified for the noise $\mathbf{n}$, *except when the performance of a least squares estimator is to be analyzed under the assumption that the noise is distributed as a multivariate normal (MVN) random vector*. That is, a least squares inversion is obtained without appeal to probability distributions for $\mathbf{x}$ and $\mathbf{n}$. Only after the inversion is complete is its performance evaluated for the case where the noise is MVN. This means some readers will wish to read sections of Appendix D on the MVN distribution to understand these evaluations, which are actually quite elementary.

2.1 The Linear Model

Throughout this chapter, we shall address the measurement model

$$\mathbf{y} = \mathbf{H}\mathbf{x} + \mathbf{n},$$

where $\mathbf{y} \in \mathbb{C}^L, \mathbf{H} \in \mathbb{C}^{L\times p}, \mathbf{x} \in \mathbb{C}^p$, and $\mathbf{n} \in \mathbb{C}^L$. It is assumed that the rank of the channel matrix is $\min(p, L)$, although the singular value decomposition (SVD) allows us to relax this assumption (see Appendix C). When $L > p$, the model is said to be over-determined, meaning there are more measurements than parameters; when $L = p$, the model is said to be determined; and when $L < p$, the model is said to be under-determined. In the under-determined case, the null space of $\mathbf{H}$ has non-zero dimension so that any proposed solution may be modified by adding a component in the null space without changing the value of $\mathbf{y} - \mathbf{H}\mathbf{x}$. Only constraints on the parameter $\mathbf{x}$ can discriminate between competing solutions.

Interpretations. Let's write the channel matrix $\mathbf{H}$ in two different ways:

$$\mathbf{H} = \begin{bmatrix}\mathbf{h}_1\ \mathbf{h}_2 \cdots \mathbf{h}_p\end{bmatrix} = \begin{bmatrix}\mathbf{c}_1^H \\ \mathbf{c}_2^H \\ \vdots \\ \mathbf{c}_L^H\end{bmatrix}.$$

Thus, the linearly modeled component $\mathbf{H}\mathbf{x}$ may be written as

$$\mathbf{H}\mathbf{x} = \sum_{k=1}^{p} \mathbf{h}_k x_k,$$

with components $[\mathbf{Hx}]_l = \mathbf{c}_l^H\mathbf{x}$. That is, $\mathbf{Hx}$ is a linear combination of response modes $\mathbf{h}_k$, each scaled by a signal component x_k, and the lth component of $\mathbf{Hx}$ is a resolution of the signal $\mathbf{x}$ onto the measurement mode $\mathbf{c}_l$ with the inner product $\mathbf{c}_l^H\mathbf{x}$. In Appendix C, it is shown that $\mathbf{H}$ may be factored as $\mathbf{H} = \mathbf{FKG}^H$. Hence, $\mathbf{y} = \mathbf{Hx}$ consists of resolutions $\mathbf{g}_i^H\mathbf{x}$ that are scaled by singular values k_i and then used to linearly combine response modes $\mathbf{f}_i$. In this way, one might say a physical forward model $\mathbf{H}$ is replaced by a mathematical forward model that illuminates the range space of $\mathbf{H}$ and its null space. These comments will be clarified in due course, but the reader may wish to refer ahead to Appendix C.

The Residual. In some applications, the residual $\mathbf{n}$ is simply a definition of the error $\mathbf{y}-\mathbf{Hx}$ that results from fitting the linear model $\mathbf{Hx}$ to measurements $\mathbf{y}$. In other applications, it is an identifiable source of *noise* in what otherwise would be an ideal measurement. Yet another interpretation is that the linear model $\mathbf{Hx}$ is modeling $\mathbf{y} - \mathbf{n}$, and not $\mathbf{y}$; so what is the sensitivity of the model to perturbations in $\mathbf{y}$? The principles of least squares and related methods do not generally exploit a model for $\mathbf{n}$ in the derivation of estimators for $\mathbf{x}$ or $\mathbf{Hx}$. This statement is relaxed in the study of the best linear unbiased estimator (BLUE), where the noise $\mathbf{n}$ is assumed to be zero mean with a known covariance matrix. However, once an estimator is derived, it is common practice to explore the performance of the estimator by assuming the noise is zero mean with a scaled identity for its covariance matrix. This is a way of understanding the sensitivity of a method to noise. Throughout this chapter, we follow this principle of performance analysis.

Sensitivity to Mismatch. A channel matrix $\mathbf{H}$ rarely models reality exactly. So even over-determined least squares solutions are vulnerable to model mismatch. This suggests a requirement for *cross-validation* of an estimator and perhaps a reduction in dimension of the estimator. For under-determined problems, where sparsity in an assumed basis is used to regularize a solution, this assumption introduces the problem of *basis mismatch*. These issues are addressed in this chapter.

Sparsity. Let's assume that the measurement model is under-determined, $L < p$. Perhaps it is hypothesized, or known a priori, that the parameters $\mathbf{x}$ are sparse in a unitary basis $\mathbf{V} \in \mathbb{C}^{p\times p}$, which is to say $\mathbf{VV}^H = \mathbf{I}_p$, and $\mathbf{V}^H\mathbf{x}$ is sparse. That is, at most $k < p$ of the coefficients in $\mathbf{V}^H\mathbf{x}$ are non-zero. For intuition, think of $\mathbf{x}$ as composed of just k DFT modes, with the index, or frequency, of these modes unknown and their complex mode weights unknown. Then the measurement model may be written as

$$\mathbf{y} = \mathbf{HVV}^H\mathbf{x} + \mathbf{n} = \mathbf{HVt} + \mathbf{n},$$

where $\mathbf{t} = \mathbf{V}^H\mathbf{x}$ is k-sparse. Now the measurement model is a sparse linear model, and this prior knowledge may be used to replace a least squares estimator for $\mathbf{x}$ by a regularized estimator of sparse $\mathbf{t}$, followed by the inverse map $\mathbf{x} = \mathbf{Vt}$.

Compression. Sometimes, an under-determined model $\mathbf{y} = \mathbf{Hx} + \mathbf{n}$ results from linearly compressing a determined or over-determined model $\mathbf{u} = \mathbf{Gx} + \mathbf{w}$ into an under-determined model. With $\mathbf{T}$ an $r \times L$ matrix, $r < L$, the under-determined model for $\mathbf{Tu}$ is

$$\mathbf{y} = \mathbf{Tu} = \mathbf{TGx} + \mathbf{n},$$

where $\mathbf{n} = \mathbf{Tw}$. Typically, the matrix $\mathbf{T}$ is a slice of a unitary matrix, which is to say $\mathbf{TT}^H = \mathbf{I}_r$. If $\mathbf{x}$ is known to be sparse in a basis $\mathbf{V}$, then the measurement model may be replaced by

$$\mathbf{y} = \mathbf{TGVV}^H\mathbf{x} + \mathbf{n} = \mathbf{TGVt} + \mathbf{n}.$$

With $\mathbf{T}$, $\mathbf{G}$, and $\mathbf{V}$ known, a regularized estimator of $\mathbf{t}$ may be extracted as suggested above, and $\mathbf{t} = \mathbf{V}^H\mathbf{x}$ may be inverted for $\mathbf{x}$ in the original model $\mathbf{u} = \mathbf{Gx} + \mathbf{w}$. Some insight is gained by considering the case where $\mathbf{V}$ is a DFT matrix, in which case the rows of $\mathbf{GV}$ are Fourier transforms of the original rows. The presumed sparse $\mathbf{t} = \mathbf{V}^H\mathbf{x}$ is then a vector of DFT coefficients, and the assumption is that just r of these p DFT coefficients are non-zero.

This is a sensitive sequence of steps and approximations: it is unlikely that the source is actually sparse in a known basis $\mathbf{V}$. That is, oscillating modes are rarely DFT modes, unless special boundary conditions force them to be; multipath components rarely lie at pre-selected delays; etc. So the assumption that the vector $\mathbf{x}$ is sparse in a known basis may introduce basis mismatch, which is to say the model $\mathbf{TGV}$ is not really the model excited by a sparse $\mathbf{t}$. The consequences of this mismatch are addressed more fully in the subsections on under-determined least squares, as these are the problems where principles of sparsity find their most important application.

2.2 Over-Determined Least Squares and Related

Consider the linear model for measurements $\mathbf{y} = \mathbf{Hx} + \mathbf{n}$. Call this the *prior* linear model for the measurements, and assume the known mode matrix $\mathbf{H}$ has full rank p, where $p \leq L$. By completing the square as in Appendix 2.A, the solution for $\mathbf{x}$ that minimizes the squared error $(\mathbf{y} - \mathbf{Hx})^H(\mathbf{y} - \mathbf{Hx})$ is found to be $\hat{\mathbf{x}} = (\mathbf{H}^H\mathbf{H})^{-1}\mathbf{H}^H\mathbf{y}$. Under the measurement model $\mathbf{y} = \mathbf{Hx} + \mathbf{n}$, this estimator decomposes as $\hat{\mathbf{x}} = \mathbf{x} + (\mathbf{H}^H\mathbf{H})^{-1}\mathbf{H}^H\mathbf{n}$. Assuming the noise $\mathbf{n}$ is zero mean with covariance matrix $E[\mathbf{nn}^H] = \sigma^2\mathbf{I}_L$, the covariance of the second term is $\sigma^2(\mathbf{H}^H\mathbf{H})^{-1}$. The estimator $\hat{\mathbf{x}}$ is said to be unbiased with error covariance $\sigma^2(\mathbf{H}^H\mathbf{H})^{-1}$. If the noise is further assumed to be MVN, then the estimator $\hat{\mathbf{x}}$ is distributed as $\hat{\mathbf{x}} \sim CN_p\left(\mathbf{x}, \sigma^2(\mathbf{H}^H\mathbf{H})^{-1}\right)$. When the model matrix is poorly conditioned, then the error covariance matrix will be poorly conditioned. The variance of the estimator is $\sigma^2 \operatorname{tr}[(\mathbf{H}^H\mathbf{H})^{-1}]$, demonstrating that small eigenvalues of the Gramian $\mathbf{H}^H\mathbf{H}$, generally corresponding to closely aligned columns of $\mathbf{H}$, contribute large components of variance.

The estimator $\mathbf{H}\hat{\mathbf{x}}$ decomposes as $\mathbf{H}\hat{\mathbf{x}} = \mathbf{H}\mathbf{x} + \mathbf{P}_{\mathbf{H}}\mathbf{n}$. This is an unbiased estimator of $\mathbf{H}\mathbf{x}$, with error covariance matrix $\sigma^2\mathbf{P}_{\mathbf{H}}$ and variance $\sigma^2\,\mathrm{tr}(\mathbf{P}_{\mathbf{H}}) = \sigma^2 p$, where $\mathbf{P}_{\mathbf{H}} = \mathbf{H}(\mathbf{H}^H\mathbf{H})^{-1}\mathbf{H}^H$ is the rank-p projector onto the p-dimensional subspace $\langle\mathbf{H}\rangle$. A plausible definition of signal-to-noise ratio is squared mean divided by variance, $\mathrm{SNR} = \mathbf{x}^H\mathbf{H}^H\mathbf{H}\mathbf{x}/\sigma^2 p$. This may be written as $\mathrm{SNR} = \frac{L}{p}\mathrm{snr}$, where

$$\mathrm{snr} = \frac{\mathbf{x}^H\mathbf{H}^H\mathbf{H}\mathbf{x}/L}{\sigma^2}.$$

In this form, the SNR may be viewed as the product of processing gain L/p times per-sample, or input, signal-to-noise ratio snr. Why is this called per-sample signal-to-noise ratio? Because the numerator of snr is the average of squared mean, averaged over the L components of $\mathbf{H}\mathbf{x}$, and σ^2 is the variance of each component of noise.

When we have occasion to compare this least squares estimator with competitors, we shall call it the ordinary least squares (LS) estimator and sometimes denote it $\hat{\mathbf{x}}_{LS}$.

The *posterior* model for measurements is

$$\mathbf{y} = \mathbf{H}\hat{\mathbf{x}} + \hat{\mathbf{n}} = \mathbf{P}_{\mathbf{H}}\mathbf{y} + (\mathbf{I}_L - \mathbf{P}_{\mathbf{H}})\mathbf{y}.$$

The geometry is this: the measurement $\mathbf{y}$ is projected onto the subspace $\langle\mathbf{H}\rangle$ for the estimator $\mathbf{H}\hat{\mathbf{x}}$. The error $\hat{\mathbf{n}}$ is orthogonal to this estimator, and together, they provide a Pythagorean decomposition of the norm-squared of the measurement: $\mathbf{y}^H\mathbf{y} = \mathbf{y}^H\mathbf{P}_{\mathbf{H}}\mathbf{y} + \mathbf{y}^H(\mathbf{I}_L - \mathbf{P}_{\mathbf{H}})\mathbf{y}$. We might say this is a *law of total power*, wherein the power in $\mathbf{y}$ is the sum of power in the estimator $\mathbf{P}_{\mathbf{H}}\mathbf{y}$ plus the power in the residual $(\mathbf{I} - \mathbf{P}_{\mathbf{H}})\mathbf{y}$. The term power is an evocative way to talk about a norm-squared like $\|\mathbf{P}_{\mathbf{H}}\mathbf{y}\|^2 = (\mathbf{P}_{\mathbf{H}}\mathbf{y})^H(\mathbf{P}_{\mathbf{H}}\mathbf{y}) = \mathbf{y}^H\mathbf{P}_{\mathbf{H}}\mathbf{y}$.

Multi-Experiment Least Squares. These results extend to the measurement model $\mathbf{Y} = \mathbf{H}\mathbf{X}$. If the channel model $\mathbf{H}$ is known, and there are no constraints on $\mathbf{X}$, then the least squares estimate of $\mathbf{X}$ is $\hat{\mathbf{X}} = (\mathbf{H}^H\mathbf{H})^{-1}\mathbf{H}^H\mathbf{Y}$, as shown Appendix 2.A. If $\mathbf{X}$ is constrained, then these constraints change the solution. For estimation in under-determined models, there are a variety of constraints one may place on the signal matrix $\mathbf{X}$ depending upon how the rows and columns are to be constrained.

Orthogonality and Cholesky Factors of a Gramian. The measurement $\mathbf{y}$ is resolved into the components, $\mathbf{y} = \mathbf{P}_{\mathbf{H}}\mathbf{y} + (\mathbf{I}_L - \mathbf{P}_{\mathbf{H}})\mathbf{y} = \hat{\mathbf{y}} + \hat{\mathbf{n}}$, where the first term may be called the estimate of $\mathbf{y}$ and the second term may be called the estimate of the noise, or the residual. These two components are orthogonal, and in fact, the estimated noise is orthogonal to the subspace $\langle\mathbf{H}\rangle$ or equivalently to every column of $\mathbf{H}$. Suppose we had asked only for orthogonality between $\mathbf{H}$ and the residual $\mathbf{y}-\mathbf{Hx}$. Then we could have concatenated the vector of measurements $\mathbf{y}$ and the channel matrix $\mathbf{H}$ to construct the matrix

$$[\mathbf{y}\ \mathbf{H}]\begin{bmatrix} 1 & \mathbf{0} \\ -\mathbf{x} & \mathbf{I}_L \end{bmatrix} = [\mathbf{n}\ \mathbf{H}]$$

and insisted that the Gramian of the matrix $[\mathbf{n}\ \mathbf{H}]$ be diagonal. That is,

$$\begin{bmatrix} \mathbf{n}^H\mathbf{n} & \mathbf{0} \\ \mathbf{0} & \mathbf{H}^H\mathbf{H} \end{bmatrix} = \begin{bmatrix} 1 & -\mathbf{x}^H \\ \mathbf{0} & \mathbf{I}_L \end{bmatrix} \begin{bmatrix} \mathbf{y}^H\mathbf{y} & \mathbf{y}^H\mathbf{H} \\ \mathbf{H}^H\mathbf{y} & \mathbf{H}^H\mathbf{H} \end{bmatrix} \begin{bmatrix} 1 & \mathbf{0} \\ -\mathbf{x} & \mathbf{I}_L \end{bmatrix}.$$

Write out the southwest element of the RHS to see that $\hat{\mathbf{x}} = (\mathbf{H}^H\mathbf{H})^{-1}\mathbf{H}^H\mathbf{y}$, and evaluate the northwest term to see that $\hat{\mathbf{n}}^H\hat{\mathbf{n}} = \mathbf{y}^H(\mathbf{I}_L - \mathbf{P_H})\mathbf{y}$. That is, the least squares solution for $\mathbf{x}$ satisfies the desired condition of orthogonality, and moreover it delivers an LDU, or Cholesky, factorization of the Gramian of $[\mathbf{y}\ \mathbf{H}]$:

$$\begin{bmatrix} \mathbf{y}^H\mathbf{y} & \mathbf{y}^H\mathbf{H} \\ \mathbf{H}^H\mathbf{y} & \mathbf{H}^H\mathbf{H} \end{bmatrix} = \begin{bmatrix} 1 & \hat{\mathbf{x}}^H \\ \mathbf{0} & \mathbf{I}_L \end{bmatrix} \begin{bmatrix} \hat{\mathbf{n}}^H\hat{\mathbf{n}} & \mathbf{0} \\ \mathbf{0} & \mathbf{H}^H\mathbf{H} \end{bmatrix} \begin{bmatrix} 1 & \mathbf{0} \\ \hat{\mathbf{x}} & \mathbf{I}_L \end{bmatrix}.$$

This is easily inverted for the LDU factorization of the inverse of this Gramian, which shows that the northwest element of the inverse of this Gramian is the inverse of the residual squared error, namely, $1/\hat{\mathbf{n}}^H\hat{\mathbf{n}}$.

More Interpretation. There is actually a little more that can be said. Rewrite the measurement model as $\mathbf{y}-\mathbf{n}-\mathbf{Hx}=\mathbf{0}$ or

$$([\mathbf{y}\ \mathbf{H}] + [-\mathbf{n}\ \mathbf{0}])\begin{bmatrix} 1 \\ -\mathbf{x} \end{bmatrix} = \mathbf{0}.$$

Evidently, without modifying the mode matrix $\mathbf{H}$, the problem is to make the minimum-norm adjustment $\begin{bmatrix}-\mathbf{n} & \mathbf{0}\end{bmatrix}$ to the matrix $\begin{bmatrix}\mathbf{y} & \mathbf{H}\end{bmatrix}$ that reduces its rank by one and forces the vector $\begin{bmatrix}1 & -\mathbf{x}^T\end{bmatrix}^T$ into the null space of the matrix $\begin{bmatrix}\mathbf{y}-\mathbf{n} & \mathbf{H}\end{bmatrix}$. Choose $\hat{\mathbf{n}} = (\mathbf{I}_L - \mathbf{P_H})\mathbf{y}$, in which case $\mathbf{y}-\hat{\mathbf{n}} = \mathbf{P_H}\mathbf{y}$ and the matrix $\begin{bmatrix}\mathbf{y}-\hat{\mathbf{n}} & \mathbf{H}\end{bmatrix}$ is $\begin{bmatrix}\mathbf{P_H}\mathbf{y} & \mathbf{H}\end{bmatrix}$. Clearly, $\mathbf{P_H}\mathbf{y}$ lies in the span of $\mathbf{H}$, making this matrix rank-deficient by one. Moreover, the estimator $\hat{\mathbf{x}} = (\mathbf{H}^H\mathbf{H})^{-1}\mathbf{H}^H\mathbf{y}$ places the vector $\begin{bmatrix}1 & -\hat{\mathbf{x}}^T\end{bmatrix}^T$ in the null space of $\begin{bmatrix}\mathbf{P_H}\mathbf{y} & \mathbf{H}\end{bmatrix}$. This insight will prove useful when we allow adjustments to the mode matrix $\mathbf{H}$ in our discussion of total least squares in Sect. 2.2.9.

Estimating a Fixed Signal from a Sequence of Measurements. This treatment of least squares applies in a straightforward way to the least squares estimation of a fixed signal $\mathbf{x}$ in a sequence of measurements $\mathbf{y}_n = \mathbf{Hx} + \mathbf{n}_n$, $n = 1, \ldots, N$, where only the noise $\mathbf{n}_n$ changes from measurement to measurement. Then the model may be written as $\mathbf{Y} = \mathbf{H}\mathbf{x}\mathbf{1}^T + \mathbf{N}$, where $\mathbf{Y} = [\mathbf{y}_1 \ \cdots \ \mathbf{y}_N]$, $\mathbf{N} = [\mathbf{n}_1 \ \cdots \ \mathbf{n}_N]$, and $\mathbf{1} = [1 \ \cdots \ 1]^T$. The problem is to minimize $\mathrm{tr}(\mathbf{N}\mathbf{N}^H)$, which is the sum of squared residuals, $\sum_{n=1}^{N} \mathbf{n}_n^H \mathbf{n}_n$. The least squares estimators of $\mathbf{x}$, $\mathbf{H}\mathbf{x}$, and $\mathbf{H}\mathbf{x}\mathbf{1}^T$ are then

$$\hat{\mathbf{x}} = (\mathbf{H}^H\mathbf{H})^{-1}\mathbf{H}^H\mathbf{Y}\mathbf{1}(\mathbf{1}^T\mathbf{1})^{-1} = (\mathbf{H}^H\mathbf{H})^{-1}\mathbf{H}^H \left(\frac{1}{N}\sum_{n=1}^{N} \mathbf{y}_n\right),$$

$$\mathbf{H}\hat{\mathbf{x}} = \mathbf{P}_\mathbf{H}\mathbf{Y}\mathbf{1}(\mathbf{1}^T\mathbf{1})^{-1} = \mathbf{P}_\mathbf{H}\left(\frac{1}{N}\sum_{n=1}^{N} \mathbf{y}_n\right),$$

and $\mathbf{H}\hat{\mathbf{x}}\mathbf{1}^T = \mathbf{P}_\mathbf{H}\mathbf{Y}\mathbf{P}_\mathbf{1}$, where $\mathbf{P}_\mathbf{1} = \mathbf{1}(\mathbf{1}^T\mathbf{1})^{-1}\mathbf{1}^T$. The interpretations are these: the columns of $\mathbf{Y}$ are averaged for an average measurement, which is then used to estimate $\mathbf{x}$ in the usual way; the estimate of $\mathbf{H}\mathbf{x}$ is the projection of the average measurement onto the subspace $\langle\mathbf{H}\rangle$; this estimate is replicated in time for the estimate of $\mathbf{H}\mathbf{x}\mathbf{1}^T$, which may be written as $\mathbf{P}_\mathbf{H}\mathbf{Y}\mathbf{P}_\mathbf{1}$. Or say it this way: squeeze the space-time matrix $\mathbf{Y}$ between pseudo-inverses of $\mathbf{H}$ and $\mathbf{1}$ for an estimator of $\mathbf{x}$; squeeze it between the projection $\mathbf{P}_\mathbf{H}$ and the pseudo-inverse of $\mathbf{1}$ for an estimator of $\mathbf{H}\mathbf{x}$; squeeze it between the spatial projector $\mathbf{P}_\mathbf{H}$ and the temporal projector $\mathbf{P}_\mathbf{1}$ for an estimator of $\mathbf{H}\mathbf{x}\mathbf{1}^T$. It is a simple matter to replace the vector $\mathbf{1}^T$ by a vector of known complex amplitudes $\mathbf{r}^H$, in which case the estimate of $\mathbf{H}\mathbf{x}\mathbf{r}^H$ is $\mathbf{H}\hat{\mathbf{x}}\mathbf{r}^H = \mathbf{P}_\mathbf{H}\mathbf{Y}\mathbf{P}_\mathbf{r}$, where the definition of $\mathbf{P}_\mathbf{r}$ is obvious. It is easy to see that $\hat{\mathbf{x}}$ is an unbiased estimator of $\mathbf{x}$. If the noises are a sequence of uncorrelated noises, then the error covariance of $\hat{\mathbf{x}}$ is $(\sigma^2/N)(\mathbf{H}^H\mathbf{H})^{-1}$. As expected, N independent copies of the same experiment reduce variance by a factor of N.

2.2.1 Linear Prediction

Suppose the model matrix $\mathbf{H} = \mathbf{Y}_{p-1} = [\mathbf{y}_1 \ \mathbf{y}_2 \ \cdots \ \mathbf{y}_{p-1}]$ is composed of a sequence of $p - 1 \leq L - 1$ measurements, and we wish to solve for the *predictor* $\mathbf{x}$ that would have best linearly predicted the next measurement $\mathbf{y}_p$ from these past measurements. Construct the data matrix $[\mathbf{y}_p \ \mathbf{Y}_{p-1}]$, and follow the prescription of the paragraph on "Orthogonality and Cholesky factors of a Gramian":

$$\begin{bmatrix} \mathbf{n}^H\mathbf{n} & \mathbf{0} \\ \mathbf{0} & \mathbf{H}^H\mathbf{H} \end{bmatrix} = \begin{bmatrix} 1 & -\mathbf{x}^H \\ \mathbf{0} & \mathbf{I}_L \end{bmatrix} \begin{bmatrix} \mathbf{y}_p^H\mathbf{y}_p & \mathbf{y}_p^H\mathbf{Y}_{p-1} \\ \mathbf{Y}_{p-1}^H\mathbf{y}_p & \mathbf{Y}_{p-1}^H\mathbf{Y}_{p-1} \end{bmatrix} \begin{bmatrix} 1 & \mathbf{0} \\ -\mathbf{x} & \mathbf{I}_L \end{bmatrix}.$$

The solution for $\hat{\mathbf{x}}$ is then $\hat{\mathbf{x}} = (\mathbf{Y}_{p-1}^H\mathbf{Y}_{p-1})^{-1}\mathbf{Y}_{p-1}^H\mathbf{y}_p$, and the solution for $\mathbf{Y}_{p-1}\hat{\mathbf{x}}$ is $\mathbf{P}_{\mathbf{Y}_{p-1}}\mathbf{y}_p$. In this case, $\hat{\mathbf{x}}$ is said to contain the coefficients of a *prediction filter*, and $\begin{bmatrix}1 \ \hat{\mathbf{x}}^T\end{bmatrix}^T$ is said to contain the coefficients of a *prediction error filter*. This language seems contrived, but the idea is that an experiment is run, or a set of experiments is run, to design the predictor $\hat{\mathbf{x}}$, which is then used unchanged on future measurements. There are a great many variations on this basis idea, among them the over-determined, reduced-rank solutions advocated by Tufts and Kumaresan in a series of important papers [208, 347, 348].

2.2.2 Order Determination

Given N measurements of the form $\mathbf{y}_n = \mathbf{H}\mathbf{x} + \mathbf{n}_n$, $n = 1, \ldots, N$, the least squares estimators of $\mathbf{x}$ and $\mathbf{H}\mathbf{x}$ are $\hat{\mathbf{x}} = (\mathbf{H}^H\mathbf{H})^{-1}\mathbf{H}^H \left(\frac{1}{N}\sum_{n=1}^{N}\mathbf{y}_n\right)$ and $\mathbf{H}\hat{\mathbf{x}} = \mathbf{P}_{\mathbf{H}}\left(\frac{1}{N}\sum_{n=1}^{N}\mathbf{y}_n\right)$. The latter may be resolved as

$$\mathbf{H}\hat{\mathbf{x}} = \mathbf{H}\mathbf{x} + \mathbf{P}_{\mathbf{H}}\left(\frac{1}{N}\sum_{n=1}^{N}\mathbf{n}_n\right).$$

Then, assuming the sequence of noises is a sequence of zero-mean uncorrelated random vectors, with common covariances $\sigma^2\mathbf{I}_L$, their average is zero mean with covariance $\frac{\sigma^2}{N}\mathbf{I}_L$. We say the estimator $\mathbf{H}\hat{\mathbf{x}}$ is an unbiased estimator of $\mathbf{H}\mathbf{x}$, with error covariance matrix $\frac{\sigma^2}{N}\mathbf{P}_{\mathbf{H}}$. The mean-squared error of this estimator is $\text{MSE}_p = \text{tr}\left(\frac{\sigma^2}{N}\mathbf{P}_{\mathbf{H}}\right) = \frac{\sigma^2 p}{N}$, and it decomposes as zero bias-squared plus variance. Perhaps there is an opportunity to introduce a small amount of bias-squared in exchange for a large savings in variance. This suggestion motivates model order reduction.

If the signal model $\mathbf{H}\mathbf{x}$ is replaced by a lower-dimensional approximating model $\mathbf{H}_r\mathbf{x}_r$, then the least squares estimator of $\mathbf{H}_r\mathbf{x}_r$ is

$$\mathbf{H}_r\hat{\mathbf{x}}_r = \mathbf{P}_{\mathbf{H}_r}\left(\frac{1}{N}\sum_{n=1}^{N}\mathbf{y}_n\right),$$

which may be resolved as

$$\mathbf{H}_r\hat{\mathbf{x}}_r = \mathbf{P}_{\mathbf{H}_r}\mathbf{H}\mathbf{x} + \mathbf{P}_{\mathbf{H}_r}\left(\frac{1}{N}\sum_{n=1}^{N}\mathbf{n}_n\right).$$

This is a biased estimator of $\mathbf{H}\mathbf{x}$, with rank-r covariance matrix $\frac{\sigma^2}{N}\mathbf{P}_{\mathbf{H}_r}$. The bias is $\mathbf{b}_r = (\mathbf{P}_{\mathbf{H}} - \mathbf{P}_{\mathbf{H}_r})\mathbf{H}\mathbf{x}$, and the mean-squared error is $\text{MSE}_r = \mathbf{b}_r^H\mathbf{b}_r + \frac{\sigma^2 r}{N}$. Evidently, variance has been reduced at the cost of bias-squared. But the bias-squared is unknown because the signal $\mathbf{x}$ is unknown. Perhaps it can be estimated.

Consider this estimator of the bias:

$$\hat{\mathbf{b}}_r = (\mathbf{P}_{\mathbf{H}} - \mathbf{P}_{\mathbf{H}_r})\frac{1}{N}\sum_{n=1}^{N}\mathbf{y}_n = \mathbf{b}_r + (\mathbf{P}_{\mathbf{H}} - \mathbf{P}_{\mathbf{H}_r})\frac{1}{N}\sum_{n=1}^{N}\mathbf{n}_n.$$

The estimator $\hat{\mathbf{b}}_r$ is an unbiased estimator of $\mathbf{b}_r$. Under the assumption that the projector $\mathbf{P}_{\mathbf{H}_r}$ is a projector onto a subspace of the subspace $\langle\mathbf{H}\rangle$, which is to say $\mathbf{P}_{\mathbf{H}}\mathbf{P}_{\mathbf{H}_r} = \mathbf{P}_{\mathbf{H}_r}$, then the covariance matrix of $\hat{\mathbf{b}}_r - \mathbf{b}_r$ is $\frac{\sigma^2}{N}(\mathbf{P}_{\mathbf{H}} - \mathbf{P}_{\mathbf{H}_r})$, and the variance of this unbiased estimator of bias is $\frac{\sigma^2}{N}(p-r)$. But in the expression for the mean-squared error MSE_r, it is $\mathbf{b}_r^H\mathbf{b}_r$ that is unknown. So we note

$$\text{E}[(\hat{\mathbf{b}}_r - \mathbf{b}_r)^H(\hat{\mathbf{b}}_r - \mathbf{b}_r)] = \text{E}[\hat{\mathbf{b}}_r^H\hat{\mathbf{b}}_r] - \mathbf{b}_r^H\mathbf{b}_r = \frac{\sigma^2}{N}(p-r).$$

It follows that $\hat{\mathbf{b}}_r^H\hat{\mathbf{b}}_r$ is a biased estimator of $\mathbf{b}_r^H\mathbf{b}_r$, with bias $\frac{\sigma^2}{N}(p-r)$.

Now, an unbiased estimator of MSE_r is obtained by replacing unknown $\mathbf{b}_r^H\mathbf{b}_r$ with its unbiased estimator:

$$\widehat{\text{MSE}}_r = \left(\hat{\mathbf{b}}_r^H\hat{\mathbf{b}}_r - \frac{\sigma^2}{N}(p-r)\right) + \frac{\sigma^2}{N}r = \hat{\mathbf{b}}_r^H\hat{\mathbf{b}}_r + \frac{\sigma^2}{N}(2r-p).$$

The order fitting rule is then to choose $\hat{\mathbf{b}}_r$ and r that minimize this estimator of mean-squared error. The penalty for large values of r comes from reasoning about unbiasedness.

Define $\mathbf{P}_{\mathbf{H}} = \mathbf{V}\mathbf{V}^H$, where $\mathbf{V} \in \mathbb{C}^{L\times p}$ is a slice of an $L \times L$ unitary matrix. Call $\frac{1}{N}\sum_{n=1}^{N}\mathbf{y}_n$ the average $\overline{\mathbf{y}}$, and order the columns of $\mathbf{V}$ according to their resolutions of $\overline{\mathbf{y}}$ onto the basis $\mathbf{V}$ as $|\mathbf{v}_1^H\overline{\mathbf{y}}|^2 > |\mathbf{v}_2^H\overline{\mathbf{y}}|^2 > \cdots > |\mathbf{v}_p^H\overline{\mathbf{y}}|^2$. Then, $\widehat{\text{MSE}}_r$ may be written as

$$\widehat{\text{MSE}}_r = \sum_{i=r+1}^{p} |\mathbf{v}_i^H\overline{\mathbf{y}}|^2 + \frac{\sigma^2}{N}(2r-p), \quad r = 0, 1, \ldots, p.$$

The winning value of r is the value that produces the minimum value of $\widehat{\text{MSE}}_r$, and this value determines $\mathbf{P}_{\mathbf{H}_r}$ to be $\mathbf{P}_{\mathbf{H}_r} = [\mathbf{v}_1 \ \cdots \ \mathbf{v}_r][\mathbf{v}_1 \ \cdots \ \mathbf{v}_r]^H$. We may say the model $\mathbf{H}$ has been replaced by the lower-dimensional model $\mathbf{P}_{\mathbf{H}_r}\mathbf{H} = [\mathbf{v}_1 \ \cdots \ \mathbf{v}_r]\mathbf{Q}_r$, where $\mathbf{Q}_r$ is an arbitrary $r \times r$ unitary matrix. Beginning at $r = p$, where $\widehat{\text{MSE}}_p = \frac{\sigma^2 p}{N}$, the rank is decreased from r to $r-1$ iff the term $|\mathbf{v}_r^H\overline{\mathbf{y}}|^2 < 2\sigma^2/N$: in other words, iff the increase in bias is smaller than the savings in variance due to the exclusion of one more dimension in the estimator of $\mathbf{Hx}$.

The formula for $\widehat{\text{MSE}}_r$ is a regularization of the discarded powers $|\mathbf{v}_i^H\overline{\mathbf{y}}|^2$ by a term that depends on $2r - p$, scaled by the variance σ^2/N. If the variance is unknown, then the regularization term serves as a Lagrange constant that depends on the order r. For each assumed value of σ^2, an optimum value of r is returned. For large values of σ^2, ranks near to 0 are promoted, whereas for small values of σ^2, ranks near to p are permitted.

These results specialize to the case where only one measurement has been made. When $N = 1$, then $\overline{\mathbf{y}} = \mathbf{y}$, and the key formula is[1]

$$\widehat{\mathrm{MSE}}_r = \sum_{i=r+1}^{p} |\mathbf{v}_i^H \mathbf{y}|^2 + \sigma^2(2r - p).$$

2.2.3 Cross-Validation

The idea behind cross-validation is to test the residual sum of squared errors, $Q = \mathbf{y}^H(\mathbf{I} - \mathbf{P_H})\mathbf{y}$, against what would have been expected had the measurements actually been drawn from the linear model $\mathbf{y} = \mathbf{Hx} + \mathbf{n}$, with $\mathbf{n}$ distributed as the MVN random vector $\mathbf{n} \sim C\mathcal{N}_L\left(\mathbf{0}, \sigma^2\mathbf{I}_L\right)$. In this case, from Cochran's theorem of Appendix F, $(2/\sigma^2)Q$ should be distributed as a chi-squared random variable with $2(L-p)$ degrees of freedom. Therefore, we may test the null hypothesis that the measurement was drawn from the distribution $\mathbf{y} \sim C\mathcal{N}_L\left(\mathbf{Hx}, \sigma^2\mathbf{I}_L\right)$ by comparing Q to a threshold η. Cross-validation fails, which is to say the model is rejected, if Q exceeds the threshold η. The probability of falsely rejecting the model is then the probability that the random variable $(2/\sigma^2)Q \sim \chi^2_{2(L-p)}$ exceeds the threshold η. We say the model is validated at confidence level $1 - Pr[(2/\sigma^2)Q > \eta]$.

There are many ways to invalidate this model: the basis $\mathbf{H}$ may be incorrect, the noise model may be incorrect, or both may be incorrect. However, if the model is validated, then at this confidence level, we have validated that the distribution of $\hat{\mathbf{x}}$ is $\hat{\mathbf{x}} \sim C\mathcal{N}_L\left(\mathbf{x}, \sigma^2(\mathbf{H}^H\mathbf{H})^{-1}\right)$. That is, we have validated at this confidence level that the estimator error is normally distributed around the true value of $\mathbf{x}$ with covariance $\sigma^2(\mathbf{H}^H\mathbf{H})^{-1}$.

What can be done when σ^2 is unknown? To address this question, the projection $\mathbf{I}_L - \mathbf{P_H}$ may be resolved into mutually orthogonal projections $\mathbf{P}_1$ and $\mathbf{P}_2$ of respective dimensions r_1 and r_2, with $r_1 + r_2 = L - p$. Define $Q_1 = \mathbf{y}^H\mathbf{P}_1\mathbf{y}$ and $Q_2 = \mathbf{y}^H\mathbf{P}_2\mathbf{y}$, so that $(2/\sigma^2)Q = (2/\sigma^2)Q_1 + (2/\sigma^2)Q_2$. From Cochran's theorem, it is known that $(2/\sigma^2)Q_1 \sim \chi^2_{2r_1}$ and $(2/\sigma^2)Q_2 \sim \chi^2_{2r_2}$ are independent random variables. Moreover, the random variable Q_1/Q is distributed as $Q_1/Q \sim \mathrm{Beta}(r_1, r_2)$. This random variable may be written as

$$\frac{Q_1}{Q} = \frac{\mathbf{y}^H\mathbf{P}_1\mathbf{y}}{\mathbf{y}^H\mathbf{P}_1\mathbf{y} + \mathbf{y}^H\mathbf{P}_2\mathbf{y}}$$

and compared with a threshold η to ensure confidence at the level $1 - Pr[Q_1/Q > \eta]$. The interpretation is that the measurement $\mathbf{y}$ is resolved into the space orthogonal to $\langle\mathbf{H}\rangle$, where its distribution is independent of $\mathbf{Hx}$. Here, its norm-squared is resolved into two components. If the cosine-squared of the angle (coherence) between $\mathbf{P}_1\mathbf{y}$ and $(\mathbf{P}_1 + \mathbf{P}_2)\mathbf{y}$, namely, Q_1/Q, is beta-distributed, then the measure-

[1] The case $N = 1$ was reported in [302]. Then, at the suggestion of B. Mazzeo, the result was extended to $N > 1$ by D. Cochran, B. Mazzeo, and LLS.

ment model is validated at confidence $1 - Pr[Q_1/Q > \eta]$. This does not validate a MVN model for the measurement, as this result holds for any spherically invariant distribution for the noise $\mathbf{n}$. But it does validate the linear model $\mathbf{y} = \mathbf{Hx} + \mathbf{n}$, at a specified confidence, for any spherically invariant noise $\mathbf{n}$.

2.2.4 Weighted Least Squares

In weighted least squares, the objective is to minimize $(\mathbf{y}-\mathbf{Hx})^H\mathbf{W}(\mathbf{y}-\mathbf{Hx})$, where the weighting matrix $\mathbf{W}$ is Hermitian positive definite. The resulting regression equation is $\mathbf{H}^H\mathbf{W}(\mathbf{y} - \mathbf{Hx})$, with solution

$$\hat{\mathbf{x}} = (\mathbf{H}^H\mathbf{WH})^{-1}\mathbf{H}^H\mathbf{Wy}.$$

To analyze the performance of this least squares estimator, we assume the measurement is $\mathbf{y} = \mathbf{Hx} + \mathbf{n}$, with $\mathbf{n}$ a zero-mean noise of covariance $\mathrm{E}[\mathbf{nn}^H] = \mathbf{R_{nn}}$. Then the estimator may be resolved as

$$\hat{\mathbf{x}} = (\mathbf{H}^H\mathbf{WH})^{-1}\mathbf{H}^H\mathbf{WHx} + (\mathbf{H}^H\mathbf{WH})^{-1}\mathbf{H}^H\mathbf{Wn} = \mathbf{x} + (\mathbf{H}^H\mathbf{WH})^{-1}\mathbf{H}^H\mathbf{Wn}.$$

This shows the estimator to be unbiased with error covariance

$$(\mathbf{H}^H\mathbf{WH})^{-1}\mathbf{H}^H\mathbf{WR_{nn}W}^H\mathbf{H}(\mathbf{H}^H\mathbf{WH})^{-1}.$$

If $\mathbf{W}$ is chosen to be $\mathbf{R}_{\mathbf{nn}}^{-1}$, then this error covariance is $(\mathbf{H}^H\mathbf{R}_{\mathbf{nn}}^{-1}\mathbf{H})^{-1}$. To assign a MVN model to $\mathbf{n}$ is to say the estimator $\hat{\mathbf{x}}$ is distributed as $\hat{\mathbf{x}} \sim CN_L\left(\mathbf{x}, (\mathbf{H}^H\mathbf{R}_{\mathbf{nn}}^{-1}\mathbf{H})^{-1}\right)$, which is a stronger statement than a statement only about the mean and covariance of the estimator.

Example 2.1 (Spectrum Analysis and Beamforming) There is an interesting special case for $\mathbf{H} = \boldsymbol{\psi}$, an $L \times 1$ vector, $\mathbf{W} = \mathbf{R}_{\mathbf{nn}}^{-1}$, and x a complex scalar. Then, proceeding according to Appendix 2.A, we have

$$(\mathbf{y} - \boldsymbol{\psi}x)^H\mathbf{R}_{\mathbf{nn}}^{-1}(\mathbf{y} - \boldsymbol{\psi}x) = \left(x - \frac{\boldsymbol{\psi}^H\mathbf{R}_{\mathbf{nn}}^{-1}\mathbf{y}}{\boldsymbol{\psi}^H\mathbf{R}_{\mathbf{nn}}^{-1}\boldsymbol{\psi}}\right)^H\left(x - \frac{\boldsymbol{\psi}^H\mathbf{R}_{\mathbf{nn}}^{-1}\mathbf{y}}{\boldsymbol{\psi}^H\mathbf{R}_{\mathbf{nn}}^{-1}\boldsymbol{\psi}}\right)$$
$$+ \mathbf{y}^H\mathbf{R}_{\mathbf{nn}}^{-1}\mathbf{y} - \frac{|\mathbf{y}^H\mathbf{R}_{\mathbf{nn}}^{-1}\boldsymbol{\psi}|^2}{\boldsymbol{\psi}^H\mathbf{R}_{\mathbf{nn}}^{-1}\boldsymbol{\psi}}.$$

The least squares estimator for x is $\hat{x} = (\boldsymbol{\psi}^H\mathbf{R}_{\mathbf{nn}}^{-1}\boldsymbol{\psi})^{-1}\boldsymbol{\psi}^H\mathbf{R}_{\mathbf{nn}}^{-1}\mathbf{y}$, and the corresponding weighted squared error is

$$(\mathbf{y} - \boldsymbol{\psi}\hat{x})^H\mathbf{R}_{\mathbf{nn}}^{-1}(\mathbf{y} - \boldsymbol{\psi}\hat{x}) = \mathbf{y}^H\mathbf{R}_{\mathbf{nn}}^{-1}\mathbf{y}(1 - \rho^2).$$

where $\rho^2 = |\mathbf{y}^H \mathbf{R}_{\mathbf{nn}}^{-1} \boldsymbol{\psi}|^2 / (\boldsymbol{\psi}^H \mathbf{R}_{\mathbf{nn}}^{-1} \boldsymbol{\psi})(\mathbf{y}^H \mathbf{R}_{\mathbf{nn}}^{-1} \mathbf{y})$ is squared coherence. When the vector $\boldsymbol{\psi}$ is the vector $[1 \; e^{j\theta} \; \cdots \; e^{j(L-1)\theta}]^T$, then this is a spectrum analyzer for the complex coefficient of the frequency component $\boldsymbol{\psi}$ at frequency or electrical angle θ. This statistic may be swept through electrical angles $-\pi < \theta \leq \pi$ to map out a complex spectrum.

2.2.5 Constrained Least Squares

When there are linear constraints $\mathbf{C}^H \mathbf{x} = \mathbf{c}$ on a solution $\mathbf{x}$, then the quadratic form is replaced by the Lagrangian $(\mathbf{y} - \mathbf{H}\mathbf{x})^H (\mathbf{y} - \mathbf{H}\mathbf{x}) + \boldsymbol{\mu}^H (\mathbf{C}^H \mathbf{x} - \mathbf{c})$, where $\mathbf{C}^H$ is an $r \times p$ constraint matrix, $r < p$, $\mathbf{c}$ is an $r \times 1$ vector, and $\boldsymbol{\mu}$ is an $r \times 1$ vector of Lagrange multipliers. We may write this Lagrangian in its dual form: $(\mathbf{x} - \hat{\mathbf{x}}_{LS})^H (\mathbf{H}^H \mathbf{H})(\mathbf{x} - \hat{\mathbf{x}}_{LS}) + \mathbf{y}^H (\mathbf{I}_L - \mathbf{P}_{\mathbf{H}})\mathbf{y} + \boldsymbol{\mu}^H (\mathbf{C}^H \mathbf{x} - \mathbf{c})$, where $\hat{\mathbf{x}}_{LS}$ is the previously derived, unconstrained, least squares estimator. Ignore the quadratic form $\mathbf{y}^H (\mathbf{I}_L - \mathbf{P}_{\mathbf{H}})\mathbf{y}$, and parameterize the unknown signal $\mathbf{x}$ as $\mathbf{x} = \hat{\mathbf{x}}_{LS} + \mathbf{t}$ to write the Lagrangian as $\mathbf{t}^H (\mathbf{H}^H \mathbf{H})\mathbf{t} + \boldsymbol{\mu}^H (\mathbf{C}^H (\hat{\mathbf{x}}_{LS} + \mathbf{t}) - \mathbf{c})$. From here, it is easy to solve for $\mathbf{t}$ as $\hat{\mathbf{t}} = -(\mathbf{H}^H \mathbf{H})^{-1} \mathbf{C} \boldsymbol{\mu}$. Enforce the constraint to solve for $\boldsymbol{\mu}$ and as a consequence $\hat{\mathbf{t}}$. The resulting solution for the constrained least squares estimate of $\mathbf{x}$ is

$$\begin{aligned}\hat{\mathbf{x}} =& \hat{\mathbf{x}}_{LS} + \hat{\mathbf{t}} = \left[\mathbf{I}_p - (\mathbf{H}^H \mathbf{H})^{-1} \mathbf{C} (\mathbf{C}^H (\mathbf{H}^H \mathbf{H})^{-1} \mathbf{C})^{-1} \mathbf{C}^H\right] \hat{\mathbf{x}}_{LS} \\ &+ (\mathbf{H}^H \mathbf{H})^{-1} \mathbf{C} (\mathbf{C}^H (\mathbf{H}^H \mathbf{H})^{-1} \mathbf{C})^{-1} \mathbf{c}.\end{aligned}$$

Condition Adjustment. Let $\mathbf{H} = \mathbf{I}_L$, so the dimension of $\mathbf{x}$ is the dimension of $\mathbf{y}$. The constrained least squares problem is to minimize $(\mathbf{y} - \mathbf{x})^H (\mathbf{y} - \mathbf{x}) + \boldsymbol{\mu}^H (\mathbf{C}^H \mathbf{x} - \mathbf{c})$. The constrained least squares solution is then

$$\hat{\mathbf{x}} = (\mathbf{I}_L - \mathbf{P}_{\mathbf{C}})\mathbf{y} + \mathbf{C}(\mathbf{C}^H \mathbf{C})^{-1} \mathbf{c},$$

where $\mathbf{P}_{\mathbf{C}} = \mathbf{C}(\mathbf{C}^H \mathbf{C})^{-1} \mathbf{C}^H$ is the projection onto the subspace $\langle \mathbf{C} \rangle$. It is easy to see that the constraint is met. Moreover, the difference between $\mathbf{y}$ and its condition adjustment $\hat{\mathbf{x}}$ is $\mathbf{P}_{\mathbf{C}} \mathbf{y} - \mathbf{C}(\mathbf{C}^H \mathbf{C})^{-1} \mathbf{c}$, which shows the difference to lie in $\langle \mathbf{C} \rangle$, the span of $\mathbf{C}$. Why is this called *condition adjustment*? Because the measurement $\mathbf{y}$ is adjusted to a constrained $\hat{\mathbf{x}}$. This is as close as one can get to the original measurement under the constraint. Condition adjustment is commonly used to smooth digitized maps. The price of smoothness is the non-zero squared error between $\mathbf{y}$ and $\mathbf{x}$.

Norm-Constrained Least Squares. When there is a norm constraint $\|\mathbf{x}\|_2^2 = t$, then the quadratic form to be minimized is replaced by the Lagrangian $(\mathbf{y} - \mathbf{H}\mathbf{x})^H (\mathbf{y} - \mathbf{H}\mathbf{x}) + \mu(\|\mathbf{x}\|_2^2 - t)$, where $\mu > 0$ is a positive real Lagrange multiplier. The solution satisfies

$$\left(\mathbf{H}^H \mathbf{H} + \mu \mathbf{I}_p\right) \mathbf{x} = \mathbf{H}^H \mathbf{y}. \tag{2.1}$$

The Gramian $\mathbf{H}^H\mathbf{H}$ is Hermitian and therefore unitarily diagonalizable. Write its eigendecomposition as $\mathbf{U}\boldsymbol{\Lambda}\mathbf{U}^H$, where $\boldsymbol{\Lambda} = \operatorname{diag}(\lambda_1, \ldots, \lambda_p)$ with eigenvalues sorted in decreasing order. Then (2.1) becomes

$$\mathbf{U}(\boldsymbol{\Lambda} + \mu\mathbf{I}_p)\mathbf{U}^H\mathbf{x} = \mathbf{H}^H\mathbf{y}.$$

Multiply both sides of this equation by $\mathbf{U}^H$ to write

$$(\boldsymbol{\Lambda} + \mu\mathbf{I}_p)\mathbf{z} = \mathbf{U}^H\mathbf{H}^H\mathbf{y} = \mathbf{b}. \tag{2.2}$$

where $\mathbf{b}$ is known and $\mathbf{z} = \mathbf{U}^H\mathbf{x}$ meets the same norm-squared constraint as $\mathbf{x}$. To solve for $\mathbf{z}$ is to solve for $\mathbf{x} = \mathbf{U}\mathbf{z}$ with the constraint met. From (2.2), it follows that $(\lambda_i+\mu)z_i = b_i$ or $z_i = b_i/(\lambda_i+\mu)$. Then the constraint $\sum_{i=1}^p |x_i|^2 = \sum_{i=1}^p |z_i|^2 = t$ is equivalent to the condition

$$g(\mu) = \sum_{i=1}^p \frac{|b_i|^2}{(\lambda_i + \mu)^2} = t.$$

The function $g(\mu)$ is continuous and monotonically decreasing for $\mu \geq -\lambda_p$, so the equation $g(\mu) = t$ has a unique solution, say μ^*, which can be easily found by bisection. Finally, the solution of the norm-constrained least squares problem is $\hat{\mathbf{x}} = \left(\mathbf{H}^H\mathbf{H} + \mu^*\mathbf{I}_p\right)^{-1}\mathbf{H}^H\mathbf{y}$. The solution to this problem can be traced back to a paper by Forsythe and Golub in 1965 [124]. An alternative proof of this result can be found in [115].

2.2.6 Oblique Least Squares

In the least squares solution, $\hat{\mathbf{x}} = (\mathbf{H}^H\mathbf{H})^{-1}\mathbf{H}^H\mathbf{y}$, the estimator $\mathbf{x}$ is computed by resolving $\mathbf{y}$ onto the columns of the channel matrix $\mathbf{H}$, and these resolutions are linearly transformed by the inverse of the Gramian $\mathbf{H}^H\mathbf{H}$ to resolve the measurement $\mathbf{y}$ into two orthogonal components, $\mathbf{P}_\mathbf{H}\mathbf{y}$ and $(\mathbf{I}_L - \mathbf{P}_\mathbf{H})\mathbf{y}$. The fitting error $(\mathbf{I}_L - \mathbf{P}_\mathbf{H})\mathbf{y}$ is orthogonal to the near point $\mathbf{P}_\mathbf{H}\mathbf{y}$ in the subspace $\langle\mathbf{H}\rangle$. Perhaps there is more insight to be had by resolving the subspace $\langle\mathbf{H}\rangle$ into a direct sum of two other lower-dimensional subspaces. To this end, parse $\mathbf{H}$ as $\mathbf{H} = [\mathbf{H}_1\ \mathbf{H}_2]$, and correspondingly parse the unknown signal as $\mathbf{x} = [\mathbf{x}_1^T\ \mathbf{x}_2^T]^T$. The prior measurement model $\mathbf{y} = \mathbf{H}\mathbf{x}$ may be written as

$$\mathbf{y} = \mathbf{H}_1\mathbf{x}_1 + \mathbf{H}_2\mathbf{x}_2 + \mathbf{n},$$

and we may interpret $\mathbf{H}_1\mathbf{x}_1$ as signal, $\mathbf{H}_2\mathbf{x}_2$ as interference, and $\mathbf{n}$ as noise. Nothing changes in the least squares solution, but with this parsing of the model, we might reasonably ask how the resulting solution for $\hat{\mathbf{x}}$ is parsed. It is not hard to show that the solutions for $\hat{\mathbf{x}}_1$, $\hat{\mathbf{x}}_2$ and $\mathbf{H}\hat{\mathbf{x}}_1$, $\mathbf{H}\hat{\mathbf{x}}_2$ are

$$\hat{\mathbf{x}}_1 = (\mathbf{H}_1^H \mathbf{P}_{\mathbf{H}_2}^{\perp} \mathbf{H}_1)^{-1} \mathbf{H}_1^H \mathbf{P}_{\mathbf{H}_2}^{\perp} \mathbf{y}, \quad \text{and} \quad \mathbf{H}_1 \hat{\mathbf{x}}_1 = \mathbf{E}_{\mathbf{H}_1\mathbf{H}_2} \mathbf{y},$$

$$\hat{\mathbf{x}}_2 = (\mathbf{H}_2^H \mathbf{P}_{\mathbf{H}_1}^{\perp} \mathbf{H}_2)^{-1} \mathbf{H}_2^H \mathbf{P}_{\mathbf{H}_1}^{\perp} \mathbf{y}, \quad \text{and} \quad \mathbf{H}_2 \hat{\mathbf{x}}_2 = \mathbf{E}_{\mathbf{H}_2\mathbf{H}_1} \mathbf{y},$$

where $\mathbf{E}_{\mathbf{H}_1\mathbf{H}_2}$ and $\mathbf{E}_{\mathbf{H}_2\mathbf{H}_1}$ are the following *oblique* projections:

$$\mathbf{E}_{\mathbf{H}_1\mathbf{H}_2} = \mathbf{H}_1 (\mathbf{H}_1^H \mathbf{P}_{\mathbf{H}_2}^{\perp} \mathbf{H}_1)^{-1} \mathbf{H}_1^H \mathbf{P}_{\mathbf{H}_2}^{\perp},$$

$$\mathbf{E}_{\mathbf{H}_2\mathbf{H}_1} = \mathbf{H}_2 (\mathbf{H}_2^H \mathbf{P}_{\mathbf{H}_1}^{\perp} \mathbf{H}_2)^{-1} \mathbf{H}_2^H \mathbf{P}_{\mathbf{H}_1}^{\perp}.$$

Evidently, $\mathbf{E}_{\mathbf{H}_1\mathbf{H}_2} + \mathbf{E}_{\mathbf{H}_2\mathbf{H}_1} = \mathbf{P}_{\mathbf{H}}$. This is a resolution of the orthogonal projection $\mathbf{P}_{\mathbf{H}}$ into two oblique projections, which are mutually orthogonal. That is, $\mathbf{E}_{\mathbf{H}_1\mathbf{H}_2}\mathbf{E}_{\mathbf{H}_1\mathbf{H}_2} = \mathbf{E}_{\mathbf{H}_1\mathbf{H}_2}$, $\mathbf{E}_{\mathbf{H}_2\mathbf{H}_1}\mathbf{E}_{\mathbf{H}_2\mathbf{H}_1} = \mathbf{E}_{\mathbf{H}_2\mathbf{H}_1}$, and $\mathbf{E}_{\mathbf{H}_1\mathbf{H}_2}\mathbf{E}_{\mathbf{H}_2\mathbf{H}_1} = \mathbf{0}$, but neither of these oblique projections is Hermitian. This replaces the two-way resolution of identity, $\mathbf{I}_L = \mathbf{P}_{\mathbf{H}} + \mathbf{P}_{\mathbf{H}}^{\perp}$, by the three-way resolution, $\mathbf{I}_L = \mathbf{E}_{\mathbf{H}_1\mathbf{H}_2} + \mathbf{E}_{\mathbf{H}_2\mathbf{H}_1} + \mathbf{P}_{\mathbf{H}}^{\perp}$. The range space of $\mathbf{E}_{\mathbf{H}_1\mathbf{H}_2}$ is $\langle \mathbf{H}_1 \rangle$, and its null space includes $\langle \mathbf{H}_2 \rangle$. Thus, $\hat{\mathbf{x}}_1$ and $\mathbf{H}_1\hat{\mathbf{x}}_1$ are unbiased estimators of $\mathbf{x}_1$ and $\mathbf{H}_1\mathbf{x}_1$, respectively.

Call $\mathbf{U}_1$ and $\mathbf{U}_2$ orthogonal spans for $\langle \mathbf{H}_1 \rangle$ and $\langle \mathbf{H}_2 \rangle$. The singular values of $\mathbf{U}_1^H\mathbf{U}_2$ determine the principal angles θ_i between the subspaces $\langle \mathbf{H}_1 \rangle$ and $\langle \mathbf{H}_2 \rangle$ (see Sect. 9.2 for a definition of the principal angles between two subspaces):

$$\sin^2 \theta_i = 1 - sv_i^2(\mathbf{U}_1^H \mathbf{U}_2).$$

These, in turn, determine the non-zero singular values of $\mathbf{E}_{\mathbf{H}_1\mathbf{H}_2}$:

$$sv_i(\mathbf{E}_{\mathbf{H}_1\mathbf{H}_2}) = \frac{1}{\sin \theta_i}.$$

The principal angles, θ_i, range from 0 to $\pi/2$, and their sines range from 0 to 1. So the singular values of the low-rank $L \times L$ oblique projection may be 0, 1, or any real value greater than 1.

What are the consequences of this result? Assume the residuals $\mathbf{n}$ have mean $\mathbf{0}$ and covariance $\sigma^2\mathbf{I}_L$. The error covariance of $\hat{\mathbf{x}}_1$ is $\sigma^2(\mathbf{H}_1^H \mathbf{P}_{\mathbf{H}_2}^{\perp} \mathbf{H}_1)^{-1}$, and the error covariance of $\mathbf{H}_1\hat{\mathbf{x}}_1$ is $\mathbf{Q} = \sigma^2 \mathbf{E}_{\mathbf{H}_1\mathbf{H}_2}\mathbf{E}_{\mathbf{H}_1\mathbf{H}_2}^H = \mathbf{H}_1 (\mathbf{H}_1^H \mathbf{P}_{\mathbf{H}_2}^{\perp} \mathbf{H}_1)^{-1} \mathbf{H}_1^H$. The eigenvalues of $\mathbf{Q}$ are the squares of the singular values of $\mathbf{E}_{\mathbf{H}_1\mathbf{H}_2}$ scaled by σ^2, namely, $\sigma^2/\sin^2\theta_i$. Thus, when the subspaces $\langle \mathbf{H}_1 \rangle$ and $\langle \mathbf{H}_2 \rangle$ are closely aligned, these eigenvalues are large. Then, for example, the trace of this error covariance matrix (the error variance) is

$$\operatorname{tr}(\mathbf{Q}) = \sigma^2 \sum_{i=1}^{r} \frac{1}{\sin^2 \theta_i} \geq r\sigma^2.$$

This squared error is the price paid for the super-resolution estimator $\mathbf{E}_{\mathbf{H}_1\mathbf{H}_2}\mathbf{y}$ that nulls the component $\mathbf{H}_2\mathbf{x}_2$ in search of the component $\mathbf{H}_1\mathbf{x}_1$. When the subspaces

$\langle \mathbf{H}_1 \rangle$ and $\langle \mathbf{H}_2 \rangle$ are nearly aligned, this price can be high, typically so high that super-resolution does not work in low to moderate signal-to-noise ratios.[2]

Example 2.2 (One-Dimensional Subspaces) When trying to resolve two closely spaced one-dimensional subspaces $\langle \mathbf{h}_1 \rangle$ and $\langle \mathbf{h}_2 \rangle$, we have

$$\mathrm{tr}(\mathbf{Q}) = \sigma^2 \frac{1}{\sin^2 \theta},$$

where $\sin^2 \theta = 1 - |\mathbf{h}_1^H \mathbf{h}_2|^2 / (\mathbf{h}_1^H \mathbf{h}_1 \mathbf{h}_2^H \mathbf{h}_2)$. When $\mathbf{h}_1$ and $\mathbf{h}_2$ are the Vandermonde modes, $\mathbf{h}_1 = [1 \; e^{j\theta_1} \; \cdots \; e^{j(L-1)\theta_1}]^T$, and $\mathbf{h}_2 = [1 \; e^{j\theta_2} \; \cdots \; e^{j(L-1)\theta_2}]^T$, then

$$\mathrm{tr}(\mathbf{Q}) = \frac{\sigma^2}{1 - \frac{1}{L^2} \frac{\sin^2(L(\theta_1 - \theta_2)/2)}{\sin^2((\theta_1 - \theta_2)/2)}},$$

which increases to ∞ without bound as $\theta_1 - \theta_2$ decreases to 0.

2.2.7 The BLUE (or MVUB or MVDR) Estimator

Although the analysis of LS for dimension reduction and cross-validation has assumed a simple zero mean and scaled identity covariance matrix for the noise $\mathbf{n}$, no model of the additive noise $\mathbf{n}$ has entered into the actual derivation of linear estimators. The best linear unbiased estimator (BLUE) changes this by assigning a zero mean and covariance $\mathbf{R}_{\mathbf{nn}}$ model to the noise. The problem is then to minimize the error covariance of a linear estimator under an unbiasedness constraint. The BLUE also goes by the names minimum variance unbiased estimator (MVUB) or minimum variance distortionless response (MVDR) estimator.

The problem is to find the estimator $\mathbf{G}^H \mathbf{y}$ that is unbiased and has minimum error variance. To say this estimator is unbiased is to say that $E[\mathbf{G}^H \mathbf{y}] = \mathbf{x}$ and to say it is best is to say no other linear estimator has smaller error variance, defined to be $Q = E[(\mathbf{G}^H \mathbf{y} - \mathbf{x})^H (\mathbf{G}^H \mathbf{y} - \mathbf{x})]$. It is assumed that the measurement model is $\mathbf{y} = \mathbf{H}\mathbf{x} + \mathbf{n}$, with the noise $\mathbf{n}$ zero mean with covariance matrix $\mathbf{R}_{\mathbf{nn}} = E[\mathbf{n}\mathbf{n}^H]$. So the problem is to minimize $\mathrm{tr}(\mathbf{G}^H \mathbf{R}_{\mathbf{nn}} \mathbf{G})$ under the constraint that $\mathbf{G}^H \mathbf{H} = \mathbf{I}_p$. It is a straightforward exercise, following the reasoning of Appendix 2.A to show that the solution for $\mathbf{G}^H$ is $\mathbf{G}^H = (\mathbf{H}^H \mathbf{R}_{\mathbf{nn}}^{-1} \mathbf{H})^{-1} \mathbf{H}^H \mathbf{R}_{\mathbf{nn}}^{-1}$. The BLUE of $\mathbf{x}$ is then $\hat{\mathbf{x}} = \mathbf{G}^H \mathbf{y}$, which resolves as $\hat{\mathbf{x}} = \mathbf{x} + \mathbf{G}^H \mathbf{n}$. This is an unbiased estimator with error covariance $\mathbf{Q} = \mathbf{G}^H \mathbf{R}_{\mathbf{nn}} \mathbf{G} = (\mathbf{H}^H \mathbf{R}_{\mathbf{nn}}^{-1} \mathbf{H})^{-1}$.

[2] Much more on the topic of oblique projections may be found in [28].

Connection with LS. When the noise covariance $\mathbf{R_{nn}} = \sigma^2\mathbf{I}_L$, then the BLUE is the LS estimator $(\mathbf{H}^H\mathbf{H})^{-1}\mathbf{H}^H\mathbf{y}$, and the error covariances for BLUE and LS are identical at $\sigma^2(\mathbf{H}^H\mathbf{H})^{-1}$. This result is sometimes called the Gauss-Markov theorem. For an arbitrary $\mathbf{R_{nn}}$, the error covariance matrix for LS is $(\mathbf{H}^H\mathbf{H})^{-1}\mathbf{H}^H\mathbf{R_{nn}}\mathbf{H}(\mathbf{H}^H\mathbf{H})^{-1}$, which produces the matrix inequality

$$(\mathbf{H}^H\mathbf{R}_{\mathbf{nn}}^{-1}\mathbf{H})^{-1} \preceq (\mathbf{H}^H\mathbf{H})^{-1}\mathbf{H}^H\mathbf{R_{nn}}\mathbf{H}(\mathbf{H}^H\mathbf{H})^{-1}.$$

When $\mathbf{H} = \boldsymbol{\psi} \in \mathbb{C}^L$, then this result is a more familiar Schwarz inequality

$$\frac{\boldsymbol{\psi}^H\boldsymbol{\psi}}{\boldsymbol{\psi}^H\mathbf{R}_{\mathbf{nn}}^{-1}\boldsymbol{\psi}} \leq \frac{\boldsymbol{\psi}^H\mathbf{R_{nn}}\boldsymbol{\psi}}{\boldsymbol{\psi}^H\boldsymbol{\psi}}.$$

In beamforming and spectrum analysis, this inequality is used to explain the sharper resolution of a Capon spectrum (the LHS) compared with the resolution of the conventional or Bartlett spectrum (the RHS).

Connection with OBLS. If in the linear model $\mathbf{y} = \mathbf{H}_1\mathbf{x}_1 + \mathbf{H}_2\mathbf{x}_2 + \mathbf{n}$, the interference term $\mathbf{H}_2\mathbf{x}_2$ is modeled as a component of noise, then the measurement model may be written as $\mathbf{y} = \mathbf{H}_1\mathbf{x}_1 + \mathbf{n}$, where the covariance matrix $\mathbf{R_{nn}}$ is structured as $\mathbf{R_{nn}} = \mathbf{H}_2\mathbf{H}_2^H + \sigma^2\mathbf{I}_L$. Then the matrix inversion lemma may be used to write

$$\sigma^2\mathbf{R}_{\mathbf{nn}}^{-1} = \mathbf{I}_L - \mathbf{H}_2(\sigma^2\mathbf{I}_p + \mathbf{H}_2^H\mathbf{H}_2)^{-1}\mathbf{H}_2^H.$$

In the limit, $\sigma^2 \to 0$, $\sigma^2\mathbf{R}_{\mathbf{nn}}^{-1} \to \mathbf{P}_{\mathbf{H}_2}^{\perp}$, and $\mathbf{G}^H \to (\mathbf{H}^H\mathbf{P}_{\mathbf{H}_2}^{\perp}\mathbf{H})^{-1}\mathbf{H}_1^H\mathbf{P}_{\mathbf{H}_2}^{\perp}$. That is, BLUE is OBLS, with error covariance matrix $\sigma^2(\mathbf{H}_1\mathbf{P}_{\mathbf{H}_2}^{\perp}\mathbf{H})^{-1}$. We might say the OBLS estimator is the low noise limit of the BLUE when the noise covariance matrix is structured as a diagonal plus a rank-r component.

The Generalized Sidelobe Canceller (GSC). The BLUE $\hat{\mathbf{x}}$ may be resolved into its components in the subspaces $\langle\mathbf{H}\rangle$ and $\langle\mathbf{H}\rangle^{\perp}$. Then

$$\begin{aligned}\hat{\mathbf{x}} &= \mathbf{G}^H(\mathbf{P_H}\mathbf{y} + \mathbf{P}_{\mathbf{H}}^{\perp}\mathbf{y}) \\ &= (\mathbf{H}^H\mathbf{H})^{-1}\mathbf{H}^H\mathbf{y} - (-\mathbf{G}^H\mathbf{P}_{\mathbf{H}}^{\perp}\mathbf{y}).\end{aligned}$$

The first term on the RHS is the LS estimate $\hat{\mathbf{x}}_{LS}$, and the second term is a filtering of $\mathbf{P}_{\mathbf{H}}^{\perp}\mathbf{y}$ by the BLUE filter $\mathbf{G}^H$. So the BLUE of $\mathbf{x}$ is the error in estimating the LS estimate of $\mathbf{x}$ by a BLUE of the component of $\mathbf{y}$ in the subspace perpendicular to $\langle\mathbf{H}\rangle$. This suggests that the BLUE filter $\mathbf{G}^H$ has minimized the quadratic form $E[(\mathbf{G}^H\mathbf{y} - \mathbf{x})^H(\mathbf{G}^H\mathbf{y} - \mathbf{x})]$ under the linear constraint $\mathbf{G}^H\mathbf{H} = \mathbf{I}_p$, or equivalently

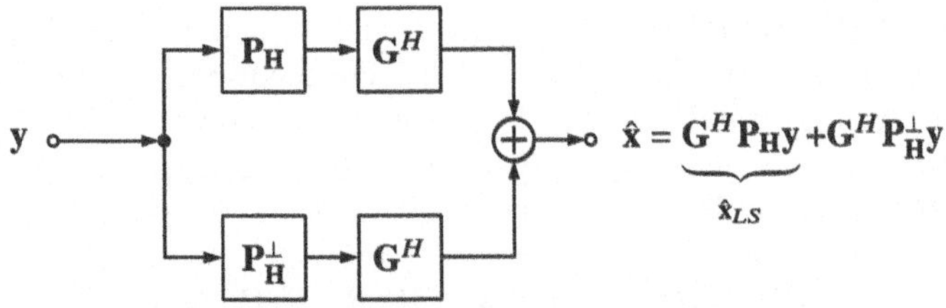

Fig. 2.1 Filtering diagram of the generalized sidelobe canceller

it has minimized the quadratic form $\mathrm{E}[(\hat{\mathbf{x}}_{LS} - (-\mathbf{G}^H\mathbf{P}_\mathbf{H}^\perp\mathbf{y}))^H(\hat{\mathbf{x}}_{LS} - (-\mathbf{G}^H\mathbf{P}_\mathbf{H}^\perp\mathbf{y}))]$, unconstrained. The filtering diagram of Fig. 2.1 is evocative. More will be said about this result in Chap. 3.

Example 2.3 (Spectrum Analysis and Beamforming) Assume $\mathbf{H} = \boldsymbol{\psi}$, where $\boldsymbol{\psi}$ is an $L \times 1$ Vandermonde vector. The signal x is a complex scalar. The constraint is that $\mathbf{g}^H\boldsymbol{\psi} = 1$. The LS estimate is $(\boldsymbol{\psi}^H\boldsymbol{\psi})^{-1}\boldsymbol{\psi}^H\mathbf{y}$. The output of the GSC channel is $(\boldsymbol{\psi}^H\mathbf{R}_{\mathbf{nn}}^{-1}\boldsymbol{\psi})^{-1}\boldsymbol{\psi}^H\mathbf{R}_{\mathbf{nn}}^{-1}\mathbf{P}_\mathbf{H}^\perp\mathbf{y}$. The BLUE is $(\boldsymbol{\psi}^H\mathbf{R}_{\mathbf{nn}}^{-1}\boldsymbol{\psi})^{-1}\boldsymbol{\psi}^H\mathbf{R}_{\mathbf{nn}}^{-1}\mathbf{y}$, with variance $(\boldsymbol{\psi}^H\mathbf{R}_{\mathbf{nn}}^{-1}\boldsymbol{\psi})^{-1}$.

2.2.8 Sequential Least Squares

Suppose the measurement equation $\mathbf{y}_{t-1} = \mathbf{H}_{t-1}\mathbf{x} + \mathbf{n}_{t-1}$ characterizes measurements recorded at time or space indices $1, 2, \ldots, t-1$. A new measurement is made at index t. Perhaps the least squares estimate of $\mathbf{x}$ can be updated. To be concrete, let's write the sequentially evolving measurement model as

$$\begin{bmatrix} y_1 \\ y_2 \\ \vdots \\ y_{t-1} \\ \hline y_t \end{bmatrix} = \begin{bmatrix} \mathbf{H}_{t-1} \\ \hline \mathbf{c}_t^H \end{bmatrix} \mathbf{x} + \begin{bmatrix} \mathbf{n}_{t-1} \\ \hline n_t \end{bmatrix}.$$

This is an explicit writing of the model $\mathbf{y}_t = \mathbf{H}_t\mathbf{x} + \mathbf{n}_t$. The least squares estimate of $\mathbf{x}$, based on measurements up to and including index t, may be written as

$$\mathbf{P}_t^{-1}\hat{\mathbf{x}}_t = \mathbf{H}_t^H\mathbf{y}_t,$$

where $\mathbf{P}_t^{-1} = \mathbf{H}_{t-1}^H\mathbf{H}_{t-1} + \mathbf{c}_t\mathbf{c}_t^H$ and $\mathbf{H}_t = [\mathbf{H}_{t-1}^H \ \mathbf{c}_t]^H$. Use the matrix inversion lemma to write $\mathbf{P}_t$ as

$$\mathbf{P}_t = \mathbf{P}_{t-1} - \gamma_t\mathbf{P}_{t-1}\mathbf{c}_t\mathbf{c}_t^H\mathbf{P}_{t-1}.$$

It is a few steps of algebra to write the solution for $\hat{\mathbf{x}}_t = \mathbf{P}_t \mathbf{H}_t^H \mathbf{y}_t$ as

$$\hat{\mathbf{x}}_t = \hat{\mathbf{x}}_{t-1} + \mathbf{k}_t (y_t - \mathbf{c}_t^H \hat{\mathbf{x}}_{t-1}),$$

where $\mathbf{P}_{t-1}^{-1}\mathbf{k}_t = \gamma_t \mathbf{c}_t$, $\gamma_t^{-1} = 1 + \mathbf{c}_t^H \mathbf{P}_{t-1}\mathbf{c}_t$, and

$$\mathbf{P}_t^{-1} = \mathbf{P}_{t-1}^{-1} + \mathbf{c}_t \mathbf{c}_t^H.$$

The key parameter is the matrix $\mathbf{P}_{t-1}^{-1} = \mathbf{H}_{t-1}^H \mathbf{H}_{t-1}$, the Gramian of $\mathbf{H}_{t-1}$. It is the inverse of the error covariance matrix for the estimator $\hat{\mathbf{x}}_{t-1}$ if the noise $\mathbf{n}_{t-1}$ is zero mean with covariance matrix $E\left[\mathbf{n}_{t-1}\mathbf{n}_{t-1}^H\right] = \mathbf{I}_{t-1}$. Here is the recursion: at index $t-1$, the estimate $\hat{\mathbf{x}}_{t-1}$, its inverse error covariance matrix $\mathbf{P}_{t-1}^{-1}$, and γ_t are computed and stored in memory. The inner product $y_{t|t-1} = \mathbf{c}_t^H \hat{\mathbf{x}}_{t-1}$ is a linear predictor of the new measurement y_t. The prediction error $y_t - y_{t|t-1}$ is scaled by the gain $\mathbf{k}_t$ to correct the previous estimate $\hat{\mathbf{x}}_{t-1}$. How is the gain computed? It is the solution to the regression equation $\mathbf{P}_{t-1}^{-1}\mathbf{k}_t = \gamma_t \mathbf{c}_t$. The inverse error covariance matrix is updated as $\mathbf{P}_t^{-1} = \mathbf{P}_{t-1}^{-1} + \mathbf{c}_t \mathbf{c}_t^H$, and the recursion continues. The computational complexity at each update is the computational complexity of solving the regression equation $\mathbf{P}_{t-1}^{-1}\mathbf{k}_t = \gamma_t \mathbf{c}_t$ for the gain $\mathbf{k}_t$ or equivalently of inverting for $\mathbf{P}_{t-1}$ to solve for γ_t and $\mathbf{k}_t$.

2.2.9 Total Least Squares

The total least squares (TLS) idea is this: deviations of measurements $\mathbf{y}$ from the model $\mathbf{Hx}$ may not be attributable only to unmodeled errors $\mathbf{n}$; rather, they might be attributable to errors in the model $\mathbf{H}$ itself. We may be led to this conclusion by unexpectedly large fitting errors, perhaps revealed by a goodness-of-fit test.

The measurement model for total least squares in the linear model is $\mathbf{y} - \mathbf{n} = (\mathbf{H} + \mathbf{E})\mathbf{x}$, which can be written as

$$\left(\begin{bmatrix}\mathbf{y} & -\mathbf{H}\end{bmatrix} + \begin{bmatrix}-\mathbf{n} & -\mathbf{E}\end{bmatrix}\right)\begin{bmatrix}1 \\ \mathbf{x}\end{bmatrix} = \mathbf{0}.$$

This constraint says the vector $\begin{bmatrix}1 & \mathbf{x}^T\end{bmatrix}^T$ lies in the null space of the adjusted matrix $\left(\begin{bmatrix}\mathbf{y} & -\mathbf{H}\end{bmatrix} + \begin{bmatrix}-\mathbf{n} & -\mathbf{E}\end{bmatrix}\right)$. The objective is to find the minimum-norm adjustment $\begin{bmatrix}-\mathbf{n} & -\mathbf{E}\end{bmatrix}$ to the model $\begin{bmatrix}\mathbf{y} & -\mathbf{H}\end{bmatrix}$, under the constraint $\mathbf{y} - \mathbf{n} = (\mathbf{H} + \mathbf{E})\mathbf{x}$, that is,

$$\begin{aligned} &\underset{\mathbf{n},\mathbf{E},\mathbf{x}}{\text{minimize}} \quad \left\|\begin{bmatrix}\mathbf{n} & \mathbf{E}\end{bmatrix}\right\|^2, \\ &\text{subject to} \quad \mathbf{y} - \mathbf{n} = (\mathbf{H} + \mathbf{E})\mathbf{x}. \end{aligned}$$

Evidently, the adjustment will reduce the rank of $\begin{bmatrix}\mathbf{y} & -\mathbf{H}\end{bmatrix}$ by one so that its null space has dimension one. The constraint forces the vector $\mathbf{y} - \mathbf{n}$ to lie in the range of the model matrix $\mathbf{H} + \mathbf{E}$.

Once more, the SVD is useful. Assume that $L \geq p+1$, and call $\mathbf{FKG}^H$ the SVD of the augmented $L \times (p+1)$ matrix $\begin{bmatrix}\mathbf{y} & -\mathbf{H}\end{bmatrix}$. Organize this SVD as

$$\mathbf{FKG}^H = \begin{bmatrix}\mathbf{F}_p & \mathbf{f}\end{bmatrix} \begin{bmatrix}\mathbf{K}_p & \mathbf{0} \\ \mathbf{0}^T & k_{p+1}\end{bmatrix} \begin{bmatrix}\mathbf{G}_p^H \\ \mathbf{g}^H\end{bmatrix},$$

where $\mathbf{F}_p \in \mathbb{C}^{L \times p}$, $\mathbf{f} \in \mathbb{C}^L$, $\mathbf{K}_p$ is a $p \times p$ diagonal matrix, $\mathbf{G}_p \in \mathbb{C}^{(p+1)\times p}$, and $\mathbf{g} = \begin{bmatrix}g_1 & \tilde{\mathbf{g}}^T\end{bmatrix}^T$, with $\tilde{\mathbf{g}} \in \mathbb{C}^p$. Assume that the multiplicity of the smallest singular value is 1. Choose $\begin{bmatrix}-\mathbf{n} & -\mathbf{E}\end{bmatrix} = -\mathbf{f}k_{p+1}\mathbf{g}^H$, which has the squared Frobenius norm k_{p+1}^2. This is the minimum-norm adjustment to $\begin{bmatrix}\mathbf{y} & -\mathbf{H}\end{bmatrix}$ that reduces its rank by one. The matrix $\left(\begin{bmatrix}\mathbf{y} & -\mathbf{H}\end{bmatrix} + \begin{bmatrix}-\mathbf{n} & -\mathbf{E}\end{bmatrix}\right)$ is then $\mathbf{F}_p\mathbf{K}_p\mathbf{G}_p^H$. Moreover, the vector $\begin{bmatrix}1 & \mathbf{x}^T\end{bmatrix}^T = (1/g_1)\mathbf{g}$ lies in its null space. The net of this procedure is that the new model $\mathbf{H} + \mathbf{E}$ is given by the p last columns of $\mathbf{FKG}^H$ with a change of sign and the adjusted measurement $\mathbf{y} - \mathbf{n}$ is the first column of $\mathbf{FKG}^H$. The solution for $\hat{\mathbf{x}}$ is given by $(1/g_1)\tilde{\mathbf{g}}$, which requires $g_1 \neq 0$.

Alternatively, the solution to the TLS problem satisfies the eigenvalue equation

$$\begin{bmatrix}\mathbf{y}^H\mathbf{y} & -\mathbf{y}^H\mathbf{H} \\ -\mathbf{H}^H\mathbf{y} & \mathbf{H}^H\mathbf{H}\end{bmatrix} \begin{bmatrix}1 \\ \hat{\mathbf{x}}\end{bmatrix} = k_{p+1}^2 \begin{bmatrix}1 \\ \hat{\mathbf{x}}\end{bmatrix}.$$

Therefore, $\hat{\mathbf{x}}$ can be expressed as

$$\hat{\mathbf{x}} = \left(\mathbf{H}^H\mathbf{H} - k_{p+1}^2\mathbf{I}_p\right)^{-1} \mathbf{H}^H\mathbf{y},$$

provided that the smallest eigenvalue of $\mathbf{H}^H\mathbf{H}$ is larger than k_{p+1}^2. Interestingly, this expression suggests that the TLS solution can be interpreted as a regularized LS solution.

The method of total least squares is discussed in Golub and Van Loan [141]. See also Chapter 12 in [142] and the monograph of Van Huffel and Vandevalle [356]. A partial SVD algorithm to efficiently compute the TLS solution in the nongeneric case of repeated smallest eigenvalue is given in [355]. There is no reason TLS cannot be extended to multi-rank adjustments to the matrix $\begin{bmatrix}\mathbf{y} & -\mathbf{H}\end{bmatrix}$, in which case the resulting null space has dimension greater than one, and there is flexibility in the choice of the estimator $\mathbf{x}$. The minimum-norm solution is one such as advocated by Tufts and Kumaresan in a series of important papers [208,347,348]. These methods are analyzed comprehensively in [218,219,354].

Extension to Multiple Measurements. A collection of $N \leq L - p$ measurements may be organized into the $L \times N$ matrix $\mathbf{Y}$. These observations correspond to the source matrix $\mathbf{X} \in \mathbb{C}^{p \times N}$, and the objective is to recover them under the previous scenario where there are also errors in the model $\mathbf{H}$. Thus, the measurement model is $\mathbf{Y}-\mathbf{N}=(\mathbf{H}+\mathbf{E})\mathbf{X}$ or, equivalently,

$$\left(\begin{bmatrix}\mathbf{Y} & -\mathbf{H}\end{bmatrix}+\begin{bmatrix}-\mathbf{N} & -\mathbf{E}\end{bmatrix}\right)\begin{bmatrix}\mathbf{I}_N \\ \mathbf{X}\end{bmatrix}=\mathbf{0}_{L\times N}.$$

This constraint now says the $(N+p)\times N$ matrix $\begin{bmatrix}\mathbf{I}_N & \mathbf{X}^T\end{bmatrix}^T$ belongs to the null space of the adjusted $L\times(N+p)$ matrix $\left(\begin{bmatrix}\mathbf{Y} & -\mathbf{H}\end{bmatrix}+\begin{bmatrix}-\mathbf{N} & -\mathbf{E}\end{bmatrix}\right)$. The problem is to find the minimum-norm adjustment $\begin{bmatrix}-\mathbf{N} & \mathbf{E}\end{bmatrix}$ to the model $\begin{bmatrix}\mathbf{Y} & -\mathbf{H}\end{bmatrix}$, under the constraint $\mathbf{Y}-\mathbf{N}=(\mathbf{H}+\mathbf{E})\mathbf{X}$. Hence, the optimization problem in this case is

$$\begin{aligned}&\underset{\mathbf{N},\mathbf{E},\mathbf{X}}{\text{minimize}} && \left\|\begin{bmatrix}\mathbf{N} & \mathbf{E}\end{bmatrix}\right\|^2,\\ &\text{subject to} && \mathbf{Y}-\mathbf{N}=(\mathbf{H}+\mathbf{E})\mathbf{X},\end{aligned}$$

and its solution reduces the rank of $\begin{bmatrix}\mathbf{Y} & -\mathbf{H}\end{bmatrix}$ by N, yielding an adjusted matrix with a null space of dimension N.

Again, the solution to TLS with N observations is based on the SVD. Assume $L\geq N+p$, and write the SVD of the $L\times(N+p)$ augmented matrix as

$$\begin{bmatrix}\mathbf{Y} & -\mathbf{H}\end{bmatrix}=\mathbf{F}\mathbf{K}\mathbf{G}^H=\begin{bmatrix}\mathbf{F}_p & \mathbf{F}_N\end{bmatrix}\begin{bmatrix}\mathbf{K}_p & \mathbf{0}\\ \mathbf{0}^T & \mathbf{K}_N\end{bmatrix}\begin{bmatrix}\mathbf{G}_p^H\\ \mathbf{G}_N^H\end{bmatrix},$$

where $\mathbf{F}_p\in\mathbb{C}^{L\times p}$, $\mathbf{F}_N\in\mathbb{C}^{L\times N}$, $\mathbf{K}_p$ is a $p\times p$ diagonal matrix, $\mathbf{K}_N$ is a $N\times N$ diagonal matrix, $\mathbf{G}_p\in\mathbb{C}^{(p+N)\times p}$, and

$$\mathbf{G}_N=\begin{bmatrix}\tilde{\mathbf{G}}_1\\ \tilde{\mathbf{G}}_2\end{bmatrix},$$

with $\tilde{\mathbf{G}}_1\in\mathbb{C}^{N\times N}$ and $\tilde{\mathbf{G}}_2\in\mathbb{C}^{p\times N}$. The adjustment is now $\begin{bmatrix}-\mathbf{N} & -\mathbf{E}\end{bmatrix}=-\mathbf{F}_N\mathbf{K}_N\mathbf{G}_N^H$, with squared Frobenius norm $\|\mathbf{K}_N\|^2$. This adjustment is the adjustment with minimum norm that reduces the rank of $\begin{bmatrix}\mathbf{Y} & -\mathbf{H}\end{bmatrix}$ by N. Moreover, the adjusted matrix $\left(\begin{bmatrix}\mathbf{Y} & -\mathbf{H}\end{bmatrix}+\begin{bmatrix}-\mathbf{N} & -\mathbf{E}\end{bmatrix}\right)$ becomes $\mathbf{F}_p\mathbf{K}_p\mathbf{G}_p^H$, and $\begin{bmatrix}\mathbf{I}_N & \mathbf{X}^T\end{bmatrix}^T=\mathbf{G}_N\tilde{\mathbf{G}}_1^{-1}$ belongs to its null space. Again, the new model $\mathbf{H}+\mathbf{E}$ is given by the p last columns of $\mathbf{F}\mathbf{K}\mathbf{G}^H$ with a change of sign, whereas the adjusted measurements $\mathbf{Y}-\mathbf{N}$ are the first N columns of $\mathbf{F}\mathbf{K}\mathbf{G}^H$. The solution for $\hat{\mathbf{X}}$ is given by $\tilde{\mathbf{G}}_2\tilde{\mathbf{G}}_1^{-1}$, which requires $\tilde{\mathbf{G}}_1$ to be a nonsingular matrix. In the very special case that $\mathbf{K}_N=k_N\mathbf{I}_N$, $\hat{\mathbf{X}}=\tilde{\mathbf{G}}_2\tilde{\mathbf{G}}_1^{-1}$ can also be rewritten as a regularized LS solution:

$$\hat{\mathbf{X}}=\left(\mathbf{H}^H\mathbf{H}-k_N^2\mathbf{I}_p\right)^{-1}\mathbf{H}^H\mathbf{Y},$$

provided that the smallest eigenvalue of $\mathbf{H}^H\mathbf{H}$ is larger than k_N^2.

2.2.10 Least Squares and Procrustes Problems for Channel Identification

The channel identification problem is to begin with synchronized measurements $\mathbf{Y}$ and signals $\mathbf{X}$, both known, and to extract a model for the channel matrix that connects them. That is, the model is $\mathbf{Y} = \mathbf{HX} + \mathbf{N}$, with $\mathbf{Y} \in \mathbb{C}^{L\times N}, \mathbf{X} \in \mathbb{C}^{p\times N}$, and $\mathbf{H} \in \mathbb{C}^{L\times p}$. Our principle will be least squares. But when the extracted channel matrix is constrained to be unitary or a slice of a unitary matrix, the problem is said to be a *Procrustes problem* [146]. When the channel matrix is constrained or parametrically modeled, then the problem is a *modal analysis problem*, to be studied in the next section.

Least Squares: $\mathbf{Y} \sim \mathbf{HX}$. The sum of squared errors between elements of $\mathbf{Y}$ and elements of $\mathbf{HX}$ is

$$\begin{aligned} V &= \operatorname{tr}\left[(\mathbf{Y}-\mathbf{HX})(\mathbf{Y}-\mathbf{HX})^H\right] \\ &= \operatorname{tr}\left(\mathbf{YY}^H - \mathbf{HXY}^H - \mathbf{YX}^H\mathbf{H}^H + \mathbf{HXX}^H\mathbf{H}^H\right). \end{aligned}$$

This is minimized at the solution $\mathbf{H} = \mathbf{YX}^H(\mathbf{XX}^H)^{-1}$, in which case $\mathbf{HX} = \mathbf{YP_X}$ and $V = \operatorname{tr}\left(\mathbf{Y}(\mathbf{I}_L - \mathbf{P_X})\mathbf{Y}^H\right)$, where $\mathbf{YP_X} = \mathbf{YX}^H(\mathbf{XX}^H)^{-1}\mathbf{X}$ is a projection of the rows of $\mathbf{Y}$ onto the subspace spanned by the rows of $\mathbf{X}$.

Procrustes: $\mathbf{Y} \sim \mathbf{HX}$, with $H^{HH} = I_p$. The problem is

$$\begin{aligned} &\underset{\mathbf{H}\in\mathbb{C}^{L\times p}}{\text{minimize}} \quad V, \\ &\text{subject to} \quad \mathbf{H}^H\mathbf{H} = \mathbf{I}_p. \end{aligned}$$

Taking into account the constraint, V becomes

$$V = \operatorname{tr}\left[\mathbf{YY}^H + \mathbf{XX}^H - 2\operatorname{Re}(\mathbf{HXY}^H)\right].$$

So the problem is to maximize the last term in this equation. Give the $p \times L$ cross-Gramian $\mathbf{XY}^H$ the SVD $\mathbf{FKG}^H$, where $\mathbf{F}$ is a $p \times p$ unitary, $\mathbf{G}$ is an $L \times p$ unitary, and $\mathbf{K}$ is a $p \times p$ matrix of non-negative singular values. Then, the problem is to maximize

$$\operatorname{Re}\left[\operatorname{tr}\left(\mathbf{KG}^H\mathbf{HF}\right)\right].$$

This is less than or equal to $\sum_{l=1}^{p} k_l$, with equality achieved at $\mathbf{H} = \mathbf{GF}^H$. The resulting channel map $\mathbf{HX}$ is then $\mathbf{HX} = \mathbf{GF}^H\mathbf{X}$, and the error is $V =$

$\mathrm{tr}\left(\mathbf{XX}^H + \mathbf{YY}^H\right) - 2\sum_{l=1}^{p} k_l$. If $\mathbf{H}$ is replaced by $\mathbf{G}_r\mathbf{F}_r^H$, where $\mathbf{F}_r$ and $\mathbf{G}_r$ are the r dominant left and right singular vectors, then the second term in the equation for V terminates at r.

Comment. Scale matters in the Procrustes problem. Had we begun with the orthogonal slices $\mathbf{U}_\mathbf{X}$ and $\mathbf{U}_\mathbf{Y}$ in place of the data matrices $\mathbf{X}$ and $\mathbf{Y}$, then the k_l would have been cosine-squareds of the principal angles between the subspaces $\langle \mathbf{U}_\mathbf{X} \rangle$ and $\langle \mathbf{U}_\mathbf{Y} \rangle$.

2.2.11 Least Squares Modal Analysis

There is a vast literature on the topic of modal analysis, as it addresses the problem of identifying *two* sets of parameters, $\mathbf{x}$ and $\boldsymbol{\theta}$, in a separable linear model $\mathbf{y} = \mathbf{H}(\boldsymbol{\theta})\mathbf{x} + \mathbf{n}$. After estimating $\mathbf{x}$, the sum of squared residuals is $V(\boldsymbol{\theta}) = \mathbf{y}^H(\mathbf{I}_L - \mathbf{P}_{\mathbf{H}(\boldsymbol{\theta})})\mathbf{y}$, with $\mathbf{P}_{\mathbf{H}(\boldsymbol{\theta})} = \mathbf{H}(\boldsymbol{\theta})[\mathbf{H}^H(\boldsymbol{\theta})\mathbf{H}(\boldsymbol{\theta})]^{-1}\mathbf{H}^H(\boldsymbol{\theta})$ the orthogonal projection onto the p-dimensional subspace $\langle \mathbf{H}(\boldsymbol{\theta}) \rangle$. The problem is made interesting by the fact that typically the modes of $\mathbf{H}(\boldsymbol{\theta})$ are nearly co-linear. Were it not for the constraint that the subspace $\langle \mathbf{H}(\boldsymbol{\theta}) \rangle$ is constrained by a parametric model, then $\mathbf{H}$ would simply be chosen to be any p-dimensional subspace that traps $\mathbf{y}$. With the constraint, the problem is to maximize the coherence

$$\frac{\mathbf{y}^H\mathbf{P}_{\mathbf{H}(\boldsymbol{\theta})}\mathbf{y}}{\mathbf{y}^H\mathbf{y}}.$$

One may construct a sequence of Newton steps, or any other numerical method, ignoring the normalization by $\mathbf{y}^H\mathbf{y}$.

There is a fairly general case that arises in modal analysis for complex exponential modes, parameterized by mode frequencies $z_k, k = 1, 2, \ldots, p$. In this case, the channel matrix is Vandermonde, $\mathbf{H}(\boldsymbol{\theta}) = [\mathbf{h}_1 \ \cdots \ \mathbf{h}_p]$, where $\mathbf{h}_k = [1 \ z_k \ \cdots \ z_k^{L-1}]^T$. The mode frequencies are zeros of a pth-order polynomial $A(z) = 1 + a_1 z + \cdots + a_p z^p$, which is to say that for any choice of $\boldsymbol{\theta} = [z_1 \ z_2 \ \cdots \ z_p]^T$, there is a corresponding $(L-p)$-dimensional subspace $\langle \mathbf{A}(\mathbf{a}) \rangle$ determined by the $(L-p)$ equations $\mathbf{A}^H(\mathbf{a})\mathbf{H}(\boldsymbol{\theta}) = \mathbf{0}$. The matrix $\mathbf{A}^H(\mathbf{a})$ is the Toeplitz matrix

$$\mathbf{A}^H(\mathbf{a}) = \begin{bmatrix} a_p & \cdots & a_1 & 1 & 0 & \cdots & 0 \\ 0 & a_p & \cdots & a_1 & 1 & \cdots & 0 \\ \vdots & \vdots & \ddots & \ddots & \ddots & \ddots & \vdots \\ 0 & \cdots & 0 & a_p & \cdots & a_1 & 1 \end{bmatrix}.$$

The projections $\mathbf{P}_{\mathbf{H}(\boldsymbol{\theta})}$ and $\mathbf{P}_\mathbf{A}$ resolve identity as $\mathbf{P}_{\mathbf{H}(\boldsymbol{\theta})} + \mathbf{P}_\mathbf{A} = \mathbf{I}_L$, so that $\mathbf{y}^H(\mathbf{I}_L - \mathbf{P}_{\mathbf{H}(\boldsymbol{\theta})})\mathbf{y}$ may be written as

$$\mathbf{y}^H\mathbf{P}_\mathbf{A}\mathbf{y} = \mathbf{a}^H\mathbf{Y}^H(\mathbf{A}^H(\mathbf{a})\mathbf{A}(\mathbf{a}))^{-1}\mathbf{Y}\mathbf{a},$$

where $\mathbf{a} = [a_p \ \cdots \ a_1 \ 1]^T$ and $\mathbf{Y}$ is the Hankel matrix

$$\mathbf{Y} = \begin{bmatrix} y_1 & y_2 & \cdots & y_{p+1} \\ y_2 & y_3 & \cdots & y_{p+2} \\ \vdots & \vdots & \ddots & \vdots \\ y_{L-p} & y_{L-p+1} & \cdots & y_L \end{bmatrix}.$$

Here, we have used the identity $\mathbf{A}^H(\mathbf{a})\mathbf{y} = \mathbf{Y}\mathbf{a}$. Call $\mathbf{a}_n$ an estimate of $\mathbf{a}$ at step n of an iteration. From it, construct the Gramian $\mathbf{A}^H(\mathbf{a}_n)\mathbf{A}(\mathbf{a}_n)$ and its inverse. Then, minimize $\mathbf{a}^H\mathbf{Y}^H(\mathbf{A}^H(\mathbf{a}_n)\mathbf{A}(\mathbf{a}_n))^{-1}\mathbf{Y}\mathbf{a}$ with respect to $\mathbf{a}$, under the constraint that its last element is 1. This is linear prediction. Call the resulting minimizer $\mathbf{a}_{n+1}$ and proceed. This algorithm may be called iterative quadratic least squares (IQLS), a variation on iterative quadratic maximum likelihood (IQML), a term used to describe the algorithm when derived in the context of maximum likelihood theory [47, 209, 237].

2.3 Under-determined Least Squares and Related

In the under-determined linear model, the measurement $\mathbf{y}$ is modeled as $\mathbf{y} = \mathbf{H}\mathbf{x}+\mathbf{n}$, where $\mathbf{H} \in \mathbb{C}^{L\times p}$, $\mathbf{x} \in \mathbb{C}^p$, and $L \leq p$. That is, the number of unknown parameters exceeds the number of measurements. The problem is to invert for $\mathbf{x}$. But once a candidate is found, an additive correction $\mathbf{x} + \mathbf{a}$, where $\mathbf{a}$ lies in the null space of $\mathbf{H}$, leaves the approximation error $\mathbf{n} = \mathbf{y} - \mathbf{H}(\mathbf{x} + \mathbf{a})$ unchanged at $\mathbf{n} = \mathbf{y} - \mathbf{H}\mathbf{x}$. So to invert for $\mathbf{x}$ with some claim to optimality or desirability, it is necessary to impose constraints on the solution. In the following subsections, we review a few common constraints. We shall assume throughout that the matrix $\mathbf{H}$ has full row rank of L.

2.3.1 Minimum-Norm Solution

The minimum-norm solution for $\mathbf{x}$ is the solution for which $\mathbf{H}\mathbf{x} = \mathbf{y}$, and the norm-squared of $\mathbf{x}$ is minimum (a tautology). One candidate is $\hat{\mathbf{x}} = \mathbf{H}^H(\mathbf{H}\mathbf{H}^H)^{-1}\mathbf{y}$. This solution lies in the range of $\mathbf{H}^H$. Any other candidate may be written as $\mathbf{x} = \alpha\hat{\mathbf{x}} + \mathbf{A}^H\boldsymbol{\beta}$, where $\mathbf{A}^H \in \mathbb{C}^{p\times(p-L)}$ is full column rank and orthogonal to $\mathbf{H}^H$, i.e., $\mathbf{A}\mathbf{H}^H = \mathbf{0}$. Then $\mathbf{H}\mathbf{x} = \alpha\mathbf{y}$, which requires $\alpha = 1$. Moreover, the norm-squared of $\mathbf{x}$ is $\mathbf{x}^H\mathbf{x} = (\hat{\mathbf{x}}^H + \mathbf{A}^H\boldsymbol{\beta})^H(\hat{\mathbf{x}} + \mathbf{A}^H\boldsymbol{\beta}) \geq \hat{\mathbf{x}}^H\hat{\mathbf{x}}$. This makes $\hat{\mathbf{x}}$ the minimum-norm inverse.

In the over-determined problem, $L \geq p$, the least squares solution for $\hat{\mathbf{x}}$ is $\hat{\mathbf{x}} = (\mathbf{H}^H\mathbf{H})^{-1}\mathbf{H}^H\mathbf{y} = \mathbf{G}\mathbf{K}^{\#}\mathbf{F}^H\mathbf{y}$, where $\mathbf{F}\mathbf{K}\mathbf{G}^H$ is the SVD of the $L \times p$ matrix $\mathbf{H}$ and $\mathbf{G}\mathbf{K}^{\#}\mathbf{F}^H$ is the pseudo-inverse of $\mathbf{H}$. In the under-determined case, $L \leq p$, the minimum-norm solution that reproduces the measurements is $\hat{\mathbf{x}} = \mathbf{H}^H(\mathbf{H}\mathbf{H}^H)^{-1}\mathbf{y} = \mathbf{G}\mathbf{K}^{\#}\mathbf{F}^H\mathbf{y}$, where $\mathbf{F}\mathbf{K}\mathbf{G}^H$ is the SVD of the $L \times p$ matrix $\mathbf{H}$. So from the SVD $\mathbf{H} = \mathbf{F}\mathbf{K}\mathbf{G}^H$, one extracts a universal pseudo-inverse $\mathbf{G}\mathbf{K}^{\#}\mathbf{F}^H$. The reader is referred to Appendix C for more details.

2.3.2 Sparse Solutions

If, in a linear model, the parameter vector $\mathbf{x}$ may be assumed to be sparse in a unitary basis, then this condition may be used to invert for an $\mathbf{x}$ that is sparse or approximately sparse in this basis. In some cases, the unitary basis is the Euclidean basis, which is to say $\mathbf{x}$ itself is sparse. For some classes of problems, such as post-processing of detected images, this is a powerful idea. The question of which basis to use is not critical. But for problems of imaging from radiated signals, as in radar, sonar, geophysics, and optics, it is rarely the case that a unitary basis in which the parameter is sparse can be known a priori. For example, radiated waves do not arrive from quantized electrical angles, broadband signals do not have line spectra at harmonic lines, and multipath copies of signals do not arrive at quantized delays. So a linear model based on such assumptions will be mismatched to the actual linear model, which is to say that the parameters in the mismatched model will not be sparse. In some cases, they will not even be compressible.[3] This issue is taken up in the subsection on basis mismatch.

The least squares problem that constrains $\mathbf{x}$ to be K sparse, which is to say its support is not larger than K, can be found as the solution to the optimization problem

$$\begin{aligned} &\underset{\mathbf{x}}{\text{minimize}} \quad \|\mathbf{y} - \mathbf{H}\mathbf{x}\|_2^2, \\ &\text{subject to} \quad \|\mathbf{x}\|_0 \leq K, \end{aligned}$$

where $\|\mathbf{x}\|_0 = \dim(\{k \mid x_k \neq 0\})$ is the ℓ_0 norm of $\mathbf{x}$. It may be written as $\|\mathbf{x}\|_0 = \sum u(|x_k|)$, where $u(a)$ is zero at $a = 0$ and one elsewhere. This problem is non-convex, and, as stated, it assumes a known bound K on the support of $\mathbf{x}$. An alternative is to replace the problem with the regularized LS problem

$$\underset{\mathbf{x}}{\text{minimize}} \quad \|\mathbf{y} - \mathbf{H}\mathbf{x}\|_2^2 + \mu \|\mathbf{x}\|_0. \tag{2.3}$$

The support is not constrained, but a large value of μ promotes small support, and a small value allows for large support. The problem remains non-convex. A convex relaxation of this non-convex problem is

$$\underset{\mathbf{x}}{\text{minimize}} \quad \|\mathbf{y} - \mathbf{H}\mathbf{x}\|_2^2 + \mu \|\mathbf{x}\|_1, \tag{2.4}$$

where $\|\mathbf{x}\|_1 = \sum |x_k|$. The magnitude $|x_k|$ may be considered an approximation to $u(|x_k|)$. The problem in (2.4) is the well-known LASSO in its Lagrangian form

[3] If $\mathbf{x}$ is sparse, then the cardinality of its support $\{k \mid x_k \neq 0\}$ is assumed to be small. If $\mathbf{x}$ is compressible, then its entries, ordered by magnitude, decay as a power law: $|x_{(k)}| \leq C_r \cdot k^{-r}$, $r > 1$ and C_r is a constant depending only on r. Then $\|\mathbf{x} - \mathbf{x}_k\|_1 \leq \sqrt{k} C_r' \cdot k^{-r+1}$, where $\mathbf{x}_k$ is the k-term approximation of $\mathbf{x}$.

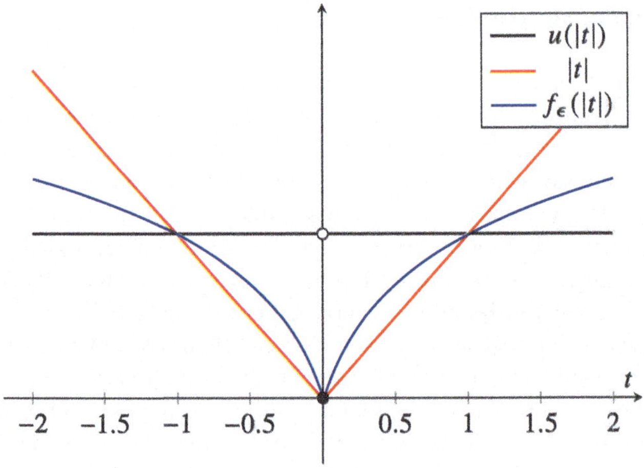

Fig. 2.2 Comparison of $u(|t|)$ and the two considered surrogates: $|t|$ and $f_\epsilon(|t|)$. Here, the step function $u(|t|)$ takes the value 0 at the origin and 1 elsewhere

[342]. The LASSO may sometimes be improved by improving on the approximation of $u(|x_k|)$, as shown in Fig. 2.2.

A typically better surrogate for the ℓ_0 norm is based on a logarithmic approximation to $u(|x_k|)$, given by [61]

$$f_\epsilon(|x_k|) = \frac{\log(1+\epsilon^{-1}|x_k|)}{\log(1+\epsilon^{-1})},$$

where the denominator ensures $f_\epsilon(1) = 1$. This surrogate, for $\epsilon = 0.2$, is depicted in Fig. 2.2, where we can see that it is a more accurate approximation of $u(\cdot)$ than is $|x|$. Using $f_\epsilon(|x_k|)$, the optimization problem in (2.3) becomes

$$\underset{\mathbf{x}}{\text{minimize}} \quad \|\mathbf{y} - \mathbf{Hx}\|_2^2 + \mu \sum_{k=1}^{p} \log(1+\epsilon^{-1}|x_k|), \tag{2.5}$$

where with some abuse of notation we have absorbed the term $\log(1+\epsilon^{-1})$ into the regularization parameter μ. However, contrary to the LASSO formulation (2.4), the optimization problem in (2.5) is no longer convex due to the logarithm. Then, to solve the optimization problem, [61] proposed an iterative approach that attains a local optimum of (2.5), which is based on a majorization-minimization (MM) algorithm [339]. The main idea behind MM algorithms is to find a majorizer of the cost function that is easy to minimize. Then, this procedure is iterated, and it can be shown that it converges to a local minimum of (2.5) [61]. Since the logarithm is a concave function, it is majorized by a first-order Taylor series. Then, applying this

Taylor series to the second term in (2.5), while keeping the first one, at each iteration the problem is to

$$\underset{\mathbf{x}}{\text{minimize}} \quad \|\mathbf{y} - \mathbf{H}\mathbf{x}\|_2^2 + \mu \sum_{k=1}^{p} w_k |x_k|, \tag{2.6}$$

where $w_k^{-1} = \left|x_k^{(i-1)}\right| + \epsilon$ and $x_k^{(i-1)}$ is the solution at the previous iteration.

The problem in (2.6) is almost identical to (2.4), but uses a re-weighted ℓ_1 norm instead of the ℓ_1 norm. If $|x_k^{(i-1)}|$ is small, then x_k will tend to be small, as w_k is large. The idea of replacing the step function by the log surrogate can be extended to other concave functions, such as atan$(\cdot)$. Moreover, approaches based on re-weighted ℓ_2 norm solutions have also been proposed in the literature (see [144] and references therein).

There are alternatives for finding a sparse solution to the under-determined LS problem. Orthogonal matching pursuit (OMP) [54,97,225,260] is an iterative greedy algorithm that selects at step $k+1$ the column of an over-complete basis that is most correlated with previous residual fitting errors of the form $(\mathbf{I}_L - \mathbf{P}_k)\mathbf{y}$ after all previously selected columns have been used to compose the orthogonal projection $\mathbf{P}_k$.

Conditions for recovery of sparse signals are described in Chap. 11, Sect. 11.1.

Bayesian Interpretation. Sparse solutions to the under-determined LS problem can also be given a maximum likelihood (ML) interpretation by positing a joint distribution for $\mathbf{y}$ and $\mathbf{x}$, with pdf $p(\mathbf{y}, \mathbf{x}) = p(\mathbf{y}|\mathbf{x})p(\mathbf{x})$. The term $p(\mathbf{y}|\mathbf{x})$ is the conditional pdf for $\mathbf{y}$, given $\mathbf{x}$, and the term $p(\mathbf{x})$ is the prior pdf for $\mathbf{x}$. For measured $\mathbf{y}$, the function $p(\mathbf{y}, \mathbf{x})$ is a likelihood function. The ML estimate of $\mathbf{x}$ maximizes log-likelihood, $\log p(\mathbf{y}|\mathbf{x}) + \log p(\mathbf{x})$.

Assume $p(\mathbf{y}|\mathbf{x})$ is the MVN pdf

$$p(\mathbf{y}|\mathbf{x}) = \frac{1}{\pi^L \sigma^{2L}} \exp\{-\|\mathbf{y} - \mathbf{H}\mathbf{x}\|_2^2/\sigma^2\}.$$

Then, the ML estimate solves the problem

$$\underset{\mathbf{x}}{\text{minimize}} \quad \|\mathbf{y} - \mathbf{H}\mathbf{x}\|_2^2 - \sigma^2 \log p(\mathbf{x}). \tag{2.7}$$

If we assume that the components of $\mathbf{x}$ are i.i.d., each with uniformly distributed phase and exponentially distributed magnitude, then $p(\mathbf{x})$ is

$$p(\mathbf{x}) = \frac{1}{(2\pi)^p} \prod_{k=1}^{p} c \exp\{-c|x_k|\}.$$

Now, plugging this prior into (2.7), the problem

$$\underset{\mathbf{x}}{\text{minimize}} \quad \|\mathbf{y} - \mathbf{H}\mathbf{x}\|^2 + \sigma^2 c \sum_{k=1}^{p} |x_k|$$

is identical to the ℓ_1-regularized LS solution in (2.4) with $\mu = \sigma^2 c$. Different priors yield different surrogates for the ℓ_0 norm.

These solutions are sometimes called MAP estimates because to maximize $p(\mathbf{y}, \mathbf{x})$ with respect to $\mathbf{x}$ is to maximize $p(\mathbf{x}|\mathbf{y}) = p(\mathbf{y}|\mathbf{x})p(\mathbf{x})/p(\mathbf{y})$, where $p(\mathbf{x}|\mathbf{y})$ is the a posteriori pdf of $\mathbf{x}$, given $\mathbf{y}$. But for measured $\mathbf{y}$, this is an a posteriori likelihood, so MAP (maximum a posteriori probability) is really MAL (maximum a posteriori likelihood). These solutions are also commonly called Bayesian solutions because they also maximize an a posteriori likelihood that is computed by Bayes rule. This is a misnomer, as Bayes rule is never actually used in the construction of the likelihood function $p(\mathbf{y}|\mathbf{x})p(\mathbf{x})$. To normalize this likelihood function by $p(\mathbf{y})$ to compute the a posteriori likelihood function $p(\mathbf{x}|\mathbf{y}) = p(\mathbf{y}|\mathbf{x})p(\mathbf{x})/p(\mathbf{y})$ would require the computation of the marginal pdf $p(\mathbf{y})$, which is computationally demanding and irrelevant for solution of the ML problem. That is, Bayes rule is not used to invert the prior pdf $p(\mathbf{x})$ for the posterior pdf $p(\mathbf{x}|\mathbf{y})$.

In spite of these reservations about terminology, these solutions continue to go by the name Bayesian, perhaps because the assignment of a prior distribution for an unknown parameter is central to Bayesian reasoning.

Dantzig Selector. The Dantzig[4] selector [59] is the solution to the optimization problem

$$\begin{aligned} &\underset{\mathbf{x}}{\text{minimize}} \quad \|\mathbf{H}^H(\mathbf{y} - \mathbf{H}\mathbf{x})\|_\infty, \\ &\text{subject to} \quad \|\mathbf{x}\|_1 \leq \kappa, \end{aligned} \tag{2.8}$$

where the ℓ_∞ norm is the largest of the absolute values of the error $\mathbf{y} - \mathbf{H}\mathbf{x}$ after resolution onto the columns of $\mathbf{H}$ or the largest of the absolute values of the gradient components of $(\mathbf{y} - \mathbf{H}\mathbf{x})^H(\mathbf{y} - \mathbf{H}\mathbf{x})$.

Interestingly, depending on the values of L, p, and κ, the Dantzig selector obtains several of the solutions derived above [158]. In over-determined scenarios ($L \geq p$) and for large κ, the solution to (2.8) is the classical LS solution presented in Sect. 2.2. This is also the case if we consider a very small μ in (2.4). In the under-determined case, the Dantzig selector achieves also a sparse solution, although not necessarily identical to the solution of (2.4) (for a properly selected μ).

[4] George Dantzig developed the simplex method for solving linear programming problems.

Basis Mismatch. Begin with an under-determined linear model $\mathbf{y} = \mathbf{Hx} + \mathbf{n}$, where $\mathbf{H} \in \mathbb{C}^{L\times p}, \mathbf{x} \in \mathbb{C}^p$, and $L \leq p$. Call $\mathbf{Hx}$ the noise-free component of the measurement $\mathbf{y}$, and ignore consideration of the noise $\mathbf{n}$ in order to gain insight into the problem of *basis mismatch.*

Suppose $\mathbf{x}$ is sparse in a unitary basis $\mathbf{V}$, which is to say $\mathbf{x} = \mathbf{Vt}$, with $\mathbf{t}$ sparse. Then $\mathbf{Hx} = \mathbf{HVt} = \mathbf{H}_0\mathbf{t}$, where $\mathbf{H}_0 = \mathbf{HV}$ is a row-transformed version of $\mathbf{H}$ by the transformation $\mathbf{V}$. That is, the signal $\mathbf{Hx}$ is sparse in the over-complete basis $\mathbf{H}_0$. One may then use an ℓ_1-regularized (or a re-weighted ℓ_1 norm) to solve the inverse problem for an approximation of sparse $\mathbf{t}$. But suppose the basis $\mathbf{V}$ is unknown and approximated with a convenient or inspired unitary dictionary $\mathbf{W}$. Then $\mathbf{x}$ is assumed to be sparse in this dictionary, $\mathbf{x} = \mathbf{Wu}$, with $\mathbf{u}$ sparse. The assumed sparse signal model is $\mathbf{HWu} = \mathbf{H}_1\mathbf{u}$, where $\mathbf{H}_1 = \mathbf{HW}$ is a row-transformed version of $\mathbf{H}$ by the transformation $\mathbf{W}$. But in this over-complete basis $\mathbf{H}_1$, the actual sparse model is $\mathbf{HVt} = \mathbf{HWW}^H\mathbf{Vt} = \mathbf{H}_1\mathbf{W}^H\mathbf{Vt}$. There are two interpretations: in the model $\mathbf{H}_1\mathbf{u}$, the vector $\mathbf{u} = \mathbf{W}^H\mathbf{Vt}$ *is not sparse*; the signal $\mathbf{Hx}$ is sparse in the basis $\mathbf{H}_1(\mathbf{W}^H\mathbf{V}) = \mathbf{H}_0$, *not the basis* $\mathbf{H}_1$. This is basis mismatch.

There are two natural questions: (1) How close is the sparsely extracted model $(\mathbf{H}_1, \mathbf{u})$ to the sparse physical model $(\mathbf{H}_0, \mathbf{t})$? (2) How close is the sparsely extracted model $(\mathbf{H}_1, \mathbf{u})$ to the non-sparse model $(\mathbf{H}_1, \mathbf{W}^H\mathbf{Vt})$? The answer to the first question cannot be known because the physical model $(\mathbf{H}_0, \mathbf{t})$ is unknown. But the difference between the sparsely extracted $\mathbf{u}$ in the assumed basis $\mathbf{H}_1$ and the actual $\mathbf{t}$ in the basis $\mathbf{H}_0$ can be bounded if the assumed basis is near enough to the physical basis, and the additive noise is small enough. The second question addresses the difference between a sparsely extracted $\mathbf{u}$ in the assumed basis $\mathbf{H}_1$ and the actual non-sparse $\mathbf{u} = \mathbf{W}^H\mathbf{Vt}$ in the basis $\mathbf{H}_1$. The answer to the second question establishes that when the over-complete basis for an under-determined problem is mismatched to the physical model, ℓ_1 regularization (or any other variant described above) is a way of *sparsifying a non-sparse parameter in the mismatched basis.* If non-sparse $\mathbf{u}$ is compressible, then this sparsification can be effective. If the mismatched basis is a DFT basis, then we are sparsifying the DFT coefficients.

These questions have motivated a number of papers on the issue of basis mismatch, and no attempt will be made to review this literature. But the early important papers to address these questions are [161] for question number (1) and [72] for question (2). Reference [312] addresses the issue for modal analysis.

2.3.3 Maximum Entropy Solution

To assume sparsity for $\mathbf{x}$ is to constrain feasible solutions and therefore to *rule out* sequences that are not sparse. Typically, this is done by relaxing the sparsity problem to a nearby problem that promotes sparsity. In many applications, this is a very important conceptual breakthrough in the solution of under-determined problems. When no such constraint is justified, then an alternative conceptual idea is to *rule in* as many sequences as possible, an important idea in statistical mechanics. This is the basis of maximum entropy inversion as demonstrated by R. Frieden [125].

To make the argument clear, let's imagine an experiment that generates a sequence of M symbols drawn with replacement from an alphabet $A = \{a_1, \ldots, a_p\}$. There are p^M possible sequences of symbols that can be generated. Each generated sequence will produce x_i for the number of symbols a_i in that sequence. The number of sequences that produce the counts $\{x_1, \ldots, x_p\}$ is $N(\mathbf{x}) = \frac{M!}{x_1! \cdots x_p!}$. This is the multinomial coefficient. For large M, this number is approximated as $2^{MH(\mathbf{x})}$, where $H(\mathbf{x}) = -\sum_{i=1}^{p}(x_i/M)\log_2(x_i/M)$. This is the entropy of a discrete random variable with pmf $\mathbf{x}/M$. It is maximized at the value $\log_2 p$ for $x_i/M = 1/p$ for $i = 1, \ldots, p$. That is, equal counts are produced by more sequences than unequal counts. (This is *not* to say equal counts are *more probable*, but that they are more typical.) If the objective were to maximize $N(\mathbf{x})$, then asymptotically in M, the solution would be to choose $x_i/M = 1/p$ for $i = 1, \ldots, p$. But our objective is to maximize the number of ways unknown $\mathbf{x}$ could have been produced, under the constraint that $\mathbf{y} = \mathbf{H}\mathbf{x}$:

$$\begin{aligned} &\underset{\mathbf{x}}{\text{maximize}} \quad \left(-\sum_{i=1}^{p} x_i \log x_i\right), \\ &\text{subject to} \quad \mathbf{y} = \mathbf{H}\mathbf{x}, \end{aligned} \tag{2.9}$$

where we have substituted $\log_2$ by log, that is, the entropy is measured in nats instead of bits.

Define the Lagrangian

$$\mathcal{L} = -\sum_{i=1}^{p} x_i \log x_i - \sum_{l=1}^{L} \lambda_l \left(y_l - \sum_{i=1}^{p} h_{li} x_i\right),$$

where h_{li} is the (l, i)-th element of $\mathbf{H}$. Set its gradients with respect to the x_i to 0 to obtain the solutions

$$\frac{\partial \mathcal{L}}{\partial x_i} = -\log x_i - 1 + \sum_{l=1}^{L} \lambda_l h_{li} = 0 \quad \Rightarrow \hat{x}_i = \exp\left\{-1 + \sum_{l=1}^{L} \lambda_l h_{li}\right\}.$$

In this maximum entropy solution, x_i is a non-negative number determined by the $L < p$ parameters $\lambda_l, l = 1, \ldots, L$. When these Lagrangian coefficients are selected to satisfy the constraints, then the resulting solutions for x_i are solutions to the under-determined problem $\mathbf{y} = \mathbf{H}\mathbf{x}$. To satisfy the constraints is to solve the equations

$$\sum_{i=1}^{p} h_{li} \exp\left\{-1 + \sum_{n=1}^{L} \lambda_n h_{ni}\right\} = y_l,\ l = 1, \ldots, L.$$

These may be written as

$$\frac{\partial}{\partial \lambda_l} Z(\lambda_1, \ldots, \lambda_L) = y_l,$$

where the partition function is $Z(\lambda_1, \ldots, \lambda_L) = \sum_{i=1}^{p} \exp\{-1+\sum_{n=1}^{L} \lambda_n h_{ni}\}$. It is now a sequence of Newton recursions to determine the λ_i, which then determine $\hat{x}_i$.

2.3.4 Minimum Mean-Squared Error Solution

Begin with the under-determined linear model $\mathbf{y} = \mathbf{H}\mathbf{x} + \mathbf{n}$, but now assign MVN distributions to $\mathbf{x}$ and $\mathbf{n}$: $\mathbf{x} \sim CN_p(\mathbf{0}, \mathbf{R}_{\mathbf{xx}})$ and $\mathbf{n} \sim CN_L(\mathbf{0}, \mathbf{R}_{\mathbf{nn}})$. In this linear model, the solution for $\mathbf{x}$ that returns the least positive definite error covariance matrix $E[(\mathbf{x} - \hat{\mathbf{x}})(\mathbf{x} - \hat{\mathbf{x}})^H]$ is $\hat{\mathbf{x}} = \mathbf{R}_{\mathbf{xx}}\mathbf{H}^H(\mathbf{H}\mathbf{R}_{\mathbf{xx}}\mathbf{H}^H + \mathbf{R}_{\mathbf{nn}})^{-1}\mathbf{y}$. The resulting minimum error covariance matrix is

$$\mathbf{Q}_{\mathbf{xx}|\mathbf{y}} = \mathbf{R}_{\mathbf{xx}} - \mathbf{R}_{\mathbf{xx}}\mathbf{H}^H(\mathbf{H}\mathbf{R}_{\mathbf{xx}}\mathbf{H}^H + \mathbf{R}_{\mathbf{nn}})^{-1}\mathbf{H}\mathbf{R}_{\mathbf{xx}}.$$

The minimum mean-squared error is the trace of this error covariance matrix. This solution is sometimes called a Bayesian solution, as it is also the mean of the posterior distribution for $\mathbf{x}$, given the measurement $\mathbf{y}$, which by the Bayes rule is $\mathbf{x} \sim CN_p(\hat{\mathbf{x}}, \mathbf{Q}_{\mathbf{xx}|\mathbf{y}})$.

In the case where $\mathbf{R}_{\mathbf{nn}} = \mathbf{0}$ and $\mathbf{R}_{\mathbf{xx}} = \mathbf{I}_p$, the minimum mean-squared error estimator is the minimum-norm solution $\mathbf{x} = \mathbf{H}^H(\mathbf{H}\mathbf{H}^H)^{-1}\mathbf{y}$. It might be said that the positive semidefinite matrices $\mathbf{R}_{\mathbf{xx}}$ and $\mathbf{R}_{\mathbf{nn}}$ determine a family of inversions. If the covariance matrices $\mathbf{R}_{\mathbf{xx}}$ and $\mathbf{R}_{\mathbf{nn}}$ are known from physical measurements or theoretical reasoning, then the solution for $\hat{\mathbf{x}}$ does what it is designed to do: minimize error covariance. If these covariance parameters are design variables, then the returned solution is of course dependent on design choices. Of course, this development holds for $p \leq L$, as well, making the solutions for $\hat{\mathbf{x}}$ and $\mathbf{Q}_{\mathbf{xx}|\mathbf{y}}$ general.

2.4 Multidimensional Scaling

Let's begin with the data matrix $\mathbf{X} \in \mathbb{C}^{L\times N}$. One interpretation is that each of the N columns of this matrix is a snapshot in time, taken by an L-element sensor array. An equivalent interpretation is that a column consists of a realization of an L-dimensional random vector. Then, each row of $\mathbf{X}$ is an N-sample random sample of the lth element of this random vector. In these interpretations, the $L \times L$ Gramian $\mathbf{G} = \mathbf{X}\mathbf{X}^H$ is a sum of rank-one outer products, or a matrix of inner products between the rows of $\mathbf{X}$, each such Euclidean inner product serving as an estimate of the Hilbert space inner product between two random variables in the L-dimensional random vector. The Gramian $\mathbf{G}$ has rank $p \leq \min(L, N)$.

Algorithm 1: From Gramian to Euclidean distance matrix and low-dimensional configuration

Input: $L \times L$ Gramian $\mathbf{G} \succ \mathbf{0}$ of rank p
Output: low-dimensional configuration of L points in $\mathbb{C}^p$, $\mathbf{X} \in \mathbb{C}^{L\times p}$, and Euclidean distance matrix $\mathbf{D}$
Compute $\mathbf{G} = \mathbf{FKF}^H$ // EVD of the Gramian
Construct $\mathbf{X} = \mathbf{FK}^{1/2}\mathbf{U}_p$ // $\mathbf{U}_p$ is an arbitrary $p \times p$ unitary matrix
Compute $\mathbf{D} = \{d_{il}\}$, where $d_{il}^2 = (\mathbf{e}_i - \mathbf{e}_l)^T\mathbf{G}(\mathbf{e}_i - \mathbf{e}_l) = (\mathbf{x}_i - \mathbf{x}_l)(\mathbf{x}_i - \mathbf{x}_l)^H$

An alternative interpretation is that each row of $\mathbf{X}$ is a point in $\mathbb{C}^N$, and there are L such points. The Gramian $\mathbf{G} = \mathbf{XX}^H$ is a matrix of inner products between these points. When $L \leq N$, then there are a small number of points in a high-dimensional space. When $L \geq N$, then there are a large number of points in a low-dimensional space. In any case, there is the suggestion that these N-dimensional points, when visualized in a lower-dimensional space, might bring insight into the structure of the data.

From Gramian to Euclidean Distance Matrix and Low-Dimensional Configuration. Is it possible, given only the non-negative definite Gramian $\mathbf{G} = \{g_{il}\} \in \mathbb{C}^{L\times L}$ of rank $p \leq \min(L, N)$, to extract a *Euclidean distance matrix*? The answer is yes, and the argument is this. Begin with $\mathbf{G} = \mathbf{FKF}^H$, and define the $L \times p$ matrix $\mathbf{X} = \mathbf{FK}^{1/2}\mathbf{U}_p$, where $\mathbf{U}_p$ is an arbitrary $p \times p$ unitary matrix, in which case $\mathbf{G} = \mathbf{XX}^H$. The matrix $\mathbf{X} \in \mathbb{C}^{L\times p}$ is a configuration of L points in $\mathbb{C}^p$ that reproduces $\mathbf{G}$. These points, which we denote by the p-dimensional row vectors $\mathbf{x}_l, l = 1, \ldots, L$, are the rows of $\mathbf{X}$. Define the standard basis vector $\mathbf{e}_l = [0 \cdots 0\, 1\, 0 \cdots 0]^T$, where the 1 appears in the lth location. Then, note that the quadratic form $(\mathbf{e}_i - \mathbf{e}_l)^T\mathbf{G}(\mathbf{e}_i - \mathbf{e}_l)$ extracts the scalar $g_{ii} - g_{il} - g_{li} + g_{ll}$. But as $\mathbf{G} = \mathbf{XX}^H$, this quadratic form is also $(\mathbf{e}_i - \mathbf{e}_l)^T\mathbf{XX}^H(\mathbf{e}_i - \mathbf{e}_l) = (\mathbf{x}_i - \mathbf{x}_l)(\mathbf{x}_i - \mathbf{x}_l)^H = d_{il}^2$. So, the terms in the Gramian $\mathbf{G}$ may be used to extract the d_{il}^2, which are the squared distances between the ith and lth rows of the configuration $\mathbf{X}$. From these d_{il}^2, which are non-negative, extract their square roots, and construct the Euclidean distance matrix $\mathbf{D} = \{d_{il}\}$, whose elements are $d_{il} = \sqrt{(\mathbf{x}_i - \mathbf{x}_l)(\mathbf{x}_i - \mathbf{x}_l)^H}$. We say the matrix $\mathbf{X} \in \mathbb{C}^{L\times p}$ is a configuration of L p-vectors that reproduces the $L \times L$, rank-p, Gramian $\mathbf{G}$ as $\mathbf{XX}^H$ and delivers a distance matrix $\mathbf{D}$ in the bargain. The program is described in Algorithm 1.

From Euclidean Distance Matrix to Gramian and Low-Dimensional Configuration. What about the other way around? If we begin with a Euclidean distance matrix $\mathbf{D} \in \mathbb{C}^{L\times L}$, can we extract a configuration $\mathbf{X} \in \mathbb{C}^{L\times p}$ and its Gramian $\mathbf{G} = \mathbf{XX}^H$? The answer is yes, and the argument is this.[5] Begin with the $L \times L$

[5] This is the original idea of multidimensional scaling (MDS). Evidently, the mathematical foundations for MDS were laid by Schoenberg [315] and by Young and Householder [390]. The theory as we describe it here was developed by Torgerson [343] and Gower [145].

distance matrix $\mathbf{D} = \{d_{il}\}$, assumed to be Euclidean for L points in $\mathbb{C}^N$, N greater or lesser than L. Necessarily, $d_{ii} = 0$ and $d_{il} = d_{li} \geq 0$. Define the modified squared distance matrix $\mathbf{A} = -(1/2)\mathbf{D} \odot \mathbf{D} \in \mathbb{C}^{L \times L}$. The configuration $\mathbf{X}$ that produces this distance matrix is unknown. But the following fundamental identities are known:

$$(\mathbf{e}_i - \mathbf{e}_l)^T \mathbf{P}_{\mathbf{1}}^{\perp} = (\mathbf{e}_i - \mathbf{e}_l)^T,$$
$$(\mathbf{e}_i - \mathbf{e}_l)^T \mathbf{A}(\mathbf{e}_i - \mathbf{e}_l) = d_{il}^2,$$
$$(\mathbf{e}_i - \mathbf{e}_l)^T \mathbf{P}_{\mathbf{1}}^{\perp} \mathbf{A} \mathbf{P}_{\mathbf{1}}^{\perp} (\mathbf{e}_i - \mathbf{e}_l) = (\mathbf{e}_i - \mathbf{e}_l)^T \mathbf{A}(\mathbf{e}_i - \mathbf{e}_l),$$

where $\mathbf{P}_{\mathbf{1}}^{\perp} = \mathbf{I}_L - \mathbf{1}(\mathbf{1}^T\mathbf{1})^{-1}\mathbf{1}^T$ is a projection onto the orthogonal complement of the subspace $\langle \mathbf{1} \rangle$. The first and third of these identities are trivial. Let's prove the second. For all pairs (i, l), $(\mathbf{e}_i - \mathbf{e}_l)^T \mathbf{A}(\mathbf{e}_i - \mathbf{e}_l) = -(1/2)(\mathbf{e}_i - \mathbf{e}_l)^T \mathbf{D} \odot \mathbf{D}(\mathbf{e}_i - \mathbf{e}_l) = (-1/2)\mathrm{tr}[(\mathbf{D} \odot \mathbf{D})(\boldsymbol{\Delta}_{ii} - \boldsymbol{\Delta}_{il} - \boldsymbol{\Delta}_{li} + \boldsymbol{\Delta}_{ll})] = -(1/2)(-d_{il}^2 - d_{li}^2) = d_{il}^2$. Here, $\boldsymbol{\Delta}_{il} = \mathbf{e}_i \mathbf{e}_l^T$ is a Kronecker matrix with 1 in location (i, l) and zeros elsewhere.

The distance matrix $\mathbf{D}$ is assumed Euclidean, which is to say $d_{il}^2 = (\mathbf{y}_i - \mathbf{y}_l)(\mathbf{y}_i - \mathbf{y}_l)^H = \mathbf{y}_i\mathbf{y}_i^H - \mathbf{y}_i\mathbf{y}_l^H - \mathbf{y}_l\mathbf{y}_i^H + \mathbf{y}_l\mathbf{y}_l^H$, for some set of row vectors $\mathbf{y}_l \in \mathbb{C}^N$. As a consequence, the matrix $\mathbf{A}$ may be written as

$$\mathbf{A} = -\frac{1}{2}\begin{bmatrix} \mathbf{y}_1\mathbf{y}_1^H \\ \mathbf{y}_2\mathbf{y}_2^H \\ \vdots \\ \mathbf{y}_L\mathbf{y}_L^H \end{bmatrix} \mathbf{1}^T - \frac{1}{2}\mathbf{1}\left[\mathbf{y}_1\mathbf{y}_1^H \; \mathbf{y}_2\mathbf{y}_2^H \; \cdots \; \mathbf{y}_L\mathbf{y}_L^H\right] + \mathrm{Re}\{\mathbf{Y}\mathbf{Y}^H\},$$

where $\mathbf{Y} \in \mathbb{C}^{L \times N}$. This matrix is not non-negative definite, but the centered matrix $\mathbf{B} = \mathbf{P}_{\mathbf{1}}^{\perp}\mathbf{A}\mathbf{P}_{\mathbf{1}}^{\perp}$ is

$$\mathbf{B} = \mathbf{P}_{\mathbf{1}}^{\perp}\mathbf{A}\mathbf{P}_{\mathbf{1}}^{\perp} = \mathrm{Re}\{\mathbf{P}_{\mathbf{1}}^{\perp}\mathbf{Y}\mathbf{Y}^H\mathbf{P}_{\mathbf{1}}^{\perp}\} \succeq \mathbf{0}.$$

Give $\mathbf{B} \in \mathbb{C}^{L \times L}$ the EVD $\mathbf{B} = \mathbf{F}\mathbf{K}\mathbf{F}^H = \mathbf{X}\mathbf{X}^H$, where $\mathbf{X} = \mathbf{F}\mathbf{K}^{1/2}\mathbf{U}_p$ is the desired configuration $\mathbf{X} \in \mathbb{C}^{L \times p}$ and $\mathbf{K}$ is a $p \times p$ diagonal matrix of non-negative scalars. It follows that

$$\begin{aligned}(\mathbf{x}_i - \mathbf{x}_l)(\mathbf{x}_i - \mathbf{x}_l)^H &= (\mathbf{e}_i - \mathbf{e}_l)^T \mathbf{X}\mathbf{X}^H(\mathbf{e}_i - \mathbf{e}_l) \\ &= (\mathbf{e}_i - \mathbf{e}_l)^T \mathbf{B}(\mathbf{e}_i - \mathbf{e}_l) \\ &= (\mathbf{e}_i - \mathbf{e}_l)^T \mathbf{A}(\mathbf{e}_i - \mathbf{e}_l) = d_{il}^2.\end{aligned}$$

Algorithm 2: MDS algorithm: From Euclidean distance matrix to low-dimensional configuration and Gramian

Input: $L \times L$ Euclidean distance matrix $\mathbf{D}$ for L points in $\mathbb{C}^N$
Output: low-dimensional configuration of L points in $\mathbb{C}^p$, $\mathbf{X} \in \mathbb{C}^{L\times p}$, and Gramian matrix $\mathbf{G}$
Construct the matrix $\mathbf{B} = \mathbf{P}_{\mathbf{1}}^{\perp}(-(1/2)\mathbf{D} \odot \mathbf{D})\mathbf{P}_{\mathbf{1}}^{\perp} = \mathbf{G}$
Compute $\mathbf{B} = \mathbf{F}\mathbf{K}\mathbf{F}^H$ // EVD of $\mathbf{B}$, with rank p
Construct $\mathbf{X} = \mathbf{F}\mathbf{K}^{1/2}\mathbf{U}_p$ // $\mathbf{U}_p$ is an arbitrary $p \times p$ unitary matrix

So the configuration $\mathbf{X}$ reproduces the Euclidean distance matrix $\mathbf{D}$, and the resulting Gramian is $\mathbf{G} = \mathbf{X}\mathbf{X}^H = \mathbf{F}\mathbf{K}\mathbf{F}^H = \mathbf{B}$, with $\mathbf{G} \succeq \mathbf{0}$. The program is described in Algorithm 2.

We may say the matrix $\mathbf{X} \in \mathbb{C}^{L\times p}$ is a configuration of L points in $\mathbb{C}^p$ that reproduces the distance matrix $\mathbf{D}$ and delivers the Gramian $\mathbf{G} = \mathbf{X}\mathbf{X}^H$ in the bargain. Moreover, this sequence of steps establishes that the matrix $\mathbf{B}$ is non-negative definite iff the distance matrix $\mathbf{D}$ is Euclidean. The original distance matrix for L points in $\mathbb{C}^N$ is reproduced by L points in $\mathbb{C}^p$. If the original points are known, then this is a dimension reduction procedure. If they are unknown, then it is a procedure that delivers points that could have produced the distance matrix. The extracted configuration is mean-centered, $\mathbf{P}_{\mathbf{1}}^{\perp}\mathbf{X} = \mathbf{X}$, and any rotation of this configuration as $\mathbf{X}\mathbf{Q}$, with $\mathbf{Q} \in U(p)$, reproduces the distance matrix $\mathbf{D}$ and the Gramian $\mathbf{G}$.

Approximation. From the identity $d_{il}^2 = (\mathbf{e}_i - \mathbf{e}_l)^T\mathbf{B}(\mathbf{e}_i - \mathbf{e}_l) = (\mathbf{e}_i - \mathbf{e}_l)^T\mathbf{F}\mathbf{K}\mathbf{F}^H(\mathbf{e}_i - \mathbf{e}_l)$, we may write $d_{il}^2 = \mathrm{tr}(\mathbf{F}\mathbf{K}\mathbf{F}^H\boldsymbol{\Delta}_{il})$. It then follows that the Frobenius norm of the distance matrix $\mathbf{D}$ may be written as

$$\|\mathbf{D}\|^2 = \sum_{i=1}^{L}\sum_{l=1}^{L} d_{il}^2 = \sum_{i=1}^{L}\sum_{l=1}^{L} \mathrm{tr}(\mathbf{F}\mathbf{K}\mathbf{F}^H\boldsymbol{\Delta}_{il}) = \mathrm{tr}(\mathbf{F}\mathbf{K}\mathbf{F}^H\boldsymbol{\Delta}),$$

where the matrix $\boldsymbol{\Delta} = \sum_{i=1}^{L}\sum_{l=1}^{L}\boldsymbol{\Delta}_{il} = 2L\mathbf{I}_L - 2\mathbf{1}\mathbf{1}^T$. As a consequence,

$$\|\mathbf{D}\|^2 = \mathrm{tr}[\mathbf{F}\mathbf{K}\mathbf{F}^H(2L\mathbf{I}_L - 2\mathbf{1}\mathbf{1}^T)] = 2L\,\mathrm{tr}(\mathbf{K}) = 2L\sum_{i=1}^{p} k_i,$$

where the last step follows from the fact that $\mathbf{1}^T\mathbf{F}\mathbf{K}\mathbf{F}^H\mathbf{1} = \mathbf{1}^T\mathbf{P}_{\mathbf{1}}^{\perp}\mathbf{A}\mathbf{P}_{\mathbf{1}}^{\perp}\mathbf{1} = 0$. This suggests that the configuration $\mathbf{X}$ of p-vectors may be approximated with the configuration $\mathbf{X}_r = \mathbf{F}\mathbf{K}_r\mathbf{U}_r$ of r-vectors, with corresponding Frobenius norm $\|\mathbf{D}_r\|^2 = 2L\sum_{i=1}^{r} k_i$. The approximation error $\|\mathbf{D}\|^2 - \|\mathbf{D}_r\|^2 = 2L\sum_{i=r+1}^{p} k_i$ is then a bulk measure of approximation, but not an element-by-element approximation error. Element by element, the approximation error is $d_{il}^2 - d_{il,r}^2 = \mathrm{tr}[\mathbf{F}(\mathbf{K} - \mathbf{K}_r)\mathbf{F}^H\boldsymbol{\Delta}_{il}]$.

Extensions to the Theory: MDS for Improper Distance Matrices. Suppose a Hermitian matrix $\mathbf{G} \in \mathbb{C}^{L\times L}$ *is not necessarily non-negative definite*. The spectral representation theorem for Hermitian matrices ensures that there is a factorization $\mathbf{G} = \mathbf{FKF}^H$, where $\mathbf{F} \in \mathbb{C}^{L\times p}$ is a slice of a unitary matrix and $\mathbf{K}$ is a $p \times p$ diagonal matrix of *real* values. Assume without loss of generality that the eigenvalues in $\mathbf{K}$ are sorted from largest to smallest, with the leading r values non-negative and the trailing $p-r$ negative. Write this diagonal matrix as $\mathbf{K} = \text{blkdiag}(\mathbf{\Lambda}_r^2, -\mathbf{\Lambda}_{p-r}^2)$, where each of $\mathbf{\Lambda}_r^2$ and $\mathbf{\Lambda}_{p-r}^2$ is a diagonal of non-negative reals. The matrix $\mathbf{K}$ may be written as

$$\mathbf{K} = \text{blkdiag}(\mathbf{\Lambda}_r, \mathbf{\Lambda}_{p-r})\,\mathbf{M}\,\text{blkdiag}(\mathbf{\Lambda}_r, \mathbf{\Lambda}_{p-r}) = \mathbf{\Lambda M \Lambda},$$

where $\mathbf{M} = \text{blkdiag}(\mathbf{I}_r, -\mathbf{I}_{p-r})$ is a Minkowski matrix with non-Hermitian factorization $\mathbf{M} = \mathbf{M}^{1/2}\mathbf{M}^{1/2}$ and $\mathbf{M}^{1/2} = \text{blkdiag}(\mathbf{I}_r, j\mathbf{I}_{p-r})$. The first block of the matrix $\mathbf{M}^{1/2}$ is Hermitian, whereas the second one is skew-Hermitian. If the configuration $\mathbf{X}$ is defined as $\mathbf{X} = \mathbf{F\Lambda}$, then the Hermitian matrix $\mathbf{G}$ is reproduced as the non-Hermitian quadratic form $\mathbf{G} = \mathbf{XMX}^H$ with non-Euclidean distances $(\mathbf{x}_i - \mathbf{x}_l)\mathbf{M}(\mathbf{x}_i - \mathbf{x}_l)^H$. If the configuration is defined as $\mathbf{X} = \mathbf{F\Lambda M}^{1/2} = \mathbf{F}\,\text{blkdiag}(\mathbf{\Lambda}_r, \mathbf{0}) + j\mathbf{F}\,\text{blkdiag}(\mathbf{0}, \mathbf{\Lambda}_{p-r})$, then the Hermitian matrix $\mathbf{G}$ is reproduced as the non-Hermitian quadratic form $\mathbf{G} = \mathbf{XX}^T$, with no change in the distances.[6]

If the matrix $\mathbf{G}$ had been real and symmetric, then the factorization would have been $\mathbf{G} = \mathbf{PKP}^T$, with $\mathbf{P}$ a slice of an orthogonal matrix and $\mathbf{K}$ a matrix of real values. The *real* configuration $\mathbf{X} = \mathbf{P\Lambda}$ reproduces the real matrix $\mathbf{G}$ as the non-Hermitian quadratic form $\mathbf{XMX}^T$. The complex configuration $\mathbf{X} = \mathbf{P\Lambda M}^{1/2} = \mathbf{P}\,\text{blkdiag}(\mathbf{\Lambda}_r, \mathbf{0}) + j\mathbf{P}\,\text{blkdiag}(\mathbf{0}, \mathbf{\Lambda}_{p-r})$ reproduces *real* $\mathbf{G}$ as the non-Hermitian quadratic form $\mathbf{G} = \mathbf{XX}^T$. So the complex field is required as an extension field to find a configuration $\mathbf{X}$ that reproduces the Gramian $\mathbf{G}$ and returns *real* pseudo-distances.

Now suppose the story had begun with an improper distance matrix $\mathbf{D}$, symmetric, $d_{ii} = 0$, $d_{il} \geq 0$, but not necessarily Euclidean. The matrix $\mathbf{B}$ may no longer be assumed non-negative definite. Still there is the spectral factorization $\mathbf{B} = \mathbf{FKF}^H$, with $\mathbf{F}$ a slice of a unitary matrix and $\mathbf{K}$ a diagonal matrix of real numbers, arranged in descending order. This matrix may be factored as before. The configuration $\mathbf{X} \in \mathbb{C}^{L\times p}$ with $\mathbf{X} = \mathbf{F\Lambda}$ reproduces the matrix $\mathbf{B}$ as the non-Hermitian quadratic form $\mathbf{XMX}^H$, with pseudo-distances $(\mathbf{x}_i - \mathbf{x}_l)\mathbf{M}(\mathbf{x}_i - \mathbf{x}_l)^H = d_{il}^2$. If the configuration is defined as $\mathbf{X} = \mathbf{F\Lambda M}^{1/2} = \mathbf{F}\,\text{blkdiag}(\mathbf{\Lambda}_r, \mathbf{0}) + j\mathbf{F}\,\text{blkdiag}(\mathbf{0}, \mathbf{\Lambda}_{p-r})$, then Hermitian $\mathbf{B}$ is reproduced as the non-Hermitian quadratic form $\mathbf{XX}^T$, and the pseudo-distances remain unchanged.

If the matrix $\mathbf{D}$ had been real and symmetric, then the factorization would have been $\mathbf{B} = \mathbf{PKP}^T$, with $\mathbf{P}$ a slice of an orthogonal matrix and $\mathbf{K}$ a matrix of real values. The *real* configuration $\mathbf{X} = \mathbf{P\Lambda}$ reproduces $\mathbf{B}$ as the non-Hermitian quadratic

[6] This reasoning is a collaboration between author LLS and Mark Blumstein. Very similar reasoning may be found in [262], and the many references therein, including Goldfarb [137].

form $\mathbf{XMX}^T$, with pseudo-distances reproducing $\mathbf{D}$. The *complex* configuration $\mathbf{X} = \mathbf{P}\,\text{blkdiag}(\mathbf{\Lambda}_r, \mathbf{0}) + j\mathbf{P}\,\text{blkdiag}(\mathbf{0}, \mathbf{\Lambda}_{p-r})$ reproduces *real* $\mathbf{B}$ as $\mathbf{B} = \mathbf{XX}^T$ and the improper distance matrix $\mathbf{D} = \{d_{il}\}$. So the complex field is required as an extension field to find a *complex* configuration $\mathbf{X}$ that reproduces the *real* improper distance matrix $\mathbf{D}$ with *real* Gramian $\mathbf{B}$. If the distance matrix had been a proper Euclidean distance matrix, then $\mathbf{M}$ would have been the identity, and the imaginary part of the complex solution for $\mathbf{X}$ would have been zero.

There are interpretations:

1. The *real* configuration $\mathbf{X}$, with $\mathbf{X} = \mathbf{P\Lambda}$ and $\mathbf{\Lambda} = \text{diag}(k_1, \ldots, k_r, |k_{r+1}|, \ldots, |k_p|)$, reproduces the improper distance matrix $\mathbf{D}$ in Minkowski space with Minkowski inner product $\mathbf{X}^T\mathbf{MX}$;
2. The *complex* configuration $\mathbf{X}$ with $\mathbf{X} = \mathbf{P\Lambda M}^{1/2}$ in complex Euclidean space reproduces the improper matrix $\mathbf{B}$ and reproduces the pseudo-distances in $\mathbf{D}$ with non-Hermitian inner product.
3. The distance $(\mathbf{x}_i - \mathbf{x}_l)\mathbf{M}(\mathbf{x}_i - \mathbf{x}_l)^T$ for the configuration $\mathbf{X} = \mathbf{P\Lambda}$ may be written as $(\mathbf{u}_i - \mathbf{u}_l)(\mathbf{u}_i - \mathbf{u}_l)^T - (\mathbf{v}_i - \mathbf{v}_l)(\mathbf{v}_i - \mathbf{v}_l)^T$, where the p-vector $\mathbf{x}_l$ is parsed into its r-dimension head and its $(p-r)$-dimensional tail as $\mathbf{x}_l = [\mathbf{u}_l\ \mathbf{v}_l]$. The first quadratic form models the Euclidean component of the matrix $\mathbf{D}$, and the second quadratic form models the non-Euclidean component. When $\mathbf{D}$ is Euclidean, then the second term vanishes.

2.5 The Johnson-Lindenstrauss Lemma

Let us state the Johnson-Lindenstrauss (JL) lemma [187] and then interpret its significance.

Lemma (Johnson-Lindenstrauss) *For any $0 < \epsilon < 1$, and any integer L, let r be a positive integer such that*

$$r \geq \frac{4}{\epsilon^2/2 - \epsilon^3/3} \log L.$$

Then, for any set V of L points in $\mathbb{R}^d$, there is a map $\mathbf{f} : \mathbb{R}^d \to \mathbb{R}^r$ such that for all $\mathbf{x}_i, \mathbf{x}_l \in V$,

$$(1-\epsilon)d_{il}^2 \leq \|\mathbf{f}(\mathbf{x}_i) - \mathbf{f}(\mathbf{x}_l)\|_2^2 \leq (1+\epsilon)d_{il}^2,$$

where $d_{il}^2 = \|\mathbf{x}_i - \mathbf{x}_l\|_2^2$.

Proof Outline. In the proof of Gupta and Dasgupta [94], it is shown that the squared length of a resolution of a d-dimensional MVN random vector of i.i.d. $\mathcal{N}_1(0, 1)$

components onto a randomly chosen subspace of dimension r is tightly concentrated around r/d times the length of the original d-dimensional random vector. Then, by constructing a moment-generating function in the difference between these two lengths, applying Markov's inequality for non-negative random variables, and using the Chernoff bound and a union bound, the lemma is proved. Moreover, Gupta and Dasgupta argue that the function $\mathbf{f}$ may be determined by resolving the original configuration onto a randomly selected r-dimensional subspace of $\mathbb{R}^d$. This randomly selected subspace will fail to serve the lemma with probability $1 - 1/L$, so this procedure may be iterated at will to achieve a probability of success equal to $1 - (1 - 1/L)^N$, which converges to 1.

Interpretation. Start with L known points in Euclidean space of dimension d. Call this a configuration V. There is no constraint on d or L. Now, specify a fidelity parameter $0 < \epsilon < 1$. With r chosen to satisfy the constraints of the JL lemma, then there is a map $\mathbf{f}$ with the property that for every pair of points $(\mathbf{x}_i, \mathbf{x}_l) \in \mathbb{R}^d$ in the original configuration, the mapped points $(\mathbf{f}(\mathbf{x}_i), \mathbf{f}(\mathbf{x}_l)) \in \mathbb{R}^r$ preserve the original pairwise squared distances to within fidelity $1-\epsilon$. Remarkably, this guarantee is universal, holding for any configuration V, and it is dependent only on the number of points in the configuration, and not the ambient dimension d of the configuration. The dimension r scales with the logarithm of the number of points in the configuration. How can this be? The proof illuminates this question and suggests that for special configurations, MDS might well improve on this universal result. This point will be elaborated shortly in a paragraph on rapprochement between MDS and the JL lemma.

The JL lemma is a universal characterization. Is the randomized polynomial time algorithm an attractive algorithm for finding a low-dimensional configuration? Beginning with a configuration V, it requires the resolution of this configuration in a randomly chosen r-dimensional subspace of $\mathbb{R}^d$, perhaps by sampling uniformly from the Grassmannian of r-dimensional subspaces. Then, for each such projection, the fidelity of pairwise distances must be checked. This requires a one-time computation of $\binom{L}{2}$ pairwise distances for the original configuration, computation of $\binom{L}{2}$ pairwise distances for each randomly projected configuration, and $\binom{L}{2}$ comparisons for fidelity. Stop at success, with the assurance that with probability $1 - (1 - 1/L)^N$, no more than N tries will be required.

So if the computation of a distance matrix would be required for an algorithmic implementation of the JL algorithm, why not begin with the distance matrix computed for the configuration V and use MDS to extract a configuration? Regardless of the ambient dimension d, MDS returns a configuration in a Euclidean space of dimension p, no greater than L, that reproduces the pairwise distances exactly when the rank of the matrix $\mathbf{B}$ is p. If this dimension is further reduced to $r < p$, then the resulting interpoint squared distances are

$$d_{il,r}^2 = d_{il}^2(1 - \epsilon_{il}),$$

where

$$\epsilon_{il} = \frac{\text{tr}[(\mathbf{B} - \mathbf{B}_r)\,\boldsymbol{\Delta}_{il}]}{d_{il}^2}.$$

Here, $\boldsymbol{\Delta}_{il} = (\mathbf{e}_i - \mathbf{e}_l)(\mathbf{e}_i - \mathbf{e}_l)^T$, $d_{il}^2 = \text{tr}\,(\mathbf{B}\boldsymbol{\Delta}_{il})$, $\mathbf{B} = \mathbf{P}_{\mathbf{1}}^{\perp}\left(-\frac{1}{2}\mathbf{D}\circ\mathbf{D}\right)\mathbf{P}_{\mathbf{1}}^{\perp}$, and $\mathbf{B}_r$ is the reduced rank version of $\mathbf{B}$. These errors ϵ_{il} depend on the pair $(\mathbf{x}_i, \mathbf{x}_l)$. So, to align our reasoning with the reasoning of the JL lemma, we define ϵ to be $\epsilon = \max_{il} \epsilon_{il}$. The question before us is how to compare an extracted MDS configuration with an extracted JL configuration.

RP vs. MDS. When attempting a comparison between the bounds of the JL lemma and the analytical results of MDS, it is important at the outset to emphasize that the JL lemma begins with a configuration of L points in an ambient Euclidean space of dimension d and replaces this configuration with a configuration of these points in a lower-dimensional Euclidean space of dimension $r \leq d$. At any choice of r larger than a constant depending on L and ϵ, the pairwise distances in the low-dimensional configuration are guaranteed to be within $\pm\epsilon$ of the original pairwise distances. The bound is universal, applying to any configuration, and it is independent of d. But any algorithm designed to meet the objectives of the JL lemma would need to begin with a configuration of points in $\mathbb{R}^d$. Moreover, an implementation of the randomized polynomial (RP) algorithm of Gupta and Dasgupta would require the computation of an $L \times L$ Euclidean distance matrix for the original configuration, so that the Euclidean distance matrix for points in each randomly selected subspace of dimension r can be tested for its distortion ϵ. This brings us to MDS, which starts only with an $L \times L$ distance matrix. The configuration of L points that may have produced this distance matrix is irrelevant, and therefore the dimension of an ambient space for these points is irrelevant. However, beginning with this Euclidean distance matrix, MDS extracts a configuration of centered points in Euclidean space of dimension $p \leq L$ whose distance matrix matches the original distance matrix *exactly*. For dimensions $r < p$, there is an algorithm for extracting an even lower-dimensional configuration and for computing the fidelity of the pairwise distances in this lower-dimensional space with the pairwise distances in the original distance matrix. There is no claim that this is the best reduced-dimension configuration for approximating the original distance matrix. The fidelity of a reduced-dimension configuration depends on the original distance matrix, or in those cases where the distance matrix comes from a configuration, on the original configuration.

So the question is this. Suppose we begin with a configuration of L points in $\mathbb{R}^d$, compute its corresponding $L \times L$ distance matrix $\mathbf{D}$, and use MDS to extract a dimension-r configuration. For each r, we compare the fidelity of the low-dimensional configuration to the original configuration by comparing pairwise distances. What can we say about the resulting fidelity, compared with the bounds of the JL lemma?

Motivated by the JL lemma, let us call ϵ *distortion* and call $D = (\epsilon^2/2 - \epsilon^3/3)/4$ *distortion measure*. Over the range of validity for the JL lemma, $0 < \epsilon < 1$, this distortion measure is bounded as $0 < D < 1/24$. For any value of D in this range, the corresponding $0 < \epsilon < 1$ may be determined. (Or, for any ϵ, D may be determined.) Define the rate R to be the dimension r, in which case according to the JL lemma, $RD > \log L$. We may restrict the range of R to $0 \leq R \leq d$, as for $R = d$, the distortion is 0. Thus, we define the rate-distortion function

$$R(D) = \min\left(d, \frac{1}{D}\log L\right).$$

Noteworthy points on this rate-distortion curve are $R = L$ at $D = \log L/L$, $R = 24 \log L$ at $D = 1/24$, and $R = d$ at $D = \log L/d$.

- *Fewer points than dimensions ($L \leq d$).* For a distortionless configuration, there is no need to consider a JL configuration at rate d, as an MDS configuration is distortionless at some rate $p \leq L \leq d$. If MDS returns a distortionless configuration at rate p, and dimension reduction is used to extract a configuration at rate $r < p$, then the distortion will increase away from 0. Will this distortion exceed the guarantee of the JL lemma? This question cannot be answered with certainty, as the MDS distortion is configuration-dependent, depending on the eigenvalues of the intermediate matrix **B**. However, it is expected that distortion will increase slowly as sub-dominant eigenvalues, and their corresponding sub-dominant coordinates, are set to zero. The smaller is p, the less likely it seems that the distortion of MDS will exceed that of the JL guarantee.
  But for some configurations it will. This imprecise reasoning is a consequence of the fact that the conclusions of the JL lemma are configuration-independent, whereas the conclusions of the MDS algorithm are configuration-dependent.
- *More points than dimensions ($L \geq d$).* For a distortionless configuration, there is no need to consider an MDS configuration, unless the MDS configuration is distortionless at rate $p < d$, in which case it is preferred over a JL configuration. If a distortionless MDS configuration at rate p is reduced in dimension to $r < d$, then the question is whether this distortion exceeds the JL guarantee. As before, this question cannot be answered with certainty, as the MDS distortion is configuration-dependent, depending on the eigenvalues of the intermediate matrix **B**. However, it is expected that distortion will increase slowly as sub-dominant eigenvalues, and their corresponding sub-dominant coordinates, are set to zero. As dimension reduction becomes more aggressive, the distortion increases dramatically. It may turn out that this method of sub-optimum dimension reduction produces distortions exceeding those of the JL guarantee. The smaller is p, the less likely it seems that the distortion of MDS will exceed that of the JL guarantee. But for some configurations it will. Again, this imprecise reasoning is a consequence of the fact that the conclusions of the JL lemma are configuration-independent, whereas the conclusions of the MDS algorithm are configuration-dependent.

Summary The net of this reasoning is that a distance matrix must be computed from a configuration of points in $\mathbb{R}^d$. From this distance matrix, an MDS configuration is extracted for all rates $0 \leq r \leq L$. For each rate, the MDS distortion is computed and compared with the distortion bound of the JL lemma at rate r. If $L \leq d$, then this comparison need only be done for rates $r \leq L$. If $L \geq d$, then this comparison is done for rates $0 \leq r \leq d$. At each rate, there will be a winner: RP or MDS.

Example 2.4 (Fewer points than dimensions) Consider a collection of $L = 500$ points in a space of dimension $d = 1000$. The vectors $\mathbf{x}_l \in \mathbb{R}^{d\times 1}$, $l = 1, \ldots, L$ have i.i.d. $\mathcal{N}_1(0, 1)$ components. The rate-distortion function determined by the JL lemma lower bounds what may be called a rate-distortion region of (r, ϵ) pairs where the guarantees of the JL lemma hold, universally. But for special data sets, and every data set is special, the rate function $r(\epsilon) = \frac{4}{\epsilon^2/2-\epsilon^3/3} \log L$ upper bounds the rate required to achieve the conclusions of the JL lemma at distortion ϵ. The rate-distortion function of the JL lemma may be written as

$$r(\epsilon) = \min\left(d, \frac{24\log(L)}{3\epsilon^2 - 2\epsilon^3}\right), \tag{2.10}$$

where $0 < \epsilon < 1$ is the distortion of the pairwise distances in the low-dimensional space, compared to the pairwise distances in the original high-dimensional space. This curve is plotted as JL in the figures to follow.

The rate computed with the random projections (RPs) of Gupta and Dasgupta is determined as follows. Begin with a configuration V of L random points $\mathbf{x}_l \in \mathbb{R}^{d\times 1}$ and a rate-distortion pair (r, ϵ) satisfying (2.10). Generate a random subspace of dimension r, $\langle \mathbf{U}\rangle \in \mathbb{G}r(r, \mathbb{R}^d)$. The RP embedding is $\mathbf{f}(\mathbf{x}_l) = \sqrt{d/r}\,\mathbf{U}^T\mathbf{x}_l$, where $\mathbf{U} \in \mathbb{R}^{d\times r}$ is an orthogonal basis for $\langle \mathbf{U}\rangle$. Check whether the randomly selected subspace satisfies the distortion conditions of the JL lemma. That is, check whether all pairwise distances satisfy

$$(1-\epsilon)\|\mathbf{x}_i - \mathbf{x}_l\|_2^2 \leq \|\mathbf{f}(\mathbf{x}_i) - \mathbf{f}(\mathbf{x}_l)\|_2^2 \leq (1+\epsilon)\|\mathbf{x}_i - \mathbf{x}_l\|_2^2.$$

If the random subspace passes the test, calculate the maximum pairwise distance distortion as

$$\hat{\epsilon} = \max_{i,l} \left| \frac{\|\mathbf{f}(\mathbf{x}_i) - \mathbf{f}(\mathbf{x}_l)\|_2^2}{\|\mathbf{x}_i - \mathbf{x}_l\|_2^2} - 1 \right|; \tag{2.11}$$

otherwise, generate another random subspace until it passes the test. For the low-dimensional embedding obtained this way, there is an $\hat{\epsilon}$ for each r. From these pairs, plot the rate-distortion function $r(\hat{\epsilon})$, and label this curve RP.

For comparison, we obtain the rate-distortion curve of an MDS embedding. Obviously, when $L \leq r \leq d$, MDS is distortionless and $\hat{\epsilon} = 0$. When $r < L$, MDS produces some distortion of the pairwise distances whose maximum can also

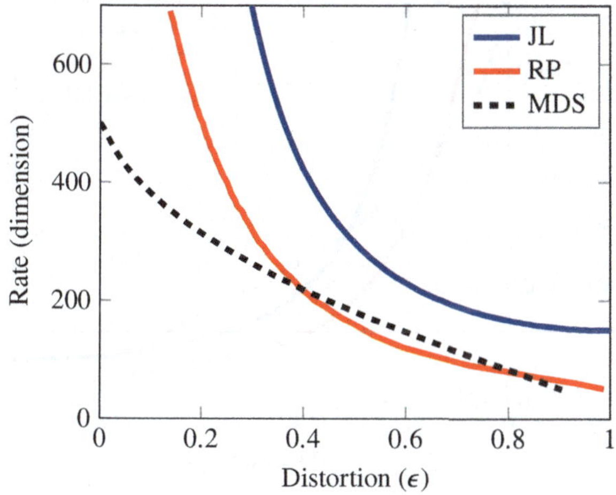

Fig. 2.3 Rate-distortion curves for random projection (RP) and MDS when $L = 500$ and $d = 1000$. The bound provided by the JL lemma is also depicted

be estimated as in (2.11). Figure 2.3 shows the results obtained by averaging 100 independent simulations where in each simulation, we generate a new collection of L points. When the reduction in dimension is not very aggressive, MDS, which is configuration dependent, provides better results than RP. For more aggressive reductions in dimension, both dimensionality reduction methods provide similar results without a clear winner.

In these experiments, the random projections are terminated at the first random projection that produces a distortion $\hat{\epsilon}$ less than ϵ. For some configurations, it may be that continued random generation of projections would further reduce distortion.

Example 2.5 (More points than dimensions) If the number of points exceeds the ambient space dimension, then $d < L$, and the question is whether the dimension r of the JL lemma can be smaller than d, leaving room for dimension reduction. That is, for a given target distortion of ϵ, is the rate guarantee less than d:

$$\frac{24}{3\epsilon^2 - 2\epsilon^3} \log L < d < L?$$

For d and L specified, a misinterpretation of the JL lemma would appear to place a bound on ϵ for which there is any potential for dimension reduction. But as our experiments show, this formula for r in the JL lemma does not actually determine what can be achieved with dimension reduction. In other words, for many special data sets, there is room for dimension reduction, even when a misinterpretation of the JL bound would suggest there is not. That is, the JL lemma simply guarantees

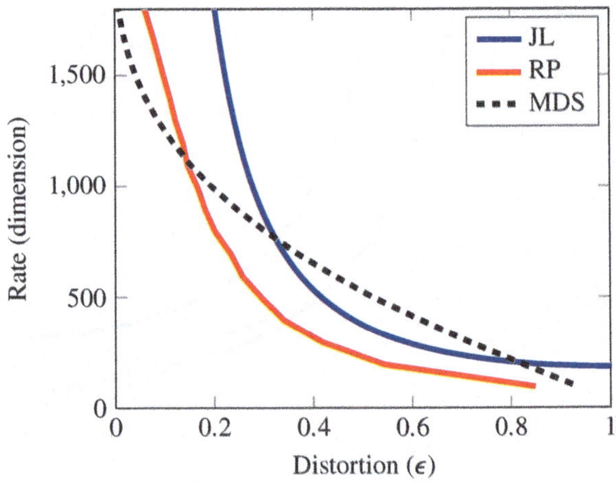

Fig. 2.4 Rate-distortion curves for random projection and MDS when $L = 2500$ and $d = 2000$. The bound provided by the JL lemma is also depicted

that for dimension greater than r, a target distortion may be achieved. It does not say that there are no dimensions smaller than r for which the distortion may be achieved.

This point is made with the following example. The ambient dimension is $d = 2000$, and the number of points in the configuration is $L = 2500$. Each point has i.i.d. $\mathcal{N}(0, 1)$ components. Figure 2.4 shows the bound provided by the JL lemma for the range of distortions where it is applicable, as well as the rate-distortion curves obtained by random projections and by MDS. In this scenario, for small distortions (for which the JL lemma is not applicable), MDS seems to be the winner, whereas for larger distortions (allowing for more aggressive reductions in dimension), random projections provide significantly better results. This seems to be the general trend when $L > d$.

These comparisons between the JL bound, and the results of random projections (RP) and MDS, run on randomly selected data sets, are illustrative. But they do not establish that RP is uniformly better than MDS, or vice versa. That is, for any distortion ϵ, the question of which method produces a smaller ambient dimension depends on the data set. So, beginning with a data set, run MDS and RP to find which returns the smaller ambient dimension. For some data sets, dimensions may be returned for values outside the range of the JL bound. This is OK: remember, the JL bound is universal; it does not speak to achievable values of rate and distortion for special data sets. And every data set is special. In many cases, the curve of rate vs. distortion will fall far below the bound suggested by the JL lemma.

2.6 Chapter Notes

Much of this chapter deals with least squares and related ideas, some of which date to the late eighteenth and early nineteenth century. But others are of more modern origin.

1. Least squares was introduced by Legendre in 1805 and independently published by Adrain in 1808 and Gauss in 1809. Gauss claimed to have discovered the essentials in 1795, and there is no reason to doubt this claim. Had Gauss and Gaspar Riche de Prony (le Baron de Prony) communicated, then Gauss's method of least squares and Prony's method of fitting damped complex exponentials to data [267] would have established the beginnings of a least squares theory of system identification more than 225 years ago.
2. Gauss discovered recursive (or sequential) least squares based on his discovery of a matrix inversion lemma in 1826. These results were re-discovered in the 1950s and 1960s by Plackett and Fagin. Sequential least squares reached its apotheosis in the Kalman filter, published in 1960. A more complete account, with references, may be found in Kamil Dedecius, "Partial forgetting in Bayesian estimation," PhD dissertation, Czech Technical University in Prague, 2010 [99].
3. The discussion of oblique least squares (OBLS) is not commonly found in books. The representation of the best linear unbiased estimator (BLUE) as a generalized sidelobe canceller is known in a few communities, but apparently unknown in many others. There are a few ideas in this chapter on reduction of model order and cross-validation that may be original.
4. There is more attention paid in this chapter to sensitivity questions in compressed sensing than is standard. Our experience is that model mismatch and compressor mismatch (physical implementations of compressors do not always conform to mathematical models for compression) should be carefully considered when compressing measurements before inverting an underdetermined problem.
5. The comparisons between MDS and the random projections proposed by Gupta and Dasgupta suggest two practical ways to reduce ambient dimension. One is deterministic, and the other is random. It is important to emphasize that random projections or MDS are likely to produce curves on the rate-distortion plane that lie *well* below the universal bound of the JL lemma. So perhaps the essential take-away is that there are two practical algorithms for reducing ambient space dimension: the sequence of random projections due to Gupta and Dasgupta and dimension reduction in MDS.

Appendices

2.A Completing the Square in Hermitian Quadratic Forms

A commonly encountered Hermitian quadratic form is $(\mathbf{y}-\mathbf{Hx})^H(\mathbf{y}-\mathbf{Hx})$. This may be written in the *completed square or dual* form as

$$(\mathbf{y}-\mathbf{Hx})^H(\mathbf{y}-\mathbf{Hx}) = (\mathbf{x}-(\mathbf{H}^H\mathbf{H})^{-1}\mathbf{H}^H\mathbf{y})^H(\mathbf{H}^H\mathbf{H})(\mathbf{x}-(\mathbf{H}^H\mathbf{H})^{-1}\mathbf{H}^H\mathbf{y})+\mathbf{y}^H(\mathbf{I}-\mathbf{P_H})\mathbf{y}.$$

This shows that the minimizing value of $\mathbf{x}$ is the least squares estimate $\hat{\mathbf{x}} = (\mathbf{H}^H\mathbf{H})^{-1}\mathbf{H}^H\mathbf{y}$, with no need to use Wirtinger calculus to differentiate a real, and consequently non-analytic, function of a complex variable. The squared error is $\mathbf{y}^H(\mathbf{I}-\mathbf{P_H})\mathbf{y}$.

This argument generalizes easily to the weighted quadratic form $(\mathbf{y}-\mathbf{Hx})^H\mathbf{W}(\mathbf{y}-\mathbf{Hx})$. Define $\mathbf{z} = \mathbf{W}^{1/2}\mathbf{y}$ and $\mathbf{G} = \mathbf{W}^{1/2}\mathbf{H}$ to rewrite this quadratic form as $(\mathbf{z}-\mathbf{Gx})^H(\mathbf{z}-\mathbf{Gx})$. This may be written as

$$(\mathbf{z}-\mathbf{Gx})^H(\mathbf{z}-\mathbf{Gx}) = (\mathbf{x}-(\mathbf{G}^H\mathbf{G})^{-1}\mathbf{G}^H\mathbf{z})^H(\mathbf{G}^H\mathbf{G})(\mathbf{x}-(\mathbf{G}^H\mathbf{G})^{-1}\mathbf{G}^H\mathbf{z})+\mathbf{z}^H(\mathbf{I}-\mathbf{P_G})\mathbf{z}.$$

This shows that the minimizing value of $\mathbf{x}$ is

$$\hat{\mathbf{x}} = (\mathbf{G}^H\mathbf{G})^{-1}\mathbf{G}^H\mathbf{z} = (\mathbf{H}^H\mathbf{W}\mathbf{H})^{-1}\mathbf{H}^H\mathbf{W}\mathbf{y},$$

with squared error $\mathbf{z}^H(\mathbf{I}-\mathbf{P_G})\mathbf{z}$.

2.A.1 Generalizing to Multiple Measurements and Other Cost Functions

This trick extends also to multiple measurements and other cost functions, which will become relevant in other parts of the book. First, it is easy to see that $(\mathbf{Y}-\mathbf{HX})^H(\mathbf{Y}-\mathbf{HX})$ can be rewritten as

$$\begin{aligned}(\mathbf{Y}-\mathbf{HX})^H&(\mathbf{Y}-\mathbf{HX})\\ &= (\mathbf{X}-(\mathbf{H}^H\mathbf{H})^{-1}\mathbf{H}^H\mathbf{Y})^H(\mathbf{H}^H\mathbf{H})(\mathbf{X}-(\mathbf{H}^H\mathbf{H})^{-1}\mathbf{H}^H\mathbf{Y})+\mathbf{Y}^H(\mathbf{I}-\mathbf{P_H})\mathbf{Y}.\end{aligned}$$

Then, for any cost function $J(\cdot) \geq 0$ that satisfies

$$J(\mathbf{Q}_1+\mathbf{Q}_2) \geq J(\mathbf{Q}_1)+J(\mathbf{Q}_2),$$

for $\mathbf{Q}_i \succeq \mathbf{0}$, the minimizing value of $\mathbf{X}$ is $\hat{\mathbf{X}} = (\mathbf{H}^H\mathbf{H})^{-1}\mathbf{H}^H\mathbf{Y}$, with residual error $J(\mathbf{Y}^H(\mathbf{I}-\mathbf{P_H})\mathbf{Y})$.

What if the roles of $\mathbf{H}$ and $\mathbf{X}$ are reversed, so that $\mathbf{X}$ is known and $\mathbf{H}$ is to be estimated. Hence, the quadratic form $(\mathbf{Y}-\mathbf{HX})(\mathbf{Y}-\mathbf{HX})^H$ may be written as

$$\begin{aligned}&(\mathbf{Y}-\mathbf{HX})(\mathbf{Y}-\mathbf{HX})^H\\&=(\mathbf{H}-\mathbf{YX}^H(\mathbf{XX}^H)^{-1})\mathbf{XX}^H(\mathbf{H}-\mathbf{YX}^H(\mathbf{XX}^H)^{-1})^H+\mathbf{Y}(\mathbf{I}-\mathbf{P_X})\mathbf{Y}^H.\end{aligned}$$

Considering a cost function with the aforementioned properties, the minimum of the cost function is $J(\mathbf{Y}(\mathbf{I}-\mathbf{P_X})\mathbf{Y}^H)$ and achieved at $\hat{\mathbf{H}}=\mathbf{YX}^H(\mathbf{XX}^H)^{-1}$.

The above results specialize to the least squares estimator, for which the cost function is $J(\cdot)=\mathrm{tr}(\cdot)$.That is, the cost function is

$$\mathrm{tr}[(\mathbf{Y}-\mathbf{HX})(\mathbf{Y}-\mathbf{HX})^H]=\mathrm{tr}[(\mathbf{Y}-\mathbf{HX})^H(\mathbf{Y}-\mathbf{HX})].$$

In this case, the LS estimator of $\mathbf{X}$ is $\hat{\mathbf{X}}=(\mathbf{H}^H\mathbf{H})^{-1}\mathbf{H}^H\mathbf{Y}$, with squared error $\mathrm{tr}(\mathbf{Y}^H(\mathbf{I}-\mathbf{P_H})\mathbf{Y})$, and the LS estimator of $\mathbf{H}$ is $\hat{\mathbf{H}}=\mathbf{YX}^H(\mathbf{XX}^H)^{-1}$, with squared error $\mathrm{tr}[\mathbf{Y}(\mathbf{I}-\mathbf{P_X})\mathbf{Y}^H]$.

2.A.2 LMMSE Estimation

In fact, this completing of the square also applies to the study of linear minimum mean-squared error (LMMSE) estimation, where the problem is to find the matrix $\mathbf{W}$ that minimizes the error covariance between the second-order random vector $\mathbf{x}$ and the filtered second-order random vector $\mathbf{y}$. This covariance matrix is

$$\mathrm{E}[(\mathbf{Wy}-\mathbf{x})(\mathbf{Wy}-\mathbf{x})^H]=\mathbf{W}^H\mathbf{R_{yy}}\mathbf{W}-\mathbf{R_{xy}}\mathbf{W}^H-\mathbf{W}\mathbf{R}_{\mathbf{xy}}^H+\mathbf{R_{xx}},$$

which may be written as

$$\mathrm{E}[(\mathbf{Wy}-\mathbf{x})(\mathbf{Wy}-\mathbf{x})^H]=(\mathbf{W}-\mathbf{R_{xy}}\mathbf{R}_{\mathbf{yy}}^{-1})^H\mathbf{R_{yy}}(\mathbf{W}-\mathbf{R_{xy}}\mathbf{R}_{\mathbf{yy}}^{-1})+\mathbf{R_{xx}}-\mathbf{R_{xy}}\mathbf{R}_{\mathbf{yy}}^{-1}\mathbf{R}_{\mathbf{xy}}^H.$$

It is now easy to show that the minimizing choice for $\mathbf{W}$ is $\mathbf{W}=\mathbf{R_{xy}}\mathbf{R}_{\mathbf{yy}}^{-1}$, yielding the error covariance matrix $\mathbf{R_{xx}}-\mathbf{R_{xy}}\mathbf{R}_{\mathbf{yy}}^{-1}\mathbf{R}_{\mathbf{xy}}^H$.

Coherence, Classical Correlations, and their Invariances

3

This chapter opens with definitions of several correlation coefficients and the distribution theory of their sampled-data estimators. Examples are worked out for Pearson's correlation coefficient, spectral coherence for wide-sense stationary (WSS) time series, and estimated signal-to-noise ratio in signal detection theory. The chapter continues with a discussion of principal component analysis (PCA) for deriving low-dimensional representations for a single channel's worth of data and then proceeds to a discussion of coherence in two and three channels. For two channels, we encounter standard correlations, multiple correlations, half-canonical correlations, and (full) canonical correlations. These may be interpreted as coherences. Half- and full-canonical coordinates serve for dimension reduction in two channels, just as principal components serve for dimension reduction in a single channel.

Canonical coordinate decomposition of linear minimum mean-squared error (LMMSE) filtering ties filtering to coherence. The role of canonical coordinates in linear minimum mean-squared error (LMMSE) estimation is explained, and these coordinates are used for dimension reduction in filtering. The Krylov subspace is introduced to illuminate the use of expanding subspaces for conjugate direction and multistage LMMSE filters. A particularly attractive feature of these filters is that they are extremely efficient to compute when the covariance matrix for the data has only a small number of distinct eigenvalues, independent of how many times each is repeated.

For the analysis of three channels worth of data, partial correlations are used to regress one channel onto two or two channels onto one. In each of these cases, partial coherence serves as a statistic for answering questions of linear dependence. When suitably normalized, they are coherences.

D. Ramírez et al., *Coherence*, https://doi.org/10.1007/978-3-031-13331-2_3

3.1 Coherence Between a Random Variable and a Random Vector

Consider a zero-mean random variable $u \in \mathbb{C}$ and a zero-mean random vector $\mathbf{v} \in \mathbb{C}^{p\times 1}$. The composite covariance matrix is defined to be

$$\mathbf{R} = \mathrm{E}\left[\begin{bmatrix} u \\ \mathbf{v} \end{bmatrix} \begin{bmatrix} u^* & \mathbf{v}^H \end{bmatrix}\right] = \begin{bmatrix} r_{uu} & \mathbf{r}_{u\mathbf{v}} \\ \mathbf{r}_{u\mathbf{v}}^H & \mathbf{R}_{\mathbf{vv}} \end{bmatrix},$$

where $r_{uu} = \mathrm{E}[uu^*]$ is a real scalar, $\mathbf{r}_{u\mathbf{v}} = \mathrm{E}[u\mathbf{v}^H]$ is $1 \times p$ complex vector, and $\mathbf{R}_{\mathbf{vv}} = \mathrm{E}[\mathbf{v}\mathbf{v}^H]$ is a $p \times p$ Hermitian matrix. Define the $(p+1) \times (p+1)$ unitary matrix $\mathbf{Q} = \mathrm{blkdiag}(q, \mathbf{Q}_p)$, with $q^*q = 1$ and $\mathbf{Q}_p^H\mathbf{Q}_p = \mathbf{I}_p$, and use it to rotate u and $\mathbf{v}$. The resulting action on $\mathbf{R}$ is

$$\mathbf{QRQ}^H = \mathrm{E}\left[\mathbf{Q}\begin{bmatrix} u \\ \mathbf{v} \end{bmatrix} \begin{bmatrix} u^* & \mathbf{v}^H \end{bmatrix}\mathbf{Q}^H\right] = \begin{bmatrix} qr_{uu}q^* & q\mathbf{r}_{u\mathbf{v}}\mathbf{Q}^H \\ \mathbf{Q}\mathbf{r}_{u\mathbf{v}}^H q^* & \mathbf{Q}\mathbf{R}_{\mathbf{vv}}\mathbf{Q}^H \end{bmatrix}.$$

Definition 3.1 The coherence between u and $\mathbf{v}$ is defined to be

$$\rho^2(\mathbf{R}) = 1 - \frac{\det(\mathbf{R})}{\det(r_{uu})\det(\mathbf{R}_{\mathbf{vv}})} = \frac{\mathbf{r}_{u\mathbf{v}}\mathbf{R}_{\mathbf{vv}}^{-1}\mathbf{r}_{u\mathbf{v}}^H}{r_{uu}}. \tag{3.1}$$

Coherence is the statistician's *multiple correlation coefficient* used in multiple regression analysis. It is invariant to unitary transformation $\mathbf{Q}$. For $p = 1$, it specializes to the standard coefficient of correlation.

Interpretations. Coherence admits several interpretations, all of them captured by Fig. 3.1:

- From a geometric perspective, coherence is the cosine-squared of the angle between the one-dimensional subspace spanned by the random variable u and the p-dimensional subspace spanned by the random variables $\mathbf{v} = [v_1 \ \cdots \ v_p]^T$.
- From an inference perspective, the mean-squared error of the estimator $\hat{u} = \mathbf{r}_{u\mathbf{v}}\mathbf{R}_{\mathbf{vv}}^{-1}\mathbf{v}$ may be written as

$$\begin{aligned} Q_{uu|\mathbf{v}} &= \mathrm{E}[(u-\hat{u})(u-\hat{u})^*] \\ &= r_{uu} - \mathbf{r}_{u\mathbf{v}}\mathbf{R}_{\mathbf{vv}}^{-1}\mathbf{r}_{u\mathbf{v}}^H \\ &= r_{uu}(1-\rho^2(\mathbf{R})). \end{aligned}$$

Equivalently, $r_{uu} = Q_{uu|\mathbf{v}} + \mathbf{r}_{u\mathbf{v}}\mathbf{R}_{\mathbf{vv}}^{-1}\mathbf{r}_{u\mathbf{v}}^H$, which is to say the variance r_{uu} decomposes into $\mathbf{r}_{u\mathbf{v}}\mathbf{R}_{\mathbf{vv}}\mathbf{r}_{u\mathbf{v}}^H$, the proportion of variance explained by $\mathbf{v}$, plus $Q_{uu|\mathbf{v}}$, the proportion unexplained by $\mathbf{v}$.

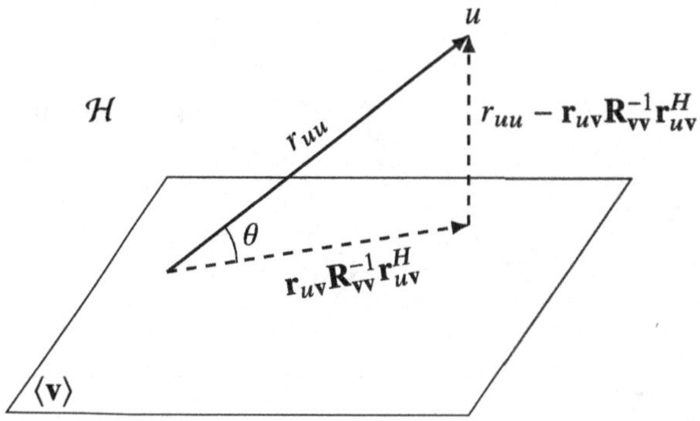

Fig. 3.1 Geometric interpretation of coherence in the Hilbert space of random variables. The subspace $\langle \mathbf{v} \rangle$ is the subspace spanned by the random variables $v_1, \ldots, v_p$

Now, suppose in place of the random variables u and $\mathbf{v}$ we have only $N > p$ i.i.d. realizations of them, u_n and $\mathbf{v}_n$, $n = 1, \ldots, N$, organized into the $1 \times N$ row vector $\mathbf{u} = [u_1 \ \cdots \ u_N]$ and the $p \times N$ matrix $\mathbf{V}$. The ith row of $\mathbf{V}$ is the $1 \times N$ row vector $\mathbf{v}_i = [v_{i1} \ \cdots \ v_{iN}]$. It is reasonable to call $\mathbf{u}$ a surrogate for the random variable u and $\mathbf{V}$ a surrogate for the random vector $\mathbf{v}$. The row vector $\mathbf{u}$ determines the one-dimensional subspace $\langle \mathbf{u} \rangle$, which is spanned by $\mathbf{u}$; the $p \times N$ matrix $\mathbf{V}$ determines the p-dimensional subspace $\langle \mathbf{V} \rangle$, which is spanned by the rows of $\mathbf{V}$. The projection of the row vector $\mathbf{u}$ onto the subspace $\langle \mathbf{V} \rangle$ is the row vector $\mathbf{u}\mathbf{V}^H(\mathbf{V}\mathbf{V}^H)^{-1}\mathbf{V}$, denoted $\mathbf{u}\mathbf{P}_\mathbf{V}$. This row vector is a linear combination of the rows of $\mathbf{V}$. The $N \times N$ matrix $\mathbf{P}_\mathbf{V} = \mathbf{V}^H(\mathbf{V}\mathbf{V}^H)^{-1}\mathbf{V}$ is Hermitian, and $\mathbf{P}_\mathbf{V}\mathbf{P}_\mathbf{V} = \mathbf{P}_\mathbf{V}$. This makes it an orthogonal projection matrix that projects row vectors onto the subspace $\langle \mathbf{V} \rangle$ by operating from the right as $\mathbf{u}\mathbf{P}_\mathbf{V}$.

Define the $(p + 1) \times N$ data matrix $\mathbf{X}$,

$$\mathbf{X} = \begin{bmatrix} \mathbf{u} \\ \mathbf{V} \end{bmatrix},$$

and its Gramian

$$\mathbf{G} = \mathbf{X}\mathbf{X}^H = \begin{bmatrix} \mathbf{u}\mathbf{u}^H & \mathbf{u}\mathbf{V}^H \\ \mathbf{V}\mathbf{u}^H & \mathbf{V}\mathbf{V}^H \end{bmatrix}.$$

This Gramian is a scaled sampled-data estimator of the covariance matrix $\mathbf{R}$. Then, $\rho^2(\mathbf{G})$ is the sample-data estimator of the coherence or multiple correlation coefficient

$$\rho^2(\mathbf{G}) = 1 - \frac{\det(\mathbf{G})}{\det(\mathbf{u}\mathbf{u}^H)\det(\mathbf{V}\mathbf{V}^H)} = \frac{\mathbf{u}\mathbf{P}_\mathbf{V}\mathbf{u}^H}{\mathbf{u}\mathbf{u}^H}. \tag{3.2}$$

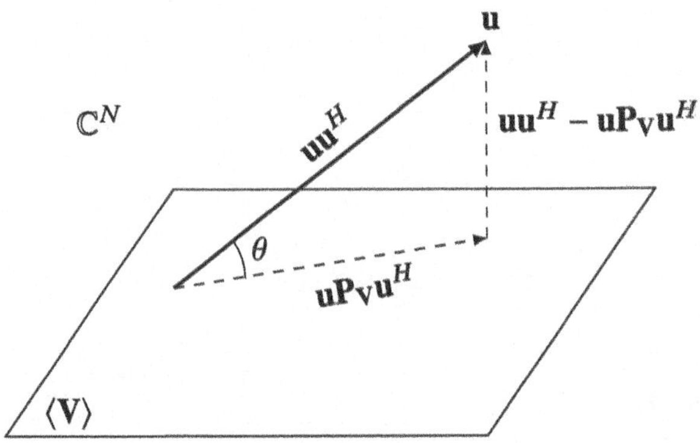

Fig. 3.2 Geometry of coherence in Euclidean space

The rightmost expression is obtained by using the Schur decomposition of determinant to write det(**G**) as

$$\det(\mathbf{G}) = \det(\mathbf{VV}^H)\det(\mathbf{uu}^H - \mathbf{uV}^H(\mathbf{VV}^H)^{-1}\mathbf{Vu}^H) = \det(\mathbf{V}^H\mathbf{V})\mathbf{u}(\mathbf{I}_N - \mathbf{P_V})\mathbf{u}^H.$$

Obviously, there is a connection:

$$\frac{\mathbf{r}_{u\mathbf{V}}\mathbf{R}_{\mathbf{VV}}^{-1}\mathbf{r}_{u\mathbf{V}}^H}{r_{uu}} \leftrightarrow \frac{\mathbf{uP_Vu}^H}{\mathbf{uu}^H}.$$

So, the sample estimator of the (population) Hilbert space coherence is an Euclidean space coherence. Euclidean space coherence may be interpreted with the help of Fig. 3.2.

Interpretations. Coherence admits several interpretations, all of them captured by Fig. 3.2:

- From a geometric perspective, coherence is the cosine-squared of the angle between the one-dimensional subspace $\langle \mathbf{u} \rangle$ spanned by the sampled-data $\mathbf{u}$ and the p-dimensional subspace $\langle \mathbf{V} \rangle$ spanned by the rows of the sampled-data matrix $\mathbf{V}$.
- From an inference perspective, the squared error of the sampled-data estimator $\hat{\mathbf{u}} - \mathbf{uP_V}$ may be written as $(\mathbf{u}-\hat{\mathbf{u}})(\mathbf{u}-\hat{\mathbf{u}})^H - \mathbf{u}(\mathbf{I}_N - \mathbf{P_V})\mathbf{u}^H$. Equivalently, $\mathbf{uu}^H$ decomposes into $\mathbf{uP_Vu}^H$, its proportion explained by $\mathbf{V}$, plus $\mathbf{u}(\mathbf{I}_N - \mathbf{P_V})\mathbf{u}^H$, its proportion unexplained by $\mathbf{V}$. It is important to note that in this discussion, the vector $\mathbf{u}$ is a row vector and the matrix $\mathbf{V}$ is an $p \times N$ matrix whose rows are $\mathbf{v}_i$. This explains the appearance of the projector $\mathbf{P_V}$ on the right of $\mathbf{u}$.

Geometry and Invariances. The geometry is, of course, the geometry of linear spaces. The population multiple correlation coefficient, or coherence, is the cosine-squared of the angle between the random variable u and the random vector $\mathbf{v}$ in the Hilbert space of second-order random variables. The sample estimator of this multiple correlation coefficient, or sample coherence, is the cosine-squared of the angle between the Euclidean vector $\mathbf{u}$ and the Euclidean subspace $\langle \mathbf{V} \rangle$.

Define the transformation $\mathbf{QXQ}_N$, where $\mathbf{Q}$ is the previously defined unitary matrix $\mathbf{Q} = \text{blkdiag}(q, \mathbf{Q}_p)$ and $\mathbf{Q}_N$ is an $N \times N$ unitary matrix. The action on $\mathbf{G}$ is $\mathbf{QGQ}^H$, which leaves $\rho^2(\mathbf{G})$ invariant.

Distribution. The null distribution ($\mathbf{r}_{u\mathbf{v}} = \mathbf{0}$) of $\rho^2(\mathbf{G})$ is this:

$$\rho^2(\mathbf{G}) \sim \text{Beta}(p, N - p),$$

where Beta$(p, N - p)$ denotes a beta random variable with density

$$f(x) = \frac{\Gamma(N)}{\Gamma(p)\Gamma(N - p)} x^{p-1}(1 - x)^{N-p-1}, \quad 0 \leq x \leq 1.$$

Some examples for various parameters (p,N) are plotted in Fig. 3.3.

This result is often derived for the case where all random variables are jointly proper complex normal (see Sect. D.6.4 for a proof when u and $\mathbf{V}$ are jointly normal). But, in fact, the result holds if

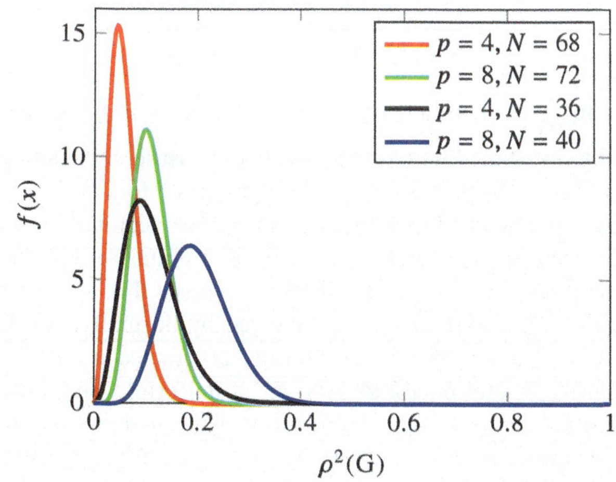

Fig. 3.3 Null distribution of coherence, Beta$(p, N - p)$, for various parameters (p,N)

- $\mathbf{u} \in \mathbb{C}^{1\times N}$ and $\mathbf{V} \in \mathbb{C}^{p\times N}$ are independently drawn, and
- One or both of $\mathbf{u}$ and $\mathbf{V}$ are invariantly distributed with respect to right unitary transformations $\mathbf{Q}_N \in U(N)$.

For example, $\mathbf{u}$ may be white Gaussian and $\mathbf{V}$ fixed, or $\mathbf{u}$ may be white Gaussian and $\mathbf{V}$ may be independently drawn, or $\mathbf{u}$ may be fixed and $\mathbf{V}$ may be drawn as p i.i.d. white Gaussian vectors. But more generally, the distribution of $\mathbf{u}$ may be spherically invariant, and the distribution of $\mathbf{V}$ may be spherically contoured.

Here is the argument. The statistic $\mathbf{u}\mathbf{P}_\mathbf{V}\mathbf{u}^H/\mathbf{u}\mathbf{u}^H$ is known to be distributed as Beta$(p, N-p)$ for $\mathbf{u} \sim \mathcal{CN}_N(\mathbf{0}, \mathbf{I}_N)$ and $\mathbf{P}_\mathbf{V}$ a fixed projection onto the p-dimensional subspace $\langle\mathbf{V}\rangle$ of $\mathbb{C}^N$ [110]. But this statistic is really only a function of the spherically invariant random vector $\mathbf{u}/\|\mathbf{u}\| \in S^{N-1}$, so this distribution applies to any spherically invariant random vector on the unit sphere. The normalized spherical normal is one such.

Now think of $\mathbf{V}$ as generated randomly, as in the construction of the sample multiple correlation coefficient. Conditionally, the multiple correlation coefficient is distributed as a beta random variable, and this distribution is invariant to $\mathbf{V}$. Therefore, its unconditional distribution is Beta$(p, N-p)$. For $p = 1$ and real random variables, the result is Sir R. A. Fisher's 1928 result for the null distribution of the correlation coefficient [119].

So how do we think about generating spherically invariant vectors $\mathbf{u}$ and spherically contoured random matrices $\mathbf{V}$? A spherically invariant random vector might as well be written as $\mathbf{u}\mathbf{Q}_N$, with $\mathbf{u}$ *deterministic* and $\mathbf{Q}_N$ uniformly distributed on $U(N)$ with respect to Haar measure. Then, the coherence statistic with fixed $\mathbf{V}$ may be written as

$$\frac{\mathbf{u}\mathbf{Q}_N\mathbf{P}_\mathbf{V}\mathbf{Q}_N^H\mathbf{u}^H}{\mathbf{u}\mathbf{Q}_N\mathbf{Q}_N^H\mathbf{u}^H} = \frac{\mathbf{u}\mathbf{P}_{\mathbf{V}\mathbf{Q}_N^H}\mathbf{u}^H}{\mathbf{u}\mathbf{u}^H}.$$

The projection $\mathbf{P}_{\mathbf{V}\mathbf{Q}_N^H}$ is distributed as a projection onto p-dimensional subspaces of $\mathbb{C}^N$, uniformly distributed with respect to Haar measure. Such subspaces may be generated by p i.i.d. realizations of spherical random vectors in $\mathbb{C}^N$.

In summary, begin with deterministic, unit-norm, vector $\mathbf{u}$ and deterministic matrix $\mathbf{V}$. Spin each of them with their respective unitary rotations, each drawn uniformly from $U(N)$. Or, as a practical alternative, this spinning action may be replaced with i.i.d. $\mathcal{CN}(0, 1)$ draws, followed by normalization for $\mathbf{u}$ and QR factorization for $\mathbf{V}$. The net of this is uniformly distributed $\mathbf{u}$ on S^{N-1}, uniformly distributed $\mathbf{V}$ on the Stiefel manifold $\mathbb{St}(p, \mathbb{C}^N)$, uniformly distributed $\langle\mathbf{V}\rangle$ on the Grassmann manifold $\mathbb{Gr}(p, \mathbb{C}^N)$, and uniformly distributed projections $\mathbf{P}_\mathbf{V}$ onto the subspaces of the Grassmannian. In all of this language, uniformly distributed means uniformly distributed with respect to Haar measure, which means invariance of distribution with respect to right unitary transformation of row vectors. Fundamentally, it is the group action of right unitary transformations on the sphere, S^{N-1}; the complex Stiefel manifold of frames, $\mathbb{St}(p, \mathbb{C}^N)$; and the complex Grassmannian of subspaces, $\mathbb{G}r(p, \mathbb{C}^N)$, that determines the meaning of uniformity and the invariance of the distribution of the angle between the subspaces $\langle \mathbf{u} \rangle$ and $\langle \mathbf{V} \rangle$.

The Case $p = 1$. When $p = 1$, coherence in (3.1) is the modulus squared of Pearson's correlation coefficient between two zero-mean complex random variables u and v

$$\rho^2(\mathbf{R}) = \frac{|r_{uv}|^2}{r_{uu} r_{vv}} = \frac{|\,\mathrm{E}[uv^*]|^2}{\mathrm{E}[|u|^2]\,\mathrm{E}[|v|^2]},$$

where in this case

$$\mathbf{R} = \begin{bmatrix} r_{uu} & r_{uv} \\ r_{uv}^* & r_{vv} \end{bmatrix}.$$

Let $\mathbf{u} = [u_1 \ \cdots \ u_N]$ and $\mathbf{v} = [v_1 \ \cdots \ v_N]$ be N i.i.d. draws of the random variables u and v, respectively; and define the Gramian

$$\mathbf{G} = \begin{bmatrix} \mathbf{u} \\ \mathbf{v} \end{bmatrix} \begin{bmatrix} \mathbf{u}^H & \mathbf{v}^H \end{bmatrix}.$$

The sample estimate of the coherence is

$$\rho^2(\mathbf{G}) = \frac{|\mathbf{u}\mathbf{v}^H|^2}{(\mathbf{u}\mathbf{u}^H)(\mathbf{v}\mathbf{v}^H)} = \frac{\left|\sum_{n=1}^N u_n v_n^*\right|^2}{\left(\sum_{n=1}^N |u_n|^2\right)\left(\sum_{n=1}^N |v_n|^2\right)}. \tag{3.3}$$

When $r_{uv} = 0$, $\rho^2(\mathbf{G})$ in (3.3) is distributed as Beta$(1, N-1)$. When the random variables are real, $\rho^2(\mathbf{G}) \sim$ Beta $\left(\frac{1}{2}, \frac{N-1}{2}\right)$. This result holds whenever u has a spherical distribution, regardless of the distribution of v.

For the sample estimate of coherence, we will sometimes use centered vectors $\bar{u}_n = u_n - \bar{u}$ and $\bar{v}_n = v_n - \bar{v}$, where $\bar{u} = \sum_{n=1}^N u_n/N$ and $\bar{v} = \sum_{n=1}^N v_n/N$ are the sample means of the vectors. The coherence between the centered vectors is

$$\rho^2(\bar{\mathbf{G}}) = \frac{|\bar{\mathbf{u}}\bar{\mathbf{v}}^H|^2}{(\bar{\mathbf{u}}\bar{\mathbf{u}}^H)(\bar{\mathbf{v}}\bar{\mathbf{v}}^H)},$$

which, under the null, is distributed as Beta$(1, N-2)$ in the complex case and as Beta $\left(\frac{1}{2}, \frac{N-2}{2}\right)$ in the real case. So centering reduces the degrees of freedom by one in the null distribution for the complex case and by $1/2$ in the real case.

In addition to deriving the null distribution of the sample coherence in his 1928 paper, Fisher considered the transformation

$$t = \frac{\rho(\mathbf{G})}{\sqrt{1-\rho^2(\mathbf{G})}}$$

and noted that when the coherence $\rho^2(\mathbf{R}) = 0$, the distribution of t is Student's distribution. In passing, he suggested the useful transformation

$$z = \frac{1}{2}\ln\left(\frac{1+\rho(\mathbf{G})}{1-\rho(\mathbf{G})}\right) = \text{artanh}(\rho(\mathbf{G})),$$

which is known as the Fisher transformation or the Fisher-z transformation. It turned out that the Fisher-z transformation was very practical since the distribution of z is approximately normal with mean $\frac{1}{2}\log\left(\frac{1+\rho(\mathbf{R})}{1-\rho(\mathbf{R})}\right)$, where $\rho(\mathbf{R})$ is the population (true) coherence, and variance $1/(N-3)$, independent of $\rho(\mathbf{R})$. This statistic is approximate normally distributed when the samples u and v are bivariate normal.

3.2 Coherence Between Two Random Vectors

The arguments of the previous section may be generalized by considering the coherence between two random vectors. To this end, consider again the Hilbert space of second-order random variables. Define the random vectors $\mathbf{u} = [u_1 \ \cdots \ u_q]^T$ and $\mathbf{v} = [v_1 \ \cdots \ v_p]^T$. In due course, these random vectors will be replaced by their sampled-data surrogates, $\mathbf{U} \in \mathbb{C}^{q\times N}$ and $\mathbf{V} \in \mathbb{C}^{p\times N}$. Then the ith row of $\mathbf{U}$ will be an N-sample version of u_i, and the lth row of $\mathbf{V}$ will be an N-sample version of v_l. So $\mathbf{u}$ and $\mathbf{v}$ are column vectors of random variables, and each row of $\mathbf{U}$ and $\mathbf{V}$ is an N-sample of one of these random variables.

The composite covariance matrix for $\mathbf{u}, \mathbf{v}$ is

$$\mathbf{R} = \begin{bmatrix} \mathbf{R}_{\mathbf{uu}} & \mathbf{R}_{\mathbf{uv}} \\ \mathbf{R}_{\mathbf{uv}}^H & \mathbf{R}_{\mathbf{vv}} \end{bmatrix},$$

where $\mathbf{R}_{\mathbf{uu}} = E[\mathbf{u}\mathbf{u}^H]$, $\mathbf{R}_{\mathbf{uv}} = E[\mathbf{u}\mathbf{v}^H]$, and $\mathbf{R}_{\mathbf{vv}} = E[\mathbf{v}\mathbf{v}^H]$ are, respectively, $q \times q$, $q \times p$, and $p \times p$.

Definition 3.2 The coherence between $\mathbf{u}$ and $\mathbf{v}$ is defined to be

$$\begin{aligned}\rho^2(\mathbf{R}) &= 1 - \frac{\det(\mathbf{R})}{\det(\mathbf{R}_{\mathbf{uu}})\det(\mathbf{R}_{\mathbf{vv}})} = 1 - \frac{\det(\mathbf{R}_{\mathbf{uu}} - \mathbf{R}_{\mathbf{uv}}\mathbf{R}_{\mathbf{vv}}^{-1}\mathbf{R}_{\mathbf{uv}}^H)}{\det(\mathbf{R}_{\mathbf{uu}})} \\ &= 1 - \det\left(\mathbf{I}_q - \mathbf{R}_{\mathbf{uu}}^{-1/2}\mathbf{R}_{\mathbf{uv}}\mathbf{R}_{\mathbf{vv}}^{-1}\mathbf{R}_{\mathbf{uv}}^H\mathbf{R}_{\mathbf{uu}}^{-1/2}\right),\end{aligned}$$

where we have used the Schur determinant identity $\det(\mathbf{R}) = \det(\mathbf{R}_{\mathbf{vv}})\det(\mathbf{R}_{\mathbf{uu}} - \mathbf{R}_{\mathbf{uv}}\mathbf{R}_{\mathbf{vv}}^{-1}\mathbf{R}_{\mathbf{uv}}^H)$ (see Appendix B). It is assumed the covariance matrices $\mathbf{R}_{\mathbf{uu}}$ and $\mathbf{R}_{\mathbf{vv}}$

are positive definite with Hermitian square roots $\mathbf{R}_{\mathbf{uu}}^{1/2}$ and $\mathbf{R}_{\mathbf{vv}}^{1/2}$, so that $\mathbf{R}_{\mathbf{uu}} = \mathbf{R}_{\mathbf{uu}}^{1/2}\mathbf{R}_{\mathbf{uu}}^{1/2}$ and $\mathbf{R}_{\mathbf{vv}} = \mathbf{R}_{\mathbf{vv}}^{1/2}\mathbf{R}_{\mathbf{vv}}^{1/2}$. Then $\mathbf{R}_{\mathbf{uu}}^{-1} = \mathbf{R}_{\mathbf{uu}}^{-1/2}\mathbf{R}_{\mathbf{uu}}^{-1/2}$ and $\mathbf{R}_{\mathbf{vv}}^{-1} = \mathbf{R}_{\mathbf{vv}}^{-1/2}\mathbf{R}_{\mathbf{vv}}^{-1/2}$.

So coherence compares the volumes of the error concentration ellipses for $\mathbf{u}$, before and after estimation of $\mathbf{u}$ from $\mathbf{v}$. This estimation is sometimes called regression or filtering.

3.2.1 Relationship with Canonical Correlations

Definition 3.3 Define the coherence matrix

$$\mathbf{C} = \mathbf{R}_{\mathbf{uu}}^{-1/2}\mathbf{R}_{\mathbf{uv}}\mathbf{R}_{\mathbf{vv}}^{-1/2},$$

with singular value decomposition

$$\mathbf{C} = \mathbf{F}\mathbf{K}\mathbf{G}^H.$$

Here, $\mathbf{F}$ and $\mathbf{G}$ are $q \times q$ and $p \times p$ unitary matrices, respectively. When $q > p$, $\mathbf{K}$ is a $q \times p$ matrix of singular values structured as

$$\mathbf{K} = \begin{bmatrix} \mathrm{diag}(k_1, \ldots, k_p) \\ \mathbf{0}_{(q-p)\times p} \end{bmatrix},$$

and for $q < p$, it is structured as

$$\mathbf{K} = \begin{bmatrix} \mathrm{diag}(k_1, \ldots, k_q) & \mathbf{0}_{q\times(p-q)} \end{bmatrix}.$$

It contains the canonical correlations between the canonical coordinates of $\mathbf{u}$ and $\mathbf{v}$ [173]. That is, each k_i is the cross-correlation between a canonical coordinate pair, $\mu_i = \mathbf{f}_i^H\mathbf{R}_{\mathbf{uu}}^{-1/2}\mathbf{u}$ and $\nu_i = \mathbf{g}_i^H\mathbf{R}_{\mathbf{vv}}^{-1/2}\mathbf{v}$.

In order to talk about coherence in a Hilbert space, it is necessary to talk about the canonical correlations between canonical coordinates. The squared canonical correlations k_i^2 are fine-grained coherences between canonical coordinates of the subspaces $\langle\mathbf{u}\rangle$ and $\langle\mathbf{v}\rangle$. That is, the coherence between $\mathbf{u}$ and $\mathbf{v}$ can be written in terms of the canonical correlations as

$$\rho^2(\mathbf{R}) = 1 - \prod_{i=1}^{\min(p,q)} \left(1 - k_i^2\right).$$

It is known that the canonical correlations form a complete set of maximal invariants under the transformation group

$$G = \left\{ g \mid g \cdot \begin{bmatrix} \mathbf{u} \\ \mathbf{v} \end{bmatrix} = \mathbf{B} \begin{bmatrix} \mathbf{u} \\ \mathbf{v} \end{bmatrix}, \mathbf{B} = \begin{bmatrix} \mathbf{B}_u & \mathbf{0} \\ 0 & \mathbf{B}_v \end{bmatrix}, \det(\mathbf{B}) \neq 0 \right\},$$

with group action $\mathbf{B}\mathbf{R}\mathbf{B}^H$.

3.2.2 The Circulant Case

Assume the covariance matrices $\mathbf{R_{uu}}$, $\mathbf{R_{uv}}$, and $\mathbf{R_{vv}}$ are circulant, in which case each has spectral representation of the form $\mathbf{R_{uv}} = \mathbf{V}_N \mathbf{D_{uv}} \mathbf{V}_N^H$, where $\mathbf{V}_N = \mathbf{F}_N/\sqrt{N}$, with $\mathbf{F}_N$ the $N \times N$ DFT matrix, and $\mathbf{D_{uv}}$ is a diagonal matrix of spectral coefficients: $\mathbf{D_{uv}} = \text{diag}(S_{\mathbf{uv}}(e^{j\theta_0}), \ldots, S_{\mathbf{uv}}(e^{j\theta_{N-1}}))$. Then the coherence matrix is

$$\begin{aligned} \mathbf{C} &= \mathbf{V}_N \mathbf{D}_{\mathbf{uu}}^{-1/2} \mathbf{D_{uv}} \mathbf{D}_{\mathbf{vv}}^{-1/2} \mathbf{V}_N^H \\ &= \mathbf{V}_N \,\text{diag}(\rho_{\mathbf{uv}}(e^{j\theta_0}), \ldots, \rho_{\mathbf{uv}}(e^{j\theta_{N-1}}))\mathbf{V}_N^H, \end{aligned}$$

where

$$\rho_{\mathbf{uv}}(e^{j\theta_k}) = \frac{S_{\mathbf{uv}}(e^{j\theta_k})}{\sqrt{S_{\mathbf{uu}}(e^{j\theta_k})\, S_{\mathbf{vv}}(e^{j\theta_k})}}.$$

Each term in the diagonal matrix, $\rho_{\mathbf{uv}}(e^{j\theta_k})$, is a *spectral coherence* at a frequency $\theta_k = 2\pi k/N$, which may be resolved as $\mathbf{V}_N^H \mathbf{C} \mathbf{V}_N$. So, in the circulant case, the SVD of $\mathbf{C}$ is $\mathbf{C} = \mathbf{F}\mathbf{K}\mathbf{G}^H$, where $\mathbf{F} = \mathbf{G} = \mathbf{V}_N$ and $\mathbf{K}$ is a diagonal matrix of spectral coherences. Moreover, each spectral coherence may be written as

$$\rho_{uv}(e^{j\theta_k}) = \frac{\mathbf{f}^H(e^{j\theta_k})\mathbf{R_{uv}}\mathbf{f}(e^{j\theta_k})}{\sqrt{\mathbf{f}^H(e^{j\theta_k})\mathbf{R_{uu}}\mathbf{f}(e^{j\theta_k})}\sqrt{\mathbf{f}^H(e^{j\theta_k})\mathbf{R_{vv}}\mathbf{f}(e^{j\theta_k})}},$$

where $\mathbf{f}(e^{j\theta_k}) = [1\, e^{-j\theta_k} \cdots e^{-j\theta_k(N-1)}]^T$ is the Fourier vector at frequency $\theta_k = 2\pi k/N$.

3.2.3 Relationship with Principal Angles

Now, suppose instead of random vectors $\mathbf{u}$ and $\mathbf{v}$, we have rectangular fat matrices $\mathbf{U} \in \mathbb{C}^{q \times N}$ and $\mathbf{V} \in \mathbb{C}^{p \times N}$, with $p, q \leq N$. The rows of $\mathbf{U}$ span the q-dimensional subspace $\langle \mathbf{U} \rangle$ of $\mathbb{C}^N$ and the rows of $\mathbf{V}$ span the p-dimensional subspace $\langle \mathbf{V} \rangle$ of $\mathbb{C}^N$.

Let us construct the Gramian (or scaled sample covariance)

$$\mathbf{G} = \begin{bmatrix} \mathbf{U}\mathbf{U}^H & \mathbf{U}\mathbf{V}^H \\ \mathbf{V}\mathbf{U}^H & \mathbf{V}\mathbf{V}^H \end{bmatrix}.$$

The Euclidean coherence between the q-dimensional subspace $\langle \mathbf{U} \rangle$ and the p-dimensional subspace $\langle \mathbf{V} \rangle$ is

$$\rho^2(\mathbf{G}) = 1 - \frac{\det(\mathbf{G})}{\det(\mathbf{U}\mathbf{U}^H)\det(\mathbf{V}\mathbf{V}^H)}$$
$$= 1 - \frac{\det(\mathbf{U}(\mathbf{I}_N - \mathbf{P}_\mathbf{V})\mathbf{U}^H)}{\det(\mathbf{U}\mathbf{U}^H)} = 1 - \prod_{i=1}^{\min(p,q)} \left(1 - \rho_i^2\right), \tag{3.4}$$

where $\mathbf{P}_\mathbf{V}$ is the projection onto $\langle \mathbf{V} \rangle$. This is a bulk definition of coherence, based on fine-grained coherences ρ_i^2. These fine-grained coherences are, in fact, cosine-squareds of principal angles θ_i between the subspaces $\langle \mathbf{U} \rangle$ and $\langle \mathbf{V} \rangle$ (see Sect. 9.2); that is, $\cos^2(\theta_i) = \rho_i^2$. When $q = 1$, then Euclidean coherence is the sample multiple correlation coefficient of (3.2). When $p = q = 1$, the squared coherence is

$$\rho^2 = \frac{\mathbf{u}\mathbf{P}_\mathbf{v}\mathbf{u}^H}{\mathbf{u}\mathbf{u}^H} = \frac{|\mathbf{u}\mathbf{v}^H|^2}{(\mathbf{u}\mathbf{u}^H)(\mathbf{v}\mathbf{v}^H)}, \tag{3.5}$$

which is the squared cosine of the angle between $\mathbf{u}$ and $\mathbf{v}$.

Let $\mathbf{k}_x = [k_{x,1} \ \cdots \ k_{x,r}]^T$ and $\mathbf{k}_y = [k_{y,1} \ \cdots \ k_{y,r}]^T$ be two vectors of canonical correlations with descending order components $k_{x,1} \geq k_{x,2} \geq \cdots \geq k_{x,r} \geq 0$. It is said that $\mathbf{k}_x$ majorizes $\mathbf{k}_y$, denoted as $\mathbf{k}_x \succeq \mathbf{k}_y$, if

$$\sum_{i=1}^{n} k_{x,i} \geq \sum_{i=1}^{n} k_{y,i}, \quad n = 1, \ldots, r-1, \quad \text{and} \quad \sum_{i=1}^{r} k_{x,i} = \sum_{i=1}^{r} k_{y,i}.$$

Majorization defines a partial ordering for vectors such that $\mathbf{k}_x \succeq \mathbf{k}_y$ if the components of $\mathbf{k}_x$ are "less spread out" than the components of $\mathbf{k}_y$. A real function $f(\cdot) : \mathcal{A} \subset \mathbb{R}^r \to \mathbb{R}$ is said to be Schur convex on $\mathcal{A}$ if

$$\mathbf{k}_x \succeq \mathbf{k}_y \text{ on } \mathcal{A} \implies f(\mathbf{k}_x) \geq f(\mathbf{k}_y).$$

Now we have the following result [317]:

Proposition 3.1 *The coherence is a Schur convex function of canonical correlations.*

So coherence is "order preserving" with respect to the partial order of majorization.

3.2.4 Distribution of Estimated Signal-to-Noise Ratio in Adaptive Matched Filtering

Begin with the measurement $\mathbf{y} \sim CN_L(\mathbf{h}x, \boldsymbol{\Sigma})$. The noise-whitened matched filter statistic is

$$\lambda = \mathbf{h}^H \boldsymbol{\Sigma}^{-1} \mathbf{y},$$

where the measurement $\mathbf{y} = \mathbf{h}x + \mathbf{n}$ consists of a scaling of a known signal $\mathbf{h} \in \mathbb{C}^L$, plus additive normal noise of zero mean and covariance $\boldsymbol{\Sigma}$. The square of the expected value of this statistic is $|x|^2(\mathbf{h}^H\boldsymbol{\Sigma}^{-1}\mathbf{h})^2$, and its variance is $\mathbf{h}^H\boldsymbol{\Sigma}^{-1}\mathbf{h}$. The output signal-to-noise ratio (SNR) for this detector statistic may then be taken to be

$$\text{SNR} = \frac{|x|^2(\mathbf{h}^H\boldsymbol{\Sigma}^{-1}\mathbf{h})^2}{\mathbf{h}^H\boldsymbol{\Sigma}^{-1}\mathbf{h}} = |x|^2\mathbf{h}^H\boldsymbol{\Sigma}^{-1}\mathbf{h}.$$

The question addressed by Reed, Mallet, and Brennan (RMB) in [281] is this: what is the distribution for an *estimate of this SNR* when known $\boldsymbol{\Sigma}$ is replaced by the sample covariance matrix $\mathbf{S} = \mathbf{Y}\mathbf{Y}^H/N$ in the matched filter. The independent columns of $\mathbf{Y} = [\mathbf{y}_1 \ \cdots \ \mathbf{y}_N]$ are drawn from the proper complex normal distribution, i.e., $\mathbf{y}_n \sim CN_L(\mathbf{0}, \boldsymbol{\Sigma})$. In fact, RMB normalized this estimate by the idealized SNR to obtain a distribution that would reveal the loss in SNR due to ignorance of $\boldsymbol{\Sigma}$. With $\boldsymbol{\Sigma}$ replaced by $\mathbf{S}$, the adaptive matched filter statistic is

$$\hat{\lambda} = \mathbf{h}^H \mathbf{S}^{-1} \mathbf{y}.$$

For fixed $\mathbf{S}$, and for $\mathbf{y}$ independent of $\mathbf{S}$, averages over the distribution of $\mathbf{y}$ produce the following squared expected value and variance of this statistic: $|x|^2(\mathbf{h}^H\mathbf{S}^{-1}\mathbf{h})^2$ and $\mathbf{h}^H\mathbf{S}^{-1}\boldsymbol{\Sigma}\mathbf{S}^{-1}\mathbf{h}$. The ratio of these is taken to be the estimated SNR:

$$\widehat{\text{SNR}} = \frac{|x|^2(\mathbf{h}^H\mathbf{S}^{-1}\mathbf{h})^2}{\mathbf{h}^H\mathbf{S}^{-1}\boldsymbol{\Sigma}\mathbf{S}^{-1}\mathbf{h}}.$$

The ratio $\rho^2 = \widehat{\text{SNR}}/\text{SNR}$ is

$$\rho^2 = \frac{(\mathbf{h}^H\mathbf{S}^{-1}\mathbf{h})^2}{(\mathbf{h}^H\boldsymbol{\Sigma}^{-1}\mathbf{h})(\mathbf{h}^H\mathbf{S}^{-1}\boldsymbol{\Sigma}\mathbf{S}^{-1}\mathbf{h})}$$

Why call this a coherence? Because by defining $\mathbf{u} = \boldsymbol{\Sigma}^{-1/2}\mathbf{h}$, and $\mathbf{v} = \boldsymbol{\Sigma}^{1/2}\mathbf{S}^{-1}\mathbf{h}$, the ratio ρ^2 may be written as a cosine-squared or coherence statistic as in (3.5):

$$\rho^2 = \frac{|\mathbf{u}^H\mathbf{v}|^2}{(\mathbf{u}^H\mathbf{u})(\mathbf{v}^H\mathbf{v})} = \frac{\mathbf{u}^H\mathbf{P}_\mathbf{v}\mathbf{u}}{\mathbf{u}^H\mathbf{u}}.$$

A few coordinate transformations allow us to rewrite the coherence statistic as

$$\rho^2 = \frac{\mathbf{e}_1^T \mathbf{W}^{-1} \mathbf{e}_1}{\mathbf{e}_1^T \mathbf{W}^{-2} \mathbf{e}_1},$$

where $\mathbf{e}_1$ is the first standard Euclidean basis vector and $\mathbf{W}$ has Wishart distribution $\mathcal{CW}_L(\mathbf{I}_L, N)$. It is now a sequence of imaginative steps to derive the celebrated Reed, Mallet, and Brennan result [281]

$$\rho^2 \sim \text{Beta}(N - L + 2, L - 1).$$

This result has formed the foundation for system designs in radar, sonar, and radio astronomy, as for a given value of L, a required value of N may be determined to ensure satisfactory output SNR with a specified confidence.

3.3 Coherence Between Two Time Series

Begin with two time series, $\{x[n]\}$ and $\{y[n]\}$, each wide-sense stationary (WSS) with correlation functions $\{r_{xx}[m], m \in \mathbb{Z}\} \longleftrightarrow \{S_{xx}(e^{j\theta}), -\pi < \theta \leq \pi\}$ and $\{r_{yy}[m], m \in \mathbb{Z}\} \longleftrightarrow \{S_{yy}(e^{j\theta}), -\pi < \theta \leq \pi\}$. The cross-correlation is assumed to be $\{r_{xy}[m], m \in \mathbb{Z}\} \longleftrightarrow \{S_{xy}(e^{j\theta}), -\pi < \theta \leq \pi\}$. The cross-correlation function $r_{xy}[m]$ is defined to be $r_{xy}[m] = \text{E}[x[n]y^*[n - m]]$, and the correlation functions are defined analogously. The two-tip arrows denote that the correlation sequence and its power spectral density are a Fourier transform pair.

The frequency-dependent squared coherence or magnitude-squared coherence (MSC) is defined as [65, 249]

$$|\rho_{xy}(e^{j\theta})|^2 = \frac{|S_{xy}(e^{j\theta})|^2}{S_{xx}(e^{j\theta})S_{yy}(e^{j\theta})}, \tag{3.6}$$

which may be interpreted as a frequency-dependent modulus-squared correlation coefficient between the two time series, thus quantifying the degree of linear dependency between them; $0 \leq |\rho_{xy}(e^{j\theta})|^2 \leq 1$. Since the pioneering work in the early 1970s by Carter, Nuttall, Knapp, and others, the MSC has found multiple applications from time delay estimation of acoustic or electromagnetic source signals [64] to the detection of evoked responses in electroencephalograms (EEG) during sensory stimulation [104, 322].

The MSC defined in (3.6) is an idealized function that is to be estimated from a finite snapshot of measurements $\mathbf{x} = [x[0] \cdots x[N-1]]^T$ and $\mathbf{y} = [y[0] \cdots y[N-1]]^T$. The correlation matrices for these snapshots are $\mathbf{R_{xy}} = \text{E}[\mathbf{xy}^H]$, $\mathbf{R_{xx}} = \text{E}[\mathbf{xx}^H]$, and $\mathbf{R_{yy}} = \text{E}[\mathbf{yy}^H]$. They are $N \times N$ Toeplitz matrices, with representations of the form

$$\mathbf{R_{xy}} = \int_{-\pi}^{\pi} \boldsymbol{\psi}(e^{j\phi}) S_{xy}(e^{j\phi}) \boldsymbol{\psi}^H(e^{j\phi}) \frac{d\phi}{2\pi},$$

where $\boldsymbol{\psi}(e^{j\theta}) = [1\ e^{j\theta}\ \cdots\ e^{j\theta(N-1)}]^T$.

Perhaps the most direct translation of the idealized definition of MSC to its finite term approximation is

$$|\hat{\rho}_{xy}(e^{j\theta})|^2 = \frac{|\boldsymbol{\psi}^H(e^{j\theta})\mathbf{R_{xy}}\boldsymbol{\psi}(e^{j\theta})|^2}{\boldsymbol{\psi}^H(e^{j\theta})\mathbf{R_{xx}}\boldsymbol{\psi}(e^{j\theta})\boldsymbol{\psi}^H(e^{j\theta})\mathbf{R_{yy}}\boldsymbol{\psi}(e^{j\theta})}.$$

That is, the term $\boldsymbol{\psi}^H(e^{j\theta})\mathbf{R_{xy}}\boldsymbol{\psi}(e^{j\theta})$ serves as an approximation to $S_{xy}(e^{j\theta})$. The approximation may be written as

$$\begin{aligned}\boldsymbol{\psi}^H(e^{j\theta})\mathbf{R_{xy}}\boldsymbol{\psi}(e^{j\theta}) &= \int_{-\pi}^{\pi} \boldsymbol{\psi}^H(e^{j\theta})\boldsymbol{\psi}(e^{j\phi})S_{xy}(e^{j\phi})\boldsymbol{\psi}^H(e^{j\phi})\boldsymbol{\psi}(e^{j\theta})\frac{d\phi}{2\pi} \\ &= \int_{-\pi}^{\pi} \left|\frac{\sin(N(\theta-\phi)/2)}{\sin((\theta-\phi)/2)}\right|^2 S_{xy}(e^{j\phi})\frac{d\phi}{2\pi}.\end{aligned}$$

This formula shows that spectral components at frequency ϕ leak through the Dirichlet kernel to contribute to the estimate of the spectrum $S_{xy}(e^{j\theta})$. This spectral leakage is called wavenumber leakage through sidelobes in array processing. It is the bane of all spectrum analysis and beamforming.

However, there are alternatives suggested by the discussion of the coherence matrix in the circulant case. If the correlation matrices $\mathbf{R_{xx}}$, etc. were circulant, then the coherence matrix would be circulant.

That is, magnitude-squared coherences $|\rho_{xy}(e^{j\theta_k})|^2$ at frequencies $\theta_k = 2\pi k/N$ can be obtained by diagonalizing the coherence matrix. This suggests an alternative to the computation of MSC. Moreover, in place of the tailoring of spectrum analysis techniques, one may consider dimension reduction of the coherence matrix by truncating canonical coherences, as a way to control spectral leakage.

There are several estimators of the MSC. A classic approach is to apply Welch's averaged periodogram method [65] as follows. Using a window of length L, we first partition the data into M possibly overlapped segments $\mathbf{x}_i$, $i = 1, \ldots, M$. The spectrum at frequency θ_k is then estimated as $\hat{S}_{xx}(e^{j\theta}) = \boldsymbol{\psi}^H(e^{j\theta})\hat{\mathbf{R}}_{\mathbf{xx}}\boldsymbol{\psi}(e^{j\theta})$, where $\hat{\mathbf{R}}_{\mathbf{xx}}$ is the sample covariance matrix estimated from the M windows or segments. Similarly, $S_{yy}(e^{j\theta})$ and $S_{xy}(e^{j\theta})$ are also estimated. The main drawback of this approach is the aforementioned spectral leakage. To address this issue, more refined MSC estimation approaches based on the use of minimum variance distortionless response (MVDR) filters [30] or the use of reduced-rank CCA coordinates for the coherence matrix [296] have been proposed. The following example demonstrates the essential ideas.

Example 3.1 (MSC spectrum) Let $s[n]$ be a complex, narrowband, WSS Gaussian time series with zero mean and unit variance. Its power spectrum is zero outside the passband $\theta \in [2\pi \cdot 0.1, 2\pi \cdot 0.15]$. This common signal is perturbed by independent additive noises $w_x[n] \sim \mathcal{N}(0, 1)$ and $w_y[n] \sim \mathcal{N}(0, 1)$ to produce the time series

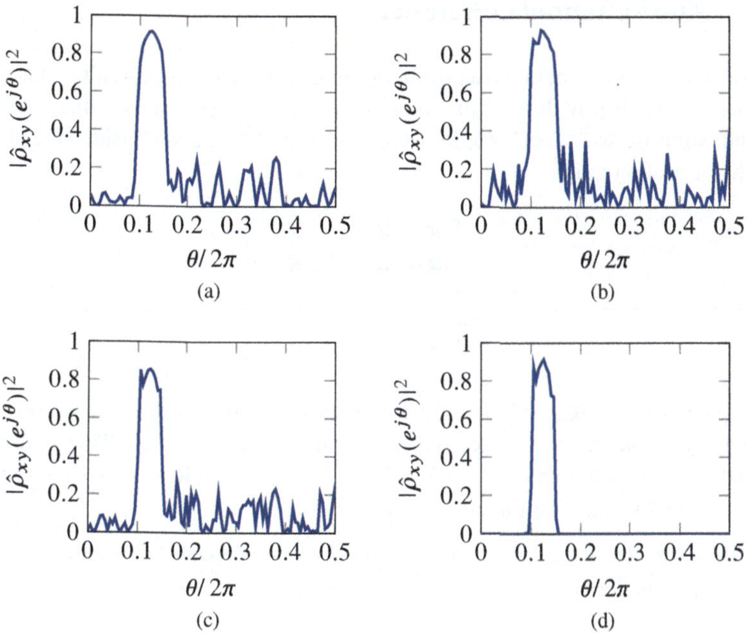

Fig. 3.4 MSC estimates for two Gaussian time series with a common narrowband signal (reprinted from [296]). (**a**) Welch (Hanning). (**b**) Welch (rectangular). (**c**) MVDR. (**d**) CCA

$x[n] = s[n] + w_x[n]$ and $y[n] = s[n] + w_y[n]$, $n = 0, \ldots, 1023$. It is easy to check that the true MSC is $|\rho_{xy}(e^{j\theta})|^2 = 0.8264$ in the band where the common narrowband signal is present and $|\rho_{xy}(e^{j\theta})|^2 = 0$ elsewhere. Figure 3.4 shows the MSC estimated by the averaged periodogram method [65] using Hanning and rectangular windows of $L = 100$ samples with 50% overlap, the MVDR approach proposed in [30] evaluated at $K = 200$ equispaced frequencies, and the reduced-rank CCA approach proposed in [296] with $p = 10$ canonical correlations.

Carter and Nuttall also studied the density of their MSC estimate. When the true MSC is zero, $|\hat{\rho}_{xy}(e^{j\theta_k})|^2$ follows a Beta$(1, N-1)$ distribution. Exact distribution results when the true MSC is not zero can be found in [64, Table 1]. It was proved in [249] that when $x[n]$ is a zero-mean Gaussian process independent of $y[n]$, the probability distribution MSC does not depend on the distribution of $y[n]$. Therefore, it is possible to set the threshold of the coherence-based detector for a specific false alarm probability independent of the statistics of the possibly non-Gaussian channel.

3.4 Multi-Channel Coherence

To analyze more general space-time problems, we need to generalize our notion of coherence to deal with multiple random vectors or vector-valued time series. To this end, suppose we have L random vectors $\mathbf{x}_l \in \mathbb{C}^{n_l \times 1}$, and consider the positive definite Hermitian matrix

$$\mathbf{R} = \begin{bmatrix} \mathbf{R}_{11} & \mathbf{R}_{12} & \cdots & \mathbf{R}_{1L} \\ \mathbf{R}_{21} & \mathbf{R}_{22} & \cdots & \mathbf{R}_{2L} \\ \vdots & \vdots & \ddots & \vdots \\ \mathbf{R}_{L1} & \mathbf{R}_{L2} & \cdots & \mathbf{R}_{LL} \end{bmatrix},$$

where each of the $\mathbf{R}_{ik}$ in this block-structured matrix is the cross-correlation between two random vectors $\mathbf{x}_i$ and $\mathbf{x}_k$ of dimensions $n_i \times n_k$. This is a matrix of puzzle pieces as shown in Fig. 3.5. If $\mathbf{R}_{11}$ is $n_1 \times n_1$ and $\mathbf{R}_{22}$ is $n_2 \times n_2$, then $\mathbf{R}_{12}$ is $n_1 \times n_2$, and so on. The puzzle pieces fit.

Definition 3.4 The multi-channel squared coherence between L random vectors $\mathbf{x}_l$, $l = 1, \ldots, L$, is defined to be

$$\rho^2(\mathbf{R}) = 1 - \frac{\det(\mathbf{R})}{\prod_{l=1}^{L} \det(\mathbf{R}_{ll})}. \tag{3.7}$$

This function is invariant to nonsingular transformation by $\mathbf{B} = \text{blkdiag}(\mathbf{B}_1, \ldots, \mathbf{B}_L)$, with group action $\mathbf{B}\mathbf{R}\mathbf{B}^H$. Moreover, the multi-channel coherence defined in (3.7) has an interesting connection with the Kullback-Leibler divergence.

Relationship with the Kullback-Leibler Divergence. Suppose we stack the column vectors $\mathbf{x}_l$ one above the other to form the $n = \sum_{l=1}^{L} n_l \times 1$ vector

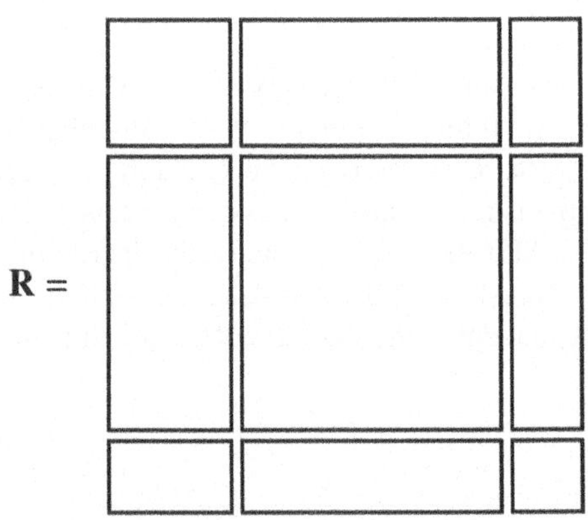

Fig. 3.5 Puzzle pieces of the block-structured covariance matrix

$$\mathbf{x} = \begin{bmatrix} \mathbf{x}_1^T & \cdots & \mathbf{x}_L^T \end{bmatrix}^T,$$

and suppose that $\mathbf{x}$ is distributed as a MVN random vector. Then the Kullback-Leibler divergence between the distribution P, which says $\mathbf{x} \sim CN_n(\mathbf{0}, \mathbf{R})$, and distribution Q, which says $\mathbf{x} \sim CN_n(\mathbf{0}, \text{blkdiag}(\mathbf{R}_{11}, \ldots, \mathbf{R}_{LL}))$, is given by

$$D_{KL}(P||Q) = -\log\left(\frac{\det(\mathbf{R})}{\prod_{l=1}^{L}\det(\mathbf{R}_{ll})}\right).$$

The connection between multiple coherence, as we have defined it, and the Kullback-Leibler divergence is then

$$\rho^2(\mathbf{R}) = 1 - e^{-D_{KL}(P||Q)}.$$

Let us now explore the use of coherence in statistical signal processing and machine learning. We first consider one-channel problems, where a low-dimensional PCA approximation maximizes coherence with the original data. Then, we move to two-channel problems where we encounter standard correlations, multiple correlations, half-canonical correlations, and (full) canonical correlations, all of which may be interpreted as coherences. Finally, for three channels, we encounter partial correlations, which when suitably normalized are coherences.

3.5 Principal Component Analysis

Begin with the zero-mean random variables, $x_l, l = 1, 2, \ldots, L$, organized into the vector $\mathbf{x} = [x_1\ x_2\ \cdots\ x_L]^T \in \mathbb{C}^L$. The covariance matrix $\mathbf{R}_{\mathbf{xx}} = E[\mathbf{x}\mathbf{x}^H]$ is Hermitian and positive semidefinite. Its diagonal elements are the variances of the x_l, and the off-diagonal elements are the cross-covariances between x_l and x_m.

The eigenvalue decomposition of this Hermitian matrix is

$$\mathbf{R}_{\mathbf{xx}} = \mathbf{U}\mathbf{\Lambda}\mathbf{U}^H,$$

where $\mathbf{U}$ is a unitary matrix and $\mathbf{\Lambda} = \text{diag}(\lambda_1, \lambda_2, \ldots, \lambda_L)$, with $\lambda_1 \geq \lambda_2 \geq \cdots \geq \lambda_L \geq 0$. The λ_l are the eigenvalues of $\mathbf{R}_{\mathbf{xx}}$. The decompositions $\mathbf{\Lambda} = \mathbf{U}^H\mathbf{R}_{\mathbf{xx}}\mathbf{U}$ and $\mathbf{R}_{\mathbf{xx}} = \mathbf{U}\mathbf{\Lambda}\mathbf{U}^H$ provide for the analysis of $\mathbf{x}$ into the random vector $\boldsymbol{\theta} = \mathbf{U}^H\mathbf{x}$, with diagonal covariance $\mathbf{\Lambda}$, and re-synthesis of $\mathbf{x}$ as $\mathbf{x} = \mathbf{U}\boldsymbol{\theta}$, with covariance $\mathbf{R}_{\mathbf{xx}}$. The total variance of $\mathbf{x}$ is preserved, as $\text{tr}(E[\boldsymbol{\theta}\boldsymbol{\theta}^H]) = \text{tr}(\mathbf{\Lambda}) = \text{tr}(\mathbf{U}^H\mathbf{R}_{\mathbf{xx}}\mathbf{U}) = \text{tr}(\mathbf{R}_{\mathbf{xx}}) = \text{tr}(E[\mathbf{x}\mathbf{x}^H])$.

Now consider a competing unitary matrix, $\mathbf{V}$, and its analysis vector $\boldsymbol{\phi} = \mathbf{V}^H\mathbf{x}$ with covariance matrix $\mathbf{V}^H\mathbf{R}_{\mathbf{xx}}\mathbf{V}$. The total variance of $\mathbf{x}$ is preserved, as $\text{tr}(\mathbf{V}^H\mathbf{R}_{\mathbf{xx}}\mathbf{V}) = \text{tr}(\mathbf{R}_{\mathbf{xx}})$. The diagonal element $(\mathbf{V}^H\mathbf{R}_{\mathbf{xx}}\mathbf{V})_{ll}$ is the variance of $(\mathbf{V}^H\mathbf{x})_l$. It is a result from the theory of majorization [230] that the eigenvalues of $\mathbf{V}^H\mathbf{R}_{\mathbf{xx}}\mathbf{V}$, which are identical to the eigenvalues of $\mathbf{R}_{\mathbf{xx}}$, majorize the diagonal elements of $\mathbf{V}^H\mathbf{R}_{\mathbf{xx}}\mathbf{V}$, which is to say

$$\sum_{l=1}^{r}\lambda_l \geq \sum_{l=1}^{r}(\mathbf{V}^H\mathbf{R}_{\mathbf{xx}}\mathbf{V})_{ll}, \qquad \text{for all } r = 1, \ldots, L,$$

with equality at $\mathbf{V} = \mathbf{U}$. So, in the competition to organize the original random variables into r linear combinations of the form $\{(\mathbf{V}^H\mathbf{x})_l, l = 1, 2, \ldots, r\}$, in order to maximize the accumulated variance for any choice of r, the winning choice is $\mathbf{V} = \mathbf{U}$. This result produces uncorrelated random variables.

We sometimes write $\mathbf{x}_r = \mathbf{U}_r\mathbf{U}_r^H\mathbf{x} = \mathbf{P}_{\mathbf{U}_r}\mathbf{x}$ as the reduced dimension version of $\mathbf{x}$, with $\mathbf{U}_r$ the $L \times r$ matrix containing the first r columns of $\mathbf{U}$. The error between $\mathbf{x}$ and $\mathbf{x}_r$ is then $(\mathbf{I}_L - \mathbf{P}_{\mathbf{U}_r})\mathbf{x}$, which is orthogonal to $\mathbf{x}_r$. There are several important properties of this decomposition:

- $\mathbf{x} = \mathbf{x}_r + (\mathbf{x} - \mathbf{x}_r)$ orthogonally decomposes $\mathbf{x}$,
- $E[(\mathbf{x} - \mathbf{x}_r)\mathbf{x}_r^H] = \mathbf{0}$ establishes the orthogonality between the approximation $\mathbf{x}_r$ and the error $\mathbf{x} - \mathbf{x}_r$,
- $E[\mathbf{x}_r\mathbf{x}^H] = \mathbf{U}\boldsymbol{\Lambda}_r\mathbf{U}^H$, where $\boldsymbol{\Lambda}_r = \operatorname{diag}(\lambda_1, \ldots, \lambda_r, 0, \ldots, 0)$, shows $\mathbf{x}_r$ to be maximally correlated with $\mathbf{x}$,
- $E[(\mathbf{x} - \mathbf{x}_r)(\mathbf{x} - \mathbf{x}_r)^H] = \mathbf{U}(\boldsymbol{\Lambda} - \boldsymbol{\Lambda}_r)\mathbf{U}^H$ is the mean-squared error matrix, with $\boldsymbol{\Lambda} - \boldsymbol{\Lambda}_r = \operatorname{diag}(0, 0, \ldots, \lambda_{r+1}, \ldots, \lambda_L)$,
- $\sum_{l=r+1}^{L}\lambda_l$ is the minimum achievable mean-squared error between $\mathbf{x}$ and $\mathbf{x}_r$,
- $\mathbf{R}_{\mathbf{xx}} = \mathbf{P}_{\mathbf{U}_r}\mathbf{R}_{\mathbf{xx}}\mathbf{P}_{\mathbf{U}_r} + (\mathbf{I}_L - \mathbf{P}_{\mathbf{U}_r})\mathbf{R}_{\mathbf{xx}}(\mathbf{I}_L - \mathbf{P}_{\mathbf{U}_r})$ is a Pythagorean decomposition into the covariance matrix of $\mathbf{x}_r$ and the covariance matrix of $(\mathbf{I}_L - \mathbf{P}_{\mathbf{U}_r})\mathbf{x}$.

So the low-dimensional approximation $\mathbf{x}_r$ maximizes coherence between $\mathbf{x}$ and $\mathbf{x}_r$ and as a consequence minimizes mean-squared error among all linear transformations to a low-dimensional approximation.

Implications for Data Analysis. Let the data matrix $\mathbf{X} = [\mathbf{x}_1\ \mathbf{x}_2\ \cdots\ \mathbf{x}_N] \in \mathbb{C}^{L\times N}$, $N \geq L$, be a random sample of the random vector $\mathbf{x} \in \mathbb{C}^L$. Each column serves as an experimental realization of $\mathbf{x}$. Or think of each column of $\mathbf{X}$ as one of N datums in $\mathbb{C}^L$, without any mention of the second-order properties of $\mathbf{x}$.

Perhaps there is a way to resolve $\mathbf{x}_n$ onto a lower-dimensional space, where coordinate values in this space can be used to resynthesize the vector $\hat{\mathbf{x}}_n$ near to the original. Let us propose the low-dimensional approximation $\hat{\mathbf{x}}_n = \mathbf{V}_r\mathbf{V}_r^H\mathbf{x}_n = \mathbf{P}_{\mathbf{V}_r}\mathbf{x}_n$, where $\mathbf{V}_r$ is an $L \times r$ slice of a unitary matrix, $r \leq L$, and $\mathbf{P}_{\mathbf{V}_r} = \mathbf{V}_r\mathbf{V}_r^H$ is the corresponding unitary projection onto the subspace $\langle\mathbf{V}_r\rangle$. The Euclidean distance between $\mathbf{x}_n$ and $\hat{\mathbf{x}}_n$ is the squared error: $(\mathbf{x}_n - \hat{\mathbf{x}}_n)^H(\mathbf{x}_n - \hat{\mathbf{x}}_n) = \mathbf{x}_n^H(\mathbf{I} - \mathbf{P}_{\mathbf{V}_r})\mathbf{x}_n$.

The total squared error in approximating the columns of $\mathbf{X}$ is

$$E = \sum_{n=1}^{N} \mathbf{x}_n^H (\mathbf{I} - \mathbf{P}_{\mathbf{V}_r}) \mathbf{x}_n = \operatorname{tr}\left((\mathbf{I} - \mathbf{P}_{\mathbf{V}_r}) \mathbf{X}\mathbf{X}^H (\mathbf{I} - \mathbf{P}_{\mathbf{V}_r}) \right),$$

where the (scaled) sample covariance (or Gramian) matrix $\mathbf{X}\mathbf{X}^H = \sum_{n=1}^{N} \mathbf{x}_n \mathbf{x}_n^H$ is non-negative definite. Give this covariance the EVD $\mathbf{X}\mathbf{X}^H = \mathbf{F}\mathbf{K}^2\mathbf{F}^H$, where $\mathbf{F}$ is $L \times L$ unitary and $\mathbf{K} = \operatorname{diag}(k_1, k_2, \ldots, k_L)$, $k_1 \geq k_2 \geq \cdots \geq k_L \geq 0$. The total squared error may be written as

$$E = \operatorname{tr}\left((\mathbf{I} - \mathbf{P}_{\mathbf{V}_r}) \mathbf{F}\mathbf{K}^2\mathbf{F}^H (\mathbf{I} - \mathbf{P}_{\mathbf{V}_r}) \right) = \operatorname{tr}\left(\mathbf{K}^2 \mathbf{F}^H (\mathbf{I} - \mathbf{P}_{\mathbf{V}_r}) \mathbf{F} \right).$$

This is minimized at the value $E = \sum_{l=r+1}^{L} k_l^2$ by aligning the subspace $\langle \mathbf{V}_r \rangle$ with the subspace spanned by the first r columns of $\mathbf{F}$. Thus, $\hat{\mathbf{X}}_r = \mathbf{F}_r\mathbf{F}_r^H\mathbf{X}$, where $\mathbf{F}_r$ is the $L \times r$ slice of $\mathbf{F}$ consisting of the first r columns of $\mathbf{F}$. This may also be written as $\hat{\mathbf{X}}_r = \mathbf{F}_r \boldsymbol{\Theta}_r$, where the columns of $\boldsymbol{\Theta}_r = \mathbf{F}_r^H \mathbf{X}$ are the coordinates of the original data in the subspace $\langle \mathbf{F}_r \rangle$.

Role of the SVD. Perhaps the SVD of $\mathbf{X}$, namely, $\mathbf{X} = \mathbf{F}\mathbf{K}\mathbf{G}^H$, lends further insight into this approximation. In this SVD, the matrix $\mathbf{F}$ is $L \times L$ unitary, $\mathbf{G}$ is $N \times N$ unitary, and $\mathbf{K}$ is $L \times N$ diagonal:

$$\mathbf{K} = \left[\operatorname{diag}(k_1, k_2, \ldots, k_L) \, \mathbf{0}_{L \times (N-L)} \right].$$

After approximation of $\mathbf{X}$, we have determined $\hat{\mathbf{X}}_r$ as $\hat{\mathbf{X}}_r = \mathbf{F}_r\mathbf{F}_r^H\mathbf{X} = \mathbf{F}_r\mathbf{F}_r^H\mathbf{F}\mathbf{K}\mathbf{G}^H = \mathbf{F}\mathbf{K}_r\mathbf{G}^H$. By the properties of the SVD, this matrix is the best rank-r approximation, in Frobenius norm, to the data matrix $\mathbf{X}$. Coming full circle, $E = \operatorname{tr}\left[(\mathbf{X} - \hat{\mathbf{X}}_r)(\mathbf{X} - \hat{\mathbf{X}}_r)^H \right] = \operatorname{tr}[\mathbf{F}(\mathbf{K} - \mathbf{K}_r)\mathbf{G}^H\mathbf{G}(\mathbf{K} - \mathbf{K}_r)\mathbf{F}^H] = \operatorname{tr}[(\mathbf{K} - \mathbf{K}_r)(\mathbf{K} - \mathbf{K}_r)] = \operatorname{tr}(\mathbf{K}^2 - \mathbf{K}_r^2) = \sum_{l=r+1}^{L} k_l^2$, which is a sum of the trailing squared singular values, corresponding to columns of $\mathbf{F}$ that are discarded in the approximation $\hat{\mathbf{X}}_r$. There are several noteworthy properties of this approximation:

1. $\hat{\mathbf{X}}_r = \mathbf{F}_r\boldsymbol{\Theta}_r$, with $\boldsymbol{\Theta}_r = \mathbf{K}_r\mathbf{G}^H$, an expansion on the basis $\mathbf{F}\mathbf{I}_r$, with coordinates $\boldsymbol{\Theta}_r$,
2. $\boldsymbol{\Theta}_r\boldsymbol{\Theta}_r^H = \mathbf{K}_r^2$, so the coordinates $\boldsymbol{\Theta} = \mathbf{F}^H\mathbf{X}$ are orthogonal with descending norm-squareds,
3. $\mathbf{X} = \hat{\mathbf{X}}_r + (\mathbf{X} - \hat{\mathbf{X}}_r)$, with $\hat{\mathbf{X}}_r(\mathbf{X} - \hat{\mathbf{X}}_r)^H = \mathbf{0}$, an orthogonality between approximants and their errors,
4. $\mathbf{X}\mathbf{X}^H = \hat{\mathbf{X}}_r\hat{\mathbf{X}}_r^H + (\mathbf{X} - \hat{\mathbf{X}}_r)(\mathbf{X} - \hat{\mathbf{X}}_r)^H$ is an orthogonal decomposition of the original covariance matrix into the estimated covariance and the error covariance, and trace of the error matrix is $\operatorname{tr}(\mathbf{K}^2 - \mathbf{K}_r^2) = \sum_{l=r+1}^{L} k_l^2$.

There is virtue in using the SVD:

1. There is no need to form the matrix $\mathbf{XX}^H$,
2. The SVD $\mathbf{X} = \mathbf{FKG}^H$ extracts the subspace $\langle \mathbf{F}_r \rangle$ for any and all $r = 1, 2, \ldots, L$,
3. The SVD extracts the coordinates of $\mathbf{X}$ in the subspace $\langle \mathbf{F}_r \rangle$ as $\mathbf{K}_r \mathbf{G}^H$, without the need to compute them as $\mathbf{F}_r^H \mathbf{X}$,
4. The sum $\sum_{l=r+1}^{L} k_l^2$ guides the choice of r.

Generalization to Accommodate Weighting of Errors. Perhaps the error should be defined as

$$E = \sum_{n=1}^{N} (\mathbf{x}_n - \hat{\mathbf{x}}_n)^H \mathbf{W}^{-1} (\mathbf{x}_n - \hat{\mathbf{x}}_n),$$

where $\mathbf{W}$ is a nonsingular Hermitian matrix. This may be written as

$$E = \sum_{n=1}^{N} (\mathbf{W}^{-1/2}\mathbf{x}_n - \mathbf{W}^{-1/2}\hat{\mathbf{x}}_n)^H (\mathbf{W}^{-1/2}\mathbf{x}_n - \mathbf{W}^{-1/2}\hat{\mathbf{x}}_n).$$

Now, all previous arguments hold, and the solution is to choose the estimator $\mathbf{W}^{-1/2}\hat{\mathbf{X}}_r = \mathbf{P}_{\mathbf{V}_r}\mathbf{W}^{-1/2}\mathbf{X}$, or $\hat{\mathbf{X}}_r = \mathbf{W}^{1/2}\mathbf{P}_{\mathbf{V}_r}\mathbf{W}^{-1/2}\mathbf{X}$, where $\mathbf{V}_r = \mathbf{F}_r$ and $\mathbf{FKF}^H$ is the EVD of the weighted Gramian $\mathbf{W}^{-1/2}\mathbf{XX}^H\mathbf{W}^{-1/2}$. It is important to note that the sequence of steps is this: 1) extract the principal subspace $\langle \mathbf{F}_r \rangle$ from the weighted Gramian $\mathbf{W}^{-1/2}\mathbf{XX}^H\mathbf{W}^{-1/2}$, 2) project the weighted data matrix $\mathbf{W}^{-1/2}\mathbf{X}$ onto this subspace, and 3) re-weight the solution by $\mathbf{W}^{1/2}$.

The SVD version of this story proceeds similarly. Give the weighted matrix $\mathbf{W}^{-1/2}\mathbf{X}$ the SVD $\mathbf{FKG}^H$. The matrix $\mathbf{F}_r\mathbf{K}_r\mathbf{G}_r^H$ is the best rank-r approximation to $\mathbf{W}^{-1/2}\mathbf{X}$ and $\mathbf{W}^{1/2}\mathbf{F}_r\mathbf{K}_r\mathbf{G}_r^H$ is the best rank-r weighted approximation to $\mathbf{X}$.

3.6 Two-Channel Correlation

Our interest is in the composite covariance matrix for the random vectors $\mathbf{x} \in \mathbb{C}^p$ and $\mathbf{y} \in \mathbb{C}^q$

$$\mathbf{R} = \mathrm{E}\left[\begin{bmatrix} \mathbf{x} \\ \mathbf{y} \end{bmatrix} \begin{bmatrix} \mathbf{x}^H & \mathbf{y}^H \end{bmatrix}\right] = \begin{bmatrix} \mathbf{R}_{\mathbf{xx}} & \mathbf{R}_{\mathbf{xy}} \\ \mathbf{R}_{\mathbf{yx}} & \mathbf{R}_{\mathbf{yy}} \end{bmatrix}. \tag{3.8}$$

This matrix is Hermitian and non-negative definite, which in some lexicons is redundant. The correlations between components of $\mathbf{x}$ and $\mathbf{y}$ are organized into the $p \times q$ matrix $\mathbf{R}_{\mathbf{xy}}$, and the normalization of this matrix by the square roots of the covariances of each gives the coherence matrix $\mathbf{C} = \mathbf{R}_{\mathbf{xx}}^{-1/2}\mathbf{R}_{\mathbf{xy}}\mathbf{R}_{\mathbf{yy}}^{-1/2}$. The eigenvalues of $\mathbf{C}$ are invariant to nonsingular transformation of $\mathbf{x}$ to $\mathbf{B}_x\mathbf{x}$, and $\mathbf{y}$ to $\mathbf{B}_y\mathbf{y}$, and these figure prominently in our treatment of canonical coordinates, and their use in model order reduction, in Sect. 3.9.

LMMSE Estimator. The linear minimum mean-squared error (LMMSE) estimator of $\mathbf{x}$ from $\mathbf{y}$ is $\hat{\mathbf{x}} = \mathbf{R}_{\mathbf{xy}}\mathbf{R}_{\mathbf{yy}}^{-1}\mathbf{y}$, and the resulting error covariance matrix is $\mathbf{Q}_{\mathbf{xx}|\mathbf{y}} = \mathrm{E}[(\mathbf{x}-\hat{\mathbf{x}})(\mathbf{x}-\hat{\mathbf{x}})^H] = \mathbf{R}_{\mathbf{xx}} - \mathbf{R}_{\mathbf{xy}}\mathbf{R}_{\mathbf{yy}}^{-1}\mathbf{R}_{\mathbf{yx}}$. We think of $\mathbf{R}_{\mathbf{xx}}$ as the covariance matrix that determines the concentration ellipse for the random vector $\mathbf{x}$ before filtering for $\hat{\mathbf{x}}$ and $\mathbf{Q}_{\mathbf{xx}|\mathbf{y}}$ as the covariance matrix that determines the concentration ellipse for the random error vector, $\mathbf{x}-\hat{\mathbf{x}}$, after filtering. When normalized by $\mathbf{R}_{\mathbf{xx}}^{-1/2}$, the error covariance matrix may be written as

$$\mathbf{R}_{\mathbf{xx}}^{-1/2}\mathbf{Q}_{\mathbf{xx}|\mathbf{y}}\mathbf{R}_{\mathbf{xx}}^{-1/2} = \mathbf{I}_p - \mathbf{R}_{\mathbf{xx}}^{-1/2}\mathbf{R}_{\mathbf{xy}}\mathbf{R}_{\mathbf{yy}}^{-1}\mathbf{R}_{\mathbf{yx}}\mathbf{R}_{\mathbf{xx}}^{-1/2} = \mathbf{I}_p - \mathbf{C}\mathbf{C}^H.$$

The term $\mathbf{R}_{\mathbf{xx}}^{-1/2}\mathbf{R}_{\mathbf{xy}}\mathbf{R}_{\mathbf{yy}}^{-1}\mathbf{R}_{\mathbf{yx}}\mathbf{R}_{\mathbf{xx}}^{-1/2}$ is a matrix-valued *multiple correlation coefficient*. It is the product of the *coherence matrix* $\mathbf{C} = \mathbf{R}_{\mathbf{xx}}^{-1/2}\mathbf{R}_{\mathbf{xy}}\mathbf{R}_{\mathbf{yy}}^{-1/2}$ and its Hermitian transpose.

The determinant of the normalized error covariance may be written as

$$\det\left(\mathbf{R}_{\mathbf{xx}}^{-1/2}\mathbf{Q}_{\mathbf{xx}|\mathbf{y}}\mathbf{R}_{\mathbf{xx}}^{-1/2}\right) = \frac{\det(\mathbf{Q}_{\mathbf{xx}|\mathbf{y}})}{\det(\mathbf{R}_{\mathbf{xx}})} = \frac{\det(\mathbf{R})}{\det(\mathbf{R}_{\mathbf{xx}})\det(\mathbf{R}_{\mathbf{yy}})}$$
$$= \prod_{i=1}^{\min(p,q)} (1 - ev_i(\mathbf{C}\mathbf{C}^H)),$$

where $ev_i(\mathbf{C}\mathbf{C}^H)$ denotes the ith eigenvalue of $\mathbf{C}\mathbf{C}^H$. A measure of bulk coherence may be written as

$$\rho^2 = 1 - \frac{\det(\mathbf{R})}{\det(\mathbf{R}_{\mathbf{xx}})\det(\mathbf{R}_{\mathbf{yy}})} = 1 - \prod_{i=1}^{\min(p,q)} (1 - ev_i(\mathbf{C}\mathbf{C}^H)).$$

This bulk coherence is near to one when the determinant of the normalized error covariance matrix is small, and this is the case where filtering for $\hat{\mathbf{x}}$ shrinks the volume of the error covariance matrix $\mathbf{Q}_{\mathbf{xx}|\mathbf{y}}$ with respect to the volume of the covariance matrix $\mathbf{R}_{\mathbf{xx}}$.

Orthogonality and Properties of the LMMSE Estimator. We should check that the estimator error $\mathbf{x} - \hat{\mathbf{x}}$ is orthogonal to the measurement $\mathbf{y}$:

$$\mathrm{E}\left[(\mathbf{x} - \hat{\mathbf{x}})\mathbf{y}^H\right] = \mathbf{R}_{\mathbf{xy}} - \mathbf{R}_{\mathbf{xy}}\mathbf{R}_{\mathbf{yy}}^{-1}\mathbf{R}_{\mathbf{yy}} = \mathbf{0}.$$

What more can be said? Write the error covariance matrix of a competing estimator $\mathbf{L}\mathbf{y}$ as

$$\begin{aligned}\mathbf{Q_L} &= \mathrm{E}\left[(\mathbf{x}-\mathbf{Ly})(\mathbf{x}-\mathbf{Ly})^H\right] \\ &= \mathbf{R_{xx}} - \mathbf{LR_{xy}} - \mathbf{R_{yx}L}^H + \mathbf{LR_{yy}L}^H \\ &= \mathbf{R_{xx}} - \mathbf{R_{xy}R_{yy}^{-1}R_{yx}} + \left(\mathbf{L} - \mathbf{R_{xy}R_{yy}^{-1}}\right)\mathbf{R_{yy}}\left(\mathbf{L} - \mathbf{R_{xy}R_{yy}^{-1}}\right)^H.\end{aligned}$$

The matrix $\mathbf{Q_L}$ may be written as $\mathbf{Q_L} = \mathbf{Q_{xx|y}} + \mathbf{M}$, where $\mathbf{M} = (\mathbf{L} - \mathbf{R_{xy}R_{yy}^{-1}})\mathbf{R_{yy}}(\mathbf{L} - \mathbf{R_{xy}R_{yy}^{-1}})^H$. The first term is determined by $\mathbf{Q_{xx|y}}$, and the other is a function of $\mathbf{L}$. The quadratic form $\mathbf{u}^H\mathbf{Q_L u}$ is minimized at $\mathbf{L} = \mathbf{W_{x|y}} = \mathbf{R_{xy}R_{yy}^{-1}}$, which is invariant to nonsingular transformation of $\mathbf{y}$. This means diagonal elements of $\mathbf{Q_L}$ are minimized, as is the trace of $\mathbf{Q_L}$. Moreover, by considering the normalized error covariance matrix $\mathbf{Q_{xx|y}^{-1/2}Q_L Q_{xx|y}^{-1/2}} = \mathbf{I}_p + \mathbf{Q_{xx|y}^{-1/2}MQ_{xx|y}^{-1/2}}$, it follows that $\det(\mathbf{Q_{xx|y}}) \le \det(\mathbf{Q_L})$, with equality iff $\mathbf{L} = \mathbf{R_{xy}R_{yy}^{-1}}$. In summary, the LMMSE estimator $\hat{\mathbf{x}} = \mathbf{R_{xy}R_{yy}^{-1}y}$ has the following properties:

1. The estimator error is orthogonal to the measurement $\mathbf{y}$ in the Hilbert space of second-order random variables, $\mathrm{E}[(\mathbf{x}-\hat{\mathbf{x}})\mathbf{y}^H] = \mathbf{0}$,
2. $\mathbf{u}^H\mathbf{Q_{xx|y}u} \le \mathbf{u}^H\mathbf{Q_L u}$, with equality iff $\mathbf{L} = \mathbf{W_{x|y}}$, that is, $\mathbf{Q_{xx|y}} \preceq \mathbf{Q_L}$ and $\mathbf{Q_{xx|y}^{-1}} \succeq \mathbf{Q_L^{-1}}$, and as a consequence, the error variances are $(\mathbf{Q_{xx|y}})_{ii} \le (\mathbf{Q_L})_{ii}$,
3. The total mean-squared error $\mathrm{E}[(\mathbf{x}-\hat{\mathbf{x}})^H(\mathbf{x}-\hat{\mathbf{x}})]$ is minimized at $\mathbf{L} = \mathbf{W_{x|y}}$, $\mathrm{tr}(\mathbf{Q_{xx|y}}) \le \mathrm{tr}(\mathbf{Q_L})$,
4. $\det(\mathbf{Q_{xx|y}}) \le \det(\mathbf{Q_L})$, and as a consequence, the volume enclosed by the concentration ellipse $\mathbf{u}^H\mathbf{Q_{xx|y}^{-1}u} = 1$ is smaller than the volume enclosed by the concentration ellipse $\mathbf{u}^H\mathbf{Q_L^{-1}u} = 1$,
5. the boundary of the concentration ellipse $\mathbf{u}^H\mathbf{Q_{xx|y}^{-1}u} = 1$ lies within the boundary of the concentration ellipse $\mathbf{u}^H\mathbf{Q_L^{-1}u} = 1$.

Signal-Plus-Noise Model. Our first idea is that the composite covariance structure $\mathbf{R}$ might be synthesized as the signal-plus-noise model $\mathbf{x} = \mathbf{x}$ and $\mathbf{y} = \mathbf{H_{y|x}x} + \mathbf{n}$, with $\mathbf{x}$ and $\mathbf{n}$ uncorrelated. In this model, $\mathbf{x}$ is interpreted to be signal, $\mathbf{H_{y|x}x}$ is considered to be signal through the channel $\mathbf{H_{y|x}}$, $\mathbf{n}$ is the channel noise, and $\mathbf{y}$ is the noisy output of the channel:

$$\begin{bmatrix}\mathbf{x}\\ \mathbf{y}\end{bmatrix} = \begin{bmatrix}\mathbf{I}_p & \mathbf{0}\\ \mathbf{H_{y|x}} & \mathbf{I}_q\end{bmatrix}\begin{bmatrix}\mathbf{x}\\ \mathbf{n}\end{bmatrix}.$$

If this signal-plus-noise model is to produce the composite covariance matrix, then

$$\begin{bmatrix}\mathbf{I}_p & \mathbf{0}\\ \mathbf{H_{y|x}} & \mathbf{I}_q\end{bmatrix}\begin{bmatrix}\mathbf{R_{xx}} & \mathbf{0}\\ \mathbf{0} & \mathbf{R_{nn}}\end{bmatrix}\begin{bmatrix}\mathbf{I}_p & \mathbf{H_{y|x}^H}\\ \mathbf{0} & \mathbf{I}_q\end{bmatrix} = \begin{bmatrix}\mathbf{R_{xx}} & \mathbf{R_{xy}}\\ \mathbf{R_{yx}} & \mathbf{R_{yy}}\end{bmatrix},$$

where $\mathbf{R_{nn}}$ is the covariance matrix of the additive noise $\mathbf{n}$ and $\mathbf{R_{xx}}$ is the covariance matrix of the signal $\mathbf{x}$. This forces the channel matrix to be $\mathbf{H_{y|x}} = \mathbf{R_{yx}}\mathbf{R_{xx}^{-1}}$ and $\mathbf{R_{nn}} = \mathbf{R_{yy}} - \mathbf{R_{yx}}\mathbf{R_{xx}^{-1}}\mathbf{R_{xy}}$. This result gives us a Cholesky or LDU factorization of $\mathbf{R}$, wherein the NW element of $\mathbf{R}$ is $\mathbf{R_{xx}}$. The additive noise covariance $\mathbf{R_{nn}}$ in the SE is the Schur complement of $\mathbf{R_{yy}}$. As a consequence, the composite covariance matrix $\mathbf{R}$ is block-diagonalized as

$$\begin{bmatrix} \mathbf{R_{xx}} & \mathbf{0} \\ \mathbf{0} & \mathbf{R_{nn}} \end{bmatrix} = \begin{bmatrix} \mathbf{I}_p & \mathbf{0} \\ -\mathbf{H_{y|x}} & \mathbf{I}_q \end{bmatrix} \begin{bmatrix} \mathbf{R_{xx}} & \mathbf{R_{xy}} \\ \mathbf{R_{yx}} & \mathbf{R_{yy}} \end{bmatrix} \begin{bmatrix} \mathbf{I}_p & -\mathbf{H}_{\mathbf{y|x}}^H \\ \mathbf{0} & \mathbf{I}_q \end{bmatrix}$$

and synthesized as the LDU Cholesky factor

$$\begin{bmatrix} \mathbf{R_{xx}} & \mathbf{R_{xy}} \\ \mathbf{R_{yx}} & \mathbf{R_{yy}} \end{bmatrix} = \begin{bmatrix} \mathbf{I}_p & \mathbf{0} \\ \mathbf{H_{y|x}} & \mathbf{I}_q \end{bmatrix} \begin{bmatrix} \mathbf{R_{xx}} & \mathbf{0} \\ \mathbf{0} & \mathbf{R_{nn}} \end{bmatrix} \begin{bmatrix} \mathbf{I}_p & \mathbf{H}_{\mathbf{y|x}}^H \\ \mathbf{0} & \mathbf{I}_q \end{bmatrix}.$$

Here, we have used for the first time the identity (see Appendix B)

$$\begin{bmatrix} \mathbf{I}_p & -\mathbf{A} \\ \mathbf{0} & \mathbf{I}_q \end{bmatrix}^{-1} = \begin{bmatrix} \mathbf{I}_p & \mathbf{A} \\ \mathbf{0} & \mathbf{I}_q \end{bmatrix}.$$

A consequence of this signal-plus-noise factorization is that the determinant of $\mathbf{R}$ may be written as $\det(\mathbf{R}) = \det(\mathbf{R_{xx}})\det(\mathbf{R_{nn}})$.

Measurement-Plus-Error Model. Our second idea is that the composite measurement vector might be decomposed into orthogonal estimator error and channel measurement:

$$\begin{bmatrix} \mathbf{e} \\ \mathbf{y} \end{bmatrix} = \begin{bmatrix} \mathbf{I}_p & -\mathbf{W_{x|y}} \\ \mathbf{0} & \mathbf{I}_q \end{bmatrix} \begin{bmatrix} \mathbf{x} \\ \mathbf{y} \end{bmatrix}.$$

If this estimator-plus-error model is to produce the composite covariance matrix, then the resulting UDL Cholesky factorization of $\mathbf{R}$ is

$$\begin{bmatrix} \mathbf{I}_p & -\mathbf{W_{x|y}} \\ \mathbf{0} & \mathbf{I}_q \end{bmatrix} \begin{bmatrix} \mathbf{R_{xx}} & \mathbf{R_{xy}} \\ \mathbf{R_{yx}} & \mathbf{R_{yy}} \end{bmatrix} \begin{bmatrix} \mathbf{I}_p & \mathbf{0} \\ -\mathbf{W}_{\mathbf{x|y}}^H & \mathbf{I}_q \end{bmatrix} = \begin{bmatrix} \mathbf{Q_{xx|y}} & \mathbf{0} \\ \mathbf{0} & \mathbf{R_{yy}} \end{bmatrix}.$$

This forces $\mathbf{W_{x|y}} = \mathbf{R_{xy}}\mathbf{R_{yy}^{-1}}$ and $\mathbf{Q_{xx|y}} = \mathbf{R_{xx}} - \mathbf{R_{xy}}\mathbf{R_{yy}^{-1}}\mathbf{R_{yx}}$. The error covariance $\mathbf{Q_{xx|y}}$ is the Schur complement of $\mathbf{R_{xx}}$. From this result, we have the following block-Cholesky factorization of the two-channel correlation matrix:

$$\begin{bmatrix} \mathbf{R_{xx}} & \mathbf{R_{xy}} \\ \mathbf{R_{yx}} & \mathbf{R_{yy}} \end{bmatrix} = \begin{bmatrix} \mathbf{I}_p & \mathbf{W_{x|y}} \\ \mathbf{0} & \mathbf{I}_q \end{bmatrix} \begin{bmatrix} \mathbf{Q_{xx|y}} & \mathbf{0} \\ \mathbf{0} & \mathbf{R_{yy}} \end{bmatrix} \begin{bmatrix} \mathbf{I}_p & \mathbf{0} \\ \mathbf{W}^H_{\mathbf{x|y}} & \mathbf{I}_q \end{bmatrix},$$

and the composite covariance matrix $\mathbf{R}^{-1}$ is therefore synthesized and block-diagonalized as

$$\begin{bmatrix} \mathbf{R_{xx}} & \mathbf{R_{xy}} \\ \mathbf{R_{yx}} & \mathbf{R_{yy}} \end{bmatrix}^{-1} = \begin{bmatrix} \mathbf{I}_p & \mathbf{0} \\ -\mathbf{W}^H_{\mathbf{x|y}} & \mathbf{I}_q \end{bmatrix} \begin{bmatrix} \mathbf{Q}^{-1}_{\mathbf{xx|y}} & \mathbf{0} \\ \mathbf{0} & \mathbf{R}^{-1}_{\mathbf{yy}} \end{bmatrix} \begin{bmatrix} \mathbf{I}_p & -\mathbf{W_{x|y}} \\ \mathbf{0} & \mathbf{I}_q \end{bmatrix}$$

and

$$\begin{bmatrix} \mathbf{I}_p & \mathbf{0} \\ \mathbf{W}^H_{\mathbf{x|y}} & \mathbf{I}_q \end{bmatrix} \begin{bmatrix} \mathbf{R_{xx}} & \mathbf{R_{xy}} \\ \mathbf{R_{yx}} & \mathbf{R_{yy}} \end{bmatrix}^{-1} \begin{bmatrix} \mathbf{I}_p & \mathbf{W_{x|y}} \\ \mathbf{0} & \mathbf{I}_q \end{bmatrix} = \begin{bmatrix} \mathbf{Q}^{-1}_{\mathbf{xx|y}} & \mathbf{0} \\ \mathbf{0} & \mathbf{R}^{-1}_{\mathbf{yy}} \end{bmatrix}.$$

When expanded, the formula for the inverse of $\mathbf{R}$ is

$$\mathbf{R}^{-1} = \begin{bmatrix} \mathbf{Q}^{-1}_{\mathbf{xx|y}} & -\mathbf{Q}^{-1}_{\mathbf{xx|y}}\mathbf{W_{x|y}} \\ -\mathbf{W}^H_{\mathbf{x|y}}\mathbf{Q}^{-1}_{\mathbf{xx|y}} & \mathbf{R}^{-1}_{\mathbf{yy}} + \mathbf{W}^H_{\mathbf{x|y}}\mathbf{Q}^{-1}_{\mathbf{xx|y}}\mathbf{W_{x|y}} \end{bmatrix}.$$

Importantly, the NW element of $\mathbf{R}^{-1}$ is $\mathbf{Q}^{-1}_{\mathbf{xx|y}}$, the inverse of the error covariance matrix, and the formula for $\det(\mathbf{R}^{-1})$ is $\det(\mathbf{R}^{-1}) = \det(\mathbf{Q}^{-1}_{\mathbf{xx|y}})\det(\mathbf{R}^{-1}_{\mathbf{yy}})$. The NE element scales with the LMMSE filter $\mathbf{W_{x|y}}$ and the SE element is a rank p inflation of the $q \times q$ inverse $\mathbf{R}^{-1}_{\mathbf{yy}}$.

Composing the Signal-Plus-Noise and Measurement-Plus-Error Models. Let us compose these two models as follows:

$$\begin{bmatrix} \mathbf{e} \\ \mathbf{y} \end{bmatrix} = \begin{bmatrix} \mathbf{I}_p & -\mathbf{W_{x|y}} \\ \mathbf{0} & \mathbf{I}_q \end{bmatrix} \begin{bmatrix} \mathbf{I}_p & \mathbf{0} \\ \mathbf{H_{y|x}} & \mathbf{I}_q \end{bmatrix} \begin{bmatrix} \mathbf{x} \\ \mathbf{n} \end{bmatrix}.$$

This establishes two connections:

$$\begin{bmatrix} \mathbf{Q_{xx|y}} & \mathbf{0} \\ \mathbf{0} & \mathbf{R_{yy}} \end{bmatrix} = \begin{bmatrix} \mathbf{I}_p & -\mathbf{W_{x|y}} \\ \mathbf{0} & \mathbf{I}_q \end{bmatrix} \begin{bmatrix} \mathbf{I}_p & \mathbf{0} \\ \mathbf{H_{y|x}} & \mathbf{I}_q \end{bmatrix} \begin{bmatrix} \mathbf{R_{xx}} & \mathbf{0} \\ \mathbf{0} & \mathbf{R_{nn}} \end{bmatrix} \begin{bmatrix} \mathbf{I}_p & \mathbf{H}^H_{\mathbf{y|x}} \\ \mathbf{0} & \mathbf{I}_q \end{bmatrix} \begin{bmatrix} \mathbf{I}_p & \mathbf{0} \\ \mathbf{W}^H_{\mathbf{x|y}} & \mathbf{I}_q \end{bmatrix} \tag{3.9}$$

and

$$\begin{bmatrix} \mathbf{Q}^{-1}_{\mathbf{xx|y}} & \mathbf{0} \\ \mathbf{0} & \mathbf{R}^{-1}_{\mathbf{yy}} \end{bmatrix} = \begin{bmatrix} \mathbf{I}_p & \mathbf{0} \\ \mathbf{W}^H_{\mathbf{x|y}} & \mathbf{I}_q \end{bmatrix} \begin{bmatrix} \mathbf{I}_p & -\mathbf{H}^H_{\mathbf{y|x}} \\ \mathbf{0} & \mathbf{I}_q \end{bmatrix} \begin{bmatrix} \mathbf{R}^{-1}_{\mathbf{xx}} & \mathbf{0} \\ \mathbf{0} & \mathbf{R}^{-1}_{\mathbf{nn}} \end{bmatrix} \begin{bmatrix} \mathbf{I}_p & \mathbf{0} \\ -\mathbf{H_{y|x}} & \mathbf{I}_q \end{bmatrix} \begin{bmatrix} \mathbf{I}_p & \mathbf{W}^H_{\mathbf{x|y}} \\ \mathbf{0} & \mathbf{I}_q \end{bmatrix}. \tag{3.10}$$

Match up the NE block of (3.9) with the SW block of (3.10), to obtain two formulas for the optimum filter $\mathbf{W}_{\mathbf{x}|\mathbf{y}}$:

$$\begin{aligned}\mathbf{W}_{\mathbf{x}|\mathbf{y}} &= \mathbf{R}_{\mathbf{xx}}\mathbf{H}_{\mathbf{y}|\mathbf{x}}^H\left(\mathbf{H}_{\mathbf{y}|\mathbf{x}}\mathbf{R}_{\mathbf{xx}}\mathbf{H}_{\mathbf{y}|\mathbf{x}}^H + \mathbf{R}_{\mathbf{nn}}\right)^{-1} \\ &= \left(\mathbf{R}_{\mathbf{xx}}^{-1} + \mathbf{H}_{\mathbf{y}|\mathbf{x}}^H\mathbf{R}_{\mathbf{nn}}^{-1}\mathbf{H}_{\mathbf{y}|\mathbf{x}}\right)^{-1}\mathbf{H}_{\mathbf{y}|\mathbf{x}}^H\mathbf{R}_{\mathbf{nn}}^{-1}.\end{aligned}$$

Then, match up the NW blocks of (3.9) and (3.10) to obtain two formulas for the error covariance matrix $\mathbf{Q}_{\mathbf{xx}|\mathbf{y}}$:

$$\begin{aligned}\mathbf{Q}_{\mathbf{xx}|\mathbf{y}} &= \mathbf{R}_{\mathbf{xx}} - \mathbf{R}_{\mathbf{xx}}\mathbf{H}_{\mathbf{y}|\mathbf{x}}^H\left(\mathbf{H}_{\mathbf{y}|\mathbf{x}}\mathbf{R}_{\mathbf{xx}}\mathbf{H}_{\mathbf{y}|\mathbf{x}}^H + \mathbf{R}_{\mathbf{nn}}\right)^{-1}\mathbf{H}_{\mathbf{y}|\mathbf{x}}\mathbf{R}_{\mathbf{xx}} \\ &= \left(\mathbf{R}_{\mathbf{xx}}^{-1} + \mathbf{H}_{\mathbf{y}|\mathbf{x}}^H\mathbf{R}_{\mathbf{nn}}^{-1}\mathbf{H}_{\mathbf{y}|\mathbf{x}}\right)^{-1}.\end{aligned}$$

These equations are Woodbury identities. It is important to note that the filter $\mathbf{W}_{\mathbf{x}|\mathbf{y}}$ does not equalize the channel filter $\mathbf{H}_{\mathbf{y}|\mathbf{x}}$. That is, $\mathbf{W}_{\mathbf{x}|\mathbf{y}}\mathbf{H}_{\mathbf{y}|\mathbf{x}} \neq \mathbf{I}_p$, but it is approximately $\mathbf{I}_p$ when $\mathbf{R}_{\mathbf{xx}}^{-1}$ is small compared with $\mathbf{H}_{\mathbf{y}|\mathbf{x}}^H\mathbf{R}_{\mathbf{nn}}^{-1}\mathbf{H}_{\mathbf{y}|\mathbf{x}}$.

Comment. The real virtue of these equations is in those cases where the problem really is a signal-plus-noise model, in which case the source covariance matrix $\mathbf{R}_{\mathbf{xx}}$, channel matrix $\mathbf{H}_{\mathbf{y}|\mathbf{x}}$, and additive noise covariance $\mathbf{R}_{\mathbf{nn}}$ are known or estimated. In such cases, these parameters are not extracted as virtual parameters that reproduce the composite covariance $\mathbf{R}$. The dimension of $\mathbf{H}_{\mathbf{y}|\mathbf{x}}$ determines which of the equations is more computationally efficient.

Law of Total Variance. The error of the linear minimum mean-squared error estimator, $\mathbf{x} - \hat{\mathbf{x}}$, is orthogonal to the estimator $\hat{\mathbf{x}}$ in a Hilbert space of second-order random variables. Much of our geometric reasoning about linear MMSE estimators generalizes to geometric reasoning about the *conditional mean* estimator. Consider the random vectors $\mathbf{x}, \mathbf{y}$, defined on the same probability space. Consider the *conditional mean* of $\mathbf{x}$, given $\mathbf{y}$, and denote it $\hat{\mathbf{x}} = E[\mathbf{x}|\mathbf{y}]$. It is easy to see that $E[\hat{\mathbf{x}}] = E[\mathbf{x}]$, which is to say that the conditional mean estimator is an unbiased estimator of $\mathbf{x}$. Moreover, $E[(\mathbf{x} - \hat{\mathbf{x}})\hat{\mathbf{x}}^H] = \mathbf{0}$, which is to say the estimator error is orthogonal to the estimator, in a Hilbert space of second-order random variables. As a consequence, from $\mathbf{x} = \hat{\mathbf{x}} + (\mathbf{x} - \hat{\mathbf{x}})$, it follows that $E[\mathbf{x}\mathbf{x}^H] = E[\hat{\mathbf{x}}\hat{\mathbf{x}}^H] + E[(\mathbf{x} - \hat{\mathbf{x}})(\mathbf{x} - \hat{\mathbf{x}})^H]$. This is a Pythagorean decomposition of correlation. Subtracting $E[\mathbf{x}]\,E[\mathbf{x}^H]$ from both sides of this equality,

$$E[\mathbf{x}\mathbf{x}^H] - E[\mathbf{x}]\,E[\mathbf{x}^H] = E[\hat{\mathbf{x}}\hat{\mathbf{x}}^H] - E[\hat{\mathbf{x}}](E[\hat{\mathbf{x}}])^H + E[(\mathbf{x} - \hat{\mathbf{x}})(\mathbf{x} - \hat{\mathbf{x}})^H].$$

This is often written as

$$\text{cov}\,\mathbf{x} = \text{cov}\,\hat{\mathbf{x}} + \text{E}[\text{cov}\,\mathbf{x}|\mathbf{y}],$$

and called the *law of total variance*. With $\mathbf{R_{xx}}$ denoting the covariance matrix of $\mathbf{x}$, the formula for normalized error covariance is now

$$\mathbf{R}_{\mathbf{xx}}^{-1/2}\,\text{E}[\text{cov}\,\mathbf{x}|\mathbf{y}]\mathbf{R}_{\mathbf{xx}}^{-1/2} = \mathbf{I}_p - \mathbf{R}_{\mathbf{xx}}^{-1/2}(\text{cov}\,\hat{\mathbf{x}})\mathbf{R}_{\mathbf{xx}}^{-1/2}.$$

In the special case that the conditional expectation is linear in $\mathbf{y}$, then $\text{E}[\text{cov}\,\mathbf{x}|\mathbf{y}] = \mathbf{Q}_{\mathbf{xx}|\mathbf{y}}$ and $\text{cov}\,\hat{\mathbf{x}} = \mathbf{R_{xy}}\mathbf{R}_{\mathbf{yy}}^{-1}\mathbf{R}_{\mathbf{xy}}^{H}$. Then, normalized error covariance is the familiar formula $\mathbf{R}_{\mathbf{xx}}^{-1/2}\mathbf{Q}_{\mathbf{xx}|\mathbf{y}}\mathbf{R}_{\mathbf{xx}}^{-1/2} = \mathbf{I}_p - \mathbf{C}\mathbf{C}^H$, with $\mathbf{C}$ the coherence matrix $\mathbf{C} = \mathbf{R}_{\mathbf{xx}}^{-1/2}\mathbf{R_{xy}}\mathbf{R}_{\mathbf{yy}}^{-1/2}$.

Distribution of the Estimators. The question before us is this: if the composite covariance matrix $\mathbf{R}$ in (3.8) is replaced by the sample covariance matrix, how are the LMMSE filter $\mathbf{W}_{\mathbf{x}|\mathbf{y}} = \mathbf{R_{xy}}\mathbf{R}_{\mathbf{yy}}^{-1}$, the error covariance matrix $\mathbf{Q}_{\mathbf{xx}|\mathbf{y}} = \mathbf{R_{xx}} - \mathbf{R_{xy}}\mathbf{R}_{\mathbf{yy}}^{-1}\mathbf{R_{yx}}$, and the measurement covariance matrix $\mathbf{R_{yy}}$ distributed?

Assume the complex proper MVN random vectors $\mathbf{x}$ and $\mathbf{y}$ are organized into the composite vector $\mathbf{z}$ as

$$\mathbf{z} = \begin{bmatrix} \mathbf{x} \\ \mathbf{y} \end{bmatrix} \sim CN_{p+q}(\mathbf{0}, \mathbf{R}).$$

Collect N i.i.d. realizations of this vector in

$$\mathbf{Z} = [\mathbf{z}_1 \ \cdots \ \mathbf{z}_N] \sim CN_{(p+q)\times N}(\mathbf{0}, \mathbf{I}_N \otimes \mathbf{R}).$$

The corresponding (scaled) sample covariance matrix, $\mathbf{S} = \mathbf{Z}\mathbf{Z}^H$, is patterned as follows:

$$\mathbf{S} = \begin{bmatrix} \mathbf{S_{xx}} & \mathbf{S_{xy}} \\ \mathbf{S_{yx}} & \mathbf{S_{yy}} \end{bmatrix} = \begin{bmatrix} \mathbf{X}\mathbf{X}^H & \mathbf{X}\mathbf{Y}^H \\ \mathbf{Y}\mathbf{X}^H & \mathbf{Y}\mathbf{Y}^H \end{bmatrix}.$$

The estimated LMMSE filter is $\widehat{\mathbf{W}}_{\mathbf{x}|\mathbf{y}} = \mathbf{S_{xy}}\mathbf{S}_{\mathbf{yy}}^{-1}$, the estimated (scaled) error covariance matrix is $\widehat{\mathbf{Q}}_{\mathbf{xx}|\mathbf{y}} = \mathbf{S_{xx}} - \mathbf{S_{xy}}\mathbf{S}_{\mathbf{yy}}^{-1}\mathbf{S_{yx}}$, and the estimated (scaled) measurement covariance matrix is $\widehat{\mathbf{R}}_{\mathbf{yy}} = \mathbf{S_{yy}}$. The distributions of these estimators are summarized as follows:

1. The sample covariance matrix is Wishart: $\mathbf{S} \sim C\mathcal{W}_{p+q}(\mathbf{R}, N)$,
2. The error covariance matrix is Wishart, $\widehat{\mathbf{Q}}_{\mathbf{xx}|\mathbf{y}} \sim C\mathcal{W}_p(\mathbf{Q}_{\mathbf{xx}|\mathbf{y}}, N-q)$, and is independent of $\mathbf{S_{xy}}$ and $\mathbf{S_{yy}}$. Equivalently, when normalized,

$$\mathbf{Q}_{\mathbf{xx}|\mathbf{y}}^{-1/2}\widehat{\mathbf{Q}}_{\mathbf{xx}|\mathbf{y}}\mathbf{Q}_{\mathbf{xx}|\mathbf{y}}^{-1/2} \sim C\mathcal{W}_p(\mathbf{I}_p, N-q),$$

3. The measurement covariance matrix is Wishart: $\mathbf{S_{yy}} \sim C\mathcal{W}_q(\mathbf{R_{yy}}, N)$,
4. Given $\mathbf{S_{yy}}$, the conditional distribution of $\mathbf{S_{xy}}$ is normal:

$$\mathbf{S_{xy}} \mid \mathbf{S_{yy}} \sim C\mathcal{N}_{p\times q}\left(\mathbf{R_{xy}}\mathbf{R_{yy}}^{-1}\mathbf{S_{yy}}, \mathbf{S_{yy}} \otimes \mathbf{Q_{xx|y}}\right),$$

5. Given $\mathbf{S_{yy}}$, the conditional distribution of $\widehat{\mathbf{W}}_{\mathbf{x|y}}$, is normal:

$$\widehat{\mathbf{W}}_{\mathbf{x|y}} \mid \mathbf{S_{yy}} \sim C\mathcal{N}_{p\times q}\left(\mathbf{W_{x|y}}, \mathbf{S_{yy}}^{-1} \otimes \mathbf{Q_{xx|y}}\right),$$

6. The unconditional distribution of $\widehat{\mathbf{W}}_{\mathbf{x|y}}$ is distributed as a matrix-t, with pdf

$$f(\widehat{\mathbf{W}}_{\mathbf{x|y}}) = \frac{\tilde{\Gamma}_q(N+p)}{\pi^{pq}\tilde{\Gamma}_q(N)}(\det(\mathbf{R_{yy}}))^{-N}(\det(\mathbf{Q_{xx|y}}))^{-q} \times \left[\det\left(\mathbf{R_{yy}}^{-1} + (\widehat{\mathbf{W}}_{\mathbf{x|y}} - \mathbf{W_{x|y}})^H \mathbf{Q_{xx|y}}^{-1}(\widehat{\mathbf{W}}_{\mathbf{x|y}} - \mathbf{W_{x|y}})\right)\right]^{-(N+p)},$$

7. The distribution of the normalized statistic $\mathbf{N} = \mathbf{Q_{xx|y}}^{-1/2}(\widehat{\mathbf{W}}_{\mathbf{x|y}} - \mathbf{W_{x|y}})\mathbf{R_{yy}}^{1/2}$ is

$$f(\mathbf{N}) = \frac{\tilde{\Gamma}_q(N+p)}{\pi^{pq}\tilde{\Gamma}_q(N)}(\det(\mathbf{I}_p + \mathbf{N}\mathbf{N}^H))^{-(N+p)},$$

where $\tilde{\Gamma}_q(x)$ is the complex multivariate gamma function

$$\tilde{\Gamma}_q(x) = \pi^{q(q-1)/2}\prod_{l=1}^{q}\Gamma(x-l+1)$$

and $\Gamma(\cdot)$ is the gamma function.

The first four of these results are proved by Muirhead in [244, Thm 3.2.10]. The last three are proved by Khatri and Rao [199] by marginalizing over the Wishart distribution of $\mathbf{S_{yy}}$.

For the case $p = q = 1$, these results specialize as follows. Let x and y be two proper complex normal random variables organized in the two-dimensional vector $\mathbf{z} = [x\ y]^T \sim C\mathcal{N}_2(\mathbf{0}, \mathbf{R})$, with covariance matrix

$$\mathbf{R} = \begin{bmatrix} \sigma_x^2 & \sigma_x\sigma_y\rho \\ \sigma_x\sigma_y\rho^* & \sigma_y^2 \end{bmatrix},$$

where $\rho = \mathrm{E}[xy^*]/(\sigma_x\sigma_y)$ denotes the complex correlation coefficient. Let us collect N i.i.d. realizations of $\mathbf{z}$ and form the row vectors $\mathbf{x} = [x_1 \ \cdots \ x_N]$ and

$\mathbf{y} = [y_1 \ \cdots \ y_N]$. The 2×2 sample covariance matrix is

$$\mathbf{S} = \begin{bmatrix} \mathbf{x}\mathbf{x}^H & \mathbf{x}\mathbf{y}^H \\ \mathbf{y}\mathbf{x}^H & \mathbf{y}\mathbf{y}^H \end{bmatrix}.$$

The estimated LMMSE filter is the scalar $\hat{w}_{x|y} = \mathbf{x}\mathbf{y}^H / \mathbf{y}\mathbf{y}^H$, and the estimated error variance is $\hat{q}_{xx|y} = \mathbf{x}\mathbf{x}^H(1 - |\hat{\rho}|^2)$, where $|\hat{\rho}|^2$ is the sample coherence. The distributions of the estimators are as follows:

1. The sample covariance matrix is Wishart: $\mathbf{S} \sim \mathcal{CW}_2(\mathbf{R}, N)$,
2. The normalized error variance is chi-squared: $\frac{2\mathbf{x}\mathbf{x}^H(1-|\hat{\rho}|^2)}{\sigma_x^2(1-|\rho|^2)} \sim \chi^2_{2N-2}$,
3. The normalized sample variance of the observations is chi-square: $2\mathbf{y}\mathbf{y}^H/\sigma_y^2 \sim \chi^2_{2N}$,
4. Given $\mathbf{y}\mathbf{y}^H$, the conditional distribution of $\mathbf{x}\mathbf{y}^H$ is normal:

$$\mathbf{x}\mathbf{y}^H \mid \mathbf{y}\mathbf{y}^H \sim \mathcal{CN}\left(\frac{\rho\sigma_x}{\sigma_y}\mathbf{y}\mathbf{y}^H, \ \sigma_x^2(1 - |\rho|^2)\mathbf{y}\mathbf{y}^H\right),$$

5. Given $\mathbf{y}\mathbf{y}^H$, the conditional distribution of $\hat{w}_{x|y}$ is normal:

$$\hat{w}_{x|y} \mid \mathbf{y}\mathbf{y}^H \sim \mathcal{CN}\left(w_{x|y}, \ \frac{\sigma_x^2(1 - |\rho|^2)}{\mathbf{y}\mathbf{y}^H}\right),$$

6. The unconditional distribution of $\hat{w}_{x|y}$ is determined by multiplying the conditional density by the marginal density for $\mathbf{y}\mathbf{y}^H$ and integrating. This is equivalent to assigning an inverse chi-squared prior for the variance of a normal distribution. The result for the unconditional density is

$$f(\hat{w}_{x|y}) = \frac{\sigma_y^2 N}{\sigma_x^2(1 - |\rho|^2)\pi} \frac{1}{\left(1 + \frac{\sigma_y^2}{\sigma_x^2(1-|\rho|^2)}|\hat{w}_{x|y} - w_{x|y}|^2\right)^{N+1}},$$

which is a scaled Student's t-distribution with $2N$ degrees of freedom.

3.7 Krylov Subspace, Conjugate Gradients, and the Multistage LMMSE Filter

The computation of the LMMSE filter $\mathbf{W}_{\mathbf{x}|\mathbf{y}} = \mathbf{R}_{\mathbf{x}\mathbf{y}}\mathbf{R}_{\mathbf{y}\mathbf{y}}^{-1}$ and the error covariance matrix $\mathbf{Q}_{\mathbf{x}\mathbf{x}|\mathbf{y}} = \mathbf{R}_{\mathbf{x}\mathbf{x}} - \mathbf{W}_{\mathbf{x}|\mathbf{y}}\mathbf{R}_{\mathbf{y}\mathbf{y}}^{-1}\mathbf{W}_{\mathbf{x}|\mathbf{y}}^H$ require inversion of the matrix $\mathbf{R}_{\mathbf{y}\mathbf{y}}$. Perhaps there is a way to decompose this computation so that a workable approximation of the LMMSE filter $\mathbf{W}_{\mathbf{x}|\mathbf{y}}$ may be computed without inverting what may be a very

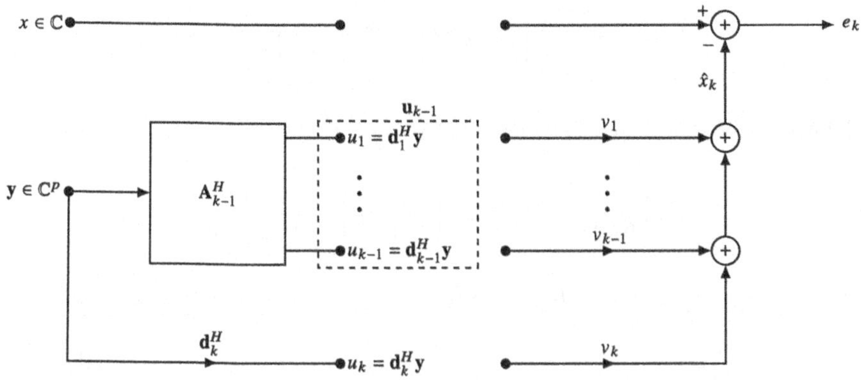

Fig. 3.6 Multistage or greedy filtering. The matrix $\mathbf{A}_k^H$ is recursively updated as $\mathbf{A}_k = [\mathbf{A}_{k-1}\ \mathbf{d}_k]$

large matrix $\mathbf{R}_{\mathbf{yy}}$. The basic idea is to transform the measurements $\mathbf{y}$ in such a way that transformed variables are diagonally correlated, as illustrated in Fig. 3.6. Then, the inverse is trivial. Of course, the EVD may be used for this purpose, but it is a non-terminating algorithm with complexity on the order of the complexity of inverting $\mathbf{R}_{\mathbf{yy}}$.

We are in search of a method, termed the method of conjugate gradients or equivalently the method of multistage LMMSE filtering.[1] The multistage LMMSE filter may be considered a greedy approximation of the LMMSE filter. But it is constructed in such a way that it converges to the LMMSE filter in a small number of steps for certain idealized, but quite common, models for $\mathbf{R}_{\mathbf{yy}}$ that arise in engineered systems. We shall demonstrate the idea for the case where the random variable x to be estimated is a complex scalar and the measurement $\mathbf{y}$ is a p-dimensional vector. The extension to vector-valued $\mathbf{x}$ is straightforward.

In Fig. 3.6, the suggestion is that the approximation of the LMMSE filter is recursively approximated as a sum of k terms, with k much smaller than p and with the computational complexity of determining each new direction vector $\mathbf{d}_k$ on the order of p^2. The net will be to replace the p^3 complexity of solving for the LMMSE filter with the kp^2 complexity of conjugate gradients for computing the direction vectors and approximating the LMMSE estimator.

According to the figure, the idea is to transform the measurements $\mathbf{y} \in \mathbb{C}^p$ into intermediate variables $\mathbf{u}_k = \mathbf{A}_k^H \mathbf{y} \in \mathbb{C}^k$, so that the LMMSE estimator $\hat{x} \in \mathbb{C}$ may be approximated as an uncoupled linear combination of the elements of $\mathbf{u}_k$. This will be a useful gambit if the number of steps in this procedure may be terminated at a number of steps k much smaller than p. From the figure, we see that the action of the linear transformation $\mathbf{A}_k^H \in \mathbb{C}^{k \times p}$ is to transform the composite covariance matrix for $[x\ \mathbf{y}^T]^T$, given by

[1] In the original, and influential, work of Goldstein and Reed, this was termed the multistage Wiener filter [140].

$$\mathrm{E}\left[\begin{bmatrix} x \\ \mathbf{y} \end{bmatrix} \begin{bmatrix} x^* & \mathbf{y}^H \end{bmatrix}\right] = \begin{bmatrix} r_{xx} & \mathbf{r}_{xy} \\ \mathbf{r}_{yx} & \mathbf{R}_{yy} \end{bmatrix},$$

to the composite covariance matrix for $[x \ (\mathbf{A}_k^H \mathbf{y})^T]^T$, given by

$$\mathrm{E}\left[\begin{bmatrix} x \\ \mathbf{A}_k^H \mathbf{y} \end{bmatrix} \begin{bmatrix} x^* & \mathbf{y}^H \mathbf{A}_k \end{bmatrix}\right] = \begin{bmatrix} r_{xx} & \mathbf{r}_{xy}\mathbf{A}_k \\ \mathbf{A}_k^H \mathbf{r}_{yx} & \mathbf{A}_k^H \mathbf{R}_{yy} \mathbf{A}_k \end{bmatrix}.$$

With $\mathbf{A}_k$ resolved as $[\mathbf{A}_{k-1} \ \mathbf{d}_k]$, the above covariance matrix may be resolved as

$$\mathrm{E}\left[\begin{bmatrix} x \\ \mathbf{A}_{k-1}^H \mathbf{y} \\ \mathbf{d}_k^H \mathbf{y} \end{bmatrix} \begin{bmatrix} x^* & \mathbf{y}^H \mathbf{A}_{k-1} & \mathbf{y}^H \mathbf{d}_k \end{bmatrix}\right] = \begin{bmatrix} r_{xx} & \mathbf{r}_{xy}\mathbf{A}_{k-1} & \mathbf{r}_{xy}\mathbf{d}_k \\ \mathbf{A}_{k-1}^H \mathbf{r}_{yx} & \mathbf{A}_{k-1}^H \mathbf{R}_{yy} \mathbf{A}_{k-1} & \mathbf{A}_{k-1}^H \mathbf{R}_{yy} \mathbf{d}_k \\ \mathbf{d}_k^H \mathbf{r}_{yx} & \mathbf{d}_k^H \mathbf{R}_{yy} \mathbf{A}_{k-1} & \mathbf{d}_k^H \mathbf{R}_{yy} \mathbf{d}_k \end{bmatrix}. \tag{3.11}$$

Our aim is to take the transformed covariance matrix in (3.11) to the form

$$\begin{bmatrix} r_{xx} & \mathbf{r}_{xy}\mathbf{A}_{k-1} & \mathbf{r}_{xy}\mathbf{d}_k \\ \mathbf{A}_{k-1}^H \mathbf{r}_{yx} & \boldsymbol{\Sigma}_{k-1}^2 & \mathbf{0} \\ \mathbf{d}_k^H \mathbf{r}_{yx} & \mathbf{0} & \sigma_k^2 \end{bmatrix},$$

where $\boldsymbol{\Sigma}_{k-1}^2$ is a diagonal $(k-1) \times (k-1)$ matrix. The motivation, of course, is to diagonalize the covariance of the transformed measurement $\mathbf{u}_k = \mathbf{A}_k^H \mathbf{y}$ so that this covariance matrix may be easily inverted. If we can achieve our aims, then the k-term approximation to the LMMSE estimator of x from $\mathbf{u}_k$ is

$$\hat{x}_k = \mathbf{r}_{xy}\mathbf{A}_{k-1}\boldsymbol{\Sigma}_{k-1}^{-2}\mathbf{u}_{k-1} + \frac{1}{\sigma_k^2}\mathbf{r}_{xy}\mathbf{d}_k u_k = \hat{x}_{k-1} + \frac{1}{\sigma_k^2}\mathbf{r}_{xy}\mathbf{d}_k u_k.$$

The trick will be to find an algorithm that keeps the computation of the *direction vectors* $\mathbf{d}_k$ alive.

To diagonalize the covariance matrix $\mathbf{A}_k^H \mathbf{R}_{yy} \mathbf{A}_k$ is to construct direction vectors $\mathbf{d}_i$ that are $\mathbf{R}_{yy}$-conjugate. That is, $\mathbf{d}_i^H \mathbf{R}_{yy} \mathbf{d}_l = \sigma_i^2 \delta[i-l]$. Perhaps these direction vectors can be recursively updated by recursively updating *gradient vectors* $\mathbf{g}_i$ that recursively Gram-Schmidt orthogonalize $\mathbf{A}_k$. We do not enforce the property that $\mathbf{G}_k = [\mathbf{g}_1 \ \mathbf{g}_2 \ \cdots \ \mathbf{g}_k]$ be unitary, but rather only that $\mathbf{G}_k^H \mathbf{G}_k = \mathrm{diag}(\kappa_1^2, \ldots, \kappa_k^2)$. That is, to say $\mathbf{g}_i^H \mathbf{g}_l = \kappa_i^2 \delta[i-l]$. The resulting algorithm is the famous conjugate gradient algorithm (CG) of Algorithm 3, first derived by Hestenes and Stiefel [162].

It is not hard to show that the direction vector $\mathbf{d}_i$ is a linear combination of the vectors $\mathbf{r}_{yx}, \mathbf{R}_{yy}\mathbf{r}_{yx}, \ldots, \mathbf{R}_{yy}^{i-1}\mathbf{r}_{yx}$. Therefore, the resulting sequence of direction vectors $\mathbf{d}_i, i = 1, 2, \ldots, k$, is a non-orthogonal basis for the *Krylov subspace*

Algorithm 3: Conjugate gradient algorithm

Initialize:

$\mathbf{d}_1 = \mathbf{r}_{\mathbf{y}x}; \quad \mathbf{g}_1 = \mathbf{r}_{\mathbf{y}x}; \quad v_1 = (\mathbf{d}_1^H \mathbf{R}_{\mathbf{yy}} \mathbf{d}_1)^{-1} \mathbf{g}_1^H \mathbf{g}_1$

$\mathbf{w}_1 = \mathbf{d}_1 v_1; \quad \hat{x}_1 = \mathbf{w}_1^H \mathbf{y}$

for $i = 2, 3, \ldots,$ until convergence **do**

$\mathbf{g}_i = \mathbf{g}_{i-1} - \mathbf{R}_{\mathbf{yy}} \mathbf{d}_{i-1} v_{i-1}$ // gradient update

$\mathbf{d}_i = \mathbf{g}_i + \mathbf{d}_{i-1} \frac{\mathbf{g}_i^H \mathbf{g}_i}{\mathbf{g}_{i-1}^H \mathbf{g}_{i-1}}$ // direction update

$v_i = \frac{\mathbf{g}_i^H \mathbf{g}_i}{\mathbf{d}_i^H \mathbf{R}_{\mathbf{yy}} \mathbf{d}_i}$ // weight update

$\mathbf{w}_i = \mathbf{w}_{i-1} + \mathbf{d}_i v_i$ // filter update

$\hat{x}_i = \mathbf{w}_i^H \mathbf{y}$ // LMMSE update

end for

$$\mathbf{K}_k = \langle \mathbf{r}_{\mathbf{y}x}, \mathbf{R}_{\mathbf{yy}} \mathbf{r}_{\mathbf{y}x}, \mathbf{R}_{\mathbf{yy}}^2 \mathbf{r}_{\mathbf{y}x}, \ldots, \mathbf{R}_{\mathbf{yy}}^{k-1} \mathbf{r}_{\mathbf{y}x} \rangle.$$

The corresponding sequence of gradients is an orthogonal basis for this subspace. So, evidently, this subspace will stop expanding in dimension, and the multistage LMMSE filter will stop growing branches, when the Krylov subspace stops expanding in dimension. The complexity of the CG algorithm is then kp^2, which may be much smaller than p^3. Can this happen?

Let us suppose the $p \times p$ Hermitian covariance $\mathbf{R}_{\mathbf{yy}}$ has just k *distinct* eigenvalues, $\lambda_1 > \lambda_2 > \cdots > \lambda_k > 0$, of respective multiplicities $r_1, r_2, \ldots, r_k$. The spectral representation of $\mathbf{R}_{\mathbf{yy}}$ is

$$\mathbf{R}_{\mathbf{yy}} = \lambda_1 \mathbf{P}_1 + \lambda_2 \mathbf{P}_2 + \cdots + \lambda_k \mathbf{P}_k,$$

where $\mathbf{P}_i$ is a rank-r_i symmetric, idempotent, projection matrix. The sum of these ranks is $\sum_{i=1}^k r_i = p$. This set of projection matrices identifies a set of mutually orthogonal subspaces, which is to say $\mathbf{P}_i \mathbf{P}_l = \mathbf{P}_i \delta[i-l]$ and $\sum_{i=1}^k \mathbf{P}_i = \mathbf{I}_p$. It follows that for any $l \geq 0$, the p-dimensional vector $\mathbf{R}_{\mathbf{yy}}^l \mathbf{r}_{\mathbf{y}x}$ may be written as

$$\mathbf{R}_{\mathbf{yy}}^l \mathbf{r}_{\mathbf{y}x} = \lambda_1^l \mathbf{P}_1 \mathbf{r}_{\mathbf{y}x} + \lambda_2^l \mathbf{P}_2 \mathbf{r}_{\mathbf{y}x} + \cdots + \lambda_k^l \mathbf{P}_k \mathbf{r}_{\mathbf{y}x}.$$

Therefore, the Krylov subspace $\mathbf{K}_k$ can have dimension no greater than k. The multistage LMMSE filter stops growing branches after k steps, and the LMMSE estimator $\hat{x}_k$ is the LMMSE estimator.

This observation explains the use of diagonal loading of the form $\mathbf{R}_{\mathbf{yy}} + \epsilon^2 \mathbf{I}_p$ as a preconditioning step in advance of conjugate gradients. Typically, an arbitrary covariance matrix will have low numerical rank, which is to say there will be $k-1$ relatively large eigenvalues, followed by $p-k+1$ relatively small eigenvalues. The addition of $\epsilon^2 \mathbf{I}_p$ only slightly biases the large eigenvalues away from their nominal values and replaces the small eigenvalues with a nearly common eigenvalue ϵ^2. The consequent number of distinct eigenvalues is essentially k, and the multistage

Table 3.1 Connection between multistage LMMSE filtering and conjugate gradients for quadratic minimization

Multistage LMMSE	CG for quadratic minimization
Subspace expansion	Iterative search
Correlation btw $x - \hat{x}_k$ and $\mathbf{y}$	Gradient vector
Analysis filter $\mathbf{d}_i$	Search direction vector
Synthesis filter v_i	Step size
Uncorrelated u_i	$\mathbf{R}_{\mathbf{yy}}$-conjugacy
Orthogonality	Zero gradient
Filter $\mathbf{w}_i$	Solution vector
Multistage LMMSE filter	Conjugate gradient algorithm

LMMSE filter may be terminated at k branches after k steps of the conjugate gradient algorithm.

The connection between the language of *multistage LMMSE filtering* and *CG for quadratic minimization* is summarized in Table 3.1.

3.8 Application to Beamforming and Spectrum Analysis

Every result to come in this section for beamforming is in fact a result for spectrum analysis. Simply replace the interpretation of $\boldsymbol{\psi} = [1\, e^{-j\phi} \cdots e^{-j(L-1)\phi}]^T/\sqrt{L}$ as a *steering vector in spatial coordinates* by its interpretation as a *steering vector in temporal coordinates*. When swept through $-\pi < \phi \le \pi$, a steering vector in spatial coordinates is an analyzer of a wavenumber spectrum; in temporal coordinates, it is an analyzer of a frequency spectrum.

Among classical and modern methods of beamforming, the *conventional* and *minimum variance distortionless response* beamformers, denoted CBF and MVDR, are perhaps the most fundamental. Of course, there are many variations on them. In this section, we use our results for estimation in two-channel models to illuminate the geometrical character of beamforming. The idea is to frame the question of beamforming as a virtual two-channel estimation problem and then derive second-order formulas that reveal the role played by coherence. A key finding is that the power out of an MVDR beamformer and the power out of a generalized sidelobe canceller (GSC) resolve the power out of a CBF beamformer.

The adaptation of the beamformers of this chapter, using a variety of rules for eigenvalue shaping, remains a topic of great interest in radar, sonar, and radio astronomy. These topics are not covered in this chapter. In fact, these rules fall more closely into the realm of the factor analysis topics treated in Chap. 5.

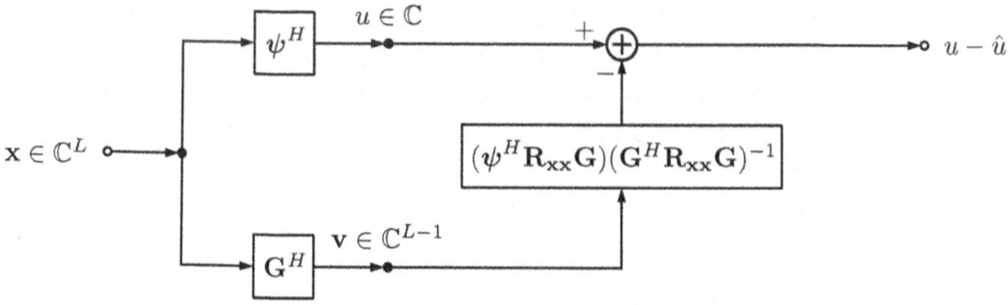

Fig. 3.7 Generalized sidelobe canceller. Top is output of conventional beamformer, bottom is output of GSC, and middle is error in estimating top from bottom

3.8.1 The Generalized Sidelobe Canceller

We begin with the generalized sidelobe canceller for beamforming or spectrum analysis, illustrated in Fig. 3.7. In this figure, the variables are defined as follows:

- $\mathbf{x} \in \mathbb{C}^L$ is the vector of measurements made at the L antenna elements of a multi-sensor array,
- $\boldsymbol{\psi} = [1\ e^{-j\phi}\ \cdots\ e^{-j(L-1)\phi}]^T/\sqrt{L}$ is the *steering vector* for a uniform linear array; ϕ is the electrical angle $\phi = 2\pi(d/\lambda)\sin\theta$, where d is the spacing between sensors, λ is the wavelength of a single-frequency propagating wave, and θ is the angle this wave makes with the perpendicular, or boresight, direction of the array; the electrical angle ϕ is the phase advance of this propagating wave as it propagates across the face of the array,
- The matrix $\mathbf{T} = [\boldsymbol{\psi}\ \mathbf{G}]$ is unitary of dimensions $L \times L$; that is, $\boldsymbol{\psi}^H\boldsymbol{\psi} = 1, \mathbf{G}^H\mathbf{G} = \mathbf{I}_{L-1}$, and $\boldsymbol{\psi}^H\mathbf{G} = \mathbf{0}$,
- $u = \boldsymbol{\psi}^H\mathbf{x} \in \mathbb{C}$ is the output of a *conventional beamformer* (CBF),
- $\mathbf{v} = \mathbf{G}^H\mathbf{x} \in \mathbb{C}^{L-1}$ is the output of a *generalized sidelobe canceller* (GSC).

Both of $\boldsymbol{\psi}$ and $\mathbf{G}$, denoted $\boldsymbol{\psi}(\phi)$ and $\mathbf{G}(\phi)$, are steered through electrical angle $-\pi < \phi \leq \pi$, to turn out bearing response patterns for the field $\mathbf{x}$ observed in an L-element array or L sample time series. At each steering angle ϕ, the steering vector $\boldsymbol{\psi}(\phi)$ determines a dimension-one subspace $\langle\boldsymbol{\psi}(\phi)\rangle$, and when scanned through the electrical angle ϕ, this set of subspaces determines the so-called array manifold. The corresponding GSC matrix $\mathbf{G}(\phi)$ may be determined by factoring the projection $\mathbf{I}_L - \boldsymbol{\psi}(\phi)\boldsymbol{\psi}^H(\phi)$ as $\mathbf{G}(\phi)\mathbf{G}^H(\phi)$. By construction, the $L \times L$ matrix $\mathbf{T} = [\boldsymbol{\psi}(\phi)\ \mathbf{G}(\phi)]$ is unitary for all ϕ.

3.8.2 Composite Covariance Matrix

We shall model the field $\mathbf{x}$ to second-order as a zero-mean vector with covariance matrix $E[\mathbf{x}\mathbf{x}^H] = \mathbf{R}_{\mathbf{xx}}$. Typically, this covariance matrix is estimated as the scaled sample covariance matrix $\mathbf{S} = \mathbf{X}\mathbf{X}^H$ from the space-time data matrix $\mathbf{X} = [\mathbf{x}_1\ \mathbf{x}_2\ \cdots\ \mathbf{x}_N] \in \mathbb{C}^{L\times N}$. Here, $\mathbf{x}_n$ is the nth temporal snapshot of $\mathbf{x}$. When there is an underlying model for this covariance, such as a low-rank signal plus noise model of the form $\mathbf{R}_{\mathbf{xx}} = \mathbf{H}\mathbf{\Lambda}\mathbf{H}^H + \mathbf{\Sigma}$, then a parametric estimator of $\mathbf{R}_{\mathbf{xx}}$ may be used in place of $\mathbf{S}$.

Since $\mathbf{T} = [\boldsymbol{\psi}\ \mathbf{G}]$, the vector $\mathbf{z} = \mathbf{T}^H\mathbf{x}$ contains in its first element the CBF output $u = \boldsymbol{\psi}^H\mathbf{x}$ and in its next $L-1$ elements the GSC output $\mathbf{v} = \mathbf{G}^H\mathbf{x}$. The composite covariance matrix for $\mathbf{z}$ is

$$\mathbf{R}_{\mathbf{zz}} = E\left[\begin{bmatrix} u \\ \mathbf{v} \end{bmatrix}\begin{bmatrix} u^* & \mathbf{v}^H \end{bmatrix}\right] = \mathbf{T}^H\mathbf{R}_{\mathbf{xx}}\mathbf{T}$$

$$= \begin{bmatrix} \boldsymbol{\psi}^H\mathbf{R}_{\mathbf{xx}}\boldsymbol{\psi} & \boldsymbol{\psi}^H\mathbf{R}_{\mathbf{xx}}\mathbf{G} \\ \mathbf{G}^H\mathbf{R}_{\mathbf{xx}}\boldsymbol{\psi} & \mathbf{G}^H\mathbf{R}_{\mathbf{xx}}\mathbf{G} \end{bmatrix} = \begin{bmatrix} \tilde{\boldsymbol{\psi}}^H\tilde{\boldsymbol{\psi}} & \tilde{\boldsymbol{\psi}}^H\tilde{\mathbf{G}} \\ \tilde{\mathbf{G}}^H\tilde{\boldsymbol{\psi}} & \tilde{\mathbf{G}}^H\tilde{\mathbf{G}} \end{bmatrix},$$

where $\tilde{\boldsymbol{\psi}} = \mathbf{R}_{\mathbf{xx}}^{1/2}\boldsymbol{\psi}$ and $\tilde{\mathbf{G}} = \mathbf{R}_{\mathbf{xx}}^{1/2}\mathbf{G}$. The LMMSE estimator of u from $\mathbf{v}$ is $\hat{u} = \tilde{\boldsymbol{\psi}}^H\tilde{\mathbf{G}}(\tilde{\mathbf{G}}^H\tilde{\mathbf{G}})^{-1}\mathbf{v}$. The Pythagorean decomposition of u is $u = \hat{u} + (u - \hat{u})$, with corresponding variance decomposition $E[|u|^2] = E[|\hat{u}|^2] + E[|u - \hat{u}|^2]$. This variance decomposition may be written as

$$\tilde{\boldsymbol{\psi}}^H\tilde{\boldsymbol{\psi}} = \tilde{\boldsymbol{\psi}}^H\mathbf{P}_{\tilde{\mathbf{G}}}\tilde{\boldsymbol{\psi}} + \tilde{\boldsymbol{\psi}}^H(\mathbf{I}_L - \mathbf{P}_{\tilde{\mathbf{G}}})\tilde{\boldsymbol{\psi}},$$

where $\mathbf{P}_{\tilde{\mathbf{G}}} = \tilde{\mathbf{G}}(\tilde{\mathbf{G}}^H\tilde{\mathbf{G}})^{-1}\tilde{\mathbf{G}}^H$. The LHS of this equation is, in fact, the power out of the conventional beamformer: $P_{CBF} = \tilde{\boldsymbol{\psi}}^H\tilde{\boldsymbol{\psi}} = \boldsymbol{\psi}^H\mathbf{R}_{\mathbf{xx}}\boldsymbol{\psi}$. The first term on the RHS is the power out of the GSC. What about the second term on the RHS? It is the error variance $Q_{uu|\mathbf{v}}$ for estimating u from $\mathbf{v}$, which may be read out of the NW element of the inverse of the composite covariance matrix. That is, $(\mathbf{R}_{\mathbf{zz}}^{-1})_{11} = Q_{uu|\mathbf{v}}^{-1}$. But by the unitarity of $\mathbf{T}$, the inverse of $\mathbf{R}_{\mathbf{zz}}$ may be written as $\mathbf{R}_{\mathbf{zz}}^{-1} = \mathbf{T}^H\mathbf{R}_{\mathbf{xx}}^{-1}\mathbf{T}$ with NW element $\boldsymbol{\psi}^H\mathbf{R}_{\mathbf{xx}}^{-1}\boldsymbol{\psi}$. The resulting important identity is

$$Q_{uu|\mathbf{v}} = \frac{1}{\boldsymbol{\psi}^H\mathbf{R}_{\mathbf{xx}}^{-1}\boldsymbol{\psi}}.$$

This is the power out of the MVDR beamformer: $P_{MVDR} = \frac{1}{\boldsymbol{\psi}^H\mathbf{R}_{\mathbf{xx}}^{-1}\boldsymbol{\psi}}$. We write these findings as

$$\boldsymbol{\psi}^H\mathbf{R}_{\mathbf{xx}}\boldsymbol{\psi} = \tilde{\boldsymbol{\psi}}^H\mathbf{P}_{\tilde{\mathbf{G}}}\tilde{\boldsymbol{\psi}} + \frac{1}{\boldsymbol{\psi}^H\mathbf{R}_{\mathbf{xx}}^{-1}\boldsymbol{\psi}}.$$

The narrative is "The power out of the MVDR beamformer and the power out of the GSC additively resolve the power out of the CBF."

Assume $\boldsymbol{\psi}$ is unit-norm. Then the Schwartz inequality shows

$$1 = |\boldsymbol{\psi}^H \boldsymbol{\psi}|^2 = |\boldsymbol{\psi}^H \mathbf{R}_{\mathbf{xx}}^{-1/2} \mathbf{R}_{\mathbf{xx}}^{1/2} \boldsymbol{\psi}|^2 \leq (\boldsymbol{\psi}^H \mathbf{R}_{\mathbf{xx}}^{-1} \boldsymbol{\psi})(\boldsymbol{\psi}^H \mathbf{R}_{\mathbf{xx}} \boldsymbol{\psi}),$$

which yields

$$\frac{1}{\boldsymbol{\psi}^H \mathbf{R}_{\mathbf{xx}}^{-1} \boldsymbol{\psi}} \leq \boldsymbol{\psi}^H \mathbf{R}_{\mathbf{xx}} \boldsymbol{\psi}.$$

This suggests better resolution for the MVDR beamformer than for the CBF.

All of this connects with our definition of coherence:

$$\rho^2(\mathbf{R}_{\mathbf{zz}}) = 1 - \frac{\det(\mathbf{R}_{\mathbf{zz}})}{\det((\mathbf{R}_{\mathbf{zz}})_{NW}) \det((\mathbf{R}_{\mathbf{zz}})_{SE})}$$
$$= 1 - \frac{\det(Q_{uu|\mathbf{v}})}{\det((\mathbf{R}_{\mathbf{zz}})_{NW})} = 1 - \frac{1/\boldsymbol{\psi}^H \mathbf{R}_{\mathbf{xx}}^{-1} \boldsymbol{\psi}}{\boldsymbol{\psi}^H \mathbf{R}_{\mathbf{xx}} \boldsymbol{\psi}}.$$

In summary, the interpretations are these:

- The output of the CBF is orthogonally decomposed as the output of the GSC and the error in estimating the output of the CBF from the output of the GSC,
- The power out of the CBF is resolved as the sum of the power out of the GSC and the power of the error in estimating the output of the CBF from the output of the GSC,
- The power out of the MVDR is less than or equal to the power out of the CBF, suggesting better resolution for MVDR,
- Coherence is one minus the ratio of the power out of the MVDR and the power out of the CBF,
- Coherence is near to one when MVDR is much smaller than CBF, suggesting that the GSC has canceled interference in sidelobes to estimate what is in the mainlobe.

Figure 3.8 illustrates this finding.

3.8.3 Distributions of the Conventional and Capon Beamformers

From a random sample $\mathbf{X} \sim CN_{L\times N}(\mathbf{0}, \mathbf{I}_N \otimes \boldsymbol{\Sigma})$ and its corresponding scaled sample covariance matrix $\mathbf{S} = \mathbf{X}\mathbf{X}^H$, typically measured from a sequence of N snapshots in an L-element array, it is a common practice in radar, sonar, and astronomy to construct the images

$$\hat{B}(\phi) = \frac{\boldsymbol{\psi}^H(\phi)\mathbf{S}\boldsymbol{\psi}(\phi)}{\boldsymbol{\psi}^H(\phi)\boldsymbol{\psi}(\phi)}, \qquad \hat{C}(\phi) = \frac{\boldsymbol{\psi}^H(\phi)\boldsymbol{\psi}(\phi)}{\boldsymbol{\psi}^H(\phi)\mathbf{S}^{-1}\boldsymbol{\psi}(\phi)},$$

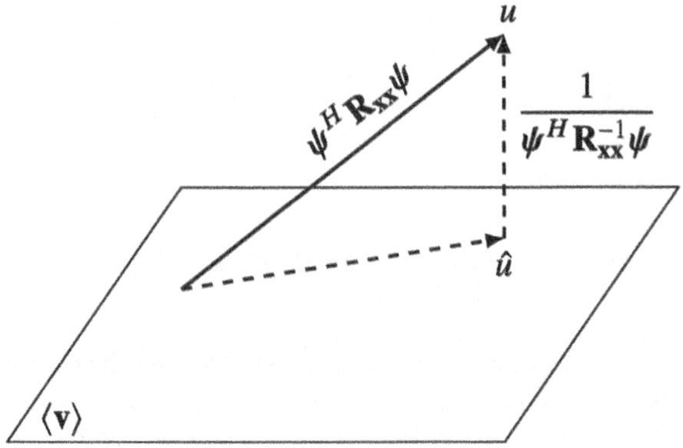

Fig. 3.8 Geometry of beamforming

with $-\pi < \phi \leq \pi$. The vector ψ is termed a steering vector, and ϕ is termed an electrical angle, as it encodes for phase delays between sensor elements in the array in response to a single-frequency propagating wave, as we saw in Sect. 3.8.1. In the case of a uniform linear array, and plane wave propagation, ψ is the Vandermonde vector $\psi = [1 \; e^{-j\phi} \; \cdots \; e^{-j(L-1)\phi}]^T / \sqrt{L}$. These images are intended to be estimators of the respective functions

$$B(\phi) = \frac{\psi^H(\phi)\boldsymbol{\Sigma}\psi(\phi)}{\psi^H(\phi)\psi(\phi)}, \qquad C(\phi) = \frac{\psi^H(\phi)\psi(\phi)}{\psi^H(\phi)\boldsymbol{\Sigma}^{-1}\psi(\phi)},$$

commonly called the conventional and Capon spectra. The Cauchy-Schwarz inequality, $(\psi^H\psi)^2 = (\psi^H\boldsymbol{\Sigma}^{-1/2}\boldsymbol{\Sigma}^{1/2}\psi)^2 \leq (\psi^H\boldsymbol{\Sigma}^{-1}\psi)(\psi^H\boldsymbol{\Sigma}\psi)$, shows

$$C(\phi) \leq B(\phi).$$

Hence, for each ϕ, the value of the Capon spectrum lies below the value of the conventional image, suggesting better resolution of closely space radiators. But more on this is to come.

The distribution of the sample covariance matrix is $\mathbf{S} \sim CW_L(\boldsymbol{\Sigma}, N)$, and the distributions of the estimated spectra $\hat{B}(\phi)$ and $\hat{C}(\phi)$ are these:

$$\hat{B}(\phi) \sim CW_1(B(\phi), N), \qquad \hat{C}(\phi) \sim CW_1(C(\phi), N - L + 1).$$

The first result follows from standard Wishart theory, and the second follows from [199, Theorem 1]. The corresponding pdfs are

$$f(\hat{B}(\phi)) = \frac{1}{\Gamma(N)B(\phi)} \left(\frac{\hat{B}(\phi)}{B(\phi)}\right)^{N-1} \operatorname{etr}\left(-\frac{\hat{B}(\phi)}{B(\phi)}\right)$$

and

$$f(\hat{C}(\phi)) = \frac{1}{\Gamma(N-L+1)C(\phi)} \left(\frac{\hat{C}(\phi)}{C(\phi)} \right)^{N-L} \operatorname{etr}\left(-\frac{\hat{C}(\phi)}{C(\phi)} \right).$$

So the random variables $\hat{B}(\phi)/B(\phi)$ and $\hat{C}(\phi)/C(\phi)$ are, respectively, distributed as χ^2_{2N} and $\chi^2_{2(N-L+1)}$ random variables. Their respective means and variances are $2N$ and $4N$; $2(N-L+1)$ and $4(N-L+1)$. It follows that the mean and variance of the estimator $\hat{B}(\phi)/2N$ are $B(\phi)$ and $(B(\phi))^2/N$. The standard deviation scales with $B(\phi)$ and inversely with $\sqrt{N}$. The mean and variance of the estimator $\hat{C}(\phi)/2(N-L+1)$ are $C(\phi)$ and $(C(\phi))^2/(N-L+1)$. The standard deviation scales with $C(\phi)$ and inversely with $\sqrt{N-L+1}$. So the higher resolution of the Capon spectrum is paid for by a loss in averaging, which is to say the effective sample size is $N-L+1$ and not N. So, in a low SNR application, the better resolution of the Capon imager is paid for by larger variance in the estimated spectrum. At low SNR, this effect is important.

These results generalize to multi-rank imagers, with the steering vector $\boldsymbol{\psi}$ replaced by an imaging matrix $\boldsymbol{\Psi} \in \mathbb{C}^{L\times r}$. Then, the matrix-valued images are $\hat{\mathbf{B}} = (\boldsymbol{\Psi}^H\boldsymbol{\Psi})^{-1/2}(\boldsymbol{\Psi}^H\mathbf{S}\boldsymbol{\Psi})(\boldsymbol{\Psi}^H\boldsymbol{\Psi})^{-1/2}$ and $\hat{\mathbf{C}} = (\boldsymbol{\Psi}^H\boldsymbol{\Psi})^{1/2}(\boldsymbol{\Psi}^H\mathbf{S}^{-1}\boldsymbol{\Psi})^{-1}(\boldsymbol{\Psi}^H\boldsymbol{\Psi})^{1/2}$. The corresponding distributions are $\mathcal{CW}_r(\mathbf{B}, N)$ and $\mathcal{CW}_r(\mathbf{C}, N-L+1)$, with obvious definitions of the matrix-valued spectra $\mathbf{B}$ and $\mathbf{C}$. It is straightforward to derive the distributions for trace or determinant of these Wishart matrices.

3.9 Canonical Correlation Analysis

Canonical correlation analysis (CCA) is to two-channel inference as principal component analysis is to one-channel inference, which is to say CCA resolves a measurement channel $\mathbf{y} \in \mathbb{C}^q$ into coordinates that carry the most information about coordinates of an unobserved message channel $\mathbf{x} \in \mathbb{C}^p$. So there will be two coordinate transformations in play, and they will be joined at the hip by a measure of coherence between the measurement and the message. As many problems in signal processing and machine learning are actual or virtual two-channel problems, CCA is of utmost importance. As examples of actual two-channel problems, we offer monopulse radar wherein detection is based on correlation between a sum beam and a difference beam; tests for wide-sense stationarity based on correlation between signals in two narrow spectral bands, passive radar and sonar where detection is based on coherence between a reference beam and a surveillance beam, and so on. As an example of a virtual two-channel problem, we offer the problem where a measured channel carries information about an unmeasured message or signal that has been transmitted through a noisy channel; or a problem where a measured time series, space series or space-time series is to be inverted for an unobserved series that could have produced it. Applications of CCA are also common in machine learning. Many data sets in practice share a common latent or semantic structure that can be described from different "viewpoints" or in different domains. For instance, an English document and its Spanish translation are different "language viewpoints" of the same semantic entity that can be learned by CCA. When more than two domains or viewpoints exist, extracting or learning a shared representation is a multi-view learning problem: a generalization of CCA to more than two datasets that will be presented in Sect. 11.2.

3.9.1 Canonical Coordinates

Let $\mathbf{x} \in \mathbb{C}^p$ be the unobserved signal to be estimated and $\mathbf{y} \in \mathbb{C}^q$ be the measurement. Without loss of generality, we may assume $p \le q$. In some cases, $\mathbf{y} = \mathbf{H}\mathbf{x} + \mathbf{n}$, but this assumption is not essential to the analysis of canonical correlations. The second-order model of the composite covariance matrix for these two channels of measurements is

$$\mathrm{E}\left[\begin{bmatrix}\mathbf{x}\\ \mathbf{y}\end{bmatrix}\begin{bmatrix}\mathbf{x}^H & \mathbf{y}^H\end{bmatrix}\right] = \begin{bmatrix}\mathbf{R}_{\mathbf{xx}} & \mathbf{R}_{\mathbf{xy}}\\ \mathbf{R}_{\mathbf{yx}} & \mathbf{R}_{\mathbf{yy}}\end{bmatrix}.$$

We may transform these variables into their *canonical coordinates* with the non-singular transformations $\mathbf{u} = \mathbf{F}^H\mathbf{R}_{\mathbf{xx}}^{-1/2}\mathbf{x}$ and $\mathbf{v} = \mathbf{G}^H\mathbf{R}_{\mathbf{yy}}^{-1/2}\mathbf{y}$ with the $p \times p$ and $q \times q$ unitary matrices $\mathbf{F}$ and $\mathbf{G}$ extracted from the SVD of the *coherence matrix* $\mathbf{C} = \mathbf{R}_{\mathbf{xx}}^{-1/2}\mathbf{R}_{\mathbf{xy}}\mathbf{R}_{\mathbf{yy}}^{-1/2}$. This coherence matrix is simply the covariance matrix for the whitened variables $\mathbf{R}_{\mathbf{xx}}^{-1/2}\mathbf{x}$ and $\mathbf{R}_{\mathbf{yy}}^{-1/2}\mathbf{y}$. That is, $\mathbf{C} = \mathrm{E}[\mathbf{R}_{\mathbf{xx}}^{-1/2}\mathbf{x}(\mathbf{R}_{\mathbf{yy}}^{-1/2}\mathbf{y})^H] = \mathbf{R}_{\mathbf{xx}}^{-1/2}\mathbf{R}_{\mathbf{xy}}\mathbf{R}_{\mathbf{yy}}^{-H/2}$. Without loss of generality, we assume the square root matrix $\mathbf{R}_{\mathbf{yy}}^{1/2}$ is Hermitian, so that $\mathbf{R}_{\mathbf{yy}}^{-H/2} = \mathbf{R}_{\mathbf{yy}}^{-1/2}$. The SVD of $\mathbf{C}$ is

$$\mathbf{C} = \mathbf{F}\mathbf{K}\mathbf{G}^H$$

where the $p \times q$ matrix $\mathbf{K} = [\mathrm{diag}(k_1, k_2, \ldots, k_p)\ \mathbf{0}_{p\times(q-p)}]$ is a matrix of non-negative *canonical correlations* bounded between zero and one. They are the correlations, or coherences, between unit-variance canonical coordinates: $\mathbf{K} = \mathrm{E}[\mathbf{u}\mathbf{v}^H]$. So this transformation produces the following composite covariance matrix for the canonical coordinates:

$$\mathrm{E}\left[\begin{bmatrix}\mathbf{u}\\ \mathbf{v}\end{bmatrix}\begin{bmatrix}\mathbf{u}^H & \mathbf{v}^H\end{bmatrix}\right] = \begin{bmatrix}\mathbf{I}_p & \mathbf{K}\\ \mathbf{K}^H & \mathbf{I}_q\end{bmatrix}.$$

The non-unity eigenvalues of this $(p+q) \times (p+q)$ matrix are $\{(1+k_i), (1-k_i), i = 1, 2, \ldots, p\}$, and its determinant is $\prod_{i=1}^{p}(1-k_i^2)$. The linear minimum mean-squared error estimator of the canonical coordinates $\mathbf{u}$ from the canonical coordinates $\mathbf{v}$ is $\hat{\mathbf{u}} = \mathbf{K}\mathbf{v}$, and the error covariance matrix is $\mathbf{I}_p - \mathbf{K}\mathbf{K}^H$. The corresponding linear minimum mean-squared error estimate of $\mathbf{x}$ from $\mathbf{y}$ is $\hat{\mathbf{x}} = \mathbf{R}_{\mathbf{xx}}^{1/2}\mathbf{F}\hat{\mathbf{u}}$, with error covariance matrix $\mathbf{R}_{\mathbf{xx}}^{1/2}\mathbf{F}(\mathbf{I}_p - \mathbf{K}\mathbf{K}^H)\mathbf{F}^H\mathbf{R}_{\mathbf{xx}}^{1/2}$.

There is much to be said about the canonical coordinates and their canonical correlations. To begin, the k_i^2 are eigenvalues of the matrix $\mathbf{C}\mathbf{C}^H$, and these eigenvalues are invariant to nonsingular linear transformations of $\mathbf{x}$ to $\mathbf{B}_x\mathbf{x}$ and of $\mathbf{y}$ to $\mathbf{B}_y\mathbf{y}$. In fact, the canonical coordinates are *maximal invariants* under these transformations. Moreover, suppose the canonical coordinates $\mathbf{u}$ and $\mathbf{v}$ are competing with white alternatives $\mathbf{w}$ and $\mathbf{z}$. The cross-covariance matrix between these variables will not be diagonal, and by the theory of majorization, the correlations ℓ_i between w_i and z_i will be majorized by the correlations k_i between u_i and v_i:

$$\sum_{i=1}^{r} k_i \geq \sum_{i=1}^{r} \ell_i, \text{ for all } r = 1, 2, \ldots, p.$$

There are a great number of problems in inference that may be framed in canonical coordinates. The reader is referred to [306].

3.9.2 Dimension Reduction Based on Canonical and Half-Canonical Coordinates

We have written the LMMSE estimator of $\mathbf{x}$ from $\mathbf{y}$ as $\hat{\mathbf{x}} = \mathbf{W}_{\mathbf{x}|\mathbf{y}}\mathbf{y}$, where $\mathbf{W}_{\mathbf{x}|\mathbf{y}} = \mathbf{R}_{\mathbf{xy}}\mathbf{R}_{\mathbf{yy}}^{-1}$. In [302], the reduced rank LMMSE estimator of rank r that minimizes the trace of the error covariance matrix is determined by minimizing the function $\mathrm{tr}((\mathbf{W}_{\mathbf{x}|\mathbf{y}} - \mathbf{W}_r)^H\mathbf{R}_{\mathbf{yy}}(\mathbf{W}_{\mathbf{x}|\mathbf{y}} - \mathbf{W}_r))$, which is the excess in the trace of $\mathbf{Q}_{\mathbf{xx}|\mathbf{y}}$ that results from dimension reduction. The solution is then $\mathbf{W}_r\mathbf{R}_{\mathbf{yy}}^{1/2} = (\mathbf{W}_{\mathbf{x}|\mathbf{y}}\mathbf{R}_{\mathbf{yy}}^{1/2})_r = (\mathbf{R}_{\mathbf{xy}}\mathbf{R}_{\mathbf{yy}}^{-1/2})_r$, where the notation $(\mathbf{A})_r$ indicates a reduced rank version of the matrix $\mathbf{A}$ obtained by setting all but the leading r singular values of $\mathbf{A}$ to zero. This makes the singular values of the half coherence matrix $\mathbf{R}_{\mathbf{xy}}\mathbf{R}_{\mathbf{yy}}^{-1/2}$ fundamental, and they are called half-canonical correlations. The reduced rank estimator of $\mathbf{x}$ from $\mathbf{y}$ is then $\hat{\mathbf{x}}_r = \mathbf{U}\boldsymbol{\Lambda}_r\mathbf{V}^H\mathbf{R}_{\mathbf{yy}}^{-1/2}\mathbf{y}$, where $\mathbf{U}\boldsymbol{\Lambda}_r\mathbf{V}^H$ is the rank-r SVD of $\mathbf{R}_{\mathbf{xy}}\mathbf{R}_{\mathbf{yy}}^{-1/2}$. The trace of the error covariance matrix for this reduced-rank Wiener filter based on half-canonical coordinates is less than or equal to the corresponding trace for the cross-spectral Wiener filter derived in [139] and reported in [101]. In [178], this problem is modified to minimize the determinant of the error covariance matrix. The solution is then $\mathbf{R}_{\mathbf{xx}}^{-1/2}\mathbf{W}_r\mathbf{R}_{\mathbf{yy}}^{1/2} = (\mathbf{R}_{\mathbf{xx}}^{-1/2}\mathbf{W}_{\mathbf{x}|\mathbf{y}}\mathbf{R}_{\mathbf{yy}}^{1/2})_r = (\mathbf{R}_{\mathbf{xx}}^{-1/2}\mathbf{R}_{\mathbf{xy}}\mathbf{R}_{\mathbf{yy}}^{-1/2})_r$, and the resulting reduced rank estimator is $\hat{\mathbf{x}}_r = \mathbf{R}_{\mathbf{xx}}^{1/2}\mathbf{U}\boldsymbol{\Lambda}_r\mathbf{V}^H\mathbf{R}_{\mathbf{yy}}^{-1/2}\mathbf{y}$, where $\mathbf{U}\boldsymbol{\Lambda}_r\mathbf{V}^H$ is the rank-r SVD of $\mathbf{R}_{\mathbf{xx}}^{-1/2}\mathbf{R}_{\mathbf{xy}}\mathbf{R}_{\mathbf{yy}}^{-1/2}$. This makes the singular values of the coherence matrix $\mathbf{R}_{\mathbf{xx}}^{-1/2}\mathbf{R}_{\mathbf{xy}}\mathbf{R}_{\mathbf{yy}}^{-1/2}$ fundamental, and these singular values are canonical correlations. For these solutions, the trace or determinant of the error covariance matrix is inflated by a non-negative function of the discarded singular values. Rank reduction of LMMSE estimators tells how to reduce the rank of $\mathbf{W}_{\mathbf{x}|\mathbf{y}}$, but does not give a principle for selecting rank.

3.10 Partial Correlation

The set-up is this: three channels produce measurements, organized into the three random vectors $\mathbf{x} \in \mathbb{C}^p$, $\mathbf{y} \in \mathbb{C}^q$, and $\mathbf{z} \in \mathbb{C}^r$, where it is assumed that $q \geq p$. The composite covariance matrix between these three is

$$\mathbf{R} = E\left[\begin{bmatrix}\mathbf{x}\\ \mathbf{y}\\ \mathbf{z}\end{bmatrix}\begin{bmatrix}\mathbf{x}^H & \mathbf{y}^H & \mathbf{z}^H\end{bmatrix}\right] = \begin{bmatrix}\mathbf{R_{xx}} & \mathbf{R_{xy}} & \mathbf{R_{xz}}\\ \mathbf{R_{yx}} & \mathbf{R_{yy}} & \mathbf{R_{yz}}\\ \mathbf{R_{zx}} & \mathbf{R_{zy}} & \mathbf{R_{zz}}\end{bmatrix}.$$

In one case to be considered, the two random vectors $\mathbf{x}$ and $\mathbf{y}$ are to be regressed onto the common random vector $\mathbf{z}$. In the other case, the random vector $\mathbf{x}$ is to be regressed onto the random vectors $\mathbf{y}$ and $\mathbf{z}$

By defining the composite vectors $\mathbf{u} = [\mathbf{x}^T\ \mathbf{y}^T]^T$ and $\mathbf{v} = [\mathbf{y}^T\ \mathbf{z}^T]^T$, the covariance matrix $\mathbf{R}$ may be parsed two ways:

$$\mathbf{R} = \begin{bmatrix}\mathbf{R_{uu}} & \mathbf{R_{uz}}\\ \mathbf{R}_{\mathbf{uz}}^H & \mathbf{R_{zz}}\end{bmatrix} = \begin{bmatrix}\mathbf{R_{xx}} & \mathbf{R_{xv}}\\ \mathbf{R}_{\mathbf{xv}}^H & \mathbf{R_{vv}}\end{bmatrix}.$$

The covariance matrix $\mathbf{R_{uu}}$ is $(p+q) \times (p+q)$, and the covariance matrix $\mathbf{R_{xx}}$ is $p \times p$. There are two useful representations for the inverse of the composite covariance matrix $\mathbf{R}$:

$$\begin{aligned}\mathbf{R}^{-1} &= \begin{bmatrix}\mathbf{Q}_{\mathbf{uu}|\mathbf{z}}^{-1} & \mathbf{Q}_{\mathbf{uu}|\mathbf{z}}^{-1}\mathbf{R_{uz}}\\ \mathbf{R}_{\mathbf{uz}}^H\mathbf{Q}_{\mathbf{uu}|\mathbf{z}}^{-1} & \mathbf{R}_{\mathbf{zz}}^{-1} + \mathbf{R}_{\mathbf{zz}}^{-1}\mathbf{R}_{\mathbf{uz}}^H\mathbf{Q}_{\mathbf{uu}|\mathbf{z}}^{-1}\mathbf{R_{uz}}\mathbf{R}_{\mathbf{zz}}^{-1}\end{bmatrix}\\ &= \begin{bmatrix}\mathbf{Q}_{\mathbf{xx}|\mathbf{v}}^{-1} & \mathbf{Q}_{\mathbf{xx}|\mathbf{v}}^{-1}\mathbf{R_{xv}}\\ \mathbf{R}_{\mathbf{xv}}^H\mathbf{Q}_{\mathbf{xx}|\mathbf{v}}^{-1} & \mathbf{R}_{\mathbf{vv}}^{-1} + \mathbf{R}_{\mathbf{vv}}^{-1}\mathbf{R}_{\mathbf{xv}}^H\mathbf{Q}_{\mathbf{xx}|\mathbf{v}}^{-1}\mathbf{R_{xv}}\mathbf{R}_{\mathbf{vv}}^{-1}\end{bmatrix}.\end{aligned} \tag{3.12}$$

The matrix $\mathbf{Q}_{\mathbf{uu}|\mathbf{z}}$ is the error covariance matrix for estimating the composite vector $\mathbf{u}$ from $\mathbf{z}$, and the matrix $\mathbf{Q}_{\mathbf{xx}|\mathbf{v}}$ is the error covariance matrix for estimating $\mathbf{x}$ from $\mathbf{v}$:

$$\begin{aligned}\mathbf{Q}_{\mathbf{uu}|\mathbf{z}} &= \mathbf{R_{uu}} - \mathbf{R_{uz}}\mathbf{R}_{\mathbf{zz}}^{-1}\mathbf{R}_{\mathbf{uz}}^H,\\ \mathbf{Q}_{\mathbf{xx}|\mathbf{v}} &= \mathbf{R_{xx}} - \mathbf{R_{xv}}\mathbf{R}_{\mathbf{vv}}^{-1}\mathbf{R}_{\mathbf{xv}}^H.\end{aligned}$$

We shall have more to say about these error covariance matrices in due course. Importantly, the inverses of each may be read out of the inverse for the composite covariance matrix $\mathbf{R}^{-1}$ of (3.12). The dimension of the error covariance $\mathbf{Q}_{\mathbf{uu}|\mathbf{z}}$ is $(p+q) \times (p+q)$, and the dimension of the error covariance $\mathbf{Q}_{\mathbf{xx}|\mathbf{v}}$ is $p \times p$.

3.10.1 Regressing Two Random Vectors onto One

The estimators of $\mathbf{x}$ and $\mathbf{y}$ from $\mathbf{z}$ and their resulting error covariance matrices are easily read out from the composite covariance matrix $\mathbf{R}$:

$$\hat{\mathbf{x}}(\mathbf{z}) = \mathbf{R}_{\mathbf{xz}}\mathbf{R}_{\mathbf{zz}}^{-1}\mathbf{z}, \qquad \mathbf{Q}_{\mathbf{xx}|\mathbf{z}} = \mathbf{R}_{\mathbf{xx}} - \mathbf{R}_{\mathbf{xz}}\mathbf{R}_{\mathbf{zz}}^{-1}\mathbf{R}_{\mathbf{xz}}^{H},$$

$$\hat{\mathbf{y}}(\mathbf{z}) = \mathbf{R}_{\mathbf{yz}}\mathbf{R}_{\mathbf{zz}}^{-1}\mathbf{z}, \qquad \mathbf{Q}_{\mathbf{yy}|\mathbf{z}} = \mathbf{R}_{\mathbf{yy}} - \mathbf{R}_{\mathbf{yz}}\mathbf{R}_{\mathbf{zz}}^{-1}\mathbf{R}_{\mathbf{yz}}^{H}.$$

The composite error covariance matrix for the errors $\mathbf{x} - \hat{\mathbf{x}}(\mathbf{z})$ and $\mathbf{y} - \hat{\mathbf{y}}(\mathbf{z})$ is the matrix

$$\mathbf{Q}_{\mathbf{uu}|\mathbf{z}} = E\left[\begin{bmatrix} \mathbf{x} - \hat{\mathbf{x}}(\mathbf{z}) \\ \mathbf{y} - \hat{\mathbf{y}}(\mathbf{z}) \end{bmatrix} \left[(\mathbf{x} - \hat{\mathbf{x}}(\mathbf{z}))^H \;\; (\mathbf{y} - \hat{\mathbf{y}}(\mathbf{z}))^H\right]\right] = \begin{bmatrix} \mathbf{Q}_{\mathbf{xx}|\mathbf{z}} & \mathbf{Q}_{\mathbf{xy}|\mathbf{z}} \\ \mathbf{Q}_{\mathbf{xy}|\mathbf{z}}^H & \mathbf{Q}_{\mathbf{yy}|\mathbf{z}} \end{bmatrix},$$

where

$$\mathbf{Q}_{\mathbf{xy}|\mathbf{z}} = E\left[(\mathbf{x} - \hat{\mathbf{x}}(\mathbf{z}))(\mathbf{y} - \hat{\mathbf{y}}(\mathbf{z}))^H\right] = \mathbf{R}_{\mathbf{xy}} - \mathbf{R}_{\mathbf{xz}}\mathbf{R}_{\mathbf{zz}}^{-1}\mathbf{R}_{\mathbf{yz}}^{H}$$

is the $p \times q$ matrix of cross-correlations between the random errors $\mathbf{x} - \hat{\mathbf{x}}(\mathbf{z})$ and $\mathbf{y} - \hat{\mathbf{y}}(\mathbf{z})$. It is called the *partial correlation between the random vectors* $\mathbf{x}$ *and* $\mathbf{y}$, *after each has been regressed onto the common vector* $\mathbf{z}$.

Equation (3.12) shows that $\mathbf{Q}_{\mathbf{uu}|\mathbf{z}}$ and its determinant may be read directly out of the $(p+q) \times (p+q)$ northwest block of the inverse of the error covariance matrix. This result was known to Harald Cramér more than 70 years ago [90] and is featured prominently in the book on graphical models by Whittaker [378].

The error covariance matrix $\mathbf{Q}_{\mathbf{uu}|\mathbf{z}}$ may be pre- and post-multiplied by the block-diagonal matrix blkdiag$(\mathbf{Q}_{\mathbf{xx}|\mathbf{z}}^{-1/2}, \mathbf{Q}_{\mathbf{yy}|\mathbf{z}}^{-1/2})$ to produce the normalized error covariance matrix and its corresponding determinant:

$$\mathbf{Q}_{\mathbf{uu}|\mathbf{z}}^{N} = \begin{bmatrix} \mathbf{I}_p & \mathbf{Q}_{\mathbf{xx}|\mathbf{z}}^{-1/2}\mathbf{Q}_{\mathbf{xy}|\mathbf{z}}\mathbf{Q}_{\mathbf{yy}|\mathbf{z}}^{-1/2} \\ \mathbf{Q}_{\mathbf{yy}|\mathbf{z}}^{-1/2}\mathbf{Q}_{\mathbf{xy}|\mathbf{z}}^{H}\mathbf{Q}_{\mathbf{xx}|\mathbf{z}}^{-1/2} & \mathbf{I}_q \end{bmatrix}, \tag{3.13}$$

$$\det(\mathbf{Q}_{\mathbf{uu}|\mathbf{z}}^{N}) = \frac{\det(\mathbf{Q}_{\mathbf{uu}|\mathbf{z}})}{\det(\mathbf{Q}_{\mathbf{xx}|\mathbf{z}})\det(\mathbf{Q}_{\mathbf{yy}|\mathbf{z}})} = \det(\mathbf{I}_p - \mathbf{C}_{\mathbf{xy}|\mathbf{z}}\mathbf{C}_{\mathbf{xy}|\mathbf{z}}^{H}).$$

The matrix $\mathbf{C}_{\mathbf{xy}|\mathbf{z}} = \mathbf{Q}_{\mathbf{xx}|\mathbf{z}}^{-1/2}\mathbf{Q}_{\mathbf{xy}|\mathbf{z}}\mathbf{Q}_{\mathbf{yy}|\mathbf{z}}^{-1/2}$ is the *partial coherence matrix*. It is noteworthy that conditioning on $\mathbf{z}$ has replaced correlation matrices with error covariance matrices in the definition of the partial coherence matrix. *Partial coherence* is then defined to be

$$\rho_{\mathbf{xy}|\mathbf{z}}^2 = 1 - \det(\mathbf{Q}_{\mathbf{uu}|\mathbf{z}}^{N}) = 1 - \frac{\det(\mathbf{Q}_{\mathbf{uu}|\mathbf{z}})}{\det(\mathbf{Q}_{\mathbf{xx}|\mathbf{z}})\det(\mathbf{Q}_{\mathbf{yy}|\mathbf{z}})}$$

$$= 1 - \det(\mathbf{I}_p - \mathbf{C}_{\mathbf{xy}|\mathbf{z}}\mathbf{C}^H_{\mathbf{xy}|\mathbf{z}}).$$

Define the SVD of the partial coherence matrix to be $\mathbf{C}_{\mathbf{xy}|\mathbf{z}} = \mathbf{FKG}^H$, where $\mathbf{F}$ is a $p \times p$ orthogonal matrix, $\mathbf{G}$ is a $q \times q$ orthogonal matrix, and $\mathbf{K}$ is a $p \times q$ diagonal matrix of partial canonical correlations. The matrix $\mathbf{K}$ may be called the *partial canonical correlation matrix.* The normalized error covariance matrix of (3.13) may be written as

$$\mathbf{Q}^N_{\mathbf{uu}|\mathbf{z}} = \begin{bmatrix} \mathbf{F} & \mathbf{0} \\ \mathbf{0} & \mathbf{G} \end{bmatrix} \begin{bmatrix} \mathbf{I}_p & \mathbf{K} \\ \mathbf{K}^H & \mathbf{I}_q \end{bmatrix} \begin{bmatrix} \mathbf{F}^H & \mathbf{0} \\ \mathbf{0} & \mathbf{G}^H \end{bmatrix}.$$

As a consequence, partial coherence may be factored as

$$\rho^2_{\mathbf{xy}|\mathbf{z}} = 1 - \det(\mathbf{I}_p - \mathbf{KK}^H) = 1 - \prod_{i=1}^{p}(1 - k_i^2).$$

The partial canonical correlations k_i are bounded between 0 and 1, as is partial coherence. When the squared partial canonical correlations k_i^2 are near to zero, then partial coherence $\rho^2_{\mathbf{xy}|\mathbf{z}}$ is near to zero, indicating linear independence of $\mathbf{x}$ and $\mathbf{y}$, given $\mathbf{z}$.

These results summarize the error analysis for linearly estimating the random vectors $\mathbf{x}$ and $\mathbf{y}$ from a common random vector $\mathbf{z}$. The only assumption is that the random vectors $(\mathbf{x}, \mathbf{y}, \mathbf{z})$ are second-order random vectors.

Example 3.2 (Partial coherence for circulant time series) Suppose the random vectors $(\mathbf{x}, \mathbf{y}, \mathbf{z})$ are of common dimension N, with every matrix in the composite covariance matrix diagonalizable by the DFT matrix $\mathbf{F}_N$. Then it is straightforward to show that each error covariance matrix of the form $\mathbf{Q}_{\mathbf{xy}|\mathbf{z}}$ may be written as $\mathbf{Q}_{\mathbf{xy}|\mathbf{z}} = \mathbf{F}_N \operatorname{diag}(Q_{\mathbf{xy}|\mathbf{z}}[0], \cdots, Q_{\mathbf{xy}|\mathbf{z}}[N-1])\mathbf{F}_N^H$, where $Q_{\mathbf{xy}|\mathbf{z}}[n]$ is a spectral representation of error covariance at frequency $2\pi n/N$. Then, $\mathbf{C}_{\mathbf{xy}|\mathbf{z}}\mathbf{C}^H_{\mathbf{xy}|\mathbf{z}} = \mathbf{F}_N\mathbf{K}^2\mathbf{F}_N^H$, where $\mathbf{K}^2 = \operatorname{diag}(k_0^2, \ldots, k_{N-1}^2)$ and $k_n^2 = |Q_{\mathbf{xy}|\mathbf{z}}[n]|^2/(Q_{\mathbf{xx}|\mathbf{z}}[n]Q_{\mathbf{yy}|\mathbf{z}}[n])$. Each term in diagonal $\mathbf{K}^2$ is a partial coherence at a frequency $2\pi n/N$. It follows that coherence is

$$\rho^2_{\mathbf{xy}|\mathbf{z}} = 1 - \prod_{n=0}^{N-1}(1 - k_n^2).$$

This may be termed broadband coherence, computed from narrowband partial coherences k_n^2. These are partial coherences computed, frequency by frequency, from the DFT coefficients.

Example 3.3 (Bivariate partial correlation coefficient) When $\mathbf{x}$ and $\mathbf{y}$ are scalar-valued, then the covariance matrix of errors $x - \hat{x}$ and $y - \hat{y}$ is

$$E\left[\begin{bmatrix} x - \hat{x} \\ y - \hat{y} \end{bmatrix} \begin{bmatrix} (x - \hat{x})^* & (y - \hat{y})^* \end{bmatrix}\right] = \begin{bmatrix} Q_{xx|\mathbf{z}} & Q_{xy|\mathbf{z}} \\ Q^*_{xy|\mathbf{z}} & Q_{yy|\mathbf{z}} \end{bmatrix}$$

where the partial correlation coefficient $Q_{xy|\mathbf{z}}$ is the scalar

$$Q_{xy|\mathbf{z}} = r_{xy} - \mathbf{r}_{x\mathbf{z}} \mathbf{R}_{\mathbf{zz}}^{-1} \mathbf{r}_{y\mathbf{z}}^H.$$

The coherence between the error in estimating x and the error in estimating y is now

$$\rho^2_{xy|\mathbf{z}} = 1 - \frac{\det\left(\begin{bmatrix} Q_{xx|\mathbf{z}} & Q_{xy|\mathbf{z}} \\ Q^*_{xy|\mathbf{z}} & Q_{yy|\mathbf{z}} \end{bmatrix}\right)}{Q_{xx|\mathbf{z}} Q_{yy|\mathbf{z}}} = \frac{|Q_{xy|\mathbf{z}}|^2}{Q_{xx|\mathbf{z}} Q_{yy|\mathbf{z}}}.$$

Invariances. The bivariate partial correlation coefficient is invariant to scaling of x and y and to nonsingular transformation of $\mathbf{z}$.

Distribution. The formula for the bivariate partial coherence is to be contrasted with the formula for the bivariate correlation coefficient. Given N independent realizations of the scalar complex random variables x and y, we have established in Sect. 3.1 that the null distribution of the sample estimator of the squared correlation coefficient is Beta$(1, N - 1)$. It is fairly straightforward to show that the net of regressing onto the third channel $\mathbf{z} \in \mathbb{C}^r$ is to replace the effective sample size from N to $N - r$, in which case the distribution of the bivariate partial correlation coefficient is Beta$(1, N - r - 1)$ [207].

3.10.2 Regressing One Random Vector onto Two

Suppose now that the random vector $\mathbf{x}$ is to be linearly regressed onto $\mathbf{v} = [\mathbf{y}^T\ \mathbf{z}^T]^T$:

$$\hat{\mathbf{x}}(\mathbf{v}) = \begin{bmatrix} \mathbf{R_{xy}} & \mathbf{R_{xz}} \end{bmatrix} \begin{bmatrix} \mathbf{R_{yy}} & \mathbf{R_{yz}} \\ \mathbf{R_{yz}}^H & \mathbf{R_{zz}} \end{bmatrix}^{-1} \begin{bmatrix} \mathbf{y} \\ \mathbf{z} \end{bmatrix}.$$

Give the matrix inverse in this equation the following block-diagonal LDU factorization:

$$\begin{bmatrix} \mathbf{R}_{\mathbf{yy}} & \mathbf{R}_{\mathbf{yz}} \\ \mathbf{R}_{\mathbf{yz}}^H & \mathbf{R}_{\mathbf{zz}} \end{bmatrix}^{-1} = \begin{bmatrix} \mathbf{I}_q & \mathbf{0} \\ -\mathbf{R}_{\mathbf{zz}}^{-1}\mathbf{R}_{\mathbf{yz}}^H & \mathbf{I}_r \end{bmatrix} \begin{bmatrix} \mathbf{Q}_{\mathbf{yy}|\mathbf{z}}^{-1} & \mathbf{0} \\ \mathbf{0} & \mathbf{R}_{\mathbf{zz}}^{-1} \end{bmatrix} \begin{bmatrix} \mathbf{I}_q & -\mathbf{R}_{\mathbf{yz}}\mathbf{R}_{\mathbf{zz}}^{-1} \\ \mathbf{0} & \mathbf{I}_r \end{bmatrix}.$$

A few lines of algebra produce this result for $\hat{\mathbf{x}}(\mathbf{v})$, the linear minimum mean-squared error estimator of $\mathbf{x}$ from $\mathbf{v}$:

$$\hat{\mathbf{x}}(\mathbf{v}) = \hat{\mathbf{x}}(\mathbf{z}) + \mathbf{Q}_{\mathbf{xy}|\mathbf{z}}\mathbf{Q}_{\mathbf{yy}|\mathbf{z}}^{-1}\left(\mathbf{y} - \hat{\mathbf{y}}(\mathbf{z})\right).$$

It is evident that the vector $\mathbf{y}$ is not used in a linear minimum mean-squared error estimator of $\mathbf{x}$ when the partial covariance $\mathbf{Q}_{\mathbf{xy}|\mathbf{z}}$ is zero. That is, the random vector $\mathbf{y}$ brings no useful second-order information to the problem of linearly estimating $\mathbf{x}$.

The error covariance matrix for estimating $\mathbf{x}$ from $\mathbf{v}$ is easily shown to be

$$\begin{aligned} \mathbf{Q}_{\mathbf{xx}|\mathbf{v}} &= E[(\mathbf{x} - \hat{\mathbf{x}}(\mathbf{v}))(\mathbf{x} - \hat{\mathbf{x}}(\mathbf{v}))^H] \\ &= \mathbf{Q}_{\mathbf{xx}|\mathbf{z}} - \mathbf{Q}_{\mathbf{xy}|\mathbf{z}}\mathbf{Q}_{\mathbf{yy}|\mathbf{z}}^{-1}\mathbf{Q}_{\mathbf{xy}|\mathbf{z}}^H. \end{aligned}$$

Thus, the error covariance $\mathbf{Q}_{\mathbf{xx}|\mathbf{z}}$ is reduced by a quadratic form depending on the covariance between the errors $\mathbf{x} - \hat{\mathbf{x}}(\mathbf{z})$ and $\mathbf{y} - \hat{\mathbf{y}}(\mathbf{z})$. If this error covariance is now normalized by the error covariance matrix achieved by regressing only on the vector $\mathbf{z}$, the result is

$$\begin{aligned} \mathbf{Q}_{\mathbf{xx}|\mathbf{v}}^N &= \mathbf{Q}_{\mathbf{xx}|\mathbf{z}}^{-1/2}\mathbf{Q}_{\mathbf{xx}|\mathbf{v}}\mathbf{Q}_{\mathbf{xx}|\mathbf{z}}^{-1/2} \\ &= \mathbf{I}_p - \mathbf{Q}_{\mathbf{xx}|\mathbf{z}}^{-1/2}\mathbf{Q}_{\mathbf{xy}|\mathbf{z}}\mathbf{Q}_{\mathbf{yy}|\mathbf{z}}^{-1}\mathbf{Q}_{\mathbf{xy}|\mathbf{z}}^H\mathbf{Q}_{\mathbf{xx}|\mathbf{z}}^{-1/2} \\ &= \mathbf{I}_p - \mathbf{C}_{\mathbf{xy}|\mathbf{z}}\mathbf{C}_{\mathbf{xy}|\mathbf{z}}^H = \mathbf{F}(\mathbf{I}_p - \mathbf{K}\mathbf{K}^H)\mathbf{F}^H. \end{aligned}$$

As in the previous subsection, $\mathbf{C}_{\mathbf{xy}|\mathbf{z}}$ is the partial coherence matrix. The determinant of this matrix measures the volume of the normalized error covariance matrix:

$$\begin{aligned} \det(\mathbf{Q}_{\mathbf{xx}|\mathbf{v}}^N) &= \frac{\det(\mathbf{Q}_{\mathbf{xx}|\mathbf{v}})}{\det(\mathbf{Q}_{\mathbf{xx}|\mathbf{z}})} \\ &= \det(\mathbf{I}_p - \mathbf{K}\mathbf{K}^H) = \prod_{i=1}^{p}(1 - k_i^2). \end{aligned}$$

As before, we may define a *partial coherence*

$$\rho_{\mathbf{x}|\mathbf{yz}}^2 = 1 - \det(\mathbf{Q}_{\mathbf{xx}|\mathbf{v}}^N) = 1 - \prod_{i=1}^{p}(1 - k_i^2).$$

When the squared partial canonical correlations k_i^2 are near to zero, then partial coherence $\rho^2_{\mathbf{x}|\mathbf{yz}}$ is near to zero, indicating linear independence of $\mathbf{x}$ on $\mathbf{y}$, given $\mathbf{z}$. Consequently, the estimator $\hat{\mathbf{x}}(\mathbf{v})$ depends only on $\mathbf{z}$, and not on $\mathbf{y}$.

These results summarize the error analysis for estimating the random vector $\mathbf{x}$ from the composite vector $\mathbf{v}$. It is notable that, except for scaling constants dependent only upon the dimensions p, q, r, the volume of the normalized error covariance matrix for estimating $\mathbf{x}$ from $\mathbf{v}$ equals the volume of the normalized error covariance matrix for estimating $\mathbf{u}$ from $\mathbf{z}$. Both of these volumes are determined by the partial canonical correlations k_i. Importantly, for answering questions of linear independence of $\mathbf{x}$ and $\mathbf{y}$, given $\mathbf{z}$, it makes no difference whether one considers $\rho^2_{\mathbf{xy}|\mathbf{z}}$ or $\rho^2_{\mathbf{x}|\mathbf{yz}}$. These two measures of coherence are identical.

Finally, the partial canonical correlations k_i are invariant to transformation of the random vector $[\mathbf{x}^T\ \mathbf{y}^T\ \mathbf{z}^T]^T$ by a block-diagonal, nonsingular, matrix $\mathbf{B} = \text{blkdiag}(\mathbf{B}_x, \mathbf{B}_y, \mathbf{B}_z)$. As a consequence, partial coherence is invariant to transformation $\mathbf{B}$. A slight variation on Proposition 10.6 in [111] shows partial canonical correlations to be maximal invariants under group action $\mathbf{B}$.

3.11 Chapter Notes

The reader is directed to Appendix D on the multivariate normal distribution and related, for a list of influential books and papers on multivariate statistics. These have guided our writing of the early parts of this chapter on coherence, which is normalized correlation, or correlation coefficient. The liberal use of unitary transformations as a device for deriving distributions is inspired by the work of James [184] and Kshirsagar [207].

1. The account of PCA for dimension reduction in a single channel reveals the central role played by the SVD and contains some geometrical insights that are uncommon.
2. The section on LMMSE filtering is based on block-structured Cholesky factorizations of two-channel correlation matrices and their inverses. The distribution theory of terms in these Cholesky factors is taken from Muirhead [244] and from Khatri and Rao [199], both must reads. The Khatri and Rao paper is not known to many researchers in signal processing and machine learning.
3. The account of the Krylov subspace and subspace expansion in the multistage Wiener filter follows [311]. But the first insights into the connection between the multistage Wiener filter and conjugate gradients were published by Weippert et al. [376]. The original derivation of the conjugate gradient algorithm is due to Hestenes and Stiefel [162], and the original derivation of the multistage Wiener filter is due to Goldstein [140].
4. In the study of beamforming, it is shown that coherence measures the ratio of power in a conventional beamformer to power in an MVDR beamformer.

The distributions of these two beamformers lend insight into their respective performances.

5. Canonical coordinates are shown to be the correct coordinate system for dimension reduction in LMMSE filtering. So canonical and half-canonical coordinates play the same role in two-channel problems as principal components play in single-channel problems.
6. Beamforming is to wavenumber localization from spatial measurements as spectrum analysis is to frequency localization from temporal measurements. The brief discussion of beamforming in this chapter does scant justice to the voluminous literature on adaptive beamforming. No comprehensive review is possible, but the reader is directed to [88, 89, 358, 361] for particularly insightful and important papers.
7. Partial coherence may be used to analyze questions of causality, questions that are fraught with ambiguity. But, nonetheless, one may propose statistical tests that are designed to reject the hypothesis of causal influence of one time series on another. The idea is to construct three time series from two, by breaking time series 1 into its past and its future. Then the question is whether time series 2 has predictive value for the future of time series 1, given the past of time series 1. This is the basis of Granger causality [147]. This question leads to the theory of partial correlations and the use of *partial coherence* or a closely related statistic as a test statistic [129, 130, 256, 309]. Partial coherence has been used to study causality in multivariable time series, neuroimages, brain scans, and marketing time series [16, 24, 25, 388].
8. Factor analysis may be said to generalize principal component analysis. It is a well-developed topic in multivariate statistics that is not covered in this chapter. However, it makes its appearance in Chaps. 5 and 7. There is a fundamental paper on factor analysis, in the notation of signal processing and machine learning, that merits special mention. In [336], Stoica and Viberg identify a regression or factor model for cases where the factor loadings are linearly dependent. This requires identification of the rank of the matrix of factor loadings, an identification that is derived in the paper. Cramér-Rao bounds are used to bound error covariances for parameter estimates in the identified factor model.

Coherence and Classical Tests in the Multivariate Normal Model

4

In this chapter, several basic results are established for inference and hypothesis testing in a multivariate normal (MVN) model. In this model, measurements are distributed as proper, complex, multivariate Gaussian random vectors. The unknown covariance matrix for these random vectors belongs to a cone. This is a common case in signal processing and machine learning. When the structured covariance matrix belongs to a cone, two important results concerning maximum likelihood (ML) estimators and likelihood ratios computed from ML estimators are reviewed. These likelihood ratios are termed generalized likelihood ratios (GLRs) in the engineering and applied sciences and ordinary likelihoods in the statistical sciences. Some basic concepts of invariance in hypothesis testing are reviewed. Equipped with these basic concepts, we then examine several classical hypothesis tests about the covariance matrix of measurements drawn from multivariate normal (MVN) models. These are the *sphericity test* that tests whether or not the covariance matrix is a scaled identity matrix with unknown scale parameter; the *Hadamard test* that tests whether or not the variables in a MVN model are independent, thus having a diagonal covariance matrix with unknown diagonal elements; and the *homogeneity test* that tests whether or not the covariance matrices of independent vector-valued MVN models are equal. We discuss the invariances and null distributions for likelihood ratios when these are known. The chapter concludes with a discussion of the expected likelihood principle for cross-validating a covariance model.

4.1 How Limiting Is the Multivariate Normal Model?

In many problems, a multivariate normal (MVN) model for measurements is justified by theoretical reasoning or empirical measurements. In others, it is a way to derive a likelihood function for a mean vector and a covariance matrix of the data. When these are estimated, then the first- and second-order moments of an underlying model are estimated. This seems quite limiting. But the key term here is *an underlying model*. When a mean vector is parameterized as a vector in a known

D. Ramírez et al., *Coherence*, https://doi.org/10.1007/978-3-031-13331-2_4

subspace or in a subspace known only by its dimension, then a solution for the mean value vector that maximizes likelihood is compelling, and it has invariances that one would be unwilling to give up. Correspondingly, when the covariance matrix is modeled to have a low-rank, or spikey, component, the solution for the covariance matrix that maximizes MVN likelihood is a compelling function of eigenvalues of the sample covariance matrix. In fact, quite generally, solutions for mean value vectors and covariance matrices that maximize MVN likelihood, under modeling constraints, produce very complicated functions of the measurements, much more complicated than the simple sample means and sample covariance matrices encountered when there are no modeling constraints. So, let us paraphrase K. J. Arrow,[1] when he says, "Simplified theory building is an absolute necessity for empirical analysis; but it is a means, not an end." We say parametric modeling in a MVN model is a means to derive what are often compelling functions of the measurements, with essential invariances and illuminating geometries. We do not say these functions are necessarily the end. In many cases, application-specific knowledge will suggest practical adjustments to these functions or experiments to assess the sensitivity of these solutions to model mismatch. Perhaps these functions become benchmarks against which alternative solutions are compared, or they form the basis for more refined model building. In summary, maximization of MVN likelihood with respect to the parameters of an underlying model is a means to *a* useful end. It may not be *the end*.

4.2 Likelihood in the Multivariate Normal Model

In the multivariate normal model, $\mathbf{x} \sim \mathcal{CN}_L(\mathbf{0}, \mathbf{R})$, the likelihood function for $\mathbf{R}$, given N i.i.d. realizations $\mathbf{X} = [\mathbf{x}_1 \ \cdots \ \mathbf{x}_N]$, is[2]

$$\ell(\mathbf{R}; \mathbf{X}) = \frac{1}{\pi^{LN} \det(\mathbf{R})^N} \exp\left\{-N \operatorname{tr}(\mathbf{R}^{-1}\mathbf{S})\right\}, \tag{4.1}$$

where $\mathbf{S} = N^{-1}\mathbf{X}\mathbf{X}^H$ is the sample covariance matrix. According to the notation used for normally distributed random matrices, introduced in Sect. D.4 and used throughout the book, the matrix $\mathbf{X}$ is distributed as $\mathbf{X} \sim \mathcal{CN}_{L\times N}(\mathbf{0}, \mathbf{I}_N \otimes \mathbf{R})$.[3] This

[1] One of the early winners of the Sveriges Riksbank Prize in Economic Sciences in Memory of Alfred Nobel, commonly referred to as the "Nobel Prize in Economics".

[2] When there is no risk of confusion, we use $\mathbf{R}$ to denote a covariance matrix that would often be denoted $\mathbf{R}_{\mathbf{xx}}$.

[3] This notation bears comment. The matrix $\mathbf{X}$ is an $L \times N$ matrix, thus explaining the subscript notation $\mathcal{CN}_{L\times N}$. The covariance matrix $\mathbf{I}_N \otimes \mathbf{R}$ is an $LN \times LN$ block-diagonal matrix with the $L \times L$ covariance matrix $\mathbf{R}$ in each of its N blocks. This is actually the covariance matrix of the $LN \times 1$ vector constructed by stacking the N columns of $\mathbf{X}$. In other words, the columns of $\mathbf{X}$ are i.i.d. proper complex normal random vectors with covariance matrix $\mathbf{R}$. Another equivalent notation replaces $\mathcal{CN}_{L\times N}$ by $\mathcal{CN}_{LN}$.

is termed a second-order statistical model for measurements, as only the second-order moments $r_{lk} = \mathrm{E}[x_l x_k^*]$ appear in the parameter $\mathbf{R} = \{r_{lk}\}$.

4.2.1 Sufficiency

Suppose $\mathbf{X}$ is a matrix whose distribution depends on the parameter $\boldsymbol{\theta}$, and let $\mathbf{t}(\mathbf{X})$ be any statistic or function of the observations. The statistic $\mathbf{t}(\mathbf{X})$, or simply $\mathbf{t}$, is said to be *sufficient* for $\boldsymbol{\theta}$ if the likelihood function $\ell(\boldsymbol{\theta}; \mathbf{X})$ factors as

$$\ell(\boldsymbol{\theta}; \mathbf{X}) = g(\boldsymbol{\theta}; \mathbf{t})h(\mathbf{X}),$$

where $h(\mathbf{X})$ is a non-negative function that does not depend on $\boldsymbol{\theta}$ and $g(\boldsymbol{\theta}; \mathbf{t})$ is a function solely of $\mathbf{t}$ and $\boldsymbol{\theta}$. In (4.1), taking $h(\mathbf{X}) = 1$, it is clear that the sample covariance $\mathbf{S}$ is a sufficient statistic for $\mathbf{R}$.

4.2.2 Likelihood

Given $\mathbf{X}$, the likelihood (4.1) is considered a function of the parameter $\mathbf{R}$, which is assumed to be unknown. The maximum likelihood (ML) estimate of $\mathbf{R}$ is the choice of $\mathbf{R}$ that maximizes the likelihood. To maximize the likelihood (4.1) with respect to an arbitrary positive definite covariance matrix is to maximize the function[4]

$$\begin{aligned} L(\mathbf{R}; \mathbf{X}) &= \log\det(\mathbf{R}^{-1}\mathbf{S}) - \mathrm{tr}(\mathbf{R}^{-1}\mathbf{S}) \\ &= \log\det(\mathbf{R}^{-1/2}\mathbf{S}\mathbf{R}^{-1/2}) - \mathrm{tr}(\mathbf{R}^{-1/2}\mathbf{S}\mathbf{R}^{-1/2}) \\ &= \sum_{l=1}^{L} \Big[\log[ev_l(\mathbf{R}^{-1/2}\mathbf{S}\mathbf{R}^{-1/2})] - ev_l(\mathbf{R}^{-1/2}\mathbf{S}\mathbf{R}^{-1/2})\Big]. \end{aligned}$$

The function $f(\alpha) = \log(\alpha) - \alpha$ has a unique maximum at $\alpha = 1$. It follows that $L(\mathbf{R}; \mathbf{X}) \leq -L$, with equality if and only if $ev_l(\mathbf{R}^{-1/2}\mathbf{S}\mathbf{R}^{-1/2}) = 1, l = 1, \ldots, L$, which is equivalent to $\mathbf{R}^{-1/2}\mathbf{S}\mathbf{R}^{-1/2} = \mathbf{I}_L$. Thus, we obtain the well-known result that the ML estimate of $\mathbf{R}$ is $\hat{\mathbf{R}} = \mathbf{S}$. Hence,

$$\ell(\mathbf{R}; \mathbf{X}) \leq \frac{1}{\pi^{LN}\det(\mathbf{S})^N} e^{-NL} = \ell(\mathbf{S}; \mathbf{X}).$$

[4] This version of log-likelihood uses the identity $-\log\det(\mathbf{R}) = \log\det(\mathbf{R}^{-1})$ and then adds a term $\log\det(\mathbf{S})$ that is independent of the parameter $\mathbf{R}$. Then $-\log\det(\mathbf{R}) + \log\det(\mathbf{S}) = \log\det(\mathbf{R}^{-1}) + \log\det(\mathbf{S}) = \log\det(\mathbf{R}^{-1}\mathbf{S})$.

It must be emphasized that this result holds for the case where the covariance matrix $\mathbf{R}$ is constrained only to be positive definite. It is not constrained by any other pattern or structure.

In many signal processing and machine learning problems of interest, $\mathbf{R}$ is a structured matrix that belongs to a given set $\mathcal{R}$. Some examples that will appear frequently in this book follow.

Example 4.1 White noise with unknown scale:

$$\mathcal{R}_1 = \{\mathbf{R} = \sigma^2 \mathbf{I}_L \mid \sigma^2 > 0\}.$$

Example 4.2 Two-channel white noises with unknown scales:

$$\mathcal{R}_2 = \left\{ \mathbf{R} = \begin{bmatrix} \sigma_1^2 \mathbf{I}_{L_1} & \mathbf{0} \\ \mathbf{0} & \sigma_2^2 \mathbf{I}_{L_2} \end{bmatrix} \mid \sigma_1^2, \sigma_2^2 > 0, \quad L_1 + L_2 = L \right\}.$$

Example 4.3 Diagonal covariance matrix with arbitrary variances:

$$\mathcal{R}_3 = \{\mathbf{R} = \text{diag}(\sigma_1^2, \ldots, \sigma_L^2) \mid \sigma_l^2 > 0, \ \forall l\}.$$

Example 4.4 Arbitrary positive definite covariance matrix:[5]

$$\mathcal{R}_4 = \{\mathbf{R} \mid \mathbf{R} \succ \mathbf{0}\}.$$

Example 4.5 Low-rank plus diagonal matrix (factor analysis):

$$\mathcal{R}_5 = \{\mathbf{R} = \mathbf{H}\mathbf{H}^H + \boldsymbol{\Sigma} \mid \mathbf{H} \in \mathbb{C}^{L \times p}, \ \boldsymbol{\Sigma} = \text{diag}(\sigma_1^2, \ldots, \sigma_L^2), \sigma_l^2 > 0, \ \forall l\}.$$

Example 4.6 Toeplitz covariance matrix:

$$\mathcal{R}_6 = \{\mathbf{R} \mid r_{l,k} = r_{l+1,k+1} = r_{l-k}\}.$$

Importantly, all structured sets $\mathcal{R}_1, \ldots, \mathcal{R}_6$ are cones. A set $\mathcal{R}$ is a cone [44] if for any $\mathbf{R} \in \mathcal{R}$ and $a > 0$, $a\mathbf{R} \in \mathcal{R}$.

The following lemma, due to Javier Vía [299], shows that when the structured set $\mathcal{R}$ is a cone, the maximizing covariance satisfies the constraint $\text{tr}(\mathbf{R}^{-1}\mathbf{S}) = L$.

---

[5] This is the set assumed for the ML estimate $\hat{\mathbf{R}} = \mathbf{S}$. It is the set for the null hypothesis in the testing problems that we will discuss in Sect. 4.3.

Lemma 4.1 *Let* $\hat{\mathbf{R}}$ *be the ML estimate for* $\mathbf{R}$ *that solves*

$$\underset{\mathbf{R}\in\mathcal{R}}{\text{maximize}} \quad \log\det(\mathbf{R}^{-1}\mathbf{S}) - \operatorname{tr}(\mathbf{R}^{-1}\mathbf{S}),$$

within a cone $\mathcal{R}$*, and let* $\mathbf{S}$ *be the sample covariance matrix. Then,*

$$\operatorname{tr}\left(\hat{\mathbf{R}}^{-1}\mathbf{S}\right) = L.$$

Proof Let $\hat{\mathbf{R}} \in \mathcal{R}$ be an estimate (not necessarily the ML estimate) of the covariance matrix within $\mathcal{R}$. Since the set $\mathcal{R}$ is a cone, we can get a new scaled estimate $\tilde{\mathbf{R}} = a\hat{\mathbf{R}}$ with $a > 0$, which also belongs to the set. The log-likelihood as a function of the scaling factor may be written as

$$\begin{aligned} g(a) = L(a\hat{\mathbf{R}}) &= \log\det\left(\frac{1}{a}\hat{\mathbf{R}}^{-1}\mathbf{S}\right) - \frac{1}{a}\operatorname{tr}\left(\hat{\mathbf{R}}^{-1}\mathbf{S}\right) \\ &= -L\log(a) + \log\det(\hat{\mathbf{R}}^{-1}\mathbf{S}) - \frac{1}{a}\operatorname{tr}\left(\hat{\mathbf{R}}^{-1}\mathbf{S}\right). \end{aligned}$$

Taking the derivative with respect to a and equating to zero, we find that the optimal scaling factor that maximizes the likelihood is

$$a^* = \frac{\operatorname{tr}\left(\hat{\mathbf{R}}^{-1}\mathbf{S}\right)}{L},$$

and thus $g(a^*) \geq g(a)$ for $a > 0$. Let $\tilde{\mathbf{R}} = a^*\hat{\mathbf{R}}$. Plugging this value into the trace term of the likelihood function, we have

$$\operatorname{tr}\left(\tilde{\mathbf{R}}^{-1}\mathbf{S}\right) = \frac{1}{a^*}\operatorname{tr}\left(\hat{\mathbf{R}}^{-1}\mathbf{S}\right) = L.$$

Since this result has been obtained *for any* estimate belonging to a cone $\mathcal{R}$, it also holds for the ML estimate, thus proving the lemma. □

Remark 4.1 The previous result extends to non-zero mean MVN models $\mathbf{X} \sim CN_{L\times N}(\mathbf{M}, \mathbf{I}_N \otimes \mathbf{R})$, with unknown $\mathbf{M}$ and $\mathbf{R}$, as long as $\mathbf{R}$ belongs to a cone $\mathcal{R}$. In this case with $\operatorname{etr}(\cdot)$ defined to be $\exp\{\operatorname{tr}(\cdot)\}$, the likelihood function is

$$\ell(\mathbf{M}, \mathbf{R}; \mathbf{X}) = \frac{1}{\pi^{LN}\det(\mathbf{R})^N}\operatorname{etr}\left(-\mathbf{R}^{-1}(\mathbf{X}-\mathbf{M})(\mathbf{X}-\mathbf{M})^H\right).$$

Any estimate $\hat{\mathbf{R}}$ of the covariance matrix $\mathbf{R}$ can be scaled to form $a\hat{\mathbf{R}}$ with $a > 0$. Repeating the steps of the proof of Lemma 4.1, the scaling factor that maximizes the likelihood makes $\operatorname{tr}\left(\hat{\mathbf{R}}^{-1}(\mathbf{X}-\mathbf{M})(\mathbf{X}-\mathbf{M})^H\right)/a^* = L$. This result holds for any estimate of the covariance, so it also holds for its ML estimate.

As a consequence of Lemma 4.1, maximum likelihood identification of $\mathbf{R}$ may be phrased as a problem of maximizing $\log\det(\mathbf{R}^{-1/2}\mathbf{S}\mathbf{R}^{-1/2})$, subject to the trace constraint $\operatorname{tr}\left(\mathbf{R}^{-1/2}\mathbf{S}\mathbf{R}^{-1/2}\right) = L$. But $\log\det(\mathbf{R}^{-1/2}\mathbf{S}\mathbf{R}^{-1/2})$ is a monotone function of $(\det(\mathbf{R}^{-1/2}\mathbf{S}\mathbf{R}^{-1/2}))^{1/L}$, so the equivalent problem is

$$\underset{\mathbf{R}\in\mathcal{R}}{\text{maximize}}\, (\det(\mathbf{T}))^{1/L} \quad \text{s.t.}\ \operatorname{tr}(\mathbf{T}) = L,$$

where $\mathbf{T} = \mathbf{R}^{-1/2}\mathbf{S}\mathbf{R}^{-1/2}$ is the sample covariance matrix for the white random vectors $\mathbf{R}^{-1/2}\mathbf{x} \sim \mathcal{CN}_L(\mathbf{0}, \mathbf{I}_L)$. The trace constraint can be removed by defining $\mathbf{T}' = \mathbf{T}/(\operatorname{tr}(\mathbf{T})/L)$, in which case $\operatorname{tr}(\mathbf{T}') = L$. Then, the problem is

$$\underset{\mathbf{R}\in\mathcal{R}}{\text{maximize}}\, (\det(\mathbf{T}'))^{1/L} = \underset{\mathbf{R}\in\mathcal{R}}{\text{maximize}}\, \frac{(\det(\mathbf{R}^{-1/2}\mathbf{S}\mathbf{R}^{-1/2}))^{1/L}}{\frac{1}{L}\operatorname{tr}(\mathbf{R}^{-1/2}\mathbf{S}\mathbf{R}^{-1/2})}.$$

So, in the multivariate normal model, maximum likelihood identification of covariance is a problem of maximizing the ratio of geometric mean of eigenvalues of $\mathbf{R}^{-1/2}\mathbf{S}\mathbf{R}^{-1/2}$ to arithmetic mean of these eigenvalues, under the constraint $\mathbf{R} \in \mathcal{R}$. This ratio is commonly taken to be a measure of spectral flatness, or whiteness, of eigenvalues. It may be called a measure of coherence. The maximum likelihood estimate of $\mathbf{R} \in \mathcal{R}$ is the covariance matrix in this constraining set that maximizes spectral flatness.

4.3 Hypothesis Testing

In this book, several hypothesis testing problems in the multivariate normal distribution are considered. Given a normal random matrix $\mathbf{X} \sim \mathcal{CN}_{L\times N}(\mathbf{0}, \mathbf{I}_N \otimes \mathbf{R})$, the basic problem for a second-order measurement model is to test

$$\begin{aligned} \mathcal{H}_1 &: \mathbf{R} \in \mathcal{R}_1, \\ \mathcal{H}_0 &: \mathbf{R} \in \mathcal{R}_0, \end{aligned}$$

where $\mathcal{R}_0$ is the structured set for the null $\mathcal{H}_0$ and $\mathcal{R}_1$ is the structured set for the alternative $\mathcal{H}_1$. The *generalized likelihood ratio* (GLR) is

$$\Lambda = \frac{\underset{\mathbf{R}_1\in\mathcal{R}_1}{\max}\, \ell(\mathbf{R}_1; \mathbf{X})}{\underset{\mathbf{R}_0\in\mathcal{R}_0}{\max}\, \ell(\mathbf{R}_0; \mathbf{X})} = \frac{\ell(\hat{\mathbf{R}}_1; \mathbf{X})}{\ell(\hat{\mathbf{R}}_0; \mathbf{X})}.$$

The GLR *test* (GLRT) is a procedure for rejecting the null hypothesis in favor of the alternative when Λ is above a predetermined threshold. When $\mathcal{R}_0$ is a covariance class of interest, then it is common to define $\mathcal{R}_1$ to be the set of positive definite covariance matrices, unconstrained by pattern or structure; then $\hat{\mathbf{R}}_1 = \mathbf{S}$. In this case, the hypothesis test is said to be a *null hypothesis test*. The null is rejected when Λ exceeds a threshold.

When $\mathcal{R}_0$ and $\mathcal{R}_1$ are cones, the following theorem establishes that the GLR for testing covariance model $\mathcal{R}_1$ vs. the covariance model $\mathcal{R}_0$ is a ratio of determinants.

Theorem 4.1 *The GLRT for the hypothesis testing problem* $\mathcal{H}_0 : \mathbf{R} \in \mathcal{R}_0$ *vs.* $\mathcal{H}_1 : \mathbf{R} \in \mathcal{R}_1$ *compares the GLR* Λ *to a threshold, with* Λ *given by*

$$\Lambda = \frac{\ell(\hat{\mathbf{R}}_1; \mathbf{X})}{\ell(\hat{\mathbf{R}}_0; \mathbf{X})} = \left(\frac{\det(\hat{\mathbf{R}}_0)}{\det(\hat{\mathbf{R}}_1)}\right)^N, \tag{4.2}$$

where

$$\hat{\mathbf{R}}_i = \arg\max_{\mathbf{R} \in \mathcal{R}_i} \log\det(\mathbf{R}^{-1}\mathbf{S}),$$

such that $\mathrm{tr}(\hat{\mathbf{R}}_i^{-1}\mathbf{S}) = L$.

Proof From Lemma 4.1, we know that the trace term of the likelihood function, when evaluated at the ML estimates, is a constant under both hypotheses. Then, substituting $\mathrm{tr}(\hat{\mathbf{R}}^{-1}\mathbf{S}) = L$ into likelihood, the result follows. □

The following remark establishes the lexicon regarding GLRs that will be used throughout the book.

Remark 4.2 The GLR Λ is a detection statistic. But in order to cast this statistic in its most illuminating light, often as a coherence statistic, we use monotone functions, like inverse, logarithm, Nth root, etc., to define a monotone function of Λ. The resulting statistic is denoted λ. For example, the GLR in (4.2) may be transformed as

$$\lambda = \frac{1}{\Lambda^{1/N}} = \frac{\det(\hat{\mathbf{R}}_1)}{\det(\hat{\mathbf{R}}_0)}.$$

If $\mathcal{H}_0$ is rejected in favor of $\mathcal{H}_1$ when $\Lambda > \eta$, then it is rejected when $\lambda < 1/\eta^{1/N}$. There is an important caution. If under $\mathcal{H}_0$, the covariance matrix $\mathbf{R}_0$ is known, then no maximization is required, and the trace term in $\ell(\mathbf{R}_0; \mathbf{X})$ becomes $\mathrm{tr}(\mathbf{R}_0^{-1}\mathbf{S})$, in which case $\exp\{N(\mathrm{tr}(\mathbf{R}_0^{-1}\mathbf{S}) - L)\}$ scales the ratio of determinants in the GLR

$$\Lambda = \left(\frac{\det(\mathbf{R}_0)}{\det(\hat{\mathbf{R}}_1)}\right)^N \exp\{N(\text{tr}(\mathbf{R}_0^{-1}\mathbf{S}) - L)\}.$$

Notice, finally, that Theorem 4.1 holds true even when the observations are non-zero mean as long as the sets of the covariance matrices under the null and the alternative hypotheses are cones. Of course, the constrained ML estimates of the covariances $\mathbf{R}_1$ and $\mathbf{R}_0$ will generally depend on the non-zero mean or an ML estimate of the mean.

4.4 Invariance in Hypothesis Testing

Many hypothesis testing problems in signal processing and machine learning have invariances that impose natural restrictions on the test statistics that may be considered for resolving hypotheses. Consider the example, taken from Chap. 3, of estimating the squared correlation coefficient or squared coherence from N i.i.d. observations of the complex random variables x_n and y_n, $n = 1, \ldots, N$. Coherence is invariant under the transformations $x_n' \to ax_n$ and $y_n' \to by_n$ for arbitrary scalars $a, b \in \mathbb{C} - \{0\}$, so it is natural to require that any statistic used for testing coherence should also be invariant to independent scalings of the two random variables.

In many hypothesis testing problems considered in this book, there is no uniformly most powerful test. However, as the correlation coefficient example shows, there is often a group of transformations with respect to which a proposed test must be invariant. In this situation, attention is restricted to test statistics that are invariant under this group of transformations. In this section, we formalize these ideas and present the transformation groups that leave the densities and the parameter spaces invariant for MVN models with structured covariance matrices. Many of these invariance arguments will appear in the chapters to follow.

Consider again the random vector $\mathbf{x} \in \mathbb{C}^L$, distributed as $\mathbf{x} \sim CN_L(\mathbf{0}, \mathbf{R})$, with $\mathbf{R} \in \mathcal{R}$. The data matrix $\mathbf{X} = [\mathbf{x}_1 \ \cdots \ \mathbf{x}_N]$ is a set of independent and identically distributed such vectors. Define the transformation group $G = \{g \mid g \cdot \mathbf{X} = \mathbf{BXQ}\}$. The group action on the measurement matrix $\mathbf{X}$ is $\mathbf{BXQ}$, where $\mathbf{B} \in GL(\mathbb{C}^L)$, with $GL(\mathbb{C}^L)$ the complex linear group of nonsingular $L \times L$ matrices, and $\mathbf{Q} \in U(N)$, with $U(N)$ the group of $N \times N$ unitary matrices. This group action leaves $\mathbf{BXQ}$ distributed as i.i.d. vectors, each distributed as $CN_L(\mathbf{0}, \mathbf{BRB}^H)$. The distribution of $\mathbf{X}$ is said to be invariant-G, and the transformation group on the parameter space induced by the transformation group G is $\overline{G} = \{\overline{g} \mid \overline{g} \cdot \mathbf{R} = \mathbf{BRB}^H\}$.

We are interested in those cases where the group G leaves a set $\mathcal{R}$ invariant-$\overline{G}$, which is to say $\overline{g} \cdot \mathcal{R} = \mathcal{R}$. We say a hypothesis test $\mathcal{H}_1 : \mathbf{R} \in \mathcal{R}_1$ vs. $\mathcal{H}_0 : \mathbf{R} \in \mathcal{R}_0$ is invariant-G when, for all $\mathbf{R} \in \mathcal{R}_i, i = 0, 1, \mathbf{BRB}^H \in \mathcal{R}_i$. That is, $\overline{g} \cdot \mathcal{R}_i = \mathcal{R}_i$. When a hypothesis testing problem is invariant-G, we shall insist that any test of it be invariant-G. That is, a detector $T(\mathbf{X})$ is invariant-G if $T(\mathbf{X}) = T(g \cdot \mathbf{X})$ for a given measurement model. It is known that the GLR will be invariant-G when the testing problem is invariant-G. That is, $\Lambda(\mathbf{X}) = \Lambda(g \cdot \mathbf{X})$. This result is proved, for example, in the discussion in [192], based on a standard result like Proposition 7.13 in [111].

The following examples are illuminating.

Example 4.7 The sets $\mathcal{R}_1 = \{\mathbf{R} \,|\, \mathbf{R} \succ 0\}$ and $\mathcal{R}_0 = \{\mathbf{R} \,|\, \mathbf{R} = \sigma^2\mathbf{I}, \sigma^2 > 0\}$ are invariant-G for group action $g \cdot \mathbf{X} = \beta\mathbf{Q}_L\mathbf{X}\mathbf{Q}_N$, where $|\beta| \neq 0$ and $\mathbf{Q}_L$ and $\mathbf{Q}_N$ are, respectively, $L \times L$ and $N \times N$ unitary matrices. The corresponding group actions on $\mathbf{R}$ are $\overline{g} \cdot \mathbf{R} = |\beta|^2\mathbf{Q}_L\mathbf{R}\mathbf{Q}_L^H \in \mathcal{R}_i$ when $\mathbf{R} \in \mathcal{R}_i$.

Example 4.8 Let us consider the hypothesis test

$$\mathcal{H}_1 : \mathbf{y} = \mathbf{H}\mathbf{x} + \mathbf{n},$$
$$\mathcal{H}_0 : \mathbf{y} = \mathbf{n},$$

where $\mathbf{H}$ is an $L \times p$ matrix that models the linear map from a p-dimensional white source $\mathbf{x} \sim CN_p(\mathbf{0}, \mathbf{I}_p)$ to the vector of observations $\mathbf{y}$ at the L sensors and $\mathbf{n} \sim CN_L(\mathbf{0}, \boldsymbol{\Sigma})$ models the noise. The data matrix $\mathbf{Y} = [\mathbf{y}_1 \ \cdots \ \mathbf{y}_N]$ is a set of independent and identically distributed observations. The sets

$$\mathcal{R}_0 = \left\{\mathbf{R} \,|\, \mathbf{R} = \sigma^2\mathbf{I}_L\right\}$$

and

$$\mathcal{R}_1 = \left\{\mathbf{R} \,|\, \mathbf{R} = \mathbf{H}\mathbf{H}^H + \sigma^2\mathbf{I}_L\right\}$$

are invariant-$\overline{G}$ for group actions $g \cdot \mathbf{X} = \beta\mathbf{Q}_L\mathbf{X}\mathbf{Q}_N$, where $\beta \neq 0$; $\mathbf{Q}_N$ and $\mathbf{Q}_L$ are unitary matrices of respective dimensions $N \times N$ and $L \times L$. The corresponding group actions on $\mathbf{R}$ are $\overline{g} \cdot \mathbf{R} = |\beta|^2\mathbf{Q}_L\mathbf{R}\mathbf{Q}_L^H \in \mathcal{R}_i$ when $\mathbf{R} \in \mathcal{R}_i$.

Example 4.9 The sets

$$\mathcal{R}_0 = \left\{\mathbf{R} \,|\, \mathbf{R} = \mathrm{diag}(\sigma_1^2, \ldots, \sigma_L^2)\right\}$$

and

$$\mathcal{R}_1 = \left\{\mathbf{R} \,|\, \mathbf{R} = \mathbf{H}\mathbf{H}^H + \mathrm{diag}(\sigma_1^2, \ldots, \sigma_L^2)\right\}$$

are invariant-$\overline{G}$ for transformation group $G = \{g \,|\, g \cdot \mathbf{X} = \mathbf{B}\mathbf{X}\mathbf{Q}_N\}$, where $\mathbf{B} = \mathrm{diag}(\beta_1, \ldots, \beta_L)$, with $\beta_l \neq 0$, and $\mathbf{Q}_N \in U(N)$. The group $\overline{G}$ is $\overline{G} = \{\overline{g} \,|\, \overline{g} \cdot \mathbf{R} = \mathbf{B}\mathbf{R}\mathbf{B}^H\}$ and $\mathbf{B}\mathbf{R}\mathbf{B}^H \in \mathcal{R}_i$ when $\mathbf{R} \in \mathcal{R}_i$.

Example 4.10 The sets

$$\mathcal{R}_0 = \{\mathbf{R} \mid \mathbf{R} = \boldsymbol{\Sigma}, \quad \boldsymbol{\Sigma} \succ \mathbf{0}\}$$

and

$$\mathcal{R}_1 = \left\{\mathbf{R} \mid \mathbf{R} = \mathbf{H}\mathbf{H}^H + \boldsymbol{\Sigma}, \quad \boldsymbol{\Sigma} \succ \mathbf{0}\right\}$$

are invariant-$\overline{G}$ for transformation group $G = \{g \mid g \cdot \mathbf{X} = \mathbf{B}\mathbf{X}\mathbf{Q}_N\}$, where $\mathbf{B}$ is a nonsingular $L \times L$ matrix. The corresponding group $\overline{G}$ is $\overline{G} = \{\overline{g} \mid \overline{g} \cdot \mathbf{R} = \mathbf{B}\mathbf{R}\mathbf{B}^H\}$ and $\mathbf{B}\mathbf{R}\mathbf{B}^H \in \mathcal{R}_i$ when $\mathbf{R} \in \mathcal{R}_i$.

4.5 Testing for Sphericity of Random Variables

The experimental set-up is this. A random sample $\mathbf{X} = [\mathbf{x}_1 \cdots \mathbf{x}_N]$ of proper complex normal i.i.d. random variables, $\mathbf{x}_n \sim CN_L(\mathbf{0}, \mathbf{R})$, with $N \geq L$ is recorded.

4.5.1 Sphericity Test: Its Invariances and Null Distribution

The hypothesis testing problem is to test the null hypothesis $\mathcal{H}_0 : \mathbf{R} = \sigma^2 \mathbf{R}_0$ vs. the alternative $\mathcal{H}_1 : \mathbf{R} \succ \mathbf{0}$, an arbitrary positive definite covariance matrix. In this experiment, the covariance matrix $\mathbf{R}_0$ is assumed to be known, but the scale constant $\sigma^2 > 0$ is unknown. Without loss of generality, it may be assumed that $\operatorname{tr}(\mathbf{R}_0) = 1$, so that the total variance $E[\mathbf{x}^H\mathbf{x}] = \operatorname{tr}(\sigma^2\mathbf{R}_0) = \sigma^2$. That is, the unknown parameter σ^2 may be regarded as the unknown mean-square value of the random variable $\mathbf{x}$.

Given the measurement matrix $\mathbf{X}$, the likelihood of the covariance matrix $\mathbf{R}$ in the MVN model $\mathbf{X} \sim CN_{L\times N}(\mathbf{0}, \mathbf{I}_N \otimes \mathbf{R})$ is (4.1), which is repeated here for convenience

$$\ell(\mathbf{R}; \mathbf{X}) = \frac{1}{\pi^{NL} \det(\mathbf{R})^N} \exp\{-N \operatorname{tr}(\mathbf{R}^{-1}\mathbf{S})\},$$

where the sample covariance $\mathbf{S} = N^{-1}\mathbf{X}\mathbf{X}^H$ is a sufficient statistic for testing $\mathcal{H}_0$ vs. $\mathcal{H}_1$. It is a straightforward exercise to show that the maximum likelihood estimator of the covariance under $\mathcal{H}_0$ is

$$\hat{\sigma}^2\mathbf{R}_0 = \frac{1}{L} \operatorname{tr}\left(\mathbf{R}_0^{-1/2}\mathbf{S}\mathbf{R}_0^{-1/2}\right)\mathbf{R}_0$$

and $\hat{\sigma}^2 = \operatorname{tr}(\mathbf{R}_0^{-1/2}\mathbf{S}\mathbf{R}_0^{-1/2})/L$. Under $\mathcal{H}_1$, the maximum likelihood estimator of $\mathbf{R}$ is the sample covariance, that is, $\hat{\mathbf{R}}_1 = \mathbf{S}$. When likelihood is evaluated at these two maximum likelihood estimates, then the GLR is

$$\lambda = \frac{1}{\Lambda^{1/N}} = \frac{\det\left(\mathbf{R}_0^{-1/2}\mathbf{S}\mathbf{R}_0^{-1/2}\right)}{\left(\frac{1}{L}\operatorname{tr}\left(\mathbf{R}_0^{-1/2}\mathbf{S}\mathbf{R}_0^{-1/2}\right)\right)^L} = \frac{\prod_{l=1}^{L} ev_l\left(\mathbf{R}_0^{-1/2}\mathbf{S}\mathbf{R}_0^{-1/2}\right)}{\left(\frac{1}{L}\sum_{l=1}^{L} ev_l\left(\mathbf{R}_0^{-1/2}\mathbf{S}\mathbf{R}_0^{-1/2}\right)\right)^L},$$

where

$$\Lambda = \frac{\ell(\hat{\mathbf{R}}_1; \mathbf{X})}{\ell(\hat{\sigma}^2\mathbf{R}_0; \mathbf{X})}.$$

Notice that $\lambda^{1/L}$ is the ratio of geometric mean to arithmetic mean of the eigenvalues of $\mathbf{R}_0^{-1/2}\mathbf{S}\mathbf{R}_0^{-1/2}$, is bounded between 0 and 1, and is invariant to scale. It is reasonably called a coherence. Under $\mathcal{H}_0$, the matrix $\mathbf{W} = \mathbf{R}_0^{-1/2}\mathbf{S}\mathbf{R}_0^{-1/2}$ is distributed as a complex Wishart matrix $\mathbf{W} \sim \mathcal{CW}_L(\mathbf{I}_L/N, N)$.

In the special case $\mathbf{R}_0 = \mathbf{I}_L$, then this likelihood ratio test is the sphericity test [236]

$$\lambda_S = \frac{\det(\mathbf{S})}{\left(\frac{1}{L}\operatorname{tr}(\mathbf{S})\right)^L} \tag{4.3}$$

and the hypothesis that the data has covariance $\sigma^2\mathbf{I}_L$ with σ^2 unknown is rejected if the sphericity statistic λ_S is below a suitably chosen threshold for a fixed probability of false rejection. This probability is commonly called a false alarm probability.

Invariances. The sphericity statistic and its corresponding hypothesis testing problem are invariant to the transformation group that composes scale and two unitary transformations, i.e., $G = \{g \mid g \cdot \mathbf{X} = \beta\mathbf{Q}_L\mathbf{X}\mathbf{Q}_N\}$, where $\beta \neq 0$, $\mathbf{Q}_L \in U(L)$, and $\mathbf{Q}_N \in U(N)$. The transformation group $\overline{G}$ is $\overline{G} = \{\overline{g} \mid \overline{g} \cdot \mathbf{R} = |\beta|^2\mathbf{Q}_L\mathbf{R}\mathbf{Q}_L^H\}$.

Notice that the sphericity test may be written as

$$\lambda_S = \frac{\prod_{l=1}^{L} ev_l(\mathbf{S})}{\left(\frac{1}{L}\sum_{l=1}^{L} ev_l(\mathbf{S})\right)^L} = \frac{\prod_{l=1}^{L-1} \frac{ev_l(\mathbf{S})}{ev_L(\mathbf{S})}}{\left[\frac{1}{L}\left(1 + \sum_{l=1}^{L-1}\frac{ev_l(\mathbf{S})}{ev_L(\mathbf{S})}\right)\right]^L}.$$

This rewriting makes the GLR a function of the statistic $ev_l(\mathbf{S})/ev_L(\mathbf{S})$, $l = 1, \ldots, L-1$. Each term in this $(L-1)$-dimensional statistic may be scaled by the common factor $ev_L(S)/\operatorname{tr}(\mathbf{S})$ to make the GLR a function of the statistic $ev_l(\mathbf{S})/\operatorname{tr}(\mathbf{S})$, $l = 1, \ldots, L-1$. This statistic is a maximal invariant statistic (see [244, Theorem 8.3.1]), and therefore, λ_S is a function of the maximal invariant statistic as any invariant test must be. Further, the probability of detection of such a test will depend on the population parameters only through the normalized eigenvalues $\frac{ev_l(\mathbf{R})}{\mathrm{tr}(\mathbf{R})}$, $l = 1, \ldots, L-1$, by a theorem of Lehmann in the theory of invariant tests [214].

Distribution Results. Under $\mathcal{H}_0$, the matrix $N\mathbf{S} = \mathbf{X}\mathbf{X}^H$ is distributed as a complex Wishart matrix $\mathcal{CW}_L(\mathbf{I}_L, N)$. For $r > 0$, the rth moment of

$$\lambda_S = \frac{\det(\mathbf{S})}{\left(\frac{1}{L}\operatorname{tr}(\mathbf{S})\right)^L} = \frac{\det\left(\mathbf{X}\mathbf{X}^H\right)}{\left(\frac{1}{L}\operatorname{tr}(\mathbf{X}\mathbf{X}^H)\right)^L}$$

is [3]

$$\mathrm{E}\left[\lambda_S^r\right] = L^{Lr}\frac{\tilde{\Gamma}_L(N+r)}{\tilde{\Gamma}_L(N)}\frac{\Gamma(LN)}{\Gamma(L(N+r))}. \tag{4.4}$$

Here, $\tilde{\Gamma}_L(N)$ is the complex multivariate gamma function

$$\tilde{\Gamma}_L(x) = \pi^{L(L-1)/2}\prod_{l=1}^{L}\Gamma(x-l+1).$$

The moments of λ_S under the null can be used to obtain exact expressions for the pdf of the sphericity test using the Mellin transform approach. In the real-valued case, the exact pdf of λ_S has been given by Consul [78] and Mathai and Rathie [235] (see also [244, pp. 341–343]). In the complex-valued case, the exact pdf of the sphericity test has been derived in [3]. The exact distributions involve Meijer's G-functions and are of limited use, so in practice one typically resorts to asymptotic distributions which can be found, for example, in [244] and [13].

It is proved in [332, Sect. 7.4] that the sphericity test λ_S in (4.3) is distributed as the product of $L-1$ independent beta random variables as $\lambda_S \stackrel{d}{=} \prod_{l=1}^{L-1} U_l$, where $U_l \sim \mathrm{Beta}\left(N-l, l\left(\frac{1}{L}+1\right)\right)$. For $L = 2$, this stochastic representation shows that the sphericity test is distributed as $\lambda_S \sim \mathrm{Beta}(N-1, 3/2)$.

For real-valued data, the sphericity statistic is distributed as the product of $L-1$ independent beta random variables with parameters $\alpha_l = (N-l)/2$ and $\beta_l = l\left(\frac{1}{L}+1/2\right)$, $l = 1, \ldots, L-1$.

4.5.2 Extensions

Sphericity Test with Known σ^2. When the variance is known, we can assume wlog that $\sigma^2 = 1$ so the problem is to test the null hypotheses $\mathcal{H}_0 : \mathbf{R} = \mathbf{I}_L$ vs. the alternative $\mathcal{H}_1 : \mathbf{R} \succ \mathbf{0}$. The generalized likelihood ratio for this test is

$$\lambda = \frac{1}{e^L \Lambda^{1/N}} = \det(\mathbf{S}) \exp\{-\operatorname{tr}(\mathbf{S})\}, \tag{4.5}$$

where $\hat{\mathbf{R}}_1 = \mathbf{S}$ and

$$\Lambda = \frac{\ell(\hat{\mathbf{R}}_1; \mathbf{X})}{\ell(\mathbf{I}_L; \mathbf{X})}.$$

The problem remains invariant under the transformation group $G = \{g \mid g \cdot \mathbf{X} = \mathbf{Q}_L \mathbf{X} \mathbf{Q}_N\}$, where $\mathbf{Q}_L \in U(L)$ and $\mathbf{Q}_N \in U(N)$. The group $\overline{G}$ is $\overline{G} = \{\overline{g} \mid \overline{g} \cdot \mathbf{R} = \mathbf{Q}_L \mathbf{R} \mathbf{Q}_L^H\}$. It can be shown that the eigenvalues of $\mathbf{S}$ are maximal invariants. Hence, any test based on the eigenvalues of $\mathbf{S}$ is invariant. In particular, the GLR in (4.5) is G-invariant, and so is the test based on the largest eigenvalue of $\mathbf{S}$, $ev_1(\mathbf{S})$, suggested by S. N. Roy [292]. The moments of (4.5) under the null and an asymptotic expression for its distribution can be found in [332, pp. 199–200].

Locally Most Powerful Invariant Test (LMPIT). In many hypothesis testing problems in multivariate normal distributions, there is no uniformly most powerful (UMP) test. This is the case for testing for sphericity of random variables. In this case, as argued in Sect. 4.4, it is sensible to restrict attention to the class of invariant tests. These are tests that are invariant to the group of transformations to which the testing problem is invariant. As we have seen, testing for the sphericity of random variables is invariant to the transformation group $G = \{g \mid g \cdot \mathbf{X} = \beta \mathbf{Q}_L \mathbf{X} \mathbf{Q}_N\}$. Any statistic that is invariant under these transformations can depend on the observations only through the eigenvalues of the matrix $\mathbf{R}_0^{-1/2} \mathbf{S} \mathbf{R}_0^{-1/2}$ or the eigenvalues of the sample covariance matrix $\mathbf{S}$ for the case where $\mathbf{R}_0 = \mathbf{I}_L$. The sphericity statistic (a generalized likelihood ratio) is one such invariant tests, but it need not to be optimal in any sense, and in some situations, better tests may exist. For example, the locally most powerful invariant (LMPI) test considers the challenging case of close hypotheses. The main idea behind the LMPIT consists in applying a Taylor series approximation of the ratio of the distributions of the maximal invariant statistic. When the lowest order term depending on the data is a monotone function of an invariant scalar statistic, the detector is locally optimal.

The LMPI test of the null hypothesis $\mathcal{H}_0 : \mathbf{R} = \sigma^2 \mathbf{I}_L$ vs. the alternative $\mathcal{H}_1 : \mathbf{R} \succ \mathbf{0}$, with σ^2 unknown, was derived by S. John in [186]. The test rejects the null when

$$\mathscr{L} = \frac{\operatorname{tr}(\mathbf{S}^2)}{(\operatorname{tr}(\mathbf{S}))^2} = \sum_{l=1}^{L} \left(\frac{ev_l(\mathbf{S})}{\operatorname{tr}(\mathbf{S})} \right)^2 \tag{4.6}$$

is larger than a threshold, determined so that the test has the required false alarm probability. Notice that (4.6) is a function of the maximal invariant $\frac{ev_l(\mathbf{S})}{\operatorname{tr}(\mathbf{S})}$, $l = 1, \ldots, L-1$, as any invariant statistic must be. Alternatively, by defining a coherence matrix as $\hat{\mathbf{C}} = \mathbf{S}/\operatorname{tr}(\mathbf{S})$, the LMPIT may be expressed as $\mathscr{L} = \|\hat{\mathbf{C}}\|^2$.

When σ^2 is known, the LMPIT does not exist. However, with $\mathrm{tr}(\mathbf{R})$ known under $\mathcal{H}_1$, the LMPIT statistic would be $\mathscr{L} = \mathrm{tr}(\mathbf{S})$. Depending on the value of $\mathrm{tr}(\mathbf{R})$, the LMPIT test would be $\mathscr{L} > \eta$, or it would be $\mathscr{L} < \eta$, with η chosen so that the test has the required false alarm probability [186].

4.6 Testing for Sphericity of Random Vectors

This section generalizes the results in the previous section to random vectors. That is, we shall consider testing for sphericity of random vectors or, as it is more commonly known, testing for block sphericity [62, 252]. Again, we are given a set of observations $\mathbf{X} = [\mathbf{x}_1 \cdots \mathbf{x}_N]$, which are i.i.d. realizations of the proper complex Gaussian random vector $\mathbf{x} \sim \mathcal{CN}_{PL}(\mathbf{0}, \mathbf{R})$. Under the null, the $PL \times 1$ random vector $\mathbf{x} = [\mathbf{u}_1^T \cdots \mathbf{u}_P^T]^T$ is composed of P independent vectors $\mathbf{u}_p$, each distributed as $\mathbf{u}_p \sim \mathcal{CN}_L(\mathbf{0}, \mathbf{R}_{\mathbf{uu}})$ with a common $L \times L$ covariance matrix $\mathbf{R}_{\mathbf{uu}}$, for $p = 1, \ldots, P$. The covariance matrix under $\mathcal{H}_1$ is $\mathbf{R} \succ \mathbf{0}$. Then, the test for sphericity of these random vectors is the test $\mathcal{H}_0 : \mathbf{R} = \mathrm{blkdiag}(\mathbf{R}_{\mathbf{uu}}, \ldots, \mathbf{R}_{\mathbf{uu}}) = \mathbf{I}_P \otimes \mathbf{R}_{\mathbf{uu}}$ vs. the alternative $\mathcal{H}_1 : \mathbf{R} \succ \mathbf{0}$. The maximum likelihood estimate of $\mathbf{R}$ under $\mathcal{H}_0$ is $\hat{\mathbf{R}} = \mathbf{I}_P \otimes \hat{\mathbf{R}}_{\mathbf{uu}}$, where $\hat{\mathbf{R}}_{\mathbf{uu}} = \frac{1}{P}\sum_{p=1}^{P} \mathbf{S}_{pp}$ and $\mathbf{S}_{pp}$ is the pth $L \times L$ block in the diagonal of $\mathbf{S} = \mathbf{X}\mathbf{X}^H/N$. The maximum likelihood estimate of $\mathbf{R}$ under $\mathcal{H}_1$ is $\hat{\mathbf{R}}_1 = \mathbf{S}$. Then, the GLR is

$$\lambda_S = \frac{1}{\Lambda^N} = \frac{\det(\mathbf{S})}{\det\left(\frac{1}{P}\sum_{p=1}^{P} \mathbf{S}_{pp}\right)^P}, \tag{4.7}$$

where

$$\Lambda = \frac{\ell(\hat{\mathbf{R}}_1; \mathbf{X})}{\ell(\mathbf{I}_P \otimes \hat{\mathbf{R}}_{\mathbf{uu}}; \mathbf{X})}.$$

The block-sphericity statistic can be written as $\lambda_S = \det(\hat{\mathbf{C}})$, where $\hat{\mathbf{C}}$ is the coherence matrix

$$\hat{\mathbf{C}} = \left(\mathbf{I}_P \otimes \hat{\mathbf{R}}_{\mathbf{uu}}\right)^{-1/2} \mathbf{S} \left(\mathbf{I}_P \otimes \hat{\mathbf{R}}_{\mathbf{uu}}\right)^{-1/2}.$$

Invariances. The statistic in (4.7) and the hypothesis test are invariant to the transformation group $G = \{g \mid g \cdot \mathbf{X} = (\mathbf{Q}_P \otimes \mathbf{B})\mathbf{X}\mathbf{Q}_N\}$, where $\mathbf{B} \in GL(\mathbb{C}^L)$, $\mathbf{Q}_P \in U(P)$, and $\mathbf{Q}_N \in U(N)$. The corresponding transformation group on the parameter space is $\overline{G} = \{\overline{g} \mid \overline{g} \cdot \mathbf{R} = (\mathbf{Q}_P \otimes \mathbf{B})\mathbf{R}(\mathbf{Q}_P \otimes \mathbf{B})^H\}$.

Distribution Results. Distribution results for the block-sphericity test are scarcer than those for the sphericity test. The null distribution for real measurements has been first studied in [62], where the authors derived the moments and the null distribution for $P = 2$ vectors, which is expressed in terms of Meijer's G-functions. Additionally, near-exact distributions are derived in [228].

In Appendix H, following along the lines in [85], a stochastic representation of the null distribution of λ_S in (4.7) is derived. This stochastic representation is

$$\lambda_S \stackrel{d}{=} P^{LP} \prod_{p=1}^{P-1} \prod_{l=1}^{L} U_{p,l} A_{p,l}^{p} \left(1 - A_{p,l}\right) B_{p,l}^{p+1},$$

where $U_{p,l} \sim \text{Beta}(N - l + 1 - pL, pL)$, $A_{p,l} \sim \text{Beta}(Np - l + 1, N - l + 1)$, and $B_{p,l} \sim \text{Beta}(N(p + 1) - 2l + 2, l - 1)$ are independent random variables.

LMPIT. The LMPIT to test the null hypothesis $\mathcal{H}_0 : \mathbf{R} = \mathbf{I}_P \otimes \mathbf{R}_{\mathbf{uu}}$ vs. the alternative $\mathcal{H}_1 : \mathbf{R} \succ \mathbf{0}$, with $\mathbf{R}_{\mathbf{uu}} \succ \mathbf{0}$, was derived in [273]. Recalling the definition of the coherence matrix $\hat{\mathbf{C}}$, the LMPIT rejects the null when

$$\mathscr{L} = \|\hat{\mathbf{C}}\|$$

is larger than a threshold, determined so that the test has the required false alarm probability.

4.7 Testing for Homogeneity of Covariance Matrices

The sphericity statistics are used to test whether a set of random vectors (or variables) are independent and identically distributed. In this section, we test only whether the random vectors are identically distributed. They are assumed independent under both hypotheses. This test is known as a homogeneity (equality) of covariance matrices and is formulated as follows [13, 382].

We are given a set of observations $\mathbf{X} = [\mathbf{x}_1 \ \cdots \ \mathbf{x}_N]$, which are i.i.d. realizations of the proper complex Gaussian $\mathbf{x} \sim \mathcal{CN}_{PL}(\mathbf{0}, \mathbf{R})$. The $PL \times 1$ random vector $\mathbf{x} = [\mathbf{u}_1^T \ \cdots \ \mathbf{u}_P^T]^T$ is composed of P independent vectors $\mathbf{u}_p$, each distributed as $\mathbf{u}_p \sim \mathcal{CN}_L(\mathbf{0}, \mathbf{R}_{\mathbf{uu}}^{(p)})$. Then the covariance matrix $\mathbf{R}$ is $\mathbf{R} = \text{blkdiag}(\mathbf{R}_{\mathbf{uu}}^{(1)}, \ldots, \mathbf{R}_{\mathbf{uu}}^{(P)})$, where each of the $\mathbf{R}_{\mathbf{uu}}^{(p)}$ is an $L \times L$ covariance matrix. The test for homogeneity of these random vectors is the test $\mathcal{H}_0 : \mathbf{R} = \text{blkdiag}(\mathbf{R}_{\mathbf{uu}}, \ldots, \mathbf{R}_{\mathbf{uu}})$ vs. the alternative $\mathcal{H}_1 : \mathbf{R} = \text{blkdiag}(\mathbf{R}_{\mathbf{uu}}^{(1)}, \ldots, \mathbf{R}_{\mathbf{uu}}^{(P)})$. The maximum likelihood estimate of $\mathbf{R}$ under $\mathcal{H}_0$ is $\hat{\mathbf{R}} = \text{blkdiag}(\hat{\mathbf{R}}_{\mathbf{uu}}, \ldots, \hat{\mathbf{R}}_{\mathbf{uu}})$, where $\hat{\mathbf{R}}_{\mathbf{uu}} = \frac{1}{P} \sum_{p=1}^{P} \mathbf{S}_{pp}$ and $\mathbf{S}_{pp}$ is the pth $L \times L$ block of $\mathbf{S} = \mathbf{X}\mathbf{X}^H / N$. The maximum likelihood estimate of $\mathbf{R}$ under $\mathcal{H}_1$ is $\hat{\mathbf{R}}_1 = \text{blkdiag}(\mathbf{S}_{11}, \ldots, \mathbf{S}_{PP})$. Then, the GLR is

$$\lambda_E = \frac{1}{\Lambda^{1/N}} = \frac{\prod_{p=1}^{P} \det\left(\mathbf{S}_{pp}\right)}{\det\left(\frac{1}{P}\sum_{p=1}^{P} \mathbf{S}_{pp}\right)^{P}}, \tag{4.8}$$

where

$$\Lambda = \frac{\ell(\hat{\mathbf{R}}_1; \mathbf{X})}{\ell(\mathbf{I}_P \otimes \hat{\mathbf{R}}_{\mathbf{uu}}; \mathbf{X})}.$$

Invariances. The statistic in (4.8) and the associated hypothesis test are invariant to the transformation group $G = \{g \mid g \cdot \mathbf{X} = (\mathbf{P}_P \otimes \mathbf{B})\mathbf{X}\mathbf{Q}_N\}$, where $\mathbf{B} \in GL(\mathbb{C}^L)$, $\mathbf{Q}_N \in U(N)$, and $\mathbf{P}_P$ is a P-dimensional permutation matrix. The corresponding transformation group on the parameter space is $\overline{G} = \{\overline{g} \mid \overline{g} \cdot \mathbf{R} = (\mathbf{P}_P \otimes \mathbf{B})\mathbf{R}(\mathbf{P}_P \otimes \mathbf{B})^H\}$.

Distribution Results. The distribution of (4.8) under each hypothesis has been studied over the past decades in [13, 244] and references therein. Moments, stochastic representations, exact distributions, and asymptotic expansions have been obtained, mainly for real observations.

Appendix H, based on the analysis for the real case in [13], presents the following stochastic representation for the null distribution of λ_E in (4.8)

$$\lambda_E \stackrel{d}{=} P^{LP} \prod_{p=1}^{P-1} \prod_{l=1}^{L} A_{p,l}^{p} \left(1 - A_{p,l}\right) B_{p,l}^{p+1},$$

where $A_{p,l} \sim \text{Beta}(Np - l + 1, N - l + 1)$ and $B_{p,l} \sim \text{Beta}(N(p+1) - 2l + 2, l - 1)$ are independent beta random variables.

The Scalar Case. When $L = 1$, we are testing equality of variances of P random variables. The GLR in (4.8) specializes to

$$\lambda_E = \frac{1}{\Lambda^{1/N}} = \frac{\prod_{p=1}^{P} s_{pp}}{\left(\frac{1}{P}\sum_{p=1}^{P} s_{pp}\right)^{P}},$$

where s_{pp} is the sample variance of the observations $u_{n,p}$, $n = 1, \ldots, N$, given by

$$s_{pp} = \frac{1}{N} \sum_{n=1}^{N} |u_{n,p}|^2.$$

Extensions. The test for homogeneity of covariance matrices can be extended for equality of power spectral density matrices, as we will discuss in Sect. 8.5. Basically, the detectors for this related problem are based on bulk coherence measures. It can be shown that no LMPIT exists for testing homogeneity of covariance matrices or equality of power spectral density matrices [275].

4.8 Testing for Independence

As usual, the experimental set-up is this. A random sample $\mathbf{X} = [\mathbf{x}_1 \cdots \mathbf{x}_N]$ of proper complex normal random variables, $\mathbf{x}_n \sim \mathcal{CN}_L(\mathbf{0}, \mathbf{R})$, is recorded. The problem is to test hypotheses about the pattern of $\mathbf{R}$.

4.8.1 Testing for Independence of Random Variables

The hypothesis testing problem is to test the hypothesis $\mathcal{H}_0 : \mathbf{R} = \text{diag}(\sigma_1^2, \ldots, \sigma_L^2)$ vs. the alternative hypothesis $\mathcal{H}_1 : \mathbf{R} \succ \mathbf{0}$, an arbitrary positive definite covariance matrix.

Given the measurement matrix $\mathbf{X}$, the likelihood of the covariance matrix $\mathbf{R}$ is given by (4.1). It is a straightforward exercise to show that the maximum likelihood estimator of the covariance $\mathbf{R}$ under $\mathcal{H}_0$ is

$$\hat{\mathbf{R}}_0 = \text{diag}(\mathbf{S}) = \text{diag}(s_{11}, \ldots, s_{LL}),$$

where s_{ll} is the lth diagonal term in the sample covariance matrix $\mathbf{S}$. Under $\mathcal{H}_1$, the maximum likelihood estimator of $\mathbf{R}$ is $\hat{\mathbf{R}}_1 = \mathbf{S}$. When likelihood is evaluated at these two maximum likelihood estimates, then the GLR is [383]

$$\lambda_I = \frac{1}{\Lambda^{1/N}} = \frac{\det(\mathbf{S})}{\det(\text{diag}(\mathbf{S}))} = \frac{\prod_{l=1}^{L} ev_l(\mathbf{S})}{\prod_{l=1}^{L} s_{ll}}, \tag{4.9}$$

where

$$\Lambda = \frac{\ell(\hat{\mathbf{R}}_1; \mathbf{X})}{\ell(\hat{\mathbf{R}}_0; \mathbf{X})}.$$

It is a result from the theory of majorization that this Hadamard ratio of eigenvalue product to diagonal product is bounded between 0 and 1, and it is invariant to individual scaling of the elements in $\mathbf{x}$. It is reasonably called a coherence.

Invariances. The hypothesis testing problem and the GLR are invariant to the transformation group $G = \{g \mid g \cdot \mathbf{X} = \mathbf{B}\mathbf{X}\mathbf{Q}_N\}$, where $\mathbf{B}$ is a nonsingular diagonal matrix $\mathbf{B} = \mathrm{diag}(\beta_1, \ldots, \beta_L)$, with $\beta_l \neq 0$, and $\mathbf{Q}_N \in U(N)$. The corresponding transformation group on the parameter space is $\overline{G} = \{\overline{g} \mid \overline{g} \cdot \mathbf{R} = \mathbf{B}\mathbf{R}\mathbf{B}^H\}$.

Distribution Results. Under the null, the random variable $\prod_{l=1}^{L} s_{ll}$ is independent of λ_I [319]. Using this result, it is shown in [319] that the rth moment of the Hadamard test, λ_I, is

$$E[\lambda_I^r] = \frac{\Gamma(N)^L}{\Gamma(N+r)^L} \frac{\prod_{l=1}^{L} \Gamma(N-L+r+l)}{\prod_{l=1}^{L} \Gamma(N-L+l)}.$$

Applying the inverse Mellin transform, the exact density of λ_I under the null is expressed in [319] as a function of Meijer's G-function. See also [3] for an alternative derivation. The exact pdf is difficult to interpret or manipulate, so one usually prefers to use either asymptotic expressions [13] or stochastic representations of the statistic.

The stochastic representation for this statistic under the null, derived in [13, 77, 201], shows that it is distributed as a product of independent beta random variables,

$$\lambda_I \stackrel{d}{=} \prod_{l=1}^{L-1} U_l,$$

where $U_l \sim \mathrm{Beta}\,(N-l, l)$. A detailed derivation of this result can be found in Appendix H.

Example 4.11 In the case of two random variables, x and y, the likelihood is an increasing monotone function of the maximal invariant statistic, which is the coherence between the two random variables. Therefore, the uniformly most powerful test for independence is

$$\lambda_I = 1 - \frac{|s_{xy}|^2}{s_{xx}s_{yy}}.$$

Under the null, $E[xy^*] = 0$, and the statistic is distributed as the beta random variable, $\lambda_I \sim \mathrm{Beta}(N-1, 1)$, so that a threshold may be set to control the probability of falsely rejecting the null.

If the covariance matrix were the covariance matrix of errors $x - \hat{x}$ and $y - \hat{y}$, as in the study of independence between random variables x and y, given $\mathbf{z}$, then this test would be

$$\lambda_I = 1 - \frac{|S_{xy|\mathbf{z}}|^2}{Q_{xx|\mathbf{z}}Q_{yy|\mathbf{z}}},$$

where the $Q_{xx|\mathbf{z}}$ and $Q_{yy|\mathbf{z}}$ are sample estimates of the error covariances when estimating x from $\mathbf{z}$ and when estimating y from $\mathbf{z}$; $S_{xy|\mathbf{z}}$ is the sample estimate of the partial correlation coefficient or equivalently the cross-covariance between these two errors. If $\mathbf{z}$ is r-dimensional, the statistic is distributed as $\lambda_I \sim \text{Beta}(N - r - 1, 1)$.

LMPIT. The LMPIT to test $\mathcal{H}_0 : \mathbf{R} = \text{diag}(\sigma_1^2, \ldots, \sigma_L^2)$ vs. $\mathcal{H}_1 : \mathbf{R} \succ \mathbf{0}$ was derived in [273]. Defining the coherence matrix

$$\hat{\mathbf{C}} = \hat{\mathbf{R}}_0^{-1/2} \mathbf{S} \hat{\mathbf{R}}_0^{-1/2},$$

with $\hat{\mathbf{R}}_0 = \text{diag}(s_{11}, \ldots, s_{LL})$, the LMPIT rejects the null when

$$\mathscr{L} = \|\hat{\mathbf{C}}\|$$

is large.

4.8.2 Testing for Independence of Random Vectors

The experimental set-up remains as before: N realizations of the proper complex random vector $\mathbf{x} \in \mathbb{C}^L$, distributed as $CN_L(\mathbf{0}, \mathbf{R})$, are organized into the data matrix $\mathbf{X} \in \mathbb{C}^{L \times N}$. However, the random variable $\mathbf{x}$ is parsed into the subsets $\mathbf{x} = [\mathbf{u}_1^T \; \mathbf{u}_2^T]^T$, $\mathbf{u}_1 \in \mathbb{C}^{L_1}, \mathbf{u}_2 \in \mathbb{C}^{L_2}$, $L_1 + L_2 = L$, with corresponding parsing of the covariance matrix and the sample covariance matrix as

$$\mathbf{R} = \begin{bmatrix} \mathbf{R}_{11} & \mathbf{R}_{12} \\ \mathbf{R}_{21} & \mathbf{R}_{22} \end{bmatrix} \quad \text{and} \quad \mathbf{S} = \begin{bmatrix} \mathbf{S}_{11} & \mathbf{S}_{12} \\ \mathbf{S}_{21} & \mathbf{S}_{22} \end{bmatrix}.$$

The hypothesis testing problem is then to test the null hypothesis $\mathcal{H}_0 : \mathbf{R} = \text{blkdiag}(\mathbf{R}_{11}, \mathbf{R}_{22})$ vs. the alternative hypothesis $\mathcal{H}_1 : \mathbf{R} \succ \mathbf{0}$, a positive definite matrix otherwise unstructured.

The maximum likelihood estimate of $\mathbf{R}$ under $\mathcal{H}_0$ is $\hat{\mathbf{R}}_0 = \text{blkdiag}(\mathbf{S}_{11}, \mathbf{S}_{22})$, and the maximum likelihood estimate under $\mathcal{H}_1$ is $\hat{\mathbf{R}}_1 = \mathbf{S}$. When likelihood is evaluated at these estimates, the resulting value of the likelihood ratio is

$$\lambda_I = \frac{1}{\Lambda^{1/N}} = \frac{\det(\mathbf{S})}{\det(\mathbf{S}_{11}) \det(\mathbf{S}_{22})} = \frac{\det(\mathbf{S}_{11} - \mathbf{S}_{12}\mathbf{S}_{22}^{-1}\mathbf{S}_{21}) \det(\mathbf{S}_{22})}{\det(\mathbf{S}_{11}) \det(\mathbf{S}_{22})}$$
$$= \det(\mathbf{I}_{L_1} - \mathbf{S}_{11}^{-1}\mathbf{S}_{12}\mathbf{S}_{22}^{-1}\mathbf{S}_{21}) = \det(\mathbf{I}_{L_1} - \hat{\mathbf{C}}\hat{\mathbf{C}}^H),$$

where

$$\Lambda = \frac{\ell(\hat{\mathbf{R}}_1; \mathbf{X})}{\ell(\hat{\mathbf{R}}_0; \mathbf{X})}.$$

The matrix $\hat{\mathbf{C}}$ is the sample coherence matrix $\hat{\mathbf{C}} = \mathbf{S}_{11}^{-1/2}\mathbf{S}_{12}\mathbf{S}_{22}^{-1/2}$. This matrix has SVD $\hat{\mathbf{C}} = \mathbf{FKG}^H$, where $\mathbf{F}$ and $\mathbf{G}$ are unitary matrices and $\mathbf{K} = \text{diag}(k_1, \ldots, k_n)$, where $n = \min(L_1, L_2)$, is a diagonal matrix of sample canonical correlations. It follows that the likelihood ratio may be written as

$$\lambda_I = \prod_{i=1}^{n}(1 - k_i^2),$$

with $0 \leq k_i \leq 1$. These sample canonical correlations are estimates of the correlations between the canonical variables $(\mathbf{F}^H\mathbf{S}_{11}^{-1/2}\mathbf{x}_1)_i$ and $(\mathbf{G}^H\mathbf{S}_{22}^{-1/2}\mathbf{x}_2)_i$. Finally, we remark that the statistic λ_I may be replaced with the statistic $1 - \prod_{i=1}^{n}(1 - k_i^2)$, which has the interpretation of a soft OR and which aligns with our previous discussion of fine-grained and bulk coherence.

This story extends to the case where the random vector $\mathbf{x}$ is parsed into P subsets,[6] $\mathbf{x} = [\mathbf{u}_1^T\ \mathbf{u}_2^T\ \cdots\ \mathbf{u}_P^T]^T$ with respective dimensions $L_1+L_2+\cdots+L_P = L$. This case is the natural extension of the test presented in Sect. 4.8.1. The hypothesis testing problem is therefore to test $\mathcal{H}_0 : \mathbf{R} = \text{blkdiag}(\mathbf{R}_{11}, \ldots, \mathbf{R}_{PP})$ vs. the alternative hypothesis $\mathcal{H}_1 : \mathbf{R} \succ \mathbf{0}$. The maximum likelihood estimates are $\hat{\mathbf{R}}_0 = \text{blkdiag}(\mathbf{S}_{11}, \ldots, \mathbf{S}_{PP})$, and $\hat{\mathbf{R}}_1 = \mathbf{S}$, which yield the GLR

$$\lambda_I = \frac{1}{\Lambda^{1/N}} = \frac{\det(\mathbf{S})}{\prod_{p=1}^{P}\det(\mathbf{S}_{pp})} = \det(\hat{\mathbf{C}}), \tag{4.10}$$

where

$$\Lambda = \frac{\ell(\hat{\mathbf{R}}_1; \mathbf{X})}{\ell(\hat{\mathbf{R}}_0; \mathbf{X})}.$$

The multiset coherence matrix is now defined as $\hat{\mathbf{C}} = \hat{\mathbf{R}}_0^{-1/2}\mathbf{S}\hat{\mathbf{R}}_0^{-1/2}$. The statistic has been called *multiple coherence* [268]. More will be said about multiple coherence when it is generalized in Chapter 8. Moreover, as shown in Chapter 7, when $P = 2$ and the deviation from the null is known a priori to be a rank-p cross-correlation matrix between $\mathbf{u}_1$ and $\mathbf{u}_2$, the statistic λ_I is modified to $\lambda_I = \prod_{i=1}^{p}(1 - k_i^2)$.

[6] The case of real random vectors is studied in [13].

Invariances. The hypothesis test and the GLR are invariant to the transformation group $G = \{g \mid g \cdot \mathbf{X} = \text{blkdiag}(\mathbf{B}_1, \ldots, \mathbf{B}_P)\mathbf{X}\mathbf{Q}_N\}$, where $\mathbf{B}_p \in GL(\mathbb{C}^{L_p})$ and $\mathbf{Q}_N \in U(N)$. The corresponding transformation group on the parameter space is $\overline{G} = \{\overline{g} \mid \overline{g} \cdot \mathbf{R} = \text{blkdiag}(\mathbf{B}_1, \ldots, \mathbf{B}_P)\, \mathbf{R}\, \text{blkdiag}(\mathbf{B}_1^H, \ldots, \mathbf{B}_P^H)\}$.

Distribution Results. As shown in Appendix H and [201], under the null, λ_I is distributed as a product of independent beta random variables

$$\lambda_I \stackrel{d}{=} \prod_{p=1}^{P-1} \prod_{l=1}^{L_{p+1}} U_{p,l},$$

where

$$U_{p,l} \sim \text{Beta}\left(N - l + 1 - \sum_{i=1}^{p} L_i, \sum_{i=1}^{p} L_i\right).$$

LMPIT. The LMPIT to test $\mathcal{H}_0 : \mathbf{R} = \text{blkdiag}(\mathbf{R}_{11}, \ldots, \mathbf{R}_{PP})$ vs. $\mathcal{H}_1 : \mathbf{R} \succ \mathbf{0}$ rejects the null when the statistic

$$\mathscr{L} = \|\hat{\mathbf{C}}\|$$

is larger than a threshold [273]. Here, we use the multiset coherence matrix $\hat{\mathbf{C}} = \hat{\mathbf{R}}_0^{-1/2}\mathbf{S}\hat{\mathbf{R}}_0^{-1/2}$, with $\hat{\mathbf{R}}_0 = \text{blkdiag}(\mathbf{S}_{11}, \ldots, \mathbf{S}_{PP})$.

4.9 Cross-Validation of a Covariance Model

In Sect. 2.2.3, we discussed the chi-squared goodness-of-fit procedure for cross-validating a least squares estimate in a first-order linear model. The essential idea was to validate that the residual squared error was consistent with what would be expected from the measurement model. That is, *one realization* of a squared error was compared against *a distribution* of squared errors, with a threshold used to control the probability that such a comparison would falsely reject a good estimate. This idea may be extended to the cross-validation of models or estimators of covariance, although the argumentation is somewhat more involved. The story is a story in MVN likelihood and how this likelihood may be used to validate a model for covariance. This idea for cross-validation of a covariance model follows from the *expected likelihood* approach proposed by Abramovich and Gorokhov in [2].

We have seen that for the problem of testing $\mathbf{X} \sim CN_{L\times N}(\mathbf{0}, \mathbf{I}_N \otimes \sigma^2\mathbf{R}_0)$, with σ^2 unknown and $\mathbf{R}_0$ known, vs. the model $\mathbf{X} \sim CN_{L\times N}(\mathbf{0}, \mathbf{I}_N \otimes \mathbf{R}), \mathbf{R} \succ \mathbf{0}$, the generalized likelihood ratio is

$$\lambda = \frac{\det(\mathbf{R}_0^{-1/2}\mathbf{S}\mathbf{R}_0^{-1/2})}{\left(\frac{1}{L}\mathrm{tr}(\mathbf{R}_0^{-1/2}\mathbf{S}\mathbf{R}_0^{-1/2})\right)^L}. \tag{4.11}$$

Under the null hypothesis that the random sample $\mathbf{X}$ was drawn from the multivariate normal model $\mathbf{X} \sim CN_{L\times N}(\mathbf{0}, \mathbf{I}_N \otimes \sigma^2\mathbf{R}_0)$, the random matrix $\mathbf{U} = \mathbf{R}_0^{-1/2}\mathbf{X}$ is distributed as $\mathbf{U} \sim CN_{L\times N}(\mathbf{0}, \mathbf{I}_N \otimes \sigma^2\mathbf{I}_L)$. Therefore, $\mathbf{W} = \mathbf{U}\mathbf{U}^H$ is an $L \times L$ matrix distributed as the complex Wishart $\mathbf{W} \sim C\mathcal{W}_L(\sigma^2\mathbf{I}_L, N)$ and λ may be written as

$$\lambda = \frac{\det(\mathbf{U}\mathbf{U}^H)}{\left(\frac{1}{L}\mathrm{tr}(\mathbf{U}\mathbf{U}^H)\right)^L}, \tag{4.12}$$

which is a stochastic representation of the sphericity statistic λ under the null. Importantly, the distribution of λ in (4.11) or (4.12) is the distribution of λ_S in (4.3), and it depends only on the parameters L and N. Moreover, it is independent of $\mathbf{R}_0$, and its support is $[0, 1]$. This null distribution for λ describes its distribution for fixed $\mathbf{R}_0$, when $\mathbf{X}$ is distributed as $\mathbf{X} \sim CN_{L\times N}(\mathbf{0}, \mathbf{I}_N \otimes \sigma^2\mathbf{R}_0)$.

A typical distribution of λ is sketched in Fig. 4.1 for $L = 12$ and $N = 24$. By comparing λ to thresholds η_1 and η_2, the null may be accepted when $\eta_1 < \lambda < \eta_2$ at confidence $1 - Pr[\eta_1 < \lambda < \eta_2]$ that the null will not be incorrectly rejected.

There is an important point to be made: the sphericity ratio λ is a random variable. One realization of $\mathbf{X}$ produces one realization of λ, and from this realization, we accept or reject the null. This acceptance or rejection is based on whether or not this one realization of λ lies within the body of the null distribution for λ.

This result suggests that λ may be computed for several candidate covariance models $\mathbf{R}$ and those that return a value of λ outside the body of the distribution may be rejected on the basis of the observation that the clairvoyant choice $\mathbf{R} = \mathbf{R}_0$ that

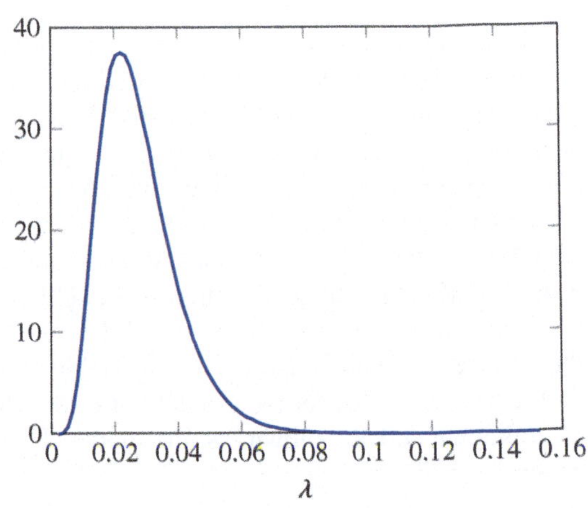

Fig. 4.1 The estimated pdf of the sphericity statistic λ in (4.11) when $L = 12$ and $N = 24$

is matched to the covariance of the measurements $\mathbf{X}$ would return such values with low probability.

Define $\lambda(\mathbf{R})$ to be the sphericity statistic for candidate $\mathbf{R}$ replacing $\mathbf{R}_0$. It may be interpreted as a normalized likelihood function for $\mathbf{R}$, given the measurement $\mathbf{X}$. Sometimes, this candidate comes from physical modeling, and sometimes, it is an estimate of covariance from an experiment. The problem is to validate it from data $\mathbf{X}$. If $\lambda(\mathbf{R})$ lies within the body of the null distribution for λ, then the measurement $\mathbf{X}$ and the model $\mathbf{R}$ have produced a normalized likelihood that would have been produced with high probability by the model $\mathbf{R}_0$. The model is said to be as likely as the model $\mathbf{R}_0$. It is cross-validated.

Given an $L \times N$ data matrix $\mathbf{X}$, it is certainly defensible to ask whether $\lambda(\mathbf{R})$ is a draw from the null distribution for the sphericity statistic λ, provided $\mathbf{R}$ comes from physical modeling or some experimental procedure that is independent of $\mathbf{X}$. For example, $\mathbf{R}$ might be computed from a data matrix $\mathbf{Y}$ that is drawn independent of $\mathbf{X}$ from the same distribution that produced $\mathbf{X}$. But what if $\lambda(\mathbf{R})$ is evaluated at an estimate of $\mathbf{R}$ that is computed from $\mathbf{X}$? For example, what if $\mathbf{R}$ is the maximum likelihood estimate of $\mathbf{R}$ when $\mathbf{R}$ is constrained to a cone class? Then, the denominator of λ is unity, and λ is the maximum of likelihood in the MVN model $\mathbf{X} \sim CN_{L\times N}(\mathbf{0}, \mathbf{I}_N \otimes \mathbf{R})$. The argument for *expected likelihood* advanced by Abramovich and Gorokhov is that this likelihood should lie within the body of the null distribution for the sphericity statistic λ, a distribution that depends only of L and N, and not on $\mathbf{R}$. If not, the estimator of $\mathbf{R}$ is deemed unreliable, which is to say that for parameter choices L and N, the candidate estimator for the $L \times L$ covariance matrix $\mathbf{R}$ is not reliable.

4.10 Chapter Notes

Inference and hypothesis testing problems in multivariate normal models are an essential part of disciplines such as statistical signal processing, data analysis, and machine learning. The brief treatment in this chapter is based mainly on the excellent and classic multivariate analysis books by Anderson [13] (whose first edition dates back to 1958); Kshirsagar [207]; Srivastava and Khatri [332]; Mardia et al. [227]; and Muirhead (1979) [244].

1. This chapter has addressed hypothesis testing problems for sphericity, independence, and homogeneity. Each of these problems is characterized by a transformation group that leaves the problem invariant.
2. Lemma 4.1 concerning ML estimates of structured covariance matrices when they belong to a cone and Theorem 4.1 showing that in this case the GLR reduces to a ratio of determinants were proved in [299].
3. The idea that maximum likelihood may lead in some circumstances to solutions whose likelihood is "too high" to be generated by the true model parameters was discussed originally in [2,3]. The sphericity test was used in these papers to reject ML estimates or other candidates that lie outside the body of the null distribution

of the sphericity test. This “expected likelihood principle” may also be used as a mechanism for cross-validating candidate estimators of covariance that come from physical modeling or from measurements that are distributed as the cross-validating measurements are distributed.

Matched Subspace Detectors 5

This chapter is addressed to signal detection from a sequence of multivariate measurements. The key idea is that the signal to be detected is constrained by a subspace model to be smooth or regular. That is, the signal component of a measurement lies in a known low-dimensional subspace or in a subspace known only by its dimension. As a consequence, these detectors may be applied to problems in beamforming, spectrum analysis, pulsed Doppler radar or sonar, synthetic aperture radar and sonar (SAR and SAS), passive localization in radar and sonar, synchronization of digital communication systems, hyperspectral imaging, and any machine learning application where data may be constrained by a subspace model.

A measurement may consist of an additive combination of signal and noise, or it may consist of noise only. The detection problem is to construct a function of the measurements (called a detector statistic or a detector score) to detect the presence of the signal. Ideally, a detector score will be invariant to a group of transformations to which the detection problem itself is invariant. It might have a claim to optimality. But this is a big ask, as it is rare to find detectors in the class of interest in this chapter that can claim optimality. Rather, they are *principled* in the sense that they are derived from a likelihood principle that is motivated by Neyman–Pearson optimality for much simpler detection problems. The principle is termed a likelihood principle in the statistics literature and a generalized likelihood principle in the engineering and applied science literature.

The principled detectors we derive have compelling geometries and invariances. In some cases, they have known null distributions that permit the setting of thresholds for control of false detections. In other cases, alternative methods that exploit the problem invariances may be used to approximate the null distribution. When these invariances yield CFAR detectors, the null distributions may be estimated using Monte Carlo simulations for fixed values of the unknown parameters under the null, and this distribution approximates the distribution for other values of the unknown parameters.

D. Ramírez et al., *Coherence*, https://doi.org/10.1007/978-3-031-13331-2_5

5.1 Signal and Noise Models

No progress can be made without a model that distinguishes signal from noise. The first modeling assumption is that in a sequence of measurements $\mathbf{y}_n, n = 1, \ldots, N$, each measurement $\mathbf{y}_n \in \mathbb{C}^L$ is a linear combination of signal and noise: $\mathbf{y}_n = \mathbf{z}_n + \mathbf{n}_n$. The sequence of noises is a sequence of independent and identically distributed random vectors, each distributed as $\mathbf{n}_n \sim CN_L(\mathbf{0}, \sigma^2 \mathbf{I}_L)$. The variance σ^2, which has the interpretation of noise power, is typically unknown, but we will also address cases where it is known or even unknown and time varying. This model is not as restrictive as it first appears. If the noise were modeled as $\mathbf{n}_n \sim CN_L(\mathbf{0}, \sigma^2 \boldsymbol{\Sigma})$, with the $L \times L$ positive definite covariance matrix $\boldsymbol{\Sigma}$ known, but σ^2 unknown, then the measurement would be whitened with the matrix $\boldsymbol{\Sigma}^{-1/2}$ to produce the noise model $\boldsymbol{\Sigma}^{-1/2}\mathbf{n}_n \sim CN_L(\mathbf{0}, \sigma^2 \mathbf{I}_L)$.[1] In the section on factor analysis, the noise covariance matrix is generalized to an unknown diagonal matrix, and in the section on the Reed-Yu detector, it is generalized to an unknown positive definite matrix.

In this chapter and the next two, it is assumed that there is a linear model for the signal, which is to say $\mathbf{z}_n = \mathbf{H}\mathbf{t}_n$. The mode weights $\mathbf{t}_n$ are unknown and unconstrained, so they might as well be modeled as $\mathbf{t}_n = \mathbf{A}\mathbf{x}_n$, where $\mathbf{A} \in GL(\mathbb{C}^p)$ is a nonsingular $p \times p$ matrix. Then, in the model $\mathbf{z}_n = \mathbf{H}\mathbf{t}_n$ it is as if $\mathbf{z}_n = \mathbf{H}\mathbf{t}_n = \mathbf{H}\mathbf{A}\mathbf{x}_n$. Without loss of generality, the matrix $\mathbf{A}$ may be parameterized as $\mathbf{A} = (\mathbf{H}^H\mathbf{H})^{-1/2}\mathbf{Q}_p$, where $\mathbf{Q}_p \in U(p)$ is an arbitrary $p \times p$ unitary matrix. The matrix $\mathbf{H}(\mathbf{H}^H\mathbf{H})^{-1/2}\mathbf{Q}_p$ is a unitary slice, so it is as if $\mathbf{z}_n = \mathbf{H}\mathbf{t}_n = \mathbf{U}\mathbf{x}_n$, where $\mathbf{U}$ is an arbitrary unitary basis for the subspace $\langle \mathbf{H} \rangle$. To be consistent with the notation employed in other chapters, we refer to this subspace as $\langle \mathbf{U} \rangle$. Moreover, many of the detectors to follow will depend on the projection matrix $\mathbf{P}_\mathbf{U} = \mathbf{U}\mathbf{U}^H$, but they may be written as $\mathbf{P}_\mathbf{H} = \mathbf{H}(\mathbf{H}^H\mathbf{H})^{-1}\mathbf{H}^H$, since $\mathbf{P}_\mathbf{U} = \mathbf{P}_\mathbf{H}$.

The evocative language is that $\mathbf{U}\mathbf{x}_n$ is a visit to the known subspace $\langle \mathbf{U} \rangle$, which is represented by the arbitrarily selected basis $\mathbf{U}$. Conditioned on $\mathbf{x}_n$, the measurement $\mathbf{y}_n$ is distributed as $\mathbf{y}_n \sim CN_L(\mathbf{U}\mathbf{x}_n, \sigma^2\mathbf{I}_L)$. These measurements may be organized into the $L \times N$ matrix $\mathbf{Y} = [\mathbf{y}_1\ \mathbf{y}_2\ \cdots\ \mathbf{y}_N]$, in which case the signal-plus-noise model is $\mathbf{Y} = \mathbf{U}\mathbf{X} + \mathbf{N}$, with $\mathbf{X}$ and $\mathbf{N}$ defined analogously to $\mathbf{Y}$.

Interpretations of the Signal Model. There are several evocative interpretations of the signal model $\mathbf{Z} = \mathbf{U}\mathbf{X}$. First, write the $L \times p$ channel matrix $\mathbf{U}$ in the following two ways:

$$\mathbf{U} = \begin{bmatrix} \mathbf{u}_1\ \mathbf{u}_2\ \cdots\ \mathbf{u}_p \end{bmatrix} = \begin{bmatrix} \boldsymbol{\rho}_1^T \\ \boldsymbol{\rho}_2^T \\ \vdots \\ \boldsymbol{\rho}_L^T \end{bmatrix}.$$

[1] In Chap. 6, we shall address the problem of unknown covariance matrix $\boldsymbol{\Sigma}$ when there is a secondary channel of measurements that carries information about it.

We may interpret the L-dimensional columns $\mathbf{u}_i, i = 1, \ldots, p$, as *modes* and the p-dimensional rows $\boldsymbol{\rho}_l^T, l = 1, \ldots, L$, as *filters*. Similarly, the $p \times N$ input matrix $\mathbf{X}$ may be written as

$$\mathbf{X} = \begin{bmatrix} \mathbf{x}_1 & \mathbf{x}_2 & \cdots & \mathbf{x}_N \end{bmatrix} = \begin{bmatrix} \boldsymbol{\xi}_1^T \\ \boldsymbol{\xi}_2^T \\ \vdots \\ \boldsymbol{\xi}_p^T \end{bmatrix}.$$

We may interpret the columns $\mathbf{x}_n, n = 1, \ldots, N$, as *spatial inputs* and the rows $\boldsymbol{\xi}_i^T, i = 1, \ldots, p$, as *temporal inputs*. The noise-free signal matrix $\mathbf{Z} = \mathbf{UX}$ is an $L \times N$ matrix written as

$$\mathbf{Z} = \begin{bmatrix} \mathbf{z}_1 & \mathbf{z}_2 & \cdots & \mathbf{z}_N \end{bmatrix} = \begin{bmatrix} \boldsymbol{\phi}_1^T \\ \boldsymbol{\phi}_2^T \\ \vdots \\ \boldsymbol{\phi}_L^T \end{bmatrix}.$$

This space-time matrix of channel responses may be written $\mathbf{Z} = \sum_{i=1}^{p} \mathbf{u}_i \boldsymbol{\xi}_i^T$, which is a sum of p rank-one outer products. The term $\mathbf{u}_i \boldsymbol{\xi}_i^T$ is the ith modal response to the ith temporal input $\boldsymbol{\xi}_i^T$. The L-dimensional column $\mathbf{z}_n$ is a spatial snapshot taken at time n, and the N-dimensional row vector $\boldsymbol{\phi}_l^T$ is a temporal snapshot taken at space l.

The spatial snapshot at time n is $\mathbf{z}_n = \mathbf{U}\mathbf{x}_n$, which is a sum of modal responses. The lth entry in this spatial snapshot is $(\mathbf{z}_n)_l = \boldsymbol{\rho}_l^T \mathbf{x}_n$, which is a filtering of the nth channel input by the lth filter. The temporal snapshot at space l is $\boldsymbol{\phi}_l^T = \boldsymbol{\rho}_l^T \mathbf{X}$, which is a collection of filterings of spatial inputs. The nth entry in this temporal snapshot is $(\boldsymbol{\phi}_l^T)_n = \boldsymbol{\rho}_l^T \mathbf{x}_n$, which is again a filtering of the nth input by the lth filter. This terminology is evocative, but not restrictive. In other applications, different terms may convey different meanings and insights. The terms space and time are then surrogates for these terms.

Conditioned on $\mathbf{X}$, the distribution of $\mathbf{Y}$ is $\mathbf{Y} \sim CN_{L\times N}(\mathbf{UX}, \mathbf{I}_N \otimes \sigma^2 \mathbf{I}_L)$. This notation for the distribution of $\mathbf{Y}$ means that if the $L \times N$ matrix $\mathbf{Y}$ is *vectorized by columns* into an $LN \times 1$ vector, this vector is distributed as multivariate normal, with mean vec($\mathbf{UX}$) and covariance matrix $\mathbf{I}_N \otimes \sigma^2 \mathbf{I}_L$.[2] If only the dimension of the subspace $\mathbf{U}$ is known, the signal matrix $\mathbf{Z} = \mathbf{UX}$ is an unknown $L \times N$ matrix of known rank p.

[2] One might wonder why the notation is not $\mathbf{Y} \sim CN_{LN}(\text{vec}(\mathbf{UX}), \mathbf{I}_N \otimes \sigma^2 \mathbf{I}_L)$. The answer is that this is convention, and as with many conventions, there is no logic.

What if the symbols $\mathbf{t}_n$ were modeled as i.i.d. random vectors, each distributed as $\mathbf{t}_n \sim \mathcal{CN}_p(\mathbf{0}, \mathbf{R}_{tt})$, with $\mathbf{R}_{tt}$ the common, but unknown, covariance matrix? Then, the covariance matrix of $\mathbf{z}_n = \mathbf{H}\mathbf{t}_n$ would be $\mathbf{R}_{zz} = \mathbf{H}\mathbf{R}_{tt}\mathbf{H}^H$, and the distribution of $\mathbf{z}_n = \mathbf{H}\mathbf{t}_n$ would be $\mathbf{z}_n \sim \mathcal{CN}_L(\mathbf{0}, \mathbf{H}\mathbf{R}_{tt}\mathbf{H}^H)$. With no constraints on the unknown covariance matrix $\mathbf{R}_{tt}$, it may be reparameterized as $\mathbf{R}_{tt} = \mathbf{A}\mathbf{R}_{xx}\mathbf{A}^H$. If $\mathbf{A}$ is chosen to be $\mathbf{A} = (\mathbf{H}^H\mathbf{H})^{-1/2}\mathbf{Q}_p$, with $\mathbf{Q}_p \in U(p)$, then this unknown covariance matrix is the rank-p covariance matrix $\mathbf{U}\mathbf{R}_{xx}\mathbf{U}^H$, with $\mathbf{U}$ an arbitrary unitary basis determined by $\mathbf{H}$ and the arbitrary unitary matrix $\mathbf{Q}_p$. It is as if $\mathbf{z}_n = \mathbf{U}\mathbf{x}_n$ with covariance matrix $\mathbf{U}\mathbf{R}_{xx}\mathbf{U}^H$. Then the evocative language is that $\mathbf{z}_n = \mathbf{U}\mathbf{x}_n$ is an unknown visit to the known subspace $\langle\mathbf{U}\rangle$, which is represented by the basis $\mathbf{U}$, with the visit constrained by the Gaussian probability law for $\mathbf{x}_n$. Finally, the distribution of $\mathbf{y}_n$ is that $\mathbf{y}_n \sim \mathcal{CN}_L(\mathbf{0}, \mathbf{U}\mathbf{R}_{xx}\mathbf{U}^H + \sigma^2\mathbf{I}_L)$, and these $\mathbf{y}_n$ are independent and identically distributed. The signal matrix $\mathbf{Z} = \mathbf{U}\mathbf{X}$ is a Gaussian matrix with distribution $\mathbf{Z} \sim \mathcal{CN}_{L\times N}(\mathbf{0}, \mathbf{I}_N \otimes \mathbf{U}\mathbf{R}_{xx}\mathbf{U}^H)$. Conditioned on $\mathbf{X}$, the measurement matrix $\mathbf{Y}$ is distributed as $\mathbf{Y} \sim \mathcal{CN}_{L\times N}(\mathbf{U}\mathbf{X}, \mathbf{I}_N \otimes \sigma^2\mathbf{I}_L)$. But with $\mathbf{U}\mathbf{X}$ distributed as $\mathbf{U}\mathbf{X} \sim \mathcal{CN}_{L\times N}(\mathbf{0}, \mathbf{I}_N \otimes \mathbf{U}\mathbf{R}_{xx}\mathbf{U}^H)$, the joint distribution of $\mathbf{Y}$ and $\mathbf{U}\mathbf{X}$ may be marginalized for the marginal distribution of $\mathbf{Y}$. The result is that $\mathbf{Y} \sim \mathcal{CN}_{L\times N}(\mathbf{0}, \mathbf{I}_N \otimes (\mathbf{U}\mathbf{R}_{xx}\mathbf{U}^H + \sigma^2\mathbf{I}_L))$. If only the dimension of the subspace $\langle\mathbf{U}\rangle$ is known, then $\mathbf{z}_n = \mathbf{U}\mathbf{x}_n$ is a Gaussian vector with covariance matrix $\mathbf{R}_{zz} = \mathbf{U}\mathbf{R}_{xx}\mathbf{U}^H$, where only the rank, p, of the covariance matrix $\mathbf{R}_{zz}$ is known.

In summary, there are four important variations on the subspace model: the subspace $\langle\mathbf{U}\rangle$ may be known, or it may be known only by its dimension p. Moreover, visits by the signal to this subspace may be given a prior distribution, or they may be treated as unknown and unconstrained by a prior distribution. When given a prior distribution, the distribution is assumed to be multivariate Gaussian. As an aid to navigating these four variations, the reader may think about points on a compass, quadrants on a map, or corners in a four-corners diagram:

NW: In the Northwest reside detectors for the case where the subspace is known, and visits to this subspace are unknown, but assigned no prior distribution. Then, conditioned on $\mathbf{x}_n$, the measurement is distributed as $\mathbf{y}_n \sim \mathcal{CN}_L(\mathbf{U}\mathbf{x}_n, \sigma^2\mathbf{I}_L), n = 1, \dots, N$, or equivalently, $\mathbf{Y} \sim \mathcal{CN}_{L\times N}(\mathbf{U}\mathbf{X}, \mathbf{I}_N \otimes \sigma^2\mathbf{I}_L)$. The matrix $\mathbf{X}$ is regarded as an unknown $p \times N$ matrix to be estimated for the construction of a generalized likelihood function.

SW: In the Southwest reside detectors for the case where the subspace is known, and visits to this subspace are unknown but assigned a prior Gaussian distribution. The marginal distribution of $\mathbf{y}_n$ is $\mathbf{y}_n \sim \mathcal{CN}_L(\mathbf{0}, \mathbf{U}\mathbf{R}_{xx}\mathbf{U}^H + \sigma^2\mathbf{I}_L), n = 1, \dots, N$, or equivalently the marginal distribution of $\mathbf{Y}$ is $\mathbf{Y} \sim \mathcal{CN}_{L\times N}(\mathbf{0}, \mathbf{I}_N \otimes (\mathbf{U}\mathbf{R}_{xx}\mathbf{U}^H + \sigma^2\mathbf{I}_L))$. The $p \times p$ covariance matrix $\mathbf{R}_{xx}$ is regarded as an unknown covariance matrix to be estimated for the construction of a generalized likelihood function.

NE: In the Northeast reside detectors for the case where only the dimension of the subspace is known, and visits to this subspace are unknown but assigned no prior distribution. Conditioned on $\mathbf{x}_n$, the measurement is distributed as $\mathbf{y}_n \sim \mathcal{CN}_L(\mathbf{U}\mathbf{x}_n, \sigma^2\mathbf{I}_L), n = 1, \dots, N$, or $\mathbf{Y} \sim \mathcal{CN}_{L\times N}(\mathbf{U}\mathbf{X}, \mathbf{I}_N \otimes \sigma^2\mathbf{I}_L)$.

With only the dimension of the subspace known, $\mathbf{Z} = \mathbf{UX}$ is an unknown $L \times N$ matrix of rank p. This matrix is regarded as an unknown matrix to be estimated for the construction of a generalized likelihood function.

SE: In the Southeast reside detectors for the case where only the dimension of the subspace is known, and visits to this subspace are unknown but assigned a prior Gaussian distribution. The marginal distribution of $\mathbf{y}_n$ is $\mathbf{y}_n \sim \mathcal{CN}_L(\mathbf{0}, \mathbf{R}_{zz} + \sigma^2\mathbf{I}_L)$, $n = 1, \ldots, N$, or equivalently, $\mathbf{Y} \sim \mathcal{CN}_{L\times N}(\mathbf{0}, \mathbf{I}_N \otimes (\mathbf{R}_{zz} + \sigma^2\mathbf{I}_L))$. The $L \times L$ covariance matrix $\mathbf{R}_{zz}$ is regarded as an unknown covariance matrix of rank p to be estimated for the construction of a generalized likelihood function.

So the western hemisphere contains detectors for signals that visit a known subspace. Our convention will be to call these *matched subspace detectors* (MSDs). The eastern hemisphere contains detectors for signals that visit a subspace known only by its dimension. Our convention will be to call these *matched direction detectors* (MDDs), as they are constructed from dominant eigenvalues of a sample covariance matrix, and these eigenvalues are associated with dominant eigenvectors (directions). The northern hemisphere contains detectors for signals that are unknown but assigned no prior distribution. These are called *first-order* detectors, as information about the signal is carried in the mean of the Gaussian measurement distribution. The southern hemisphere contains detectors for signals that are constrained by a Gaussian prior distribution. These are called *second-order detectors*, as information about the signal is carried in the covariance matrix of the Gaussian measurement distribution. In our navigation of these cases we begin in the NW and proceed to SW, NE, and SE, in that order. Table 5.1 summarizes the four kinds of detectors and the signal models are illustrated in Fig. 5.1, where panel (a) accounts for the NW and NE, and panel (b) accounts for the SW and SE.

Are the detector scores for signals that are constrained by a prior distribution *Bayesian detectors*? Perhaps, but not in our lexicon, or the standard lexicon of statistics. They are *marginal detectors*, where the measurement density is obtained

Table 5.1 First-order and second-order detectors for known subspace and unknown subspace of known dimension. In the NW corner, the signal matrix $\mathbf{X}$ is unknown; in the SW corner, the $p \times p$ signal covariance matrix $\mathbf{R}_{xx}$ is unknown; in the NE corner, the $L \times N$ rank-p signal matrix $\mathbf{Z} = \mathbf{UX}$ is unknown; and in the SE corner, the $L \times L$ rank-p signal covariance matrix $\mathbf{R}_{zz} = \mathbf{UR}_{xx}\mathbf{U}^H$ is unknown

	Known subspace	Unknown subspace of known dimension
First-order detectors	$\mathbf{Y} \sim \mathcal{CN}_{L\times N}(\mathbf{UX}, \mathbf{I}_N \otimes \sigma^2\mathbf{I}_L)$	$\mathbf{Y} \sim \mathcal{CN}_{L\times N}(\mathbf{Z}, \mathbf{I}_N \otimes \sigma^2\mathbf{I}_L)$
Second-order detectors	$\mathbf{Y} \sim \mathcal{CN}_{L\times N}(\mathbf{0}, \mathbf{I}_N \otimes (\mathbf{UR}_{xx}\mathbf{U}^H + \sigma^2\mathbf{I}_L))$	$\mathbf{Y} \sim \mathcal{CN}_{L\times N}(\mathbf{0}, \mathbf{I}_N \otimes (\mathbf{R}_{zz} + \sigma^2\mathbf{I}_L))$

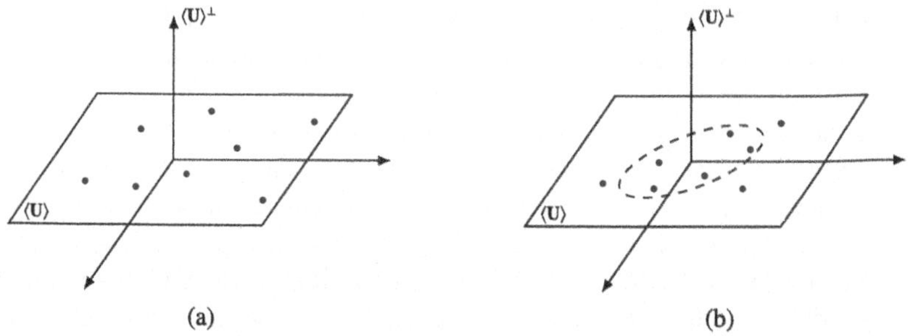

Fig. 5.1 Subspace signal models. In (**a**), the signal $\mathbf{x}_n$, unconstrained by a prior distribution, visits a subspace $\langle \mathbf{U} \rangle$ that is known or known only by its dimension. In (**b**), the signal $\mathbf{x}_n$, constrained by a prior MVN distribution, visits a subspace $\langle \mathbf{U} \rangle$ that is known or known only by its dimension

by marginalizing a conditional Gaussian density with respect to a Gaussian prior for the signal component. We reserve the term Bayesian for those statistical procedures that use Bayes rule to invert a prior distribution for a posterior distribution. Of course, a Gaussian prior to the signal is not the only possibility. Sirianunpiboon, Howard, and Cochran have marginalized with respect to non-Gaussian priors that are motivated by Haar measure on the Stiefel manifold [323, 324].

Throughout we use the abbreviation GLR for generalized likelihood ratio. The GLR is a ratio of two generalized likelihood functions. In each generalized likelihood, unknown parameters are replaced by their maximum likelihood estimates. A GLR is a *detector score* that may be used in a *detector* that compares the GLR to a threshold. The resulting detector is called a generalized likelihood ratio test (GLRT). If the threshold is exceeded it is concluded (not determined) that the signal is present in the measurement. Otherwise it is concluded that the signal is absent. It is awkward to call the GLR a detector score, so we often call it a detector, with the understanding that it will be used in a GLRT.

5.2 The Detection Problem and Its Invariances

The detection problem is the hypothesis test

$$\begin{aligned} \mathcal{H}_1 &: \mathbf{y}_n = \mathbf{U}\mathbf{x}_n + \mathbf{n}_n, \quad && n = 1, 2, \ldots, N, \\ \mathcal{H}_0 &: \mathbf{y}_n = \mathbf{n}_n, && n = 1, 2, \ldots, N, \end{aligned}$$

which may be written

$$\begin{aligned} \mathcal{H}_1 &: \mathbf{Y} = \mathbf{U}\mathbf{X} + \mathbf{N}, \\ \mathcal{H}_0 &: \mathbf{Y} = \mathbf{N}. \end{aligned} \tag{5.1}$$

If $\mathbf{X}$ is given no distribution, then a more precise statement of the hypothesis test is to say $\mathcal{H}_1$ denotes the set of distributions $\mathbf{Y} \sim CN_{L\times N}(\mathbf{UX}, \mathbf{I}_N \otimes \sigma^2\mathbf{I}_L)$, with $\sigma^2 > 0$ and $\mathbf{X} \in \mathbb{C}^{p\times N}$; $\mathcal{H}_0$ denotes the set of distributions $\mathbf{Y} \sim CN_{L\times N}(\mathbf{0}, \mathbf{I}_N \otimes \sigma^2\mathbf{I}_L)$, with $\sigma^2 > 0$. If $\mathbf{X}$ is given a Gaussian distribution, then $\mathcal{H}_1$ in (5.1) denotes the set of distributions $\mathbf{Y} \sim CN_{L\times N}(\mathbf{0}, \mathbf{I}_N \otimes (\mathbf{U}\mathbf{R}_{\mathbf{xx}}\mathbf{U}^H + \sigma^2\mathbf{I}_L))$, with $\sigma^2 > 0$ and $\mathbf{R}_{\mathbf{xx}} \succ \mathbf{0}$; $\mathcal{H}_0$ denotes the set of distributions $\mathbf{Y} \sim CN_{L\times N}(\mathbf{0}, \mathbf{I}_N \otimes \sigma^2\mathbf{I}_L)$, with $\sigma^2 > 0$.

The invariances of this detection problem depend on which parameters are known and which are unknown. For example, when σ^2 is unknown and $\langle\mathbf{U}\rangle$ is known, the hypothesis testing problem is invariant to the transformation group

$$G = \{g \mid g \cdot \mathbf{Y} = \beta\mathbf{V}_L\mathbf{Y}\mathbf{Q}_N\}, \tag{5.2}$$

where $\beta \neq 0$, $\mathbf{Q}_N \in U(N)$ is an arbitrary $N \times N$ unitary matrix, and $\mathbf{V}_L = \mathbf{U}\mathbf{Q}_p\mathbf{U}^H + \mathbf{P}_{\mathbf{U}}^{\perp}$, with $\mathbf{Q}_p \in U(p)$, $\mathbf{P}_{\mathbf{U}}^{\perp} = \mathbf{I}_L - \mathbf{P}_{\mathbf{U}}$, and $\mathbf{P}_{\mathbf{U}} = \mathbf{U}\mathbf{U}^H$. The matrix $\mathbf{V}_L$ is a rotation matrix. Each element of G leaves the distribution of $\mathbf{Y}$ multivariate normal, but with transformed coordinates. That is, conditioned on $\mathbf{X}$, the distribution of $g \cdot \mathbf{Y}$ is $g \cdot \mathbf{Y} \sim CN_{L\times N}(\beta\mathbf{U}\mathbf{Q}_p\mathbf{X}\mathbf{Q}_N, \mathbf{I}_N \otimes |\beta|^2\sigma^2\mathbf{I}_L)$. The parameterization changes from $(\mathbf{X}, \sigma^2)$ to $(\beta\mathbf{Q}_p\mathbf{X}\mathbf{Q}_N, |\beta|^2\sigma^2)$, but the distribution of $g \cdot \mathbf{Y}$ remains in the set of distributions corresponding to $\mathcal{H}_1$. If $\mathbf{X}$ is given a Gaussian prior distribution, then the marginal distribution of $\mathbf{Y}$ is $\mathbf{Y} \sim CN_{L\times N}(\mathbf{0}, \mathbf{I}_N \otimes (|\beta|^2\mathbf{U}\mathbf{Q}_p\mathbf{R}_{\mathbf{xx}}\mathbf{Q}_p^H\mathbf{U}^H + |\beta|^2\sigma^2\mathbf{I}_L))$. The parameterization changes from $(\mathbf{R}_{\mathbf{xx}}, \sigma^2)$ to $(|\beta|^2\mathbf{Q}_p\mathbf{R}_{\mathbf{xx}}\mathbf{Q}_p^H, |\beta|^2\sigma^2)$, but the distribution of $g \cdot \mathbf{Y}$ remains in the set of distributions corresponding to $\mathcal{H}_1$. Under $\mathcal{H}_0$, similar arguments hold. The invariance set for each measurement is a double cone, consisting of a vertical cone perched at the origin of the subspace $\langle\mathbf{U}\rangle$, and its reflection through the subspace.

We shall insist that each of the detectors we derive is invariant to the transformation group that leaves the hypothesis testing problem invariant. The detectors will be invariant to scale, which means their distributions under $\mathcal{H}_0$ will be invariant to scaling. Such detectors are said to be scale-invariant, or CFAR, which means a scale-invariant threshold may be set to ensure that the detector has constant false alarm rate (CFAR).

5.3 Detectors in a First-Order Model for a Signal in a Known Subspace

The detection problem for a first-order signal model in a known subspace (NW quadrant in Table 5.1) is

$$\begin{aligned} \mathcal{H}_1 &: \mathbf{Y} \sim CN_{L\times N}(\mathbf{UX}, \mathbf{I}_N \otimes \sigma^2\mathbf{I}_L), \\ \mathcal{H}_0 &: \mathbf{Y} \sim CN_{L\times N}(\mathbf{0}, \mathbf{I}_N \otimes \sigma^2\mathbf{I}_L), \end{aligned} \tag{5.3}$$

with $\mathbf{X} \in \mathbb{C}^{p \times N}$ and $\sigma^2 > 0$ unknown parameters of the distribution for $\mathbf{Y}$ under $\mathcal{H}_1$, and $\sigma^2 > 0$ an unknown parameter of the distribution under $\mathcal{H}_0$. The subspace $\langle \mathbf{U} \rangle$ is known, with arbitrarily chosen basis $\mathbf{U}$. This hypothesis testing problem is invariant to the transformation group of (5.2).

5.3.1 Scale-Invariant Matched Subspace Detector

From the multivariate Gaussian distribution for $\mathbf{Y}$, the likelihood of the parameters $\mathbf{X}$ and σ^2 under the alternative $\mathcal{H}_1$ is

$$\ell(\mathbf{X}, \sigma^2; \mathbf{Y}) = \frac{1}{\pi^{LN} \sigma^{2LN}} \operatorname{etr}\left\{ -\frac{1}{\sigma^2} (\mathbf{Y} - \mathbf{U}\mathbf{X})(\mathbf{Y} - \mathbf{U}\mathbf{X})^H \right\},$$

where $\operatorname{etr}\{\cdot\}$ stands for $\exp\{\operatorname{tr}(\cdot)\}$. Under the hypothesis $\mathcal{H}_0$, this likelihood function is

$$\ell(\sigma^2; \mathbf{Y}) = \frac{1}{\pi^{LN} \sigma^{2LN}} \operatorname{etr}\left\{ -\frac{1}{\sigma^2} \mathbf{Y}\mathbf{Y}^H \right\}.$$

The GLR is then the ratio of maximized likelihoods

$$\Lambda_1 = \frac{\ell(\hat{\mathbf{X}}, \hat{\sigma}_1^2; \mathbf{Y})}{\ell(\hat{\sigma}_0^2; \mathbf{Y})},$$

where $\hat{\sigma}_i^2$ is the ML estimate of the noise variance under $\mathcal{H}_i$ and $\hat{\mathbf{X}}$ is the ML estimate of $\mathbf{X}$ under $\mathcal{H}_1$. Under $\mathcal{H}_0$, the ML estimate of σ^2 is

$$\hat{\sigma}_0^2 = \frac{1}{NL} \operatorname{tr}\left(\mathbf{Y}^H \mathbf{Y} \right).$$

This average of the squares of each entry in $\mathbf{Y}$ is an average of powers. The ML estimates of $\mathbf{X}$ and σ^2 under $\mathcal{H}_1$ are $\hat{\mathbf{X}} = \mathbf{U}^H \mathbf{Y}$, and

$$\hat{\sigma}_1^2 = \frac{1}{NL} \operatorname{tr}\left(\mathbf{Y}^H \mathbf{P}_{\mathbf{U}}^{\perp} \mathbf{Y} \right).$$

The estimator $\hat{\mathbf{X}}$ is the resolution of $\mathbf{Y}$ onto the basis for the subspace $\langle \mathbf{U} \rangle$ and the ML estimator $\mathbf{U}\hat{\mathbf{X}} = \mathbf{P}_{\mathbf{U}} \mathbf{Y}$ is a projection of the measurement onto the subspace $\langle \mathbf{U} \rangle$. The ML estimator of the noise variance is an average of all squares in the components of $\mathbf{Y}$ that lie outside the subspace $\langle \mathbf{U} \rangle$. This is an average of powers in the so-called orthogonal subspace, where there is no signal component. The GLR is then

$$\lambda_1 = 1 - \frac{1}{\Lambda_1^{1/NL}} = \frac{\mathrm{tr}\left(\mathbf{Y}^H\mathbf{P}_U\mathbf{Y}\right)}{\mathrm{tr}\left(\mathbf{Y}^H\mathbf{Y}\right)} = \sum_{n=1}^{N} \tilde{\mathbf{y}}_n^H \mathbf{P}_U \tilde{\mathbf{y}}_n, \tag{5.4}$$

where

$$\tilde{\mathbf{y}}_n = \frac{\mathbf{y}_n}{\sqrt{\sum_{m=1}^{N} \mathbf{y}_m^H \mathbf{y}_m}}$$

is a normalized measurement.

The GLR in (5.4) is a coherence detector that measures the fraction of the energy that lies in the subspace $\langle \mathbf{U} \rangle$. In fact, it is an average coherence between the normalized measurements and the subspace $\langle \mathbf{U} \rangle$. This GLR, proposed in [307], is a multipulse generalization of the CFAR matched subspace detector [303] and we will refer to it as the scale-invariant matched subspace detector.

Invariances. The scale-invariant matched subspace detector is invariant to the group G defined in (5.2). As a consequence, the detector and its distribution are invariant to scale, which makes its false alarm rate invariant to scale (or CFAR to scale).

Null Distribution. To compute the distribution of λ_1 in (5.4) under $\mathcal{H}_0$, let us rewrite it as

$$\lambda_1 = \frac{\mathrm{tr}\left(\mathbf{Y}^H\mathbf{P}_U\mathbf{Y}\right)}{\mathrm{tr}\left(\mathbf{Y}^H\mathbf{P}_U\mathbf{Y}\right) + \mathrm{tr}\left(\mathbf{Y}^H\mathbf{P}_U^{\perp}\mathbf{Y}\right)},$$

and note that each of the traces is a sum of quadratic forms of Gaussian variables. Then, using the results in Appendix F, under $\mathcal{H}_0$, $2\,\mathrm{tr}\left(\mathbf{Y}^H\mathbf{P}_U\mathbf{Y}\right) \sim \chi^2_{2Np}$ and $2\,\mathrm{tr}\left(\mathbf{Y}^H\mathbf{P}_U^{\perp}\mathbf{Y}\right) \sim \chi^2_{2N(L-p)}$. These are independent random variables, so $\lambda_1 \sim \mathrm{Beta}(Np, N(L-p))$. The transformation

$$\frac{L-p}{p}\frac{\lambda_1}{1-\lambda_1} = \frac{L-p}{p}\frac{\mathrm{tr}\left(\mathbf{Y}^H\mathbf{P}_U\mathbf{Y}\right)}{\mathrm{tr}\left(\mathbf{Y}^H\mathbf{P}_U^{\perp}\mathbf{Y}\right)}$$

is distributed as $F_{2Np,2N(L-p)}$ under $\mathcal{H}_0$. These distributions only depend on known parameters: N, L, and p. They do not depend on σ^2.

5.3.2 Matched Subspace Detector

The measurement model under $\mathcal{H}_1$ is now $\mathbf{Y} \sim \mathcal{CN}_{L\times N}(\mathbf{UX}, \mathbf{I}_N \otimes \sigma^2\mathbf{I}_L)$, where σ^2 is known. Following the steps of Sect. 5.3.1, and replacing estimates of the noise variance with the known variance σ^2, the GLR is the matched subspace detector

$$\lambda_1 = \sigma^2 \log \Lambda_1 = \operatorname{tr}\left(\mathbf{Y}^H \mathbf{P}_{\mathrm{U}} \mathbf{Y}\right) = \sum_{n=1}^{N} \mathbf{y}_n^H \mathbf{P}_{\mathrm{U}} \mathbf{y}_n, \tag{5.5}$$

with

$$\Lambda_1 = \frac{\ell(\hat{\mathbf{X}}, \sigma^2; \mathbf{Y})}{\ell(\sigma^2; \mathbf{Y})},$$

and $\hat{\mathbf{X}} = \mathbf{U}^H \mathbf{Y}$. This might be termed a multipulse or multi-snapshot generalization of the matched subspace detector [303].

Invariances. The matched subspace detector score in (5.5) is invariant to all actions of G defined in (5.2), except for scale. This is the transformation group $G = \{g \mid g \cdot \mathbf{Y} = \mathbf{V}_L \mathbf{Y} \mathbf{Q}_N\}$.

Distribution. The null distribution of $2\lambda_1$ in (5.5) is χ^2_{2Np} and, under $\mathcal{H}_1$, the mean of $\mathbf{P}_{\mathrm{U}}\mathbf{Y}$ is $\mathbf{UX}$, so the non-null distribution of $2\lambda_1$ is noncentral $\chi^2_{2Np}(\delta)$, where the noncentrality parameter is $\delta = 2\operatorname{tr}(\mathbf{X}^H\mathbf{X})$.

5.4 Detectors in a Second-Order Model for a Signal in a Known Subspace

An alternative to estimating the $p \times N$ signal matrix $\mathbf{X}$ is to model it as a random matrix with a specified probability distribution (SW quadrant in Table 5.1). For example, it may be given the distribution $\mathbf{X} \sim CN_{p\times N}(0, \mathbf{I}_N \otimes \mathbf{R}_{\mathbf{xx}})$. Then, the marginal distribution of $\mathbf{Y}$ is $\mathbf{Y} \sim CN_{L\times N}(\mathbf{0}, \mathbf{I}_N \otimes (\mathbf{U}\mathbf{R}_{\mathbf{xx}}\mathbf{U}^H + \sigma^2\mathbf{I}_L))$. In this model, the $p \times p$ covariance $\mathbf{R}_{\mathbf{xx}}$ and the noise variance are unknown. The net effect is that the construction of a GLR requires the replacement of pN unknown parameters of $\mathbf{X}$ by the p^2 unknown parameters of the Hermitian matrix $\mathbf{R}_{\mathbf{xx}}$. The impact of this modeling assumption on performance is not a simple matter of comparing parameter counts, as these parameters are estimated in different probability distributions for $\mathbf{Y}$. This hypothesis testing problem is invariant to the transformation group of (5.2).

5.4.1 Scale-Invariant Matched Subspace Detector

The hypothesis test is

$$\begin{aligned} \mathcal{H}_1 &: \mathbf{Y} \sim CN_{L\times N}(\mathbf{0}, \mathbf{I}_N \otimes (\mathbf{U}\mathbf{R}_{\mathbf{xx}}\mathbf{U}^H + \sigma^2\mathbf{I}_L)), \\ \mathcal{H}_0 &: \mathbf{Y} \sim CN_{L\times N}(\mathbf{0}, \mathbf{I}_N \otimes \sigma^2\mathbf{I}_L), \end{aligned}$$

where $\mathbf{R}_{\mathbf{xx}} \succ \mathbf{0}$ and $\sigma^2 > 0$ are unknown parameters. In other words, there are two competing models for the covariance matrix. Denote these covariance matrices by $\mathbf{R}_i$ to write the likelihood function as

$$\ell(\mathbf{R}_i; \mathbf{Y}) = \frac{1}{\pi^{LN} \det(\mathbf{R}_i)^N} \operatorname{etr}\left\{-N\mathbf{R}_i^{-1}\mathbf{S}\right\},$$

where $\mathbf{S}$ is the sample covariance matrix

$$\mathbf{S} = \frac{1}{N}\mathbf{Y}\mathbf{Y}^H = \frac{1}{N}\sum_{n=1}^{N} \mathbf{y}_n\mathbf{y}_n^H.$$

Under $\mathcal{H}_1$, the covariance matrix $\mathbf{R}_1$ is an element of the set $\mathcal{R}_1 = \{\mathbf{R} \mid \mathbf{R} = \mathbf{U}\mathbf{R}_{\mathbf{xx}}\mathbf{U}^H + \sigma^2\mathbf{I}_L,\ \sigma^2 > 0,\ \mathbf{R}_{\mathbf{xx}} \succ \mathbf{0}\}$. Under $\mathcal{H}_0$, the covariance matrix $\mathbf{R}_0$ is an element of the set $\mathcal{R}_0 = \{\mathbf{R} \mid \mathbf{R} = \sigma^2\mathbf{I}_L,\ \sigma^2 > 0\}$. Both sets are cones. Then, according to Lemma 4.1, the GLR simplifies to a ratio of determinants

$$\lambda_2 = \Lambda_2^{1/N} = \frac{\det(\hat{\mathbf{R}}_0)}{\det(\hat{\mathbf{R}}_1)}, \tag{5.6}$$

where $\hat{\mathbf{R}}_1$ is the ML estimate of $\mathbf{R}_1$, $\hat{\mathbf{R}}_0$ is the ML estimate of $\mathbf{R}_0$, and

$$\Lambda_2 = \frac{\ell(\hat{\mathbf{R}}_1; \mathbf{Y})}{\ell(\hat{\mathbf{R}}_0; \mathbf{Y})}.$$

The ML estimate of the covariance matrix under the null hypothesis is $\hat{\mathbf{R}}_0 = \hat{\sigma}_0^2\mathbf{I}_L$, where

$$\hat{\sigma}_0^2 = \frac{1}{L}\operatorname{tr}(\mathbf{S}),$$

and the corresponding solution for $\det(\hat{\mathbf{R}}_0)$ is

$$\det(\hat{\mathbf{R}}_0) = \hat{\sigma}_0^{2L} = \left(\frac{1}{L}\operatorname{tr}(\mathbf{S})\right)^L.$$

The ML estimate of $\hat{\mathbf{R}}_1$ is much more involved. It was first obtained by Bresler [46], and later used by Ricci in [282] to derive the GLR. The solution given in [282] for (5.6) is

$$\lambda_2 = \frac{\left(\frac{1}{L}\operatorname{tr}(\mathbf{S})\right)^L}{\left[\frac{1}{L-q}\left(\operatorname{tr}(\mathbf{S}) - \sum_{l=1}^{q} ev_l(\mathbf{U}^H\mathbf{S}\mathbf{U})\right)\right]^{L-q} \prod_{l=1}^{q} ev_l(\mathbf{U}^H\mathbf{S}\mathbf{U})}. \tag{5.7}$$

In this formula, the $ev_l(\mathbf{U}^H\mathbf{S}\mathbf{U})$ are eigenvalues of the sample covariance matrix resolved onto an arbitrary basis $\mathbf{U}$ for the known subspace $\langle\mathbf{U}\rangle$. These eigenvalues are invariant to right unitary transformation of $\mathbf{U}$ as $\mathbf{U}\mathbf{Q}_p$, which is another arbitrary basis for $\langle\mathbf{U}\rangle$. As shown in [46], the integer q is the unique integer satisfying

$$ev_{q+1}(\mathbf{U}^H\mathbf{S}\mathbf{U}) \leq \underbrace{\frac{1}{L-q}\left(\operatorname{tr}(\mathbf{S}) - \sum_{l=1}^{q} ev_l(\mathbf{U}^H\mathbf{S}\mathbf{U})\right)}_{\hat{\sigma}_1^2} < ev_q(\mathbf{U}^H\mathbf{S}\mathbf{U}). \tag{5.8}$$

The term sandwiched between $ev_{q+1}(\mathbf{U}^H\mathbf{S}\mathbf{U})$ and $ev_q(\mathbf{U}^H\mathbf{S}\mathbf{U})$ is in fact the ML estimate of σ^2 under the alternative $\mathcal{H}_1$. The basic idea of the algorithm is thus to sweep q from 0 to p, evaluate $\hat{\sigma}_1^2$ for each q, and keep the one that fulfills (5.8). In this sweep, initial and final conditions are set as $ev_0(\mathbf{U}^H\mathbf{S}\mathbf{U}) = \infty$ and $ev_{p+1}(\mathbf{U}^H\mathbf{S}\mathbf{U}) = 0$. This solution is derived in the appendix to this chapter, in Section 5.B, following the derivation in [46]. An alternative solution based on a sequence of alternating maximizations was presented in [301].

Equivalence of the GLRs for First- and Second-Order Models when the Subspace is One-Dimensional. The following lemma establishes the equivalence between the GLRs for first- and second-order models when the subspace is one-dimensional and the noise variance is unknown.

Lemma (Remark 1 in [301]) *For* $p = 1$*, the GLR* λ_1 *in* (5.4) *and the GLR* λ_2 *in* (5.7) *are related as*

$$\lambda_2 = \begin{cases} \frac{1}{L}\left(1 - \frac{1}{L}\right)^{L-1} \frac{1}{\lambda_1(1-\lambda_1)^{L-1}}, & \lambda_1 > \frac{1}{L}, \\ 1, & \lambda_1 \leq \frac{1}{L}. \end{cases}$$

Hence, λ_2 *is a monotone transformation of* λ_1 *(or vice versa), making the two GLRs statistically equivalent, with the same performance.*

Invariances. The hypothesis testing problem, and the resulting GLR, are invariant to the transformation group of (5.2).

Null Distribution. The distribution under $\mathcal{H}_0$ of λ_2 in (5.7) is intractable. Of course, there is an exception for $p = 1$ since λ_1 (5.4) and λ_2 (5.7) are statistically equivalent. Nonetheless, since the detection problem, and therefore the GLR, are invariant to scalings, it is possible to approximate the null distribution for fixed L, p, and N using Monte Carlo simulations for $\sigma^2 = 1$. The approximation is valid for any other noise variance.

5.4.2 Matched Subspace Detector

When the noise variance σ^2 is known, a minor modification of Section 5.B in Appendix shows that the ML estimate for $\mathbf{R}_1$ is

$$\begin{aligned}\hat{\mathbf{R}}_1 &= \mathbf{U}\hat{\mathbf{R}}_{\mathbf{xx}}\mathbf{U}^H + \sigma^2\mathbf{I}_L \\ &= \mathbf{U}\mathbf{W}\,\mathrm{diag}\left(\max(ev_1(\mathbf{U}^H\mathbf{S}\mathbf{U}), \sigma^2), \ldots, \max(ev_p(\mathbf{U}^H\mathbf{S}\mathbf{U}), \sigma^2)\right)\mathbf{W}^H\mathbf{U}^H + \sigma^2\mathbf{P}_{\mathbf{U}}^{\perp},\end{aligned}$$

where $\mathbf{W}$ is a matrix containing the eigenvectors of $\mathbf{U}^H\mathbf{S}\mathbf{U}$. Then,

$$\lambda_2 = \frac{1}{N}\log\Lambda_2 = \sum_{l=1}^{q}\frac{ev_l(\mathbf{U}^H\mathbf{S}\mathbf{U})}{\sigma^2} - \sum_{l=1}^{q}\log\left(\frac{ev_l(\mathbf{U}^H\mathbf{S}\mathbf{U})}{\sigma^2}\right) - q, \tag{5.9}$$

where

$$\Lambda_2 = \frac{\ell(\hat{\mathbf{R}}_1; \mathbf{Y})}{\ell(\sigma^2; \mathbf{Y})},$$

and q is the integer that fulfills $ev_{q+1}(\mathbf{U}^H\mathbf{S}\mathbf{U}) \leq \sigma^2 < ev_q(\mathbf{U}^H\mathbf{S}\mathbf{U})$.

Equivalence of the GLRs for First- and Second-Order Models when the Subspace is One-Dimensional. The following lemma establishes the equivalence between the GLRs for first- and second-order models when the subspace is one-dimensional and the noise variance is known.

Lemma *For $p = 1$, the GLR λ_1 in (5.5) and the GLR λ_2 in (5.9) are related as*

$$\lambda_2 = \begin{cases}\dfrac{\lambda_1}{N\sigma^2} - \log\left(\dfrac{\lambda_1}{N\sigma^2}\right) - 1, & \dfrac{\lambda_1}{N} > \sigma^2, \\ 0, & \dfrac{\lambda_1}{N} \leq \sigma^2,\end{cases}$$

which is a monotone transformation of λ_1. Then, the GLRs are statistically equivalent.

Invariances. The hypothesis testing problem, and the resulting GLR, are invariant to the transformation group $G = \{g \mid g \cdot \mathbf{Y} = \mathbf{V}_L\mathbf{Y}\mathbf{Q}_N\}$. The invariance to scale is lost.

Null Distribution. The distribution under $\mathcal{H}_0$ of λ_2 in (5.9) is intractable. Of course, there is an exception for $p = 1$ since the GLRs λ_1 in (5.5) and λ_2 in (5.9) are statistically equivalent. Nonetheless, it is possible to approximate the null distribution for fixed L, p, N, and σ^2 using Monte Carlo simulations.

This concludes our discussion of matched subspace detectors for signals in a known subspace. We now turn to a discussion of subspace detectors for signals in an unknown subspace of known dimension.

5.5 Detectors in a First-Order Model for a Signal in a Subspace Known Only by its Dimension

The hypothesis testing problem for an unknown deterministic signal $\mathbf{X}$ and an unknown subspace known only by its dimension (NE quadrant in Table 5.1) is

$$\begin{aligned}\mathcal{H}_1 &: \mathbf{Y} \sim CN_{L\times N}(\mathbf{U}\mathbf{X}, \mathbf{I}_N \otimes \sigma^2\mathbf{I}_L),\\ \mathcal{H}_0 &: \mathbf{Y} \sim CN_{L\times N}(\mathbf{0}, \mathbf{I}_N \otimes \sigma^2\mathbf{I}_L),\end{aligned}$$

with $\mathbf{U}\mathbf{X} \in \mathbb{C}^{L\times N}$ and $\sigma^2 > 0$ unknown parameters of the distribution for $\mathbf{Y}$ under $\mathcal{H}_1$, and $\sigma^2 > 0$ an unknown parameter of the distribution under $\mathcal{H}_0$. Importantly, with the subspace $\langle\mathbf{U}\rangle$ known only by its dimension, $\mathbf{U}\mathbf{X}$ is now an unknown $L \times N$ matrix of known rank p.

As in the case of a known subspace, this detection problem is invariant to scalings and right multiplication of the data matrix by a unitary matrix. However, since $\mathbf{U}$ is unknown, the rotation invariance is more general, making the detection problem also invariant to left multiplication by a unitary matrix. Then, the invariance group is

$$G = \{g \mid g \cdot \mathbf{Y} = \beta\mathbf{Q}_L\mathbf{Y}\mathbf{Q}_N\}, \tag{5.10}$$

with $\beta \neq 0$, $\mathbf{Q}_L \in U(L)$, and $\mathbf{Q}_N \in U(N)$. The invariance to unitary transformation $\mathbf{Q}_L$ is more general than the invariance to rotation $\mathbf{V}_L$, because there is now no constraint on the subspace. As this detection problem is also scale invariant, its GLR will be CFAR with respect to measurement scaling.

5.5.1 Scale-Invariant Matched Direction Detector

When the subspace $\langle\mathbf{U}\rangle$ is known, the GLR is given in (5.4). For the subspace known only by its dimension p, there is one additional maximization of the likelihood under $\mathcal{H}_1$. That is, there is one more maximization of the numerator of the GLR. The maximizing subspace may be obtained by matching a basis $\mathbf{U}$ to the first p eigenvectors of $\mathbf{YY}^H$. The maximizing choice for $\mathbf{P_U}$ is $\hat{\mathbf{P}}_\mathbf{U} = \mathbf{W}_p\mathbf{W}_p^H$, where $\mathbf{W} = [\mathbf{W}_p\ \mathbf{W}_{L-p}]$ is the matrix of eigenvectors of $\mathbf{YY}^H$, and $\mathbf{W}_p$ is the $L \times p$ matrix of eigenvectors corresponding to the largest p eigenvalues of $\mathbf{YY}^H$. Consequently, $\mathrm{tr}(\mathbf{Y}^H\hat{\mathbf{P}}_\mathbf{U}\mathbf{Y}) = \sum_{l=1}^{p} ev_l(\mathbf{YY}^H)$. The resulting GLR is

$$\lambda_1 = 1 - \frac{1}{\Lambda_1^{1/NL}} = \frac{\sum_{l=1}^{p} ev_l(\mathbf{YY}^H)}{\sum_{l=1}^{L} ev_l(\mathbf{YY}^H)}, \tag{5.11}$$

where $ev_l(\mathbf{YY}^H)$ are the eigenvalues of $\mathbf{YY}^H$, and

$$\Lambda_1 = \frac{\ell(\hat{\mathbf{U}}, \hat{\mathbf{X}}, \hat{\sigma}^2; \mathbf{Y})}{\ell(\hat{\sigma}^2; \mathbf{Y})}.$$

The GLRs in (5.4) and in (5.11) are both coherence detectors. In one case, the subspace is known, and in the other, the unknown subspace of known dimension p is estimated to be the subspace spanned by the dominant eigenvectors of $\mathbf{YY}^H$. Of course, these are also the dominant left singular vectors of $\mathbf{Y}$. The GLR λ_1 in (5.11) may be called the scale-invariant matched direction detector. It is the extension of the one-dimensional matched direction detector [34] reported in [323].

Invariances. The scale-invariant matched direction detector is invariant to the transformation group of (5.10).

Null Distribution. The null distribution for $p = 1$ and $L = 2$ was derived in [34]:

$$f(\lambda_1) = \frac{\Gamma(2N)}{\Gamma(N)\Gamma(N-1)}\lambda_1^{N-2}(1-\lambda_1)^{N-2}(2\lambda_1 - 1)^2, \quad \frac{1}{2} \le \lambda_1 \le 1. \tag{5.12}$$

However, for other choices of p and L, the null distribution is not known. Nevertheless, exploiting the problem invariances and for fixed L, p, and N, the null distribution may be estimated by using Monte Carlo simulations for $\sigma^2 = 1$, which is valid for other values of the noise variance. Alternatively, for $p > 1$, the false alarm probability of λ_1 can be determined from the joint density of the L ordered eigenvalues of $\mathbf{YY}^H \sim \mathcal{CW}_L(\mathbf{I}_L, N)$ as given in [184, Equation (95)] combined with the importance sampling method of [183].

5.5.2 Matched Direction Detector

When the noise variance σ^2 is known, then the likelihood under $\mathcal{H}_0$ is known, and there is no maximization with respect to the noise variance. The resulting GLR is the matched direction detector [33]

$$\lambda_1 = \sigma^2 \log \Lambda_1 = \sum_{l=1}^{p} ev_l(\mathbf{Y}\mathbf{Y}^H), \tag{5.13}$$

where $ev_l(\mathbf{Y}\mathbf{Y}^H)$ are the eigenvalues of $\mathbf{Y}\mathbf{Y}^H$ and

$$\Lambda_1 = \frac{\ell(\hat{\mathbf{U}}, \hat{\mathbf{X}}, \sigma^2; \mathbf{Y})}{\ell(\sigma^2; \mathbf{Y})}.$$

This amounts to an alignment of the subspace $\langle \mathbf{U} \rangle$ with the p principal eigenvectors, or orthogonal directions, of the sample covariance matrix.

Invariances. The invariance group for unknown subspace and known variance is $G = \{g \mid g \cdot \mathbf{Y} = \mathbf{Q}_L \mathbf{Y} \mathbf{Q}_N\}$, where $\mathbf{Q}_N \in U(N)$ and $\mathbf{Q}_L \in U(L)$. The invariance to scale is lost.

Null Distribution. The null distribution of λ_1 in (5.13) is not known, apart from the case $p = 1$, where it is the distribution of the largest eigenvalue of a Wishart matrix [197, Theorem 2]. However, for $p > 1$, the null distribution may be determined numerically from the joint distribution of all the L ordered eigenvalues of $\mathbf{Y}\mathbf{Y}^H \sim C\mathcal{W}_L(\mathbf{I}_L, N)$ as given in [184, Equation (95)]. Alternatively, one may compute the false alarm probability using the importance sampling scheme developed in [183].

5.6 Detectors in a Second-Order Model for a Signal in a Subspace Known Only by its Dimension

The hypothesis testing problem is to detect a Gaussian signal $\mathbf{X}$ in an unknown subspace $\langle \mathbf{U} \rangle$ of known dimension (SE quadrant in Table 5.1). The covariance matrix of the signal $\mathbf{Z} = \mathbf{U}\mathbf{X}$ is $\mathbf{I}_N \otimes \mathbf{R}_{\mathbf{zz}}$, with $\mathbf{R}_{\mathbf{zz}}$ an unknown non-negative definite matrix of rank p. The detection problem is

$$\begin{aligned} \mathcal{H}_1 &: \mathbf{Y} \sim C\mathcal{N}_{L\times N}(\mathbf{0}, \mathbf{I}_N \otimes (\mathbf{R}_{\mathbf{zz}} + \sigma^2 \mathbf{I}_L)), \\ \mathcal{H}_0 &: \mathbf{Y} \sim C\mathcal{N}_{L\times N}(\mathbf{0}, \mathbf{I}_N \otimes \sigma^2 \mathbf{I}_L), \end{aligned}$$

where $\mathbf{R}_{\mathbf{zz}} \succ \mathbf{0}$ and $\sigma^2 > 0$ are unknown. This detection problem is invariant to the transformation group in (5.10), $G = \{g \mid g \cdot \mathbf{Y} = \beta \mathbf{Q}_L \mathbf{Y} \mathbf{Q}_N\}$, with $\beta \neq 0$, $\mathbf{Q}_L \in U(L)$, and $\mathbf{Q}_N \in U(N)$.

5.6.1 Scale-Invariant Matched Direction Detector

Under $\mathcal{H}_1$, the covariance matrix $\mathbf{R}_1$ is an element of the set $\mathcal{R}_1 = \{\mathbf{R} \mid \mathbf{R} = \mathbf{R}_{\mathbf{zz}} + \sigma^2 \mathbf{I}_L,\ \sigma^2 > 0, \mathbf{R}_{\mathbf{zz}} \succ \mathbf{0}\}$. Under $\mathcal{H}_0$, the covariance matrix $\mathbf{R}_0$ is an element of the set $\mathcal{R}_0 = \{\mathbf{R} \mid \mathbf{R} = \sigma^2 \mathbf{I}_L,\ \sigma^2 > 0\}$. Since both sets are cones, Lemma 4.1 establishes that the GLR is the ratio of determinants

$$\lambda_2 = \Lambda_2^{1/N} = \frac{\det(\hat{\mathbf{R}}_0)}{\det(\hat{\mathbf{R}}_1)},$$

where $\hat{\mathbf{R}}_1$ is the ML estimate of $\mathbf{R}_1 = \mathbf{R}_{\mathbf{zz}} + \sigma^2 \mathbf{I}_L$, $\hat{\mathbf{R}}_0$ is the ML estimate of $\mathbf{R}_0 = \sigma^2 \mathbf{I}$, and

$$\Lambda_2 = \frac{\ell(\hat{\mathbf{R}}_1; \mathbf{Y})}{\ell(\hat{\mathbf{R}}_0; \mathbf{Y})}.$$

The ML estimate of the covariance matrix under the null hypothesis is again $\hat{\mathbf{R}}_0 = \hat{\sigma}_0^2 \mathbf{I}_L$, where

$$\hat{\sigma}_0^2 = \frac{1}{L} \operatorname{tr}(\mathbf{S}).$$

Let $\mathbf{W}$ be the matrix of eigenvectors of $\mathbf{S}$, which are ordered according to the eigenvalues $ev_l(\mathbf{S})$, with $ev_l(\mathbf{S}) \geq ev_{l+1}(\mathbf{S})$. Then, the fundamental result of Anderson [14] shows that the ML estimate of $\hat{\mathbf{R}}_1$ is

$$\hat{\mathbf{R}}_1 = \hat{\mathbf{R}}_{\mathbf{zz}} + \hat{\sigma}_1^2 \mathbf{I}_L = \mathbf{W} \operatorname{diag}(ev_1(\mathbf{S}), \ldots, ev_p(\mathbf{S}), \hat{\sigma}_1^2, \ldots, \hat{\sigma}_1^2) \mathbf{W}^H,$$

where the ML estimate $\hat{\sigma}_1^2$ is

$$\hat{\sigma}_1^2 = \frac{1}{L-p} \sum_{l=p+1}^{L} ev_l(\mathbf{S}).$$

Note that the elements of $\operatorname{diag}(ev_1(\mathbf{S}), \ldots, ev_p(\mathbf{S}), \hat{\sigma}_1^2, \ldots, \hat{\sigma}_1^2)$ are non-negative, and the first p of them are larger than the trailing $L - p$, which are constant at $\hat{\sigma}_1^2$. There are two observations to be made about this ML estimate of $\mathbf{R}_1$: (1) the ML estimate of $\mathbf{R}_{\mathbf{zz}}$ is

$$\hat{\mathbf{R}}_{\mathbf{zz}} = \mathbf{W}_p \operatorname{diag}(ev_1(\mathbf{S}) - \hat{\sigma}_1^2, \ldots, ev_p(\mathbf{S}) - \hat{\sigma}_1^2) \mathbf{W}_p^H,$$

where $\mathbf{W}_p$ is the $L \times p$ matrix of eigenvectors corresponding to the largest p eigenvalues, meaning the dominant eigenspace of $\mathbf{S}$ determines the rank-p covariance matrix $\mathbf{R}_{\mathbf{zz}}$; (2) if the ML estimate of $\mathbf{R}_1^{-1/2}$ were used to whiten the sample covariance matrix, the result would be

$$\hat{\mathbf{R}}_1^{-1/2}\mathbf{S}\hat{\mathbf{R}}_1^{-1/2} = \mathbf{W}\ \mathrm{diag}(1,\ldots,1, ev_{p+1}(\mathbf{S})/\hat{\sigma}_1^2, \ldots, ev_L(\mathbf{S})/\hat{\sigma}_1^2)\mathbf{W}^H.$$

This is a whitening of $\mathbf{S}$, under a constraint on $\mathbf{R}_1$.

Using these ML estimates, the GLR is the following function of the eigenvalues of the sample covariance matrix $\mathbf{S}$:

$$\lambda_2 = \frac{\left(\frac{1}{L}\sum_{l=1}^{L} ev_l(\mathbf{S})\right)^L}{\left(\frac{1}{L-p}\sum_{l=p+1}^{L} ev_l(\mathbf{S})\right)^{L-p}\prod_{l=1}^{p} ev_l(\mathbf{S})}. \tag{5.14}$$

The GLR in (5.14) was proposed in [270]. As this reference shows, (5.14) is the GLR only when the covariance matrix $\mathbf{R}_1$ is a non-negative definite matrix of rank-p plus a scaled identity, and $p < L - 1$. For $p \geq L - 1$, $\mathbf{R}_1$ is a positive definite matrix without further structure, and the GLR is the sphericity test (see Sect. 4.5).

Equivalence of the GLRs for First- and Second-Order Models when the Subspace is One-Dimensional. As in the case of a known subspace, the next lemma shows the equivalence of the GLRs for first- and second-order models for rank-1 signals.

Lemma (Remark 2 in [301]) *For $p = 1$, the GLR λ_1 in* (5.11) *and the GLR λ_2 in* (5.14) *are related as*

$$\lambda_2 = \frac{1}{L}\left(1 - \frac{1}{L}\right)^{L-1} \frac{1}{\lambda_1 (1 - \lambda_1)^{L-1}}.$$

Therefore, both have the same performance in this particular case since this transformation is monotone in $[1/L, 1]$, which is the support of λ_1.

Invariances. The GLR in (5.14) is invariant to the transformation group of (5.10).

Null Distribution. The null distribution of (5.14) is not known, with the exception of $p = 1$ and $L = 2$. Then, taking into account the previous lemma, the distribution is given by (5.12). For other cases, the null distribution may be approximated using Monte Carlo simulations.

Locally Most Powerful Invariant Test. For an unknown subspace of known dimension, the locally most powerful invariant test (LMPIT) is

$$\mathscr{L} = \|\hat{\mathbf{C}}\|, \tag{5.15}$$

where

$$\hat{\mathbf{C}} = \frac{\mathbf{S}}{\mathrm{tr}\,(\mathbf{S})}.$$

This LMPIT to test the null hypothesis $\mathcal{H}_0 : \mathbf{R}_0 = \sigma^2\mathbf{I}_L$ vs. the alternative $\mathcal{H}_1 : \mathbf{R}_1 \succ \mathbf{0}$ with unknown σ^2 was first derived by S. John in [186]. This is the LMPIT for testing sphericity (cf. (4.6) in Chap. 4). Later, it was shown in [273] that (5.15) is also the LMPIT when $\mathbf{R}_1$ is a low-rank plus a scaled identity.

A Geometric Interpretation. In the model $\mathbf{Y} \sim \mathcal{CN}_{L\times N}(\mathbf{0}, \mathbf{I}_N \otimes (\mathbf{R}_{\mathbf{zz}} + \sigma^2\mathbf{I}_L))$, it is as if the signal matrix $\mathbf{Z} = \mathbf{UX}$ is a Gaussian random matrix with covariance matrix $\mathbf{I}_N \otimes \mathbf{R}_{\mathbf{zz}}$. This random matrix may be factored as $\mathbf{Z} = \mathbf{R}_{\mathbf{zz}}^{1/2}\mathbf{V}$, where $\mathbf{V}$ is an $L \times N$ matrix of independent $\mathcal{CN}_1(0, 1)$ random variables. The matrix $\mathbf{V}$ may be given an LQ factorization $\mathbf{V} = \mathbf{LQ}$, where $\mathbf{L}$ is lower triangular and $\mathbf{Q}$ is an $L \times N$ row-slice of an $N \times N$ unitary matrix. Then, the signal matrix is $\mathbf{R}_{\mathbf{zz}}^{1/2}\mathbf{LQ}$. In this factorization, it is as if the random unitary matrix $\mathbf{Q}$ is drawn from the Stiefel manifold, filtered by a random channel $\mathbf{L}$, and transformed by unknown $\mathbf{R}_{\mathbf{zz}}^{1/2}$ to produce the rank-p Gaussian matrix $\mathbf{Z}$. There is a lot that can be said:

- The unitary matrix $\mathbf{Q}$ visits the compact Stiefel manifold $\mathbb{S}t(p, \mathbb{C}^N)$ uniformly with respect to Haar measure. That is, the distribution of $\mathbf{Q}$ is invariant to right unitary transformation. This statement actually requires only that the entries in $\mathbf{V}$ are spherically invariant.
- The random unitary $\mathbf{Q}$ and lower triangular $\mathbf{L}$ are statistically independent.
- The matrix $\mathbf{LL}^H$ is an LU factorization of the Gramian $\mathbf{VV}^H$.
- The matrix $\mathbf{L}$ is distributed as a matrix of independent random variables: $l_{ii} \sim \sqrt{\frac{1}{2}\chi^2_{2(N-i+1)}}$, $l_{ik} \sim \mathcal{CN}_1(0, 1)$, $i > k$. This is Bartlett's decomposition of the Wishart matrix $\mathbf{VV}^H \sim \mathcal{CW}_L(\mathbf{I}_L, N)$ (cf. Appendix G).

We might say the second-order Gaussian model for $\mathbf{Z} = \mathbf{UX} \sim \mathcal{CN}_{L\times N}(\mathbf{0}, \mathbf{I}_N \otimes \mathbf{R}_{\mathbf{zz}})$ actually codes for a uniform draw of an $L \times N$ matrix $\mathbf{Q}$ from the Stiefel manifold, followed by random filtering by the lower-triangular matrix $\mathbf{L}$ and unknown linear transformation by $\mathbf{R}_{\mathbf{zz}}^{1/2}$. This is the geometry.

5.6.2 Matched Direction Detector

When the noise variance σ^2 is known, then there is no estimator of it under the two hypotheses. The estimator of $\mathbf{R}_1$ is

$$\hat{\mathbf{R}}_1 = \hat{\mathbf{R}}_{\mathbf{zz}} + \sigma^2 \mathbf{I}_L = \mathbf{W}\ \mathrm{diag}(ev_1(\mathbf{S}), \ldots, ev_p(\mathbf{S}), \sigma^2, \ldots, \sigma^2)\mathbf{W}^H,$$

where $\mathbf{W}$ is the matrix that contains the eigenvectors of $\mathbf{S}$ and $ev_p(\mathbf{S})$ must exceed σ^2. Otherwise, this ML estimate would be incompatible with the assumptions; i.e., the data do not support the assumptions about dimension and variance for the experiment.

Assuming that the data support the assumptions, the GLR is

$$\lambda_2 = \frac{1}{N} \log \Lambda_2 = \sum_{l=1}^{p} \frac{ev_l(\mathbf{S})}{\sigma^2} - \sum_{l=1}^{p} \log \left(\frac{ev_l(\mathbf{S})}{\sigma^2} \right) - p, \tag{5.16}$$

where

$$\Lambda_2 = \frac{\ell(\hat{\mathbf{R}}_1; \mathbf{Y})}{\ell(\sigma^2; \mathbf{Y})}.$$

The following identities are noteworthy and intuitive:

$$\det(\hat{\mathbf{R}}_1) = \sigma^{2(L-p)} \prod_{l=1}^{p} ev_l(\mathbf{S}), \qquad \mathrm{tr}\left(\hat{\mathbf{R}}_1^{-1}\mathbf{S}\right) = p + \sum_{l=p+1}^{L} \frac{ev_l(\mathbf{S})}{\sigma^2}.$$

Equivalence of the GLRs for First- and Second-Order Models when the Subspace is One-Dimensional. As with previous GLRs, one can show equivalence for first- and second-order models when the subspace is one-dimensional. This result is presented in the next lemma.

Lemma *For $p = 1$, the GLR λ_1 in* (5.13) *and the GLR λ_2 in* (5.16) *are related as*

$$\lambda_2 = \frac{\lambda_1}{N\sigma^2} - \log\left(\frac{\lambda_1}{N\sigma^2}\right) - 1.$$

The GLR λ_2 is a monotone transformation of λ_1 for $\lambda_1/N \geq \sigma^2$ or, equivalently, for $ev_1(\mathbf{S}) > \sigma^2$, which is required to make $\hat{\mathbf{R}}_1$ compatible with the assumptions. Then, if the data support the assumptions of the second-order GLR, the first- and second-order GLRs are statistically equivalent.

Invariances. The invariance group for unknown subspace and known variance is $G = \{g \mid g \cdot \mathbf{Y} = \mathbf{Q}_L \mathbf{Y} \mathbf{Q}_N\}$, where $\mathbf{Q}_N \in U(N)$ and $\mathbf{Q}_L \in U(L)$. The invariance to scale is lost.

Null Distribution. The null distribution of λ_2 in (5.16) is not known, apart from the case $p = 1$, where it is the distribution of the largest eigenvalue of a Wishart matrix [197, Theorem 2]. Similarly to the GLR for the first-order model, for $p > 1$, the null distribution may be determined numerically from the joint distribution of all the L ordered eigenvalues of $\mathbf{Y}\mathbf{Y}^H \sim \mathcal{CW}_L(\mathbf{I}_L, N)$ as given in [184, Equation (95)]. Alternatively, one may compute the false alarm probability using the importance sampling scheme developed in [183]. Finally, since the distribution under $\mathcal{H}_0$ does not have unknown parameters, the null distribution can be approximated using Monte Carlo simulations.

5.7 Factor Analysis

There is one more generalization of the theory of second-order subspace detection, which is based on factor analysis (FA) [330]. The aims of factor analysis are to fit a low-rank-plus-diagonal covariance model to measurements [212, 213, 330].

When adapted to detection theory, FA is a foundation for detecting a random signal that lies in a low-dimensional subspace known only by its dimension. The covariance matrix of this signal is modeled by the low-rank covariance matrix, $\mathbf{R}_{\mathbf{zz}}$. The covariance matrix of independent additive noise is modeled by a positive definite diagonal matrix $\boldsymbol{\Sigma}$. Neither $\mathbf{R}_{\mathbf{zz}}$ nor $\boldsymbol{\Sigma}$ is known. This model is more general than the white noise model assumed in previous sections, but it forces iterative maximization for an approximation to the GLR.

The detection problem is

$$\begin{aligned} \mathcal{H}_1 &: \mathbf{Y} \sim \mathcal{CN}_{L\times N}(\mathbf{0}, \mathbf{I}_N \otimes (\mathbf{R}_{\mathbf{zz}} + \boldsymbol{\Sigma})), \\ \mathcal{H}_0 &: \mathbf{Y} \sim \mathcal{CN}_{L\times N}(\mathbf{0}, \mathbf{I}_N \otimes \boldsymbol{\Sigma}), \end{aligned}$$

where $\boldsymbol{\Sigma}$ is an unknown diagonal covariance matrix and $\mathbf{R}_{\mathbf{zz}}$ is an unknown positive definite covariance matrix of known rank p. The set of covariance matrices under each hypothesis is a cone, which allows us to write the GLR as

$$\lambda_2 = \Lambda_2^{1/N} = \frac{\det(\hat{\mathbf{R}}_0)}{\det(\hat{\mathbf{R}}_1)}, \tag{5.17}$$

where

$$\Lambda_2 = \frac{\ell(\hat{\mathbf{R}}_1; \mathbf{Y})}{\ell(\hat{\mathbf{R}}_0; \mathbf{Y})}.$$

The matrix $\hat{\mathbf{R}}_1$ is the ML estimate of $\mathbf{R}_1 = \mathbf{R}_{\mathbf{zz}} + \boldsymbol{\Sigma}$ and $\hat{\mathbf{R}}_0$ is the ML estimate of $\mathbf{R}_0 = \boldsymbol{\Sigma}$. An iterative solution for this GLR was derived in [270]. In this section of the book, we present an alternative to that solution, based on block minorization-maximization.

Under $\mathcal{H}_0$, the ML estimate of the covariance matrix is

$$\hat{\mathbf{R}}_0 = \text{diag}(\mathbf{S}),$$

which is just the main diagonal of the sample covariance matrix $\mathbf{S}$. Under $\mathcal{H}_1$, there is no closed-form solution for the ML estimates and numerical procedures are necessary, such as [188, 189, 270]. Here, we use block minorization-maximization (BMM) [279]. The method described in [196] considers two blocks: the low-rank term $\mathbf{R}_{\mathbf{zz}}$ and the noise covariance matrix $\boldsymbol{\Sigma}$. Fixing one of these two blocks, BMM aims to find the solution that maximizes a minorizer of the likelihood, which ensures that the likelihood is increased. Then, alternating between the optimization of each of these blocks, BMM guarantees that the solution converges to a stationary point of the likelihood $\ell(\mathbf{R}_1; \mathbf{Y})$.

Start by fixing $\boldsymbol{\Sigma}$. In this case, there is no need for a minorizer as it is possible to find the solution for $\mathbf{R}_{\mathbf{zz}}$ in closed-form. Compute the whitened sample covariance matrix $\mathbf{S}_{\boldsymbol{\Sigma}} = \boldsymbol{\Sigma}^{-1/2}\mathbf{S}\boldsymbol{\Sigma}^{-1/2}$. Denote the eigenvectors of this whitened sample covariance matrix by $\mathbf{W}$ and the eigenvalues by $ev_l(\mathbf{S}_{\boldsymbol{\Sigma}})$, with $ev_l(\mathbf{S}_{\boldsymbol{\Sigma}}) \geq ev_{l+1}(\mathbf{S}_{\boldsymbol{\Sigma}})$. The solution for $\mathbf{R}_{\mathbf{zz}}$ that maximizes likelihood is again a variation on the Anderson result [14]

$$\hat{\mathbf{R}}_{\mathbf{zz}} = \boldsymbol{\Sigma}^{1/2}\mathbf{W}\,\text{diag}\left((ev_1(\mathbf{S}_{\boldsymbol{\Sigma}}) - 1)^+, \ldots, (ev_p(\mathbf{S}_{\boldsymbol{\Sigma}}) - 1)^+, 0, \ldots, 0\right)\mathbf{W}^H\boldsymbol{\Sigma}^{1/2},$$

where $(x)^+ = \max(x, 0)$. When $\mathbf{R}_{\mathbf{zz}}$ is fixed, and using a minorizer based on a linearization of the log-likelihood, the solution that maximizes this minorizer is

$$\hat{\boldsymbol{\Sigma}} = \text{diag}\left(\mathbf{S} - \mathbf{R}_{\mathbf{zz}}\right).$$

A fixed point of likelihood is obtained by alternating between these two solutions, and the resulting estimate of $\mathbf{R}_1$ is taken to be an approximation to the ML estimate. Once estimates of the covariance matrix under both hypotheses are available, the GLR is given by the ratio of determinants in (5.17). However, and similar to the GLR in (5.14), the derivation in this section is only valid when $\mathbf{R}_1$ is rank-p plus diagonal, which requires $p < L - \sqrt{L}$. Otherwise, the model is not identifiable and the GLR is the Hadamard ratio of Sect. 4.8.1 [270].

Invariances. The invariance group for this problem is $G = \{g \mid g \cdot \mathbf{Y} = \mathbf{B}\mathbf{Y}\mathbf{Q}_N\}$, where $\mathbf{Q}_N \in U(N)$ and $\mathbf{B} = \text{diag}(\beta_1, \ldots, \beta_L)$, with $\beta_l \neq 0$. That is, the detection problem is invariant to arbitrary and independent scalings of the components of $\mathbf{y}_n$. This invariance contrasts with most of the previously considered tests, which were invariant to common scalings. This invariance makes the detector CFAR, which means a threshold may be set for a fixed probability of false alarm.

Null Distribution. The null distribution of the GLR is not known, but taking into account the invariance to independent scalings, the null distribution may be obtained using Monte Carlo simulations for a given choice of L, p, and N.

Locally Most Powerful Invariant Test. For the detection problem considered in this section, the locally most powerful invariant test (LMPIT) statistic is

$$\mathscr{L} = \|\hat{\mathbf{C}}\|,$$

where $\hat{\mathbf{C}}$ is the following coherence matrix

$$\hat{\mathbf{C}} = (\text{diag}(\mathbf{S}))^{-1/2}\mathbf{S}(\text{diag}(\mathbf{S}))^{-1/2}.$$

This LMPIT was derived in [273], where it was also shown to be the LMPIT for testing independence of random variables (see Sect. 4.8.1).

5.8 A MIMO Version of the Reed-Yu Detector and its Connection to the Wilks Lambda and the Hotelling T^2 Statistics

The roles of channel and symbol may be reversed to consider the model $\mathbf{HX}$, with the symbol (or weight) matrix $\mathbf{X}$ *known* and the channel $\mathbf{H}$ *unknown*. The first contribution in the engineering literature to this problem was made by Reed and Yu [280], who derived the probability distribution for a generalized likelihood ratio in the case of a single-input multiple-output (SIMO) channel. Their application was optical pattern detection with unknown spectral distribution, so measurements were real. Bliss and Parker [39] generalized this result for synchronization in a complex multiple-input multiple-output (MIMO) communication channel. In this section, it is shown that the generalized likelihood ratio (GLR) for the Reed-Yu problem, as generalized by Bliss and Parker, is a Wilks Lambda statistic that generalizes the Hotelling T^2 statistic [51, 259].

The detection problem is to test the hypotheses

$$\begin{aligned} \mathcal{H}_1 &: \mathbf{Y} \sim CN_{L\times N}(\mathbf{HX}, \mathbf{I}_N \otimes \boldsymbol{\Sigma}), \\ \mathcal{H}_0 &: \mathbf{Y} \sim CN_{L\times N}(\mathbf{0}, \mathbf{I}_N \otimes \boldsymbol{\Sigma}). \end{aligned} \tag{5.18}$$

The symbol matrix $\mathbf{X}$ is a known, full-rank, $p \times N$ symbol matrix, but the channel $\mathbf{H}$ and noise covariance matrix $\boldsymbol{\Sigma}$ are unknown parameters of the distribution for

$\mathbf{Y}$ under $\mathcal{H}_1$. The covariance matrix $\boldsymbol{\Sigma}$ is the only unknown parameter of the distribution under $\mathcal{H}_0$. Interestingly, for this reversal of roles between channel and signal, the problem of signal detection in noise of unknown positive definite covariance matrix is well-posed. This generalization is not possible in the case where the subspace is known but the signal is unknown. As usual, it is assumed that $N > L$ and $L > p$, but it is assumed also that $N > L + p$. For $p > L$, there is a parallel development in [51], but this case is not included here.

The likelihood function under $\mathcal{H}_1$ and $\mathcal{H}_0$ are, respectively,

$$\ell(\mathbf{H}, \boldsymbol{\Sigma}; \mathbf{Y}) = \frac{1}{\pi^{LN} \det(\boldsymbol{\Sigma})^N} \operatorname{etr}\left\{-\boldsymbol{\Sigma}^{-1}(\mathbf{Y} - \mathbf{H}\mathbf{X})(\mathbf{Y} - \mathbf{H}\mathbf{X})^H\right\},$$

and

$$\ell(\boldsymbol{\Sigma}; \mathbf{Y}) = \frac{1}{\pi^{LN} \det(\boldsymbol{\Sigma})^N} \operatorname{etr}\left\{-\boldsymbol{\Sigma}^{-1}\mathbf{Y}\mathbf{Y}^H\right\}.$$

Under $\mathcal{H}_0$, the ML estimate of the covariance matrix $\boldsymbol{\Sigma}$ is

$$\hat{\boldsymbol{\Sigma}}_0 = \frac{1}{N}\mathbf{Y}\mathbf{Y}^H.$$

Similarly, under $\mathcal{H}_1$, the ML estimates of unknown parameters are $\hat{\mathbf{H}} = \mathbf{Y}\mathbf{X}^H(\mathbf{X}\mathbf{X}^H)^{-1}$ and

$$\hat{\boldsymbol{\Sigma}}_1 = \frac{1}{N}(\mathbf{Y} - \hat{\mathbf{H}}\mathbf{X})(\mathbf{Y} - \hat{\mathbf{H}}\mathbf{X})^H = \frac{1}{N}\mathbf{Y}(\mathbf{I}_N - \mathbf{P}_\mathbf{X})\mathbf{Y}^H,$$

with $\mathbf{P}_\mathbf{X} = \mathbf{X}^H(\mathbf{X}\mathbf{X}^H)^{-1}\mathbf{X}$ a rank-p projection onto the subspace $\langle \mathbf{X} \rangle$, spanned by the rows of the $p \times N$ matrix $\mathbf{X}$. The projection $\hat{\mathbf{H}}\mathbf{X} = \mathbf{Y}\mathbf{X}^H(\mathbf{X}\mathbf{X}^H)^{-1}\mathbf{X} = \mathbf{Y}\mathbf{P}_\mathbf{X}$ projects rows of $\mathbf{Y}$ onto the subspace $\langle \mathbf{X} \rangle$ spanned by the rows of $\mathbf{X}$.

The GLR is then

$$\lambda_1 = \Lambda_1^{1/N} = \frac{\det(\mathbf{Y}\mathbf{Y}^H)}{\det(\mathbf{Y}(\mathbf{I}_N - \mathbf{P}_\mathbf{X})\mathbf{Y}^H)}, \tag{5.19}$$

where

$$\Lambda_1 = \frac{\ell(\hat{\mathbf{H}}, \hat{\boldsymbol{\Sigma}}_1; \mathbf{Y})}{\ell(\hat{\boldsymbol{\Sigma}}_0; \mathbf{Y})}.$$

Connection to the Wilks Lambda. To establish the connection between (5.19) and the Wilks Lambda, write λ_1^{-1} as

$$\lambda_1^{-1} = \frac{\det(\mathbf{Y}\mathbf{P}_\mathbf{X}^\perp\mathbf{Y}^H)}{\det(\mathbf{Y}\mathbf{Y}^H)} = \frac{\det(\mathbf{Y}\mathbf{P}_\mathbf{X}^\perp\mathbf{Y}^H)}{\det(\mathbf{Y}\mathbf{P}_\mathbf{X}\mathbf{Y}^H + \mathbf{Y}\mathbf{P}_\mathbf{X}^\perp\mathbf{Y}^H)},$$

where $\mathbf{P}_\mathbf{X}^\perp = \mathbf{I}_N - \mathbf{P}_\mathbf{X}$ is a rank-$(N-p)$ projection matrix onto the subspace orthogonal to $\langle\mathbf{X}\rangle$. Define the $p \times N$ matrix $\mathbf{V}_1$ to be an arbitrary basis for $\langle\mathbf{X}\rangle$ and $\mathbf{V}_2$ to be a basis for the orthogonal subspace, in which case $\mathbf{P}_\mathbf{X} = \mathbf{V}_1^H\mathbf{V}_1$ and $\mathbf{P}_\mathbf{X}^\perp = \mathbf{V}_2^H\mathbf{V}_2$. Then,

$$\lambda_1^{-1} = \frac{\det(\mathbf{Y}_2\mathbf{Y}_2^H)}{\det(\mathbf{Y}_1\mathbf{Y}_1^H + \mathbf{Y}_2\mathbf{Y}_2^H)},$$

where $\mathbf{Y}_1 = \mathbf{Y}\mathbf{V}_1^H$ and $\mathbf{Y}_2 = \mathbf{Y}\mathbf{V}_2^H$, making λ_1^{-1} the same as the Wilks Lambda [382]. This statistic may be written in several equivalent forms:

$$\begin{aligned}\lambda_1^{-1} &= \frac{1}{\det(\mathbf{I}_L + (\mathbf{Y}_2\mathbf{Y}_2^H)^{-1/2}\mathbf{Y}_1\mathbf{Y}_1^H(\mathbf{Y}_2\mathbf{Y}_2^H)^{-1/2})}\\ &= \frac{1}{\det(\mathbf{I}_p + \mathbf{Y}_1^H(\mathbf{Y}_2\mathbf{Y}_2^H)^{-1}\mathbf{Y}_1)}\\ &= \det((\mathbf{I}_p + \mathbf{F})^{-1}) = \det(\mathbf{B}).\end{aligned}$$

The matrix $\mathbf{F} = \mathbf{Y}_1^H(\mathbf{Y}_2\mathbf{Y}_2^H)^{-1}\mathbf{Y}_1$ is distributed as a matrix F-statistic, and the matrix $\mathbf{B} = (\mathbf{I}_p + \mathbf{F})^{-1}$ is distributed as a matrix Beta statistic.

Connection to the Hotelling T^2. In the special case $p = 1$, with $\mathbf{X} = \mathbf{1}^T$, a constant symbol sequence, the resolutions $\mathbf{Y}_1$ and $\mathbf{Y}_2$ may be written $\mathbf{Y}_1 = \mathbf{Y}\mathbf{1}/\sqrt{N} = \sqrt{N}\bar{\mathbf{y}}$ and $\mathbf{Y}_2\mathbf{Y}_2^H = \mathbf{Y}\mathbf{Y}^H - N\bar{\mathbf{y}}\bar{\mathbf{y}}^H$, where the vector $\bar{\mathbf{y}} = \mathbf{Y}\mathbf{1}/N$ replaces $\mathbf{Y}$ by its row averages. Then, $\mathbf{F}$ is the scalar-valued statistic

$$F = \sqrt{N}\bar{\mathbf{y}}^H\left(\mathbf{Y}\mathbf{Y}^H - N\bar{\mathbf{y}}\bar{\mathbf{y}}^H\right)^{-1}\sqrt{N}\bar{\mathbf{y}}.$$

Thus, F and λ_1^{-1} are monotone increasing in Hotelling's T^2 statistic. For further emphasis, the GLR may be written

$$\begin{aligned}\lambda_1^{-1} &= \frac{\det(\mathbf{Y}_2\mathbf{Y}_2^H)}{\det(\mathbf{Y}\mathbf{Y}^H)} = \frac{\det(\mathbf{Y}\mathbf{Y}^H - N\bar{\mathbf{y}}\bar{\mathbf{y}}^H)}{\det(\mathbf{Y}\mathbf{Y}^H)} = \frac{(1 - N\bar{\mathbf{y}}^H(\mathbf{Y}\mathbf{Y})^{-1}\bar{\mathbf{y}})\det(\mathbf{Y}\mathbf{Y}^H)}{\det(\mathbf{Y}\mathbf{Y}^H)}\\ &= 1 - N\bar{\mathbf{y}}^H(\mathbf{Y}\mathbf{Y})^{-1}\bar{\mathbf{y}}.\end{aligned}$$

The monotone function $N(1 - \lambda_1^{-1}) = (\sum_{n=1}^N \mathbf{y}_n)^H(\mathbf{Y}\mathbf{Y}^H)^{-1}(\sum_{n=1}^N \mathbf{y}_n)$ is Hotelling's T^2 statistic.

So the multi-rank version of the Reed-Yu problem is a generalization of Hotelling's problem, where Hotelling's unknown $\mathbf{h}$ is replaced by a sequence of unknown $\mathbf{H}\mathbf{x}_n$, with the linear combining weights $\mathbf{x}_n$ known, but the common channel matrix $\mathbf{H}$ unknown.

Related Tests. The hypothesis test may also be addressed by using three other competing test statistics as alternatives to $\lambda_1^{-1} = \det(\mathbf{B})$. For the case $p = 1$, all four tests reduce to the use of the Hotelling T^2 test statistic, which is uniformly most powerful (UMP) invariant. For the case $p > 1$, however, no single test can be expected to dominate the others in terms of power. The three other tests use the Bartlett-Nanda-Pillai trace statistic, $\text{tr}(\mathbf{B})$, the Lawley-Hotelling trace statistic, $\text{tr}(\mathbf{F}) = \text{tr}(\mathbf{B}^{-1}(\mathbf{I}_p - \mathbf{B}))$, and Roy's maximum root statistic, $ev_1(\mathbf{B})$.

Invariances. The hypothesis testing problem and the GLR are invariant to the transformation group $G = \{g \mid g \cdot \mathbf{Y} = \mathbf{B}\mathbf{Y}\}$, for $\mathbf{B} \in GL(\mathbb{C}^L)$ any $L \times L$ nonsingular complex matrix. This transformation is more general than the transformation $\mathbf{V}_L$ in the case where $\mathbf{U}$ is known (c.f. (5.2)). The invariance to right unitary transformation is lost because the symbol matrix is known, rather than unknown as in previous examples. One particular nonsingular transformation is the noise covariance matrix, so the GLR is CFAR.

For $p = 1$, the test based on F is UMP invariant test among tests for $\mathcal{H}_0$ versus $\mathcal{H}_1$ at fixed false alarm probability. The uniformity is over all non-zero values of the $1 \times N$ symbol matrix $\mathbf{X}$. Starting with sufficient statistics $\mathbf{Y}_1$ and $\mathbf{Y}_2\mathbf{Y}_2^H$, it is easily shown that F is the maximal invariant. Since the noncentral F distribution is known to possess the monotone likelihood ratio property [215, p. 307, Problem 7.4], it is concluded that the GLRT that accepts the hypothesis $\mathcal{H}_1$ for large values of the GLR, λ_1, is UMP invariant [132, Theorem 3.2].

Null Distribution. Under $\mathcal{H}_0$, the inverse GLR $\lambda_1^{-1} = \det(\mathbf{B})$ is distributed as a central complex Beta, denoted $CBeta_p(N - L, L)$ [51, 259]. The corresponding stochastic representation is

$$\lambda_1^{-1} \sim \prod_{i=1}^{p} b_i, \tag{5.20}$$

where the $b_i \sim \text{Beta}(N - L - i + 1, L)$ are independent beta-distributed random variables. This stochastic representation affords a numerically stable stochastic simulation of the GLR, without simulation of the detector statistic itself, which may involve determinants of large matrices. Moreover, from this representation, the moment generating function (mgf) of the GLR may be derived. In [51], saddle point inversions of $Z = \log\det(\mathbf{B})$ of this mgf are treated. Much more on the distribution of the GLR may be found in [51], where comparisons are made with the large random matrix approximations of [163, 164].

Under the alternative, the distribution of λ_1^{-1} is the distribution of a noncentral Wilks Lambda. For the case $p = 1$, it is the distribution derived in [280, 332]. In [51, 259], the distribution is derived for arbitrary values of N, L, and p. Hiltunen, *et al.* [164] show that in the case where the number of receiving antennas L and the number of snapshots N are large and of the same order of magnitude, but the number of transmitting antennas p remains fixed, a standardized version of $\log(\lambda_1)$ converges to a normal distribution under $\mathcal{H}_0$ and $\mathcal{H}_1$. Then, pragmatic approximations for the distribution are derived for large p.

A Numerical Result. Under the null, the distribution of λ_1^{-1} may be obtained using the stochastic representation in (5.20), given by a product of independent beta-distributed random variables. Or, its mgf may be inverted with the method of saddle points or inverted exactly from its rational mgf. These methods may be used to predict the probability of false alarm with precision, without asymptotic approximation. These methods may be compared with the asymptotic approximations of [163, 164]. The purpose is not to call into question asymptotic results, which for some parameter choices and false alarm probabilities may be quite accurate. Rather, it is to show that asymptotic approximations are just that: approximations that are to be used with caution in non-asymptotic regimes.

In Fig. 5.2, false alarm probabilities are predicted from the stochastic representation in (5.20) (labeled Stoch. Rep.), from saddle point approximation of

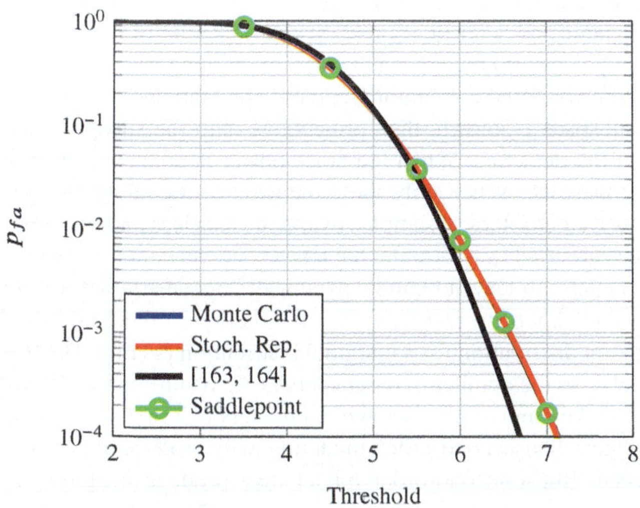

Fig. 5.2 Probability of false alarm (p_{fa}) on log-scale for a scenario with $p = 5$ sources, $L = 10$ antenna elements, and $N = 20$ snapshots

the null distribution of λ_1^{-1}, and from the large random matrix approximation in [163, 164], using their approximation (b). These are compared with the false alarm probabilities predicted from simulation of λ_1 itself (labeled Monte Carlo). These latter are invisible, as they lie exactly under the predictions from the stochastic representation (5.20) and from saddle point inversion of the moment generating function. The figure demonstrates that when the asymptotic approximation to the probability of false alarm is predicted to be 10^{-4}, the actual probability of false alarm is 10^{-3}. For some applications, such as radar and communications, this has consequences. For other applications, it may not. For much larger values of L and N, the asymptotic approximations become more accurate.

5.9 Chapter Notes

The common theme in this chapter is that the signal component of a measurement is assumed to lie in a known low-dimensional subspace, or in a subspace known only by its dimension. This modeling assumption generalizes the matched filter model, where the subspace dimension is one. In many branches of engineering and applied science, this kind of model arises from physical modeling of signal sources. But in other branches, the model arises as a tractable way to enforce smoothness or regularity on a component of a measurement that differs from additive noise or interference. This makes the subspace model applicable to a wide range of problems in signal processing and machine learning.

1. Many of the detectors in this chapter have been, and continue to be, applied to problems in beamforming, spectrum analysis, pulsed Doppler radar or sonar, synthetic aperture radar and sonar (SAR and SAS), passive localization of electromagnetic and acoustic sources, synchronization of digital communication systems, hyperspectral imaging, and machine learning. We have made no attempt to review the voluminous literature on these applications.
2. When a subspace is known, then projections onto the subspace are a common element of the detectors. When the noise power is unknown, then the detectors measure coherence. When only the dimension of the subspace is known, then detectors use eigenvalues of sample covariance matrices and, in some cases, these eigenvalues are used in a formula that has a coherence interpretation.
3. Which is better? To leave unknown parameters unconstrained (as in a first-order statistical models), or to assign a prior distribution to them and marginalize the resulting joint distribution for a marginal distribution (as in a second-order statistical model)? As the number of parameters in the resulting second-order model is smaller than the number of unknown parameters in a first-order model, intuition would suggest that second-order modeling will produce detectors with better performance. But a second-order model may produce a marginal distribution that does not accurately model measurements. This is the mismatch problem. In fact the question has no unambiguous answer. For a detailed empirical study

we refer the reader to [301], which shows that the answer to the question depends on what is known about the signal subspace. For a subspace known only by its dimension, this study suggests that second-order detectors outperform first-order detectors for a MVN prior on unknown parameters, and for all choices of the parameters (L, p, N, and SNR) considered in the study. Nevertheless, when the subspace is known, the conclusions are not clearcut. The performance of the first-order GLR is rather insensitive to the channel eigenvalue spread, measured by the spectral flatness, whereas the performance of the second-order GLR is not. The first-order GLR performs better than the second-order detector for spectrally flat channels, but this ordering of performance is reversed for non-flat channels. As for the comparison between the GLR and the LMPIT (when it exists) we point the reader to [272] and [271]. Both papers consider the case of a second-order model with unknown subspace of known dimension. The first considers white noise of unknown variance (c.f. Sect. 5.6.1), whereas the second considers the case of an unknown diagonal covariance matrix for the noise (c.f. Sect. 5.7). In both cases, the LMPIT outperforms the GLR for low and moderate SNRs.

Appendices

5.A Variations on Matched Subspace Detectors in a First-Order Model for a Signal in a Known Subspace

This appendix contains variations on the matched subspace detectors (MSDs) in a first-order model for a signal in a known subspace.

5.A.1 Scale-Invariant, Geometrically Averaged, Matched Subspace Detector

When the noise variance varies from time-to-time, or snapshot-to-snapshot, the measurement model is $\mathbf{Y} \sim CN_{L\times N}(\mathbf{UX}, \text{diag}(\sigma_1^2, \ldots, \sigma_N^2) \otimes \mathbf{I}_L)$, where $\mathbf{X} = \mathbf{0}$ under $\mathcal{H}_0$ and, under $\mathcal{H}_1$, $\mathbf{X} \in \mathbb{C}^{p\times N}$ is unknown; $\sigma_n^2 > 0$ are unknown parameters under both hypotheses. This means a sequence of noise variances must be estimated. Under the null, the ML estimates of the noise variances are

$$\hat{\sigma}_{n,0}^2 = \frac{1}{L}\mathbf{y}_n^H \mathbf{y}_n.$$

Under the alternative, the estimates of $\mathbf{X}$ and σ_n^2 are $\hat{\mathbf{X}} = \mathbf{U}^H\mathbf{Y}$, and

$$\hat{\sigma}_{n,1}^2 = \frac{1}{L}\mathbf{y}_n^H \mathbf{P}_{\mathbf{U}}^{\perp}\mathbf{y}_n.$$

Then, the GLR is

$$\lambda_1 = 1 - \Lambda_1^{-1/L} = 1 - \prod_{n=1}^{N} \frac{\mathbf{y}_n^H \mathbf{P}_{\mathbf{U}}^{\perp} \mathbf{y}_n}{\mathbf{y}_n^H \mathbf{y}_n} = 1 - \prod_{n=1}^{N} \left(1 - \frac{\mathbf{y}_n^H \mathbf{P}_{\mathbf{U}} \mathbf{y}_n}{\mathbf{y}_n^H \mathbf{y}_n}\right), \tag{5.21}$$

where

$$\Lambda_1 = \frac{\ell(\hat{\mathbf{X}}, \hat{\sigma}_{1,1}^2, \ldots, \hat{\sigma}_{N,1}^2; \mathbf{Y})}{\ell(\hat{\sigma}_{1,0}^2, \ldots, \hat{\sigma}_{N,0}^2; \mathbf{Y})}.$$

Then, λ_1 in (5.21) is a *bulk coherence statistic in a product of coherences*. That is, the time-dependent function within the product is a time-dependent coherence, and one minus this product is a coherence. It is equivalent to say that

$$1 - \frac{\mathbf{y}_n^H \mathbf{P}_{\mathbf{U}} \mathbf{y}_n}{\mathbf{y}_n^H \mathbf{y}_n}$$

is the sine-squared of the angle between the measurement $\mathbf{y}_n$ and the subspace $\langle \mathbf{U} \rangle$. Then, one minus a product of such sine-squared is itself a kind of bulk cosine-squared.

This detector has been derived independently in [1] and [258, 307], using different means. In [1], a Gamma distributed prior was assigned to the sequence of unknown variances, and a resulting Bessel function was approximated for large L. In [258, 307], the detector was derived as a GLR as outlined above.

Invariances. This detection problem is invariant to time-varying rotations in $\langle \mathbf{U} \rangle$, and non-zero scalings of the $\mathbf{y}_n, n = 1, \ldots, N$.

Null Distribution. In [258, 307], it is shown that the null distribution of (5.21) is the distribution of the random variable $1 - \prod_{n=1}^{N} b_n$, where the random variables b_n are independent random variables distributed as $b_n \sim \text{Beta}(L - p, p)$.

5.A.2 Refinement: Special Signal Sequences

If the sequence $\mathbf{x}_n$ is the constant sequence $\mathbf{x}_n = \mathbf{x}$, where $\mathbf{x}$ is a single unknown vector, then every function of the form $\sum_{n=1}^{N} \mathbf{y}_n^H (\mathbf{I}_L - \mathbf{P}_{\mathbf{U}}) \mathbf{y}_n$ is replaced by the function $\sum_{n=1}^{N} (\mathbf{y}_n - \mathbf{P}_{\mathbf{U}} \overline{\mathbf{y}})^H (\mathbf{y}_n - \mathbf{P}_{\mathbf{U}} \overline{\mathbf{y}})$. The statistic $\overline{\mathbf{y}} = (1/N) \sum_{n=1}^{N} \mathbf{y}_n$ is a *coherent average* of measurements, distributed under the null as $\overline{\mathbf{y}} \sim CN_L(\mathbf{0}, (1/N)\mathbf{I}_L)$. A corresponding likelihood ratio may then be written $2N\overline{\mathbf{y}}^H \mathbf{P}_{\mathbf{U}} \overline{\mathbf{y}} = 2(\sqrt{N}\overline{\mathbf{y}})^H \mathbf{P}_{\mathbf{U}} (\sqrt{N}\overline{\mathbf{y}})$, rather than $2 \sum_{n=1}^{N} \mathbf{y}_n \mathbf{P}_{\mathbf{U}} \mathbf{y}_n$. This amounts to replacing this *non-coherent* sum of N quadratic forms with one quadratic form in a *coherent* sum of N measurements. Instead of having a sum of N independent χ_{2p}^2 random variables, to produce a

χ^2_{2Np} random variable, there is one χ^2_{2p} random variable. This is the net, under the null hypothesis, of *matched filter combining* versus *diversity combining* of measurements.

If the sequence $\mathbf{x}_n$ is factored as $\mathbf{x}_n = \alpha_n \mathbf{f}_n$ and the sequence of signals $\mathbf{f}_n$ is known, then the subspace $\langle \mathbf{U} \rangle$ is replaced by a sequence of subspaces $\langle \mathbf{g}_n \rangle$, with $\mathbf{g}_n = \mathbf{U}\mathbf{f}_n$, and the function $2\sum_{n=1}^{N} \mathbf{y}_n(\mathbf{I}_L - \mathbf{P}_\mathbf{U})\mathbf{y}_n$, distributed as $\chi^2_{2N(L-p)}$, is replaced by $2\sum_{n=1}^{N} \mathbf{y}_n^H(\mathbf{I}_L - \mathbf{P}_{\mathbf{g}_n})\mathbf{y}_n$. This is a sum of N independent $\chi^2_{2(L-1)}$ random variables, so the distribution of the sum is $\chi^2_{2N(L-1)}$. If the signal sequence is constant at $\mathbf{f}_n = \mathbf{f}$, then there is a single subspace $\langle \mathbf{g} \rangle$, with $\mathbf{g} = \mathbf{U}\mathbf{f}$, and the sum is $2\sum_{n=1}^{N} \mathbf{y}_n^H(\mathbf{I}_L - \mathbf{P}_\mathbf{g})\mathbf{y}_n$, distributed as $\chi^2_{2N(L-1)}$. There is no change in the functional form of the detector statistics. Only the quadratic form $\sum_{n=1}^{N} \mathbf{y}_n^H(\mathbf{I}_L - \mathbf{P}_\mathbf{U})\mathbf{y}_n$ is replaced by $\sum_{n=1}^{N} \mathbf{y}_n^H(\mathbf{I}_L - \mathbf{P}_{\mathbf{g}_n})\mathbf{y}_n$ or $\sum_{n=1}^{N} \mathbf{y}_n^H(\mathbf{I}_L - \mathbf{P}_\mathbf{g})\mathbf{y}_n$.

5.A.3 Rapprochement

The matched subspace detector in (5.5), the scale-invariant matched subspace detector in (5.4), and the scale-invariant, geometrically averaged, matched subspace detector in (5.21) apply to these respective cases: σ^2 is known, σ^2 is unknown but constant for all n, and σ_n^2 is unknown and variable with n. These cases produce a family of three detectors for three assumptions about the noise $\mathbf{n}_n \sim CN_L(\mathbf{0}, \sigma_n^2 \mathbf{I}_L)$. The detectors may be written in a common format:

1. Matched subspace detector:

$$\operatorname{tr}(\mathbf{Y}^H \mathbf{P}_\mathbf{U} \mathbf{Y}) = \operatorname{tr}(\mathbf{Y}^H \mathbf{Y}) \left(1 - \frac{\operatorname{tr}(\mathbf{Y}^H \mathbf{P}_\mathbf{U}^\perp \mathbf{Y})}{\operatorname{tr}(\mathbf{Y}^H \mathbf{Y})}\right).$$

2. Scale-invariant matched subspace detector:

$$\frac{\operatorname{tr}(\mathbf{Y}^H \mathbf{P}_\mathbf{U} \mathbf{Y})}{\operatorname{tr}(\mathbf{Y}^H \mathbf{Y})} = 1 - \frac{\operatorname{tr}(\mathbf{Y}^H \mathbf{P}_\mathbf{U}^\perp \mathbf{Y})}{\operatorname{tr}(\mathbf{Y}^H \mathbf{Y})}.$$

3. Scale-invariant, geometrically averaged, matched subspace detector:

$$1 - \prod_{n=1}^{N} \frac{\mathbf{y}_n^H \mathbf{P}_\mathbf{U}^\perp \mathbf{y}_n}{\mathbf{y}_n^H \mathbf{y}_n} = 1 - \prod_{n=1}^{N} \left(1 - \frac{\mathbf{y}_n^H \mathbf{P}_\mathbf{U} \mathbf{y}_n}{\mathbf{y}_n^H \mathbf{y}_n}\right).$$

The first of these detector statistics accumulates the total power resolved into the subspace $\langle \mathbf{U} \rangle$. The second sums the cosine-squared of angles between normalized measurements and the subspace $\langle \mathbf{U} \rangle$. The third computes one minus the product of sine-squared of angles between measurements and the subspace $\langle \mathbf{U} \rangle$. Each of these detector statistics is coordinate-free, which is to say every statistic depends only on the known subspace $\langle \mathbf{U} \rangle$, and not on any particular basis for the subspace.

5.B Derivation of the Matched Subspace Detector in a Second-Order Model for a Signal in a Known Subspace

The original proof of [46] may be adapted to our problem as follows. The covariance matrix $\mathbf{R}_1 = \mathbf{U}\mathbf{R}_{\mathbf{xx}}\mathbf{U}^H + \sigma^2\mathbf{I}_L$ may be written

$$\begin{aligned}\mathbf{R}_1 &= \sigma^2(\mathbf{U}\mathbf{Q}_{\mathbf{xx}}\mathbf{U}^H + \mathbf{I}_L)\\ &= \left[\mathbf{U}\ \mathbf{U}^{\perp}\right] \operatorname{blkdiag}\left(\sigma^2(\mathbf{Q}_{\mathbf{xx}} + \mathbf{I}_p), \sigma^2\mathbf{I}_{L-p}\right) \left[\mathbf{U}\ \mathbf{U}^{\perp}\right]^H,\end{aligned}$$

where $\mathbf{U}^{\perp}$ is a unitary matrix orthogonal to $\mathbf{U}$, $\mathbf{U}^H\mathbf{U}^{\perp} = \mathbf{0}$, and $\mathbf{Q}_{\mathbf{xx}} = \mathbf{R}_{\mathbf{xx}}/\sigma^2$. Then,

$$\begin{aligned}\mathbf{R}_1^{-1} &= \left[\mathbf{U}\ \mathbf{U}^{\perp}\right] \operatorname{blkdiag}\left(\sigma^{-2}(\mathbf{Q}_{\mathbf{xx}} + \mathbf{I}_p)^{-1}, \sigma^{-2}\mathbf{I}_{L-p}\right) \left[\mathbf{U}\ \mathbf{U}^{\perp}\right]^H\\ &= \sigma^{-2}\mathbf{U}(\mathbf{Q}_{\mathbf{xx}} + \mathbf{I}_p)^{-1}\mathbf{U}^H + \sigma^{-2}\mathbf{U}^{\perp}(\mathbf{U}^{\perp})^H,\end{aligned}$$

and $\det(\mathbf{R}_1) = \sigma^{2L}\det(\mathbf{Q}_{\mathbf{xx}} + \mathbf{I}_p)$.

It is a few algebraic steps to write the likelihood function as

$$\ell(\mathbf{R}_1; \mathbf{Y}) = \frac{1}{\pi^{LN}\det(\mathbf{R}_1)^N} \operatorname{etr}\left\{-N\mathbf{R}_1^{-1}\mathbf{S}\right\} = \ell_1(\sigma^2; \mathbf{Y}) \cdot \ell_2(\mathbf{Q}_{\mathbf{xx}}, \sigma^2; \mathbf{Y}),$$

where

$$\ell_1(\sigma^2; \mathbf{Y}) = \frac{1}{\pi^{(L-p)N}\sigma^{2(L-p)N}} \operatorname{etr}\left\{-N\frac{(\mathbf{U}^{\perp})^H\mathbf{S}\mathbf{U}^{\perp}}{\sigma^2}\right\},$$

and

$$\ell_2(\mathbf{Q}_{\mathbf{xx}}, \sigma^2; \mathbf{Y}) = \frac{1}{\pi^{pN}\sigma^{2pN}\det(\mathbf{Q}_{\mathbf{xx}} + \mathbf{I}_p)^N} \operatorname{etr}\left\{-N(\mathbf{Q}_{\mathbf{xx}} + \mathbf{I}_p)^{-1}\frac{\mathbf{U}^H\mathbf{S}\mathbf{U}}{\sigma^2}\right\}.$$

That is, likelihood decomposes into the product of a Gaussian likelihood with covariance matrix $\sigma^2\mathbf{I}_{L-p}$ and sample covariance matrix $(\mathbf{U}^{\perp})^H\mathbf{S}\mathbf{U}^{\perp}$ and another Gaussian likelihood with covariance matrix $\mathbf{Q}_{\mathbf{xx}} + \mathbf{I}_p$ and sample covariance matrix $\mathbf{U}^H\mathbf{S}\mathbf{U}/\sigma^2$.

For fixed σ^2, the maximization of $\ell(\mathbf{R}_1; \mathbf{Y})$ simplifies to the maximization of $\ell_2(\mathbf{Q}_{\mathbf{xx}}, \sigma^2; \mathbf{Y})$, which is an application of the fundamental result of Anderson [14]. Denote the eigenvectors of the resolved covariance matrix $\mathbf{U}^H\mathbf{S}\mathbf{U}$ by $\mathbf{W}$ and its eigenvalues by $ev_l(\mathbf{U}^H\mathbf{S}\mathbf{U})$, with $ev_l(\mathbf{U}^H\mathbf{S}\mathbf{U}) \geq ev_{l+1}(\mathbf{U}^H\mathbf{S}\mathbf{U})$. Apply Anderson's result to find the ML estimate of $\mathbf{Q}_{\mathbf{xx}}$:

$$\hat{\mathbf{Q}}_{\mathbf{xx}} = \mathbf{W}\,\mathrm{diag}\left(\left(\frac{ev_1(\mathbf{U}^H\mathbf{S}\mathbf{U})}{\sigma^2} - 1\right)^+, \ldots, \left(\frac{ev_p(\mathbf{U}^H\mathbf{S}\mathbf{U})}{\sigma^2} - 1\right)^+\right)\mathbf{W}^H, \tag{5.22}$$

which depends on σ^2, yet to be estimated. For fixed $\mathbf{Q}_{\mathbf{xx}}$, the ML estimate of σ^2 is

$$\hat{\sigma}^2 = \frac{1}{L}\left[\mathrm{tr}\left((\mathbf{U}^\perp)^H\mathbf{S}\mathbf{U}^\perp\right) + \mathrm{tr}\left((\mathbf{Q}_{\mathbf{xx}} + \mathbf{I})^{-1}\,\mathbf{U}^H\mathbf{S}\mathbf{U}\right)\right], \tag{5.23}$$

which depends on $\mathbf{Q}_{\mathbf{xx}}$.

The estimates of (5.22) and (5.23) are coupled: the ML estimate of σ^2 depends of $\mathbf{Q}_{\mathbf{xx}}$ and the ML estimate of $\mathbf{Q}_{\mathbf{xx}}$ depends of σ^2. Substitute $\hat{\mathbf{Q}}_{\mathbf{xx}}$ to write (5.23) as

$$L\hat{\sigma}^2 = \mathrm{tr}\left((\mathbf{U}^\perp)^H\mathbf{S}\mathbf{U}^\perp\right) + \sum_{l=1}^{p} \min\left(ev_l(\mathbf{U}^H\mathbf{S}\mathbf{U}), \hat{\sigma}^2\right), \tag{5.24}$$

which is a non-linear equation with no closed-form solution. However, it is very easy to find a solution based on a simple algorithm. First, define the following functions:

$$f_1(x) = Lx - \mathrm{tr}\left((\mathbf{U}^\perp)^H\mathbf{S}\mathbf{U}^\perp\right), \qquad f_2(x) = \sum_{l=1}^{p} \min\left(ev_l(\mathbf{U}^H\mathbf{S}\mathbf{U}), x\right).$$

Equipped with these two function, (5.24) may be re-cast as $f_1(\hat{\sigma}^2) = f_2(\hat{\sigma}^2)$, which is the intersection between the affine function $f_1(\cdot)$ and the piecewise-linear function $f_2(\cdot)$. It can be shown that there exists just one intersection between $f_1(\cdot)$ and $f_2(\cdot)$, which translates into a unique solution for $\hat{\sigma}^2$. To obtain this solution, denote by q the integer for which

$$ev_{q+1}(\mathbf{U}^H\mathbf{S}\mathbf{U}) \leq \hat{\sigma}^2 < ev_q(\mathbf{U}^H\mathbf{S}\mathbf{U}), \tag{5.25}$$

where $ev_0(\mathbf{U}^H\mathbf{S}\mathbf{U})$ is set to ∞ and $ev_{p+1}(\mathbf{U}^H\mathbf{S}\mathbf{U})$ is set to 0. Therefore, (5.24) becomes

$$L\hat{\sigma}^2 = \mathrm{tr}\left((\mathbf{U}^\perp)^H\mathbf{S}\mathbf{U}^\perp\right) + \sum_{l=q+1}^{p} ev_l(\mathbf{U}^H\mathbf{S}\mathbf{U}) + q\hat{\sigma}^2,$$

or

$$\hat{\sigma}^2 = \frac{1}{L-q}\left[\mathrm{tr}\left((\mathbf{U}^\perp)^H\mathbf{S}\mathbf{U}^\perp\right) + \sum_{l=q+1}^{p} ev_l(\mathbf{U}^H\mathbf{S}\mathbf{U})\right]. \tag{5.26}$$

The parameter q is the unique natural number satisfying (5.25). The basic idea of the algorithm is thus to sweep q from 0 to p, compute (5.26) for each q and keep the one fulfilling (5.25). Once this estimate is available, it can be used in (5.22) to obtain $\hat{\mathbf{Q}}_{\mathbf{xx}}$. The determinant of $\mathbf{R}_1$ required for the GLR is

$$\begin{aligned}\det\left(\hat{\mathbf{R}}_1\right) &= \det\left[\hat{\sigma}^2\left(\mathbf{U}\hat{\mathbf{Q}}_{\mathbf{xx}}\mathbf{U}^H + \mathbf{I}\right)\right] \\ &= \frac{1}{(L-q)^{L-q}}\left[\operatorname{tr}\left((\mathbf{U}^{\perp})^H\mathbf{S}\mathbf{U}^{\perp}\right) + \sum_{l=q+1}^{p} ev_l(\mathbf{U}^H\mathbf{S}\mathbf{U})\right]^{L-q} \prod_{l=1}^{q} ev_l(\mathbf{U}^H\mathbf{S}\mathbf{U}) \\ &= \frac{1}{(L-q)^{L-q}}\left[\operatorname{tr}(\mathbf{S}) - \sum_{l=1}^{q} ev_l(\mathbf{U}^H\mathbf{S}\mathbf{U})\right]^{L-q} \prod_{l=1}^{q} ev_l(\mathbf{U}^H\mathbf{S}\mathbf{U}).\end{aligned}$$

5.C Variations on Matched Direction Detectors in a Second-Order Model for a Signal in a Subspace Known Only by its Dimension

There are two variations on the matched direction detector (MDD) in a second-order model for a signal in a subspace known only by its dimension (SE): (1) the dimension of the unknown subspace is unknown, but the noise variance is known and (2) the dimension and the noise variance are both unknown. The detection problem remains

$$\begin{aligned}\mathcal{H}_1 &: \mathbf{Y} \sim \mathcal{CN}_{L\times N}(\mathbf{0}, \mathbf{I}_N \otimes (\mathbf{R}_{\mathbf{zz}} + \sigma^2\mathbf{I}_L)), \\ \mathcal{H}_0 &: \mathbf{Y} \sim \mathcal{CN}_{L\times N}(\mathbf{0}, \mathbf{I}_N \otimes \sigma^2\mathbf{I}_L),\end{aligned}$$

where $\mathbf{R}_{\mathbf{zz}}$ is the unknown, rank p, covariance matrix for visits to an unknown subspace.

Known Noise Variance, but Unknown Subspace Dimension. The GLR is

$$\Lambda_2 = \frac{\ell(\hat{\mathbf{R}}_1; \mathbf{Y})}{\ell(\sigma^2; \mathbf{Y})},$$

where the ML estimate of $\hat{\mathbf{R}}_1$ is to be determined. Following a derivation similar to that in Sect. 5.6.1, the ML estimate of $\mathbf{R}_1$ is

$$\hat{\mathbf{R}}_1 = \hat{\mathbf{R}}_{\mathbf{zz}} + \sigma^2\mathbf{I}_L = \mathbf{W}\operatorname{diag}(ev_1(\mathbf{S}), \ldots, ev_{\hat{p}}(\mathbf{S}), \sigma^2, \ldots, \sigma^2)\mathbf{W}^H, \tag{5.27}$$

where $\mathbf{W}$ is a matrix that contains the eigenvectors of $\mathbf{S}$, $ev_l(\mathbf{S})$ are the corresponding eigenvalues, and $\hat{p}$ is the integer that satisfies $ev_{\hat{p}+1}(\mathbf{S}) \leq \sigma^2 < ev_{\hat{p}}(\mathbf{S})$. That is, the noise variance determines the identified rank, $\hat{p}$, of the low-rank covariance $\hat{\mathbf{R}}_{\mathbf{zz}}$. Using (5.27), a few lines of algebra show that

$$\det(\hat{\mathbf{R}}_1) = \sigma^{2(L-\hat{p})} \prod_{l=1}^{\hat{p}} ev_l(\mathbf{S}),$$

and

$$\operatorname{tr}\left(\hat{\mathbf{R}}_1^{-1}\mathbf{S}\right) = \hat{p} + \sum_{l=\hat{p}+1}^{L} \frac{ev_l(\mathbf{S})}{\sigma^2}.$$

Then,

$$\lambda_2 = \frac{1}{N}\log\Lambda_2 = \sum_{l=1}^{\hat{p}} \frac{ev_l(\mathbf{S})}{\sigma^2} - \sum_{l=1}^{\hat{p}} \log\left(\frac{ev_l(\mathbf{S})}{\sigma^2}\right) - \hat{p}.$$

This detection problem and the GLR are invariant to the transformation defined in (5.2), without the invariance to scale.

Unknown Noise Variance and Unknown Subspace Dimension. When the dimension p and noise variance σ^2 are both unknown, then the ML estimate of p is $\hat{p} = L$ and the ML estimate of $\mathbf{R}_1$ is simply the sample covariance matrix: $\hat{\mathbf{R}}_1 = \mathbf{S}$. Therefore, the GLR is the sphericity test (see Sect. 4.5). To return an estimate of p less than L requires the use of an order selection rule, such as the Akaike information criterion (AIC) or minimum description length (MDL) rules described in the paper of Wax and Kailath [374].

Adaptive Subspace Detectors

6

Adaptive subspace detectors address the problem of detecting subspace signals in noise of *unknown covariance*. Typically, it is assumed that a secondary channel of signal-free measurements may be used in an estimate of this unknown covariance. The question addressed in this chapter is how to fuse these signal-free measurements from a secondary channel with measurements from a primary channel, to determine whether the primary channel contains a subspace signal. Once this question is answered, then the *matched subspace detectors* (MSDs) of Chap. 5 become the *adaptive subspace detectors* (ASDs) of this chapter.

The theory of adaptive subspace detectors originated in the radar literature. But make no mistake: the theory is so general and well developed that it applies to any application where a signal may be modeled as a visit to a known subspace or an unknown subspace of known dimension. Our economical language is that such signals are *subspace signals*. They might be said to be smooth or constrained. Importantly, only the subspace or its dimension is known, not a particular basis for the subspace. Consequently, every detector statistic derived in this chapter is invariant to right unitary transformations of an arbitrary unitary basis for the subspace.

Adaptive subspace detectors have found widespread application in radar, sonar, digital communication, medical imaging, hyperspectral imaging, vibration analysis, remote sensing, and many problems of applied science.

6.1 Introduction

As with so much of adaptive detection theory, the story begins with the late greats, Ed Kelly and Irving Reed. In the early days of the theory, attention was paid largely to the problem of detecting what we would now call dimension-one signals in Gaussian noise of unknown covariance. But as the story has evolved, it has become a story in the detection of multidimensional signals in noise of unknown covariance, when there is secondary training data that may be used to estimate this unknown

D. Ramírez et al., *Coherence*, https://doi.org/10.1007/978-3-031-13331-2_6

covariance matrix. The pioneering papers were [69,70,193,194]. The paper by Kelly and Forsythe [194] laid the groundwork for much of the work that was to follow.

The innovation of [193] was to introduce a *homogeneous* secondary channel of signal-free measurements whose unknown covariance matrix was equal to the unknown covariance matrix of primary measurements. Likelihood theory was then used to derive what is now called the Kelly detector. In [194], adaptive subspace detection was formulated in terms of the generalized multivariate analysis of variance for complex variables. These papers were followed by the adaptive detectors of [70,289]. Then, in 1991 and 1994, a scale-invariant MSD was introduced [302,303], and in 1995 and 1996, a scale-invariant ASD was introduced [80, 240, 305]. The corresponding adaptive detector statistic is now commonly called ACE, as it is an *adaptive coherence estimator*. In [80], this detector was derived as an asymptotic approximation to the generalized likelihood ratio (GLR) for detecting a coherent signal in compound Gaussian noise, and in [240, 305], it was derived as an estimate and plug (EP) version of the scale-invariant MSD [303].

In [204], the authors showed that ACE was a likelihood ratio statistic for a *non-homogeneous* secondary channel of measurements whose unknown covariance matrix was an unknown scaling of the unknown covariance matrix of the primary channel. ACE was extended to multidimensional subspace signals in [205]. Then, in [206], ACE was shown to be a uniformly most powerful invariant (UMPI) detector. In subsequent years, there has been a flood of important papers on adaptive subspace detectors. Among published references on adaptive detection, we cite here [20, 21, 41, 42, 50, 82, 185] and references therein.

All of this work is addressed to adaptive detection in what might be called a first-order statistical model for measurements. That is, the measurements in the primary channel may contain a subspace signal plus Gaussian noise of unknown covariance, but no prior distribution is assigned to the location of the signal in the subspace. These results were first derived for the case where there were $N_S > L$ secondary snapshots composing the $L \times N_S$ secondary channel of measurements and just one primary snapshot composing the $L \times 1$ primary channel of measurements. The dimension of the subspace was one. Then, in [21, 82], the authors extended ASDs to multiple measurements in the primary channel and compared them to EP adaptations. The first attempt to replace this first-order model by a second-order model was made in [282], where the authors used a Gaussian model for the signal, and a result of [46], to derive the *second-order matched subspace detector* residing in the SW of Table 5.1 in Chap. 5. An EP adaptation from secondary measurements was proposed. In [35], the GLR for a second-order statistical model of a dimension-one signal was derived.

The full development of ASDs for first- and second-order models of multidimensional signals, and multiple measurements in the primary channel, is contained in [255] and [6].

Organization of the Chapter. This chapter begins with *estimate and plug* (EP) adaptations of the MSD statistics on the NW, NE, SW, and SE points of the compass in Chap. 5. The noise covariance matrix that was assumed known in

Chap. 5 is replaced by the sample covariance matrix of signal-free measurements in a secondary channel. The resulting detector statistics are adaptive, but they are not generalized likelihood ratio (GLR) statistics.

The rest of the chapter is devoted to the derivation of GLRs for ASDs in the NW only, beginning with the Kelly and ACE detectors and continuing to their generalizations for multidimensional subspace signals and multiple measurements in the primary channel. These generalizations were first reported in [82] and [21]. The GLRs in the NE, SW, and SE are now known [255], but they are not included in this chapter. The reader is directed to [255] for a comprehensive account of adaptive subspace detectors in the first- and second-order statistical models in the NW, NE, SW, and SE, in homogeneous and partially homogeneous problems.

As in Chap. 5, a first-order statistical model for a multidimensional subspace signal assumes the signal modulates the mean value of a multivariate normal distribution. In a second-order statistical model, the signal modulates the covariance matrix of the multivariate normal model. In each of these models, the signal may visit a known subspace, or it may visit a subspace known only by its dimension. So there are four variations on the signal model of the primary data. The secondary measurements introduced in this chapter may be *homogeneous* with the primary data, which is to say they are scaled as the primary data is scaled, or they may be *partially homogeneous*, which is to say the primary and secondary data are unequally scaled by an unknown positive factor.

6.2 Adaptive Detection Problems

The problem is to detect a subspace signal in MVN noise of unknown covariance matrix when there is a secondary channel of signal-free measurements to be fused with measurements in a primary channel that may or may not carry a signal. The notation throughout this chapter will be that N_P primary measurements in an L-element array of sensors are organized into an $L \times N_P$ matrix $\mathbf{Y}_P = [\mathbf{y}_1 \cdots \mathbf{y}_{N_P}]$ and N_S secondary measurements in this or another L-element array are organized into an $L \times N_S$ matrix $\mathbf{Y}_S = [\mathbf{y}_{N_P+1} \cdots \mathbf{y}_{N_P+N_S}]$. The total number of measurements is $N = N_P + N_S$. The measurements in these two matrices are independent, but they share a common noise covariance matrix, a point to be clarified in due course.

6.2.1 Signal Models

There are four variations on the multidimensional signal model, corresponding to the points NW, NE, SW, and SE on the compass of Chap. 5:

NW: The signal visits a known subspace, unconstrained by a prior distribution. This is a first-order statistical model, as the signal appears as a low-rank component in the mean of a multivariate Gaussian distribution for the measurements. When there is only one measurement in the primary channel, then the GLRs are those of [80, 193, 204–206, 240, 305]. For multiple measurements, the results are those of [21, 82].

SW: The signal visits a known subspace, constrained by a Gaussian prior distribution. This is a second-order statistical model, as the signal model appears as a low-rank component in the covariance matrix of a multivariate Gaussian distribution for the measurements. EP statistics have been derived in [185, 282]. The GLR results are those of [35] in the rank-one case and [255] in the multi-rank case.

NE: The signal visits an unknown subspace of known dimension, unconstrained by a prior distribution. This a first-order statistical model. The results are those of [255].

SE: The signal visits an unknown subspace of known dimension, constrained by a Gaussian prior distribution; this is a second-order statistical model. The estimated low-rank covariance matrix for the subspace signal may be called an adaptive factor model. The results are those of [255].

These signal models are illustrated in Fig. 6.1, where panel (a) accounts for the NW and NE and panel (b) accounts for the SW and SE.

6.2.2 Hypothesis Tests

In the NW and NE where the measurement model is a first-order MVN model, the adaptive detection problem is the following test of hypothesis $\mathcal{H}_0$ vs. alternative $\mathcal{H}_1$:

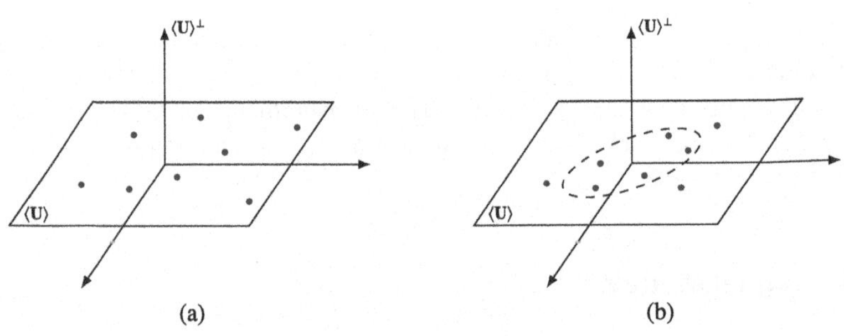

Fig. 6.1 Subspace signal models. In (**a**), the signal $\mathbf{x}_n$, unconstrained by a prior distribution, visits a subspace $\langle \mathbf{U} \rangle$ that is known or known only by its dimension. In (**b**), the signal $\mathbf{x}_n$, constrained by a prior MVN distribution, visits a subspace $\langle \mathbf{U} \rangle$ that is known or known only by its dimension

$$\mathcal{H}_0 : \begin{cases} \mathbf{Y}_P \sim CN_{L\times N_P}(\mathbf{0}, \mathbf{I}_{N_P} \otimes \sigma^2\boldsymbol{\Sigma}), \\ \mathbf{Y}_S \sim CN_{L\times N_S}(\mathbf{0}, \mathbf{I}_{N_S} \otimes \boldsymbol{\Sigma}), \end{cases}$$
$$\mathcal{H}_1 : \begin{cases} \mathbf{Y}_P \sim CN_{L\times N_P}(\mathbf{UX}, \mathbf{I}_{N_P} \otimes \sigma^2\boldsymbol{\Sigma}), \\ \mathbf{Y}_S \sim CN_{L\times N_S}(\mathbf{0}, \mathbf{I}_{N_S} \otimes \boldsymbol{\Sigma}), \end{cases}$$

where $\mathbf{U} \in \mathbb{C}^{L\times p}$ is either a known arbitrary basis for a known subspace $\langle\mathbf{U}\rangle$ or an unknown basis with known rank $p \leq L$; $\mathbf{X} = [\mathbf{x}_1 \ \cdots \ \mathbf{x}_{N_P}]$ is the $p \times N_P$ matrix of unknown signal coordinates, $\boldsymbol{\Sigma}$ is an $L \times L$ unknown positive definite covariance matrix, and $\sigma^2 > 0$ is a scale parameter that is known in the homogeneous case and unknown in the partially homogeneous case. The notation $CN_{L\times N_S}(\mathbf{0}, \mathbf{I}_{N_S} \otimes \boldsymbol{\Sigma})$ denotes the complex normal, or Gaussian, distribution of a matrix, which when vectorized by columns would be an LN_S-variate normal random vector with mean $\mathbf{0}$ and block-diagonal covariance matrix $\mathbf{I}_{N_S} \otimes \boldsymbol{\Sigma}$. This is a matrix of common $L \times L$ blocks $\boldsymbol{\Sigma}$ on each of its N_S diagonals.

In the following, we suppose that $N_S \geq L$ and $N_P \geq 1$ and, without loss of generality, it is assumed that $\mathbf{U}$ is a slice of a unitary matrix. This is the model assumed in the generalizations of the Kelly and ACE detectors in [21, 82].

In the SW and SE, where the measurement model is a second-order MVN model, a prior Gaussian distribution is assumed for the matrix $\mathbf{X}$, namely, $\mathbf{X} \sim CN_{p\times N_P}(\mathbf{0}, \mathbf{I}_{N_P} \otimes \mathbf{R}_{\mathbf{xx}})$. The $p \times p$ covariance matrix $\mathbf{R}_{\mathbf{xx}}$ models correlations $E[\mathbf{x}_i\mathbf{x}_k^H] = \mathbf{R}_{\mathbf{xx}}\delta[i-k]$. The joint distribution of $\mathbf{Y}_P$ and $\mathbf{X}$ is marginalized for $\mathbf{Y}_P$, with the result that, under the alternative $\mathcal{H}_1$, $\mathbf{Y}_P \sim CN_{L\times N_P}(\mathbf{0}, \mathbf{I}_{N_P} \otimes (\mathbf{U}\mathbf{R}_{\mathbf{xx}}\mathbf{U}^H + \boldsymbol{\Sigma}))$. The adaptive detection problem is the following test of hypothesis $\mathcal{H}_0$ vs. alternative $\mathcal{H}_1$:

$$\mathcal{H}_0 : \begin{cases} \mathbf{Y}_P \sim CN_{L\times N_P}(\mathbf{0}, \mathbf{I}_{N_P} \otimes \sigma^2\boldsymbol{\Sigma}), \\ \mathbf{Y}_S \sim CN_{L\times N_S}(\mathbf{0}, \mathbf{I}_{N_S} \otimes \boldsymbol{\Sigma}), \end{cases}$$
$$\mathcal{H}_1 : \begin{cases} \mathbf{Y}_P \sim CN_{L\times N_P}(\mathbf{0}, \mathbf{I}_{N_P} \otimes (\mathbf{U}\mathbf{R}_{\mathbf{xx}}\mathbf{U}^H + \sigma^2\boldsymbol{\Sigma})), \\ \mathbf{Y}_S \sim CN_{L\times N_S}(\mathbf{0}, \mathbf{I}_{N_S} \otimes \boldsymbol{\Sigma}), \end{cases}$$

where the $L \times p$ matrix $\mathbf{U}$ is either an arbitrary unitary basis for $\langle\mathbf{U}\rangle$ or an unknown matrix with known rank $p \leq L$, $\boldsymbol{\Sigma}$ is an unknown positive definite covariance matrix, and $\sigma^2 > 0$ is known in homogeneous problems and unknown in partially homogeneous problems. This is the model assumed in the generalizations of [255] for adaptive subspace detection in a second-order statistical model.

In the following section, estimate and plug (EP) solutions are given for fusing measurements from a secondary channel with measurements in a primary channel. These are solutions for all four points on the compass, NW, NE, SW, and SE. They are *not* GLR solutions.

6.3 Estimate and Plug (EP) Solutions for Adaptive Subspace Detection

The results from Chap. 5 may be re-worked for the case where the noise covariance matrix $\sigma^2\mathbf{I}_L$ is replaced by $\sigma^2\boldsymbol{\Sigma}$, with $\boldsymbol{\Sigma}$ a known positive definite covariance matrix and σ^2 an unknown, positive scale constant. In a first-order statistical model, where the subspace $\langle\mathbf{U}\rangle$ is known, the measurement matrix $\mathbf{Y}$, now denoted $\mathbf{Y}_P$, is replaced by its whitened version $\boldsymbol{\Sigma}^{-1/2}\mathbf{Y}_P \sim CN_{L\times N_P}(\boldsymbol{\Sigma}^{-1/2}\mathbf{UX}, \mathbf{I}_{N_P} \otimes \sigma^2\mathbf{I}_L)$. The subspace $\langle\mathbf{U}\rangle$ is replaced by the subspace $\langle\boldsymbol{\Sigma}^{-1/2}\mathbf{U}\rangle$, and the GLR is determined as in Chap. 5. When the subspace $\langle\mathbf{U}\rangle$ is known only by its dimension, this dimension is assumed unchanged by the whitening. Similarly, in a second-order statistical model, the measurement matrix $\mathbf{Y}_P$ is replaced by its whitened version $\boldsymbol{\Sigma}^{-1/2}\mathbf{Y}_P \sim CN_{L\times N_P}(\mathbf{0}, \mathbf{I}_{N_P} \otimes (\boldsymbol{\Sigma}^{-1/2}\mathbf{U}\mathbf{R}_{\mathbf{xx}}\mathbf{U}^H\boldsymbol{\Sigma}^{-1/2} + \sigma^2\mathbf{I}_L))$. When the subspace $\langle\mathbf{U}\rangle$ is known only by its dimension, the matrix $\boldsymbol{\Sigma}^{-1/2}\mathbf{U}\mathbf{R}_{\mathbf{xx}}\mathbf{U}^H\boldsymbol{\Sigma}^{-1/2}$ is an unknown and unconstrained $L \times L$ matrix of known rank p. This makes the results of Chap. 5 more general than they might appear at first reading.

But what if the noise covariance matrix $\boldsymbol{\Sigma}$ is unknown? One alternative is to estimate the unknown covariance matrix $\boldsymbol{\Sigma}$ as $\hat{\boldsymbol{\Sigma}} = \mathbf{S}_S = \mathbf{Y}_S\mathbf{Y}_S^H/N_S$ and use this estimate in place of $\boldsymbol{\Sigma}$ in the whitenings $\boldsymbol{\Sigma}^{-1/2}\mathbf{Y}_P$ and $\boldsymbol{\Sigma}^{-1/2}\mathbf{U}$. This gambit returns EP versions of the various detectors of Chap. 5. These EP adaptations are not generally GLR statistics, although in a few special and important cases [204], they are. A comprehensive comparison of EP and GLR detectors is carried out in [6].

This raises the question "what are the GLR statistics for the case where the measurements are $\mathbf{Y}_P$ and $\mathbf{Y}_S$, with $\mathbf{Y}_S \sim CN_{L\times N_S}(\mathbf{0}, \mathbf{I}_{N_S} \otimes \boldsymbol{\Sigma})$ and $\mathbf{Y}_P$ distributed according to one of the four possible subspace signal models at the NW, NE, SW, or SE points of the compass in Chap. 5?" When there is only a single snapshot ($N_P = 1$) in $\mathbf{Y}_P$, and the subspace signal model is the first-order statistical model of the NW, the statistics are the Kelly and ACE statistics. In Sect. 6.4 of this chapter, the GLRs for the NW are derived, following [82] and [21]. The derivations for the NE, SW, and SE are recently reported in [255], and not described in this chapter.

6.3.1 Detectors in a First-Order Model for a Signal in a Known Subspace

In the notation of this chapter, the hypothesis test in the NW corner of Table 5.1 in Chap. 5 (cf. (5.3)) is

$$\begin{aligned}\mathcal{H}_1 &: \mathbf{Y}_P \sim CN_{L\times N_P}(\mathbf{UX}, \mathbf{I}_{N_P} \otimes \sigma^2\mathbf{I}_L),\\ \mathcal{H}_0 &: \mathbf{Y}_P \sim CN_L(\mathbf{0}, \mathbf{I}_{N_P} \otimes \sigma^2\mathbf{I}_L),\end{aligned}$$

with $\mathbf{X}$ and σ^2 unknown parameters of the distribution for $\mathbf{Y}_P$ under $\mathcal{H}_1$ and σ^2 an unknown parameter of the distribution under $\mathcal{H}_0$. The subspace $\langle\mathbf{U}\rangle$ is known by its arbitrary basis $\mathbf{U}$, and the question is whether the mean of $\mathbf{Y}_P$ carries visits to this subspace.

The scale-invariant matched subspace detector of Chap. 5 may be written as

$$\lambda_1 = 1 - \Lambda_1^{-N_P L} = \frac{\operatorname{tr}\left(\mathbf{Y}_P^H \mathbf{P}_{\mathbf{U}} \mathbf{Y}_P\right)}{\operatorname{tr}\left(\mathbf{Y}_P^H \mathbf{Y}_P\right)},$$

which is a coherence statistic that measures the fraction of energy that lies in the subspace $\langle \mathbf{U} \rangle$.

Suppose the noise covariance model $\mathbf{I}_{N_P} \otimes \sigma^2 \mathbf{I}_L$ is replaced by the model $\mathbf{I}_{N_P} \otimes \sigma^2 \boldsymbol{\Sigma}$, with $\boldsymbol{\Sigma}$ a known $L \times L$ positive definite covariance matrix. Then the measurement $\mathbf{Y}_P$ may be whitened as $\boldsymbol{\Sigma}^{-1/2}\mathbf{Y}_P$, which is then distributed as $\mathbf{Y}_P \sim CN_{L\times N_P}(\boldsymbol{\Sigma}^{-1/2}\mathbf{UX}, \mathbf{I}_{N_P} \otimes \sigma^2 \mathbf{I}_L)$. The hypothesis testing problem may be phrased as a hypothesis testing problem on $\boldsymbol{\Sigma}^{-1/2}\mathbf{Y}_P$, and the GLR remains essentially unchanged, with $\boldsymbol{\Sigma}^{-1/2}\mathbf{Y}_P$ replacing $\mathbf{Y}_P$ and $\boldsymbol{\Sigma}^{-1/2}\mathbf{U}$ replacing $\mathbf{U}$:

$$\lambda_1(\boldsymbol{\Sigma}) = \frac{\operatorname{tr}\left((\boldsymbol{\Sigma}^{-1/2}\mathbf{Y}_P)^H \mathbf{P}_{\boldsymbol{\Sigma}^{-1/2}\mathbf{U}} (\boldsymbol{\Sigma}^{-1/2}\mathbf{Y}_P)\right)}{\operatorname{tr}\left((\boldsymbol{\Sigma}^{-1/2}\mathbf{Y}_P)^H (\boldsymbol{\Sigma}^{-1/2}\mathbf{Y})\right)}. \tag{6.1}$$

If there is a secondary channel of measurements, distributed as $\mathbf{Y}_S \sim CN_{L\times N_S}(\mathbf{0}, \mathbf{I}_{N_S} \otimes \boldsymbol{\Sigma})$, then with no assumed parametric model for $\boldsymbol{\Sigma}$, its ML estimate from the secondary channel only is $\hat{\boldsymbol{\Sigma}} = \mathbf{S}_S = \mathbf{Y}_S \mathbf{Y}_S^H / N_S$. This estimator may be inserted into (6.1) to obtain the EP adaptation of the scale-invariant MSD in a first-order signal model,

$$\begin{aligned}
\lambda_1(\mathbf{S}_S) &= \frac{\operatorname{tr}\left((\mathbf{S}_S^{-1/2}\mathbf{Y}_P)^H \mathbf{P}_{\mathbf{G}} (\mathbf{S}_S^{-1/2}\mathbf{Y}_P)\right)}{\operatorname{tr}\left((\mathbf{S}_S^{-1/2}\mathbf{Y}_P)^H (\mathbf{S}_S^{-1/2}\mathbf{Y}_P)\right)} \\
&= \frac{\operatorname{tr}(\mathbf{P}_{\mathbf{G}} \mathbf{T}_P)}{\operatorname{tr}(\mathbf{T}_P)},
\end{aligned}$$

where $\mathbf{G} = \mathbf{S}_S^{-1/2}\mathbf{U}$, $\mathbf{P}_{\mathbf{G}} = \mathbf{G}(\mathbf{G}^H\mathbf{G})^{-1}\mathbf{G}^H$, and

$$\mathbf{T}_P = \frac{1}{N_P}\mathbf{S}_S^{-1/2}\mathbf{Y}_P\mathbf{Y}_P^H\mathbf{S}_S^{-1/2} = \mathbf{S}_S^{-1/2}\mathbf{S}_P\mathbf{S}_S^{-1/2}.$$

The statistic $\mathbf{T}_P$ is a compression of the measurements that will figure prominently throughout this chapter.

This EP statistic is not a GLR because the estimate of $\boldsymbol{\Sigma}$ uses only secondary measurements, and not a fusing of secondary and primary measurements.

6.3.2 Detectors in a Second-Order Model for a Signal in a Known Subspace

In the notation of this chapter, the hypothesis test in the SW corner of the compass in Chap. 5 is

$$\begin{aligned}\mathcal{H}_1 &: \mathbf{Y}_P \sim CN_{L\times N_P}(\mathbf{0}, \mathbf{I}_{N_P} \otimes (\mathbf{U}\mathbf{R}_{\mathbf{xx}}\mathbf{U}^H + \sigma^2\mathbf{I}_L)),\\ \mathcal{H}_0 &: \mathbf{Y}_P \sim CN_{L\times N_P}(\mathbf{0}, \mathbf{I}_{N_P} \otimes \sigma^2\mathbf{I}_L),\end{aligned}$$

with the $p \times p$ matrix $\mathbf{R}_{\mathbf{xx}} \succ \mathbf{0}$ and σ^2 unknown parameters under $\mathcal{H}_1$ and σ^2 unknown under $\mathcal{H}_0$. The GLR is

$$\lambda_2 = \Lambda_2^{1/N_P} = \frac{\left(\frac{1}{L}\operatorname{tr}(\mathbf{S}_P)\right)^{1/L}}{\left[\frac{1}{L-q}\left(\operatorname{tr}(\mathbf{S}_P) - \sum_{l=1}^{q} ev_l(\mathbf{U}^H\mathbf{S}_P\mathbf{U})\right)\right]^{L-q} \prod_{l=1}^{q} ev_l(\mathbf{U}^H\mathbf{S}_P\mathbf{U})},$$

where q is the integer satisfying

$$ev_{q+1}(\mathbf{U}^H\mathbf{S}_P\mathbf{U}) < \frac{1}{L-q}\left(\operatorname{tr}(\mathbf{S}_P) - \sum_{l=1}^{q} ev_l(\mathbf{U}^H\mathbf{S}_P\mathbf{U})\right) < ev_q(\mathbf{U}^H\mathbf{S}_P\mathbf{U}). \tag{6.2}$$

The term sandwiched between $ev_{q+1}(\mathbf{U}^H\mathbf{S}_P\mathbf{U})$ and $ev_q(\mathbf{U}^H\mathbf{S}_P\mathbf{U})$ is the ML estimate of the noise variance under $\mathcal{H}_1$. This result was derived in [282].

If now the noise covariance matrix is $\mathbf{I}_{N_P} \otimes \sigma^2\boldsymbol{\Sigma}$, the GLR remains essentially unchanged, with $\boldsymbol{\Sigma}^{-1/2}\mathbf{S}_P\boldsymbol{\Sigma}^{-1/2}$ replacing $\mathbf{S}_P$ and $(\boldsymbol{\Sigma}^{-1/2}\mathbf{U})^H\boldsymbol{\Sigma}^{-1/2}\mathbf{S}_P\boldsymbol{\Sigma}^{-1/2}(\boldsymbol{\Sigma}^{-1/2}\mathbf{U})$ replacing $\mathbf{U}^H\mathbf{S}_P\mathbf{U}$. If there is a signal-free secondary channel of measurements, an EP adaptation of the scale-invariant MSD in a second-order signal model is obtained by replacing $\boldsymbol{\Sigma}$ by its ML estimator $\hat{\boldsymbol{\Sigma}} = \mathbf{S}_S$.

6.3.3 Detectors in a First-Order Model for a Signal in a Subspace Known Only by its Dimension

In the notation of this chapter, the hypothesis test in the NE corner of Chap. 5 is

$$\begin{aligned}\mathcal{H}_1 &: \mathbf{Y}_P \sim CN_{L\times N_P}(\mathbf{U}\mathbf{X}, \mathbf{I}_{N_P} \otimes \sigma^2\mathbf{I}_L),\\ \mathcal{H}_0 &: \mathbf{Y}_P \sim CN_{L\times N_P}(\mathbf{0}, \mathbf{I}_{N_P} \otimes \sigma^2\mathbf{I}_L),\end{aligned}$$

with $\mathbf{U}\mathbf{X}$ and σ^2 unknown parameters of the distribution for $\mathbf{Y}_P$ under $\mathcal{H}_1$ and σ^2 an unknown parameter of the distribution under $\mathcal{H}_0$. With the subspace $\langle\mathbf{U}\rangle$ unknown,

$\mathbf{UX}$ is now a factorization of an unknown $L \times N_P$ matrix of rank p. The question is whether the measurement $\mathbf{Y}_P$ carries such an unknown matrix of rank p in its mean. As derived in Chap. 5, the GLR is (cf. (5.11))

$$\lambda_1 = 1 - \Lambda_1^{1/N_P L} = \frac{\sum_{l=1}^{p} ev_l(\mathbf{S}_P)}{\sum_{l=1}^{L} ev_l(\mathbf{S}_P)},$$

where the $ev_l(\mathbf{S}_P)$ are the positive eigenvalues of the sample covariance matrix $\mathbf{S}_P = \mathbf{Y}_P\mathbf{Y}_P^H/N_P$.

Repeating the reasoning of the preceding sections, if the noise covariance model $\mathbf{I}_{N_P} \otimes \sigma^2\mathbf{I}_L$ is replaced by the model $\mathbf{I}_{N_P} \otimes \sigma^2\boldsymbol{\Sigma}$, then the measurement $\mathbf{Y}_P$ may be whitened as $\boldsymbol{\Sigma}^{-1/2}\mathbf{Y}_P$, and the GLR remains essentially unchanged, with $\boldsymbol{\Sigma}^{-1/2}\mathbf{S}_P\boldsymbol{\Sigma}^{-1/2}$ replacing $\mathbf{S}_P$:

$$\lambda_1(\boldsymbol{\Sigma}) = \frac{\sum_{l=1}^{p} ev_l(\boldsymbol{\Sigma}^{-1/2}\mathbf{S}_P\boldsymbol{\Sigma}^{-1/2})}{\sum_{l=1}^{L} ev_l(\boldsymbol{\Sigma}^{-1/2}\mathbf{S}_P\boldsymbol{\Sigma}^{-1/2})}. \tag{6.3}$$

If there is a signal-free secondary channel of measurements, distributed as $\mathbf{Y}_S \sim CN_{L\times N_S}(\mathbf{0}, \mathbf{I}_{N_S} \otimes \boldsymbol{\Sigma})$, then the estimator $\hat{\boldsymbol{\Sigma}} = \mathbf{S}_S$ may be inserted into (6.3) to obtain an EP version of the scale-invariant adaptive subspace detector for a first-order model of a signal in an unknown subspace of known dimension

$$\lambda_1(\mathbf{S}_S) = \frac{\sum_{l=1}^{p} ev_l(\mathbf{T}_P)}{\sum_{l=1}^{L} ev_l(\mathbf{T}_P)},$$

where $\mathbf{T}_P = \mathbf{S}_S^{-1/2}\mathbf{S}_P\mathbf{S}_S^{-1/2}$ is a compression of the measurements into a secondarily whitened sample covariance matrix.

6.3.4 Detectors in a Second-Order Model for a Signal in a Subspace Known Only by its Dimension

In the notation of this chapter, the hypothesis test in the SE corner of Table 5.1 in Chap. 5 is

$$\begin{aligned}
\mathcal{H}_1 &: \mathbf{Y}_P \sim CN_{L\times N_P}(\mathbf{0}, \mathbf{I}_{N_P} \otimes (\mathbf{R}_{\mathbf{zz}} + \sigma^2\mathbf{I}_L)),\\
\mathcal{H}_0 &: \mathbf{Y}_P \sim CN_{L\times N_P}(\mathbf{0}, \mathbf{I}_{N_P} \otimes \sigma^2\mathbf{I}_L),
\end{aligned}$$

where the rank-p covariance $\mathbf{R}_{\mathbf{zz}}$ and scale σ^2 are unknown. The scale-invariant matched direction detector derived in Chap. 5 is

$$\lambda_2 = \Lambda_2^{1/N_P} = \frac{\left[\frac{1}{L}\sum_{l=1}^{L} ev_l(\mathbf{S}_P)\right]^L}{\left[\frac{1}{L-p}\sum_{l=p+1}^{L} ev_l(\mathbf{S}_P)\right]^{L-p} \prod_{l=1}^{p} ev_l(\mathbf{S}_P)}.$$

When the noise covariance model $\mathbf{I}_{N_P} \otimes \sigma^2 \mathbf{I}_L$ is replaced by $\mathbf{I}_{N_P} \otimes \sigma^2 \boldsymbol{\Sigma}$, with $\boldsymbol{\Sigma}$ known, the GLR remains unchanged, with $\boldsymbol{\Sigma}^{-1/2}\mathbf{S}_P\boldsymbol{\Sigma}^{-1/2}$ replacing $\mathbf{S}_P$. If there is a signal-free secondary channel of measurements, distributed as $\mathbf{Y}_S \sim C\mathcal{N}_{L\times N_S}(\mathbf{0}, \mathbf{I}_{N_S} \otimes \boldsymbol{\Sigma})$, the EP adaptation of the GLR is

$$\lambda_2(\mathbf{S}_S) = \frac{\left[\frac{1}{L}\sum_{l=1}^{L} ev_l(\mathbf{T}_P)\right]^L}{\left[\frac{1}{L-p}\sum_{l=p+1}^{L} ev_l(\mathbf{T}_P)\right]^{L-p} \prod_{l=1}^{p} ev_l(\mathbf{T}_P)},$$

where $\mathbf{T}_P = \mathbf{S}_S^{-1/2}\mathbf{S}_P\mathbf{S}_S^{-1/2}$.

6.4 GLR Solutions for Adaptive Subspace Detection

For the noise covariance $\boldsymbol{\Sigma}$ unknown and the scale σ^2 known or unknown, the GLR statistics for all four points on the compass are now known [21,82,255]. These GLRs generalize previous adaptations by assuming the signal model is multidimensional and by allowing for $N_P \geq 1$ measurements in the primary channel. The results of [255] may be said to be a general theory of adaptive subspace detectors.

In the remainder of this chapter, we address only the GLRs for the NW. These GLRs are important generalizations of the Kelly and ACE statistics, which number among the foundational results for adaptive subspace detection.

As usual, the procedure will be to define a multivariate normal likelihood function under the alternative and null hypotheses, to maximize likelihood with respect to unknown parameters, and then to form a likelihood ratio. A monotone function of this likelihood ratio is the detector statistic, sometimes called the detector score. This procedure has no claims to optimality, but it is faithful to the philosophy of Neyman-Pearson hypothesis testing, and the resulting detector statistics have desirable invariances.

6.4.1 The Kelly and ACE Detector Statistics

The original Kelly and ACE detector statistics were derived for the case of $N_S \geq L$ secondary measurements and a single primary measurement. That is, $N_P = 1$. Moreover, the subspace signal was modeled as a dimension-one signal. Hence, the primary measurement was distributed as $\mathbf{y}_P \sim CN_L(\mathbf{u}x, \sigma^2\boldsymbol{\Sigma})$, and the secondary measurements were distributed as $\mathbf{Y}_S \sim CN_{L\times N_S}(\mathbf{0}, \mathbf{I}_{N_S} \otimes \boldsymbol{\Sigma})$. The parameter σ^2 was assumed equal to 1 by Kelly, but it was assumed unknown to model scale mismatch between the primary channel and the secondary channels in [80,204,305]. The one-dimensional subspace was considered known with representative basis $\mathbf{u}$, but the location $\mathbf{u}x$ of the signal in this subspace was unknown. In other words, x is unknown.

Under the alternative $\mathcal{H}_1$, the joint likelihood of the primary and secondary measurements is

$$\ell(x, \boldsymbol{\Sigma}, \sigma^2; \mathbf{y}_P, \mathbf{Y}_S) = \frac{1}{\pi^L \sigma^{2L} \det(\boldsymbol{\Sigma})} \exp\left\{-\frac{1}{\sigma^2} \operatorname{tr}(\boldsymbol{\Sigma}^{-1}(\mathbf{y}_P - \mathbf{u}x)(\mathbf{y}_P - \mathbf{u}x)^H)\right\}$$
$$\times \frac{1}{\pi^{LN_S} \det(\boldsymbol{\Sigma})^{N_S}} \operatorname{etr}\left\{-N_S \boldsymbol{\Sigma}^{-1}\mathbf{S}_S\right\},$$

where $\mathbf{S}_S = \mathbf{Y}_S\mathbf{Y}_S^H/N_S$ is the sample covariance matrix for the measurements in the secondary channel.

Kelly [193] assumed $\sigma^2 = 1$, maximized likelihood with respect to x and $\boldsymbol{\Sigma}$ under $\mathcal{H}_1$ and with respect to $\boldsymbol{\Sigma}$ under $\mathcal{H}_0$, and obtained the GLR

$$\lambda_{Kelly} = 1 - \frac{1}{\Lambda_{Kelly}^{1/N}} = \frac{|\mathbf{u}^H\mathbf{S}_S^{-1}\mathbf{y}_P|^2}{\mathbf{u}^H\mathbf{S}_S^{-1}\mathbf{u}(N_S + \mathbf{y}_P^H\mathbf{S}_S^{-1}\mathbf{y}_P)} = \frac{\mathbf{z}^H\mathbf{P}_\mathbf{g}\mathbf{z}}{N_S + \mathbf{z}^H\mathbf{z}},$$

where $\mathbf{z} = \mathbf{S}_S^{-1/2}\mathbf{y}_P$, $\mathbf{g} = \mathbf{S}_S^{-1/2}\mathbf{u}$, $\mathbf{P}_\mathbf{g} = \mathbf{g}(\mathbf{g}^H\mathbf{g})^{-1}\mathbf{g}^H$, and

$$\Lambda_{Kelly} = \frac{\ell(\hat{x}, \hat{\boldsymbol{\Sigma}}, \sigma^2 = 1; \mathbf{y}_P, \mathbf{Y}_S)}{\ell(\hat{\boldsymbol{\Sigma}}, \sigma^2 = 1; \mathbf{y}_P, \mathbf{Y}_S)}.$$

This detector is invariant to common scaling of $\mathbf{y}_P$ and each of the secondary measurement in $\mathbf{Y}_S$.

In [204], the authors maximized likelihood over unknown σ^2 to derive the ACE statistic

$$\lambda_{ACE} = 1 - \frac{1}{\Lambda_{ACE}^{1/N}} = \frac{|\mathbf{u}^H\mathbf{S}_S^{-1}\mathbf{y}_P|^2}{(\mathbf{u}^H\mathbf{S}_S^{-1}\mathbf{u})(\mathbf{y}_P^H\mathbf{S}_S^{-1}\mathbf{y}_P)} = \frac{\mathbf{z}^H\mathbf{P}_\mathbf{g}\mathbf{z}}{\mathbf{z}^H\mathbf{z}},$$

where

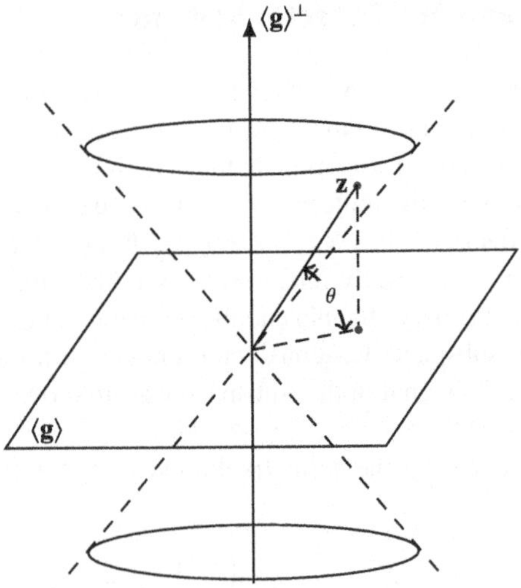

Fig. 6.2 The ACE statistic $\lambda_{ACE} = \cos^2(\theta)$ is invariant to scale and rotation in $\langle \mathbf{g} \rangle$ and $\langle \mathbf{g}^{\perp} \rangle$. This is the double cone illustrated

$$\Lambda_{ACE} = \frac{\ell(\hat{x}, \hat{\boldsymbol{\Sigma}}, \hat{\sigma}^2; \mathbf{y}_P, \mathbf{Y}_S)}{\ell(\hat{\boldsymbol{\Sigma}}, \hat{\sigma}^2; \mathbf{y}_P, \mathbf{Y}_S)}.$$

This form shows that the ACE statistic is invariant to rotation of the whitened measurement $\mathbf{z}$ in the subspaces $\langle \mathbf{g} \rangle$ and $\langle \mathbf{g} \rangle^{\perp}$ and invariant to *uncommon* scaling of $\mathbf{y}_P$ and $\mathbf{Y}_S$. These invariances define a double cone of invariances, as described in [204] and illustrated in Fig. 6.2. The ACE statistic is a *coherence statistic* that measures the cosine-squared of the angle that the whitened measurement makes with a whitened subspace. In [204], ACE was shown to be a GLR; in [205], it was generalized to multidimensional subspace signals; and in [206], it was shown to be uniformly most powerful invariant (UMPI). The detector statistic λ_{ACE} was derived in [80] as an asymptotic statistic for detecting a signal in compound Gaussian noise. In [305], ACE was proposed as an EP version of the scale-invariant matched subspace detector [302, 303].

Rapproachment: The AMF, Kelly, and ACE Detectors. When the noise covariance $\boldsymbol{\Sigma}$ and scaling σ^2 are both known, the so-called non-coherent matched filter statistic is

$$\lambda_{MF} = \log \Lambda_{MF} = \frac{|\mathbf{u}^H \boldsymbol{\Sigma}^{-1} \mathbf{y}_P|^2}{\sigma^2 \mathbf{u}^H \boldsymbol{\Sigma}^{-1} \mathbf{u}},$$

where $\hat{x} = \frac{\mathbf{u}^H \boldsymbol{\Sigma}^{-1} \mathbf{y}_P}{\mathbf{u}^H \boldsymbol{\Sigma}^{-1} \mathbf{u}}$ and

$$\Lambda_{MF} = \frac{\ell(\hat{x}, \mathbf{u}, \boldsymbol{\Sigma}, \sigma^2; \mathbf{y}_P)}{\ell(\boldsymbol{\Sigma}, \sigma^2; \mathbf{y}_P)}.$$

An EP adaptation replaces $\boldsymbol{\Sigma}$ by the sample covariance matrix $\mathbf{S}_S$ of the secondary channel. Then, assuming the scaling $\sigma^2 = 1$, the *adaptive matched filter* [70, 289] is

$$\lambda_{AMF} = \frac{|\mathbf{u}^H \mathbf{S}_S^{-1} \mathbf{y}_P|^2}{\mathbf{u}^H \mathbf{S}_S^{-1} \mathbf{u}} = \mathbf{z}^H \mathbf{P}_{\mathbf{g}} \mathbf{z},$$

where as before $\mathbf{z} = \mathbf{S}_S^{-1/2} \mathbf{y}_P$ and $\mathbf{g} = \mathbf{S}_S^{-1/2} \mathbf{u}$. This detector statistic is *not* a generalized likelihood ratio. The Kelly statistic [193] is

$$\lambda_{Kelly} = \frac{\mathbf{z}^H \mathbf{P}_{\mathbf{g}} \mathbf{z}}{N_S + \mathbf{z}^H \mathbf{z}}.$$

Compare these two detector statistics with the ACE statistic:

$$\lambda_{ACE} = \frac{\mathbf{z}^H \mathbf{P}_{\mathbf{g}} \mathbf{z}}{\mathbf{z}^H \mathbf{z}}.$$

The Kelly GLR is invariant to common scaling of $\mathbf{y}_P$ and $\mathbf{Y}_S$. It is not invariant to uncommon scaling, as the ACE statistic is. The geometric interpretation of ACE is compelling as the cosine-squared of the angle between a whitened measurement and a whitened subspace.

The generalization of the Kelly statistic to multidimensional subspace signals was derived in [194], and the generalization of the ACE statistic to multidimensional subspace signals was derived in [205]. The generalization of the Kelly and ACE statistics for dimension-one subspaces and multiple measurements in the primary channel was derived in [82]. The generalization of these detectors to multidimensional subspaces and multiple measurements in the primary channel was derived in [21]. It is one of these generalizations that is treated in the next section.

In [205], the AMF, Kelly, and ACE detectors are given stochastic representations in terms of several independent random variables. These stochastic representations characterize the distribution of these detectors.

6.4.2 Multidimensional and Multiple Measurement GLR Extensions of the Kelly and ACE Detector Statistics

In the NW corner, the signal subspace is known. The signal model is multidimensional, and the number of measurements in the primary channel may be greater than one. Visits to this subspace are unconstrained, which is to say the measurement model is a first-order MVN model where information about the signal is carried in the mean matrix of the measurements. The resulting GLRs are those of [82] and [21], although the expressions for these GLRs found in this section differ somewhat from the forms found in these references.

Under the alternative $\mathcal{H}_1$, the joint likelihood of primary and secondary measurements is

$$\ell(\mathbf{X}, \boldsymbol{\Sigma}, \sigma^2; \mathbf{Y}_P, \mathbf{Y}_S) = \frac{1}{\pi^{L(N_S+N_P)}\sigma^{2LN_P}\det(\boldsymbol{\Sigma})^{N_S+N_P}} \operatorname{etr}\left\{-\boldsymbol{\Sigma}^{-1}\mathbf{Y}_S\mathbf{Y}_S^H\right\}$$
$$\times \operatorname{etr}\left\{-\frac{1}{\sigma^2}\boldsymbol{\Sigma}^{-1}\left(\mathbf{Y}_P - \mathbf{U}\mathbf{X}\right)\left(\mathbf{Y}_P - \mathbf{U}\mathbf{X}\right)^H\right\}.$$

This can be rewritten as

$$\ell(\mathbf{X}, \boldsymbol{\Sigma}, \sigma^2; \mathbf{Y}_P, \mathbf{Y}_S) = \frac{1}{\pi^{LN}\sigma^{2LN_P}\det(\boldsymbol{\Sigma})^{N}}$$
$$\times \operatorname{etr}\left\{-\boldsymbol{\Sigma}^{-1}\left(\mathbf{Y}_S\mathbf{Y}_S^H + \frac{1}{\sigma^2}\left(\mathbf{Y}_P - \mathbf{U}\mathbf{X}\right)\left(\mathbf{Y}_P - \mathbf{U}\mathbf{X}\right)^H\right)\right\},$$

where $N = N_S + N_P$. Under the hypothesis $\mathcal{H}_0$, the joint likelihood is

$$\ell(\boldsymbol{\Sigma}, \sigma^2; \mathbf{Y}_P, \mathbf{Y}_S) = \frac{1}{\pi^{LN}\sigma^{2LN_P}\det(\boldsymbol{\Sigma})^{N}} \operatorname{etr}\left\{-\boldsymbol{\Sigma}^{-1}\left(\mathbf{Y}_S\mathbf{Y}_S^H + \frac{1}{\sigma^2}\mathbf{Y}_P\mathbf{Y}_P^H\right)\right\}.$$

The Case of Known σ^2. Under $\mathcal{H}_1$, the likelihood is maximized by the maximum likelihood (ML) estimates of $\boldsymbol{\Sigma}$ and $\mathbf{X}$. For fixed $\mathbf{X}$, the ML estimate of $\boldsymbol{\Sigma}$ is

$$N\hat{\boldsymbol{\Sigma}} = \mathbf{Y}_S\mathbf{Y}_S^H + \frac{1}{\sigma^2}\left(\mathbf{Y}_P - \mathbf{U}\mathbf{X}\right)\left(\mathbf{Y}_P - \mathbf{U}\mathbf{X}\right)^H$$
$$= \mathbf{S}_S^{1/2}\left[N_S\mathbf{I}_L + \frac{1}{\sigma^2}\left(\mathbf{S}_S^{-1/2}\mathbf{Y}_P - \mathbf{G}\mathbf{X}\right)\left(\mathbf{S}_S^{-1/2}\mathbf{Y}_P - \mathbf{G}\mathbf{X}\right)^H\right]\mathbf{S}_S^{1/2},$$

where $\mathbf{G} = \mathbf{S}_S^{-1/2}\mathbf{U}$. The ML estimate of $\mathbf{X}$ is obtained by plugging $\hat{\boldsymbol{\Sigma}}$ into the likelihood function and maximizing this compressed likelihood with respect to $\mathbf{X}$. This is equivalent to minimizing the determinant of $\hat{\boldsymbol{\Sigma}}$ with respect to $\mathbf{X}$, yielding

$$\mathbf{S}_S^{-1/2}\mathbf{Y}_P - \mathbf{G}\hat{\mathbf{X}} = \mathbf{P}_{\mathbf{G}}^{\perp}\mathbf{S}_S^{-1/2}\mathbf{Y}_P.$$

This result is proved in Sect. B.9.2, cf. (B.15). Therefore, we have

$$N\hat{\boldsymbol{\Sigma}} = \mathbf{S}_S^{1/2}\left(N_S\mathbf{I}_L + \frac{N_P}{\sigma^2}\mathbf{P}_{\mathbf{G}}^{\perp}\mathbf{T}_P\mathbf{P}_{\mathbf{G}}^{\perp}\right)\mathbf{S}_S^{1/2},$$

and the compressed likelihood becomes

$$\ell(\hat{\mathbf{X}}, \hat{\boldsymbol{\Sigma}}, \sigma^2; \mathbf{Y}_P, \mathbf{Y}_S) = \frac{N^{LN}}{(e\pi)^{LN}\sigma^{2LN_P}} \frac{1}{\det^N(\mathbf{S}_S)} \frac{1}{\det^N(N_S\mathbf{I}_L + \frac{N_P}{\sigma^2}\mathbf{P}_{\mathbf{G}}^{\perp}\mathbf{T}_P\mathbf{P}_{\mathbf{G}}^{\perp})}. \tag{6.4}$$

It is straightforward to show that compressed likelihood under $\mathcal{H}_0$ is

$$\ell(\hat{\boldsymbol{\Sigma}}, \sigma^2; \mathbf{Y}_P, \mathbf{Y}_S) = \frac{N^{LN}}{(e\pi)^{LN}\sigma^{2LN_P}} \frac{1}{\det^N(\mathbf{S}_S)} \frac{1}{\det^N(N_S\mathbf{I}_L + \frac{N_P}{\sigma^2}\mathbf{T}_P)}. \tag{6.5}$$

The GLR in the homogeneous case, $\sigma^2 = 1$, and $p < L$ may be written as the Nth root of the ratio of these generalized likelihoods

$$\lambda_1 = \Lambda_1^{1/N} = \frac{\det\left(\mathbf{I}_L + \frac{N_P}{N_S}\mathbf{T}_P\right)}{\det\left(\mathbf{I}_L + \frac{N_P}{N_S}\mathbf{P}_{\mathbf{G}}^{\perp}\mathbf{T}_P\mathbf{P}_{\mathbf{G}}^{\perp}\right)}, \tag{6.6}$$

where

$$\Lambda_1 = \frac{\ell(\hat{\mathbf{X}}, \hat{\boldsymbol{\Sigma}}, \sigma^2 = 1; \mathbf{Y}_P, \mathbf{Y}_S)}{\ell(\hat{\boldsymbol{\Sigma}}, \sigma^2 = 1; \mathbf{Y}_P, \mathbf{Y}_S)}.$$

When the subspace $\langle \mathbf{U} \rangle$ is one dimensional, and $N_P = 1$, then this GLR is within a monotone function of the Kelly statistic. So the result of [21, 82] is a full generalization of the original Kelly result.

The GLR statistic in (6.6) illuminates the role of the *secondarily whitened primary data* $\mathbf{S}_S^{-1/2}\mathbf{Y}_P$, its corresponding whitened sample covariance $\mathbf{T}_P$, and the sample covariance of whitened measurements after their projection onto the subspace $\mathbf{P}_{\mathbf{G}}^{\perp}$. The GLR is a function only of the eigenvalues of $\mathbf{T}_P$ and the eigenvalues of $\mathbf{P}_{\mathbf{G}}^{\perp}\mathbf{T}_P\mathbf{P}_{\mathbf{G}}^{\perp}$. With just a touch of license, the inverse of the GLR statistic may be called a coherence statistic. For $p = L$ and $\sigma^2 = 1$, the GLR reduces to $\det\left(\mathbf{I}_L + \frac{N_P}{N_S}\mathbf{T}_P\right)$.

This GLR is derived for $\sigma^2 = 1$, but generalization to any known value of σ^2 is straightforward: the primary data may be normalized by the square root of σ^2, to produce a homogeneous case. So, without loss of generality, it may be assumed that $\sigma^2 = 1$ when σ^2 is known.

The Case of Unknown σ^2. Determining the GLR for a partially homogeneous case requires one more maximization of the likelihoods in (6.4) and (6.5) with respect to σ^2. For $p = L$, the likelihood under $\mathcal{H}_1$ is unbounded with respect to $\sigma^2 > 0$, and, hence, the GLR does not exist. Therefore, we assume $p < L$.

When the scale parameter σ^2 is unknown, then each of the compressed likelihoods in (6.4) and (6.5) must be maximized with respect to σ^2. The function to be *minimized* may be written $(1/\sigma^{2LN_S})\det^N(\sigma^2\mathbf{I}_L + \frac{N_P}{N_S}\mathbf{M})$, where $\mathbf{M} = \mathbf{P}_{\mathbf{G}}^{\perp}\mathbf{T}_P\mathbf{P}_{\mathbf{G}}^{\perp}$, under $\mathcal{H}_1$, and $\mathbf{M} = \mathbf{T}_P$, under $\mathcal{H}_0$. The Nth root of this function may be written as

$$\sigma^{2LN_P/N-2t}\prod_{l=1}^{t}\left(\sigma^2 + \frac{N_P}{N_S}ev_l(\mathbf{M})\right),$$

where t is the rank of $\mathbf{M}$ and $ev_l(\mathbf{M}), l = 1, \ldots, t$, are the non-zero eigenvalues of $\mathbf{M}$, ordered from largest to smallest. The rank of the matrix $\mathbf{M}$ is $t_1 = \min(L-p, N_P)$ under $\mathcal{H}_1$ and $t_0 = \min(L, N_P)$ under $\mathcal{H}_0$. To minimize this function is to minimize its logarithm, which is to minimize

$$\left(\frac{LN_P}{N} - t\right)\log\sigma^2 + \sum_{l=1}^{t}\log\left(\sigma^2 + \frac{N_P}{N_S}ev_l(\mathbf{M})\right).$$

Differentiate with respect to σ^2 and equate to zero to find the condition for the minimizing σ^2:

$$t - \frac{LN_P}{N} = \sum_{l=1}^{t}\frac{1}{1 + \frac{1}{\sigma^2}\frac{N_P}{N_S}ev_l(\mathbf{M}))}. \tag{6.7}$$

There can be no positive solution for σ^2 unless $t > LN_P/N$. Under $\mathcal{H}_0$, the condition $\min(L, N_P) > LN_P/N$ is always satisfied. Under $\mathcal{H}_1$, the condition is $\min(L-p, N_P) > LN_P/N$. For $L - p \geq N_P$, the condition is satisfied, but for $L-p < N_P$, the condition is $L-p > LN_P/N$ or, equivalently, $pN_P < (L-p)N_S$. For fixed L and p, this imposes a constraint on the fraction N_S/N_P, given by $N_S/N_P > \frac{p}{L-p}$. So the constraint is $N_S > N_P\frac{p}{L-p}$. Furthermore, recall that $N_S \geq L$.

Call $\hat{\sigma}_1^2$ the solution to (6.7) when $\mathbf{M} = \mathbf{P}_{\mathbf{G}}^{\perp}\mathbf{T}_P\mathbf{P}_{\mathbf{G}}^{\perp}$, and $\hat{\sigma}_0^2$ the solution when $\mathbf{M} = \mathbf{T}_P$. In general, there is no closed-form solution to (6.7). Then, the GLR for detecting a subspace signal in a first-order signal model is

$$\lambda_1 = \Lambda_1^{1/N} = \frac{\hat{\sigma}_0^{2LN_P/N}}{\hat{\sigma}_1^{2LN_P/N}}\frac{\det\left(\mathbf{I}_L + \frac{N_P}{N_S\hat{\sigma}_0^2}\mathbf{T}_P\right)}{\det\left(\mathbf{I}_L + \frac{1}{\hat{\sigma}_1^2}\frac{N_P}{N_S}\mathbf{P}_{\mathbf{G}}^{\perp}\mathbf{T}_P\mathbf{P}_{\mathbf{G}}^{\perp}\right)}$$

$$= \frac{\hat{\sigma}_1^{2LN_S/N}}{\hat{\sigma}_0^{2LN_S/N}}\frac{\det\left(\hat{\sigma}_0^2\mathbf{I}_L + \frac{N_P}{N_S}\mathbf{T}_P\right)}{\det\left(\hat{\sigma}_1^2\mathbf{I}_L + \frac{N_P}{N_S}\mathbf{P}_{\mathbf{G}}^{\perp}\mathbf{T}_P\mathbf{P}_{\mathbf{G}}^{\perp}\right)},$$

where

$$\Lambda_1 = \frac{\ell(\hat{\mathbf{X}}, \hat{\boldsymbol{\Sigma}}, \hat{\sigma}_1^2; \mathbf{Y}_P, \mathbf{Y}_S)}{\ell(\hat{\boldsymbol{\Sigma}}, \hat{\sigma}_0^2; \mathbf{Y}_P, \mathbf{Y}_S)},$$

provided $N_S/N_P > \frac{p}{L-p}$. With just a touch of license, the inverse of this GLR may be interpreted as a coherence statistic. For $p = 1$ and $N_P = 1$, this GLR is within a monotone function of the original ACE statistic. So the result of [82] is a full generalization of the original GLR derivation of ACE [204].

Rapproachment between the EP and GLR Statistics in the NW. The EP adaptation, repeated here for convenience, is

$$\lambda_1(\mathbf{S}_S) = \frac{\mathrm{tr}\,(\mathbf{T}_P)}{\mathrm{tr}\,\left(\mathbf{P}_{\mathbf{G}}^{\perp}\mathbf{T}_P\mathbf{P}_{\mathbf{G}}^{\perp}\right)}.$$

This estimate and plug solution stands in contrast to the GLR solution depending, as it does, only on sums of eigenvalues of $\mathbf{P}_{\mathbf{G}}^{\perp}\mathbf{T}_P\mathbf{P}_{\mathbf{G}}^{\perp}$ and $\mathbf{T}_P$.

This concludes our treatment of adaptive subspace detectors. The EP adaptations cover each point of the compass: NW, NE, SW, and SE. The GLRs cover only the NW. In [255], all four points are covered for EPs and GLRs. In [6, 255], the performances of the EP and GLR solutions are evaluated and compared. The reader is directed to these papers.

6.5 Chapter Notes

The literature on matched and adaptive subspace detectors is voluminous. The reader is referred to [205] for an account of the provenance for first-order adaptive subspace detectors. This provenance includes the original work of Kelly [193]; Kelly and Forsythe [194]; Chen and Reed [70]; Robey et al. [289]; Conte et al. [80, 81]; and Kraut et al. [204–206, 240, 305].

1. References [205, 206] establish that the ACE statistic of [80, 305] is a uniformly most powerful invariant (UMPI) detector of multidimensional subspace signals. Its invariances, optimalities, and performances are well understood.
2. Bandiera, Besson, Conte, Lops, Orlando, Ricci, and their collaborators continue to advance the theory of ASDs with the extension of ASDs to multi-snapshot primary data in first- and second-order signal models [20, 21, 31, 33, 35, 82, 83, 238, 255, 282].
3. The work of Besson and collaborators [31–34] addresses several variations on the NE problem of detecting signals in unknown dimension-one subspaces for homogeneous and partially homogeneous problems.
4. In [284, 285] and subsequent work, Richmond has analyzed the performance of many adaptive detectors for multi-sensor array processing.
5. When a model is imposed on the unknown covariance matrix, such as Toeplitz or persymmetric, then estimate and plug solutions may be modified to accommodate these models, and GLRs may be approximated.

Two-Channel Matched Subspace Detectors 7

This chapter is addressed to the problem of detecting a common subspace signal in two multi-sensor channels. In passive detection problems, for instance, we use observations from a reference channel where a noisy and linearly distorted version of the signal of interest is always present and from a surveillance channel that contains either noise or signal-plus-noise. Following the structure of Chap. 5, we study second-order detectors where the unknown transmitted signal is modeled as a zero-mean Gaussian and averaged out or marginalized and first-order detectors where the unknown transmitted signal appears in the mean of the observations with no prior distribution assigned to it. The signal subspaces at the two sensor arrays may be known, or they may be unknown with known dimension.

Adhering to the nomenclature introduced in Chap. 5, when the subspaces are known, the resulting detectors are termed *matched subspace detectors* with different variations in what is assumed to be known or unknown. When the subspaces are unknown, they are termed *matched direction detectors*. In some cases, the GLR admits a closed-form solution, while, in others, numerical optimization approaches, mainly in the form of alternating optimization techniques, must be applied. We study in this chapter different noise models, ranging from spatially white noises with identical variances to arbitrarily correlated Gaussian noises (but independent across channels). For each noise and signal model, the invariances of the hypothesis testing problem and its GLR are established. Maximum likelihood estimation of unknown signal and noise parameters leads to a variety of coherence statistics. In some specific cases, we also present the locally most powerful invariant test, which is also a coherence statistic.

7.1 Signal and Noise Models for Two-Channel Problems

The problem considered is the detection of electromagnetic or acoustic sources from their radiated fields, measured passively at two spatially separated sensor arrays. In one of the sensor arrays, known as the reference channel, a noisy and linearly

D. Ramírez et al., *Coherence*, https://doi.org/10.1007/978-3-031-13331-2_7

distorted version of the signal of interest is always present, and the problem is to detect whether or not the signal of interest is present at the other sensor array, known as the surveillance channel.

We follow the framework established in Chap. 5 and consider first- and second-order multivariate normal measurements at the surveillance and reference channels, each consisting of L sensors that record N measurements. The nth measurement is

$$\begin{bmatrix} \mathbf{y}_{s,n} \\ \mathbf{y}_{r,n} \end{bmatrix} = \begin{bmatrix} \mathbf{H}_s \\ \mathbf{H}_r \end{bmatrix} \mathbf{x}_n + \begin{bmatrix} \mathbf{n}_{s,n} \\ \mathbf{n}_{r,n} \end{bmatrix}, \quad n = 1, \ldots, N,$$

where $\mathbf{y}_{s,n} \in \mathbb{C}^L$ and $\mathbf{y}_{r,n} \in \mathbb{C}^L$ are the surveillance and reference measurements; $\mathbf{x}_n \in \mathbb{C}^p$ contains the unknown transmitted signal; $\mathbf{H}_s \in \mathbb{C}^{L\times p}$ and $\mathbf{H}_r \in \mathbb{C}^{L\times p}$ represent the $L \times p$ channels from the transmitter(s) to the surveillance and reference multiantenna receivers, respectively; and the vectors $\mathbf{n}_{s,n}$ and $\mathbf{n}_{r,n}$ model the additive noise. For notational convenience, the signals, noises, and channel matrices may be stacked as $\mathbf{y}_n = [\mathbf{y}_{s,n}^T\ \mathbf{y}_{r,n}^T]^T$, $\mathbf{n}_n = [\mathbf{n}_{s,n}^T\ \mathbf{n}_{r,n}^T]^T$, and $\mathbf{H} = [\mathbf{H}_s^T\ \mathbf{H}_r^T]^T$.

The two-channel passive detection problem is to test the hypothesis that the surveillance channel contains no signal, versus the alternative that it does:

$$\mathcal{H}_0 : \mathbf{y}_n = \begin{bmatrix} \mathbf{0} \\ \mathbf{H}_r \end{bmatrix} \mathbf{x}_n + \mathbf{n}_n, \quad n = 1, \ldots, N,$$

$$\mathcal{H}_1 : \mathbf{y}_n = \begin{bmatrix} \mathbf{H}_s \\ \mathbf{H}_r \end{bmatrix} \mathbf{x}_n + \mathbf{n}_n, \quad n = 1, \ldots, N.$$

The observations may be organized into the $2L \times N$ matrix $\mathbf{Y} = [\mathbf{y}_1 \cdots \mathbf{y}_N]$. The measurement model under $\mathcal{H}_1$ is then

$$\mathbf{Y} = \begin{bmatrix} \mathbf{H}_s \\ \mathbf{H}_r \end{bmatrix} \mathbf{X} + \mathbf{N}, \tag{7.1}$$

with the $p \times N$ transmit signal matrix, $\mathbf{X}$, and the $2L \times N$ noise matrix, $\mathbf{N}$, defined analogously to $\mathbf{Y}$. The sample covariance matrix is

$$\mathbf{S} = \frac{1}{N}\mathbf{Y}\mathbf{Y}^H = \begin{bmatrix} \mathbf{S}_{ss} & \mathbf{S}_{sr} \\ \mathbf{S}_{sr}^H & \mathbf{S}_{rr} \end{bmatrix},$$

where $\mathbf{S}_{ss}$ is the sample covariance matrix of the surveillance channel and the other blocks are defined similarly.

The signal matrix $\mathbf{X}$ is the matrix

$$\mathbf{X} = \begin{bmatrix} \mathbf{x}_1 & \cdots & \mathbf{x}_N \end{bmatrix} = \begin{bmatrix} \boldsymbol{\rho}_1^T \\ \vdots \\ \boldsymbol{\rho}_p^T \end{bmatrix}.$$

That is, $\mathbf{X}$ consists of a sequence of N column vectors $\mathbf{x}_n \in \mathbb{C}^p$, each of which is a $p \times 1$ vector of source transmissions at time n, or $\mathbf{X}$ consists of p row vectors $\boldsymbol{\rho}_l^T \in \mathbb{C}^N$, each of which is a $1 \times N$ vector of transmissions from source l. When $\boldsymbol{\rho}$ is pronounced "row," then it should be a reminder that $\boldsymbol{\rho}^T$ is a row vector.

7.1.1 Noise Models

The additive noise is assumed to be temporally white and distributed as a sequence of proper complex zero-mean Gaussian random vectors, each with covariance matrix $\boldsymbol{\Sigma}$. That is, the noise matrix is distributed as $\mathbf{N} \sim \mathcal{CN}_{2L\times N}(\mathbf{0}, \mathbf{I}_N \otimes \boldsymbol{\Sigma})$. Moreover, the noises at the surveillance and reference channels are assumed to be uncorrelated, so $\boldsymbol{\Sigma}$ is a block-diagonal matrix

$$\boldsymbol{\Sigma} = \begin{bmatrix} \boldsymbol{\Sigma}_{ss} & \mathbf{0} \\ \mathbf{0} & \boldsymbol{\Sigma}_{rr} \end{bmatrix} \in \mathcal{E},$$

where $\mathcal{E}$ is the set of block-diagonal covariance matrices. Examples of these structured sets of interest in single-channel and multi-channel detection problems have been presented in Sect. 4.2. In this chapter, we consider the following noise models:

- White noises (i.i.d.) with identical variance at both channels. $\boldsymbol{\Sigma}_{ss} = \boldsymbol{\Sigma}_{rr} = \sigma^2 \mathbf{I}_L$:

$$\mathcal{E}_1 = \left\{ \boldsymbol{\Sigma} \succ 0 \mid \boldsymbol{\Sigma} = \sigma^2 \mathbf{I}_{2L},\ \sigma^2 > 0 \right\} \tag{7.2}$$

- White noises but with different variances at the surveillance and reference channels. $\boldsymbol{\Sigma}_{ss} = \sigma_s^2 \mathbf{I}_L$, $\boldsymbol{\Sigma}_{rr} = \sigma_r^2 \mathbf{I}_L$:

$$\mathcal{E}_2 = \left\{ \boldsymbol{\Sigma} \succ 0 \mid \boldsymbol{\Sigma} = \begin{bmatrix} \sigma_s^2 \mathbf{I}_L & \mathbf{0} \\ \mathbf{0} & \sigma_r^2 \mathbf{I}_L \end{bmatrix},\ \sigma_s^2 > 0, \sigma_r^2 > 0 \right\} \tag{7.3}$$

- Uncorrelated noises across antennas. $\boldsymbol{\Sigma}_{ss}$ and $\boldsymbol{\Sigma}_{rr}$ are diagonal matrices with unknown positive elements along their diagonals:

$$\mathcal{E}_3 = \left\{ \boldsymbol{\Sigma} \succ 0 \mid \boldsymbol{\Sigma} = \begin{bmatrix} \operatorname{diag}(\sigma_{s,1}^2, \dots, \sigma_{s,L}^2) & \mathbf{0} \\ \mathbf{0} & \operatorname{diag}(\sigma_{r,1}^2, \dots, \sigma_{r,L}^2) \end{bmatrix},\ \sigma_{s,l}^2, \sigma_{r,l}^2 > 0 \right\} \tag{7.4}$$

- Noises with arbitrary spatial correlation. $\boldsymbol{\Sigma}_{ss}$ and $\boldsymbol{\Sigma}_{rr}$ are arbitrary positive definite matrices:

$$\mathcal{E}_4 = \left\{ \boldsymbol{\Sigma} \succ 0 \mid \boldsymbol{\Sigma} = \begin{bmatrix} \boldsymbol{\Sigma}_{ss} & \mathbf{0} \\ \mathbf{0} & \boldsymbol{\Sigma}_{rr} \end{bmatrix},\ \boldsymbol{\Sigma}_{ss} \succ 0,\ \boldsymbol{\Sigma}_{rr} \succ 0 \right\}$$

For the noise models considered in this chapter, it is easy to check that, for unknown parameters, the structured parameter sets under both hypotheses are cones. Therefore, Lemma 4.1 in Chap. 4 can be used to show the trace term of the likelihood function for first-order or second-order models, when evaluated at the ML estimates, is a constant under both hypotheses. Consequently, the GLR tests reduce to a ratio of determinants.

7.1.2 Known or Unknown Subspaces

The reference and surveillance channels may be decomposed as

$$\mathbf{H}_s = \mathbf{U}_s \mathbf{A}_s, \qquad \mathbf{H}_r = \mathbf{U}_r \mathbf{A}_r,$$

where $\mathbf{U}_s$ and $\mathbf{U}_r$ are $L \times p$ matrices whose columns form a unitary basis for the subspaces $\langle \mathbf{U}_s \rangle$ and $\langle \mathbf{U}_r \rangle$, respectively, and $\mathbf{A}_s \in GL(\mathbb{C}^p)$ and $\mathbf{A}_r \in GL(\mathbb{C}^p)$ are arbitrary $p \times p$ invertible matrices. Analogously to the subspace detectors studied for the single channel case in Chap. 5, in some cases the subspaces for the reference and surveillance channels are known, while in others only their dimension p is known.

Conditioned on $\mathbf{X}$, the observations under $\mathcal{H}_1$ with known subspaces under a first-order measurement model are distributed as

$$\mathbf{Y} \sim CN_{2L \times N}\left(\begin{bmatrix} \mathbf{U}_s \mathbf{A}_s \\ \mathbf{U}_r \mathbf{A}_r \end{bmatrix} \mathbf{X},\ \mathbf{I}_N \otimes \boldsymbol{\Sigma} \right).$$

As the source signal $\mathbf{X}$ and the $p \times p$ matrices $\mathbf{A}_s$ and $\mathbf{A}_r$ are unknown, without loss of generality, this model may be rewritten as the model

$$\mathbf{Y} \sim CN_{2L \times N}\left(\begin{bmatrix} \mathbf{U}_s \mathbf{A} \\ \mathbf{U}_r \end{bmatrix} \mathbf{X},\ \mathbf{I}_N \otimes \boldsymbol{\Sigma} \right),$$

with $\mathbf{A}$ and $\mathbf{X}$ unknown. When the subspaces $\langle \mathbf{U}_r \rangle$ and $\langle \mathbf{U}_s \rangle$ are known, detectors for signals in these models will be called *matched subspace detectors* in a first-order statistical model. When these subspaces are unknown, then $\mathbf{Y} \sim CN_{2L \times N}(\mathbf{Z}, \mathbf{I}_N \otimes \boldsymbol{\Sigma})$, where $\mathbf{Z}$ is an unknown $2L \times N$ matrix of rank p. The detectors will be called *matched direction detectors* in a first-order statistical model.

When a Gaussian prior distribution is assigned to $\mathbf{x} \sim CN_p(\mathbf{0}, \mathbf{R}_{\mathbf{xx}})$, the signal model (7.1) can be marginalized with respect to $\mathbf{X}$, resulting in the covariance matrix for the measurements

$$\mathbf{R}_{\mathbf{yy}} = \begin{bmatrix} \mathbf{U}_s \mathbf{A}_s \mathbf{R}_{\mathbf{xx}} \mathbf{A}_s^H \mathbf{U}_s^H & \mathbf{U}_s \mathbf{A}_s \mathbf{R}_{\mathbf{xx}} \mathbf{A}_r^H \mathbf{U}_r^H \\ \mathbf{U}_r \mathbf{A}_r \mathbf{R}_{\mathbf{xx}} \mathbf{A}_s^H \mathbf{U}_s^H & \mathbf{U}_r \mathbf{A}_r \mathbf{R}_{\mathbf{xx}} \mathbf{A}_r^H \mathbf{U}_r^H \end{bmatrix} + \begin{bmatrix} \boldsymbol{\Sigma}_{ss} & \mathbf{0} \\ \mathbf{0} & \boldsymbol{\Sigma}_{rr} \end{bmatrix}. \tag{7.5}$$

Since the $p \times p$ covariance matrix $\mathbf{R}_{\mathbf{xx}}$ and the linear mappings $\mathbf{A}_s$ and $\mathbf{A}_r$ are unknown, the covariance matrix (7.5) can be written as

$$\mathbf{R_{yy}} = \begin{bmatrix} \mathbf{U}_s\mathbf{Q}_{ss}\mathbf{U}_s^H & \mathbf{U}_s\mathbf{Q}_{sr}\mathbf{U}_r^H \\ \mathbf{U}_r\mathbf{Q}_{rs}\mathbf{U}_r^H & \mathbf{U}_r\mathbf{Q}_{rr}\mathbf{U}_r^H \end{bmatrix} + \begin{bmatrix} \mathbf{\Sigma}_{ss} & \mathbf{0} \\ \mathbf{0} & \mathbf{\Sigma}_{rr} \end{bmatrix},$$

where $\mathbf{Q}_{ss}$ and $\mathbf{Q}_{rr}$ are unknown positive definite matrices and $\mathbf{Q}_{sr} = \mathbf{Q}_{rs}^H$ is an unknown $p \times p$ matrix. Together with the noise covariance matrix $\mathbf{\Sigma} = \text{blkdiag}(\mathbf{\Sigma}_{ss}, \mathbf{\Sigma}_{rr})$, these are the variables to be estimated in an ML framework. The marginal distribution for $\mathbf{Y}$ under $\mathcal{H}_1$ for a second-order model is then

$$\mathbf{Y} \sim CN_{2L\times N}\left(\mathbf{0}, \mathbf{I}_N \otimes \left(\begin{bmatrix} \mathbf{U}_s\mathbf{Q}_{ss}\mathbf{U}_s^H & \mathbf{U}_s\mathbf{Q}_{sr}\mathbf{U}_r^H \\ \mathbf{U}_r\mathbf{Q}_{rs}\mathbf{U}_r^H & \mathbf{U}_r\mathbf{Q}_{rr}\mathbf{U}_r^H \end{bmatrix} + \mathbf{\Sigma}\right)\right).$$

Adhering to the convention established in Chap. 5, the detectors for signals in known subspaces $\langle \mathbf{U}_r \rangle$ and $\langle \mathbf{U}_s \rangle$ will be called *matched subspace detectors* in a second-order statistical model. If only the dimension of the subspaces, p, is known, any special structure in $\mathbf{R_{xx}}$ will be washed out by $\mathbf{H}_s$ and $\mathbf{H}_r$. Therefore, without loss of generality, the transmit signal covariance can be absorbed into these factor loadings, and thus we assume $\mathbf{R_{xx}} = \mathbf{I}_p$. The marginal distribution for $\mathbf{Y}$ is

$$\mathbf{Y} \sim CN_{2L\times N}\left(\mathbf{0}, \mathbf{I}_N \otimes \left(\mathbf{H}\mathbf{H}^H + \mathbf{\Sigma}\right)\right),$$

where $\mathbf{H}$ is an unknown $2L \times p$ matrix. Detectors derived from this model will be called *matched direction detectors* in a second-order statistical model.

The four variations of the two-channel detectors considered in this chapter are summarized in Table 7.1. To derive detectors for the 4 measurement models of Table 7.1, for the 4 covariance sets $\mathcal{E}_1$ through $\mathcal{E}_4$, would be to derive 32 different detectors: 16 for the cases where the noise covariance matrix is known and 16 for the cases where the noise covariance is unknown. In fact, some of these combinations

Table 7.1 First-order and second-order detectors for known subspace and unknown subspace of known dimension. In the NW corner, the signal $\mathbf{X}$ and the $p \times p$ matrix $\mathbf{A}$ are unknown; in the SW corner, the $p \times p$ matrices $\mathbf{Q}_{ss}$, $\mathbf{Q}_{rr}$, and $\mathbf{Q}_{sr} = \mathbf{Q}_{rs}^H$ are unknown; in the NE corner, the $2L \times N$ rank-p signal matrix $\mathbf{Z}$ is unknown with $\mathbf{Z} = \mathbf{H}\mathbf{X}$; and in the SE corner, the $2L \times 2L$ rank-p signal covariance matrix $\mathbf{H}\mathbf{H}^H$ is unknown. In each of the corners, the noise covariance matrix $\mathbf{\Sigma}$ may be known, or it may be an unknown covariance matrix in one of the covariance sets $\mathcal{E}_m$, $m = 1, \ldots, 4$

	Known subspace	Unknown subspace of known dimension
First-order detectors	$\mathbf{Y} \sim CN_{2L\times N}\left(\begin{bmatrix} \mathbf{U}_s\mathbf{A} \\ \mathbf{U}_r \end{bmatrix}\mathbf{X}, \mathbf{I}_N \otimes \mathbf{\Sigma}\right)$	$\mathbf{Y} \sim CN_{2L\times N}(\mathbf{Z}, \mathbf{I}_N \otimes \mathbf{\Sigma})$
Second-order detectors	$\mathbf{Y} \sim CN_{2L\times N}\left(\mathbf{0}, \mathbf{I}_N \otimes \left(\begin{bmatrix} \mathbf{U}_s\mathbf{Q}_{ss}\mathbf{U}_s^H & \mathbf{U}_s\mathbf{Q}_{sr}\mathbf{U}_r^H \\ \mathbf{U}_r\mathbf{Q}_{rs}\mathbf{U}_r^H & \mathbf{U}_r\mathbf{Q}_{rr}\mathbf{U}_r^H \end{bmatrix} + \mathbf{\Sigma}\right)\right)$	$\mathbf{Y} \sim CN_{2L\times N}(\mathbf{0}, \mathbf{I}_N \otimes (\mathbf{H}\mathbf{H}^H + \mathbf{\Sigma}))$

are ill-posed GLR problems, and others pose intractable optimization problems in the ML identification of unknown parameters. Therefore, in the sections and subsections to follow, we select for study a small subset of the most interesting combinations of signal model and noise model.

7.2 Detectors in a First-Order Model for a Signal in a Known Subspace

The detection problem for a first-order signal model in known surveillance and reference subspaces (NW quadrant in Table 7.1) is

$$\begin{aligned}\mathcal{H}_0 &: \mathbf{Y} \sim \mathcal{CN}_{2L\times N}\left(\begin{bmatrix}\mathbf{0}\\ \mathbf{U}_r\end{bmatrix}\mathbf{X},\ \mathbf{I}_N \otimes \boldsymbol{\Sigma}\right),\\ \mathcal{H}_1 &: \mathbf{Y} \sim \mathcal{CN}_{2L\times N}\left(\begin{bmatrix}\mathbf{U}_s\mathbf{A}\\ \mathbf{U}_r\end{bmatrix}\mathbf{X},\ \mathbf{I}_N \otimes \boldsymbol{\Sigma}\right),\end{aligned} \tag{7.6}$$

where $\mathbf{U}_s$ and $\mathbf{U}_r$ are arbitrary bases for the known p-dimensional subspaces of the surveillance and reference channels, respectively. The signal $\mathbf{X}$ and the block diagonal noise covariance matrix $\boldsymbol{\Sigma}$ are unknown, and so is $\mathbf{A}$. In the following subsection, we consider the case where the dimension of the known subspaces is $p = 1$.

For the noise model, we consider the set $\mathcal{E}_1$ in (7.2), in which case noises are spatially white with identical variances in both channels, $\boldsymbol{\Sigma} = \sigma^2\mathbf{I}_{2L}$. Other cases call for numerical optimization to obtain the ML estimates of the unknowns [313].

7.2.1 Scale-Invariant Matched Subspace Detector for Equal and Unknown Noise Variances

When $p = 1$ and $\boldsymbol{\Sigma} = \sigma^2\mathbf{I}_{2L}$, the detection problem (7.6) reduces to

$$\begin{aligned}\mathcal{H}_0 &: \mathbf{Y} \sim \mathcal{CN}_{2L\times N}\left(\begin{bmatrix}\mathbf{0}\\ \mathbf{u}_r\end{bmatrix}\boldsymbol{\rho}^T,\ \mathbf{I}_N \otimes \sigma^2\mathbf{I}_{2L}\right),\\ \mathcal{H}_1 &: \mathbf{Y} \sim \mathcal{CN}_{2L\times N}\left(\begin{bmatrix}a\mathbf{u}_s\\ \mathbf{u}_r\end{bmatrix}\boldsymbol{\rho}^T,\ \mathbf{I}_N \otimes \sigma^2\mathbf{I}_{2L}\right),\end{aligned}$$

where $\boldsymbol{\rho}^T$ is now an unknown $1 \times N$ row vector. The matrices $a\mathbf{u}_s\boldsymbol{\rho}^T$ and $\mathbf{u}_r\boldsymbol{\rho}^T$ are $L \times N$ matrices of rank 1. The noise variance σ^2 and a are unknown.

Under $\mathcal{H}_0$, the likelihood function is

$$\ell(\boldsymbol{\rho}, \sigma^2; \mathbf{Y}) = \frac{1}{\pi^{2LN}(\sigma^2)^{2LN}} \operatorname{etr}\left(-\frac{N\mathbf{S}_{ss}}{\sigma^2}\right) \operatorname{etr}\left(-\frac{(\mathbf{Y}_r - \mathbf{u}_r\boldsymbol{\rho}^T)(\mathbf{Y}_r - \mathbf{u}_r\boldsymbol{\rho}^T)^H}{\sigma^2}\right).$$

The ML estimate of the source signal is $\hat{\boldsymbol{\rho}}^T = \mathbf{u}_r^H \mathbf{Y}_r$, and therefore the compressed likelihood as a function solely of σ^2 is

$$\ell(\hat{\boldsymbol{\rho}}, \sigma^2; \mathbf{Y}) = \frac{1}{\pi^{2LN}(\sigma^2)^{2LN}} \operatorname{etr}\left(-\frac{N\mathbf{S}_{ss}}{\sigma^2}\right) \operatorname{etr}\left(-\frac{N\mathbf{P}_{\mathbf{u}_r}^{\perp}\mathbf{S}_{rr}}{\sigma^2}\right),$$

where $\mathbf{P}_{\mathbf{u}_r}^{\perp} = \mathbf{I}_L - \mathbf{u}_r\mathbf{u}_r^H$. Equating the derivative of $\ell(\hat{\mathbf{x}}, \sigma^2; \mathbf{Y})$ w.r.t. σ^2 to zero, we obtain the ML estimate

$$\hat{\sigma}_0^2 = \frac{\operatorname{tr}(\mathbf{S}_{ss}) + \operatorname{tr}(\mathbf{P}_{\mathbf{u}_r}^{\perp}\mathbf{S}_{rr})}{2L}.$$

Under $\mathcal{H}_1$, the likelihood is

$$\ell(a, \boldsymbol{\rho}, \sigma^2; \mathbf{Y}) = \frac{1}{\pi^{2LN}(\sigma^2)^{2LN}} \operatorname{etr}\left\{-\frac{1}{\sigma^2}\left(\mathbf{Y} - \mathbf{v}(a)\boldsymbol{\rho}^T\right)\left(\mathbf{Y} - \mathbf{v}(a)\boldsymbol{\rho}^T\right)^H\right\}.$$

where $\mathbf{v}(a)$ is the $2L \times 1$ vector $\mathbf{v}(a) = [a\mathbf{u}_s^T \; \mathbf{u}_r^T]^T$. For any fixed a, the maximizing solution for $\mathbf{v}(a)\boldsymbol{\rho}^T$ is $\mathbf{P}_{\mathbf{v}}(a)\mathbf{Y}$, where $\mathbf{P}_{\mathbf{v}}(a)$ is the rank one projection matrix $\mathbf{v}(a)(\mathbf{v}^H(a)\mathbf{v}(a))^{-1}\mathbf{v}^H(a)$. It is then easy to show that the ML estimate of σ^2 is

$$\hat{\sigma}_1^2(a) = \frac{\operatorname{tr}(\mathbf{P}_{\mathbf{v}}^{\perp}(a)\mathbf{S})}{2L}.$$

The compressed likelihood is now a function of $\hat{\sigma}_1^2(a)$, and this function is maximized by minimizing $\operatorname{tr}(\mathbf{P}_{\mathbf{v}}^{\perp}(a)\mathbf{S})$, or maximizing $\operatorname{tr}(\mathbf{P}_{\mathbf{v}}(a)\mathbf{S})$ with respect to a. To this end, write $\operatorname{tr}(\mathbf{P}_{\mathbf{v}}(a)\mathbf{S}) = (|a|^2+1)^{-1}(|a|^2\alpha_{ss} + 2\operatorname{Re}\{a^*\alpha_{sr}\} + \alpha_{rr})$, where $\alpha_{sr} = \mathbf{u}_s^H\mathbf{S}_{sr}\mathbf{u}_r$, $\alpha_{ss} = \operatorname{tr}(\mathbf{P}_{\mathbf{u}_s}\mathbf{S}_{ss})$, and $\alpha_{rr} = \operatorname{tr}(\mathbf{P}_{\mathbf{u}_r}\mathbf{S}_{rr})$. It is a few steps of algebra to parameterize a as $a = \xi e^{j\theta}$ and show that the maximizing values of θ and ξ are $\hat{\theta} = \arg(\alpha_{sr})$ and $\hat{\xi} = \gamma_{rs}/2 + \sqrt{\gamma_{rs}^2 + 1/2}$, where $\gamma_{rs} = (\alpha_{rr} - \alpha_{ss})/|\alpha_{sr}|$. This determines the variance estimator $\hat{\sigma}_1^2$.

As is common throughout this book, the GLR is a ratio of determinants,

$$\lambda_1 = \Lambda_1^{\frac{1}{2LN}} = \frac{\hat{\sigma}_0^2}{\hat{\sigma}_1^2}$$

where

$$\Lambda_1 = \frac{\ell(\hat{a}, \hat{\boldsymbol{\rho}}, \hat{\sigma}_1^2; \mathbf{Y})}{\ell(\hat{\boldsymbol{\rho}}, \hat{\sigma}_0^2; \mathbf{Y})}.$$

Invariances. The GLR, and the corresponding detection problem, are invariant to the transformation group $G = \{g \mid g \cdot \mathbf{Y} = \beta\mathbf{Y}\mathbf{Q}_N\}$, where $\beta \neq 0$ and $\mathbf{Q}_N \in U(N)$

an arbitrary $N \times N$ unitary matrix. That is, the detector is invariant to a common scaling of the surveillance and reference channels and a right unitary transformation of the measurement matrix $\mathbf{Y}$. Hence, the GLR is CFAR with respect to common scalings.

7.2.2 Matched Subspace Detector for Equal and Known Noise Variances

When the common noise variance σ^2 is known, then without loss of generality it may be taken to be $\sigma^2 = 1$. Under $\mathcal{H}_0$, the compressed likelihood function is

$$\ell(\hat{\rho}, \sigma^2 = 1; \mathbf{Y}) = \frac{1}{\pi^{2LN}} \operatorname{etr}\left(-N\mathbf{S}_{ss} - N\mathbf{P}_{\mathbf{u}_r}^{\perp}\mathbf{S}_{rr}\right),$$

Under $\mathcal{H}_1$ the compressed likelihood is

$$\ell(\hat{a}, \hat{\rho}, \sigma^2 = 1; \mathbf{Y}) = \frac{1}{\pi^{2LN}} \operatorname{etr}\left(-N\mathbf{P}_{\mathbf{v}}^{\perp}(\hat{a})\mathbf{S}\right),$$

where $\hat{a}$ is the solution derived in the previous subsection. The GLR is the log-likelihood ratio

$$\lambda_1 = \log \Lambda_1 = N \operatorname{tr}\left(\mathbf{P}_{\mathbf{v}}(\hat{a})\mathbf{S} - \mathbf{P}_{\mathbf{u}_r}\mathbf{S}_{rr}\right),$$

where

$$\Lambda_1 = \frac{\ell(\hat{a}, \hat{\rho}, \sigma^2 = 1; \mathbf{Y})}{\ell(\hat{\rho}, \sigma^2 = 1; \mathbf{Y})}.$$

Invariances. The GLR, and the corresponding detection problem, are invariant to the transformation group $G = \{g \mid g \cdot \mathbf{Y} = \mathbf{Y}\mathbf{Q}_N\}$, where $\mathbf{Q}_N \in U(N)$ is an arbitrary $N \times N$ unitary matrix. That is, the detector is invariant to a right unitary transformation of the measurement matrix $\mathbf{Y}$. The GLR is *not* CFAR with respect to scalings.

7.3 Detectors in a Second-Order Model for a Signal in a Known Subspace

When the signal is assigned a Gaussian prior, the joint distribution of $\mathbf{Y}$ and $\mathbf{X}$ may be marginalized for the marginal MVN distribution of $\mathbf{Y}$. The resulting measurement model is given in the SW quadrant in Table 7.1. We restrict ourselves to the case $p = 1$, since the multi-rank case requires the use of optimization techniques to obtain ML estimates of the unknown parameters. For the noise model, we consider the set $\mathcal{E}_1$ in (7.2) (i.i.d. white noise) and $\mathcal{E}_2$ in (7.3) (white noise of different variance in each channel).

The detection problem for a second-order signal model in known surveillance and reference subspaces of dimension $p = 1$ is

$$\mathcal{H}_0 : \mathbf{Y} \sim \mathcal{CN}_{2L\times N}\left(\mathbf{0}, \mathbf{I}_N \otimes \left(\begin{bmatrix} \mathbf{0} & \mathbf{0} \\ \mathbf{0} & \mathbf{u}_r q_{rr} \mathbf{u}_r^H \end{bmatrix} + \mathbf{\Sigma}\right)\right),$$
$$\mathcal{H}_1 : \mathbf{Y} \sim \mathcal{CN}_{2L\times N}\left(\mathbf{0}, \mathbf{I}_N \otimes \left(\begin{bmatrix} \mathbf{u}_s q_{ss} \mathbf{u}_s^H & \mathbf{u}_s q_{sr} \mathbf{u}_r^H \\ \mathbf{u}_r q_{sr}^* \mathbf{u}_s^H & \mathbf{u}_r q_{rr} \mathbf{u}_r^H \end{bmatrix} + \mathbf{\Sigma}\right)\right),$$

where $\mathbf{u}_s \in \mathbb{C}^L$ and $\mathbf{u}_r \in \mathbb{C}^L$ are known unitary bases for the one-dimensional subspaces $\langle \mathbf{u}_s \rangle$ and $\langle \mathbf{u}_r \rangle$.

7.3.1 Scale-Invariant Matched Subspace Detector for Equal and Unknown Noise Variances

We consider the case of white noises with identical unknown variance at both channels: $\mathbf{\Sigma} = \sigma^2 \mathbf{I}_{2L}$. The known unitary basis for the surveillance channel, $\mathbf{u}_s$, can be completed with its orthogonal complement to form the unitary matrix $\mathbf{U}_s = [\mathbf{u}_s \, \mathbf{u}_{s1}^{\perp} \cdots \mathbf{u}_{s(L-1)}^{\perp}]$. Similarly, we form the $L \times L$ unitary matrix $\mathbf{U}_r$ for the reference channel. The powers of the observations after projection into the one-dimensional surveillance and reference subspaces are denoted as $\alpha_{ss} = \mathbf{u}_s^H \mathbf{S}_{ss} \mathbf{u}_s = \operatorname{tr}(\mathbf{P}_{\mathbf{u}_s} \mathbf{S}_{ss})$ and $\alpha_{rr} = \mathbf{u}_r^H \mathbf{S}_{rr} \mathbf{u}_r = \operatorname{tr}(\mathbf{P}_{\mathbf{u}_r} \mathbf{S}_{rr})$. These values are positive real constants, with probability one. The complex cross-correlation between the surveillance and reference signals after projection is denoted $\alpha_{sr} = \mathbf{u}_s^H \mathbf{S}_{sr} \mathbf{u}_r$, which is in general complex.

Under $\mathcal{H}_0$, the covariance matrix is structured as

$$\mathbf{R}_0 = \begin{bmatrix} \sigma^2 \mathbf{I}_L & \mathbf{0} \\ \mathbf{0} & \mathbf{u}_r q_{rr} \mathbf{u}_r^H + \sigma^2 \mathbf{I}_L \end{bmatrix},$$

with unknown parameters $\xi_r = q_{rr} + \sigma^2$ and σ^2 to be estimated under a maximum likelihood framework. It is a simple exercise to show that their ML estimates are

$$\hat{\sigma}_0^2 = \frac{\operatorname{tr}(\mathbf{S}_{ss} + \mathbf{P}_{\mathbf{u}_r}^{\perp} \mathbf{S}_{rr})}{2L - 1},$$
$$\hat{\xi}_r = \hat{q}_{rr} + \hat{\sigma}_0^2 = \max(\operatorname{tr}(\mathbf{P}_{\mathbf{u}_r} \mathbf{S}_{rr}), \hat{\sigma}_0^2).$$

The resulting determinant (assuming for simplicity $\operatorname{tr}(\mathbf{P}_{\mathbf{u}_r} \mathbf{S}_{rr}) \geq \hat{\sigma}_0^2$, meaning that the power after projection in the reference channel is larger than or equal to the estimated noise variance) is

$$\det(\hat{\mathbf{R}}_0) = \frac{\left(\mathrm{tr}(\mathbf{S}_{ss} + \mathbf{P}_{\mathbf{u}_r}^{\perp}\mathbf{S}_{rr})\right)^{2L-1} \mathrm{tr}(\mathbf{P}_{\mathbf{u}_r}\mathbf{S}_{rr})}{(2L-1)^{2L-1}}.$$

Under $\mathcal{H}_1$, the covariance matrix is patterned as

$$\mathbf{R}_1 = \begin{bmatrix} \mathbf{u}_s q_{ss}\mathbf{u}_s^H + \sigma^2\mathbf{I}_L & \mathbf{u}_s q_{sr}\mathbf{u}_r^H \\ \mathbf{u}_r q_{sr}^*\mathbf{u}_s^H & \mathbf{u}_r q_{rr}\mathbf{u}_r^H + \sigma^2\mathbf{I}_L \end{bmatrix} = \begin{bmatrix} \mathbf{R}_{ss} & \mathbf{R}_{sr} \\ \mathbf{R}_{sr}^H & \mathbf{R}_{rr} \end{bmatrix}. \tag{7.7}$$

The northeast (southwest) block $\mathbf{R}_{sr} = \mathbf{u}_s q_{sr}\mathbf{u}_r^H$ ($\mathbf{R}_{sr}^H = \mathbf{u}_r q_{sr}^*\mathbf{u}_s^H$) is a rank-one matrix. The inverse of the patterned matrix in (7.7) is (see Sect. B.4)

$$\begin{bmatrix} \mathbf{R}_{ss} & \mathbf{R}_{sr} \\ \mathbf{R}_{sr}^H & \mathbf{R}_{rr} \end{bmatrix}^{-1} = \begin{bmatrix} \mathbf{M}_{rr}^{-1} & -\mathbf{R}_{ss}^{-1}\mathbf{R}_{sr}\mathbf{M}_{ss}^{-1} \\ -\mathbf{R}_{rr}^{-1}\mathbf{R}_{sr}^H\mathbf{M}_{rr}^{-1} & \mathbf{M}_{ss}^{-1} \end{bmatrix},$$

where $\mathbf{M}_{ss} = \mathbf{R}_{rr} - \mathbf{R}_{sr}^H\mathbf{R}_{ss}^{-1}\mathbf{R}_{sr}$ and $\mathbf{M}_{rr} = \mathbf{R}_{ss} - \mathbf{R}_{sr}\mathbf{R}_{rr}^{-1}\mathbf{R}_{sr}^H$ are the Schur complements of the blocks in the diagonal of $\mathbf{R}_1$. Defining $\xi_s = q_{ss} + \sigma^2$ and $\xi_r = q_{rr} + \sigma^2$, we get

$$\mathbf{M}_{ss}^{-1} = \mathbf{U}_r \begin{bmatrix} \frac{\xi_s}{\xi_s\xi_r - |q_{sr}|^2} & 0 & 0 & \dots & 0 \\ 0 & \frac{1}{\sigma^2} & 0 & \dots & 0 \\ \vdots & \vdots & \ddots & \vdots & \vdots \\ 0 & 0 & \dots & 0 & \frac{1}{\sigma^2} \end{bmatrix} \mathbf{U}_r^H,$$

$$\mathbf{M}_{rr}^{-1} = \mathbf{U}_s \begin{bmatrix} \frac{\xi_r}{\xi_s\xi_r - |q_{sr}|^2} & 0 & 0 & \dots & 0 \\ 0 & \frac{1}{\sigma^2} & 0 & \dots & 0 \\ \vdots & \vdots & \ddots & \vdots & \vdots \\ 0 & 0 & \dots & 0 & \frac{1}{\sigma^2} \end{bmatrix} \mathbf{U}_s^H,$$

and

$$-\mathbf{R}_{ss}^{-1}\mathbf{R}_{sr}\mathbf{M}_{ss}^{-1} = (-\mathbf{R}_{rr}^{-1}\mathbf{R}_{sr}^H\mathbf{M}_{rr}^{-1})^H = \mathbf{u}_s \frac{q_{sr}}{\xi_s\xi_r - |q_{sr}|^2}\mathbf{u}_r^H.$$

The northeast and southwest blocks of $\mathbf{R}_1^{-1}$ are rank-one matrices. From these results, we obtain

$$\det(\mathbf{R}_1) = (\sigma^2)^{2(L-1)}\left(\xi_s\xi_r - |q_{sr}|^2\right),$$

$$\mathrm{tr}(\mathbf{R}_1^{-1}\mathbf{S}) = \frac{\xi_s\alpha_{rr} + \xi_r\alpha_{ss} - 2\mathrm{Re}\{q_{sr}\alpha_{sr}^*\}}{\xi_s\xi_r - |q_{sr}|^2} + \frac{\mathrm{tr}(\mathbf{S}) - \alpha_{rr} - \alpha_{ss}}{\sigma^2},$$

and it must be satisfied that $\xi_s\xi_r - |q_{rs}|^2 > 0$ for the covariance matrix to be positive definite. Taking derivatives of the log-likelihood function and equating them to zero, it is easy to check that the ML estimates are $\hat{\xi}_r = \alpha_{rr}$, $\hat{\xi}_s = \alpha_{ss}$, $\hat{q}_{sr} = \alpha_{sr}$, and

$$\hat{\sigma}_1^2 = \frac{\operatorname{tr}(\mathbf{P}_{\mathbf{u}_s}^{\perp}\mathbf{S}_{ss}) + \operatorname{tr}(\mathbf{P}_{\mathbf{u}_r}^{\perp}\mathbf{S}_{rr})}{2(L-1)}.$$

Substituting these estimates and discarding constant terms, the GLR for this problem is

$$\lambda_2 = \Lambda_2^{1/N} = \frac{\left(\operatorname{tr}(\mathbf{S}_{ss} + \mathbf{P}_{\mathbf{u}_r}^{\perp}\mathbf{S}_{rr})\right)^{2L-1}\operatorname{tr}(\mathbf{P}_{\mathbf{u}_r}\mathbf{S}_{rr})}{\left(\operatorname{tr}(\mathbf{P}_{\mathbf{u}_r}^{\perp}\mathbf{S}_{rr} + \mathbf{P}_{\mathbf{u}_s}^{\perp}\mathbf{S}_{ss})\right)^{2(L-1)}\left(\operatorname{tr}(\mathbf{P}_{\mathbf{u}_s}\mathbf{S}_{ss})\operatorname{tr}(\mathbf{P}_{\mathbf{u}_r}\mathbf{S}_{rr}) - |\alpha_{sr}|^2\right)},$$

where $\lambda_2 = \det(\hat{\mathbf{R}}_0)/\det(\hat{\mathbf{R}}_1)$ and

$$\Lambda_2 = \frac{\ell(\hat{q}_{ss}, \hat{q}_{rr}, \hat{q}_{sr}, \hat{\sigma}_1^2; \mathbf{Y})}{\ell(\hat{q}_{rr}, \hat{\sigma}_0^2; \mathbf{Y})} = \frac{\ell(\hat{\xi}_s, \hat{\xi}_r, \hat{q}_{sr}, \hat{\sigma}_1^2; \mathbf{Y})}{\ell(\hat{\xi}_r, \hat{\sigma}_0^2; \mathbf{Y})}.$$

Invariances. This second-order scale-invariant MSD in white noises of equal variance is invariant to the transformation group $G = \{g \mid g \cdot \mathbf{Y} = \beta\mathbf{Y}\mathbf{Q}_N\}$, where $\beta \neq 0$ and $\mathbf{Q}_N \in U(N)$ is an arbitrary $N \times N$ unitary matrix.

7.3.2 Scale-Invariant Matched Subspace Detector for Unequal and Unknown Noise Variances

Repeating the steps of the previous section, when the noise at each channel is white but with different variance, $\boldsymbol{\Sigma}_{ss} = \sigma_s^2\mathbf{I}_L$ and $\boldsymbol{\Sigma}_{rr} = \sigma_r^2\mathbf{I}_L$, the determinant of the ML estimate of the covariance matrix under $\mathcal{H}_0$ is

$$\det(\hat{\mathbf{R}}_0) = \left(\frac{1}{L}\operatorname{tr}(\mathbf{S}_{ss})\right)^L \left(\frac{1}{L-1}\operatorname{tr}(\mathbf{P}_{\mathbf{u}_r}^{\perp}\mathbf{S}_{rr})\right)^{(L-1)} \operatorname{tr}(\mathbf{P}_{\mathbf{u}_r}\mathbf{S}_{rr}).$$

The covariance matrix under $\mathcal{H}_1$ is patterned as (7.7) with σ_s^2 replacing σ^2 in its northwest block and σ_r^2 replacing σ^2 in its southeast block. The ML estimates for the unknowns are $\hat{\xi}_r = \alpha_{rr}$, $\hat{\xi}_s = \alpha_{ss}$, $\hat{q}_{sr} = \alpha_{sr}$, and

$$\hat{\sigma}_{s,1}^2 = \frac{\operatorname{tr}(\mathbf{P}_{\mathbf{u}_s}^{\perp}\mathbf{S}_{ss})}{L-1}, \qquad \hat{\sigma}_{r,1}^2 = \frac{\operatorname{tr}(\mathbf{P}_{\mathbf{u}_r}^{\perp}\mathbf{S}_{rr})}{L-1},$$

so the determinant of $\hat{\mathbf{R}}_1$ is

$$\det(\hat{\mathbf{R}}_1) = \left(\frac{\mathrm{tr}(\mathbf{P}_{\mathbf{u}_s}^{\perp}\mathbf{S}_{ss})}{L-1}\right)^{(L-1)} \left(\frac{\mathrm{tr}(\mathbf{P}_{\mathbf{u}_r}^{\perp}\mathbf{S}_{rr})}{L-1}\right)^{(L-1)} \left(\mathrm{tr}(\mathbf{P}_{\mathbf{u}_s}\mathbf{S}_{ss})\,\mathrm{tr}(\mathbf{P}_{\mathbf{u}_r}\mathbf{S}_{rr}) - |\alpha_{sr}|^2\right).$$

The GLR is the ratio of determinants

$$\lambda_2 = \Lambda_2^{1/N} = \frac{\det(\hat{\mathbf{R}}_0)}{\det(\hat{\mathbf{R}}_1)} \tag{7.8}$$

where

$$\Lambda_2 = \frac{\ell(\hat{q}_{ss}, \hat{q}_{rr}, \hat{q}_{sr}, \hat{\sigma}_{s,1}^2, \hat{\sigma}_{r,1}^2; \mathbf{Y})}{\ell(\hat{q}_{rr}, \hat{\sigma}_{s,0}^2, \hat{\sigma}_{r,0}^2; \mathbf{Y})}.$$

Invariances. This second-order scale-invariant matched subspace detector in white noises of unequal variances (7.8) is invariant to the transformation group $G = \{g \mid g \cdot \mathbf{Y} = \mathrm{blkdiag}(\beta_s \mathbf{I}_L, \beta_r \mathbf{I}_L)\mathbf{Y}\mathbf{Q}_N\}$, where $\beta_s, \beta_r \neq 0$, and $\mathbf{Q}_N \in U(N)$ is an arbitrary $N \times N$ unitary matrix.

7.4 Detectors in a First-Order Model for a Signal in a Subspace Known Only by its Dimension

In the NE quadrant of Table 7.1, the signal model is $\mathbf{Y} = \mathbf{HX} + \mathbf{N}$, where $\mathbf{Y}$ is a $2L \times N$ matrix of measurements, $\mathbf{X}$ is a $p \times N$ matrix of unknown source signals, $\mathbf{N}$ is a $2L \times N$ matrix of noises, and $\mathbf{H}$ is an unknown $2L \times p$ channel matrix. As a consequence, the signal matrix $\mathbf{Z} = \mathbf{HX}$ is an unknown $2L \times N$ matrix of known rank p, and $\mathbf{HX}$ may be taken to be a general factorization of $\mathbf{Z}$. The channel under $\mathcal{H}_0$ is structured as

$$\mathbf{H} = \begin{bmatrix} \mathbf{0} \\ \mathbf{H}_r \end{bmatrix};$$

under $\mathcal{H}_1$ the matrix $\mathbf{H}$ is an arbitrary unknown $2L \times p$ matrix.

Under a first-order model for the measurements, the number of deterministic unknowns in $\mathbf{Z}$ increases linearly with N. The generalized likelihood approach leads to ill-posed problems except when the noise covariance matrices in the surveillance and reference channels are scaled identities, possibly with different scale factors. This can be seen as follows. Suppose the noise covariance matrix for the reference channel is diagonal with unknown variances $\boldsymbol{\Sigma}_{rr} = \mathrm{diag}(\sigma_{r,1}^2, \ldots, \sigma_{r,L}^2)$. The likelihood function for the reference channel is

$$\ell(\mathbf{H}_r, \mathbf{X}, \boldsymbol{\Sigma}_{rr}; \mathbf{Y}_r) = \frac{1}{\pi^{2LN}\det(\boldsymbol{\Sigma}_{rr})^N} \,\mathrm{etr}\left\{-\boldsymbol{\Sigma}_{rr}^{-1}(\mathbf{Y}_r - \mathbf{H}_r\mathbf{X})(\mathbf{Y}_r - \mathbf{H}_r\mathbf{X})^H\right\}.$$

Choose $\hat{\mathbf{X}}$ to be a basis for the row space of the first p rows of $\mathbf{Y}_r$ and choose $\hat{\mathbf{H}}_r = \mathbf{Y}_r\hat{\mathbf{X}}^H$. Then,

$$\mathbf{Y}_r - \hat{\mathbf{H}}_r\hat{\mathbf{X}} = \mathbf{Y}_r - \mathbf{Y}_r\hat{\mathbf{X}}^H\hat{\mathbf{X}}^H = \mathbf{Y}_r(\mathbf{I}_N - \hat{\mathbf{X}}^H\hat{\mathbf{X}}) = \begin{bmatrix} \mathbf{0} \\ \vdots \\ \mathbf{0} \\ \boldsymbol{\nu}_{p+1}^T \\ \vdots \\ \boldsymbol{\nu}_L^T \end{bmatrix},$$

where $\boldsymbol{\nu}_l^T$ denotes the lth row of $\mathbf{Y}_r$. Choosing these estimates for the source matrix and the channel, the compressed likelihood for the noise variances is

$$\ell(\hat{\mathbf{H}}_r, \hat{\mathbf{X}}, \boldsymbol{\Sigma}_{rr}; \mathbf{Y}_r) = \frac{1}{\pi^{2LN}\left(\prod_{l=1}^{L}\sigma_{r,l}^2\right)^N} \exp\left(-\sum_{l=p+1}^{L}\frac{\|\boldsymbol{\nu}_l\|^2}{\sigma_{r,l}^2}\right).$$

It is now possible to make the likelihood arbitrarily large by letting one of more of the $\sigma_{r,l}^2 \to 0$ for $l = 1, \ldots, p$. This was first pointed out in [324]. For this reason, under a first-order model for the measurements, only the noise models $\boldsymbol{\Sigma} = \sigma^2\mathbf{I}_{2L}$ (white noises) and $\boldsymbol{\Sigma} = \text{blkdiag}(\sigma_s^2\mathbf{I}_L, \sigma_r^2\mathbf{I}_L)$ (white noises but with different variances at the surveillance and reference channels) yield well-posed problems.

7.4.1 Scale-Invariant Matched Direction Detector for Equal and Unknown Noise Variances

When the noise is white with unknown scale, the noise covariance matrix $\boldsymbol{\Sigma}$ belongs to the cone $\mathcal{E}_1 = \{\boldsymbol{\Sigma} = \sigma^2\mathbf{I}_{2L} \mid \sigma^2 > 0\}$, which was defined in (7.2). We may reproduce the arguments of Lemma 4.1 in Chap. 4 to show that the trace term in a likelihood function evaluated at the ML estimate for σ^2 is a constant equal to the dimension of the observations, in this case $2L$. Since this argument holds under both hypotheses, it follows that the GLR is

$$\lambda_1 = \Lambda_1^{\frac{1}{2LN}} = \frac{\hat{\sigma}_0^2}{\hat{\sigma}_1^2},$$

where

$$\Lambda_1 = \frac{\ell(\hat{\mathbf{H}}_s, \hat{\mathbf{H}}_r, \hat{\mathbf{X}}, \hat{\sigma}_1^2; \mathbf{Y})}{\ell(\hat{\mathbf{H}}_r, \hat{\mathbf{X}}, \hat{\sigma}_0^2; \mathbf{Y})}.$$

The $\hat{\sigma}_i^2$, $i = 0, 1$, are the ML estimates of the noise variance under $\mathcal{H}_i$. It remains to find these estimates.

Under $\mathcal{H}_0$, the likelihood is

$$\ell(\mathbf{H}_r, \mathbf{X}, \sigma^2; \mathbf{Y}) = \frac{1}{(\pi\sigma^2)^{2LN}} \operatorname{etr}\left\{-\frac{1}{\sigma^2}\mathbf{Y}_s\mathbf{Y}_s^H\right\} \operatorname{etr}\left\{-\frac{1}{\sigma^2}(\mathbf{Y}_r - \mathbf{H}_r\mathbf{X})(\mathbf{Y}_r - \mathbf{H}_r\mathbf{X})^H\right\}.$$

The ML estimate of $\mathbf{H}_r\mathbf{X}$ is the best rank-p approximation of $\mathbf{Y}_r$ according to the Frobenius norm. If the $L \times N$ matrix of observations for the reference channel has singular value decomposition

$$\mathbf{Y}_r = \mathbf{F}_r \left[\operatorname{diag}\left(ev_1(\mathbf{S}_{rr}^{1/2}), \ldots, ev_L(\mathbf{S}_{rr}^{1/2})\right) \; \mathbf{0}_{L\times(N-L)}\right] \mathbf{G}_r^H,$$

with singular values $ev_1(\mathbf{S}_{rr}^{1/2}) \geq \cdots \geq ev_L(\mathbf{S}_{rr}^{1/2})$. Then the value of $\mathbf{H}_r\mathbf{X}$ that maximizes the likelihood is

$$\hat{\mathbf{H}}_r\hat{\mathbf{X}} = \mathbf{F}_r \left[\operatorname{diag}\left(ev_1(\mathbf{S}_{rr}^{1/2}), \ldots, ev_p(\mathbf{S}_{rr}^{1/2}), 0, \ldots, 0\right) \; \mathbf{0}_{L\times(N-L)}\right] \mathbf{G}_r^H.$$

Plugging these ML estimates for $\mathbf{H}_r\mathbf{X}$ and discarding constant terms, the ML estimate of the noise variance under the null is derived as

$$\hat{\sigma}_0^2 = \frac{1}{2L}\left(\operatorname{tr}(\mathbf{S}) - \sum_{l=1}^{p} ev_l(\mathbf{S}_{rr})\right).$$

Under the alternative, the ML estimate of $\mathbf{HX}$ is the best rank-p approximation of $\mathbf{Y}$, and the ML estimate of the noise variance is

$$\hat{\sigma}_1^2 = \frac{1}{2L}\left(\operatorname{tr}(\mathbf{S}) - \sum_{l=1}^{p} ev_l(\mathbf{S})\right).$$

Now, substituting these ML estimates in the GLR, the test statistic is

$$\lambda_1 = \frac{\operatorname{tr}(\mathbf{S}) - \sum_{l=1}^{p} ev_l(\mathbf{S}_{rr})}{\operatorname{tr}(\mathbf{S}) - \sum_{l=1}^{p} ev_l(\mathbf{S})}. \tag{7.9}$$

This result extends the one-channel multipulse CFAR matched direction detector derived in Chap. 5 to a two-channel passive detection problem.

Invariances. The detector statistic is invariant to common scaling of the surveillance and reference channels and to independent transformations $\mathbf{Q}_s\mathbf{Y}_s$ and $\mathbf{Q}_r\mathbf{Y}_r$, where $\mathbf{Q}_s \in U(L)$ and $\mathbf{Q}_r \in U(L)$. It is invariant to a right multiplication by an $N \times N$ unitary matrix $\mathbf{Q}_N \in U(N)$. That is, the invariant transformation group for the GLR in (7.9), and the corresponding detection problem, is $G =$

$\{g \mid g \cdot \mathbf{Y} = \beta \, \text{blkdiag}(\mathbf{Q}_s, \mathbf{Q}_r)\mathbf{Y}\mathbf{Q}_N\}$, where $\beta \neq 0$, $\mathbf{Q}_s, \mathbf{Q}_r \in U(L)$, and $\mathbf{Q}_N \in U(N)$.

7.4.2 Matched Direction Detector for Equal and Known Noise Variances

When the common noise variance σ^2 is known, it may be assumed without loss of generality that $\sigma^2 = 1$. Under $\mathcal{H}_0$, the likelihood is

$$\ell(\mathbf{H}_r, \mathbf{X}; \mathbf{Y}) = \frac{1}{\pi^{2LN}} \text{etr}(-N\mathbf{S}_{ss}) \, \text{etr}\left\{-(\mathbf{Y}_r - \mathbf{H}_r\mathbf{X})(\mathbf{Y}_r - \mathbf{H}_r\mathbf{X})^H\right\}.$$

Discarding constant terms, it is easy to check that the maximum of the log-likelihood under the null is

$$\log \ell(\hat{\mathbf{H}}_r, \hat{\mathbf{X}}; \mathbf{Y}) = -N \, \text{tr}(\mathbf{S}_{ss}) - N \sum_{l=p+1}^{L} ev_l(\mathbf{S}_{rr}) = -N \, \text{tr}(\mathbf{S}) + N \sum_{l=1}^{p} ev_l(\mathbf{S}_{rr}).$$

Following a similar procedure, the maximum of the log-likelihood under $\mathcal{H}_1$ is

$$\log \ell(\hat{\mathbf{H}}_s, \hat{\mathbf{H}}_r, \hat{\mathbf{X}}; \mathbf{Y}) = -N \sum_{l=p+1}^{2L} ev_l(\mathbf{S}) = -N \, \text{tr}(\mathbf{S}) + N \sum_{l=1}^{p} ev_l(\mathbf{S}).$$

Then, the GLR is

$$\lambda_1 = \frac{1}{N} \log \Lambda_1 = \sum_{l=1}^{p} \left(ev_l(\mathbf{S}) - ev_l(\mathbf{S}_{rr}) \right), \tag{7.10}$$

where

$$\Lambda_1 = \frac{\ell(\hat{\mathbf{H}}_s, \hat{\mathbf{H}}_r, \hat{\mathbf{X}}; \mathbf{Y})}{\ell(\hat{\mathbf{H}}_r, \hat{\mathbf{X}}; \mathbf{Y})}.$$

For $p = 1$, this is the result by Hack et al. in [153]. The GLR (7.10) generalizes the detector in [153] to an arbitrary p.

Invariances. Compared to the scale-invariant matched direction detector in (7.9), the GLR in (7.10) loses the invariance to scale. Hence, the invariant transformation group is $G = \{g \mid g \cdot \mathbf{Y} = \text{blkdiag}(\mathbf{Q}_s, \mathbf{Q}_r)\mathbf{Y}\mathbf{Q}_N\}$, where $\mathbf{Q}_s, \mathbf{Q}_r \in U(L)$, and $\mathbf{Q}_N \in U(N)$.

7.4.3 Scale-Invariant Matched Direction Detector in Noises of Different and Unknown Variances

The GLR is again a ratio of determinants, which in this case reads

$$\lambda_1 = \Lambda_1^{1/LN} = \frac{\hat{\sigma}_{s,0}^2 \hat{\sigma}_{r,0}^2}{\hat{\sigma}_{s,1}^2 \hat{\sigma}_{r,1}^2}, \tag{7.11}$$

where $\hat{\sigma}_{s,i}^2$ and $\hat{\sigma}_{r,i}^2$, $i = 0, 1$ are respectively the ML estimates of the noise variance of the surveillance and reference channels under $\mathcal{H}_i$, and

$$\Lambda_1 = \frac{\ell(\hat{\mathbf{H}}_s, \hat{\mathbf{H}}_r, \hat{\mathbf{X}}, \hat{\sigma}_{s,1}^2, \hat{\sigma}_{r,1}^2; \mathbf{Y})}{\ell(\hat{\mathbf{H}}_r, \hat{\mathbf{X}}, \hat{\sigma}_{s,0}^2, \hat{\sigma}_{r,0}^2; \mathbf{Y})}.$$

Under $\mathcal{H}_0$, the likelihood is

$$\ell(\mathbf{H}_r, \mathbf{X}, \sigma_s^2, \sigma_r^2; \mathbf{Y}) = \frac{1}{\pi^{2LN}(\sigma_s^2)^{LN}(\sigma_r^2)^{LN}} \operatorname{etr}\left(-\frac{N}{\sigma_s^2}\mathbf{S}_{ss}\right) \times \operatorname{etr}\left\{-\frac{1}{\sigma_r^2}(\mathbf{Y}_r - \mathbf{H}_r\mathbf{X})(\mathbf{Y}_r - \mathbf{H}_r\mathbf{X})^H\right\}.$$

The ML estimates of the noise variances are

$$\hat{\sigma}_{s,0}^2 = \frac{\operatorname{tr}(\mathbf{S}_{ss})}{L}, \qquad \hat{\sigma}_{r,0}^2 = \frac{\operatorname{tr}(\mathbf{S}_{rr}) - \sum_{l=1}^{p} ev_l(\mathbf{S}_{rr})}{L},$$

which can be derived using results from previous subsections.

Under $\mathcal{H}_1$, we need an iterative procedure to obtain the ML estimates of the noise variances, $\sigma_{s,1}^2$ and $\sigma_{r,1}^2$, and the rank-p signal component $\mathbf{HX}$. Let $\mathbf{S}_{\boldsymbol{\Sigma}}$ be the noise-whitened sample covariance matrix $\mathbf{S}_{\boldsymbol{\Sigma}} = \boldsymbol{\Sigma}^{-1/2}\mathbf{S}\boldsymbol{\Sigma}^{-1/2}$, with EVD $\mathbf{S}_{\boldsymbol{\Sigma}} = \mathbf{W}\operatorname{diag}(ev_1(\mathbf{S}_{\boldsymbol{\Sigma}}), \ldots, ev_{2L}(\mathbf{S}_{\boldsymbol{\Sigma}}))\,\mathbf{W}^H$. Then, for a given matrix $\boldsymbol{\Sigma} = \operatorname{blkdiag}(\sigma_s^2\mathbf{I}_L, \sigma_r^2\mathbf{I}_L)$, the value of $\mathbf{HX}$ that maximizes the likelihood is

$$\hat{\mathbf{H}}\hat{\mathbf{X}} = \boldsymbol{\Sigma}^{1/2}\mathbf{W}\mathbf{D}^{1/2},$$

where $\mathbf{D} = \operatorname{diag}\left(ev_1(\mathbf{S}_{\boldsymbol{\Sigma}}), \ldots, ev_p(\mathbf{S}_{\boldsymbol{\Sigma}}), 0, \ldots, 0\right)$. This fixes the values of $\mathbf{H}_s\mathbf{X}$ and $\mathbf{H}_r\mathbf{X}$. The noise variances that maximize the likelihood are

$$\hat{\sigma}_{s,1}^2 = \frac{1}{NL}\operatorname{tr}\left((\mathbf{Y}_s - \mathbf{H}_s\mathbf{X})(\mathbf{Y}_s - \mathbf{H}_s\mathbf{X})^H\right),$$

$$\hat{\sigma}_{r,1}^2 = \frac{1}{NL}\operatorname{tr}\left((\mathbf{Y}_r - \mathbf{H}_r\mathbf{X})(\mathbf{Y}_r - \mathbf{H}_r\mathbf{X})^H\right).$$

Iterating between these convergent steps, we obtain the ML estimates $\hat{\sigma}_{s,1}^2$ and $\hat{\sigma}_{r,1}^2$. Substituting the final estimates into (7.11) yields the GLR for this model.

An approximate closed-form GLR can be obtained by estimating the noise variances under $\mathcal{H}_1$ directly from the surveillance and reference channels as

$$\hat{\sigma}_{s,1}^2 = \frac{1}{L} \sum_{l=p+1}^{L} ev_l(\mathbf{S}_{ss}), \qquad \hat{\sigma}_{r,1}^2 = \frac{1}{L} \sum_{l=p+1}^{L} ev_l(\mathbf{S}_{rr}).$$

With these approximations, the GLR may be approximated as

$$\lambda_{app} = \exp(-2)\, \Lambda_{app}^{1/LN} = \frac{\operatorname{tr}(\mathbf{S}_{ss})}{\operatorname{tr}(\mathbf{S}_{ss}) - \sum_{l=1}^{p} ev_l(\mathbf{S}_{ss})} \exp\left(-\frac{1}{L} \sum_{l=p+1}^{2L} ev_l\left(\mathbf{S}_{\hat{\boldsymbol{\Sigma}}}\right) \right), \tag{7.12}$$

where

$$\Lambda_{app} = \frac{\ell(\hat{\mathbf{H}}_s, \hat{\mathbf{H}}_r, \hat{\mathbf{X}}, \hat{\sigma}_{s,1}^2, \hat{\sigma}_{r,1}^2; \mathbf{Y})}{\ell(\hat{\mathbf{H}}_r, \hat{\mathbf{X}}, \hat{\sigma}_{s,0}^2, \hat{\sigma}_{r,0}^2; \mathbf{Y})}.$$

The whitened covariance matrix is approximated as

$$\mathbf{S}_{\hat{\boldsymbol{\Sigma}}} = \begin{bmatrix} \frac{\mathbf{S}_{ss}}{\hat{\sigma}_{s,1}^2} & \frac{\mathbf{S}_{sr}}{\hat{\sigma}_{s,1}\hat{\sigma}_{r,1}} \\ \frac{\mathbf{S}_{sr}^H}{\hat{\sigma}_{s,1}\hat{\sigma}_{r,1}} & \frac{\mathbf{S}_{rr}}{\hat{\sigma}_{r,1}^2} \end{bmatrix}.$$

The leading ratio term in the detector (7.12) is a GLR for the surveillance channel; the exponential term takes into account the coupling between the two channels.

Invariances. The GLR in (7.11) is invariant to the transformation group $G = \{g \mid g \cdot \mathbf{Y} = \operatorname{blkdiag}(\beta_s \mathbf{Q}_s, \beta_r \mathbf{Q}_r)\mathbf{Y}\mathbf{Q}_N\}$, where $\beta_s, \beta_r \neq 0$, $\mathbf{Q}_s, \mathbf{Q}_r \in U(L)$, and $\mathbf{Q}_N \in U(N)$.

7.4.4 Matched Direction Detector in Noises of Known but Different Variances

The likelihood function under $\mathcal{H}_0$ is maximized with respect to $\mathbf{H}_r\mathbf{X}$ at

$$\ell(\hat{\mathbf{H}}_r, \hat{\mathbf{X}}, \sigma_s^2, \sigma_r^2; \mathbf{Y}) = \frac{1}{\pi^{2LN}(\sigma_s^2\sigma_r^2)^{LN}} \exp\left\{ -\frac{N}{\sigma_s^2} \sum_{l=1}^{L} ev_l(\mathbf{S}_{ss}) - \frac{N}{\sigma_r^2} \sum_{l=p+1}^{L} ev_l(\mathbf{S}_{rr}) \right\}.$$

The likelihood function under $\mathcal{H}_1$ is maximized with respect to **HX** at

$$\ell(\hat{\mathbf{H}}_s, \hat{\mathbf{H}}_r, \hat{\mathbf{X}}, \sigma_s^2, \sigma_r^2; \mathbf{Y}) = \frac{1}{\pi^{2LN}(\sigma_s^2\sigma_r^2)^{LN}} \exp\left\{-N \sum_{l=p+1}^{L} ev_l(\mathbf{S}_{\boldsymbol{\Sigma}})\right\},$$

where $\mathbf{S}_{\boldsymbol{\Sigma}}$ is the whitened sample covariance $\mathbf{S}_{\boldsymbol{\Sigma}} = \boldsymbol{\Sigma}^{-1/2}\mathbf{S}\boldsymbol{\Sigma}^{-1/2}$ and $\boldsymbol{\Sigma} = \text{blkdiag}(\sigma_s^2\mathbf{I}_L, \sigma_r^2\mathbf{I}_L)$.

7.5 Detectors in a Second-Order Model for a Signal in a Subspace Known Only by its Dimension

In a second-order model, the signal sequence $\{\mathbf{x}_n\}$ is assumed to be a sequence of proper, complex Gaussian, random vectors with zero mean and unknown covariance matrix $E[\mathbf{x}\mathbf{x}^H] = \mathbf{R}_{\mathbf{xx}}$. From the joint distribution of the measurement and signal, the marginal distribution of the measurement is determined by integrating this joint distribution over $\mathbf{x} \in \mathbb{C}^p$. Since the subspaces are unknown, $\mathbf{R}_{\mathbf{xx}}$ may be absorbed into the unknown channels, and thus it may be assumed $\mathbf{R}_{\mathbf{xx}} = \mathbf{I}_p$. The signal model corresponds to the SE quadrant of Table 7.1. The detection problem becomes a hypothesis testing problem on the structure of the covariance matrix for the measurements.

For the covariance of the noise component, we consider the four different models presented in Sect. 7.1. The detection problem for a second-order signal model in unknown surveillance and reference subspaces of known dimension is

$$\begin{aligned}
\mathcal{H}_0 &: \mathbf{Y} \sim \mathcal{CN}_{2L\times N}\left(\mathbf{0}, \mathbf{I}_N \otimes \left(\begin{bmatrix} \mathbf{0} & \mathbf{0} \\ \mathbf{0} & \mathbf{H}_r\mathbf{H}_r^H \end{bmatrix} + \boldsymbol{\Sigma}\right)\right), \\
\mathcal{H}_1 &: \mathbf{Y} \sim \mathcal{CN}_{2L\times N}\left(\mathbf{0}, \mathbf{I}_N \otimes \left(\begin{bmatrix} \mathbf{H}_s\mathbf{H}_s^H & \mathbf{H}_s\mathbf{H}_r^H \\ \mathbf{H}_r\mathbf{H}_s^H & \mathbf{H}_r\mathbf{H}_r^H \end{bmatrix} + \boldsymbol{\Sigma}\right)\right).
\end{aligned}$$

This second-order detection problem essentially amounts to testing between the two different structures for the composite covariance matrix under the null hypothesis and alternative hypothesis. There are two possible interpretations of this model: (1) it is a one-channel factor model with special constraints on the loadings under $\mathcal{H}_0$, or (2) it is a two-channel factor model with common factors in the two channels under $\mathcal{H}_1$ and no loadings of the surveillance channel under $\mathcal{H}_0$.

The sets defining the structured covariance matrices under each of the two hypotheses are

$$\mathcal{R}_0 = \left\{\begin{bmatrix} \mathbf{0} & \mathbf{0} \\ \mathbf{0} & \mathbf{H}_r\mathbf{H}_r^H \end{bmatrix} + \boldsymbol{\Sigma}, \quad \text{for } \boldsymbol{\Sigma} \in \mathcal{E}\right\},$$

$$\mathcal{R}_1 = \left\{\begin{bmatrix} \mathbf{H}_s\mathbf{H}_s^H & \mathbf{H}_s\mathbf{H}_r^H \\ \mathbf{H}_r\mathbf{H}_s^H & \mathbf{H}_r\mathbf{H}_r^H \end{bmatrix} + \boldsymbol{\Sigma}, \quad \text{for } \boldsymbol{\Sigma} \in \mathcal{E}, \right\},$$

where $\mathcal{E}$ indicates any one of the noise covariance models described in Sect. 7.1. Since these sets are cones, the resulting GLR is a ratio of determinants

$$\lambda_2 = \Lambda_2^{1/N} = \frac{\det(\hat{\mathbf{R}}_0)}{\det(\hat{\mathbf{R}}_1)}, \tag{7.13}$$

where

$$\Lambda_2 = \frac{\ell(\hat{\mathbf{R}}_1; \mathbf{Y})}{\ell(\hat{\mathbf{R}}_0; \mathbf{Y})},$$

and $\hat{\mathbf{R}}_i$ is the ML estimate of the covariance matrix under $\mathcal{H}_i$ with the required structure.

Optimization Problems for ML Estimation. In this case the ML estimates of covariance may be obtained by solving the following optimization problem:

$$\begin{array}{lll} \text{Problem 1:} & \underset{\mathbf{R}\in\mathcal{R}_i}{\text{maximize}} & \log\det\left(\mathbf{R}^{-1}\mathbf{S}\right), \\ & \text{subject to} & \operatorname{tr}\left(\mathbf{R}^{-1}\mathbf{S}\right) = 2L. \end{array} \tag{7.14}$$

The following theorem illuminates the problem of determining $\mathbf{R}$ in Problem 1 and leads also to an alternative formulation to be given in Problem 2.

Theorem 7.1 *For a given block-diagonal noise covariance $\boldsymbol{\Sigma}$, define the noise-whitened sample covariance matrix and its eigenvalue decomposition*

$$\mathbf{S}_{\boldsymbol{\Sigma}} = \boldsymbol{\Sigma}^{-1/2}\mathbf{S}\boldsymbol{\Sigma}^{-1/2} = \begin{bmatrix} \mathbf{S}_{\boldsymbol{\Sigma},ss} & \mathbf{S}_{\boldsymbol{\Sigma},sr} \\ \mathbf{S}_{\boldsymbol{\Sigma},sr}^H & \mathbf{S}_{\boldsymbol{\Sigma},rr} \end{bmatrix} = \mathbf{W}\,\mathrm{diag}\left(ev_1(\mathbf{S}_{\boldsymbol{\Sigma}}), \ldots, ev_{2L}(\mathbf{S}_{\boldsymbol{\Sigma}})\right)\mathbf{W}^H, \tag{7.15}$$

with $ev_1(\mathbf{S}_{\boldsymbol{\Sigma}}) \geq \cdots \geq ev_{2L}(\mathbf{S}_{\boldsymbol{\Sigma}})$. Similarly, the southeast block has eigenvalue decomposition

$$\mathbf{S}_{\boldsymbol{\Sigma},rr} = \mathbf{W}_{rr}\,\mathrm{diag}\left(ev_1(\mathbf{S}_{\boldsymbol{\Sigma},rr}), \ldots, ev_L(\mathbf{S}_{\boldsymbol{\Sigma},rr})\right)\mathbf{W}_{rr}^H,$$

with $ev_1(\mathbf{S}_{\boldsymbol{\Sigma},rr}) \geq \cdots \geq ev_L(\mathbf{S}_{\boldsymbol{\Sigma},rr})$. Then, under the alternative $\mathcal{H}_1$, the value of $\mathbf{H}\mathbf{H}^H$ that maximizes the likelihood is

$$\hat{\mathbf{H}}\hat{\mathbf{H}}^H = \boldsymbol{\Sigma}^{1/2}\mathbf{W}\mathbf{D}\mathbf{W}^H\boldsymbol{\Sigma}^{1/2},$$

where $\mathbf{D} = \mathrm{diag}\left(d_1, \ldots, d_p, 0, \ldots, 0\right)$, and $d_l = (ev_l(\mathbf{S}_{\boldsymbol{\Sigma}}) - 1)^+$.

For a given noise covariance matrix $\boldsymbol{\Sigma}_{rr}$ in the reference channel, the value of $\mathbf{H}_r\mathbf{H}_r^H$ that maximizes the likelihood under the null is

$$\hat{\mathbf{H}}_r\hat{\mathbf{H}}_r^H = \boldsymbol{\Sigma}_{rr}^{1/2}\mathbf{W}_{rr}\mathbf{D}_{rr}\mathbf{W}_{rr}^H\boldsymbol{\Sigma}_{rr}^{1/2}, \tag{7.16}$$

where $\mathbf{D}_{rr} = \operatorname{diag}\left(d_{rr,1}, \dots, d_{rr,p}, 0, \dots, 0\right)$, *and* $d_{rr,l} = (ev_l(\mathbf{S}_{\boldsymbol{\Sigma},rr}) - 1)^+$.

Proof The proof for $\mathcal{H}_1$ is identical to Theorem 9.4.1 in [227] (cf. pages 264–265). The proof for $\mathcal{H}_0$ is straightforward after we rewrite the log-likelihood function using the block-wise decomposition in (7.15) and use the fact that the noise covariance $\boldsymbol{\Sigma}$ is block diagonal. □

Theorem 7.1 can be used to derive Problem 2 for the ML estimate of covariance, under the alternative $\mathcal{H}_1$. For a given $\boldsymbol{\Sigma}$, Theorem 7.1 gives the value of $\mathbf{H}\mathbf{H}^H$ that maximizes the log-likelihood function with respect to $\mathbf{R} = \mathbf{H}\mathbf{H}^H + \boldsymbol{\Sigma}$. Thus, we have the solution $\mathbf{R} = \boldsymbol{\Sigma}^{1/2}\mathbf{W}\mathbf{D}\mathbf{W}^H\boldsymbol{\Sigma}^{1/2} + \boldsymbol{\Sigma}$. Straightforward calculation shows that $\det(\mathbf{R}^{-1}\mathbf{S}) = \prod_{l=1}^{p}\min(ev_l(\mathbf{S}_{\boldsymbol{\Sigma}}), 1)\prod_{l=p+1}^{2L} ev_l(\mathbf{S}_{\boldsymbol{\Sigma}})$ and $\operatorname{tr}(\mathbf{R}^{-1}\mathbf{S}) = \sum_{l=1}^{p}\min(ev_l(\mathbf{S}_{\boldsymbol{\Sigma}}), 1) + \sum_{l=p+1}^{2L} ev_l(\mathbf{S}_{\boldsymbol{\Sigma}})$. Therefore, Problem 1 may be rewritten as

$$\begin{aligned} \text{Problem 2:}\quad & \underset{\boldsymbol{\Sigma}\in\mathcal{E}}{\text{maximize}} \left(\prod_{l=1}^{p}\min(ev_l(\mathbf{S}_{\boldsymbol{\Sigma}}), 1)\prod_{l=p+1}^{2L} ev_l(\mathbf{S}_{\boldsymbol{\Sigma}})\right)^{\frac{1}{2L}}, \\ & \text{subject to}\ \frac{1}{2L}\left(\sum_{l=1}^{p}\min(ev_l(\mathbf{S}_{\boldsymbol{\Sigma}}), 1) + \sum_{l=p+1}^{2L} ev_l(\mathbf{S}_{\boldsymbol{\Sigma}})\right) = 1. \end{aligned} \tag{7.17}$$

Recall that $ev_1(\mathbf{S}_{\boldsymbol{\Sigma}}) \geq \cdots \geq ev_{2L}(\mathbf{S}_{\boldsymbol{\Sigma}}) \geq 0$ is the set of ordered eigenvalues of the noise-whitened sample covariance matrix. Thus, the trace constraint in (7.17) directly implies $ev_l(\mathbf{S}_{\boldsymbol{\Sigma}}) \geq 1$ for $l = 1, \dots, p$. In consequence, Problem 2 can be written more compactly as

$$\begin{aligned} \text{Problem 2}:\quad & \underset{\boldsymbol{\Sigma}\in\mathcal{E}}{\text{maximize}} \left(\prod_{l=p+1}^{2L} ev_l(\mathbf{S}_{\boldsymbol{\Sigma}})\right)^{\frac{1}{2L-p}}, \\ & \text{subject to}\ \frac{1}{2L-p}\sum_{l=p+1}^{2L} ev_l(\mathbf{S}_{\boldsymbol{\Sigma}}) = 1. \end{aligned} \tag{7.18}$$

That is, the ML estimation problem under the alternative hypothesis comes down to finding the noise covariance matrix with the required structure that maximizes the geometric mean of the trailing eigenvalues of the noise-whitened sample covariance matrix, subject to the constraint that the arithmetic mean of these trailing eigenvalues is 1. For some specific structures, Problem 2 may significantly simplify the derivation of the ML solution, as shown later.

The GLR may now be derived for different noise models. For white noises with identical variances at both channels, or for noises with arbitrary correlation, the GLRs admit a closed-form expression. For white noises with different variances at the surveillance and reference channels, or for diagonal noise covariance matrices, closed-form GLRs do not exist, and one resorts to iterative algorithms to approximate the ML estimates of the unknown parameters. One of these iterative algorithms that are particularly efficient is the alternating optimization method presented later in this chapter.

7.5.1 Scale-Invariant Matched Direction Detector for Equal and Unknown Noise Variances

For $\boldsymbol{\Sigma} = \sigma^2\mathbf{I}_{2L}$, with unknown variance σ^2, the GLR may be called a scale-invariant matched direction detector. We assume that $p < L-1$, since otherwise the covariance matrices would not be modeled as the sum of a low-rank non-negative definite matrix plus a scaled identity.

Suppose the sample covariance matrices have these eigenvalue decompositions:

$$\begin{aligned}\mathbf{S} &= \mathbf{W}\,\mathrm{diag}(ev_1(\mathbf{S}),\ldots,ev_{2L}(\mathbf{S}))\mathbf{W}^H,\\ \mathbf{S}_{ss} &= \mathbf{W}_{ss}\,\mathrm{diag}(ev_1(\mathbf{S}_{ss}),\ldots,ev_L(\mathbf{S}_{ss}))\mathbf{W}_{ss}^H,\\ \mathbf{S}_{rr} &= \mathbf{W}_{rr}\,\mathrm{diag}(ev_1(\mathbf{S}_{rr}),\ldots,ev_L(\mathbf{S}_{rr}))\mathbf{W}_{rr}^H,\end{aligned}$$

all with eigenvalues ordered as $ev_1(\mathbf{S}) \geq ev_2(\mathbf{S}) \geq \cdots \geq ev_{2L}(\mathbf{S})$ taking $\mathbf{S}$ as an example. When $\boldsymbol{\Sigma} = \sigma^2\mathbf{I}_{2L}$, Problem 2 in (7.18) directly gives the ML solution for σ^2 under the alternative hypothesis $\mathcal{H}_1$ by realizing that

$$\frac{1}{2L-p}\sum_{l=p+1}^{2L} ev_l(\mathbf{S}_{\boldsymbol{\Sigma}}) = \frac{1}{2L-p}\sum_{l=p+1}^{2L}\frac{1}{\sigma^2}ev_l(\mathbf{S}),$$

which returns the ML estimate

$$\hat{\sigma}_1^2 = \frac{1}{2L-p}\sum_{l=p+1}^{2L} ev_l(\mathbf{S}). \tag{7.19}$$

Therefore, the ML estimate of the covariance matrix under the alternative $\mathcal{H}_1$ is

$$\hat{\mathbf{R}}_1 = \mathbf{W}\mathbf{D}\mathbf{W}^H + \hat{\sigma}_1^2\mathbf{I}_{2L},$$

where $\mathbf{D} = \text{diag}(d_1, \ldots, d_p, 0, \ldots, 0)$ is an $2L \times 2L$ diagonal matrix with $d_l = ev_l(\mathbf{S}) - \hat{\sigma}_1^2$; $d_l \geq 0$ by virtue of the eigenvalue ordering.

Under the null hypothesis, for a given $\mathbf{\Sigma}_{rr} = \sigma_0^2 \mathbf{I}_L$, the result in Theorem 7.1 gives the value of $\mathbf{H}_r \mathbf{H}_r^H$ that maximizes the likelihood. Then, $\mathbf{R}_0$ is a function solely of σ_0^2,

$$\mathbf{R}_0 = \begin{bmatrix} \mathbf{0} & \mathbf{0} \\ \mathbf{0} & \hat{\mathbf{H}}_r \hat{\mathbf{H}}_r^H \end{bmatrix} + \sigma_0^2 \mathbf{I}_{2L}, \tag{7.20}$$

where $\hat{\mathbf{H}}_r \hat{\mathbf{H}}_r^H = \mathbf{W}_{rr} \mathbf{D}_{rr} \mathbf{W}_{rr}^H$, $\mathbf{D}_{rr} = \text{diag}(d_{rr,1}, \ldots, d_{rr,p}, 0, \ldots, 0)$ is an $L \times L$ diagonal matrix and $d_{rr,l} = (ev_l(\mathbf{S}_{rr}) - \sigma_0^2)^+$. Taking the inverse of (7.20), it is straightforward to show that the trace constraint is

$$\text{tr}(\mathbf{R}_0^{-1}\mathbf{S}) = p_r + \frac{1}{\sigma_0^2} \sum_{l=p_r+1}^{L} ev_l(\mathbf{S}_{rr}) + \frac{1}{\sigma_0^2} \text{tr}(\mathbf{S}_{ss}) = 2L,$$

where $p_r = \min(p, p_0)$ and p_0 is the number of eigenvalues satisfying $ev_l(\mathbf{S}_{rr}) \geq \sigma_0^2$. Therefore, the ML estimate of the noise variance is

$$\hat{\sigma}_0^2 = \frac{1}{2L - p_r} \left(\sum_{l=1}^{L} ev_l(\mathbf{S}_{ss}) + \sum_{l=p_r+1}^{L} ev_l(\mathbf{S}_{rr}) \right),$$

and the covariance matrix under the null is

$$\hat{\mathbf{R}}_0 = \begin{bmatrix} \mathbf{0} & \mathbf{0} \\ \mathbf{0} & \hat{\mathbf{H}}_r \hat{\mathbf{H}}_r^H \end{bmatrix} + \hat{\sigma}_0^2 \mathbf{I}_{2L}.$$

Plugging the ML estimates, $\hat{\mathbf{R}}_0$ and $\hat{\mathbf{R}}_1$, into (7.13), the GLR for white noises with identical unknown variance is given by

$$\lambda_2 = \frac{(\hat{\sigma}_0^2)^{2L-p_r} \prod_{l=1}^{p_r} ev_l(\mathbf{S}_{rr})}{(\hat{\sigma}_1^2)^{2L-p} \prod_{l=1}^{p} ev_l(\mathbf{S})}, \tag{7.21}$$

where, recall, p_r is the largest value of l between 1 and p such that $ev_l(\mathbf{S}_{rr}) \geq \hat{\sigma}_0^2$.

In practice, the procedure for obtaining the ML estimate of σ_0^2 starts with $p_r = p$ and then checks whether the candidate solution satisfies $ev_{p_r}(\mathbf{S}_{rr}) \geq \hat{\sigma}_0^2$. If the condition is not satisfied, the rank of the signal subspace is decreased to $p_r = p - 1$, which implies in turn a decrease in the estimate of the noise variance until the condition $ev_{p_r}(\mathbf{S}_{rr}) \geq \hat{\sigma}_0^2$ is satisfied. The intuition behind this behavior is clear. If the assumed dimension of the signal subspace is not compatible with the estimated noise variance $\hat{\sigma}_0^2$, that is, if the number of signal mode powers above the estimated noise level, $\hat{\sigma}_0^2$, is lower than expected, then the dimension of the signal subspace is reduced, and the noise variance is estimated based on a lower-dimensional signal subspace and correspondingly a larger-dimensional noise subspace. Thus, the potential solutions for the ML estimates under the null range from the case $p_r = p$ (meaning that it is possible to estimate a signal subspace of dimension p in the reference channel) to the case $p_r = 0$ when the sample variance in the surveillance channel is larger than the sample variance in the reference channel, which leads to the conclusion that all the energy in the reference channel is due to noise.

Invariances. As in the analogous problem for first-order models, the detector statistic is invariant to the transformation group $G = \{g \mid g \cdot \mathbf{Y} = \beta \,\text{blkdiag}(\mathbf{Q}_s, \mathbf{Q}_r)\mathbf{Y}\mathbf{Q}_N\}$, where $\beta \neq 0$, $\mathbf{Q}_s, \mathbf{Q}_r \in U(L)$, and $\mathbf{Q}_N \in U(N)$.

7.5.2 Matched Direction Detector for Equal and Known Noise Variances

When the noise variance σ^2 is known, the ML estimate of the covariance under the alternative is

$$\hat{\mathbf{R}}_1 = \mathbf{W}\mathbf{D}\mathbf{W}^H + \sigma^2\mathbf{I}_{2L}, \tag{7.22}$$

where $\mathbf{D} = \text{diag}\left(ev_1(\mathbf{S}) - \sigma^2, \ldots, ev_{p_a}(\mathbf{S}) - \sigma^2, 0, \ldots, 0\right)$ and p_a is the largest value of l between 1 and p such that $ev_l(\mathbf{S}) \geq \sigma^2$. Likewise, the ML estimate of the covariance matrix under the null when σ^2 is known is

$$\hat{\mathbf{R}}_0 = \begin{bmatrix} \mathbf{0} & \mathbf{0} \\ \mathbf{0} & \mathbf{W}_{rr}\mathbf{D}_{rr}\mathbf{W}_{rr}^H \end{bmatrix} + \sigma^2\mathbf{I}_{2L}, \tag{7.23}$$

where $\mathbf{D}_{rr} = \text{diag}\left(ev_1(\mathbf{S}_{rr}) - \sigma^2, \ldots, ev_{p_n}(\mathbf{S}_{rr}) - \sigma^2, 0, \ldots, 0\right)$, with p_n the largest value of l between 1 and p such that $ev_l(\mathbf{S}_{rr}) \geq \sigma^2$.

Using the ML estimates in (7.22) and (7.23), straightforward algebraic steps show that

$$\det(\hat{\mathbf{R}}_1) = \sigma^{2(2L-p_a)} \prod_{l=1}^{p_a} ev_l(\mathbf{S}), \quad \text{tr}(\hat{\mathbf{R}}_1^{-1}\mathbf{S}) = \frac{1}{\sigma^2}\text{tr}(\mathbf{S}) - \frac{1}{\sigma^2}\sum_{l=1}^{p_a} ev_l(\mathbf{S}),$$

and

$$\det(\hat{\mathbf{R}}_0) = \sigma^{2(2L-p_n)} \prod_{l=1}^{p_n} ev_l(\mathbf{S}_{rr}), \quad \text{tr}(\hat{\mathbf{R}}_0^{-1}\mathbf{S}) = \frac{1}{\sigma^2}\text{tr}(\mathbf{S}) - \frac{1}{\sigma^2}\sum_{l=1}^{p_n} ev_l(\mathbf{S}_{rr}).$$

Hence, the GLR under white noise with known variance is

$$\lambda_2 = \Lambda_2^{1/N} = \frac{\prod_{l=1}^{p_n} ev_l(\mathbf{S}_{rr})}{\prod_{l=1}^{p_a} ev_l(\mathbf{S})} \exp\left(\frac{1}{\sigma^2}\sum_{l=1}^{p_a} ev_l(\mathbf{S}) - \frac{1}{\sigma^2}\sum_{l=1}^{p_n} ev_l(\mathbf{S}_{rr})\right)\sigma^{2(p_a-p_n)},$$

where

$$\Lambda_2 = \frac{\ell(\hat{\mathbf{R}}_1; \mathbf{Y})}{\ell(\hat{\mathbf{R}}_0; \mathbf{Y})}.$$

Invariances. The detector statistic is invariant to the transformation group $G = \{g \mid g \cdot \mathbf{Y} = \text{blkdiag}(\mathbf{Q}_s, \mathbf{Q}_r)\mathbf{Y}\mathbf{Q}_N\}$, where $\mathbf{Q}_s, \mathbf{Q}_r \in U(L)$, and $\mathbf{Q}_N \in U(N)$. That is, the invariance to scale is lost.

7.5.3 Scale-Invariant Matched Direction Detector for Uncorrelated Noises Across Antennas (or White Noises with Different Variances)

When $\boldsymbol{\Sigma}$ is structured as (7.3) or (7.4), closed-form GLRs do not exist, and one resorts to numerical methods. An important property of the sets of structured covariance matrices considered in this chapter, which allows us to obtain relatively simple ML estimation algorithms, is given in the following proposition.

Proposition 7.1 *The structure of the sets $\mathcal{E}$ considered in this chapter is preserved under matrix inversion. That is,*

$$\boldsymbol{\Sigma} \in \mathcal{E} \quad \Leftrightarrow \quad \boldsymbol{\Sigma}^{-1} \in \mathcal{E}.$$

Proof The result directly follows from the (block)-diagonal structure of the matrices in the sets $\mathcal{E}$. □

In order to obtain a simple iterative algorithm, we rely on the following property of the sets of inverse covariance or precision matrices associated with $\mathcal{R}_i$, $i = 0, 1$.

Proposition 7.2 *The sets of inverse covariance matrices $\mathcal{P}_i = \{\mathbf{R}^{-1} \mid \mathbf{R} \in \mathcal{R}_i\}$, $i = 0, 1$, can be written as*

$$\mathcal{P}_i = \left\{\mathbf{D} - \mathbf{G}\mathbf{G}^H \mid \mathbf{D} \in \mathcal{E} \quad \text{and} \quad \mathbf{D} \succeq \mathbf{G}\mathbf{G}^H\right\}.$$

In particular, $\mathbf{D} = \boldsymbol{\Sigma}^{-1}$ and $\mathbf{G}\mathbf{G}^H = \boldsymbol{\Sigma}^{-1}\mathbf{H}\left(\mathbf{I}_p + \mathbf{H}^H\boldsymbol{\Sigma}^{-1}\mathbf{H}\right)^{-1}\mathbf{H}^H\boldsymbol{\Sigma}^{-1}$, or equivalently, $\boldsymbol{\Sigma} = \mathbf{D}^{-1}$ and $\mathbf{H}\mathbf{H}^H = \mathbf{D}^{-1/2}\mathbf{F}\left(\boldsymbol{\Lambda}^{-1} - \mathbf{I}_{2L}\right)^{-1}\mathbf{F}^H\mathbf{D}^{-1/2}$, where $\mathbf{F}$ and $\boldsymbol{\Lambda}$ are the eigenvector and eigenvalue matrices in the EV decomposition $\mathbf{D}^{-1/2}\mathbf{G}\mathbf{G}^H\mathbf{D}^{-1/2} = \mathbf{F}\boldsymbol{\Lambda}\mathbf{F}^H$.

Proof Applying the matrix inversion lemma (see Sect. B.4.2), we can write

$$\mathbf{R}^{-1} = \left(\mathbf{H}\mathbf{H}^H + \boldsymbol{\Sigma}\right)^{-1} = \boldsymbol{\Sigma}^{-1} - \boldsymbol{\Sigma}^{-1}\mathbf{H}\left(\mathbf{I}_p + \mathbf{H}^H\boldsymbol{\Sigma}^{-1}\mathbf{H}\right)^{-1}\mathbf{H}^H\boldsymbol{\Sigma}^{-1},$$

which allows us to identify $\mathbf{D} = \boldsymbol{\Sigma}^{-1}$ and $\mathbf{G}\mathbf{G}^H = \boldsymbol{\Sigma}^{-1}\mathbf{H}(\mathbf{I}_p + \mathbf{H}^H\boldsymbol{\Sigma}^{-1}\mathbf{H})^{-1}\mathbf{H}^H\boldsymbol{\Sigma}^{-1}$. In order to recover $\mathbf{H}$ from $\mathbf{D}$ and $\mathbf{G}$, let us write $\tilde{\mathbf{H}} = \mathbf{D}^{1/2}\mathbf{H}$, which yields

$$\mathbf{D}^{-1/2}\mathbf{G}\mathbf{G}^H\mathbf{D}^{-1/2} = \mathbf{F}\boldsymbol{\Lambda}\mathbf{F}^H = \tilde{\mathbf{H}}\left(\mathbf{I}_p + \tilde{\mathbf{H}}^H\tilde{\mathbf{H}}\right)^{-1}\tilde{\mathbf{H}}^H,$$

where the first equality is the EV decomposition of $\mathbf{D}^{-1/2}\mathbf{G}\mathbf{G}^H\mathbf{D}^{-1/2}$. Finally, writing the EV decomposition of $\tilde{\mathbf{H}}\tilde{\mathbf{H}}^H$ as $\mathbf{F}_{\tilde{\mathbf{H}}}\boldsymbol{\Lambda}_{\tilde{\mathbf{H}}}\mathbf{F}_{\tilde{\mathbf{H}}}^H$ allows us to identify

$$\mathbf{F}_{\tilde{\mathbf{H}}} = \mathbf{F}, \qquad \boldsymbol{\Lambda}_{\tilde{\mathbf{H}}} = \left(\boldsymbol{\Lambda}^{-1} - \mathbf{I}_{2L}\right)^{-1},$$

which obviously requires $\mathbf{I}_{2L} \succeq \boldsymbol{\Lambda}$, or equivalently $\mathbf{D} \succeq \mathbf{G}\mathbf{G}^H$. □

Thanks to Proposition 7.2, the ML estimation problem can be formulated in terms of the matrices $\mathbf{D}$ and $\mathbf{G}$ as

$$\begin{aligned} \underset{\mathbf{D},\mathbf{G}}{\text{maximize}} \quad & \log\det(\mathbf{D} - \mathbf{G}\mathbf{G}^H) - \operatorname{tr}\left[(\mathbf{D} - \mathbf{G}\mathbf{G}^H)\mathbf{S}\right], \\ \text{subject to} \quad & \mathbf{D} - \mathbf{G}\mathbf{G}^H \succeq \mathbf{0}, \\ & \mathbf{D} \in \mathcal{E}. \end{aligned} \tag{7.24}$$

Although this problem is non-convex, it is formulated in a form suitable for applying the alternating optimization approach. Thus, for a fixed inverse noise covariance matrix $\mathbf{D} = \boldsymbol{\Sigma}^{-1}$, the problem of finding the optimal $\mathbf{G}$ reduces to

$$\underset{\mathbf{G}}{\text{maximize}} \quad \log\det(\mathbf{I}_{2L} - \mathbf{D}^{-1/2}\mathbf{G}\mathbf{G}^H\mathbf{D}^{-1/2}) + \operatorname{tr}\left(\mathbf{G}\mathbf{G}^H\mathbf{S}\right),$$

or, in terms of $\tilde{\mathbf{G}} = \mathbf{D}^{-1/2}\mathbf{G}$ and $\mathbf{S}_{\mathbf{D}^{-1}} = \mathbf{D}^{1/2}\mathbf{S}\mathbf{D}^{1/2} = \mathbf{S}_{\boldsymbol{\Sigma}}$,

$$\underset{\tilde{\mathbf{G}}}{\text{maximize}} \quad \log\det(\mathbf{I}_{2L} - \tilde{\mathbf{G}}\tilde{\mathbf{G}}^H) + \operatorname{tr}\left(\tilde{\mathbf{G}}\tilde{\mathbf{G}}^H\mathbf{S}_{\boldsymbol{\Sigma}}\right). \tag{7.25}$$

The solution of (7.25) can be found in a straightforward manner and is given by any $\tilde{\mathbf{G}}$ of the form

$$\tilde{\mathbf{G}} = \mathbf{W}_p \operatorname{diag}(d_1, \ldots, d_p)\mathbf{Q},$$

where $d_l = \left(\sqrt{1 - 1/ev_l(\mathbf{S}_\boldsymbol{\Sigma})}\right)^+$; $\mathbf{W}_p$ is a matrix containing the p principal eigenvectors of $\mathbf{S}_\boldsymbol{\Sigma}$, with $ev_l(\mathbf{S}_\boldsymbol{\Sigma}), l = 1, \ldots, p$, the corresponding eigenvalues; and $\mathbf{Q} \in U(p)$ is an arbitrary unitary matrix. Finally, using Proposition 7.2, the optimal matrix $\mathbf{H}$ satisfies

$$\hat{\mathbf{H}}\hat{\mathbf{H}}^H = \boldsymbol{\Sigma}^{1/2}\mathbf{W}_p \operatorname{diag}\left((ev_1(\mathbf{S}_\boldsymbol{\Sigma}) - 1)^+, \ldots, (ev_p(\mathbf{S}_\boldsymbol{\Sigma}) - 1)^+\right) \mathbf{W}_p^H \boldsymbol{\Sigma}^{1/2}.$$

Fixing the matrix $\mathbf{G}$, the optimization problem in (7.24) reduces to

$$\underset{\mathbf{D} \in \mathcal{E}}{\text{maximize}} \quad \log\det(\mathbf{D} - \mathbf{G}\mathbf{G}^H) - \operatorname{tr}(\mathbf{D}\mathbf{S}), \tag{7.26}$$

which is a convex optimization problem. Taking the constrained gradient of (7.26) with respect to $\mathbf{D}$ yields

$$\nabla_\mathbf{D} = \boldsymbol{\Theta}\left[(\mathbf{D} - \mathbf{G}\mathbf{G}^H)^{-1} - \mathbf{S}\right],$$

where $\boldsymbol{\Theta}[\cdot]$ is an operator imposing the structure of the set $\mathcal{E}$. Noting that $(\mathbf{D} - \mathbf{G}\mathbf{G}^H)^{-1} = \mathbf{H}\mathbf{H}^H + \boldsymbol{\Sigma}$, we conclude that the gradient is zero under the alternative hypothesis when $\boldsymbol{\Theta}\left[\mathbf{H}\mathbf{H}^H + \boldsymbol{\Sigma} - \mathbf{S}\right] = \mathbf{0}_{2L}$. For instance, when $\mathcal{E} = \mathcal{E}_3$ is the set of diagonal matrices with positive elements, then the optimal $\boldsymbol{\Sigma}$ is

$$\hat{\boldsymbol{\Sigma}} = \operatorname{diag}\left(\mathbf{S} - \hat{\mathbf{H}}\hat{\mathbf{H}}^H\right).$$

On the other hand, when $\mathcal{E} = \mathcal{E}_2$ is the set of matrices structured as in (7.3), the optimal $\boldsymbol{\Sigma}$ is

$$\hat{\boldsymbol{\Sigma}} = \begin{bmatrix} \frac{1}{L}\operatorname{tr}\left(\mathbf{S}_{ss} - \hat{\mathbf{H}}_s\hat{\mathbf{H}}_s^H\right)\mathbf{I}_L & \mathbf{0} \\ \mathbf{0} & \frac{1}{L}\operatorname{tr}\left(\mathbf{S}_{rr} - \hat{\mathbf{H}}_r\hat{\mathbf{H}}_r^H\right)\mathbf{I}_L \end{bmatrix}.$$

Finally, this overall alternating optimization approach for the ML estimation of $\mathbf{H}$ and $\boldsymbol{\Sigma}$ (when the noises are uncorrelated across antennas) is summarized in Algorithm 4. Since at each step the value of the objective function can only increase, the method is guaranteed to converge to a (possibly local) maximum. However, this alternating minimization approach does not guarantee that the global maximizer of the log-likelihood has been found.

This alternating optimization approach can readily be extended to other noise models with covariance matrices in a cone for which closed-form ML estimates do not exist. For instance, it was generalized in [276] to a multichannel factor analysis (MFA) problem where each of the channels carries measurements that share factors with all other channels but also contains factors that are unique to the channel. As is usual in factor analysis, each channel carries an additive noise whose covariance is

Algorithm 4: GLR for noises with diagonal covariance matrix

Input: Sample covariance matrix $\mathbf{S}$ with blocks $\mathbf{S}_{ss}$, $\mathbf{S}_{rr}$, $\mathbf{S}_{sr} = \mathbf{S}_{rs}^H$, and rank p
Output: GLR statistic λ_2
`/* Obtain ML estimates under` $\mathcal{H}_1$ `*/`
Initialize $\hat{\boldsymbol{\Sigma}}_1 = \mathbf{I}_{2L}$
repeat

Compute SVD of the noise-whitened sample covariance matrix

$$\mathbf{S}_{\hat{\boldsymbol{\Sigma}}_1} = \hat{\boldsymbol{\Sigma}}_1^{-1/2}\mathbf{S}\hat{\boldsymbol{\Sigma}}_1^{-1/2} = \mathbf{W}\,\mathrm{diag}\left(ev_1(\mathbf{S}_{\hat{\boldsymbol{\Sigma}}_1}), \ldots, ev_{2L}(\mathbf{S}_{\hat{\boldsymbol{\Sigma}}_1})\right)\mathbf{W}^H$$

Compute new channel estimate as

$$\hat{\mathbf{H}}_1\hat{\mathbf{H}}_1^H = \hat{\boldsymbol{\Sigma}}_1^{1/2}\mathbf{W}_p\,\mathrm{diag}\left(\left(ev_1(\mathbf{S}_{\hat{\boldsymbol{\Sigma}}_1}) - 1\right)^+, \ldots, \left(ev_p(\mathbf{S}_{\hat{\boldsymbol{\Sigma}}_1}) - 1\right)^+\right)\mathbf{W}_p^H\hat{\boldsymbol{\Sigma}}_1^{1/2}$$

Estimate new noise covariance matrix as $\hat{\boldsymbol{\Sigma}}_1 = \mathrm{diag}\left(\mathbf{S} - \hat{\mathbf{H}}_1\hat{\mathbf{H}}_1^H\right)$

until *Convergence*
ML estimate $\hat{\mathbf{R}}_1 = \hat{\mathbf{H}}_1\hat{\mathbf{H}}_1^H + \hat{\boldsymbol{\Sigma}}_1$
`/* Obtain ML estimates under` $\mathcal{H}_0$ `*/`
Initialize $\hat{\boldsymbol{\Sigma}}_0 = \mathrm{blkdiag}(\hat{\boldsymbol{\Sigma}}_{ss}, \hat{\boldsymbol{\Sigma}}_{rr}) = \mathbf{I}_{2L}$
repeat

Compute SVD of the noise-whitened sample covariance matrix for the reference channel

$$\mathbf{S}_{\hat{\boldsymbol{\Sigma}},rr} = \hat{\boldsymbol{\Sigma}}_{rr}^{-1/2}\mathbf{S}_{rr}\hat{\boldsymbol{\Sigma}}_{rr}^{-1/2} = \mathbf{W}_{rr}\,\mathrm{diag}\left(ev_1(\mathbf{S}_{\hat{\boldsymbol{\Sigma}},rr}), \ldots, ev_L(\mathbf{S}_{\hat{\boldsymbol{\Sigma}},rr})\right)\mathbf{W}_{rr}^H$$

Compute new channel estimates as

$$\hat{\mathbf{H}}_r\hat{\mathbf{H}}_r^H = \hat{\boldsymbol{\Sigma}}_{rr}^{1/2}\mathbf{W}_{rr,p}\,\mathrm{diag}\left(\left(ev_1(\mathbf{S}_{\hat{\boldsymbol{\Sigma}},rr}) - 1\right)^+, \ldots, \left(ev_p(\mathbf{S}_{\hat{\boldsymbol{\Sigma}},rr}) - 1\right)^+\right)\mathbf{W}_{rr,p}^H\hat{\boldsymbol{\Sigma}}_{rr}^{1/2}$$

$$\hat{\mathbf{H}}_0\hat{\mathbf{H}}_0^H = \mathrm{blkdiag}\left(\mathbf{0}, \hat{\mathbf{H}}_r\hat{\mathbf{H}}_r^H\right)$$

Estimate new noise covariance matrices

$$\hat{\boldsymbol{\Sigma}}_{ss} = \mathrm{diag}(\mathbf{S}_{ss}) \quad \hat{\boldsymbol{\Sigma}}_{rr} = \mathrm{diag}\left(\mathbf{S}_{rr} - \hat{\mathbf{H}}_{rr}\hat{\mathbf{H}}_{rr}^H\right) \quad \hat{\boldsymbol{\Sigma}}_0 = \mathrm{blkdiag}\left(\hat{\boldsymbol{\Sigma}}_{ss}, \hat{\boldsymbol{\Sigma}}_{rr}\right)$$

until *Convergence*
ML estimate $\hat{\mathbf{R}}_0 = \hat{\mathbf{H}}_0\hat{\mathbf{H}}_0^H + \hat{\boldsymbol{\Sigma}}_0$
Obtain GLR as $\lambda_2 = \frac{\det(\hat{\mathbf{R}}_0)}{\det(\hat{\mathbf{R}}_1)}$

diagonal but is otherwise unknown. In this scenario, the unique-factors-plus-noise covariance matrix is a block-diagonal matrix, and each block admits a low-rank-plus-diagonal matrix decomposition. The alternating optimization method presented in this chapter can readily be adapted to obtain ML estimates for this MFA problem.

Invariances. For white noises with different variances, c.f. (7.3), the detector statistic is invariant to the transformation group $G = \{g \mid g \cdot \mathbf{Y} = \text{blkdiag}(\beta_s \mathbf{Q}_s, \beta_r \mathbf{Q}_r) \mathbf{Y}\mathbf{Q}_N\}$, where $\beta_s, \beta_r \neq 0$, $\mathbf{Q}_s, \mathbf{Q}_r \in U(L)$, and $\mathbf{Q}_N \in U(N)$.

When the noise covariance matrix is diagonal, as in (7.4), the detector statistic is invariant to the transformation group

$$G = \left\{g \mid g \cdot \mathbf{Y} = \text{diag}(\beta_{s,1}, \ldots, \beta_{s,L}, \beta_{r,1}, \ldots, \beta_{r,L})\mathbf{Y}\mathbf{Q}_N\right\},$$

where $\beta_{s,l}, \beta_{r,l} \neq 0$ and $\mathbf{Q}_N \in U(N)$.

7.5.4 Transformation-Invariant Matched Direction Detector for Noises with Arbitrary Spatial Correlation

When the noises in each channel have arbitrary positive definite spatial covariance matrices, the ML estimate of the covariance matrix under the null is $\hat{\mathbf{R}}_0 = \text{blkdiag}(\mathbf{S}_{ss}, \mathbf{S}_{rr})$.

Under the alternative, the ML estimate has been derived in [337, 370].[1] To present this result, let $\hat{\mathbf{C}} = \mathbf{S}_{ss}^{-1/2}\mathbf{S}_{sr}\mathbf{S}_{rr}^{-1/2}$ be the sample coherence matrix between the surveillance and reference channels, and let $\hat{\mathbf{C}} = \mathbf{F}\mathbf{K}\mathbf{G}^H$ be its singular value decomposition, where the matrix $\mathbf{K} = \text{diag}\,(k_1, \ldots, k_L)$ contains the sample canonical correlations $1 \geq k_1 \geq \cdots \geq k_L \geq 0$ along its diagonal. The ML estimate of the covariance matrix under $\mathcal{H}_1$ is

$$\hat{\mathbf{R}}_1 = \begin{bmatrix} \mathbf{S}_{ss} & \mathbf{S}_{ss}^{1/2}\hat{\mathbf{C}}_p\mathbf{S}_{rr}^{1/2} \\ \mathbf{S}_{rr}^{1/2}\hat{\mathbf{C}}_p^H\mathbf{S}_{ss}^{1/2} & \mathbf{S}_{rr} \end{bmatrix}, \tag{7.27}$$

where $\hat{\mathbf{C}}_p = \mathbf{F}\mathbf{K}_p\mathbf{G}^H$, with $\mathbf{K}_p = \text{diag}\left(k_1, \ldots, k_p, 0, \ldots, 0\right)$ a rank-p truncation of $\mathbf{K}$. Plugging the ML estimates into (7.13), and using the decomposition

$$\hat{\mathbf{R}}_1 = \begin{bmatrix} \mathbf{S}_{ss}^{1/2} & \mathbf{0} \\ \mathbf{0} & \mathbf{S}_{rr}^{1/2} \end{bmatrix} \begin{bmatrix} \mathbf{F} & \mathbf{0} \\ \mathbf{0} & \mathbf{G} \end{bmatrix} \begin{bmatrix} \mathbf{I}_L & \mathbf{K}_p \\ \mathbf{K}_p & \mathbf{I}_L \end{bmatrix} \begin{bmatrix} \mathbf{F}^H & \mathbf{0} \\ \mathbf{0} & \mathbf{G}^H \end{bmatrix} \begin{bmatrix} \mathbf{S}_{ss}^{1/2} & \mathbf{0} \\ \mathbf{0} & \mathbf{S}_{rr}^{1/2} \end{bmatrix},$$

it is easy to check that the GLR in a second-order model for a signal in an unknown subspace of known dimension, when the channel noises have arbitrary unknown covariance matrices, is

$$\lambda_2 = \frac{1}{\det(\mathbf{I}_L - \mathbf{K}_p^2)} = \prod_{l=1}^{p} \frac{1}{(1 - k_l^2)}, \tag{7.28}$$

---

[1] An alternative derivation for the particular case of $p = 1$ can be found in [297].

where k_l is the lth sample canonical correlation between the surveillance and reference channels. Interestingly, $1 - \lambda_2{}^{-1}$ is the coherence statistic, $0 \leq 1 - \prod_{l=1}^{p}(1 - k_l^2) \leq 1$, based on squared canonical correlations for a rank-p signal.

If the covariance matrix under $\mathcal{H}_1$ were assumed an arbitrary positive definite matrix, instead of rank-p (which happens for sufficiently large p), the GLR statistic would be the following generalized Hadamard ratio

$$\frac{\det(\mathbf{S}_{ss})\det(\mathbf{S}_{rr})}{\det(\mathbf{S})} = \prod_{l=1}^{L} \frac{1}{(1 - k_l^2)}, \tag{7.29}$$

which is the statistic to test for independence between two MVN random vectors, as in Sect. 4.8.2. Notice also that $1 - \prod_{l=1}^{L}(1 - k_l^2)$ is the generalized coherence (GC) originally defined in [77]. So, for noises with arbitrary covariance matrices, the net of prior knowledge of the signal dimension p is to replace L by p in the coherence.

From the identified model for $\hat{\mathbf{R}}_1$ in (7.27), it is a standard result in the theory of MMSE estimation that the estimator of a measurement $\mathbf{y}_s$ in the surveillance channel can be orthogonally decomposed as $\mathbf{y}_s = \hat{\mathbf{y}}_s + \hat{\mathbf{e}}_s$, where

$$\hat{\mathbf{y}}_s = \mathbf{S}_{ss}^{1/2}\mathbf{F}\mathbf{K}_p\mathbf{G}^H\mathbf{S}_{rr}^{-1/2}\mathbf{y}_r \sim \mathcal{CN}_L(\mathbf{0}, \mathbf{S}_{ss}^{1/2}\mathbf{F}\mathbf{K}_p\mathbf{K}_p\mathbf{F}^H\mathbf{S}_{ss}^{1/2}),$$

and

$$\hat{\mathbf{e}}_s \sim \mathcal{CN}_L(\mathbf{0}, \mathbf{S}_{ss}^{1/2}(\mathbf{I}_L - \mathbf{F}\mathbf{K}_p\mathbf{K}_p\mathbf{F}^H)\mathbf{S}_{ss}^{1/2}).$$

The matrix $\mathbf{S}_{ss}^{1/2}\mathbf{F}\mathbf{K}_p\mathbf{G}^H\mathbf{S}_{rr}^{-1/2}$ is the MMSE filter in canonical coordinates, and the matrix $\mathbf{S}_{ss}^{1/2}(\mathbf{I}_L - \mathbf{F}\mathbf{K}_p\mathbf{K}_p\mathbf{F}^H)\mathbf{S}_{ss}^{1/2}$ is the error covariance matrix in canonical coordinates. The matrix $\mathbf{K}_p$ is the MMSE filter for estimating the canonical coordinates $\mathbf{F}^H\mathbf{S}_{ss}^{-1/2}\mathbf{x}_s$ from the canonical coordinates $\mathbf{G}^H\mathbf{S}_{rr}^{-1/2}\mathbf{y}_r$, and the matrix $\mathbf{I}_L - \mathbf{F}\mathbf{K}_p\mathbf{K}_p\mathbf{F}^H$ is the error covariance matrix when doing so. As a consequence, we may interpret the coherence or canonical coordinate detector λ_2^{-1} as the volume of the error concentration ellipse when predicting the canonical coordinates of the surveillance channel signal from the canonical coordinates of the reference channel signal. When the channels are highly correlated, then this prediction is accurate, the volume of the error concentration ellipse is small, and $1 - \lambda_2^{-1}$ is near to one, indicating a detection.

Invariances. Independent transformation of the surveillance channel by a non-singular transformation $\mathbf{B}_s$ and the reference channel by a non-singular transformation $\mathbf{B}_r$ leaves the coherence matrix $\hat{\mathbf{C}}$ invariant. Consequently, its singular values k_l are invariant, and as a consequence the detector (7.28) is also invariant. Additionally, we can also permute surveillance and reference channels without modifying the structure of the covariance matrix under each hypothesis. That is, the GLR is invariant under transformations in the group $G =$

$\{g \mid g \cdot \mathbf{Y} = (\mathbf{P} \otimes \mathbf{I}_L)\,\mathrm{blkdiag}(\mathbf{B}_s, \mathbf{B}_r)\mathbf{Y}\mathbf{Q}_N\}$, where $\mathbf{B}_s, \mathbf{B}_r \in GL(\mathbb{C}^L)$, $\mathbf{Q}_N \in U(N)$ and $\mathbf{P}$ is a 2×2 permutation matrix. As a special case, λ_2 is CFAR with respect to noise power in the surveillance channel and signal-plus-noise power in the reference channel.

Comment. This detector is quite general. But, how is it that the rank-p covariance matrices $\mathbf{H}_s\mathbf{H}_s^H$ and $\mathbf{H}_r\mathbf{H}_r^H$ can be identified in noises of arbitrary unknown positive definite covariance matrices, when no such identification is possible in standard factor analysis? The answer is that in this two-channel problem the sample covariance matrix $\mathbf{S}_{sr}$ brings information about $\mathbf{H}_s\mathbf{H}_r^H$ and this information is used with $\mathbf{S}_{ss}$ and $\mathbf{S}_{rr}$ to identify the covariance models $\mathbf{H}_s\mathbf{H}_s^H + \boldsymbol{\Sigma}_{ss}$ in the surveillance channel and $\mathbf{H}_r\mathbf{H}_r^H + \boldsymbol{\Sigma}_{rr}$ in the reference channel.

Locally Most Powerful Invariant Test. When the noise vectors in the surveillance and reference channels are uncorrelated with each other, and the covariance matrix for each is an arbitrary unknown covariance matrix, then $\mathbf{R}_0$, the covariance matrix under $\mathcal{H}_0$, is a block-diagonal matrix with positive definite blocks and no further structure. Under $\mathcal{H}_1$, the covariance matrix $\mathbf{R}_1$ is the sum of a rank-p signal covariance matrix and a block-diagonal matrix with positive definite blocks and no further structure. Hence, the results in [273] apply, and the LMPIT statistic is

$$\mathscr{L}_2 = \|\hat{\mathbf{C}}\|,$$

where

$$\hat{\mathbf{C}} = \mathrm{blkdiag}(\mathbf{S}_{ss}^{-1/2}, \mathbf{S}_{rr}^{-1/2})\,\mathbf{S}\,\mathrm{blkdiag}(\mathbf{S}_{ss}^{-1/2}, \mathbf{S}_{rr}^{-1/2}).$$

This, too, is a coherence statistic. It may be written

$$\hat{\mathbf{C}} = \begin{bmatrix} \mathbf{I}_L & \mathbf{S}_{ss}^{-1/2}\mathbf{S}_{sr}\mathbf{S}_{rr}^{-1/2} \\ \mathbf{S}_{rr}^{-1/2}\mathbf{S}_{rs}\mathbf{S}_{ss}^{-1/2} & \mathbf{I}_L \end{bmatrix},$$

where the northeast block is the sample coherence matrix between the surveillance and reference channels and the southwest block is its Hermitian transpose. With some abuse of notation, we can write the square of the LMPI statistic as

$$\mathscr{L}_2^2 = \|\hat{\mathbf{C}}\|^2 = 2\|\mathbf{S}_{ss}^{-1/2}\mathbf{S}_{sr}\mathbf{S}_{rr}^{-1/2}\|^2 + 2L = \sum_{l=1}^{L} k_l^2,$$

where k_l is the lth sample canonical correlation between the surveillance and reference channels; that is, k_l is a singular value of $\mathbf{S}_{ss}^{-1/2}\mathbf{S}_{sr}\mathbf{S}_{rr}^{-1/2}$. Two comments are in order. First, the statistic $(1/L)\sum_{l=1}^{L} k_l^2$ is coherence. Second, the LMPIT considers all L canonical correlations, contrary to the GLR in (7.28). As shown in [273], the low-rank structure is locally irrelevant, and the LMPIT is identical to the case where the covariance matrix under $\mathcal{H}_1$ is assumed to be an arbitrary positive definite matrix.

7.6 Chapter Notes

This chapter has addressed the problem of detecting a subspace signal when in addition to the surveillance sensor array there is a reference sensor array in which a distorted and noisy version of the signal to be detected is received. The problem is to determine if there are complex demodulations and synchronizations that bring signals in the surveillance sensors into coherence with signals in the reference sensors. This approach forms the basis of passive detectors typically used in radar, sonar, and other detection and localization problems in which it is possible to take advantage of the signal transmitted by a non-cooperative illuminator of opportunity.

1. Passive radar systems have been studied for several decades due to their simplicity and low cost of implementation in comparison to systems with dedicated transmitters [150, 151]. The conventional approach for passive detection uses the cross-correlation between the data received in the reference and surveillance channels as the test statistic. In [222] the authors study the performance of the cross-correlation (CC) detector for rank-one signals and known noise variance.
2. The literature of passive sensing for detection and localization of sources is developing so rapidly that a comprehensive review of the literature is impractical. But a cursory review up to about 2019 would identify the following papers and their contributions. Passive MIMO target detection with a noisy reference channel has been considered in [153], where the transmitted waveform is considered to be deterministic, but unknown. The authors of [153] derive the generalized likelihood ratio test (GLRT) for this deterministic target model under spatially white noise of *known* variance. The work in [92] derives the GLRT in a passive radar problem that models the received signal as a deterministic rank-one waveform scaled by an unknown single-input single-output (SISO) channel. The noise is white of either known or unknown variance. In another line of work, a passive detector that exploits the subspace structure of the received signal has been proposed in [135]. Instead of computing the cross-correlation between the surveillance and reference channel measurements, the ad hoc detector proposed in [135] cross-correlates the dominant left singular vectors of the matrices containing the observations acquired at both channels. Passive MIMO target detection under a second-order measurement model has been addressed in [299], where GLR statistics under different noise models have been derived.
3. The null distributions for most of the detection statistics derived in this chapter are unknown or intractable. When the number of observations grows, the Wilks approximation, which states that the test statistic $2 \log \Lambda$ converges to a chi-squared distribution with degrees of freedom equal to the difference in

dimensionality of the parameters in $\mathcal{H}_1$ and $\mathcal{H}_0$, is often accurate. Alternatively, by taking advantage of the invariances of the problem, which carry over to the respective GLRs, it is possible to approximate the null distribution by Monte Carlo simulations for some parameter of the problem (e.g., the noise variance), such approximation being valid for other values of that parameter. In some particular cases, the distribution is known: when the Gaussian noises in the two channels have arbitrary spatial correlation matrices, and the signal lies in a one-dimensional subspace, the GLR is the largest sample canonical correlation between the two channels. The distribution is known, but it is complicated. Applying random matrix theory results, it was shown in [299] that, after an appropriate transformation, the distribution of the largest canonical correlation under the null converges to a Tracy-Widom law of order 2.

Detection of Spatially Correlated Time Series 8

This chapter is addressed to several problems concerning the independence of measurements recorded in a network of L sensors or measuring instruments. It is common to label measurements at each instrument by a time index and to label instruments by a space index. Measurements are then space-time measurements, and a test for independence among multiple time series is a test for *spatial* independence. Is the measurement at sensor l independent of the measurement at sensor m for all pairs (l, m)? When measurements are assumed to be MVN, then a test for independence is a test of spatial correlation between measurements. Without the assumption of normality, this test for correlation may be said to be a test for *linear* independence of measurements.

In the simplest case, the measurement at each sensor is a complex normal random variable, and the joint distribution of these L random variables is the distribution of a MVN random vector. At the next level of complexity, the measurement at each sensor is itself a MVN random vector. The joint distribution of these L random vectors is the distribution of a MVN random vector that is a concatenation of random vectors. In the most general case, the measurement at each sensor is a Gaussian time series, which is to say the joint distribution of any finite collection of samples from the time series is distributed as a MVN random vector. The collection of L time series is a multivariate Gaussian time series. Any finite collection of samples from this multivariate time series is distributed as a MVN random vector. Therefore, a test of independence between time series is a test for independence between random vectors. However, when the time series are jointly wide-sense stationary (WSS), a limiting argument in the number of samples taken from each time series may be used to replace a space-time statistic for independence by a space-frequency statistic. This replacement leads to a definition of *broadband multi-channel coherence* for time series.

Testing for spatial independence in space-time measurements is actually more general than it might appear. For example, it is shown that the problem of testing for cyclostationarity of a time series may be reformulated as a test for spatial independence in a virtual space-time problem. More generally, any problem that is,

D. Ramírez et al., *Coherence*, https://doi.org/10.1007/978-3-031-13331-2_8

or may be reformulated as, a problem of independence testing in multiple datasets may be evocatively termed a space-time problem.

8.1 Introduction

The problem of testing for independence among several *real* normal random variables has a provenance beginning with Wilks [383], who used likelihood in a real MVN model to derive the Hadamard ratio and its null distribution. Anderson [13] extended the Wilks results to several real MVN random vectors by deriving a generalized Hadamard ratio and its null distribution. The Hadamard ratio for *complex* random variables was derived by geometrical means in [76,77,133], where a complex MVN assumption was then used to derive the null distribution for the Hadamard ratio. The Hadamard ratio was derived for the complex case also in [216] based on likelihood theory, where an approximation was given that has turned out to be the locally most powerful invariant test of independence in the case of complex normal random variables. The reader is referred to Chap. 4 for more details on these detectors.

The extension of these results to several time series amounts to adapting the generalized Hadamard ratio for vectors to a finite sampling of each time series and then using limiting arguments in the number of samples to derive a generalized Hadamard ratio. This is the program of [268], where it is shown that for wide-sense stationary (WSS) time series, the limiting form of the generalized Hadamard ratio has a spectral form that estimates what might be called *multi-channel broadband coherence*.

In [201], the authors extend the results of [268] to the case where each of the sensors in a network of sensors is replaced by a multi-sensor array. The stochastic representation of the Hadamard ratio as a product of independent beta-distributed random variables extends the Anderson result to complex random vectors. In [202], the authors use the method of saddle points to accurately compute the probability that the test statistic will exceed a threshold. These results are used to set a threshold that controls the probability of false alarm for the GLR of [201].

The Hadamard ratio, and its generalization to time series, has inspired a large body of literature on spectrum sensing and related problems. The work in [270] studies the problem of detecting a WSS communication signal that is common to several sensors. The authors of [269], [290], and [363] specialize the reasoning of the Hadamard ratio to the case where potentially dependent time series at each sensor have known, or partially known, space-time covariance structure. A variation on the Hadamard ratio is derived in [7] for the case where the space-time structure of potentially dependent random vectors is known to be separable and persymmetric.

The detection of cyclostationarity has a long tradition that dates to the original work of Gardner [314] and Cochran [116]. The more recent work of [266,274,314] and [175,325] reformulates the problem of detecting cyclostationarity as a problem of testing for coherence in a virtual space-time problem.

8.2 Testing for Independence of Multiple Time Series

The lth element in an L-element sensor array records N samples of a time series $\{x_l[n]\}$. These may be called space-time measurements. The resulting samples are organized into the time vectors $\mathbf{x}_l = [x_l[0] \cdots x_l[N-1]]^T \in \mathbb{C}^N$. The question to be answered is whether the random time vectors $\mathbf{x}_l$ are mutually uncorrelated, i.e., whether they are spatially uncorrelated. In the case of MVN random vectors, this is a question of mutual independence.

To begin, we shall assume the random variables $x_l[n], l = 1, \ldots, L, n = 0, \ldots, N-1$, in the time vectors $\mathbf{x}_l, l = 1, \ldots, L$, have arbitrary but unknown auto-covariances and cross-covariances, but when a limiting form of the test statistic is derived for large N, it will be assumed that the time series from which these measurements are made are jointly wide sense stationary (WSS). Then, the results of this chapter extend the results in Sect. 4.8 from random variables and random vectors to time series.

When the random vectors $\mathbf{x}_l, l = 1, \ldots, L$, are assumed to be zero-mean complex normal random vectors, then their concatenation is an $LN \times 1$ time-space vector $\mathbf{z} = [\mathbf{x}_1^T \cdots \mathbf{x}_L^T]^T$, distributed as $\mathbf{z} \sim CN_{LN}(\mathbf{0}, \mathbf{R})$. The $LN \times LN$ covariance matrix $\mathbf{R}$ is structured as

$$\mathbf{R} = \begin{bmatrix} \mathbf{R}_{11} & \mathbf{R}_{12} & \cdots & \mathbf{R}_{1L} \\ \mathbf{R}_{21} & \mathbf{R}_{22} & \cdots & \mathbf{R}_{2L} \\ \vdots & \vdots & \ddots & \vdots \\ \mathbf{R}_{L1} & \mathbf{R}_{L2} & \cdots & \mathbf{R}_{LL} \end{bmatrix}. \tag{8.1}$$

For $l = m$, the $N \times N$ matrix $\mathbf{R}_{ll} = \mathrm{E}\left[\mathbf{x}_l \mathbf{x}_l^H\right], l = 1, \ldots, L$, is an auto-covariance matrix for the measurement vector $\mathbf{x}_l$. For $l \neq m$, the $N \times N$ matrix $\mathbf{R}_{lm} = \mathrm{E}\left[\mathbf{x}_l \mathbf{x}_m^H\right], l, m = 1, \ldots, L$, is a cross-covariance matrix between the measurement vectors $\mathbf{x}_l$ and $\mathbf{x}_m$.

8.2.1 The Detection Problem and its Invariances

Assume a random sample consisting of M independent and identically distributed (i.i.d.) samples of the $LN \times 1$ vector $\mathbf{z}$, organized into the $LN \times M$ data matrix $\mathbf{Z} = [\mathbf{z}_1 \cdots \mathbf{z}_M]$. This data matrix is distributed as $\mathbf{Z} \sim CN_{LN \times M}(\mathbf{0}, \mathbf{I}_M \otimes \mathbf{R})$. Under the null hypothesis $\mathcal{H}_0$, the time series are spatially uncorrelated, and therefore $\mathbf{R}_{lm} = \mathbf{0}_N$ for $l \neq m$. The structure of $\mathbf{R}$ is then

$$\mathbf{R}_0 = \mathrm{blkdiag}\left(\mathbf{R}_{11}, \mathbf{R}_{22}, \ldots, \mathbf{R}_{LL}\right).$$

Under the alternative $\mathcal{H}_1$, the covariance matrix is defined as in (8.1) with the $LN \times LN$ covariance matrix $\mathbf{R}$ constrained only to be a positive definite Hermitian covariance matrix. In this case, the covariance matrix is denoted $\mathbf{R}_1$.

A null hypothesis test for spatial independence of the time vectors $\mathbf{x}_l, l = 1, \ldots, L$, based on the random sample $\mathbf{Z}$ is then

$$\begin{aligned} \mathcal{H}_1 &: \mathbf{Z} \sim CN_{LN \times M}(\mathbf{0}, \mathbf{I}_M \otimes \mathbf{R}_1), \\ \mathcal{H}_0 &: \mathbf{Z} \sim CN_{LN \times M}(\mathbf{0}, \mathbf{I}_M \otimes \mathbf{R}_0). \end{aligned} \tag{8.2}$$

This hypothesis testing problem is invariant to the transformation group $G = \{g \mid g \cdot \mathbf{Z} = \mathbf{PBZQ}_M\}$, where $\mathbf{P} = \mathbf{P}_L \otimes \mathbf{I}_N$ is an $LN \times LN$ block-structured permutation matrix that reorders the sensor elements, $\mathbf{B} = \text{blkdiag}(\mathbf{B}_1, \ldots, \mathbf{B}_L)$ is a block-structured matrix of non-singular $N \times N$ matrices $\mathbf{B}_l \in GL(\mathbb{C}^N)$, and $\mathbf{Q}_M \in U(M)$ is an $M \times M$ unitary matrix.

8.2.2 Test Statistic

The unknown covariance matrices $\mathbf{R}_1$ and $\mathbf{R}_0$ are elements of the cones $\mathcal{R}_1 = \{\mathbf{R} \mid \mathbf{R} \succ \mathbf{0}\}$ and $\mathcal{R}_0 = \{\mathbf{R} \mid \mathbf{R} = \text{blkdiag}(\mathbf{R}_{11}, \ldots, \mathbf{R}_{LL}), \mathbf{R}_{ll} \succ \mathbf{0}\}$. By Lemma 4.1, a monotone function of the GLR is

$$\lambda = \frac{1}{\Lambda^{1/M}} = \frac{\det(\hat{\mathbf{R}}_1)}{\det(\hat{\mathbf{R}}_0)},$$

where

$$\Lambda = \frac{\ell(\hat{\mathbf{R}}_1; \mathbf{Z})}{\ell(\hat{\mathbf{R}}_0; \mathbf{Z})}.$$

As usual, $\ell(\hat{\mathbf{R}}_i; \mathbf{Z})$ is the likelihood of the ith hypothesis when the covariance matrix $\mathbf{R}_i$ is replaced by its ML estimate $\hat{\mathbf{R}}_i$. These ML estimates are $\hat{\mathbf{R}}_1 = \mathbf{S}$, and $\hat{\mathbf{R}}_0 = \text{blkdiag}(\mathbf{S}_{11}, \ldots, \mathbf{S}_{LL})$, where the sample covariance matrix is

$$\mathbf{S} = \frac{1}{M} \sum_{m=1}^{M} \mathbf{z}_m \mathbf{z}_m^H = \begin{bmatrix} \mathbf{S}_{11} & \cdots & \mathbf{S}_{1L} \\ \vdots & \ddots & \vdots \\ \mathbf{S}_{L1} & \cdots & \mathbf{S}_{LL} \end{bmatrix}.$$

The resulting GLR is

$$\lambda = \frac{\det(\mathbf{S})}{\prod_{l=1}^{L} \det(\mathbf{S}_{ll})}, \tag{8.3}$$

where $\mathbf{S}_{ll}$ is the lth $N \times N$ block on the diagonal of $\mathbf{S}$. Following the nomenclature in Sect. 4.8.2, the expression for the GLR in (8.3) is a *generalized* Hadamard ratio.

The GLR in (8.3) may be rewritten as the determinant of a coherence matrix,

$$\lambda = \det(\hat{\mathbf{C}}), \tag{8.4}$$

where the estimated coherence matrix is

$$\hat{\mathbf{C}} = \hat{\mathbf{R}}_0^{-1/2}\hat{\mathbf{R}}_1\hat{\mathbf{R}}_0^{-1/2}. \tag{8.5}$$

This coherence statistic, a generalized Hadamard ratio, was derived in [268]. The generalized Hadamard ratio for testing independence of *real* random vectors was first derived in [13].

Invariances. The GLR shares the invariances of the hypothesis testing problem. That is, $\lambda(g \cdot \mathbf{Z}) = \lambda(\mathbf{Z})$ for g in the transformation group $G = \{g | g \cdot \mathbf{Z} = \mathbf{PBZQ}_M\}$.

Null Distribution. The results of [201], re-derived in Appendix H, provide the following stochastic representation for the GLR λ under the null (see also Sect. 4.8.2),

$$\lambda \stackrel{d}{=} \prod_{l=1}^{L-1} \prod_{n=0}^{N-1} U_{l,n},$$

where $\stackrel{d}{=}$ denotes equality in distribution. The $U_{l,n}$ are independent beta-distributed random variables:

$$U_{l,n} \sim \text{Beta}(M - lN - n, lN).$$

LMPIT. The LMPIT for the hypothesis test in (8.2) rejects the null when the statistic

$$\mathscr{L} = \|\hat{\mathbf{C}}\|,$$

exceeds a threshold [273]. The coherence matrix $\hat{\mathbf{C}}$ is defined in (8.5).

Frequency-Domain Representation of the GLR. The GLR in (8.4) may be computed in the frequency domain by exploiting the problem invariances. To this end, rewrite the GLR as

$$\lambda = \det\left((\mathbf{F}_N \otimes \mathbf{I}_L)\hat{\mathbf{C}}(\mathbf{F}_N \otimes \mathbf{I}_L)^H\right),$$

where $\mathbf{F}_N$ is the N-dimensional Fourier matrix. Constant multiplicative terms have been ignored. Now, by a simple permutation of the rows, and columns of the matrix

$(\mathbf{F}_N \otimes \mathbf{I}_L)\hat{\mathbf{C}}(\mathbf{F}_N \otimes \mathbf{I}_L)^H$, which leaves the determinant unchanged, the GLR may be expressed in the frequency domain as

$$\lambda = \det\left(\begin{bmatrix} \hat{\mathbf{C}}(e^{j\theta_0}) & \hat{\mathbf{C}}(e^{j\theta_0}, e^{j\theta_1}) & \cdots & \hat{\mathbf{C}}(e^{j\theta_0}, e^{j\theta_{N-1}}) \\ \hat{\mathbf{C}}(e^{j\theta_1}, e^{j\theta_0}) & \hat{\mathbf{C}}(e^{j\theta_1}) & \cdots & \hat{\mathbf{C}}(e^{j\theta_1}, e^{j\theta_{N-1}}) \\ \vdots & \vdots & \ddots & \vdots \\ \hat{\mathbf{C}}(e^{j\theta_{N-1}}, e^{j\theta_0}) & \hat{\mathbf{C}}(e^{j\theta_{N-1}}, e^{j\theta_1}) & \cdots & \hat{\mathbf{C}}(e^{j\theta_{N-1}}) \end{bmatrix}\right). \tag{8.6}$$

The $L \times L$ spectral coherence matrix $\hat{\mathbf{C}}(e^{j\theta_k}, e^{j\theta_n})$ is defined by its (l, m) element as

$$(\hat{\mathbf{C}}(e^{j\theta_k}, e^{j\theta_n}))_{l,m} = \mathbf{f}^H(e^{j\theta_k})\mathbf{S}_{ll}^{-1/2}\mathbf{S}_{lm}\mathbf{S}_{mm}^{-1/2}\mathbf{f}(e^{j\theta_n}),$$

with $\mathbf{f}(e^{j\theta_k})$ the Fourier vector at frequency $\theta_k = 2\pi k/N$ and $\hat{\mathbf{C}}(e^{j\theta_k})$ is a shorthand for $\hat{\mathbf{C}}(e^{j\theta_k}, e^{j\theta_k})$.

8.3 Approximate GLR for Multiple WSS Time Series

The data matrix $\mathbf{Z}$ is a random sample of L time vectors $\mathbf{x}_l, l = 1, \ldots, L$, each an N-variate time vector. When these time vectors are measurements from an L-variate WSS time series, then the covariance matrix $\mathbf{R}$ is Hermitian with Toeplitz blocks under $\mathcal{H}_1$ and Hermitian and block-diagonal with Toeplitz blocks under $\mathcal{H}_0$. A direct attack on the GLR would require an iterative algorithm for estimating Toeplitz matrices from measurements. So the GLR derived so far for finite L, N, M is not truly the GLR when the L-variate time series is WSS and the corresponding covariance matrix $\mathbf{R}$ has Toeplitz blocks. To approximate the true GLR for the asymptotic regime, there are two approaches. The first is to *relax* the Toeplitz constraint by requiring only that the covariance matrix $\mathbf{R}$ is Hermitian and positive definite, in which case the GLR is (8.3), or (8.6) in its frequency-domain version. This GLR may then be *analyzed* in the asymptotic regime of large N, M, under the assumption that the covariance matrix *is* Toeplitz. The second approach is to *approximate* the Toeplitz constraint by a circulant constraint and use Whittle's likelihood for large N [379, 380]. In the subsections to follow, each of these approaches is outlined as a way to approximate the GLR from M i.i.d. realizations of the $LN \times 1$ vector $\mathbf{z}$.

8.3.1 Limiting Form of the Nonstationary GLR for WSS Time Series

The GLR of (8.6) assumes only that the covariance matrix $\mathbf{R}$ is Hermitian and positive definite. It may be called the GLR for a nonstationary L-variate time series. The GLR decomposes as $\lambda = \lambda_{WSS}\lambda_{NS}$, where

$$\lambda_{WSS} = \prod_{k=0}^{N-1} \det\left(\hat{\mathbf{C}}(e^{j\theta_k})\right),$$

and

$$\lambda_{NS} = \det\left(\begin{bmatrix} \mathbf{I}_L & \hat{\mathbf{Q}}(e^{j\theta_0}, e^{j\theta_1}) & \cdots & \hat{\mathbf{Q}}(e^{j\theta_0}, e^{j\theta_{N-1}}) \\ \hat{\mathbf{Q}}(e^{j\theta_1}, e^{j\theta_0}) & \mathbf{I}_L & \cdots & \hat{\mathbf{Q}}(e^{j\theta_1}, e^{j\theta_{N-1}}) \\ \vdots & \vdots & \ddots & \vdots \\ \hat{\mathbf{Q}}(e^{j\theta_{N-1}}, e^{j\theta_0}) & \hat{\mathbf{Q}}(e^{j\theta_{N-1}}, e^{j\theta_1}) & \cdots & \mathbf{I}_L \end{bmatrix}\right).$$

The $L \times L$ spectral matrices in λ_{NS} are defined as

$$\hat{\mathbf{Q}}(e^{j\theta_k}, e^{j\theta_n}) = \hat{\mathbf{C}}^{-1/2}(e^{j\theta_k})\hat{\mathbf{C}}(e^{j\theta_k}, e^{j\theta_n})\hat{\mathbf{C}}^{-1/2}(e^{j\theta_n}).$$

The monotone function $(1/N)\log\lambda$ may be written

$$\frac{1}{N}\log\lambda = \frac{1}{2\pi}\sum_{k=0}^{N-1}\log\det(\hat{\mathbf{C}}(e^{j\theta_k})\frac{2\pi}{N} + \frac{1}{N}\log\lambda_{NS},$$

where $\theta_k = 2\pi k/N$. As $M \to \infty$, the sample covariance matrix $\mathbf{S}$ converges almost surely to the covariance matrix $\mathbf{R}$, which has Toeplitz blocks for WSS time series. Hence, using Szegö's theorem [149], it can be shown that λ_{NS} converges to 0 when $M, N \to \infty$. Exploiting these results, it is shown in [268] that the log-GLR may be approximated for large N, M as

$$\begin{aligned}\frac{1}{N}\log\lambda &= \frac{1}{2\pi}\sum_{k=0}^{N-1}\log\det(\hat{\mathbf{C}}(e^{j\theta_k})\frac{2\pi}{N} \\ &= \frac{1}{2\pi}\sum_{k=0}^{N-1}\log\frac{\det(\hat{\mathbf{S}}(e^{j\theta_k})}{\prod_{l=1}^{L}(\hat{\mathbf{S}}(e^{j\theta_k}))_{ll}}\frac{2\pi}{N},\end{aligned} \tag{8.7}$$

where $(\hat{\mathbf{S}}(e^{j\theta_k}))_{l,m} = \hat{S}_{lm}(e^{j\theta_k}) = \mathbf{f}^H(e^{j\theta_k})\mathbf{S}_{lm}\mathbf{f}(e^{j\theta_k})$ is a quadratic estimator of the power spectral density (PSD) at radian frequency θ_k. The result of (8.7) is a broadband spectral coherence, composed of the logarithms of the narrowband spectral coherences $\det(\hat{\mathbf{S}}(e^{j\theta_k}))/\prod_{l=1}^{L}(\hat{\mathbf{S}}(e^{j\theta_k}))_{ll}$, each of which is a Hadamard ratio.

For intuition, (8.7) may be said to be an approximation of

$$\int_{-\pi}^{\pi}\log\left[\frac{\det\left(\mathbf{S}(e^{j\theta})\right)}{\prod_{l=1}^{L}S_{ll}(e^{j\theta})}\right]\frac{d\theta}{2\pi}, \tag{8.8}$$

with the understanding that no practical implementation would estimate $\mathbf{S}(e^{j\theta})$ for every $\theta \in (-\pi, \pi]$. The observation that $1/N \log \lambda_{NS}$ approaches zero suggests its use as a measure of the degree of nonstationarity of the multiple time series.

A Generalized MSC. As demonstrated in [268], the integrand of (8.8) is a function of the magnitude-squared coherence (MSC) in the case of bivariate time series. That is, for $L = 2$,

$$\frac{\det\left(\mathbf{S}(e^{j\theta})\right)}{\prod_{l=1}^{L} S_{ll}(e^{j\theta})} = \frac{S_{11}(e^{j\theta})S_{22}(e^{j\theta}) - |S_{12}(e^{j\theta})|^2}{S_{11}(e^{j\theta})S_{22}(e^{j\theta})} = 1 - |\rho_{12}(e^{j\theta})|^2,$$

where

$$|\rho_{12}(e^{j\theta})|^2 = \frac{|S_{12}(e^{j\theta})|^2}{S_{11}(e^{j\theta})S_{22}(e^{j\theta})},$$

is the MSC defined in (3.6). More generally, $1 - \det\left(\mathbf{S}(e^{j\theta})\right) / \prod_{l=1}^{L} S_{ll}(e^{j\theta})$ may be called a generalized MSC [268].

Relationship with Mutual Information. In the case of L random variables, a reasonable generalization of mutual information is computed as the Kullback-Leibler (KL) divergence between their joint pdf and the product of the L marginal pdfs [91]. This natural generalization of mutual information [87], called the *multi-information*, captures more than just pairwise relations in an information-theoretic framework. The KL divergence may be rewritten as the sum of the L marginal entropies minus the joint entropy of the L random variables. Hence, for L time series, the marginal and joint entropy rates [87] may be substituted for the marginal and joint entropies. The multi-information is then

$$\begin{aligned} I(\{x_1[n]\}, \ldots, \{x_L[n]\}) = &\sum_{l=1}^{L} \lim_{N\to\infty} \frac{1}{N} H(x_l[0], \ldots, x_l[N-1]) \\ &- \lim_{N\to\infty} \frac{1}{N} H(\mathbf{x}[0], \ldots, \mathbf{x}[N-1]), \end{aligned}$$

where the first term on the right-hand side is the sum of the marginal entropy rates of the time series $\{x_l[n], l = 1, \ldots, L\}$ and the second term is the joint entropy rate of $\{\mathbf{x}[n]\}$, where $\mathbf{x}[n] = [x_1[n] \cdots x_L[n]]^T$. For jointly proper complex Gaussian WSS processes, this is

$$I(\{x_1[n]\}, \ldots, \{x_L[n]\}) = -\int_{-\pi}^{\pi} \log\left[\frac{\det\left(\mathbf{S}(e^{j\theta})\right)}{\prod_{l=1}^{L} S_{ll}(e^{j\theta})}\right] \frac{d\theta}{2\pi}.$$

Then, comparing $I(\{x_1[n]\}, \ldots, \{x_L[n]\})$ with (8.7), it can be seen that the log-GLR is an approximation of minus the mutual information among L Gaussian time series.

8.3.2 GLR for Multiple Circulant Time Series and an Approximate GLR for Multiple WSS Time Series

When the L time series $\{x_l[n], l = 1, \ldots, L\}$, are jointly WSS, each of the covariance blocks in $\mathbf{R}$ is Toeplitz. There is no closed-form expression or terminating algorithm for estimating these blocks. So in the previous subsection, the GLR was computed for multiple nonstationary time series, and its limiting form was used to approximate the GLR for multiple WSS time series. An alternative is to compute the GLR using a multivariate extension of Whittle's likelihood [379, 380], which is based on Szegö's spectral formulas [149]. The basic idea is that the likelihood of a block-Toeplitz covariance matrix converges in mean squared error to that of a block-circulant matrix [274], which is easily block-diagonalized with the Fourier matrix.

In contrast to the derivation in the previous subsections, we shall now arrange the space-time measurements into the L-dimensional space vectors $\mathbf{x}[n] = [x_1[n] \cdots x_L[n]]^T, n = 0, \ldots, N-1$. These are then stacked into the $LN \times 1$ space-time vector $\mathbf{y} = [\mathbf{x}^T[0] \cdots \mathbf{x}^T[N-1]]^T$. This vector is distributed as $\mathbf{y} \sim \mathcal{CN}_{LN}(\mathbf{0}, \mathbf{R})$, where the covariance matrix is restructured as

$$\mathbf{R}_1 = \begin{bmatrix} \mathbf{R}_1[0] & \mathbf{R}_1[-1] & \cdots & \mathbf{R}_1[-N+1] \\ \mathbf{R}_1[1] & \mathbf{R}_1[0] & \cdots & \mathbf{R}_1[-N+2] \\ \vdots & \vdots & \ddots & \vdots \\ \mathbf{R}_1[N-1] & \mathbf{R}_1[N-2] & \cdots & \mathbf{R}_1[0] \end{bmatrix},$$

under $\mathcal{H}_1$, and as

$$\mathbf{R}_0 = \begin{bmatrix} \mathbf{R}_0[0] & \mathbf{R}_0[-1] & \cdots & \mathbf{R}_0[-N+1] \\ \mathbf{R}_0[1] & \mathbf{R}_0[0] & \cdots & \mathbf{R}_0[-N+2] \\ \vdots & \vdots & \ddots & \vdots \\ \mathbf{R}_0[N-1] & \mathbf{R}_0[N-2] & \cdots & \mathbf{R}_0[0] \end{bmatrix},$$

under $\mathcal{H}_0$. The $L \times L$ covariance matrix $\mathbf{R}_1[m]$ is $\mathbf{R}_1[m] = \mathrm{E}[\mathbf{x}[n]\mathbf{x}^H[n-m]]$. This covariance sequence under $\mathcal{H}_1$ has no further structure, but $\mathbf{R}_0[m] = \mathrm{E}[\mathbf{x}[n]\mathbf{x}^H[n-m]]$, the covariance sequence under $\mathcal{H}_0$, is diagonal because the time series are spatially uncorrelated. That is, $\mathbf{R}_1$ is a block-Toeplitz matrix with $L \times L$ arbitrary blocks, whereas $\mathbf{R}_0$ is block-Toeplitz but with $L \times L$ diagonal blocks.

To avoid the block-Toeplitz structure, as outlined above, we use a multivariate extension of Whittle's likelihood [379, 380]. Then, defining the transformation

$$\mathbf{w} = (\mathbf{F}_N \otimes \mathbf{I}_L)\mathbf{y},$$

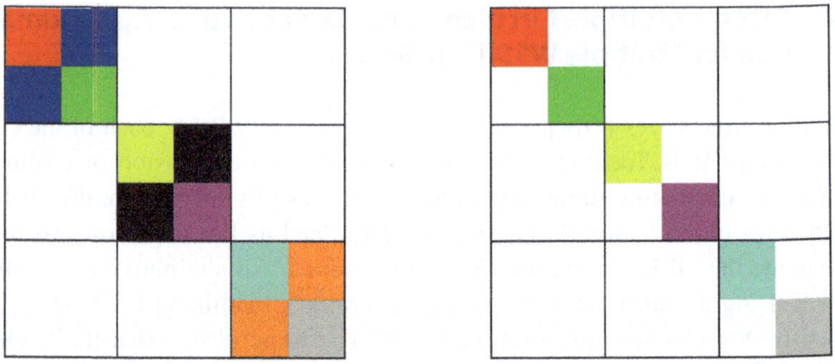
(a)
(b)

Fig. 8.1 Structure of the covariance matrices of $\mathbf{w}$ for $N = 3$ and $L = 2$ under both hypotheses for WSS processes. Each square represents a scalar. (**a**) Spatially correlated. (**b**) Spatially uncorrelated

which contains samples of the discrete Fourier transform (DFT) of $\mathbf{x}[n]$, the test for spatial correlation is approximated as

$$\begin{aligned} \mathcal{H}_1 &: \mathbf{w} \sim \mathcal{CN}_{LN}(\mathbf{0}, \mathbf{D}_1), \\ \mathcal{H}_0 &: \mathbf{w} \sim \mathcal{CN}_{LN}(\mathbf{0}, \mathbf{D}_0). \end{aligned} \tag{8.9}$$

Here, the frequency-domain covariance matrix $\mathbf{D}_1$ is block-diagonal with block size L and $\mathbf{D}_0$ is diagonal with positive elements, as depicted in Fig. 8.1. The accuracy of this approximation improves as N grows. In fact, [274] proves the convergence in mean squared error of the likelihood of $\mathbf{D}_i$ to the likelihood of $\mathbf{R}_i$ as $N \to \infty$, for $i = 0, 1$.

The $L \times L$ blocks of the covariance matrices are given by the power spectral density matrix of $\mathbf{x}[n]$ at frequencies $\theta_k = 2\pi k/N, k = 0, \ldots, N-1$, and for $\mathbf{D}_0$ these blocks only contain along their diagonals the PSDs of each vector component. To summarize, the test in (8.9) is again a test for the covariance structure of the observations: block-diagonal vs. diagonal. However, it is important to note that (8.9) is only an approximate (or nearby) detection problem for finite N and it is this approximating problem that is addressed in the following, based on the observations $\mathbf{W} = [\mathbf{w}_1 \cdots \mathbf{w}_M]$, where $\mathbf{w}_m, m = 1, \ldots, M$, are i.i.d.

Invariances. The test for spatial correlation for WSS processes in (8.9) is invariant to the transformation group $G = \{g \mid g \cdot \mathbf{W} = \mathbf{P}\,\mathrm{diag}(\beta_1, \ldots, \beta_{LN})\mathbf{W}\mathbf{Q}_M\}$, where $\mathbf{Q}_M \in U(M)$ is an arbitrary unitary matrix, $\beta_l \neq 0$, and $\mathbf{P} = \mathbf{P}_N \otimes \mathbf{P}_L$, with $\mathbf{P}_N$ and $\mathbf{P}_L$ permutation matrices of sizes N and L, respectively. Interestingly, the multiplication by a diagonal matrix represents an independent linear filtering of each time series, $\{x_l[n]\}$, implemented in the frequency domain (a circular convolution). Moreover, the permutation $\mathbf{P}_N$ represents an arbitrary reordering of the DFT frequencies, and the permutation $\mathbf{P}_L$ applies a reordering of the L channels. These invariances make sense since the matrix-valued PSD is arbitrary and unknown. Hence, modifying the PSD by permuting the frequencies, arbitrarily changing the shape of the PSD of each component or exchanging channels, does not modify the test.

GLR Test. The GLR for the detection problem in (8.9) is

$$\lambda = \frac{\det(\hat{\mathbf{D}}_1)}{\det(\hat{\mathbf{D}}_0)}, \tag{8.10}$$

where $\hat{\mathbf{D}}_1 = \text{blkdiag}_L(\mathbf{S})$ and $\hat{\mathbf{D}}_0 = \text{diag}(\mathbf{S})$, with sample covariance matrix

$$\mathbf{S} = \frac{1}{M}\sum_{m=1}^{M} \mathbf{w}_m \mathbf{w}_m^H.$$

Defining now the coherence matrix

$$\hat{\mathbf{C}} = \hat{\mathbf{D}}_0^{-1/2}\hat{\mathbf{D}}_1\hat{\mathbf{D}}_0^{-1/2}, \tag{8.11}$$

the GLR (8.10) may be rewritten as

$$\lambda = \det(\hat{\mathbf{C}}) = \prod_{k=1}^{N} \det(\hat{\mathbf{C}}_k),$$

where $\hat{\mathbf{C}}_k$ is the kth $L \times L$ block on the diagonal of $\hat{\mathbf{C}}$. Taking into account that $\mathbf{w}$ contains samples of the DFT of $\mathbf{x}[n]$, the $L \times L$ blocks on the diagonal of $\mathbf{S}$ are given by Bartlett-type estimates of the PSD, i.e.,

$$\hat{\mathbf{S}}(e^{j\theta_k}) = \frac{1}{M}\sum_{m=1}^{M} \mathbf{x}_m(e^{j\theta_k})\mathbf{x}_m^H(e^{j\theta_k}),$$

where

$$\mathbf{x}_m(e^{j\theta_k}) = \sum_{n=0}^{N-1} \mathbf{x}_m[n]e^{-j\theta_k n},$$

with $\theta_k = 2\pi k/N$ and $\mathbf{x}_m[n]$ being the nth sample of the mth realization of the multivariate time series. Then, we can write the log-GLR as

$$\log\lambda = \sum_{k=0}^{N-1} \log\left[\frac{\det(\hat{\mathbf{S}}(e^{j\theta_k}))}{\prod_{l=1}^{L} \hat{S}_{ll}(e^{j\theta_k})}\right] = \sum_{k=0}^{N-1} \log\det(\hat{\mathbf{C}}(e^{j\theta_k})), \tag{8.12}$$

where $\hat{S}_{ll}(e^{j\theta})$ is the PSD estimate of the lth process, i.e., the lth diagonal element of $\hat{\mathbf{S}}(e^{j\theta})$, and the spectral coherence is

$$\hat{\mathbf{C}}(e^{j\theta}) = \hat{\mathbf{D}}^{-1/2}(e^{j\theta})\hat{\mathbf{S}}(e^{j\theta})\hat{\mathbf{D}}^{-1/2}(e^{j\theta}), \tag{8.13}$$

with $\hat{\mathbf{D}}(e^{j\theta}) = \operatorname{diag}\left(\hat{\mathbf{S}}(e^{j\theta})\right)$.

The statistic in (8.7) is a limiting form of a GLR, derived for a nonstationary process, when M and N are large, and the time series are WSS; the statistic in (8.12) is the exact GLR for a circulant time series, which for large N approximates a WSS time series.

The GLR λ in (8.12) is a product of N independent Hadamard ratios for each of which the distribution under the null is stochastically equivalent to a product of betas. Therefore, the stochastic representation for the null distribution of λ is $\prod_{n=0}^{N-1}\prod_{l=1}^{L-1} U_l^{(n)}$, where $U_l^{(n)} \sim \text{Beta}(M-l, l)$. Note that the distribution of $U_l^{(n)}, n = 0, \ldots, N-1$, does not depend on n. Hence, these are just N different realizations of the same random variable.

LMPIT. The LMPIT for the test in (8.9) is given by

$$\mathscr{L} = \sum_{k=0}^{N-1} \|\hat{\mathbf{C}}(e^{j\theta_k})\|^2, \tag{8.14}$$

where the spectral coherence $\hat{\mathbf{C}}(e^{j\theta})$ is defined in (8.13). Again, this test statistic is a measure of broadband coherence that is obtained by fusing fine-grained spectral coherences.

8.4 Applications

Detecting correlation among time series applies to sensor networks [393], cooperative networks with multiple relays using the amplify-and-forward (AF) scheme [120, 211, 242], and MIMO radar [217]. Besides these applications, there are two that are particularly important: (1) the detection of primary user transmissions in cognitive radio and (2) testing for impropriety of time series. These are analyzed in more detail in the following two subsections.

8.4.1 Cognitive Radio

Cognitive radio (CR) is a mature communications paradigm that could potentially boost spectrum usage [52, 68, 243]. In interweave CR, which is one of the three main techniques in CR [138], the opportunistic access of the so-called "cognitive" or "secondary" users to a given channel is allowed when the primary user (the licensed user of the channel) is not transmitting. Thus, every cognitive user must detect when a channel is idle, which is known as spectrum sensing and is a key ingredient of interweave CR [18].

Spectrum sensing can be formulated as the hypothesis test:

$$\begin{aligned} \mathcal{H}_1 &: \mathbf{x}[n] = (\mathbf{H} * \mathbf{s})[n] + \mathbf{n}[n], \\ \mathcal{H}_0 &: \mathbf{x}[n] = \mathbf{n}[n], \end{aligned} \tag{8.15}$$

where $\mathbf{x}[n] \in \mathbb{C}^L$ is the received signal at the cognitive user's array; $\mathbf{n}[n] \in \mathbb{C}^L$ is spatially uncorrelated WSS noise, which is Gaussian distributed with zero mean and arbitrary PSDs; $\mathbf{H}[n] \in \mathbb{C}^{L\times p}$ is a time-invariant and frequency-selective MIMO channel; and $\mathbf{s}[n] \in \mathbb{C}^p$ is the signal transmitted by a primary user equipped with p antennas. Among the different features that may be used to derive statistics for the detection problem (8.15) [18], it is possible to exploit the spatial correlation induced by the transmitted signal on the received signal at the cognitive user's array. That is, due to the term $(\mathbf{H} * \mathbf{s})[n]$ and the spatially uncorrelated noise, the received signal $\mathbf{x}[n]$ is spatially correlated under $\mathcal{H}_1$, but it is uncorrelated under $\mathcal{H}_0$. Based on this observation, the (approximate) GLRT and LMPIT for the CR detection problem in (8.15) are (8.12) and (8.14), respectively.

8.4.2 Testing for Impropriety in Time Series

Another important application of correlation detection in time series is the problem of testing whether a zero-mean univariate WSS complex time series is proper or improper. The case of multivariate processes is straightforwardly derived from the results in this chapter. It is well known [318] that the complex time series $\{x[n]\}$ is proper if and only if it is uncorrelated with its conjugate, namely, the complementary covariance function is $E[x[n]x[n-m]] = 0, \forall m, n$ (see Appendix E). Hence, when the detectors of this chapter are applied to the bivariate time series $\{\mathbf{x}[n]\}$, with $\mathbf{x}[n] = [x[n] \; x^*[n]]^T$, they become tests for impropriety. In this way, the (approximate) log-GLR is given by (8.12) and the LMPIT by (8.14). After some algebra, both test statistics become

$$\log \lambda = \sum_{k=0}^{N-1} \log(1 - |\hat{C}(e^{j\theta_k})|^2),$$

and

$$\mathscr{L} = \sum_{k=0}^{N-1} |\hat{C}(e^{j\theta_k})|^2,$$

where the spectral coherence is

$$|\hat{C}(e^{j\theta})|^2 = \frac{|\hat{\tilde{S}}(e^{j\theta})|^2}{\hat{S}(e^{j\theta})\hat{S}(e^{-j\theta})},$$

with $\hat{S}(e^{j\theta})$ and $\hat{\tilde{S}}(e^{j\theta})$ the estimated PSD and complementary PSD, respectively. These detectors are frequency-resolved versions of those developed for testing whether a random variable is proper [318]. For a more detailed analysis, the reader is referred to [66] and references therein.

8.5 Extensions

In the previous sections, we have assumed that the spatial correlation is arbitrary; that is, no correlation model has been assumed. Nevertheless, there are some scenarios where this knowledge is available and can be exploited. For instance, the detection problem in (8.15) may have additional structure. Among all possible models that can be considered for the spatial structure, those in Chap. 5 are of particular interest.

For instance, when measurements are taken from a WSS time series, the approximate GLR for the second-order model with unknown subspace of known dimension p and unknown variance (see Sect. 5.6.1, Eq. (5.14)) is [270]

$$\log \lambda = \sum_{k=0}^{N-1} \log \left\{ \frac{\left[\frac{1}{L}\sum_{l=1}^{L} ev_l\left(\hat{\mathbf{S}}(e^{j\theta_k})\right)\right]^L}{\left[\frac{1}{L-p}\sum_{l=p+1}^{L} ev_l\left(\hat{\mathbf{S}}(e^{j\theta_k})\right)\right]^{L-p}\prod_{l=1}^{p} ev_l\left(\hat{\mathbf{S}}(e^{j\theta_k})\right)} \right\}.$$

Note that this is the GLR only when $p < L - 1$, as otherwise the structure induced by the low-rank transmitted signal disappears [270]. The asymptotic LMPIT for the models in Chap. 5 can be derived in a similar manner. However, as shown in [273], the LMPIT is not modified by the rank-p signal, regardless of the value of p, and only the noise covariance matters. Hence, for spatially uncorrelated noises, the asymptotic LMPIT is still given by (8.14).

This chapter has addressed the question of whether or not a set of L univariate time series are correlated. The work in [201] develops an extension of this problem to a set of P multivariate time series. Assuming wide-sense stationarity in both time and space, the log-GLR is asymptotically approximated by

$$\begin{aligned}\log \lambda &= \sum_{k=0}^{N-1}\sum_{l=0}^{L-1} \log\left[\frac{\det(\hat{\mathbf{S}}(e^{j\theta_k}, e^{j\phi_l}))}{\prod_{p=1}^{P} \hat{S}_{pp}(e^{j\theta_k}, e^{j\phi_l})}\right] \\ &= \sum_{k=0}^{N-1}\sum_{l=0}^{L-1} \log\det(\hat{\mathbf{C}}(e^{j\theta_k}, e^{j\phi_l})), \end{aligned} \tag{8.16}$$

where $\theta_k = 2\pi k/N$, $\phi_l = 2\pi l/L$, $\hat{\mathbf{S}}(e^{j\theta}, e^{j\phi})$ is the PSD estimate in the frequency/wavenumber domain. The spectral coherence is

$$\hat{\mathbf{C}}(e^{j\theta}, e^{j\phi}) = \hat{\mathbf{D}}^{-1/2}(e^{j\theta}, e^{j\phi})\hat{\mathbf{S}}(e^{j\theta}, e^{j\phi})\hat{\mathbf{D}}^{-1/2}(e^{j\theta}, e^{j\phi}),$$

with $\hat{\mathbf{D}}(e^{j\theta}, e^{j\phi}) = \operatorname{diag}\left(\hat{\mathbf{S}}(e^{j\theta}, e^{j\phi})\right)$. Note that (8.16) is the sum in wavenumber of a wavenumber-resolved version of (8.7).

A different, yet related, problem is that of testing whether a set of P independent L-variate time series have the same power spectral density, which can be seen as an extension of the problem in Sect. 4.7. There are many approaches for addressing this problem, such as [75, 102, 103, 123, 349], all of which assume $P = 2$. In [275], following the developments of this chapter, the log-GLR is approximated for arbitrary P by

$$\log \lambda = \sum_{p=1}^{P} \sum_{k=0}^{N-1} \log \det(\hat{\mathbf{C}}_p(e^{j\theta_k})),$$

where now the coherence matrix is defined as

$$\hat{\mathbf{C}}_p(e^{j\theta}) = \left(\frac{1}{P}\sum_{m=1}^{P} \hat{\mathbf{S}}_m(e^{j\theta})\right)^{-1/2} \hat{\mathbf{S}}_p(e^{j\theta}) \left(\frac{1}{P}\sum_{m=1}^{P} \hat{\mathbf{S}}_m(e^{j\theta})\right)^{-1/2},$$

with $\hat{\mathbf{S}}_p(e^{j\theta})$ the estimate of the PSD matrix of the pth multivariate time series at frequency θ. However, the LMPIT does not exist, as the local approximation to the ratio of the distributions of the maximal invariant statistic depends on unknown parameters.

8.6 Detection of Cyclostationarity

A multivariate zero-mean random process $\{\mathbf{u}[n]\}$ is (second-order) cyclostationary if the matrix-valued covariance function, defined as

$$\mathbf{R}_{\mathbf{uu}}[n, m] = \mathrm{E}[\mathbf{u}[n]\mathbf{u}^H[n - m]],$$

is periodic in n: $\mathbf{R}_{\mathbf{uu}}[n, m] = \mathbf{R}_{\mathbf{uu}}[n + P, m]$. The period P is a natural number larger than one; if $P = 1$, the process is WSS. The period P is called the *cycle* period of the process. Hence, CS processes can be used to model phenomena generated by periodic effects in communications, meteorology and climatology, oceanography, astronomy, and economics. For a very detailed review of the bibliography of CS processes, which also includes the aforementioned applications and others, the reader is referred to [320].

There is a spectral theory of CS processes. However, since the covariance function depends on two time indexes, there are two Fourier transforms to be taken. The Fourier series expansion of $\mathbf{R}_{\mathbf{uu}}[n,m]$ may be written

$$\mathbf{R}_{\mathbf{uu}}[n,m] = \sum_{c=0}^{P-1} \mathbf{R}_{\mathbf{uu}}^{(c)}[m] e^{j2\pi cn/P}.$$

The cth coefficient of the Fourier series is

$$\mathbf{R}_{\mathbf{uu}}^{(c)}[m] = \frac{1}{P} \sum_{n=0}^{P-1} \mathbf{R}_{\mathbf{uu}}[n,m] e^{-j2\pi cn/P},$$

which is known as the *cyclic* covariance function at *cycle frequency* $2\pi c/P$. The Fourier transform (in m) of this cyclic covariance function is the *cyclic* power spectral density, given by

$$\mathbf{S}_{\mathbf{uu}}^{(c)}(e^{j\theta}) = \sum_{m} \mathbf{R}_{\mathbf{uu}}^{(c)}[m] e^{-j\theta m}.$$

The cyclic PSD is related to the Loève (or 2D) spectrum as

$$\mathbf{S}_{\mathbf{uu}}(e^{j\theta_1}, e^{j\theta_2}) = \sum_{c=0}^{P-1} \mathbf{S}_{\mathbf{uu}}^{(c)}(e^{j\theta_1}) \delta\left(\theta_1 - \theta_2 - \frac{2\pi c}{P}\right),$$

where $\mathbf{S}_{\mathbf{uu}}(e^{j\theta_1}, e^{j\theta_2})$ is the Loève spectrum [223]. That is, the support of the Loève spectrum for CS processes is the lines $\theta_1 - \theta_2 = 2\pi c/P$, which are harmonics of the fundamental cycle frequency $2\pi/P$. Additionally, for $c = 0$, the cyclic PSD reduces to the PSD, and the line $\theta_1 = \theta_2$ is therefore known as the *stationary manifold*.

Gladyshev's Representation of a CS Process. Yet another representation of CS processes was introduced by Gladyshev [134]. The representation is given by the time series $\{\mathbf{x}[n]\}$, where

$$\mathbf{x}[n] = \left[\mathbf{u}^T[nP]\ \mathbf{u}^T[nP+1] \cdots \mathbf{u}^T[(n+1)P-1]\right]^T \in \mathbb{C}^{LP}.$$

This is the stack of P samples of the L-variate random vector $\mathbf{u}[n]$. Gladyshev proved that $\{\mathbf{x}[n]\}$ is a vector-valued WSS process when $\{\mathbf{u}[n]\}$ is CS with cycle period P. That is, its covariance function only depends on the time lag

$$\mathbf{R}_{\mathbf{xx}}[n,m] = E[\mathbf{x}[n]\mathbf{x}^H[n-m]] = \mathbf{R}_{\mathbf{xx}}[m].$$

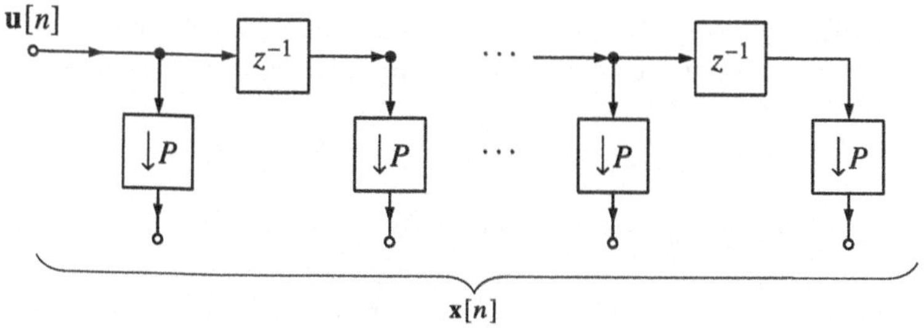

Fig. 8.2 Gladyshev's representation of a CS process

Figure 8.2 depicts Gladyshev's characterization of the CS process $\{\mathbf{u}[n]\}$, suggesting that the components of the WSS process $\{\mathbf{x}[n]\}$ can be interpreted as a polyphase representation of the signal $\mathbf{u}[n]$. In the figure the down arrows denote subsampling of $\{\mathbf{u}[n]\}$ and its one-sample delays, by a factor of P. The outputs of these subsamplers are the so-called polyphase components of $\{\mathbf{u}[n]\}$.

Detectors of Cyclostationarity. There are roughly three categories of cyclostationarity detectors:

1. Techniques based on the Loève spectrum [48, 49, 182]: These methods compare the energy that lies on the lines $\theta_1 - \theta_2 = \frac{2\pi c}{P}$ to the energy in the rest of the 2D frequency plane $[\theta_1\ \theta_2]^T \in \mathbb{R}^2$.
2. Techniques based on testing for non-zero cyclic covariance function or cyclic spectrum [17, 93]: These approaches test whether $\mathbf{R}_{\mathbf{uu}}^{(c)}(e^{j\theta})$ (or $\mathbf{S}_{\mathbf{uu}}^{(c)}[m]$) are zero for $c > 1$.
3. Techniques based on testing for correlation between the process and its frequency-shifted version [116, 314, 353]: In a CS process, $\mathbf{u}[n]$ is correlated with $\mathbf{v}[n] = \mathbf{u}[n]e^{j2\pi nc/P}$, whereas in a WSS it is not. Hence, this family of techniques tests for correlation between $\mathbf{u}[n]$ and $\mathbf{v}[n]$.

In the remainder of this section, the problem of detecting CS is formulated as a virtual detection problem for spatial correlation, which allows us to use all the machinery presented in this chapter. Additionally, it is shown that the derived detectors have interpretations in all three categories.

8.6.1 Problem Formulation and Its Invariances

The problem of detecting a cyclostationary signal, with *known* cycle period P, contaminated by WSS noise can be formulated in its most general form as

$$\begin{aligned} \mathcal{H}_1 &: \{\mathbf{u}[n]\} \text{ is CS with cycle period } P, \\ \mathcal{H}_0 &: \{\mathbf{u}[n]\} \text{ is WSS,} \end{aligned} \tag{8.17}$$

where $\{\mathbf{u}[n]\} \in \mathbb{C}^L$ is an L-variate complex time series, which we take as a zero-mean proper Gaussian. Given NP samples of $\mathbf{u}[n]$, which we arrange into the vector

$$\mathbf{y} = \left[\mathbf{u}^T[0]\ \mathbf{u}^T[1]\ \mathbf{u}^T[2] \cdots \mathbf{u}^T[NP-1]\right]^T,$$

the test in (8.17) boils down to a test for the covariance structure of $\mathbf{y}$:

$$\begin{aligned} \mathcal{H}_1 &: \mathbf{y} \sim \mathcal{CN}_{LNP}(\mathbf{0}, \mathbf{R}_1), \\ \mathcal{H}_0 &: \mathbf{y} \sim \mathcal{CN}_{LNP}(\mathbf{0}, \mathbf{R}_0), \end{aligned} \tag{8.18}$$

where $\mathbf{R}_i \in \mathbb{C}^{LNP}$ is the covariance matrix of $\mathbf{y}$ under the ith hypothesis. Thus, as in previous sections, we have to determine the structure of the covariance matrices.

The covariance under $\mathcal{H}_0$, i.e., $\{\mathbf{u}[n]\}$ is WSS, is easy to derive taking into account that $\mathbf{y}$ is the stack of NP samples of a multivariate WSS process:

$$\mathbf{R}_0 = \begin{bmatrix} \mathbf{R}_{\mathbf{uu}}[0] & \mathbf{R}_{\mathbf{uu}}[-1] & \cdots & \mathbf{R}_{\mathbf{uu}}[-NP+1] \\ \mathbf{R}_{\mathbf{uu}}[1] & \mathbf{R}_{\mathbf{uu}}[0] & \cdots & \mathbf{R}_{\mathbf{uu}}[-NP+2] \\ \vdots & \vdots & \ddots & \vdots \\ \mathbf{R}_{\mathbf{uu}}[NP-1] & \mathbf{R}_{\mathbf{uu}}[NP-2] & \cdots & \mathbf{R}_{\mathbf{uu}}[0] \end{bmatrix},$$

where $\mathbf{R}_{\mathbf{uu}}[m] = E[\mathbf{u}[n]\mathbf{u}^H[n-m]] \in \mathbb{C}^{L \times L}$ is the matrix-valued covariance sequence under $\mathcal{H}_0$. The covariance matrix $\mathbf{R}_0$ is a block-Toeplitz matrix with block size L. It is important to point out that only the structure of $\mathbf{R}_0$ is known, but the particular values are not. That is, the matrix-valued covariance function $\mathbf{R}_{\mathbf{uu}}[m]$ is unknown.

When $\mathbf{u}[n]$ is cyclostationary, under $\mathcal{H}_1$, we can use Gladyshev's representation of a CS process to write

$$\mathbf{y} = \left[\mathbf{x}^T[0]\ \mathbf{x}^T[1]\ \mathbf{x}^T[2] \cdots \mathbf{x}^T[N-1]\right]^T,$$

where

$$\mathbf{x}[n] = \left[\mathbf{u}^T[nP]\ \mathbf{u}^T[nP+1] \cdots \mathbf{u}^T[(n+1)P-1]\right]^T \in \mathbb{C}^{LP},$$

is WSS. Then, the covariance matrix $\mathbf{R}_1$ becomes

$$\mathbf{R}_1 = \begin{bmatrix} \mathbf{R}_{\mathbf{xx}}[0] & \mathbf{R}_{\mathbf{xx}}[-1] & \cdots & \mathbf{R}_{\mathbf{xx}}[-N+1] \\ \mathbf{R}_{\mathbf{xx}}[1] & \mathbf{R}_{\mathbf{xx}}[0] & \cdots & \mathbf{R}_{\mathbf{xx}}[-N+2] \\ \vdots & \vdots & \ddots & \vdots \\ \mathbf{R}_{\mathbf{xx}}[N-1] & \mathbf{R}_{\mathbf{xx}}[N-2] & \cdots & \mathbf{R}_{\mathbf{xx}}[0] \end{bmatrix},$$

Fig. 8.3 Structure of the covariance matrices of $\mathbf{y}$ for $N = 3$ and $P = 2$ under both hypotheses. Each square represents an $L \times L$ matrix. (**a**) Stationary signal. (**b**) Cyclostationary signal

where $\mathbf{R}_{\mathbf{xx}}[m] = \mathrm{E}[\mathbf{x}[n]\mathbf{x}^H[n-m]] \in \mathbb{C}^{LP \times LP}$ is the matrix-valued covariance sequence under $\mathcal{H}_1$. That is, $\mathbf{R}_1$ is a block-Toeplitz matrix with block size LP, and each block has no further structure beyond being positive definite. The test in (8.17), under the Gaussian assumption, may therefore be formulated as a test for the covariance structure of the observations. Specifically, we are testing two block-Toeplitz covariance matrices with different block sizes: LP under $\mathcal{H}_1$ and L under $\mathcal{H}_0$, as shown in Fig. 8.3.

The block-Toeplitz structure of the covariance matrices in (8.18) precludes the derivation of closed-form expressions for both the GLR and LMPIT [274]. To overcome this issue, and derive closed-form detectors, [274] solves an approximate problem in the frequency domain as done in Sect. 8.3.2. First, let us define the vector

$$\mathbf{z} = (\mathbf{L}_{NP,N} \otimes \mathbf{I}_L)(\mathbf{F}_{NP} \otimes \mathbf{I}_L)\mathbf{y},$$

where $\mathbf{L}_{NP,N}$ is the commutation matrix. Basically, $\mathbf{z}$ contains samples of the discrete Fourier transform (DFT) of $\mathbf{u}[n]$ arranged in a particular order. Then, the test in (8.18) can be approximated as

$$\begin{aligned} \mathcal{H}_1 &: \mathbf{z} \sim \mathcal{CN}_{LNP}(\mathbf{0}, \mathbf{D}_1), \\ \mathcal{H}_0 &: \mathbf{z} \sim \mathcal{CN}_{LNP}(\mathbf{0}, \mathbf{D}_0). \end{aligned} \tag{8.19}$$

Here, $\mathbf{D}_0$ is a block-diagonal matrix with block size L, and $\mathbf{D}_1$ is also block-diagonal but with block size LP, as shown in Fig. 8.4. Thus, the problem of detecting a cyclostationary signal contaminated by WSS noise boils down to testing whether the covariance matrix of $\mathbf{z}$ is block-diagonal with block size LP or L.

Interestingly, if we focus on each of the N blocks of size $LP \times LP$ in the diagonal of $\mathbf{D}_1$ and $\mathbf{D}_0$, we would be testing whether each block is just positive definite or block-diagonal with positive definite blocks. That is, in each block, the problem is that of testing for spatial correlation (c.f. (8.2)). Alternatively, it could

Fig. 8.4 Structure of the covariance matrices of $\mathbf{z}$ for $N = 3$ and $P = 2$ under both hypotheses. Each square represents an $L \times L$ matrix. (**a**) Stationary signal. (**b**) Cyclostationary signal

also be interpreted as a generalization of (8.9). In particular, if we consider $L = 1$ in (8.19), both problems would be equivalent. This explains why we called this a virtual problem of detecting spatial correlation.

Given the observations $\mathbf{Z} = [\mathbf{z}_1 \cdots \mathbf{z}_M]$, we can obtain the invariances of the detection problem in (8.19), which are instrumental in the development of the detectors, and gain insight into the problem. It is clear that multiplying the observations $\mathbf{z}$ by a block diagonal matrix with block size L does not modify the structure of $\mathbf{D}_1$ and $\mathbf{D}_0$. This invariance is interesting as it represents a multiple-input-multiple-output (MIMO) filtering in the frequency domain (a circular convolution). Additionally, we may permute $LP \times LP$ blocks of $\mathbf{D}_1$ and $\mathbf{D}_0$ without varying their structure, and within these blocks, permuting $L \times L$ blocks also leaves the structure unchanged. Thus, these permutations result in a particular reordering of the frequencies of the DFT of $\mathbf{u}[n]$. Finally, as always, the problem is invariant to a right multiplication by a unitary matrix. Hence, the invariances are captured by the invariance group

$$G = \{g \mid g \cdot \mathbf{Z} = \mathbf{PBZQ}_M\},$$

where $\mathbf{B}$ is an invertible block-diagonal matrix with block size L, $\mathbf{Q}_M \in U(M)$, and $\mathbf{P} = \mathbf{P}_N \otimes (\mathbf{P}_P \otimes \mathbf{I}_L)$, with $\mathbf{P}_N$ and $\mathbf{P}_P$ being permutation matrices of sizes N and P, respectively.

8.6.2 Test Statistics

Taking into account the reformulation as a virtual problem, it is easy to show that the (approximate) GLR for detecting cyclostationarity is given by [274]

$$\lambda = \frac{1}{\Lambda^{1/M}} = \frac{\det(\hat{\mathbf{D}}_1)}{\det(\hat{\mathbf{D}}_0)}, \tag{8.20}$$

where

$$\Lambda = \frac{\ell(\hat{\mathbf{D}}_1; \mathbf{Z})}{\ell(\hat{\mathbf{D}}_0; \mathbf{Z})},$$

and $\ell(\hat{\mathbf{D}}_i; \mathbf{Z})$ is the likelihood of the ith hypothesis where the covariance matrix $\mathbf{D}_i$ has been replaced by its ML estimate $\hat{\mathbf{D}}_i$. Under the alternative, the ML estimate of the covariance matrix is

$$\hat{\mathbf{D}}_1 = \text{blkdiag}_{LP}(\mathbf{S}),$$

with the sample covariance matrix given by

$$\mathbf{S} = \frac{1}{M} \sum_{m=1}^{M} \mathbf{z}_m \mathbf{z}_m^H.$$

That is, $\hat{\mathbf{D}}_1$ is a block-diagonal matrix obtained from the $LP \times LP$ blocks of $\mathbf{S}$. Similarly, under the null, the ML estimate $\hat{\mathbf{D}}_0$ is given by $\hat{\mathbf{D}}_0 = \text{blkdiag}_L(\mathbf{S})$. The GLR in (8.20) can alternatively be rewritten as

$$\lambda = \det(\hat{\mathbf{C}}) = \prod_{k=1}^{N} \det(\hat{\mathbf{C}}_k),$$

where the coherence matrix is

$$\hat{\mathbf{C}} = \hat{\mathbf{D}}_0^{-1/2} \hat{\mathbf{D}}_1 \hat{\mathbf{D}}_0^{-1/2}, \tag{8.21}$$

and $\hat{\mathbf{C}}_k$ is the kth $LP \times LP$ block on its diagonal. This expression provides a more insightful interpretation of the GLR, which, as we will see in Sect. 8.6.3, is a measure of bulk coherence that can be resolved into fine-grained spectral coherences.

Null Distribution. Appendix H shows that the GLR in (8.20), under the null, is stochastically equivalent to $\prod_{n=0}^{N-1} \prod_{l=0}^{L-1} \prod_{p=0}^{P-1} U_{l,p}^{(n)}$, where $U_{l,p}^{(n)} \sim \text{Beta}(M - (Lp + l), Lp)$.

LMPIT. The LMPIT for the approximate problem in (8.19) is given by [274]

$$\mathscr{L} = \|\hat{\mathbf{C}}\|^2 = \sum_{k=1}^{N} \|\hat{\mathbf{C}}_k\|^2, \tag{8.22}$$

where $\hat{\mathbf{C}}$ is defined in (8.21).

8.6.3 Interpretation of the Detectors

In previous sections, we have presented the GLR and the LMPIT for detecting a cyclostationary signal in WSS noise, which has an arbitrary spatiotemporal structure. One common feature of both detectors is that they are given by (different) functions of the same coherence matrix. In this section, we will show that these detectors are also functions of a spectral coherence, which is related to the cyclic PSD and the PSD. This, of course, sheds some light on the interpretation of the detectors and allows for a more comprehensive comparison with the different categories of cyclostationary detectors presented before.

The GLR and the LMPIT are functions of the coherence matrix $\hat{\mathbf{C}}$ in (8.21). In [274] it is shown that the blocks of this matrix are given by a spectral coherence, defined as

$$\hat{\mathbf{C}}^{(c)}(e^{j\theta_k}) = \hat{\mathbf{S}}^{-1/2}(e^{j\theta_k})\hat{\mathbf{S}}^{(c)}(e^{j\theta_k})\hat{\mathbf{S}}^{-1/2}\left(e^{j(\theta_k - 2\pi c/P)}\right), \tag{8.23}$$

where $\hat{\mathbf{S}}^{(c)}(e^{j\theta_k})$ is an estimate of the cyclic PSD of $\{\mathbf{u}[n]\}$ at frequency $\theta_k = 2\pi k/NP$ and cycle frequency $2\pi c/P$, and $\hat{\mathbf{S}}(e^{j\theta_k})$ is an estimate of the PSD of $\{\mathbf{u}[n]\}$ at frequency θ_k, i.e., $\hat{\mathbf{S}}(e^{j\theta_k}) = \hat{\mathbf{S}}^{(0)}(e^{j\theta_k})$. Based on the spectral coherence (8.23), the LMPIT in (8.22) may be more insightfully rewritten as

$$\mathscr{L} = \sum_{c=1}^{P-1} \sum_{k=0}^{(P-c)N-1} \left\| \hat{\mathbf{C}}^{(c)}(e^{j\theta_k}) \right\|^2. \tag{8.24}$$

Unfortunately, due to the nonlinearity of the determinant, an expression similar to (8.24) for the GLR is only possible for $P = 2$ and is given by

$$\log \lambda = \sum_{k=0}^{NP-1} \log \det \left[\mathbf{I}_L - \hat{\mathbf{C}}^{(1)H}(e^{j\theta_k})\hat{\mathbf{C}}^{(1)}(e^{j\theta_k}) \right].$$

For other values of P, the GLR is still a function of $\hat{\mathbf{C}}^{(c)}(e^{j\theta_k})$, albeit with no closed-form expression.

A further interpretation comes from considering the spectral representation of $\{\mathbf{u}[n]\}$ [318]:

$$\mathbf{u}[n] = \int_{-\pi}^{\pi} d\boldsymbol{\xi}(e^{j\theta})e^{j\theta n},$$

where $d\boldsymbol{\xi}(e^{j\theta})$ is an increment of the spectral process $\{\boldsymbol{\xi}(e^{j\theta})\}$. Based on this representation, we may express the cyclic PSD as [318]

$$\mathbf{S}^{(c)}(e^{j\theta})d\theta = E\left[d\boldsymbol{\xi}(e^{j\theta})d\boldsymbol{\xi}^H\left(e^{j(\theta+2\pi c/P)}\right)\right],$$

which clearly shows that $\hat{\mathbf{C}}^{(c)}(\theta)$ is an estimate of the coherence matrix of $d\boldsymbol{\xi}(\theta)$ and its frequency-shifted version.

8.7 Chapter Notes

- Testing for independence among random vectors is a basic problem in multi-sensor signal processing, where the typical problem is to detect a correlated effect in measurements recorded at several sensors. In some cases, this common effect may be attributed to a common source.
- In this chapter, we have assumed that there are M i.i.d. realizations of the space-time snapshot. In some applications, this assumption is justified. More commonly, M realizations are obtained by constructing space-time snapshots from consecutive windows of a large space-time realization. These windowings are not i.i.d. realizations, but in many applications, they are approximately so.
- There are many variations on the problem of testing for spatial correlation. For example: Is the time series at sensor 1 linearly independent of the time series at sensors 2 through L, without regard for whether these time series are linearly independent? This kind of question arises in many contexts, including the construction of links in graphs of measurement nodes [22].
- Testing the correlation between $L = 2$ stationary time series is a problem that has been studied prior to the references cited in the introduction to this chapter. For instance, [159] proposed two test statistics obtained as different functions of several lags of the cross-covariance function normalized by the standard deviations of the prewhitened time series. The subsequent work in [169] presented an improved test statistic. In both cases, the test statistics can be interpreted as coherence detectors.
- The results in Sect. 8.6 may be extended in several ways. For example, in [266] the WSS process has further structure: it can be spatially uncorrelated, temporally white, or temporally white and spatially uncorrelated. For the three cases, it is possible to derive the corresponding GLRs; however, the LMPIT only exists in the case of spatially uncorrelated processes. For temporally white processes, spatially correlated or uncorrelated, [266] showed that the LMPIT does not exist and used this proof to propose *LMPIT-inspired* detectors.
- Cyclostationarity may be exploited in passive detection of communication signals or signals radiated from rotating machinery. In [172], the authors derived GLR and LMPIT-inspired tests for passively detecting cyclostationary signals.

Subspace Averaging and its Applications 9

All distances between subspaces are functions of the principal angles between them and thus can ultimately be interpreted as measures of coherence between pairs of subspaces, as we have seen throughout this book. In this chapter, we first review the geometry and statistics of the Grassmann and Stiefel manifolds, in which q-dimensional subspaces and q-dimensional frames live, respectively. Then, we pay particular attention to the problem of subspace averaging using the projection (a.k.a. chordal) distance. Using this metric, the average of orthogonal projection matrices turns out to be the central quantity that determines, through its eigendecomposition, both the central subspace and its dimension. The dimension is determined by thresholding the eigenvalues of an average of projection matrices, while the corresponding eigenvectors form a basis for the central subspace. We discuss applications of subspace averaging to subspace clustering and to source enumeration in array processing.

9.1 The Grassmann and Stiefel Manifolds

We present a brief introduction to Riemannian geometry, focusing on the Grassmann manifold of q-dimensional subspaces in $\mathbb{R}^n$, which is denoted as $\mathbb{G}r(q, \mathbb{R}^n)$, and the Stiefel manifold of q-frames in $\mathbb{R}^n$, denoted as $\mathbb{S}t(q, \mathbb{R}^n)$. The main ideas carry over naturally to a complex ambient space $\mathbb{C}^n$. A more detailed account of Riemannian manifolds and, in particular, Grassmann and Stiefel manifolds can be found in [4, 74, 114].

Let us begin with basic concepts. A manifold is a topological space that is locally similar to a Euclidean space. Every point on the manifold has a neighborhood for which there exists a homeomorphism (i.e., a bijective continuous mapping) mapping the neighborhood to $\mathbb{R}^n$. For differentiable manifolds, it is possible to define the derivatives of curves on the manifold. The derivatives at a point $\mathbf{V}$ on the manifold lie in a vector space $\mathcal{T}_{\mathbf{V}}$, which is the tangent space at that point. A Riemannian manifold $\mathcal{M}$ is a differentiable manifold for which each tangent space has an inner

D. Ramírez et al., *Coherence*, https://doi.org/10.1007/978-3-031-13331-2_9

product that varies smoothly from point to point. The inner product induces a norm for tangent vectors in the tangent space.

The Stiefel and Grassmann manifolds are compact smooth Riemannian manifolds, with an inner product structure. This inner product determines distance functions, which are required to compute averages or to perform optimization tasks on the manifold.

The Stiefel Manifold. The Stiefel manifold $\mathbb{S}t(q, \mathbb{R}^n)$ is the space of q-frames in $\mathbb{R}^n$, where a set of q orthonormal vectors in $\mathbb{R}^n$ is called a q-frame. The Stiefel manifold is represented by the set of $n \times q$ matrices, $\mathbf{V} \in \mathbb{R}^{n \times q}$, such that $\mathbf{V}^T\mathbf{V} = \mathbf{I}_q$. The orthonormality of $\mathbf{V}$ enforces $q(q+1)/2$ independent conditions on the nq elements of $\mathbf{V}$, hence $\dim(\mathbb{S}t(q, \mathbb{R}^n)) = nq - q(q+1)/2$. Since $\text{tr}(\mathbf{V}^T\mathbf{V}) = \sum_{i,k=1}^{n} v_{ik}^2 = q$, the Stiefel is also a subset of a sphere of radius $\sqrt{q}$ in $\mathbb{R}^{nq}$. The Stiefel is invariant to left-orthogonal transformations

$$\mathbf{V} \to \mathbf{Q}\mathbf{V}, \quad \text{for any} \quad \mathbf{Q} \in O(n),$$

where $O(n)$ is the orthogonal group of $n \times n$ matrices. That is, the orthogonal matrix $\mathbf{Q} \in O(n)$ acts transitively on the elements of the Stiefel manifold, which is to say the left transformation $\mathbf{QV}$ is another q-frame in $\mathbb{S}t(q, \mathbb{R}^n)$.

Taking a representative

$$\mathbf{V}_0 = \begin{bmatrix} \mathbf{I}_q \\ \mathbf{0} \end{bmatrix} \in \mathbb{S}t(q, \mathbb{R}^n),$$

the matrix $\mathbf{Q}$ that leaves $\mathbf{V}_0$ invariant must be of the form

$$\mathbf{Q} = \begin{bmatrix} \mathbf{I}_q & \mathbf{0} \\ \mathbf{0} & \mathbf{Q}_{n-q} \end{bmatrix},$$

where $\mathbf{Q}_{n-q} \in O(n-q)$. This shows that $\mathbb{S}t(q, \mathbb{R}^n)$ may be thought of as a quotient space $O(n)/O(n-q)$. Alternatively, one may say: begin with an $n \times n$ orthogonal matrix from the orthogonal group $O(n)$, extract the first q columns, and you have a q-dimensional frame from $\mathbb{S}t(q, \mathbb{R}^n)$. The extraction is invariant to rotation of the last $n - q$ columns of $O(n)$, and this accounts for the mod notation $/O(n-q)$. Likewise, one can define the complex Stiefel manifold of q-frames in $\mathbb{C}^n$, denoted as $\mathbb{S}t(q, \mathbb{C}^n)$, which is a compact manifold of dimension $2nq - q^2$.

The notion of a directional derivative in a vector space can be generalized to Riemannian manifolds by replacing the increment $\mathbf{V} + t\Delta\mathbf{V}$ in the definition of the directional derivative

$$\lim_{t \to 0} \frac{f(\mathbf{V} + t\Delta\mathbf{V}) - f(\mathbf{V})}{t},$$

by a smooth curve $\gamma(t)$ on the manifold that passes through $\mathbf{V}$ (i.e., $\gamma(0) = \mathbf{V}$). This yields a well-defined directional derivative $\frac{d(f(\gamma(t)))}{dt}|_{t=0}$ and a well-defined tangent vector to the manifold at a point $\mathbf{V}$. The *tangent space* to the manifold $\mathcal{M}$ at $\mathbf{V}$, denoted as $T_{\mathbf{V}}\mathcal{M}$, is the set of all tangent vectors to $\mathcal{M}$ at $\mathbf{V}$. The tangent space is a vector space that provides a local approximation of the manifold in the same way that the derivative of a real-valued function provides a local linear approximation of the function. The dimension of the tangent space is the dimension of the manifold.

The tangent space of the Stiefel manifold at a point $\mathbf{V} \in \mathbb{S}t(q, \mathbb{R}^n)$ is easily obtained by differentiating $\mathbf{V}^T\mathbf{V} = \mathbf{I}_q$, yielding

$$T_{\mathbf{V}}\mathbb{S}t(q,\mathbb{R}^n) = \left\{ \Delta\mathbf{V} \in \mathbb{R}^{n\times q} \mid \left.\frac{d(\mathbf{V}+t\Delta\mathbf{V})^T(\mathbf{V}+t\Delta\mathbf{V})}{dt}\right|_{t=0} = \mathbf{0} \right\}$$
$$= \left\{ \Delta\mathbf{V} \in \mathbb{R}^{n\times q} \mid (\Delta\mathbf{V})^T\mathbf{V} + \mathbf{V}^T(\Delta\mathbf{V}) = \mathbf{0} \right\}.$$

From $(\Delta\mathbf{V})^T\mathbf{V} + \mathbf{V}^T(\Delta\mathbf{V}) = \mathbf{0}$, it follows that $\mathbf{V}^T(\Delta\mathbf{V})$ is a $q \times q$ skew-symmetric matrix. This imposes $q(q+1)/2$ constraints on $\Delta\mathbf{V}$ so the tangent space has dimension $nq - q(q+1)/2$, the same as that of $\mathbb{S}t(q, \mathbb{R}^n)$. Since $\Delta\mathbf{V}$ is an $n \times q$ full-rank matrix, we may write the following alternative characterization of the elements of the tangent plane

$$\Delta\mathbf{V} = \mathbf{V}\mathbf{A} + \mathbf{V}_\perp\mathbf{B}, \tag{9.1}$$

where $\mathbf{V}_\perp$ is an $(n-q)$ frame for the orthogonal complement to $\mathbf{V}$ such that $\mathbf{V}\mathbf{V}^T + \mathbf{V}_\perp\mathbf{V}_\perp^T = \mathbf{I}_n$, $\mathbf{B}$ is $(n-q)\times q$ and $\mathbf{A}$ is $q \times q$. Using (9.1) in $(\Delta\mathbf{V})^T\mathbf{V} + \mathbf{V}^T(\Delta\mathbf{V}) = \mathbf{0}$ yields $\mathbf{A}^T + \mathbf{A} = \mathbf{0}$, so $\mathbf{A}$ is a skew-symmetric matrix. In (9.1), $\mathbf{VA}$ belongs to the vertical space $\mathcal{V}_{\mathbf{V}}\mathbb{S}t$ and $\mathbf{V}_\perp\mathbf{B}$ to the horizontal space $\mathcal{H}_{\mathbf{V}}\mathbb{S}t$ at the point $\mathbf{V}$. This language is clarified in the paragraphs below.

The Grassmann Manifold. The Grassmann manifold $\mathbb{G}r(q, \mathbb{R}^n)$ is a space whose elements are q-dimensional subspaces of the ambient n-dimensional vector space $\mathbb{R}^n$. For $q = 1$, the Grassmannian $\mathbb{G}r(1, \mathbb{R}^n)$ is the space of lines through the origin in $\mathbb{R}^n$, so it is the projective space of dimension $n-1$. Points on $\mathbb{G}r(q, \mathbb{R}^n)$ are equivalence classes of $n \times q$ matrices, where the orthogonal bases for the subspaces $\langle\mathbf{V}_1\rangle$ and $\langle\mathbf{V}_2\rangle$ are equivalent if $\mathbf{V}_2 = \mathbf{V}_1\mathbf{Q}$ for some $\mathbf{Q} \in O(q)$. Therefore, the Grassmann manifold may be defined as a quotient space $\mathbb{S}t(q, \mathbb{R}^n)/O(q) = O(n)/O(q) \times O(n-q)$, with canonical projection $\pi : \mathbb{S}t \to \mathbb{G}r$, so that the equivalence class of $\mathbf{V} \in \mathbb{S}t$ is the fiber $\pi^{-1}(\langle\mathbf{V}\rangle) \in \mathbb{S}t$. Computations on Grassmann manifolds are performed using orthonormal matrix representatives for the points, that is, points on the Stiefel. However, the matrix representative for points in the Grassmann manifold is not unique. Any basis for the subspace can be a representative for the class of equivalence. Alternatively, any full rank $n \times q$ matrix $\mathbf{X}$ may be partitioned as

$$\mathbf{X} = \begin{bmatrix} \mathbf{X}_1 \\ \mathbf{X}_2 \end{bmatrix},$$

where $\mathbf{X}_1$ is $q \times q$ and $\mathbf{X}_2$ is $(n-q) \times q$. Since $\mathbf{X}$ is full rank, we have $\mathbf{X}^T\mathbf{X} \succ \mathbf{0}$, which implies $\mathbf{X}_1^T\mathbf{X}_1 \succ \mathbf{0}$ and $\mathbf{X}_2^T\mathbf{X}_2 \succ \mathbf{0}$. Therefore, the column span of $\mathbf{X}$ is the same as the column span of $\mathbf{X}\mathbf{X}_1^{-1}$, meaning that we can pick a matrix of the form

$$\mathbf{G} = \begin{bmatrix} \mathbf{I}_q \\ \mathbf{X}_2\mathbf{X}_1^{-1} \end{bmatrix} = \begin{bmatrix} \mathbf{I}_q \\ \mathbf{F} \end{bmatrix}$$

as representative for a point in the Grassmannian, where $\mathbf{F} \in \mathbb{R}^{(n-q)\times q}$. From this parametrization, it is clear that $\mathbb{G}r(q, \mathbb{R}^n)$ is of dimension $q(n-q)$. In the complex case, the dimensions are doubled.

For points on the Grassmannian, we will find it convenient in this chapter to distinguish between the subspace, $\langle \mathbf{V} \rangle \in \mathbb{G}r(q, \mathbb{R}^n)$, and its non-unique q-frame $\mathbf{V} \in \mathbb{S}t(q, \mathbb{R}^n)$. When the meaning is clear, we may sometimes say that $\mathbf{V}$ is a point on the Grassmannian, or that $\{\mathbf{V}_r\}_{r=1}^R$ is a collection or a random sample of subspaces of size R.

Yet another useful representation for a point on the Grassmann manifold is given by its orthogonal projection matrix $\mathbf{P}_\mathbf{V} = \mathbf{V}\mathbf{V}^T$. $\mathbf{P}_\mathbf{V}$ is the idempotent orthogonal projection onto $\langle \mathbf{V} \rangle$ and is a unique representative of $\langle \mathbf{V} \rangle$. In fact, the Grassmannian $\mathbb{G}r(q, \mathbb{R}^n)$ may be identified as the set of rank-q projection matrices, denoted here as $\mathbb{P}r(q, \mathbb{R}^n)$:

$$\mathbb{P}r(q, \mathbb{R}^n) = \{\mathbf{P} \in \mathbb{R}^{n\times n} \mid \mathbf{P}^T = \mathbf{P},\ \mathbf{P}^2 = \mathbf{P},\ \mathrm{tr}(\mathbf{P}) = q\}.$$

The frame $\mathbf{V}$ determines the subspace $\langle \mathbf{V} \rangle$, but not vice versa. However, the subspace $\langle \mathbf{V} \rangle$ does determine the subspace $\langle \mathbf{V} \rangle^\perp$, and a frame $\mathbf{V}_\perp$ may be taken as a basis for this subspace. The tangent space to $\langle \mathbf{V} \rangle$ is defined to be

$$T_\mathbf{V}\mathbb{G}r(q, \mathbb{R}^n) = \left\{\Delta\mathbf{V} \in T_\mathbf{V}\mathbb{S}t(q, \mathbb{R}^n) \mid \Delta\mathbf{V} \perp \mathbf{V}\mathbf{A}, \quad \forall \mathbf{A} \text{ skew-symmetric}\right\}.$$

From $\Delta\mathbf{V} = \mathbf{V}\mathbf{A} + \mathbf{V}_\perp\mathbf{B}$, it follows that $\Delta\mathbf{V} = \mathbf{V}_\perp$. That is, the skew-symmetric matrix $\mathbf{A} = \mathbf{0}$, and the tangent space to the Grassmannian at subspace $\langle \mathbf{V} \rangle$ is the subspace $\langle \mathbf{V}_\perp \rangle$. Note that this solution for $\Delta\mathbf{V}$ depends only on the subspace $\langle \mathbf{V} \rangle$, and not on a specific choice for the frame $\mathbf{V}$. Then, we may identify the tangent space $T_\mathbf{V}\mathbb{G}r(q, \mathbb{R}^n)$ with the horizontal space of the Stiefel

$$T_\mathbf{V}\mathbb{G}r(q, \mathbb{R}^n) = \left\{\left(\mathbf{I}_n - \mathbf{V}\mathbf{V}^T\right)\mathbf{B} \mid \mathbf{B} \in \mathbb{R}^{n\times q}\right\} \cong \mathcal{H}_\mathbf{V}\mathbb{S}t.$$

For intuition, this subspace is the linear space of vectors $(\mathbf{I}_n - \mathbf{V}\mathbf{V}^T)\mathbf{B}$, which shows that the Grassmannian may be thought of as the set of orthogonal projections $\mathbf{V}\mathbf{V}^T$, with tangent spaces $\langle \mathbf{V}_\perp \rangle$. The geometry is illustrated in Fig. 9.1.

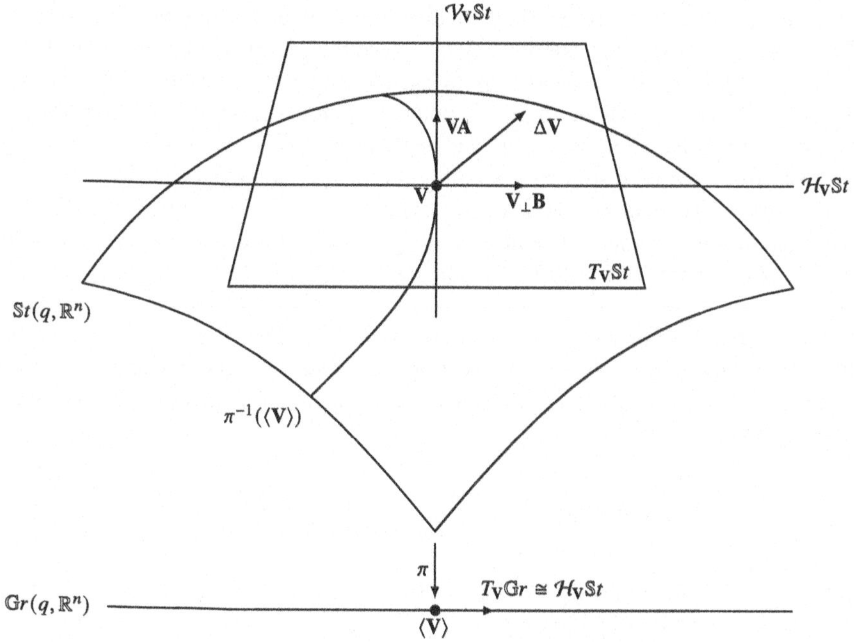

Fig. 9.1 The Stiefel manifold (the fiber bundle) is here represented as a surface in the Euclidean ambient space of all matrices, and the quotient by the orthogonal group action, which generates orbits on each matrix $\mathbf{V}$ (the fibers), is the Grassmannian manifold (the base manifold), represented as a straight line below. The idea is that every point on that bottom line represents a fiber, drawn there as a "curve" in the Stiefel "surface." Then, each of the three manifolds mentioned has its own tangent space, the Grassmannian tangent space represented by a horizontal arrow at the bottom, the tangent space to the Stiefel as a plane tangent to the surface, and the tangent to the fiber/orbit as a vertical line in that plane. The perpendicular horizontal line is thus orthogonal to the fiber curve at the point and is called the Horizontal space of the Stiefel at that matrix. It is clear from the figure then that moving from a fiber to a nearby fiber, i.e., moving on the Grassmannian, can only be measured by horizontal tangent vectors (as, e.g., functions on the Grassmannian are functions on the Stiefel that are constant along such curves); thus the Euclidean orthogonality in the ambient spaces yields the formula for the representation of vectors in the Grassmannian tangent space

9.1.1 Statistics on the Grassmann and Stiefel Manifolds

In many problems, it is useful to assume that there is an underlying distribution on the Stiefel or the Grassmann manifolds from which random samples are drawn independently. In this way, we may generate a collection of orthonormal frames $\{\mathbf{V}_r\}_{r=1}^R$ on the Stiefel manifold, or a collection of subspaces $\{\langle\mathbf{V}_r\rangle\}_{r=1}^R$, or projection matrices $\{\mathbf{P}_r = \mathbf{V}_r\mathbf{V}_r^T\}_{r=1}^R$, on the Grassmann manifold. Strictly speaking, distributions are different on the Stiefel and on the Grassmann manifolds. For example, in $\mathbb{R}^2$, the classic von Mises distribution is a distribution on the Stiefel $\mathbb{S}t(1,\mathbb{R}^2)$ that accounts for directions and hence has support $[0, 2\pi)$. The corresponding distribution on the Grassmann $\mathbb{G}r(1,\mathbb{R}^2)$, whose points are lines in

$\mathbb{R}^2$ instead of directions, is the Bingham distribution, which has antipodal symmetry, i.e., $f(\mathbf{v}) = f(-\mathbf{v})$ and support $[0, \pi)$. Nevertheless, there is a mapping from $\mathbb{St}(q, \mathbb{R}^n)$ to $\mathbb{Gr}(q, \mathbb{R}^n)$ that assigns to each q-frame the subspace that it spans: $\mathbf{V} \rightarrow \langle \mathbf{V} \rangle \cong \mathbf{P_V}$. Then, for most practical purposes, we can concentrate on distributions on the Stiefel manifold and apply the mapping, $\mathbf{V} \rightarrow \mathbf{V}\mathbf{V}^T$, to generate samples from the corresponding distribution on the Grassmannian. Uniform and non-uniform distributions on $\mathbb{St}(q, \mathbb{R}^n)$ and $\mathbb{Gr}(q, \mathbb{R}^n)$ or, equivalently, on the manifold of rank-q projection matrices, $\mathbb{Pr}(q, \mathbb{R}^n)$, have been extensively discussed in [74]. We review some useful results below.

Uniform Distributions on $\mathbb{St}(q, \mathbb{R}^n)$ and $\mathbb{Gr}(q, \mathbb{R}^n)$. The Haar measure, $(d\mathbf{V})$, on the Stiefel manifold is invariant under transformation $\mathbf{Q}_1\mathbf{V}\mathbf{Q}_2$, where $\mathbf{Q}_1 \in O(n)$ and $\mathbf{Q}_2 \in O(q)$. The integral of this measure on the manifold gives the total volume of $\mathbb{St}(q, \mathbb{R}^n)$:

$$\mathrm{Vol}(\mathbb{St}(q, \mathbb{R}^n)) = \int_{\mathbb{St}(q,\mathbb{R}^n)} (d\mathbf{V}) = \frac{2^q \pi^{qn/2}}{\Gamma_q\left(\frac{n}{2}\right)},$$

where $\Gamma_q(x)$ is the multivariate Gamma function. This function is defined as (see also Appendix D, Eq. (D.7))

$$\Gamma_q(x) = \int_{\mathbf{A}\succ \mathbf{0}} \mathrm{etr}(-\mathbf{A}) \det(\mathbf{A})^{x-\frac{(q+1)}{2}} d\mathbf{A} = \pi^{q(q-1)/4} \prod_{i=1}^{q} \Gamma\left(x - \frac{(i-1)}{2}\right),$$

where $\mathbf{A}$ is a $q \times q$ positive definite matrix.

Example 9.1 As an example, consider $\mathbb{St}(1, \mathbb{R}^2)$. For

$$\mathbf{h} = \begin{bmatrix} \cos(\theta) \\ \sin(\theta) \end{bmatrix} \in \mathbb{St}(1, \mathbb{R}^2),$$

with $\theta \in [0, 2\pi)$, choose

$$\mathbf{h}_\perp = \begin{bmatrix} -\sin(\theta) \\ \cos(\theta) \end{bmatrix},$$

such that $\mathbf{H} = [\mathbf{h}\ \mathbf{h}_\perp] \in O(2)$. Then, the differential form for the invariant measure on $\mathbb{St}(1, \mathbb{R}^2)$ is

$$(d\mathbf{V}) = \mathbf{h}_\perp^T d\mathbf{h} = \left[-\sin(\theta)\ \cos(\theta)\right] \begin{bmatrix} -\sin(\theta) d\theta \\ \cos(\theta) d\theta \end{bmatrix} = d\theta,$$

and hence

$$\text{Vol}(\mathbb{S}t(1, \mathbb{R}^2)) = \int_{\mathbb{S}t(1,\mathbb{R}^2)} (d\mathbf{V}) = \int_0^{2\pi} d\theta = 2\pi.$$

The invariant measure on $\mathbb{G}r(q, \mathbb{R}^n)$ or $\mathbb{P}r(q, \mathbb{R}^n)$ is invariant to the transformation $\mathbf{P} \rightarrow \mathbf{Q}\mathbf{P}\mathbf{Q}^T$ for $\mathbf{Q} \in O(n)$. Its integral on the Grassmann manifold yields [177]

$$\text{Vol}(\mathbb{G}r(q, \mathbb{R}^n)) = \int_{\mathbb{G}r(q,\mathbb{R}^n)} (d\mathbf{P}) = \frac{\pi^{q(n-q)/2}\Gamma_q\left(\frac{q}{2}\right)}{\Gamma_q\left(\frac{n}{2}\right)},$$

which is the volume of the Grassmannian. Note that $(d\mathbf{V})$ and $(d\mathbf{P})$ are unnormalized probability measures that do not integrate to one. It is also common to express the densities on the Stiefel or the Grassmann manifolds in terms of normalized invariant measures defined as

$$[d\mathbf{V}] = \frac{(d\mathbf{V})}{\text{Vol}(\mathbb{S}t(q, \mathbb{R}^n))}, \quad \text{and} \quad [d\mathbf{P}] = \frac{(d\mathbf{P})}{\text{Vol}(\mathbb{G}r(q, \mathbb{R}^n))},$$

which integrate to one on the respective manifolds. In this chapter, we express the densities with respect to the normalized invariant measures.

For sampling from uniform distributions, the basic experiment is this: generate $\mathbf{X}$ as a random $n \times q$ tall matrix $(n > q)$ with i.i.d. $\mathcal{N}(0, 1)$ random variables. Perform a QR decomposition of this random matrix as $\mathbf{X} = \mathbf{T}\mathbf{R}$. Then, the matrix $\mathbf{T}$ is uniformly distributed on $\mathbb{S}t(q, \mathbb{R}^n)$, and $\mathbf{T}\mathbf{T}^T$ is uniformly distributed on $\mathbb{G}r(q, \mathbb{R}^n) \cong \mathbb{P}r(q, \mathbb{R}^n)$. Remember that points on $\mathbb{G}r(q, \mathbb{R}^n)$ are equivalence classes of $n \times q$ matrices, where $\mathbf{T}_1 \sim \mathbf{T}_2$ if $\mathbf{T}_1 = \mathbf{T}_2\mathbf{Q}$, for some $\mathbf{Q} \in O(q)$.

Alternatively, given $\mathbf{X} \sim \mathcal{N}_{n\times q}(\mathbf{0}, \mathbf{I}_q \otimes \mathbf{I}_n)$, its unique polar decomposition is defined as

$$\mathbf{X} = \mathbf{T}\mathbf{R}, \qquad \text{with} \quad \mathbf{T} = \mathbf{X}(\mathbf{X}^T\mathbf{X})^{-1/2} \qquad \text{and} \quad \mathbf{R} = (\mathbf{X}^T\mathbf{X})^{1/2},$$

where $(\mathbf{X}^T\mathbf{X})^{1/2}$ denotes the unique square root of the matrix $\mathbf{X}^T\mathbf{X}$. In the polar decomposition, $\mathbf{T}$ is usually called the *orientation* of the matrix. The random matrix $\mathbf{T} = \mathbf{X}(\mathbf{X}^T\mathbf{X})^{-1/2}$ is uniformly distributed on $\mathbb{S}t(q, \mathbb{R}^n)$, and $\mathbf{P} = \mathbf{T}\mathbf{T}^T$ is uniformly distributed on $\mathbb{G}r(q, \mathbb{R}^n)$ or $\mathbb{P}r(q, \mathbb{R}^n)$.

The Matrix Langevin Distribution. Let us begin with a random normal matrix $\mathbf{X} \sim \mathcal{N}_{n\times q}(\mathbf{M}, \mathbf{\Sigma} \otimes \mathbf{I}_n)$ where $\mathbf{\Sigma}$ is a $q \times q$ positive definite matrix. Its density is

$$f(\mathbf{X}) \propto \text{etr}\left(-\frac{(\mathbf{X}-\mathbf{M})\mathbf{\Sigma}^{-1}(\mathbf{X}-\mathbf{M})^T}{2}\right)$$

$$\propto \text{etr}\left(-\frac{\mathbf{\Sigma}^{-1}\mathbf{X}^T\mathbf{X} + \mathbf{\Sigma}^{-1}\mathbf{M}^T\mathbf{M} - 2\mathbf{\Sigma}^{-1}\mathbf{M}^T\mathbf{X}}{2}\right).$$

Imposing the condition $\mathbf{X}^T\mathbf{X} = \mathbf{I}_q$ and defining $\mathbf{H} = \mathbf{M}\boldsymbol{\Sigma}^{-1}$, we get a distribution of the form $f(\mathbf{X}) \propto \text{etr}(\mathbf{H}^T\mathbf{X})$. The normalizing constant is

$$\int_{\mathbb{S}t(q,n)} \text{etr}(\mathbf{H}^T\mathbf{X})[d\mathbf{X}] = {}_0F_1\left(\frac{n}{2}, \frac{1}{4}\mathbf{H}^T\mathbf{H}\right),$$

where ${}_0F_1$ is a hypergeometric function of matrix argument (see [74, Appendix A.6]). Therefore, $\mathbf{X} \in \mathbb{S}t(q, \mathbb{R}^n)$ is said to have the matrix Langevin distribution $\mathcal{L}_{n\times q}(\mathbf{H})$ if its density has the form [107]

$$f(\mathbf{X}) = \frac{1}{{}_0F_1\left(\frac{n}{2}, \frac{1}{4}\mathbf{H}^T\mathbf{H}\right)} \text{etr}(\mathbf{H}^T\mathbf{X}), \tag{9.2}$$

where $\mathbf{H}$ is an $n \times q$ matrix. Write the SVD of the matrix $\mathbf{H}$ as $\mathbf{H} = \mathbf{F}\boldsymbol{\Lambda}\mathbf{G}^T$, where $\mathbf{F} \in \mathbb{S}t(q, \mathbb{R}^n)$, $\mathbf{G} \in O(q)$, and $\boldsymbol{\Lambda} = \text{diag}(\lambda_1, \ldots, \lambda_q)$. The singular values, λ_i, which are assumed to be different, are the concentration parameters, and $\mathbf{H}_0 = \mathbf{F}\mathbf{G}^T$ is the orientation matrix, which is the mode of the distribution. The distribution is rotationally symmetric around the orientation matrix $\mathbf{H}_0$. For $\mathbf{H} = \mathbf{0}$, we recover the uniform distribution on the Stiefel.

For $q = 1$, we have $\mathbf{h} = \lambda\mathbf{f}$, and the matrix Langevin density (9.2) reduces to the von Mises-Fisher distribution for $\mathbf{x} \in \mathbb{S}t(1, \mathbb{R}^n)$

$$f(\mathbf{x}) = \frac{1}{a_n(\lambda)} \exp(\lambda\mathbf{f}^T\mathbf{x}), \quad \mathbf{x}^T\mathbf{x} = 1,$$

where $\lambda \geq 0$ is the concentration parameter, $\mathbf{f} \in \mathbb{S}t(1, \mathbb{R}^n)$, and the normalizing constant is

$$a_n(\lambda) = (2\pi)^{n/2} I_{n/2-1}(\lambda)\lambda^{-\frac{n}{2}+1},$$

with $I_\nu(x)$ the modified Bessel function of the first kind and order ν. The distribution is unimodal with mode $\mathbf{f}$. The higher the λ, the higher the concentration around the mode direction $\mathbf{f}$. When $n = 2$, the vectors in $\mathbb{S}t(1, \mathbb{R}^2)$ may be parameterized as $\mathbf{x} = [\cos(\theta)\ \sin(\theta)]^T$ and $\mathbf{f} = [\cos(\phi)\ \sin(\phi)]^T$; the density becomes

$$f(\theta) = \frac{e^{\lambda\cos(\theta-\phi)}}{2\pi I_0(\lambda)}, \qquad -\pi < \theta \leq \pi.$$

So the distribution is clustered around the angle ϕ; the larger the λ, the more concentrated the distribution is around ϕ.

As suggested in [74], to generate samples from $\mathcal{L}_{n\times q}(\mathbf{H})$, we might use a rejection sampling mechanism with the uniform as proposal density. Rejection sampling, however, can be very inefficient for large n and $q > 1$. More efficient sampling algorithms have been proposed in [168].

The Matrix Bingham Distribution. Begin with a random normal matrix $\mathbf{X} \sim \mathcal{N}_{n\times q}(\mathbf{0}, \mathbf{I}_q \otimes \boldsymbol{\Sigma})$, where $\boldsymbol{\Sigma}$ is an $n \times n$ positive definite matrix. Its density is

$$f(\mathbf{X}) \propto \text{etr}\left(-\frac{\mathbf{X}^T\boldsymbol{\Sigma}^{-1}\mathbf{X}}{2}\right).$$

Imposing the condition $\mathbf{X}^T\mathbf{X} = \mathbf{I}_q$ and defining $\mathbf{H} = -\boldsymbol{\Sigma}^{-1}/2$, we get a distribution of the form $f(\mathbf{X}) \propto \text{etr}(\mathbf{X}^T\mathbf{H}\mathbf{X})$, where $\mathbf{H}$ is now an $n \times n$ symmetric matrix. Calculating the normalizing constant for this density, we get the matrix Bingham distribution with parameter $\mathbf{H}$, which we denote as $\mathcal{B}_{n\times q}(\mathbf{H})$:

$$f(\mathbf{X}) = \frac{1}{{}_1F_1\left(\frac{k}{2}, \frac{n}{2}, \mathbf{H}\right)}\,\text{etr}(\mathbf{X}^T\mathbf{H}\mathbf{X}), \quad \mathbf{X}^T\mathbf{X} = \mathbf{I}_q. \tag{9.3}$$

Let $\mathbf{H} = \mathbf{F}\boldsymbol{\Lambda}\mathbf{F}^T$, with $\mathbf{F} \in O(n)$ and $\boldsymbol{\Lambda} = \text{diag}(\lambda_1, \ldots, \lambda_n)$. The distribution has multimodal orientations $\mathbf{H}_0 = \mathbf{F}_q\mathbf{G}$, where $\mathbf{F}_q$ contains the q columns of $\mathbf{F}$ that correspond to the largest eigenvalues of $\mathbf{H}$, and $\mathbf{G} \in O(q)$. The density (9.3) is invariant under right-orthogonal transformations $\mathbf{X} \to \mathbf{X}\mathbf{Q}$ for $\mathbf{Q} \in O(q)$. For $q = 1$, it is the Bingham distribution on the $(n-1)$-dimensional sphere [37].

Note that the matrix Bingham distribution can be viewed as a distribution on the Grassmann manifold or the manifold of rank-q projection matrices. In fact, we can rewrite (9.3) as

$$f(\mathbf{P}) = \frac{1}{{}_1F_1\left(\frac{k}{2}, \frac{n}{2}, \mathbf{H}\right)}\,\text{etr}(\mathbf{H}\mathbf{P}), \tag{9.4}$$

where $\mathbf{P} = \mathbf{X}\mathbf{X}^T \in \mathbb{P}r(q, \mathbb{R}^n)$. This distribution has mode $\mathbf{P}_0 = \mathbf{F}_q\mathbf{F}_q^T$, which is the closest idempotent rank-q matrix to $\mathbf{H}$. If we take $\mathbf{H} = \mathbf{0}$ in (9.3) or in (9.4), we recover the uniform distributions on $\mathbb{S}t(q, \mathbb{R}^n)$ and $\mathbb{G}r(q, \mathbb{R}^n)$, respectively. The distributions (9.2) and (9.3) can be combined to form the family of matrix Langevin-Bingham distributions with density $f(\mathbf{X}) \propto \text{etr}\left(\mathbf{H}_1^T\mathbf{X} + \mathbf{X}^T\mathbf{H}_2\mathbf{X}\right)$.

The Matrix Angular Central Gaussian Distribution. Begin with a random normal matrix $\mathbf{X} \sim \mathcal{N}_{n\times q}(\mathbf{0}, \mathbf{I}_q \otimes \boldsymbol{\Sigma})$ and write its unique polar decomposition as $\mathbf{X} = \mathbf{T}\mathbf{R}$. Then, it is proved in [73] that the distribution of $\mathbf{T} = \mathbf{X}(\mathbf{X}^T\mathbf{X})^{-1/2}$ follows the matrix angular central Gaussian distribution with parameter $\boldsymbol{\Sigma}$, which we denote as $MACG(\boldsymbol{\Sigma})$ with density

$$f(\mathbf{T}) = \frac{1}{\text{Vol}(\mathbb{S}t(q, \mathbb{R}^n))}\det(\boldsymbol{\Sigma})^{-q/2}\det\left(\mathbf{T}^T\boldsymbol{\Sigma}^{-1}\mathbf{T}\right)^{-n/2}, \quad \mathbf{T}^T\mathbf{T} = \mathbf{I}_q.$$

For $q = 1$, we have the angular central Gaussian distribution with parameter $\boldsymbol{\Sigma}$. It is denoted as $AGC(\boldsymbol{\Sigma})$, and its density is

$$f(\mathbf{t}) = \frac{\Gamma\left(\frac{n}{2}\right)}{2\pi^{n/2}} \det(\boldsymbol{\Sigma})^{-1/2} \left(\mathbf{t}^T \boldsymbol{\Sigma}^{-1} \mathbf{t}\right)^{-n/2}, \quad \mathbf{t}^T\mathbf{t} = 1,$$

where the normalizing constant $2\pi^{n/2}/\Gamma\left(\frac{n}{2}\right)$ is the volume of $\mathbb{S}t(1, \mathbb{R}^n)$. The $ACG(\boldsymbol{\Sigma})$ distribution is an alternative to the Bingham distribution for modeling antipodally symmetric data on $\mathbb{S}t(1, \mathbb{R}^n)$, and its statistical theory has been studied by Tyler in [352]. In particular, a method developed by Tyler shows that if $\mathbf{T}_m \sim MACG(\boldsymbol{\Sigma})$, $m = 1, \ldots, M$, is a random i.i.d. sample of size M from the $MACG(\boldsymbol{\Sigma})$ distribution, then the maximum likelihood estimator $\hat{\boldsymbol{\Sigma}}$ of $\boldsymbol{\Sigma}$ is the fixed-point solution of the equation

$$\hat{\boldsymbol{\Sigma}} = \frac{n}{qM} \sum_{m=1}^{M} \mathbf{T}_m \left(\mathbf{T}_m^T \hat{\boldsymbol{\Sigma}}^{-1} \mathbf{T}_m\right)^{-1} \mathbf{T}_m^T.$$

The following property shows that the $MACG(\boldsymbol{\Sigma})$ distribution can be transformed to uniformity by a simple linear transformation. There is no known simple transformation to uniformity for any other antipodal symmetric distribution on $\mathbb{S}t(q, \mathbb{R}^n)$.

Property 9.1 Let $\mathbf{X} \sim \mathcal{N}_{n\times q}(\mathbf{0}, \mathbf{I}_q \otimes \boldsymbol{\Sigma})$ with $\mathbf{T}_X = \mathbf{X}\left(\mathbf{X}^T\mathbf{X}\right)^{-1/2} \sim MACG(\boldsymbol{\Sigma})$. We consider the linear transformation $\mathbf{Y} = \mathbf{B}\mathbf{X}$ with orientation matrix $\mathbf{T}_Y = \mathbf{Y}\left(\mathbf{Y}^T\mathbf{Y}\right)^{-1/2}$, where $\mathbf{B}$ is an $n \times n$ nonsingular matrix. Then,

- $\mathbf{T}_Y \sim MACG(\mathbf{B}\boldsymbol{\Sigma}\mathbf{B}^T)$.
- In particular, if $\mathbf{T}_X$ is uniformly distributed on $\mathbb{S}t(q, \mathbb{R}^n)$ (i.e., $\mathbf{T}_X \sim MACG(\mathbf{I}_n)$), then $\mathbf{T}_Y \sim MACG(\mathbf{B}\mathbf{B}^T)$.
- If $\mathbf{T}_X \sim MACG(\boldsymbol{\Sigma})$ and $\mathbf{B}$ is chosen such that $\mathbf{B}\boldsymbol{\Sigma}\mathbf{B}^T = \mathbf{I}_n$, then $\mathbf{T}_Y$ is uniformly distributed on $\mathbb{S}t(q, n)$.

A Discrete Distribution on Projection Matrices. It is sometimes useful to define discrete distributions over finite sets of projection matrices of different ranks. The following example was proposed in [128]. Let $\mathbf{U} = [\mathbf{u}_1 \cdots \mathbf{u}_n] \in O(n)$ be an arbitrary orthogonal basis of the ambient space, and let $\boldsymbol{\alpha} = [\alpha_1 \cdots \alpha_n]^T$, with $0 \leq \alpha_i \leq 1$. The α_i are ordered from largest to smallest, but they need not sum to 1. We define a discrete distribution on the set of random projection matrices $\mathbf{P} = \mathbf{V}\mathbf{V}^H$ (or, equivalently, the set of random subspaces $\langle \mathbf{V} \rangle$, or set of frames $\mathbf{V}$) with parameter vector $\boldsymbol{\alpha}$ and orientation matrix $\mathbf{U}$. The distribution of $\mathbf{P}$ will be denoted $\mathbf{P} \sim \mathcal{D}(\mathbf{U}, \boldsymbol{\alpha})$.

To shed some light on this distribution, let us explain the experiment that determines $\mathcal{D}(\mathbf{U}, \boldsymbol{\alpha})$. Draw 1 includes $\mathbf{u}_1$ with probability α_1 and excludes it with probability $(1 - \alpha_1)$. Draw 2 includes $\mathbf{u}_2$ with probability α_2 and excludes it with probability $(1 - \alpha_2)$. Continue in this way until draw n includes $\mathbf{u}_n$ with probability α_n and excludes it with probability $(1 - \alpha_n)$. We may call the string $i_1, i_2, \ldots, i_n$, the indicator sequence for the draws. That is, $i_k = 1$, if $\mathbf{u}_k$ is drawn on draw k, and $i_k = 0$ otherwise. In this way, Pascal's triangle shows that the probability of drawing the subspace $\langle \mathbf{V} \rangle$ is $Pr[\langle \mathbf{V} \rangle] = \prod_I \alpha_i \prod_{\bar{I}}(1-\alpha_j)$, where the index set I is the set of indices k for which $i_k = 1$ in the construction of $\mathbf{V}$. This is also the probability law on frames $\mathbf{V}$ and projections $\mathbf{P}$. For example, the probability of drawing an empty frame is $\prod_{i=1}^n (1 - \alpha_i)$, the probability of drawing the dimension-1 frame $\mathbf{u}_i \mathbf{u}_i^H$ is $\alpha_i \prod_{j \neq i}^n (1 - \alpha_j)$, and so on. It is clear from this distribution on the 2^n frames that all probabilities lie between 0 and 1 and that they sum to 1.

Property 9.2 Let $\mathbf{P}_r \sim \mathcal{D}(\mathbf{U}, \boldsymbol{\alpha})$, $r = 1, \ldots, R$, be a sequence of i.i.d. draws from the distribution $\mathcal{D}(\mathbf{U}, \boldsymbol{\alpha})$ and let $\overline{\mathbf{P}} = \sum_{r=1}^R \mathbf{P}_r / R$ be its sample mean with decreasing eigenvalues $k_1, \ldots, k_n$. Then, we have the following properties:

1. $\mathrm{E}[\mathbf{P}_r] = \mathbf{U} \operatorname{diag}(\boldsymbol{\alpha}) \mathbf{U}^T$.
2. $\mathrm{E}\left[\operatorname{tr}(\mathbf{P}_r)\right] = \sum_{i=1}^n \alpha_i$.
3. $\mathrm{E}\left[k_i\right] = \alpha_i$.

These properties follow directly from the definition of $\mathcal{D}(\mathbf{U}, \boldsymbol{\alpha})$. In fact, the definition for this distribution takes an average matrix $\mathbf{P}_0 = \mathbf{U} \operatorname{diag}(\boldsymbol{\alpha}) \mathbf{U}^T$ (a symmetric matrix with eigenvalues between 0 and 1) and then defines a discrete distribution such that the mathematical expectation of a random draw from this distribution coincides with $\mathbf{P}_0$ (this is Property 1 above).

Remark 9.1 The α_i's control the *concentrations* or probabilities in the directions determined by the orthogonal basis $\mathbf{U}$. For instance, if $\alpha_i = 1$ all random subspaces contain direction $\mathbf{u}_i$, whereas if $\alpha_i = 0$, the angle between $\mathbf{u}_i$ and all random subspaces drawn from that distribution will be $\pi/2$.

Example 9.2 Suppose $\mathbf{U} = [\mathbf{u}_1\ \mathbf{u}_2\ \mathbf{u}_3]$ is the standard basis in $\mathbb{R}^3$ and let $\boldsymbol{\alpha}$ be a three-dimensional vector with elements $\alpha_1 = 3/4$ and $\alpha_2 = \alpha_3 = 1/4$. The discrete distribution $\mathbf{P} \sim \mathcal{D}(\mathbf{U}, \boldsymbol{\alpha})$ has an alphabet of $2^3 = 8$ subspaces with the following probabilities:

- $\Pr(\mathbf{P} = \mathbf{0}) = 9/64$
- $\Pr\left(\mathbf{P} = \mathbf{u}_1 \mathbf{u}_1^T\right) = 27/64$
- $\Pr\left(\mathbf{P} = \mathbf{u}_2 \mathbf{u}_2^T\right) = 3/64$
- $\Pr\left(\mathbf{P} = \mathbf{u}_3 \mathbf{u}_3^T\right) = 3/64$
- $\Pr\left(\mathbf{P} = \mathbf{u}_1 \mathbf{u}_1^T + \mathbf{u}_2 \mathbf{u}_2^T\right) = 9/64$
- $\Pr\left(\mathbf{P} = \mathbf{u}_1 \mathbf{u}_1^T + \mathbf{u}_3 \mathbf{u}_3^T\right) = 9/64$
- $\Pr\left(\mathbf{P} = \mathbf{u}_2 \mathbf{u}_2^T + \mathbf{u}_3 \mathbf{u}_3^T\right) = 1/64$
- $\Pr(\mathbf{P} = \mathbf{I}_3) = 3/64$

The distribution is unimodal with mean

$$E[\mathbf{P}] = \mathbf{U}\,\mathrm{diag}(\boldsymbol{\alpha})\mathbf{U}^T.$$

and expected dimension $E[\mathrm{tr}(\mathbf{P})] = 5/4$. Given R draws from the distribution $\mathbf{P} \sim D(\mathbf{U}, \boldsymbol{\alpha})$, the eigenvalues of the sample average of projections $\overline{\mathbf{P}} = \sum_{r=1}^{R} \mathbf{P}_r/R$, k_i, converge to α_i as R grows. It is easy to check that the probability of drawing a dimension-1 subspace for this example is $33/64$.

As we will see in in Sect. 9.7, the generative model underlying $D(\mathbf{U}, \boldsymbol{\alpha})$ is useful for the application of subspace averaging techniques to array processing.

9.2 Principal Angles, Coherence, and Distances Between Subspaces

Let us consider two subspaces $\langle \mathbf{V} \rangle \in \mathbb{G}r(p, \mathbb{R}^n)$ and $\langle \mathbf{U} \rangle \in \mathbb{G}r(q, \mathbb{R}^n)$. Let $\mathbf{V} \in \mathbb{R}^{n \times p}$ be a matrix whose columns form an orthogonal basis for $\langle \mathbf{V} \rangle$. Then $\mathbf{V}^T\mathbf{V} = \mathbf{I}_p$, and $\mathbf{P}_\mathbf{V} = \mathbf{V}\mathbf{V}^T$ is the idempotent orthogonal projection onto $\langle \mathbf{V} \rangle$. Recall that $\mathbf{P}_\mathbf{V}$ is a unique representation of $\langle \mathbf{V} \rangle$, whereas $\mathbf{V}$ is not unique. In a similar way, we define $\mathbf{U}$ and $\mathbf{P}_\mathbf{U}$ for the subspace $\langle \mathbf{U} \rangle$.

Principal Angles. To measure the distance between two subspaces, we need the concept of principal angles, which is introduced in the following definition [142].

Definition 9.1 Let $\langle \mathbf{V} \rangle$ and $\langle \mathbf{U} \rangle$ be subspaces of $\mathbb{R}^n$ whose dimensionality satisfy $\dim(\langle \mathbf{V} \rangle) = p \geq \dim(\langle \mathbf{U} \rangle) = q \geq 1$. The principal angles $\theta_1, \ldots, \theta_q \in [0, \pi/2]$, between $\langle \mathbf{V} \rangle$ and $\langle \mathbf{U} \rangle$, are defined recursively by

$$\begin{aligned} \cos(\theta_k) = \; & \max_{\mathbf{u} \in \langle \mathbf{U} \rangle} \max_{\mathbf{v} \in \langle \mathbf{V} \rangle} \mathbf{u}^T\mathbf{v} = \mathbf{u}_k^T\mathbf{v}_k \\ \text{subject to} \quad & \|\mathbf{u}\|_2 = \|\mathbf{v}\|_2 = 1, \\ & \mathbf{u}^T\mathbf{u}_i = 0, \quad i = 1, \ldots, k-1, \\ & \mathbf{v}^T\mathbf{v}_i = 0, \quad i = 1, \ldots, k-1, \end{aligned}$$

for $k = 1, \ldots, q$.

The smallest principal angle θ_1 is the minimum angle formed by a pair of unit vectors $(\mathbf{u}_1, \mathbf{v}_1)$ drawn from $\langle \mathbf{U} \rangle \times \langle \mathbf{V} \rangle$. That is,

$$\theta_1 = \min_{\mathbf{u} \in \langle \mathbf{U} \rangle, \mathbf{v} \in \langle \mathbf{V} \rangle} \arccos\left(\langle \mathbf{u}, \mathbf{v} \rangle\right), \tag{9.5}$$

subject to $\|\mathbf{u}\|_2 = \|\mathbf{v}\|_2 = 1$. The second principal angle θ_2 is defined as the smallest angle attained by a pair of unit vectors $(\mathbf{u}_2, \mathbf{v}_2)$ that is orthogonal to the first pair and so on. The sequence of principal angles is nondecreasing, and it is contained in the range $\theta_i \in [0, \pi/2]$.

A more computational definition of the principal angles is presented in [38]. Suppose that $\mathbf{U}$ and $\mathbf{V}$ are two matrices whose columns form orthonormal bases for $\langle\mathbf{U}\rangle$ and $\langle\mathbf{V}\rangle$. Then, the singular values of $\mathbf{U}^T\mathbf{V}$ are $\cos(\theta_1), \ldots, \cos(\theta_q)$. This definition of the principal angles is most convenient numerically because singular value decompositions can be computed efficiently with standard linear algebra software packages. Note also that this definition of the principal angles does not depend on the choice of bases that represent the two subspaces.

Coherence. The Euclidean squared coherence between subspaces was defined in Chap. 3 as

$$\rho^2(\langle\mathbf{V}\rangle, \langle\mathbf{U}\rangle) = 1 - \frac{\det\left(\mathbf{U}^T(\mathbf{I}_n - \mathbf{P_V})\mathbf{U}\right)}{\det\left(\mathbf{U}^T\mathbf{U}\right)}.$$

Using the definition of principal angles from SVD, the squared coherence can now be written as

$$\rho^2(\langle\mathbf{V}\rangle, \langle\mathbf{U}\rangle) = 1 - \prod_{k=1}^{q}(1 - \cos^2\theta_k) = 1 - \prod_{k=1}^{q}\sin^2\theta_k.$$

The geometry of the squared cosines has been discussed in Chap. 3.

Distances Between Subspaces. The principal angles induce several distance metrics, which can be used in subspace averaging [229], subspace packing [100], or subspace clustering [366]. Note that computations on Grassmann manifolds are performed using orthonormal (or unitary in the complex case) matrix representatives for the points, so any measure of distance must be orthogonally invariant. The following are the most widely used [114] (we assume for the definitions that both subspaces have dimension q):

1. **Geodesic distance**:

$$d_{geo}(\langle\mathbf{U}\rangle, \langle\mathbf{V}\rangle) = \left(\sum_{r=1}^{q}\theta_r^2\right)^{1/2}. \tag{9.6}$$

This distance takes values between zero and $\sqrt{q}\pi/2$. It measures the geodesic distance between two subspaces on the Grassmann manifold. This distance function has the drawback of not being differentiable everywhere. For example, consider the case of $\mathbb{Gr}(1, \mathbb{R}^2)$ (lines passing through the origin) and hold one line $\mathbf{u}$ fixed while the other line $\mathbf{v}$ rotates. As $\mathbf{v}$ rotates, the principal angle θ_1 increases from 0 to $\pi/2$ ($\mathbf{u}^T\mathbf{v} = 0$) and then decreases to zero as the angle between the two lines increases to π. Then, the geodesic distance function is nondifferentiable at $\theta = \pi/2$ [84]. As another drawback, there is no way to isometrically embed $\mathbb{G}r(q, \mathbb{R}^n)$ into an Euclidean space of any dimension so that the geodesic distance d_{geo} is the Euclidean distance in that space [84].

2. **Chordal distance**: The Grassmannian $\mathbb{G}r(q, \mathbb{R}^n)$ can be embedded into a Euclidean space of dimension $(n-q)q$, or higher, such that the distance between subspaces is represented by the distance in that space. Some of these embeddings map the subspaces to points on a sphere so that the straight-line distance between points on the sphere is the chord between them, which naturally explains the name *chordal distance* for such a metric. Different embeddings are possible, and therefore one may find different "chordal" distances defined in the literature. However, the most common embedding that defines a chordal distance is given by the projection matrices. We already know that it is possible to associate each subspace $\langle \mathbf{V} \rangle \in \mathbb{G}r(q, \mathbb{R}^n)$ with its corresponding projection matrix $\mathbf{P_V} = \mathbf{V}\mathbf{V}^T$. $\mathbf{P_V}$ is a symmetric, idempotent, matrix whose trace is q. Therefore, $\|\mathbf{P_V}\|^2 = \mathrm{tr}(\mathbf{P}_\mathbf{V}^T \mathbf{P_V}) = \mathrm{tr}(\mathbf{P_V}) = q$. Defining the traceless part of $\mathbf{P_V}$ as $\tilde{\mathbf{P}}_\mathbf{V} = \mathbf{P_V} - \frac{q}{n}\mathbf{I}_n$, we have $\mathrm{tr}(\tilde{\mathbf{P}}_\mathbf{V}^T \tilde{\mathbf{P}}_\mathbf{V}) = q(n-q)/n$, which allows us to embed the Grassmannian $\mathbb{G}r(q, \mathbb{R}^n)$ on a sphere of radius $\sqrt{q(n-q)/n}$ in $\mathbb{R}^D$, with $D = (n-1)(n+2)/2$ [84]. Then, the chordal distance between two subspaces is given by the Frobenius norm of the difference between the respective projection matrices, which is the straight line (chord) between two points embedded on the sphere:

$$d_c(\langle \mathbf{U} \rangle, \langle \mathbf{V} \rangle) = \frac{1}{\sqrt{2}} \|\mathbf{P_U} - \mathbf{P_V}\| = \left(q - \left\|\mathbf{U}^T\mathbf{V}\right\|^2\right)^{1/2}$$
$$= \left(\sum_{r=1}^{q} \left(1 - \cos^2\theta_r\right)\right)^{1/2} = \left(\sum_{r=1}^{q} \sin^2\theta_r\right)^{1/2}. \tag{9.7}$$

This is the metric referred to as chordal distance in the majority of works on this subject [29, 84, 100, 136, 160], although it might as well be well called projection distance or projection F-norm, as in [114], or simply extrinsic distance as in [331]. In this chapter, we will use the terms "chordal," "projection," or "extrinsic" interchangeably to refer to the distance in (9.7), which will be the fundamental metric used in this chapter for the computation of an average of subspaces. In any case, to avoid confusion, the reader should remember that other embeddings are possible, in Euclidean spaces of different dimensions. When the elements of $\mathbb{G}r(q, \mathbb{R}^n)$ are mapped to points on a sphere, the resulting distances may also be properly called chordal distances. An example is the distance

$$d_{c'}(\langle \mathbf{U} \rangle, \langle \mathbf{V} \rangle) = 2\left(\sum_{r=1}^{q} \sin^2\left(\frac{\theta_r}{2}\right)\right)^{1/2} = 2\left(\sum_{r=1}^{q} \left(\frac{1-\cos\theta_r}{2}\right)\right)^{1/2}$$

$$= \sqrt{2}\left(q - \sum_{r=1}^{q} \cos\theta_r\right)^{1/2} = \sqrt{2}\left(q - \|\mathbf{U}^T\mathbf{V}\|_*\right)^{1/2}, \qquad (9.8)$$

where $\|\mathbf{X}\|_* = \sum_r sv_r(\mathbf{X})$ denotes the nuclear norm of $\mathbf{X}$. Removing the $\sqrt{2}$ in the above expression gives the so-called Procrustes distance, frequently used in shape analysis [74, Chapter 9]. The Procrustes distance for the Grassmannian is defined as the smallest Euclidean distance between any pair of matrices in the two corresponding equivalence classes. The value of the chordal or scaled Procrustes distance defined as in (9.8) ranges between 0 and $\sqrt{2q}$, whereas the value of the chordal or projection distance defined as in (9.7) ranges between 0 and $\sqrt{q}$. The following example illustrates the difference between the different distance measures.

Example 9.3 Let us consider the points $\mathbf{u} = [1\ 0]^T$ and $\mathbf{v} = [\cos(\pi/4)\ \sin(\pi/4)]^T$ on the Grassmannian $\mathbb{G}r(1, \mathbb{R}^2)$. They have a single principal angle of $\pi/4$. The geodesic distance is $d_{geo} = \pi/4$. The chordal distance as defined in (9.8) is the length of the chord joining the points embedded on the unit sphere in $\mathbb{R}^2$, given by $d_{c'} = 2\sin(\pi/8)$. The chordal or projection distance as defined in (9.7) is

$$d_c = \frac{1}{\sqrt{2}} \left\| \underbrace{\begin{bmatrix} 1 & 0 \\ 0 & 0 \end{bmatrix}}_{\mathbf{P_u}} - \underbrace{\begin{bmatrix} 1/2 & 1/2 \\ 1/2 & 1/2 \end{bmatrix}}_{\mathbf{P_v}} \right\| = \frac{1}{\sqrt{2}},$$

which is the chord between the projection matrices when viewed as points on the unit sphere on $\mathbb{R}^3$, but it is the length of the projection from $\mathbf{u}$ to $\mathbf{v}$ if we consider the points embedded on $\mathbb{R}^2$. As pointed out in [114], a distance defined in a higher dimensional ambient space tends to be shorter, since in a space of higher dimensionality, it is possible to take a shorter path (we may "cut corners" in measuring the distance between two points, as explained in [114]). In this example,

$$d_c = \frac{1}{\sqrt{2}} < d_{c'} = 2\sin(\pi/8) < d_{geo} = \pi/4.$$

Note that the definition of the chordal distance in (9.7) can be extended to subspaces of different dimension. If $\dim(\langle\mathbf{V}\rangle) = q_V \geq \dim(\langle\mathbf{U}\rangle) = q_U$, then the squared projection distance is

$$d_c^2\left(\langle \mathbf{U} \rangle, \langle \mathbf{V} \rangle\right) = \frac{1}{2}\left\|\mathbf{P}_\mathbf{U} - \mathbf{P}_\mathbf{V}\right\|^2$$

$$= \left(q_U - \sum_{r=1}^{q_U} \cos^2\theta_r\right) + \frac{1}{2}\left(q_V - q_U\right). \tag{9.9}$$

The first term in the last expression of (9.9) measures the chordal distance defined by the principal angles, whereas the second term accounts for projection matrices of different ranks. Note that the second term may dominate the first one when $q_V \gg q_U$. If $q_V = q_U = q$, then (9.9) reduces to (9.7).

There are arguments in favor of the chordal distance. Among them is its computational simplicity, as it requires the Frobenius norm of a difference of projection matrices, in contrast to other metrics that depend on the singular values of $\mathbf{U}^T\mathbf{V}$. Unlike the geodesic distance, the chordal distance is differentiable everywhere and can be isometrically embedded into a Euclidean space. It is also possible to define a Grassmann kernel based on the chordal distance, thus enabling the application of data-driven kernel methods [156, 392]. In addition, the chordal distance is related to the squared error in resolving the standard basis for the ambient space, $\{\mathbf{e}_i\}_{i=1}^n$, onto the subspace $\langle \mathbf{V} \rangle$ as opposed to the subspace $\langle \mathbf{U} \rangle$. Let $\{\mathbf{e}_i\}_{i=1}^n$ denote the standard basis for the ambient space $\mathbb{R}^n$. Then, the error in resolving $\mathbf{e}_i$ onto the subspace $\langle \mathbf{V} \rangle$ as opposed to the subspace $\langle \mathbf{U} \rangle$ is $(\mathbf{P}_\mathbf{V} - \mathbf{P}_\mathbf{U})\mathbf{e}_i$, and the squared error computed over the basis $\{\mathbf{e}_i\}_{i=1}^n$ is

$$\sum_{i=1}^n \mathbf{e}_i^T(\mathbf{P}_\mathbf{U} - \mathbf{P}_\mathbf{V})^T(\mathbf{P}_\mathbf{U} - \mathbf{P}_\mathbf{V})\mathbf{e}_i = \operatorname{tr}\left[(\mathbf{P}_\mathbf{U} - \mathbf{P}_\mathbf{V})^T(\mathbf{P}_\mathbf{U} - \mathbf{P}_\mathbf{V})\right]$$

$$= \left\|\mathbf{P}_\mathbf{U} - \mathbf{P}_\mathbf{V}\right\|^2$$

$$= 2d_c^2\left(\langle \mathbf{U} \rangle, \langle \mathbf{V} \rangle\right).$$

A final argument in favor of the chordal or projection distance is that projections operate in the ambient space and it is here that we wish to measure error.

3. **Other subspace distances**: Other subspace distances proposed in the literature are the *spectral distance*

$$d_{spec}^2\left(\langle \mathbf{U} \rangle, \langle \mathbf{V} \rangle\right) = \sin^2\theta_1 = 1 - \left\|\mathbf{U}^T\mathbf{V}\right\|_2^2,$$

where θ_1 is the smallest principal angle given in (9.5), and $\|\mathbf{X}\|_2 = sv_1(\mathbf{X})$ is the ℓ_2 (or spectral) norm of $\mathbf{X}$. It takes values between 0 and 1. The *Fubini-Study distance*

$$d_{FS}\left(\langle \mathbf{U} \rangle, \langle \mathbf{V} \rangle\right) = \arccos\left(\prod_{k=1}^q \cos(\theta_k)\right) = \arccos\left(\det(\mathbf{U}^T\mathbf{V})\right),$$

which takes values between 0 and $\pi/2$, and the *Binet-Cauchy distance* [387]

$$d_{BC}\left(\langle \mathbf{U}\rangle, \langle \mathbf{V}\rangle\right) = \left(1 - \prod_{k=1}^{q} \cos^2(\theta_k)\right)^{1/2} = \left(1 - \det{}^2(\mathbf{U}^T\mathbf{V})\right)^{1/2},$$

which takes values between 0 and 1.

9.3 Subspace Averages

In many applications of statistical signal processing, pattern recognition, computer vision, and machine learning, high-dimensional data exhibits a low-dimensional structure that is revealed by a subspace representation. In computer vision, for instance, the set of images under different illuminations can be represented by a low-dimensional subspace [26]. And subspaces appear also as invariant representations of signals geometrically deformed under a set of affine transformations [154]. There are many more applications where low-dimensional subspaces capture the intrinsic geometry of the problem, ranging from array processing, motion segmentation, subspace clustering, spectrum sensing for cognitive radio, or noncoherent multiple-input multiple-output (MIMO) wireless communications [29, 136].

When input data are modeled as subspaces, a fundamental problem is to compute an average or central subspace. Let us consider a collection of measured subspaces $\langle \mathbf{V}_r\rangle \in \mathbb{G}r(q, \mathbb{R}^n)$, $r = 1, \ldots, R$. The subspace averaging problem is stated as follows

$$\langle \mathbf{V}^* \rangle = \underset{\langle \mathbf{V}\rangle \in \mathbb{G}r(q,\mathbb{R}^n)}{\arg\min} \frac{1}{R}\sum_{r=1}^{R} d^2\left(\langle \mathbf{V}\rangle, \langle \mathbf{V}_r\rangle\right),$$

where $d\left(\langle \mathbf{V}\rangle, \langle \mathbf{V}_r\rangle\right)$ could be any of the metrics presented in Sect. 9.2. We shall focus on the computation of averages using the geodesic distance (9.6) and the projection or chordal distance (9.7).

9.3.1 The Riemannian Mean

The Riemannian mean or center of mass, also known as Karcher or Frechet mean [190], of a collection of subspaces is the point on $\mathbb{G}r(q, \mathbb{R}^n)$ that minimizes the sum of squared geodesic distances:

$$\langle \mathbf{V}^* \rangle = \underset{\langle \mathbf{V}\rangle \in \mathbb{G}r(q,\mathbb{R}^n)}{\arg\min} \frac{1}{R}\sum_{r=1}^{R} d_{geo}^2\left(\langle \mathbf{V}\rangle, \langle \mathbf{V}_r\rangle\right). \tag{9.10}$$

Algorithm 5: Riemannian mean on the Grassmann manifold

Input: Basis for the subspaces $\{\mathbf{V}_r\}_{r=1}^R$, tolerance for convergence ϵ, and step-size δ
Output: Riemannian mean $\mathbf{V}^*$
Initialize: $\mathbf{V}^* = \mathbf{V}_i$, usually obtained by picking one of the elements of $\{\mathbf{V}_r\}_{r=1}^R$ at random.

1. Compute the Log map for each subspace at the current estimate for the mean: $\text{Log}_{\mathbf{V}^*}(\mathbf{V}_r)$, $r = 1, \ldots, R$
2. Compute the average tangent vector
$$\Delta\mathbf{V} = \frac{1}{R}\sum_{r=1}^{R} \text{Log}_{\mathbf{V}^*}(\mathbf{V}_r)$$
3. If $||\Delta\mathbf{V}|| < \epsilon$ stop, else go to step 4
4. Update $\mathbf{V}^* = \text{Exp}_{\mathbf{V}^*}(\delta\Delta\mathbf{V})$ moving δ along the geodesic on the direction of $\Delta\mathbf{V}$

The Karcher mean is most commonly found by using an iterative algorithm that exploits the matrix Exp and Log maps to move the data to and from the tangent space of a single point at each step. The Exp map is a "pullback" map that takes points on the tangent plane and pulls them onto the manifold in a manner that preserves distances: $\text{Exp}_{\mathbf{V}}(\mathbf{W}) : \mathbf{W} \in T_{\mathbf{V}}\mathcal{M} \to \mathcal{M}$. We can think of a vector $\mathbf{W} \in T_{\mathbf{V}}\mathcal{M}$ as a velocity for a geodesic curve in $\mathcal{M}$. This defines a natural bijective correspondence between points in $T_{\mathbf{V}}\mathcal{M}$ and points in $\mathcal{M}$ in a small ball around $\mathbf{V}$ such that points along the same tangent vector will be mapped along the same geodesic. The function inverse of the Exp map is the Log map, which maps a point $\mathbf{V} \in \mathcal{M}$ in the manifold to the tangent plane at $\mathbf{V}'$: $\text{Log}_{\mathbf{V}'}(\mathbf{V}) : \mathbf{V} \in \mathcal{M} \to T_{\mathbf{V}'}\mathcal{M}$. That is, $\text{Exp}_{\mathbf{V}'}\left(\text{Log}_{\mathbf{V}'}(\mathbf{V})\right) = \mathbf{V}$.

It is then straightforward to see in the case of the sphere that the Riemannian mean between the north pole and south pole is not unique since any point on the equator qualifies as a Riemannian mean. More formally, if the collection of subspaces is spread such that the Exp and Log maps are no longer bijective, then the Riemannian or Karcher mean is no longer unique. A unique optimal solution is guaranteed for data that lives within a convex ball on the Grassmann manifold, but in practice not all datasets satisfy this criterion. When this criterion is satisfied, a convergent iterative algorithm, proposed in [351], to compute the Riemannian mean (9.10) is summarized in Algorithm 5. Figure 9.2 illustrates the steps involved in the algorithm to compute the average of a cloud of points on a circle. To compute the Exp and Log maps for the Grassmannian, the reader is referred to [4].

Although the number of iterations needed to find the Riemannian mean depends on the diameter of the dataset [229], the iterative Algorithm 5 is in general computationally costly.

Finally, note that the average of the geodesic distances to the Riemannian mean $\mathbf{V}^*$, given by

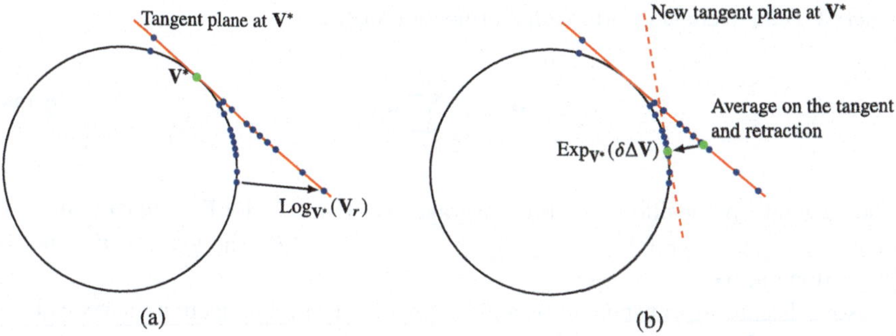

Fig. 9.2 Riemannian mean iterations on a circle. (**a**) Log map. (**b**) Exp map

$$\sigma_{geo}^2 = \sum_{r=1}^{R} d_{geo}^2\left(\langle \mathbf{V}^* \rangle, \langle \mathbf{V}_r \rangle\right),$$

may be called the Karcher or Riemannian variance.

9.3.2 The Extrinsic or Chordal Mean

Srivastava and Klassen proposed the so-called extrinsic mean, which uses the projection or chordal distance as a metric, as an alternative to the Riemannian mean in [331]. In this chapter, we shall refer to this mean as the extrinsic or chordal mean. Given a set of points on $\mathbb{G}r(q, \mathbb{R}^n)$, the chordal mean is the point

$$\langle \mathbf{V}^* \rangle = \underset{\langle \mathbf{V} \rangle \in \mathbb{G}r(q,\mathbb{R}^n)}{\arg\min} \quad \frac{1}{R} \sum_{r=1}^{R} d_c^2\left(\langle \mathbf{V} \rangle, \langle \mathbf{V}_r \rangle\right).$$

Using the definition of the chordal distance as the Frobenius norm of the difference of projection matrices, the solution may be written as

$$\mathbf{P}^* = \underset{\mathbf{P} \in \mathbb{P}r(q,\mathbb{R}^n)}{\arg\min} \quad \frac{1}{2R} \sum_{r=1}^{R} \|\mathbf{P} - \mathbf{P}_r\|^2, \tag{9.11}$$

where $\mathbf{P}_r = \mathbf{V}_r \mathbf{V}_r^T$ is the orthogonal projection matrix onto the rth subspace and $\mathbb{P}r(q, \mathbb{R}^n)$ denotes the set of all idempotent projection matrices of rank q. In contrast to the Riemannian mean, the extrinsic mean can be found analytically, as it is shown next. Let us begin by expanding the cost function in (9.11) as

$$\underset{\mathbf{P} \in \mathbb{P}r(q,\mathbb{R}^n)}{\text{minimize}} \quad \frac{1}{2} \operatorname{tr}\left(\mathbf{P}(\mathbf{I} - 2\overline{\mathbf{P}}) + \overline{\mathbf{P}}\right), \tag{9.12}$$

where $\overline{\mathbf{P}}$ is an average of orthogonal projection matrices:

$$\overline{\mathbf{P}} = \frac{1}{R}\sum_{r=1}^{R} \mathbf{P}_r. \tag{9.13}$$

The eigendecomposition of this average is $\overline{\mathbf{P}} = \mathbf{F}\mathbf{K}\mathbf{F}^T$, where $\mathbf{K} = \text{diag}(k_1, \ldots, k_n)$, with $1 \geq k_1 \geq k_2 \geq \cdots \geq k_n \geq 0$. The average $\bar{\mathbf{P}}$ is not a projection matrix.

Now, discarding constant terms and writing the desired projection matrix as $\mathbf{P} = \mathbf{U}\mathbf{U}^T$, where $\mathbf{U}$ is an orthogonal $n \times q$ matrix, problem (9.12) may be rewritten as

$$\underset{\mathbf{U} \in \mathbb{S}t(q,\mathbb{R}^n)}{\text{maximize}} \quad \text{tr}\left(\mathbf{U}^T\overline{\mathbf{P}}\mathbf{U}\right). \tag{9.14}$$

The solution to (9.14) is given by any orthogonal matrix whose column space is the same as the subspace spanned by the q principal eigenvectors of $\mathbf{F}$

$$\mathbf{U}^* = \left[\mathbf{f}_1\ \mathbf{f}_2\ \cdots\ \mathbf{f}_q\right] = \mathbf{F}_q,$$

and $\mathbf{P}^* = \mathbf{U}^*(\mathbf{U}^*)^T$. So the average subspace according the extrinsic distance is the subspace determined by the q eigenvectors of the average projection matrix corresponding to its q largest eigenvalues.

In fact, the eigenvectors of $\overline{\mathbf{P}}$ provide a flag or a nested sequence of central subspaces of increasing dimension $\langle \mathbf{V}_1 \rangle \subset \langle \mathbf{V}_2 \rangle \subset \cdots \subset \langle \mathbf{V}_q \rangle$, where $\dim(\langle \mathbf{V}_r \rangle) = r$. The flag is central in the sense that the kth subspace within the flag is the best k-dimensional representation of the data with respect to a cost function based on the Frobenius norm of the difference of projection matrices [108].

9.4 Order Estimation

The subspace averaging problem of Sect. 9.3 begins with a collection of q-dimensional subspaces in $\mathbb{R}^n$, and hence its average, $\langle \mathbf{V}^* \rangle$, is also a q-dimensional subspace in $\mathbb{G}r(q, \mathbb{R}^n)$. In some applications, the input subspaces may have different dimensions, which raises the question of estimating the dimension of an average subspace. This section addresses this order estimation problem, whose solution provides a simple order fitting rule based on thresholding the eigenvalues of the average of projection matrices. The proposed rule appears to be particularly well suited to problems involving high-dimensional data and low sample support, such as the determination of the number of sources with a large array of sensors, the so-called source enumeration problem, as discussed in [128]. The order fitting rule for subspace averaging was first published in [298], and an unrefined application to source enumeration was presented in [300].

Let us consider a collection of measured subspaces $\{\langle \mathbf{V}_r \rangle\}_{r=1}^{R}$ of $\mathbb{R}^n$, each with respective dimension $\dim(\langle \mathbf{V}_r \rangle) = q_r < n$. Each subspace $\langle \mathbf{V}_r \rangle$ is a point on the Grassmann manifold $\mathbb{G}r(q_r, \mathbb{R}^n)$, and the collection of subspaces lives on a disjoint union of Grassmannians. Without loss of generality, the dimension of the union of all subspaces is assumed to be the ambient space dimension n.

Using the chordal distance between subspaces, an order estimation criterion for the central subspace that "best approximates" the collection is

$$\left(s^*, \mathbf{P}_s^*\right) = \underset{\substack{s \in \{0,1,\ldots,n\} \\ \mathbf{P} \in \mathbb{P}r(s,\mathbb{R}^n)}}{\arg\min} \quad \frac{1}{2R} \sum_{r=1}^{R} \|\mathbf{P} - \mathbf{P}_r\|^2 .$$

For completeness, we also accept solutions $\mathbf{P} = \mathbf{0}$ with rank $s = 0$, meaning that there is no central "signal subspace" shared by the collection of input subspaces. Repeating the steps in Sect. 9.3, we find that the optimal order s^* is the number of negative eigenvalues of the matrix

$$\mathbf{S} = \mathbf{I}_n - 2\overline{\mathbf{P}},$$

or, equivalently, the number of eigenvalues of $\overline{\mathbf{P}}$ larger than 1/2, which is the order fitting rule proposed in [298]. The proposed rule may be written alternatively as

$$s^* = \underset{s \in \{0,1,\ldots,n\}}{\arg\min} \quad \sum_{i=1}^{s} (1 - k_i) + \sum_{i=s+1}^{n} k_i .$$

A similar rule was developed in [167] for the problem of designing optimum time-frequency subspaces with a specified time-frequency pass region.

Once the optimal order s^* is known, a basis for the average subspace is given by any unitary matrix whose column space is the same as the subspace spanned by the s^* principal eigenvectors of $\mathbf{F}$. So the average subspace is constructed by quantizing the eigenvalues of the average projection matrix at 0 or 1.

Example 9.4 We first generate a covariance matrix

$$\boldsymbol{\Sigma} = \mathbf{V}_c \mathbf{V}_c^T + \sigma^2 \mathbf{I}_n,$$

where $\mathbf{V}_c \in \mathbb{R}^{n \times q}$ is a matrix whose columns form an orthonormal basis for the central subspace $\langle \mathbf{V}_c \rangle \in \mathbb{G}r(q, \mathbb{R}^n)$ and the value of σ^2 determines the signal-to-noise ratio of the experiment, which is defined here as $\mathrm{SNR} = 10 \log_{10}\left(\frac{q}{n\sigma^2}\right)$. The covariance matrix $\boldsymbol{\Sigma}$ generated this way is the parameter of a matrix angular central Gaussian distribution $MACG(\boldsymbol{\Sigma})$.

We now generate R perturbed versions, possibly of different dimensions, of the central subspace as

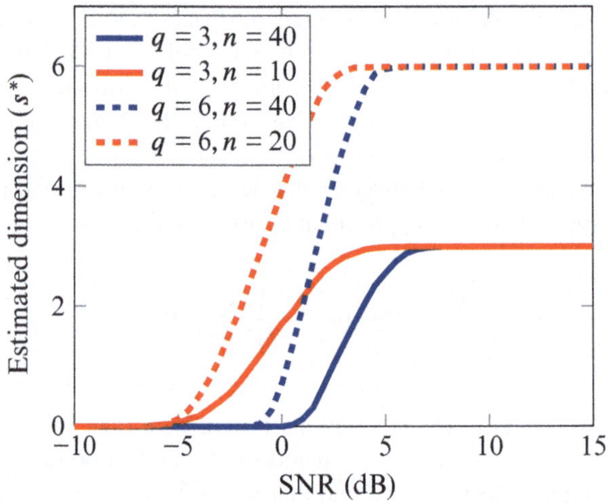

Fig. 9.3 Estimated dimension of the subspace average as a function of the SNR for different values of q (dimension of the true central subspace) and n (dimension of the ambient space). The number of averaged subspaces is $R = 50$

$$q_r \sim \mathcal{U}(q-1, q+1), \qquad \mathbf{X}_r \in \mathbb{R}^{n \times q_r} \sim MACG(\mathbf{\Sigma}),$$

for $r = 1, \ldots, R$. So we first sample the subspace dimension q_r from a uniform distribution $\mathcal{U}(q-1, q+1)$ and then sample a q_r-dimensional subspace from a $MACG(\mathbf{\Sigma})$ distribution. Let us recall that sampling from $MACG(\mathbf{\Sigma})$ amounts to sampling from a normal distribution and then extracting the orientation matrix of its $n \times q_r$ polar decomposition, i.e.,

$$\mathbf{Z}_r \sim \mathcal{N}_{n \times q_r}(\mathbf{0}, \mathbf{I}_{q_r} \otimes \mathbf{\Sigma}), \qquad \mathbf{X}_r = \mathbf{Z}_r(\mathbf{Z}_r^T \mathbf{Z}_r)^{-1/2}.$$

Figure 9.3 shows the estimated dimension of the subspace average as a function of the SNR for different values of q (dimension of the true central subspace) and n (dimension of the ambient space). The number of averaged subspaces is $R = 50$. The curves represent averaged results of 500 independent simulations. As demonstrated, there is transition behavior between an estimated order of $s^* = 0$ (no central subspace) and the correct order $s^* = q$, in the vicinity of SNR $= 0$ dB.

9.5 The Average Projection Matrix

When the chordal distance is used to measure the pairwise dissimilarity between subspaces, the average of the corresponding orthogonal projection matrices plays a central role in determining the subspace average and its dimension. It is therefore of interest to review some of its properties.

Property 9.3 Let us consider a set of subspaces $\langle \mathbf{V}_r \rangle \in \mathbb{G}r(q_r, \mathbb{R}^n)$ with respective projection matrices $\mathbf{P}_r$, for $r = 1, \ldots, R$. Each projection matrix is idempotent with $\text{rank}(\mathbf{P}_r) = q_r$. The average of projection matrices

$$\overline{\mathbf{P}} = \frac{1}{R} \sum_{r=1}^{R} \mathbf{P}_r,$$

with eigendecomposition $\overline{\mathbf{P}} = \mathbf{F}\mathbf{K}\mathbf{F}^T$, is not a projection matrix itself and therefore is not idempotent. However, it has the following properties:

(P1) $\overline{\mathbf{P}}$ is a symmetric matrix. This is trivially proved by noticing that $\overline{\mathbf{P}}$ is an average of symmetric matrices.

(P2) Its eigenvalues are real and satisfy $0 \leq k_i \leq 1$. To prove this, take without loss of generality the ith eigenvalue-eigenvector pair $(k_i, \mathbf{f}_i)$. Then

$$k_i = \mathbf{f}_i^H \overline{\mathbf{P}} \mathbf{f}_i = \frac{1}{R} \sum_{r=1}^{R} \mathbf{f}_i^T \mathbf{P}_r \mathbf{f}_i \overset{(1)}{=} \frac{1}{R} \sum_{r=1}^{R} \mathbf{f}_i^T \mathbf{P}_r^2 \mathbf{f}_i = \frac{1}{R} \sum_{r=1}^{R} ||\mathbf{P}_r \mathbf{f}_i||^2 \leq 1,$$

where (1) holds because all $\mathbf{P}_r$ are idempotent and the inequality follows from the fact that each term $||\mathbf{P}_r \mathbf{f}_i||^2$ is the squared norm of the projection of a unit norm vector, $\mathbf{f}_i$, onto the subspace $\langle \mathbf{V}_r \rangle$ and therefore $||\mathbf{P}_r \mathbf{f}_i||^2 \leq 1$, with equality only if the eigenvector belongs to the subspace.

(P3) The trace of the average of projections satisfies

$$\text{tr}(\overline{\mathbf{P}}) = \frac{1}{R} \sum_{r=1}^{R} q_r.$$

Therefore, when all subspaces have the same dimension, q, the trace of the average projection is $\text{tr}(\overline{\mathbf{P}}) = q$.

The previous properties hold for arbitrary sets of subspaces $\{\langle \mathbf{V}_r \rangle\}_{r=1}^R$. When the subspaces are i.i.d. realizations of some distribution on the Grassmann manifold, the average of projections, which could also be called in this case the sample mean of the projections, is a random matrix whose expectation can be sometimes analytically characterized. A first result is the following. Let $\{\langle \mathbf{V}_r \rangle\}_{r=1}^R$ be a random sample of size R *uniformly distributed* in $\mathbb{G}r(q, \mathbb{R}^n)$. Equivalently, each $\mathbf{P}_r$ is a rank-q projection uniformly distributed in $\mathbb{P}r(q, \mathbb{R}^n)$. Then, it is immediate to prove that (see [74, p. 29])

$$\mathrm{E}[\overline{\mathbf{P}}] = \mathrm{E}[\mathbf{P}_r] = \frac{q}{n}\mathbf{I}_n,$$

so all eigenvalues of the expected value of the average of projections are identical to $k_i = q/n$, $i = 1, \ldots, n$, indicating no preference for any particular direction. So, asymptotically, for uniformly distributed subspaces, the order fitting rule of Sect. 9.4 will return 0 if $q < n/2$, and n otherwise, in both cases suggesting there is no central low-dimensional subspace. This result is the basis of the Bingham test for uniformity, which rejects uniformity if the average of projection matrices, $\overline{\mathbf{P}}$, is far from its expected value $(q/n)\mathbf{I}_n$.

For non-uniform distributions on the Grassmannian, the expectation of a projection matrix is in general difficult to obtain. Nevertheless, for the angular central Gaussian distribution defined on the projective space $\mathbb{G}r(1, \mathbb{R}^2)$, the following example is illustrative.

Example 9.5 Let $\tilde{\mathbf{x}} \sim ACG(\mathbf{F}\boldsymbol{\Sigma}\mathbf{F}^T)$ be a unit-norm random vector in $\mathbb{R}^2$ having the angular central Gaussian distribution with dispersion matrix $\mathbf{F}\boldsymbol{\Sigma}\mathbf{F}^T$, where $\boldsymbol{\Sigma} = \mathrm{diag}(\sigma_1^2, \sigma_2^2)$, and let $\mathbf{P} = \tilde{\mathbf{x}}\tilde{\mathbf{x}}^T$ be the corresponding 2×2 random rank-1 projection matrix. Then, $\mathrm{E}[\mathbf{P}] = \mathbf{F}\boldsymbol{\Lambda}\mathbf{F}^T$ where

$$\boldsymbol{\Lambda} = \frac{\boldsymbol{\Sigma}^{1/2}}{\mathrm{tr}(\boldsymbol{\Sigma}^{1/2})} = \begin{bmatrix} \frac{\sigma_1}{\sigma_1+\sigma_2} & 0 \\ 0 & \frac{\sigma_2}{\sigma_1+\sigma_2} \end{bmatrix}.$$

Recall that if $\mathbf{x} \sim \mathcal{N}_2(\mathbf{0}, \boldsymbol{\Sigma})$, then

$$\tilde{\mathbf{x}} = \frac{\mathbf{F}\mathbf{x}}{||\mathbf{F}\mathbf{x}||} = \mathbf{F}\frac{\mathbf{x}}{||\mathbf{x}||} \sim AG(\mathbf{F}\boldsymbol{\Sigma}\mathbf{F}^T).$$

The expectation of the projection matrix is

$$\mathrm{E}[\mathbf{P}] = \mathrm{E}[\tilde{\mathbf{x}}\tilde{\mathbf{x}}^T] = \mathbf{F}\,\mathrm{E}\left[\frac{\mathbf{x}\mathbf{x}^T}{||\mathbf{x}||^2}\right]\mathbf{F}^T,$$

so the problem reduces to calculating $\mathrm{E}\left[\frac{\mathbf{x}\mathbf{x}^T}{||\mathbf{x}||^2}\right]$ when $\mathbf{x} \sim \mathcal{N}_2(\mathbf{0}, \boldsymbol{\Sigma})$. The result is

$$\mathrm{E}\left[\frac{\mathbf{x}\mathbf{x}^T}{||\mathbf{x}||^2}\right] = \begin{bmatrix} \mathrm{E}\left[\frac{x_1^2}{x_1^2+x_2^2}\right] & \mathrm{E}\left[\frac{x_1x_2}{x_1^2+x_2^2}\right] \\ \mathrm{E}\left[\frac{x_1x_2}{x_1^2+x_2^2}\right] & \mathrm{E}\left[\frac{x_2^2}{x_1^2+x_2^2}\right] \end{bmatrix}.$$

The off-diagonal terms of this matrix are calculated as

$$\mathrm{E}\left[\frac{x_1x_2}{x_1^2+x_2^2}\right] = K\int_{-\infty}^{\infty}\int_{-\infty}^{\infty}\frac{x_1x_2}{x_1^2+x_2^2}e^{-\left(\frac{x_1^2}{2\sigma_1^2}+\frac{x_2^2}{2\sigma_2^2}\right)}dx_1dx_2,$$

where $K^{-1} = (2\pi)\det(\mathbf{\Sigma})^{1/2}$. Transforming from Euclidean to polar coordinates, $x_1 = r\cos\theta$, $x_2 = r\sin\theta$, with Jacobian $J(x_1, x_2 \to r, \theta) = r$, we have

$$\begin{aligned}\mathrm{E}\left[\frac{x_1x_2}{x_1^2+x_2^2}\right] &= K\int_0^{2\pi}\sin\theta\cos\theta\int_0^{\infty} re^{-r^2\left(\frac{\cos^2\theta}{2\sigma_1^2}+\frac{\sin^2\theta}{2\sigma_2^2}\right)}drd\theta\\ &= K\int_0^{2\pi}\frac{\sin\theta\cos\theta}{\frac{\cos^2\theta}{\sigma_1^2}+\frac{\sin^2\theta}{\sigma_2^2}}d\theta = 0,\end{aligned}$$

where the last equality follows from the fact that the integrand is a zero-mean periodic function with period π.

Similarly, the Northwest diagonal term of $\mathrm{E}[\mathbf{P}]$ is

$$\begin{aligned}\mathrm{E}\left[\frac{x_1^2}{x_1^2+x_2^2}\right] &= K\int_0^{2\pi}\frac{\cos^2\theta}{\frac{\cos^2\theta}{\sigma_1^2}+\frac{\sin^2\theta}{\sigma_2^2}}d\theta\\ &= K\sigma_2^2\int_0^{2\pi}\frac{1}{\frac{\sigma_2^2}{\sigma_1^2}+\tan^2\theta}d\theta.\end{aligned} \tag{9.15}$$

The last integral can be solved analytically, yielding

$$\int_0^{2\pi}\frac{1}{\frac{\sigma_2^2}{\sigma_1^2}+\tan^2\theta}d\theta = 2\pi\frac{\sigma_1^2}{\sigma_2(\sigma_1+\sigma_2)}.$$

Substituting this result in (9.15), we obtain

$$\mathrm{E}\left[\frac{x_1^2}{x_1^2+x_2^2}\right] = \frac{\sigma_1}{\sigma_1+\sigma_2}.$$

Therefore,

$$\mathrm{E}\left[\frac{\mathbf{x}\mathbf{x}^T}{||\mathbf{x}||^2}\right] = \begin{bmatrix}\frac{\sigma_1}{\sigma_1+\sigma_2} & 0\\ 0 & \frac{\sigma_2}{\sigma_1+\sigma_2}\end{bmatrix} = \frac{\mathbf{\Sigma}^{1/2}}{\mathrm{tr}(\mathbf{\Sigma}^{1/2})},$$

thus proving that $\mathrm{E}[\mathbf{P}] = \mathbf{F}\mathbf{\Lambda}\mathbf{F}^T$ with $\mathbf{\Lambda} = \mathbf{\Sigma}^{1/2}/\,\mathrm{tr}(\mathbf{\Sigma}^{1/2})$.

From the perspective of subspace averaging, the previous example may be interpreted as follows. Let $\{\langle \mathbf{V}_r \rangle\}_{r=1}^R$ be a random sample of size R of one-dimensional subspaces (lines) in $\mathbb{G}r(1, \mathbb{R}^2)$ with angular central Gaussian distribution $MACG(\mathbf{F}\boldsymbol{\Sigma}\mathbf{F}^T)$. Matrix $\mathbf{F}$ gives the orientation of the distribution, and $\boldsymbol{\Sigma} = \mathrm{diag}(\sigma_1^2, \sigma_2^2)$ are the concentration parameters. Equivalently, the projection matrix onto the random subspace $\langle \mathbf{V}_r \rangle$ may be formed by sampling a two-dimensional Gaussian vector $\mathbf{x}_r \sim \mathcal{N}_2(\mathbf{0}, \mathbf{F}\boldsymbol{\Sigma}\mathbf{F}^T)$ and then forming $\mathbf{P}_r = \frac{1}{||\mathbf{x}_r||^2}\mathbf{x}_r\mathbf{x}_r^T$. When the sample size grows, the sample mean converges to the expectation

$$\overline{\mathbf{P}} = \frac{1}{R}\sum_{r=1}^{R} \mathbf{P}_r \overset{R\to\infty}{\longrightarrow} \frac{1}{\mathrm{tr}(\boldsymbol{\Sigma}^{1/2})}\mathbf{F}\boldsymbol{\Sigma}^{1/2}\mathbf{F}^T.$$

The net of this result is that for a sufficiently large collection of subspaces, as long as the distribution has some directionality, i.e., $\sigma_1 > \sigma_2$, the subspace averaging procedure will output as central subspace the eigenvector corresponding to the largest eigenvalue of the matrix $\mathbf{F}\boldsymbol{\Sigma}\mathbf{F}^T$, as one would expect. For isotropic data, $\sigma_1 = \sigma_2$, the eigenvalues of the average of projection matrices converge to 1/2 as $R \to \infty$, suggesting in this case that there is no central subspace.

9.6 Application to Subspace Clustering

Given a collection of subspaces, we have shown in previous sections of this chapter how to determine an average or central subspace according to a projection distance measure and how to determine the dimension of this average. The eigenvectors and eigenvalues of the average projection matrix play a central role in determining, respectively, the central subspace and its dimension. In some applications, the input data could be drawn from a mixture of distributions on a union of Grassmann manifolds and hence can be better modeled by multiple clusters of subspaces with centroids of different dimensions. For instance, it has been demonstrated that there exist low-dimensional representations for a set of images of a fixed object or subject under variations in illumination conditions or pose [26]. A collection of images of K different subjects under different illumination or pose conditions should be modeled by K different clusters of subspaces. The goal of subspace clustering is to determine the number of clusters K, their central subspaces or centroids $\{\langle \mathbf{M}_k \rangle\}_{k=1}^K$ (this amounts to determining their dimensions and the subspace bases $\{\mathbf{M}_k\}_{k=1}^K$), and the segmentation of the input subspaces into clusters. In this section, we address this problem by leveraging the averaging procedure and order fitting rule described in Sects. 9.3 and 9.4, respectively.

The standard subspace clustering problem begins with a collection of data points $\{\mathbf{x}_r\}_{r=1}^R$ drawn from an unknown union of K subspaces. The points in the kth subspace are modeled as $\mathbf{x}_{rk} = \mathbf{M}_k\mathbf{y}_{rk} + \mathbf{n}_{rk}$, for $r = 1, \ldots, R_k$, where $\mathbf{M}_k \in \mathbb{S}t(q_k, \mathbb{R}^L)$ is a basis for the kth centroid of unknown dimension q_k; $\mathbf{y}_{rk} \in \mathbb{R}^{q_k}$; and $\mathbf{n}_{rk}$ models the noise in the observations [366, 367]. The total number of data points is $R = \sum_{k=1}^{K} R_k$. We consider a different formulation of the subspace clustering problem in which we begin with a collection of subspaces $\{\langle \mathbf{V}_r \rangle\}_{r=1}^{R}$, and the goal is to find the number of clusters K and the segmentation or assignment of subspaces to clusters. Each subspace in the collection may have a different dimension, q_r, but all of them live in an ambient space of dimension L. Notice that once the number of clusters has been found and the segmentation problem has been solved, we can fit a central subspace to each group by the averaging procedure described in Sect. 9.3. For each group, the dimension of the central subspace is the number of eigenvalues of the average of projection matrices larger than 1/2, and the corresponding eigenvectors form a basis for that centroid.

For a fixed number of clusters, K, the subspace clustering problem can be formulated using projection matrices as follows

$$\begin{aligned} &\underset{\{q_k\},\{\mathbf{P}_{\mathbf{M}_k}\},\{w_{rk}\}}{\arg\min} && \sum_{k=1}^{K} \frac{1}{2R_k} \sum_{r=1}^{R} w_{rk} \left\| \mathbf{P}_{\mathbf{M}_k} - \mathbf{P}_{\mathbf{V}_r} \right\|^2, \\ &\text{subject to} && w_{rk} \in \{0, 1\} \quad \text{and} \quad \sum_{r=1}^{R} w_{rk} = R_k. \end{aligned} \tag{9.16}$$

The binary assignment variables are $w_{rk} = 1$ if subspace $\mathbf{V}_r$ with projection matrix $\mathbf{P}_{\mathbf{V}_r} = \mathbf{V}_r \mathbf{V}_r^T$ belongs to cluster k and $w_{rk} = 0$ otherwise. Notice that $\sum_{r=1}^{R} w_{rk} = R_k$ is the number of subspaces in cluster k and hence $\sum_{k=1}^{K} R_k = R$.

Given the orthogonal projection matrices $\{\mathbf{P}_{\mathbf{M}_k}\}_{k=1}^{K}$ that represent K centroids, the optimal values for w_{rk} assign each subspace to its closest centroid. Given the segmentation variables, Problem (9.16) is decoupled into K problems, each of which can be solved by performing the SVD of the average projection matrix of the R_k subspaces assigned to the kth cluster. By iterating these two steps, a refined solution for the subspace clustering can be obtained. Obviously, this is just a variant of the K-means algorithm [109] applied to subspaces, similar in spirit to the K-planes algorithm [45] or the K-subspaces algorithm [346]. This clustering algorithm is very simple, but its convergence to the global optimum depends on a good initialization. In fact, the cost function (9.16) has many local minima, and the iterative procedure may easily get trapped into one of them. Therefore, many random initializations may be needed to find a good clustering solution. A more efficient alternative for solving the segmentation problem is described below.

Segmentation via MDS Embedding. Subspaces may be embedded into a low-dimensional Euclidean space by applying the multidimensional scaling (MDS) procedure (cf. Sect. 2.4 of Chap. 2). Then, standard K-means may be used to obtain the segmentation variables w_{rk}. In this way, the K-means algorithm works with low-dimensional vectors in a Euclidean space, and, therefore, it is less prone to local minima. The MDS procedure begins by building an $R \times R$ squared Euclidean distance matrix $\mathbf{D}$. As a distance metric, we choose the projection subspace distance

$$(\mathbf{D})_{i,l} = \frac{1}{2} \left\| \mathbf{P}_{\mathbf{V}_i} - \mathbf{P}_{\mathbf{V}_l} \right\|^2,$$

where $\mathbf{P}_{\mathbf{V}_i}$ and $\mathbf{P}_{\mathbf{V}_l}$ are the orthogonal projection matrices into the subspaces $\langle \mathbf{V}_i \rangle$ and $\langle \mathbf{V}_l \rangle$, respectively. The goal of MDS is to find a configuration of points in a low-dimensional subspace such that their pairwise distances reproduce (or approximate) the original distance matrix. Let $d_{MDS} < L$ be the dimension of the configuration of points and recall that L is the dimension of the ambient space for all subspaces. Then, $\mathbf{X} \in \mathbb{R}^{R \times d_{MDS}}$ and $(\mathbf{x}_i - \mathbf{x}_l)(\mathbf{x}_i - \mathbf{x}_l)^T \approx (\mathbf{D})_{i,l}$, where $\mathbf{x}_i$ is the ith row of $\mathbf{X}$. The MDS procedure computes the non-negative definite centered matrix

$$\mathbf{B} = -\frac{1}{2} \mathbf{P}_{\mathbf{1}}^{\perp} \mathbf{D} \mathbf{P}_{\mathbf{1}}^{\perp},$$

where $\mathbf{P}_{\mathbf{1}}^{\perp} = \mathbf{I}_R - \mathbf{1} \left(\mathbf{1}^T \mathbf{1} \right)^{-1} \mathbf{1}^T$ is a projection matrix onto the orthogonal space of subspace $\langle \mathbf{1} \rangle$. From the EVD of $\mathbf{B} = \mathbf{F} \mathbf{K} \mathbf{F}^T$, we can extract a configuration $\mathbf{X} = \mathbf{F}_{d_{MDS}} \mathbf{K}_{d_{MDS}}$, where $\mathbf{F}_{d_{MDS}} = [\mathbf{f}_1 \cdots \mathbf{f}_{d_{MDS}}]$ and $\mathbf{K}_{d_{MDS}} = \operatorname{diag}(k_1, \ldots, k_{d_{MDS}})$ contain the d_{MDS} largest eigenvectors and eigenvalues of $\mathbf{B}$, respectively. We can now cluster the rows of $\mathbf{X}$ with the K-means algorithm (or any other clustering method). Since the R points $\mathbf{x}_i \in \mathbb{R}^{d_{MDS}}$ belong to a low-dimensional Euclidean space, the convergence of the K-means is faster and requires fewer random initializations to converge to the global optimum. The low-dimensional embedding of subspaces via MDS allows us to determine the number of clusters using standard clustering validity indices proposed in the literature, namely, the Davies-Bouldin index [96], Calinski-Harabasz index [55], or the silhouette index [291]. The example below assesses their performance in a subspace clustering problem with MDS embedding.

Example 9.6 (Determining the number of clusters) A fundamental question that needs to be addressed in any subspace clustering problem is that of determining how many clusters are actually present. Related to this question is that of the validity of the clusters formed. Many clustering validity indices have been proposed since the 1970s, including the Davies-Bouldin index [96], the Calinski-Harabasz index [55], and the silhouette index [291]. All these indices are functions of the within and between cluster scatter matrices obtained for different numbers of clusters. The value of K that maximizes these indices is considered to be the correct number of clusters.

Consider an example with $K = 3$ clusters formed by subspaces in $\mathbb{R}^L$, with $L = 50$ (ambient space dimension). The subspaces belonging to the kth cluster are generated by sampling from a $MACG(\boldsymbol{\Sigma}_k)$ distribution with parameter $\boldsymbol{\Sigma}_k = \mathbf{M}_k \mathbf{M}_k^T + \sigma^2 \mathbf{I}_L$, where $\mathbf{M}_k \in \mathbb{St}(q_k, \mathbb{R}^L)$ is an orthogonal basis for the central subspace (or centroid) of the kth cluster and σ^2 is the variance of an isotropic perturbation that determines the cluster spread. The signal-to-noise ratio in dB is defined as

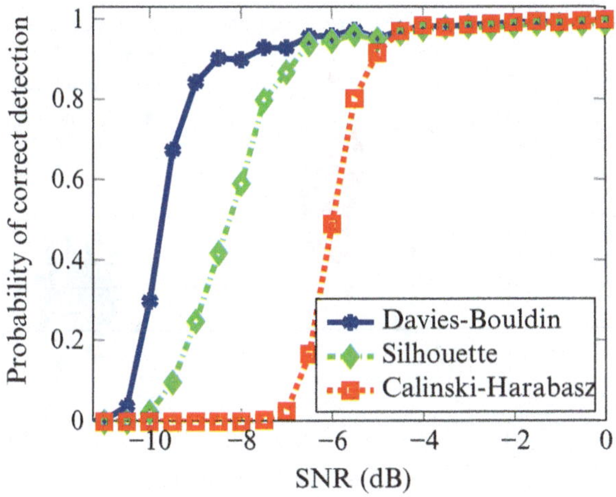

Fig. 9.4 Probability of detection of the correct number of clusters for different cluster validity indices

$$\mathrm{SNR} = -10\log_{10}(L\sigma^2).$$

Therefore, for the kth cluster, we generate $\mathbf{V}_{rk} \sim MACG(\boldsymbol{\Sigma}_k)$, $r = 1, \ldots, R_k$. The dimensions of the central subspaces are $q_1 = 3$, $q_2 = 3$, and $q_3 = 5$. The bases for the central subspaces are generated such that $\mathbf{M}_2$ and $\mathbf{M}_3$ intersect in one dimension. That is, $\dim(\langle \mathbf{M}_2 \rangle \cup \langle \mathbf{M}_3 \rangle) = 7$. This makes the subspace clustering problem much harder to solve. The number of subspaces in each cluster is $R_1 = 50$, $R_2 = 100$, and $R_3 = 50$. MDS is applied to obtain a low-dimensional embedding of the $R = 200$ subspaces into a $d_{MDS} = 5$ dimensional Euclidean space. The membership function w_{rk} of pattern $\mathbf{x}_r \in \mathbb{R}^{d_{MDS}}$ to cluster $\mathbf{M}_k$ is obtained by the standard K-means algorithm. The process is repeated for different values of K, and the value that optimizes the corresponding validity index is considered to be the correct number of clusters.

Figure 9.4 shows the probability of determining the correct number of clusters versus the SNR for the Davies-Bouldin index, the Calinski-Harabasz index, and the silhouette index. The Davies-Bouldin index provides the best result. Therefore, we select this criterion for determining the number of clusters. Figure 9.5 shows the final clustering obtained in an example with SNR $= 0$ dB. To represent the clusters in a bidimensional space, we have used the first two MDS components. Cluster 3, which is formed by subspaces in $\mathbb{G}r(5, \mathbb{R}^{50})$, is well separated from Clusters 1 and 2 in $\mathbb{G}r(3, \mathbb{R}^{50})$, even though Clusters 2 and 3 intersect in one dimension. Once the membership function and the number of clusters have been determined, the subspace averaging procedure equipped with the order estimation rule returns the correct dimensions for the central subspaces. The complete subspace clustering algorithm is shown in Algorithm 6.

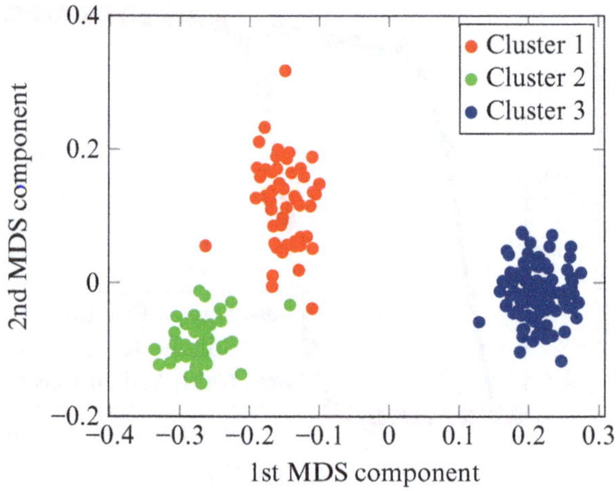

Fig. 9.5 Subspace clustering example. The clusters are depicted in a bidimensional Euclidean space formed by the first and second MDS components

9.7 Application to Array Processing

In this section, we apply the order fitting rule for subspace averaging described in Sect. 9.4 to the problem of estimating the number of signals received by a sensor array, which is referred to in the related literature as source enumeration. This is a classic and well-researched problem in radar, sonar, and communications [302,358], and numerous criteria have been proposed over the last decades to solve it, most of which are given by functions of the eigenvalues of the sample covariance matrix [191,224,374,389,394]. These methods tend to underperform when the number of antennas is large and/or the number of snapshots is relatively small in comparison to the number of antennas, the so-called small-sample regime [245], which is the situation of interest here.

The proposed method to solve this problem forms a collection of subspaces based on the array geometry and sampling from the discrete distribution $D(\mathbf{U}, \boldsymbol{\alpha})$ presented in Sect. 9.1.1. Then, the order fitting rule for averages of projections described in Sect. 9.4 can be used to enumerate the sources. This method is particularly effective when the dimension of the input space is large (high-dimensional arrays), and we have only a few snapshots, which is when the eigenvalues of sample covariance matrices are poorly estimated and methods based on functions of these eigenvalues underperform design objectives.

Source Enumeration in Array Processing. Let us consider K narrowband signals impinging on a large, uniform, half-wavelength linear array with M antennas, as depicted in Fig. 9.6. The received signal is

Algorithm 6: Subspace clustering algorithm

Input: Subspaces $\{\langle \mathbf{V}_r \rangle\}_{r=1}^R$ or, equivalently, orthogonal projection matrices $\{\mathbf{P}_r\}_{r=1}^R$ and MDS dimension d_{MDS}

Output: Number of clusters K, bases (and dimensions) of the central subspaces $\mathbf{M}_k \in \mathbb{S}t(q_k, \mathbb{R}^L)$, $k = 1, \ldots, K$, and membership function w_{rk}

```
/* Euclidean embedding via MDS */
```

Generate squared extrinsic distance matrix $\mathbf{D}$ with $(\mathbf{D})_{i,l} = \frac{1}{2} \left\| \mathbf{P}_{\mathbf{V}_i} - \mathbf{P}_{\mathbf{V}_l} \right\|^2$

Obtain $\mathbf{B} = -\frac{1}{2}\mathbf{P}_1^\perp \mathbf{D} \mathbf{P}_1^\perp$ and perform its EVD $\mathbf{B} = \mathbf{F}\mathbf{K}\mathbf{F}^T$

MDS embedding $\mathbf{X} = \mathbf{F}_{d_{MDS}} \mathbf{K}_{d_{MDS}} \in \mathbb{R}^{R \times d_{MDS}}$

/* Determine K and w_{rk} */

for $k = 1, \ldots, K_{max}$ **do**

  Cluster the rows of $\mathbf{X}$ with K-means into clusters $\mathbf{M}_1, \ldots, \mathbf{M}_k$

  Calculate the Davies-Bouldin index $DB(k)$

Find the number of clusters as $K = \underset{k \in \{1,\ldots,K_{max}\}}{\arg\min} \; DB(k)$ and recover the corresponding membership function w_{rk} obtained with K-means

```
/* Determine the dimensions and bases of the central
   subspaces */
```

for $k = 1, \ldots, K$ **do**

  Average projection matrix for cluster $\mathbf{M}_k$

$$\overline{\mathbf{P}}_k = \frac{1}{\sum_{r=1}^R w_{rk}} \sum_{r=1}^R w_{rk} \mathbf{P}_r$$

  Find $\overline{\mathbf{P}}_k = \mathbf{F}\mathbf{K}\mathbf{F}^T$

  Estimate q_k as the number of eigenvalues of $\overline{\mathbf{P}}_k$ larger than 1/2

  Find a basis for the central subspace as

$$\mathbf{M}_k = [\mathbf{f}_1 \;\cdots\; \mathbf{f}_{q_k}]$$

$$\mathbf{x}[n] = [\mathbf{a}(\theta_1) \quad \cdots \quad \mathbf{a}(\theta_K)]\,\mathbf{s}[n] + \mathbf{n}[n] = \mathbf{A}\mathbf{s}[n] + \mathbf{n}[n], \tag{9.17}$$

where $\mathbf{s}[n] = [s_1[n] \cdots s_K[n]]^T$ is the transmit signal; $\mathbf{A} \in \mathbb{C}^{M \times K}$ is the steering matrix, whose kth column $\mathbf{a}(\theta_k) = [1 \; e^{-j\theta_k} \; e^{-j\theta_k(M-1)}]^T$ is the complex array response to the kth source; and θ_k is the unknown electrical angle for the kth source. In the case of narrowband sources, free space propagation, and a uniform linear array (ULA) with inter-element spacing d, the spatial frequency or electrical angle is

$$\theta_k = \frac{2\pi}{\lambda} d \sin(\phi_k),$$

where λ is the wavelength and ϕ_k is the direction of arrival (DOA). We will refer to θ_k as the DOA of source k. Note that for a half-wavelength ULA $\theta_k = \pi \sin(\phi_k)$,

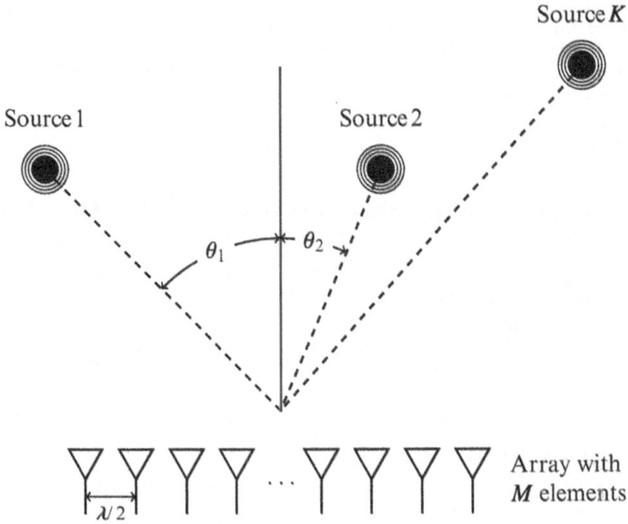

Fig. 9.6 Source enumeration problem in large scale arrays: estimating the number of sources K in a ULA with a large number of antenna elements M

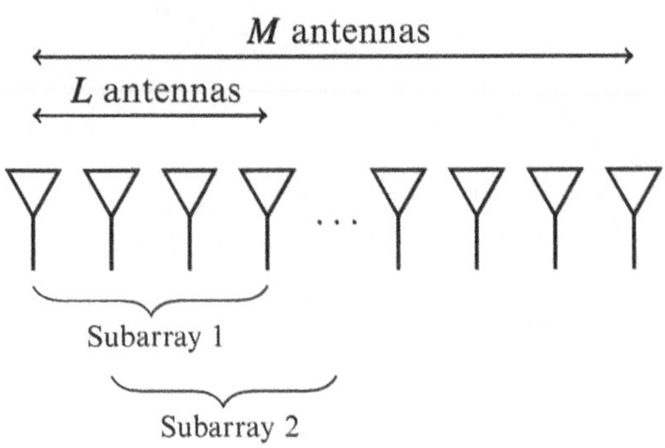

Fig. 9.7 L-dimensional subarrays extracted from a uniform linear array with $M > L$ elements

and the spatial frequency varies between $-\pi$ and π when the direction of arrival varies between $-90°$ and $90°$, with $0°$ being the broadside direction.

The signal and noise vectors are modeled as $\mathbf{s}[n] \sim \mathcal{CN}_K(\mathbf{0}, \mathbf{R}_{ss})$ and $\mathbf{n}[n] \sim \mathcal{CN}_M(\mathbf{0}, \sigma^2\mathbf{I}_M)$, respectively. From the signal model (9.17), the covariance matrix of the measurements is

$$\mathbf{R} = \mathrm{E}\left[\mathbf{x}[n]\mathbf{x}^H[n]\right] = \mathbf{A}\mathbf{R}_{ss}\mathbf{A}^H + \sigma^2\mathbf{I}_M.$$

We assume there are N snapshots collected in the data matrix $\mathbf{X} = [\mathbf{x}[1] \ \cdots \ \mathbf{x}[N]]$. The source enumeration problem consists of estimating K from $\mathbf{X}$.

Shift Invariance. When uniform linear arrays are used, a property called shift invariance holds, which forms the basis of the ESPRIT (estimation of signal parameters via rotational invariance techniques) method [261, 293] and its many variants. Let $\mathbf{A}_l$ be the $L \times K$ matrix with rows $l, \ldots, l+L-1$ extracted from the steering matrix $\mathbf{A}$. This steering matrix for the lth subarray is illustrated in Fig. 9.7. Then, from (9.17) it is readily verified that

$$\mathbf{A}_l \operatorname{diag}(e^{-j\theta_1}, \ldots, e^{-j\theta_K}) = \mathbf{A}_{l+1}, \quad l = 1, \ldots, M-L+1,$$

which is the shift invariance property. In this way, $\mathbf{A}_l$ and $\mathbf{A}_{l+1}$ are related by a nonsingular rotation matrix,

$$\mathbf{Q} = \operatorname{diag}(e^{-j\theta_1}, \ldots, e^{-j\theta_K}),$$

and therefore they span the same subspace. That is, $\langle \mathbf{A}_l \rangle = \langle \mathbf{A}_{l+1} \rangle$, with $\dim(\langle \mathbf{A}_l \rangle) = K < L$. In ESPRIT, two subarrays of dimension $L = M-1$ are considered, and thus we have $\mathbf{A}_1\mathbf{Q} = \mathbf{A}_2$, where $\mathbf{A}_1$ and $\mathbf{A}_2$ select, respectively, the first and the last $M-1$ rows of $\mathbf{A}$.

When noise is present, however, the shift-invariance property does not hold for the main eigenvectors extracted from the sample covariance matrix. The optimal subspace estimation (OSE) technique proposed by Vaccaro et al. obtains an improved estimate of the signal subspace with the required structure (up to the first order) [219, 354]. Nevertheless, the OSE technique requires the dimension of the signal subspace to be known in advance and, therefore, does not apply directly to the source enumeration problem.

From the $L \times 1$ ($L > K$) subarray snapshots $\mathbf{x}_l[n]$, we can estimate an $L \times L$ sample covariance as

$$\mathbf{S}_l = \frac{1}{N} \sum_{n=1}^{N} \mathbf{x}_l[n]\mathbf{x}_l^H[n].$$

Note that each $\mathbf{S}_l$ block corresponds to an $L \times L$ submatrix of the full sample covariance $\mathbf{S}$ extracted along its diagonal.

Due to the shift invariance property of ULAs, the noiseless signal subspaces of all submatrices $\mathbf{R}_l = E[\mathbf{x}_l[n]\mathbf{x}_l^H[n]]$ are identical. Since there are M sensors and we extract L-dimensional subarrays, there are $J = M - L + 1$ different sample covariance estimates $\mathbf{S}_l$, $l = 1, \ldots, J$. For each $\mathbf{S}_l$ we compute its eigendecomposition $\mathbf{S}_l = \mathbf{U}_l \boldsymbol{\Lambda}_l \mathbf{U}_l^H$, where $\Lambda_l = \operatorname{diag}\left(\lambda_{l,1}, \ldots, \lambda_{l,L}\right)$, $\lambda_{l,1} \geq \cdots \geq \lambda_{l,L}$. For each $\mathbf{S}_l$, we define a discrete distribution $\mathcal{D}(\mathbf{U}_l, \boldsymbol{\alpha}_l)$, as defined in Sect. 9.1.1, from which to draw random projections: $\mathbf{P}_{lt}, t = 1, \ldots, T$. Obviously, a key point for the success of the SA method is to determine a good distribution $\mathcal{D}(\mathbf{U}_l, \boldsymbol{\alpha}_l)$ and a good sampling procedure to draw random subspaces. This is described in the following.

Random Generation of Subspaces. To describe the random sampling procedure for subspace generation, let us take for simplicity the full $M \times M$ sample covariance matrix $\mathbf{S} = \mathbf{U}\boldsymbol{\Lambda}\mathbf{U}^H$, where $\boldsymbol{\Lambda} = \text{diag}(\lambda_1, \ldots, \lambda_M)$, $\lambda_1 \geq \cdots \geq \lambda_M$. Each random subspace $\langle \mathbf{V} \rangle$ has dimension $\dim(\langle \mathbf{V} \rangle) = k_{max}$, where $k_{max} < \min(M, N)$ is an overestimate of the maximum number of sources that we expect in our problem. The subspace is iteratively constructed as follows:

1. Initialize $\langle \mathbf{V} \rangle = \emptyset$
2. While $\text{rank}(\mathbf{V}) \leq k_{max}$ do
   (a) Generate a random draw $\langle \mathbf{G} \rangle \sim \mathcal{D}(\mathbf{U}, \boldsymbol{\alpha})$
   (b) $\langle \mathbf{V} \rangle = \langle \mathbf{V} \rangle \cup \langle \mathbf{G} \rangle$

The orientation matrix $\mathbf{U}$ of the distribution $\mathcal{D}$ is given by the eigenvectors of the sample covariance matrix. On the other hand, the concentration parameters should be chosen so that the signal subspace directions are selected more often than the noise subspace directions, and, consequently, they should be a function of the eigenvalues of the sample covariance λ_m. The following concentration parameters for $\mathcal{D}(\mathbf{U}, \boldsymbol{\alpha})$ were used in [128]

$$\alpha_m = \frac{\Delta\lambda_m}{\sum_k \Delta\lambda_k}, \tag{9.18}$$

where

$$\Delta\lambda_m = \begin{cases} \lambda_m - \lambda_{m+1}, & m = 1, \ldots, M-1, \\ 0, & m = M. \end{cases} \tag{9.19}$$

With this choice for $\mathcal{D}(\mathbf{U}, \boldsymbol{\alpha})$, the probability of picking the mth direction from $\mathbf{U}$ is proportional to $\lambda_m - \lambda_{m+1}$, thus placing more probability on jumps of the eigenvalue profile. Notice also that whenever $\lambda_m = \lambda_{m+1}$, then $\alpha_m = 0$, which means that $\mathbf{u}_m$ will never be chosen in any random draw. We take the convention that if $\Delta\lambda_m = 0$, $\forall m$, then we do not apply the normalization in (9.18), and hence the concentration parameters are also all zero: $\alpha_m = 0, \forall m$. A summary of the algorithm is shown in Algorithm 7.

Source Enumeration in Array Processing Through Subspace Averaging. For each subarray sample covariance matrix, we can generate T random subspaces according to the procedure described above. Since we have J subarray matrices, we get a total of JT subspaces. The SA approach simply finds the average of projection matrices

$$\overline{\mathbf{P}} = \frac{1}{JT} \sum_{l=1}^{J} \sum_{t=1}^{T} \mathbf{P}_{lt},$$

Algorithm 7: Generation of a random subspace

Input: $\mathbf{S} = \mathbf{U}_0 \mathbf{\Lambda} \mathbf{U}_0^H$, k_{max}
Output: Unitary basis for a random subspace $\mathbf{V}$
Initialization: $\mathbf{U} = \mathbf{U}_0$, $\boldsymbol{\lambda} = \text{diag}(\mathbf{\Lambda})$, and $\mathbf{V} = \emptyset$
while rank($\mathbf{V}$) $\leq k_{max}$ **do**

```
/* Generate concentration parameters α                 */
```
$M = |\boldsymbol{\lambda}|$
$\alpha_m = \frac{\Delta\lambda_m}{\sum_i \Delta\lambda_i}$, $m = 1, \ldots, M$, with $\Delta\lambda_m$ given by (9.19)
```
/* Sample from D(U, α)                                 */
```
$\mathbf{g} = [g_1 \cdots g_M]^T$, with $g_m \sim \mathcal{U}(0, 1)$
$\mathcal{I} = \{m \mid g_m \leq \alpha_m\}$
$\mathbf{G} = \mathbf{U}(:, \mathcal{I})$
```
/* Append new subspace                                 */
```
$\mathbf{V} = \begin{bmatrix}\mathbf{V} & \mathbf{G}\end{bmatrix}$
```
/* Eliminate selected directions                       */
```
$\mathbf{U} = \mathbf{U}(:, \overline{\mathcal{I}})$
$\boldsymbol{\lambda} = \boldsymbol{\lambda}(\overline{\mathcal{I}})$

Algorithm 8: Subspace averaging (SA) criterion

Input: $\mathbf{S}$, L, T and k_{max};
Output: Order estimate $\hat{k}_{SA}$
for $l = 1, \ldots, J$ **do**
Extract $\mathbf{S}_l$ from $\mathbf{S}$ and obtain $\mathbf{S}_l = \mathbf{U}_l \mathbf{\Sigma}_l \mathbf{U}_l^H$
Generate T random subspaces from $\mathbf{S}_l$ using Algorithm 7
Compute the projection matrices $\mathbf{P}_{lt} = \mathbf{V}_{lt} \mathbf{V}_{lt}^H$

Compute $\overline{\mathbf{P}}$ and its eigenvalues $(k_1, \ldots, k_L)$
Estimate $\hat{k}_{SA}$ as the number of eigenvalues of $\overline{\mathbf{P}}$ larger than 1/2

to which the order estimation method described in Sect. 9.4 may be applied. Note that the only parameters in the method are the dimension of the subarrays, L; the dimension of the extracted subspaces, k_{max}; and the number T of random subspaces extracted from each subarray. A summary of the proposed algorithm is shown in Algorithm 8.

Numerical Results. We consider a scenario with $K = 3$ narrowband incoherent unit-power signals, with DOAs separated by $\Delta = 10°$, impinging on a uniform linear array with $M = 100$ antennas and half-wavelength element separation as shown in Fig. 9.6. The number of snapshots is $N = 60$, thus yielding a rank-deficient sample covariance matrix. The Rayleigh limit for this scenario is $2\pi/M \approx 3.6°$, so in this example the sources are well separated. The proposed SA method uses subarrays of size $L = M - 5$, so the total number of subarrays is $J = 6$. From the sample covariance matrix of each subarray, we generate $T = 20$ random subspaces of dimension $k_{max} = \lfloor M/5 \rfloor$, which gives us a total of 120 subspaces on the Grassmann manifold $\mathbb{G}r(k_{max}, \mathbb{R}^L)$ to compute the average of projection matrices $\overline{\mathbf{P}}$. For the examples in this section, we define the signal-to-noise ratio as $\text{SNR} = 10\log_{10}(1/\sigma^2)$, which is the input or per-sample SNR. The SNR at the output of the array is $20\log_{10} M$ dBs higher.

Some representative methods for source enumeration with high-dimensional data and few snapshots have been selected for comparison. They exploit random matrix results and are specifically designed to operate in this regime. Further, all of them are functions of the eigenvalues $\lambda_1 \geq \cdots \geq \lambda_M$ of the sample covariance matrix $\mathbf{S}$. We now present a brief description of the methods under comparison.

- LS-MDL criterion in [179]: The standard MDL method proposed by Wax and Kailath in [374], based on a fundamental result of Anderson [14], is

$$\hat{k}_{MDL} = \underset{0\leq k\leq M-1}{\text{argmin}} \; (M-k)N\log\left(\frac{a(k)}{g(k)}\right) + \frac{1}{2}k(2M-k)\log N, \tag{9.20}$$

where $a(k)$ and $g(k)$ are the arithmetic and the geometric mean, respectively, of the $M-k$ smallest eigenvalues of $\mathbf{S}$. When the number of snapshots is smaller than the number of sensors or antennas ($N < M$), the sample covariance becomes rank-deficient and (9.20) cannot be applied directly. The LS-MDL method proposed by Huang and So in [179] replaces the noise eigenvalues λ_m in the MDL criterion by a linear shrinkage, calculated as

$$\rho_m^{(k)} = \beta^{(k)}a(k) + (1-\beta^{(k)})\lambda_m, \quad m = k+1, \ldots, M,$$

where $\beta^{(k)} = \min(1, \alpha^{(k)})$, with

$$\alpha^{(k)} = \frac{\sum_{m=k+1}^{M} \lambda_m^2 + (M-k)^2 a(k)^2}{(N+1)\left(\sum_{m=k+1}^{M} \lambda_m^2 - (M-k)a(k)^2\right)}.$$

- NE criterion in [245]: The method proposed by Nadakuditi and Edelman in [245], which we refer to as the NE criterion, is given by

$$\hat{k}_{NE} = \underset{0\leq k\leq M-1}{\text{argmin}} \left\{\frac{1}{2}\left(\frac{Nt_k}{M}\right)^2\right\} + 2(k+1),$$

where

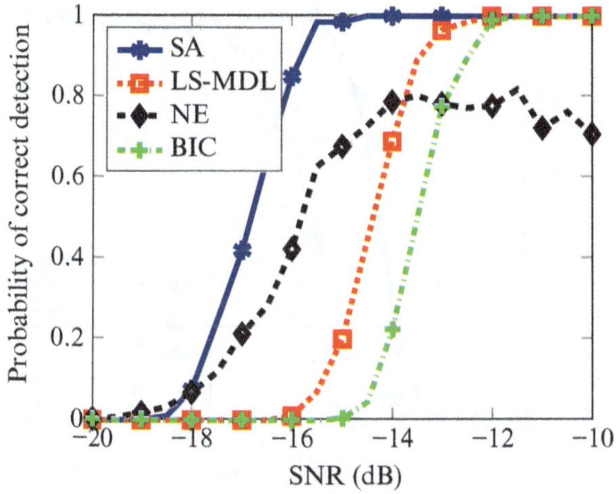

Fig. 9.8 Probability of correct detection vs. SNR for all methods. In this experiment, there are $K = 3$ sources separated by $\Delta_\theta = 10°$, the number of antennas is $M = 100$, and the number of snapshots is $N = 60$ and $L = \lfloor M - 5 \rfloor$

$$t_k = \left[\frac{\sum_{m=k+1}^{M} \lambda_m^2}{a(k)^2(M-k)} - \left(1 + \frac{M}{N}\right) \right] M.$$

- BIC method for large-scale arrays in [180]: The variant of the Bayesian information criterion (BIC) [224] for large-scale arrays proposed in [180] is

$$\hat{k}_{BIC} = \underset{0 \le k \le M-1}{\text{argmin}} \; 2(M-k)N \log\left(\frac{a(k)}{g(k)}\right) + P(k, M, N),$$

where

$$P(k, M, N) = Mk \left(\log(2N) - \frac{1}{k} \sum_{m=1}^{k} \log\left(\frac{\lambda_m}{a(k)}\right) \right).$$

Figure 9.8 shows the probability of correct detection vs. the signal-to-noise ratio (SNR) for the methods under comparison. Increasing the number of snapshots to $N = 150$ and keeping fixed the rest of the parameters, we obtain the results shown in Fig. 9.9. For this scenario, where source separations are roughly three times the Rayleigh limit, the SA method outperforms competing methods. Other examples may be found in [128].

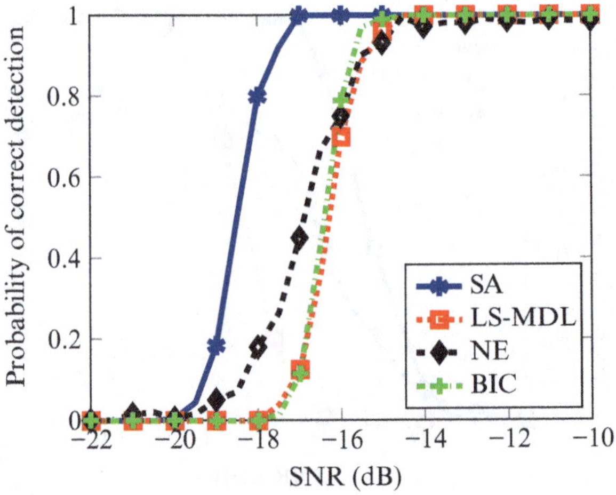

Fig. 9.9 Probability of correct detection vs. SNR for all methods. In this experiment, there are $K = 3$ sources separated by $\Delta_\theta = 10°$, the number of antennas is $M = 100$, and the number of snapshots is $N = 150$ and $L = \lfloor M - 5 \rfloor$

9.8 Chapter Notes

1. A good review of the Grassmann and Stiefel manifolds, including how to develop optimization algorithms on these Riemannian manifolds, is given in the classic paper by Edelman, Arias, and Smith [114]. A more detailed treatment of the topic can be found in the book on matrix optimization algorithms on manifolds by Absil, Mahony, and Sepulchre [4].
2. A rigorous treatment of distributions on the Stiefel and Grassmann manifolds is the book by Yasuko Chikuse [73]. Much of the material in Sect. 9.1.1 of this chapter is based on that book.
3. The application of subspace averaging techniques for order determination in array processing problems has been discussed in [128, 298, 300].
4. A robust formulation of the subspace averaging problem (9.11) is described in [128]. It uses a smooth concave increasing function of the chordal distance that saturates for large distance values so that outliers or subspaces far away from the average have a limited effect on the average. An efficient majorization-minimization algorithm [339] is proposed in [128] for solving the resulting nonconvex optimization problem.

Performance Bounds and Uncertainty Quantification

10

This chapter is addressed to performance bounding and uncertainty quantification when estimating parameters from measured data. The assumption is that measurements are drawn from a probability distribution within a known class. The actual distribution within this class is unknown because one or more parameters of the distribution are unknown.

While parameter estimation may be the goal, it seems intuitively clear that the quality of a parameter estimator will depend on the resolvability of one probability distribution from another, within the known class. It is not so clear that one should be able to bound the performance of a parameter estimator, or speak to the resolvability of distributions, without ever committing to the estimator to be used. But, in fact, this is possible, as first demonstrated by Calyampudi Radhakrishna Rao and Harald Cramér.

In 1945, C. R. Rao published his classic paper on *information and accuracy attainable in the estimation of statistical parameters* [277]. In 1946, H. Cramér independently derived some of the same results [90]. Important extensions followed in [67, 155]. The bounds on the error covariance matrix of parameter estimators first derived in [90, 277] have since become known as Cramér-Rao bounds and typically denoted CRBs. These bounds are derived by reasoning about the covariance matrix of *Fisher score*, and as a consequence, the bounds depend on *Fisher information*. But Fisher score may be replaced by other measurement scores to produce other bounds. This raises the question of what constitutes a good score. Certainly, the Fisher score is one, but there are others.

Once a score is chosen, then there are several geometries that emerge: two Hilbert space geometries and, in the case of multivariate normal (MVN) measurements, a Euclidean geometry. For Fisher score, there is an insightful information geometry.

There is the question of compression of measurements and the effect of compression on performance bounds. As expected, there is a loss of information and an increase in bounds. This issue is touched upon briefly in the chapter notes (Sect. 10.7), where the reader is directed to a paper that quantifies this loss in a representative model for measurements.

D. Ramírez et al., *Coherence*, https://doi.org/10.1007/978-3-031-13331-2_10

The story told in this chapter is a *frequentist story*, which is to say no prior distribution is assigned to unknown parameters and therefore no Bayes rule may be used to compute a posterior distribution that might be used to estimate the unknown parameters according to a loss function such as mean-squared error. Rather, the point of view is a frequentist view, where the only connection between measurements and parameters is carried in a pdf $p(\mathbf{x}; \boldsymbol{\theta})$, not in a joint pdf $p(\mathbf{x}, \boldsymbol{\theta}) = p(\mathbf{x}; \boldsymbol{\theta})p(\boldsymbol{\theta})$. The consequence is that an estimate of $\boldsymbol{\theta}$, call it $\mathbf{t}(\mathbf{x})$, amounts to a *principle of inversion* of a likelihood function $p(\mathbf{x}; \boldsymbol{\theta})$ for the parameter $\boldsymbol{\theta}$. A comprehensive account of Bayesian bounds may be found in [359], and a comparison of Bayesian and frequentist bounds may be found in [318].

We shall assume throughout this chapter that measurements are real and parameters are real. But with straightforward modifications, the story extends to complex measurements and with a little more work to complex parameters [318].

10.1 Conceptual Framework

The conceptual framework is this. A real parameter $\boldsymbol{\theta} \in \boldsymbol{\Theta} \subseteq \mathbb{R}^r$ determines a probability law $P_{\boldsymbol{\theta}}$ from which a measurement $\mathbf{x} \in \mathbb{R}^n$ is drawn. This is economical language for the statement that there is a probability space $(\mathbb{R}^n, \mathcal{B}, P_{\boldsymbol{\theta}})$ and the identity map $\mathbf{X} : \mathbb{R}^n \longrightarrow \mathbb{R}^n$ for which $Pr[\mathbf{X} \in B] = \int_B dP_{\boldsymbol{\theta}}$ for all Borel sets $B \in \mathcal{B}$. Here $\mathcal{B}$ is the Borel set of open and closed cylinder sets in $\mathbb{R}^n$. We shall assume $dP_{\boldsymbol{\theta}} = p(\mathbf{x}; \boldsymbol{\theta})d\mathbf{x}$ and interpret $p(\mathbf{x}; \boldsymbol{\theta})d\mathbf{x}$ as the probability that the random variable $\mathbf{X}$ lies in the cylinder $(\mathbf{x}, \mathbf{x} + d\mathbf{x}]$.

Fix $\boldsymbol{\theta}$. Then $p(\mathbf{x}; \boldsymbol{\theta})$ is the pdf for $\mathbf{X}$, given the parameter choice $\boldsymbol{\theta}$. Alternatively, fix a measurement $\mathbf{x}$. Then $p(\mathbf{x}; \boldsymbol{\theta})$ is termed the *likelihood* of the parameter $\boldsymbol{\theta}$, given the measurement $\mathbf{x}$. Each realization of the random variable $\mathbf{X}$ determines a different likelihood for the unknown parameter $\boldsymbol{\theta}$, so we shall speak of the *likelihood random variable* $p(\mathbf{X}; \boldsymbol{\theta})$ and its realizations $p(\mathbf{x}; \boldsymbol{\theta})$. In fact, as the story of performance bounding develops, we shall find that it is the *log-likelihood random variable* $\log p(\mathbf{X}; \boldsymbol{\theta})$ and its corresponding *Fisher score random variable* $\frac{\partial}{\partial \boldsymbol{\theta}} \log p(\mathbf{X}; \boldsymbol{\theta})$ that play the starring roles.[1]

The pdf $p(\mathbf{x}; \boldsymbol{\theta})$ may be called a *synthesis or forward model* for how the model parameters $\boldsymbol{\theta}$ determine which measurements $\mathbf{x}$ are likely and which are unlikely. The *analysis or inverse* problem is to invert a measurement $\mathbf{x}$ for the value of $\boldsymbol{\theta}$, or really the pdf $p(\mathbf{x}; \boldsymbol{\theta})$, from which the measurement was likely drawn. The principle of *maximum likelihood* takes this qualitative reasoning to what seems to be its quantitative absurdity: "if $\mathbf{x}$ is observed, then it must have been likely, and therefore let's estimate $\boldsymbol{\theta}$ to be the value that would have made $\mathbf{x}$ most likely. That is, let's

[1] In the other chapters and appendices of the book, when there was little risk of confusion, no distinction was made between a random variable X and its realization x, both usually denoted as x for scalars, $\mathbf{x}$ for vectors, and $\mathbf{X}$ for matrices. In this chapter, however, we will be more meticulous and distinguish the random variable from its realization to emphasize the fact that when dealing with bounds, such as the CRB, it is the random variables $p(\mathbf{X}; \theta)$, $\log p(\mathbf{X}; \theta)$, and $\frac{\partial}{\partial \theta_i} \log p(\mathbf{X}; \theta)$ that play a primary role.

estimate $\boldsymbol{\theta}$ to be the value that maximizes the likelihood $p(\mathbf{x}; \boldsymbol{\theta})$." As it turns out, this principle is remarkably useful as a principle for inference. It sometimes produces unbiased estimators, sometimes efficient estimators, etc. A typical application of this analysis-synthesis problem begins with $\mathbf{x}$ as a measured time series, space series, or space-time series and $\boldsymbol{\theta}$ as a set of physical parameters that account for what is measured. There is no deterministic map from $\boldsymbol{\theta}$ to $\mathbf{x}$, but there is a probability statement about which measurements are likely and which are unlikely for each candidate value of $\boldsymbol{\theta}$. This probability law is the only known connection between parameters and measurements, and from this probability law, one aims to invert a measurement for a parameter.

10.2 Fisher Information and the Cramér-Rao Bound

There is no better introduction to the general topic of performance bounds than a review of Rao's reasoning, which incidentally anticipated what is now called information geometry [11, 12, 248].

With a few variations on the notation in [277], consider a pdf $p(\mathbf{x}; \boldsymbol{\theta})$, a probability density function for the variables $x_1, \ldots, x_n$, parameterized by the unknown parameters $\theta_1, \ldots, \theta_r$. The problem is to observe the vector $\mathbf{x} = [x_1 \ \cdots \ x_n]^T$ and from it estimate the unknown parameter vector $\boldsymbol{\theta} = [\theta_1 \ \cdots \ \theta_r]^T \in \boldsymbol{\Theta}$. This estimate then becomes an estimate of the pdf from which the observations were drawn. After several important qualifiers about the role of minimum variance unbiased estimators, Rao directs his attention to unbiased estimators and asks what can be said about their efficiency, a term he associates with error covariance. We shall assume, as Rao did, that bias is zero and then account for it in a later section to obtain the more general results that remain identified as CRBs in the literature.

Consider an estimate of the parameters $\boldsymbol{\theta} \in \boldsymbol{\Theta}$, organized into the $r \times 1$ vector $\mathbf{t}(\mathbf{x}) \in \mathbb{R}^r$. The ith element of $\mathbf{t}(\mathbf{x})$, denoted $t_i(\mathbf{x})$, is the estimate of θ_i. The mean of the random variable $\mathbf{t}(\mathbf{X})$ is defined to be the vector $\mathbf{m}(\boldsymbol{\theta}) = \mathrm{E}[\mathbf{t}(\mathbf{X})]$, and this is assumed to be $\boldsymbol{\theta}$. So the estimator is unbiased. The notation $\mathrm{E}[\cdot]$ here stands for expectation with respect to the pdf $p(\mathbf{x}; \boldsymbol{\theta})$. To say the bias is zero is to say

$$\mathbf{b}(\boldsymbol{\theta}) = \int_{\mathbb{R}^n} \mathbf{t}(\mathbf{x}) p(\mathbf{x}; \boldsymbol{\theta}) d\mathbf{x} - \boldsymbol{\theta} = \mathbf{0}.$$

Differentiate the bias vector with respect to $\boldsymbol{\theta}$ to produce the $r \times r$ matrix of sensitivities

$$\frac{\partial}{\partial \boldsymbol{\theta}} \mathbf{b}(\boldsymbol{\theta}) = \int_{\mathbb{R}^n} \mathbf{t}(\mathbf{x}) \frac{\frac{\partial}{\partial \boldsymbol{\theta}} p(\mathbf{x}; \boldsymbol{\theta})}{p(\mathbf{x}; \boldsymbol{\theta})} p(\mathbf{x}; \boldsymbol{\theta}) d\mathbf{x} - \mathbf{I}_r = \mathbf{0}. \tag{10.1}$$

The derivative of $p(\mathbf{x}; \boldsymbol{\theta})$ is the $1 \times r$ vector

$$\frac{\partial}{\partial \boldsymbol{\theta}} p(\mathbf{x}; \boldsymbol{\theta}) = \left[\frac{\partial}{\partial \theta_1} p(\mathbf{x}; \boldsymbol{\theta}) \ \cdots \ \frac{\partial}{\partial \theta_r} p(\mathbf{x}; \boldsymbol{\theta}) \right].$$

The normalized $1 \times r$ vector of partial derivatives is called the *Fisher score* and denoted $\mathbf{s}^T(\mathbf{x}; \boldsymbol{\theta})$:

$$\mathbf{s}^T(\mathbf{x}; \boldsymbol{\theta}) = \frac{1}{p(\mathbf{x}; \boldsymbol{\theta})} \left[\frac{\partial}{\partial \theta_1} p(\mathbf{x}; \boldsymbol{\theta}) \cdots \frac{\partial}{\partial \theta_r} p(\mathbf{x}; \boldsymbol{\theta}) \right].$$

10.2.1 Properties of Fisher Score

Fisher score may be written as

$$\begin{aligned} \mathbf{s}^T(\mathbf{x}; \boldsymbol{\theta}) &= \left[\frac{\partial}{\partial \theta_1} \log p(\mathbf{x}; \boldsymbol{\theta}) \cdots \frac{\partial}{\partial \theta_r} \log p(\mathbf{x}; \boldsymbol{\theta}) \right] \\ &= \left[s_1(\mathbf{x}; \boldsymbol{\theta}) \cdots s_r(\mathbf{x}; \boldsymbol{\theta}) \right]. \end{aligned}$$

The term $\frac{\partial}{\partial \theta_i} \log p(\mathbf{x}; \boldsymbol{\theta}) d\theta_i$ measures the fractional change in $p(\mathbf{x}; \boldsymbol{\theta})$ due to an infinitesimal change in θ_i.

Equation (10.1) may now be written $\mathrm{E}[\mathbf{t}(\mathbf{X})\mathbf{s}^T(\mathbf{X}; \boldsymbol{\theta})] = \mathbf{I}_r$, which is to say $\mathrm{E}[t_i(\mathbf{X}) s_l(\mathbf{X}; \boldsymbol{\theta})] = \delta[i - l]$. The ith estimator is correlated only with the ith measurement score. Denoting the variance of $t_i(\mathbf{X})$ as $Q_{ii}(\boldsymbol{\theta})$ and the variance of $s_i(\mathbf{X}; \boldsymbol{\theta})$ as $J_{ii}(\boldsymbol{\theta})$, the coherence between these two random variables is $1/(Q_{ii}(\boldsymbol{\theta}) J_{ii}(\boldsymbol{\theta})) \leq 1$, and therefore $Q_{ii}(\boldsymbol{\theta}) \geq 1/J_{ii}(\boldsymbol{\theta})$. But, as we shall see, this is not generally a tight bound.

The Fisher score $\mathbf{s}(\mathbf{X}; \boldsymbol{\theta})$ is a zero-mean random vector:

$$\mathrm{E}[\mathbf{s}^T(\mathbf{X}; \boldsymbol{\theta})] = \int_{\mathbb{R}^n} \frac{\frac{\partial}{\partial \boldsymbol{\theta}} p(\mathbf{x}; \boldsymbol{\theta})}{p(\mathbf{x}; \boldsymbol{\theta})} p(\mathbf{x}; \boldsymbol{\theta}) d\mathbf{x} = \frac{\partial}{\partial \boldsymbol{\theta}} \int_{\mathbb{R}^n} p(\mathbf{x}; \boldsymbol{\theta}) d\mathbf{x} = \mathbf{0}_{1 \times r}.$$

So, in fact, $\mathrm{E}[(\mathbf{t}(\mathbf{X}) - \boldsymbol{\theta})\mathbf{s}^T(\mathbf{X}; \boldsymbol{\theta})] = \mathbf{I}_r$. The covariance of the Fisher score, denoted $\mathbf{J}(\boldsymbol{\theta})$, is defined to be $\mathrm{E}[\mathbf{s}(\mathbf{X}; \boldsymbol{\theta})\mathbf{s}^T(\mathbf{X}; \boldsymbol{\theta})]$, and it may be written as[2]

$$\mathbf{J}(\boldsymbol{\theta}) = \mathrm{E}[\mathbf{s}(\mathbf{X}; \boldsymbol{\theta})\mathbf{s}^T(\mathbf{X}; \boldsymbol{\theta})] = -\mathrm{E}\left[\frac{\partial^2}{\partial \boldsymbol{\theta}^2} \log p(\mathbf{X}; \boldsymbol{\theta}) \right].$$

This $r \times r$ matrix is called the Fisher information matrix and abbreviated as FIM.

[2] To show this, use

$$\begin{aligned} \frac{\partial^2 \log p(\mathbf{x}; \boldsymbol{\theta})}{\partial \theta_i \partial \theta_l} &= \frac{\partial}{\partial \theta_i} \left[\frac{1}{p(\mathbf{x}; \boldsymbol{\theta})} \frac{\partial p(\mathbf{x}; \boldsymbol{\theta})}{\partial \theta_l} \right] \\ &= \frac{1}{p(\mathbf{x}; \boldsymbol{\theta})} \frac{\partial^2 p(\mathbf{x}; \boldsymbol{\theta})}{\partial \theta_i \partial \theta_l} - \frac{1}{p(\mathbf{x}; \boldsymbol{\theta})} \frac{\partial p(\mathbf{x}; \boldsymbol{\theta})}{\partial \theta_i} \frac{1}{p(\mathbf{x}; \boldsymbol{\theta})} \frac{\partial p(\mathbf{x}; \boldsymbol{\theta})}{\partial \theta_l}. \end{aligned}$$

Then, assume that the order of integration and differentiation may be freely arranged and integrate with respect to $p(\mathbf{x}; \boldsymbol{\theta}) d\mathbf{x}$.

10.2.2 The Cramér-Rao Bound

Figure 10.1 describes a virtual two-channel experiment, where the *error score* $\mathbf{e}(\mathbf{X};\boldsymbol{\theta}) = \mathbf{t}(\mathbf{X}) - \boldsymbol{\theta}$ is considered a *message* and $\mathbf{s}(\mathbf{X};\boldsymbol{\theta})$ is considered a *measurement*. Each of these is a zero-mean random vector. The composite covariance matrix of these two scores is

$$\mathbf{C}(\boldsymbol{\theta}) = E\left[\begin{bmatrix}\mathbf{e}(\mathbf{X};\boldsymbol{\theta})\\ \mathbf{s}(\mathbf{X};\boldsymbol{\theta})\end{bmatrix}\begin{bmatrix}\mathbf{e}^T(\mathbf{X};\boldsymbol{\theta}) & \mathbf{s}^T(\mathbf{X};\boldsymbol{\theta})\end{bmatrix}\right] = \begin{bmatrix}\mathbf{Q}(\boldsymbol{\theta}) & \mathbf{I}_r\\ \mathbf{I}_r & \mathbf{J}(\boldsymbol{\theta})\end{bmatrix},$$

where $\mathbf{Q}(\boldsymbol{\theta}) = E[\mathbf{e}(\mathbf{X};\boldsymbol{\theta})\mathbf{e}^T(\mathbf{X};\boldsymbol{\theta})]$ is the covariance matrix of the zero-mean estimator error $\mathbf{e}(\mathbf{X};\boldsymbol{\theta})$. The Fisher information matrix $\mathbf{J}(\boldsymbol{\theta})$ is assumed to be positive definite. Therefore, this covariance matrix is non-negative definite iff the Schur complement $\mathbf{Q}(\boldsymbol{\theta}) - \mathbf{J}^{-1}(\boldsymbol{\theta}) \succeq \mathbf{0}$. That is,

$$\mathbf{Q}(\boldsymbol{\theta}) \succeq \mathbf{J}^{-1}(\boldsymbol{\theta}),$$

with equality iff $\mathbf{e}(\mathbf{X};\boldsymbol{\theta}) = \mathbf{J}^{-1}(\boldsymbol{\theta})\mathbf{s}(\mathbf{X},\boldsymbol{\theta})$. No assumption has been made about the estimator $\mathbf{t}(\mathbf{X})$, except that it is unbiased. The term $\mathbf{J}^{-1}(\boldsymbol{\theta})\mathbf{s}(\mathbf{X},\boldsymbol{\theta})$ is in fact the LMMSE estimator of the random error $\mathbf{e}(\mathbf{X};\boldsymbol{\theta})$ from the zero-mean random score $\mathbf{s}(\mathbf{X};\boldsymbol{\theta})$.

Efficiency and Maximum Likelihood. An estimator is said to be efficient with respect to Fisher score if it is unbiased and $\mathbf{Q}(\boldsymbol{\theta}) = \mathbf{J}^{-1}(\boldsymbol{\theta})$, which is to say $\mathbf{e}(\mathbf{X};\boldsymbol{\theta}) = \mathbf{J}^{-1}(\boldsymbol{\theta})\mathbf{s}(\mathbf{X};\boldsymbol{\theta})$ for all $\boldsymbol{\theta}$. That is, the covariance of an (unbiased) efficient estimator equals the CRB. Assume there exists such an estimator. Certainly, under some regularity assumptions (the fundamental assumption is that $p(\mathbf{x};\boldsymbol{\theta})$ is differentiable as a function of $\boldsymbol{\theta}$, with a derivative that is jointly continuous in $\mathbf{x}$ and $\boldsymbol{\theta}$), the condition holds at the maximum likelihood estimator of $\boldsymbol{\theta}$, denoted $\mathbf{t}_{ML}(\mathbf{X})$, where $\mathbf{s}(\mathbf{X}, \mathbf{t}_{ML}) = \mathbf{0}$. That is, $\mathbf{t}(\mathbf{X}) - \mathbf{t}_{ML}(\mathbf{X}) = \mathbf{0}$, demonstrating that if an efficient estimator exists, it is an ML estimator. It does not follow that an ML estimator is necessarily efficient.

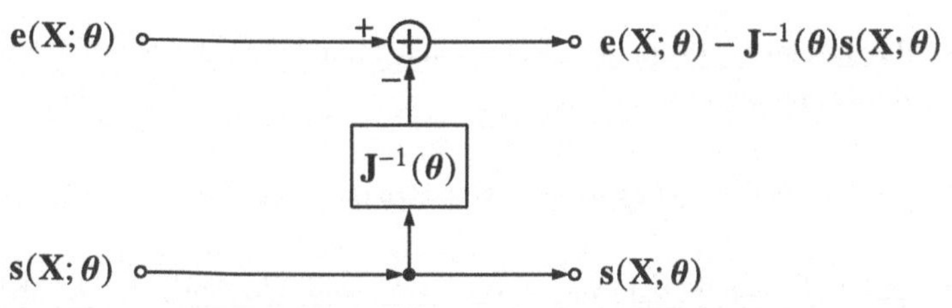

Fig. 10.1 A virtual two-channel experiment for deriving the CRB. The estimator $\mathbf{J}^{-1}(\boldsymbol{\theta})\mathbf{s}(\mathbf{X};\boldsymbol{\theta})$ is the LMMSE estimator of the error score $\mathbf{e}(\mathbf{X};\boldsymbol{\theta})$ from the Fisher score $\mathbf{s}(\mathbf{X};\boldsymbol{\theta})$

Example 10.1 Let $\mathbf{x} = [x_1 \cdots x_N]^T$ be independent samples of a Gaussian random variable with pdf $N(\theta, \sigma^2)$. The Fisher score is

$$s(\mathbf{X}; \theta) = \frac{\partial}{\partial \theta} \log p(\mathbf{X}; \theta) = \frac{N(\bar{x} - \theta)}{\sigma^2},$$

where $\bar{x} = \sum_{n=1}^{N} x_n / N$. The CRB is $J(\theta) = N/\sigma^2$, so the variance of any unbiased estimator of the mean θ is

$$\mathrm{Var}(\hat{\theta}) \geq \frac{\sigma^2}{N}.$$

The ML estimate of the mean is $\hat{\theta}_{ML} = \bar{x}$ with variance $\mathrm{Var}(\hat{\theta}_{ML}) = \frac{\sigma^2}{N}$. So CRB equality is achieved; the ML estimator is efficient.

Invariances. The CRB is invariant to transformation of Fisher score by the non-singular transformation $\mathbf{T}$. The Fisher score $\mathbf{Ts}(\mathbf{X}; \boldsymbol{\theta})$ remains zero mean with transformed covariance matrix $\mathbf{TJ}(\boldsymbol{\theta})\mathbf{T}^T$. The cross-covariance $\mathrm{E}[\mathbf{e}(\mathbf{X}; \boldsymbol{\theta})\mathbf{s}^T(\mathbf{X}; \boldsymbol{\theta})]$ transforms to $\mathbf{T}^T$, so that $\mathbf{Q}(\boldsymbol{\theta}) \succeq \mathbf{T}^T(\mathbf{TJ}(\boldsymbol{\theta})\mathbf{T}^T)^{-1}\mathbf{T} = \mathbf{J}^{-1}(\boldsymbol{\theta})$.

Estimating Functions of Parameters. Suppose it is the parameters $\mathbf{w} = \mathbf{f}(\boldsymbol{\theta})$, and not $\boldsymbol{\theta}$, that are to be estimated. Assume an inverse $\boldsymbol{\theta} = \mathbf{g}(\mathbf{w})$. The original Fisher score may be written as

$$\mathbf{s}^T(\mathbf{X}; \boldsymbol{\theta}) = \mathbf{s}^T(\mathbf{X}; \mathbf{w})\mathbf{H}; \quad \mathbf{H} = \frac{\partial}{\partial \boldsymbol{\theta}} \mathbf{w}.$$

The (i, l)th element of $\mathbf{H}$ is $(\mathbf{H})_{il} = \frac{\partial}{\partial \theta_l} w_i$. The connection between Fisher informations is $\mathbf{J}(\boldsymbol{\theta}) = \mathbf{H}^T\mathbf{J}(\mathbf{w})\mathbf{H}$. Assume $\mathbf{t_w}(\mathbf{X})$ is an unbiased estimator of $\mathbf{w}$. Then, the CRB on the covariance of the error $\mathbf{t_w}(\mathbf{X}) - \mathbf{w}$ is $\mathbf{Q}(\mathbf{w}) \succeq \mathbf{HJ}^{-1}(\boldsymbol{\theta})\mathbf{H}^T$ at $\mathbf{w} = \mathbf{f}(\boldsymbol{\theta})$. This result actually extends to maps from $\mathbb{R}^r$ to $\mathbb{R}^q$, $q < r$, provided the log-likelihood $\log p(\mathbf{x}; \mathbf{w}) = \max_{\boldsymbol{\theta} \in \mathbf{g}(\mathbf{w})} \log p(\mathbf{x}; \boldsymbol{\theta})$ is a continuous, bounded mapping from $\mathbb{R}^r$ to $\mathbb{R}^q$.

Nuisance Parameters. From the CRB we may bound the variance of any unbiased estimator of one parameter θ_i as

$$Q_{ii} = \boldsymbol{\delta}_i^T \mathbf{Q}(\boldsymbol{\theta})\boldsymbol{\delta}_i \geq \boldsymbol{\delta}_i^T \mathbf{J}^{-1}(\boldsymbol{\theta})\boldsymbol{\delta}_i = (\mathbf{J}^{-1})_{ii},$$

where $\boldsymbol{\delta}_i$ is the ith standard basis vector in $\mathbb{R}^r$ and Q_{ii} and $(\mathbf{J}^{-1})_{ii}$ denote the ith element on the diagonal of $\mathbf{Q}(\boldsymbol{\theta})$ and $\mathbf{J}^{-1}(\boldsymbol{\theta})$, respectively. We would like to show that this variance is larger than $(J_{ii})^{-1}$, which would be the variance bound if only the parameter θ_i were unknown. To this end, consider the Schwarz inequality

$(\mathbf{y}^T\mathbf{J}(\boldsymbol{\theta})\mathbf{x})^2 \le (\mathbf{y}^T\mathbf{J}(\boldsymbol{\theta})\mathbf{y})(\mathbf{x}^T\mathbf{J}(\boldsymbol{\theta})\mathbf{x})$. Choose $\mathbf{y} = \mathbf{J}^{-1}(\boldsymbol{\theta})\mathbf{x}$ and $\mathbf{x} = \boldsymbol{\delta}_i$. Then

$$1 \le (\boldsymbol{\delta}_i^T\mathbf{J}^{-1}(\boldsymbol{\theta})\boldsymbol{\delta}_i)(\boldsymbol{\delta}_i^T\mathbf{J}(\boldsymbol{\theta})\boldsymbol{\delta}_i) = (\mathbf{J}^{-1})_{ii}J_{ii},$$

or $(\mathbf{J}^{-1})_{ii} \ge (1/J_{ii})$. This means unknown nuisance parameters interfere with the estimation of the parameter θ_i.

This argument generalizes. Parse $\mathbf{J}$ and its inverse as follows:[3]

$$\mathbf{J}(\boldsymbol{\theta}) = \begin{bmatrix} \mathbf{J}_{11} & \mathbf{J}_{12} \\ \mathbf{J}_{12}^T & \mathbf{J}_{22} \end{bmatrix}$$

$$\mathbf{J}^{-1}(\boldsymbol{\theta}) = \begin{bmatrix} (\mathbf{J}_{11} - \mathbf{J}_{12}\mathbf{J}_{22}^{-1}\mathbf{J}_{12}^T)^{-1} & * \\ * & (\mathbf{J}_{22} - \mathbf{J}_{12}^T\mathbf{J}_{11}^{-1}\mathbf{J}_{12})^{-1} \end{bmatrix}.$$

$\mathbf{J}_{11}$ is the $q \times q$ Fisher matrix for a subset of the parameters of interest, $\theta_1, \ldots, \theta_q$, and $\mathbf{J}_{22}$ is the Fisher matrix for the remaining parameters, $\theta_{q+1}, \ldots, \theta_r$, which we label nuisance parameters. If only the parameters of interest corresponding to $\mathbf{J}_{11}$ were unknown, then the Fisher information matrix would be $\mathbf{J}_{11}$, and the covariance bound for estimating these parameters would be $\mathbf{J}_{11}^{-1}$. With the nuisance parameters unknown as well, the Fisher matrix is $\mathbf{J}(\boldsymbol{\theta})$, and the covariance bound is

$$\mathbf{Q}_{11}(\boldsymbol{\theta}) \succeq (\mathbf{J}_{11} - \mathbf{J}_{12}\mathbf{J}_{22}^{-1}\mathbf{J}_{12}^T)^{-1} \succeq (\mathbf{J}_{11})^{-1},$$

where $\mathbf{Q}_{11}(\boldsymbol{\theta})$ is the $q \times q$ NW corner of $\mathbf{Q}(\boldsymbol{\theta})$. Nuisance parameters increase the CRB. The increase depends on $\mathbf{J}_{11} - \mathbf{J}_{12}\mathbf{J}_{22}^{-1}\mathbf{J}_{12}^T$, which is the error covariance in estimating the primary scores from the nuisance scores, a point that is further clarified in the next section on geometry.

A Sequence of N Independent, Identically Distributed Measurements. It should be clear that when an experiment returns a sequence of N i.i.d. measurements, the governing pdf random variable is the product $\prod_{i=1}^{n} p(\mathbf{X}_i; \boldsymbol{\theta})$, in which case log-likelihoods, scores, and Fisher matrices add. Thus, the Fisher matrix is $N\mathbf{J}(\boldsymbol{\theta})$, and the CRB is $\mathbf{Q}(\boldsymbol{\theta}) \succeq \frac{1}{N}\mathbf{J}^{-1}(\boldsymbol{\theta})$. This additivity of Fisher information is one of the many arguments in its favor.

10.2.3 Geometry

There are two related geometries to be developed. The first is the geometry of error and measurement scores, and the second is the geometry of the Fisher scores.

[3] To avoid unnecessary clutter, when there is no risk of confusion, we shall sometimes write a term like $\mathbf{J}_{12}(\boldsymbol{\theta})$ as $\mathbf{J}_{12}$, suppressing the dependence on θ.

Projection of Error Scores onto the Subspace Spanned by Measurement Scores. To begin, consider one element of the unbiased error, $e_1(\mathbf{X};\boldsymbol{\theta}) = t_1(\mathbf{X}) - \theta_1$, and the Fisher scores $\mathbf{s}^T(\mathbf{X};\boldsymbol{\theta}) = [s_1(\mathbf{X};\boldsymbol{\theta}) \cdots s_r(\mathbf{X},\boldsymbol{\theta})]$. Each of these random variables may be called a vector in the Hilbert space of second-order random variables. The span of the scores, denoted as the linear space $\langle \mathbf{S} \rangle = \langle s_1(\mathbf{X};\boldsymbol{\theta}), \ldots, s_r(\mathbf{X},\boldsymbol{\theta})\rangle$, is illustrated by a plane in Fig. 10.2. The random variable $e_1(\mathbf{X};\boldsymbol{\theta})$ is illustrated as a vector lying off the plane. The ambient space is a Hilbert space of second-order random variables, where inner products between random variables are computed as expectations with respect to the pdf $p(\mathbf{x};\boldsymbol{\theta})$.

The composite covariance matrix for the error score and the measurement score is

$$\mathrm{E}\left[\begin{bmatrix} e_1(\mathbf{X};\boldsymbol{\theta}) \\ \mathbf{s}(\mathbf{X};\boldsymbol{\theta}) \end{bmatrix} \begin{bmatrix} e_1(\mathbf{X};\boldsymbol{\theta}) & \mathbf{s}^T(\mathbf{X};\boldsymbol{\theta}) \end{bmatrix}\right] = \begin{bmatrix} Q_{11} & \boldsymbol{\delta}_1^T \\ \boldsymbol{\delta}_1 & \mathbf{J}(\boldsymbol{\theta}) \end{bmatrix}.$$

The projection of the error $e_1(\mathbf{X};\boldsymbol{\theta})$ onto the span of the measurement scores is $\boldsymbol{\delta}_1^T \mathbf{J}^{-1}(\boldsymbol{\theta})\mathbf{s}(\mathbf{X};\boldsymbol{\theta})$ as illustrated in Fig. 10.2. It is easily checked that the error between $e_1(\mathbf{X};\boldsymbol{\theta})$ and its estimate in the subspace spanned by the scores is orthogonal to the subspace $\langle s_1(\mathbf{X};\boldsymbol{\theta}), \ldots, s_r(\mathbf{X},\boldsymbol{\theta})\rangle$ and the variance of this error is

$$Q_{11} - \boldsymbol{\delta}_1^T \mathbf{J}^{-1}(\boldsymbol{\theta})\boldsymbol{\delta}_1 = Q_{11} - (\mathbf{J}^{-1})_{11}.$$

The cosine-squared of the angle between the error and the subspace is $(\mathbf{J}^{-1})_{11}/Q_{11} \le 1$. The choice of parameter θ_1 is arbitrary. So the conclusion is that $(\mathbf{J}^{-1})_{ii} \le Q_{ii}$, and $(\mathbf{J}^{-1})_{ii}/Q_{ii}$ is the cosine-squared of the angle, or *coherence*, between the error score $e_i(\mathbf{X};\boldsymbol{\theta})$ and the subspace spanned by the Fisher scores.

This argument generalizes. Define the whitened error $\mathbf{u}(\mathbf{X};\boldsymbol{\theta}) = \mathbf{Q}^{-1/2}(\boldsymbol{\theta})\mathbf{e}(\mathbf{X};\boldsymbol{\theta})$. That is, $\mathrm{E}[\mathbf{u}(\mathbf{X};\boldsymbol{\theta})\mathbf{u}^T(\mathbf{X};\boldsymbol{\theta})] = \mathbf{I}_r$. The components of $\mathbf{u}(\mathbf{X};\boldsymbol{\theta})$ may be considered an orthogonal basis for the subspace $\langle \mathbf{U} \rangle = \langle u_1(\mathbf{X};\boldsymbol{\theta}), \ldots, u_r(\mathbf{X};\boldsymbol{\theta})\rangle$. Similarly, define the whitened score $\mathbf{v}(\mathbf{X};\boldsymbol{\theta}) = \mathbf{J}^{-1/2}(\boldsymbol{\theta})\mathbf{s}(\mathbf{X};\boldsymbol{\theta})$. The components of $\mathbf{v}(\mathbf{X};\boldsymbol{\theta})$ may be considered an orthogonal basis for the subspace $\langle \mathbf{V} \rangle = \langle v_1(\mathbf{X};\boldsymbol{\theta}), \ldots, v_r(\mathbf{X};\boldsymbol{\theta})\rangle$. Then, $\mathrm{E}[\mathbf{u}\mathbf{v}^T] = \mathbf{Q}(\boldsymbol{\theta})^{-1/2}\mathbf{J}(\boldsymbol{\theta})^{-1/2}$, and the SVD of this cross-correlation is $\mathbf{F}(\boldsymbol{\theta})\mathbf{K}(\boldsymbol{\theta})\mathbf{G}^T(\boldsymbol{\theta})$, with $\mathbf{F}(\boldsymbol{\theta})$ and $\mathbf{G}(\boldsymbol{\theta})$ unitary. The $r \times r$ diagonal matrix $\mathbf{K}(\boldsymbol{\theta}) = \mathrm{diag}(k_1(\boldsymbol{\theta}), \ldots, k_r(\boldsymbol{\theta}))$ extracts the $k_i(\boldsymbol{\theta})$ as cosines of the principal angles between the subspaces $\langle \mathbf{U} \rangle$ and $\langle \mathbf{V} \rangle$. The cosine-squareds, or coherences, are extracted as the eigenvalues of

$$\mathbf{Q}^{-1/2}(\boldsymbol{\theta})\mathbf{J}^{-1}(\boldsymbol{\theta})\mathbf{Q}^{-1/2}(\boldsymbol{\theta}) = \mathbf{F}(\boldsymbol{\theta})\mathbf{K}^2(\boldsymbol{\theta})\mathbf{F}^T(\boldsymbol{\theta}).$$

From the CRB, $\mathbf{Q}(\boldsymbol{\theta}) \succeq \mathbf{J}^{-1}(\boldsymbol{\theta})$, it follows that $\mathbf{Q}^{-1/2}(\boldsymbol{\theta})\mathbf{J}^{-1}(\boldsymbol{\theta})\mathbf{Q}^{-1/2}(\boldsymbol{\theta}) \preceq \mathbf{I}_r$, which is to say these cosine-squareds are less than or equal to one. Figure 10.1 may be redrawn as in Fig. 10.3. In this figure, the random variables $\boldsymbol{\mu}(\mathbf{X};\boldsymbol{\theta}) = \mathbf{F}^T(\boldsymbol{\theta})\mathbf{u}(\mathbf{X};\boldsymbol{\theta})$ and $\boldsymbol{\nu}(\mathbf{X};\boldsymbol{\theta}) = \mathbf{G}^T(\boldsymbol{\theta})\mathbf{v}(\mathbf{X};\boldsymbol{\theta})$ are canonical coordinates and

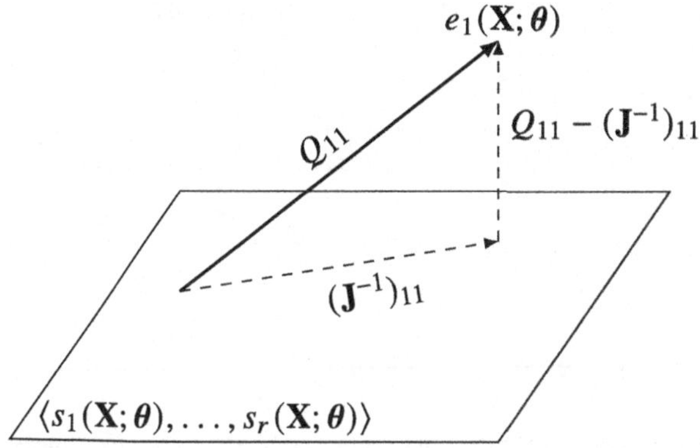

Fig. 10.2 Projection of the error score $e_1(\mathbf{X};\boldsymbol{\theta})$ onto the subspace $\langle s_1(\mathbf{X};\boldsymbol{\theta}), \ldots, s_r(\mathbf{X};\boldsymbol{\theta})\rangle$ spanned by the measurement scores. The labelings illustrate the Pythagorean decomposition of variance for the error score, Q_{11}, into its components $(\mathbf{J}^{-1})_{11}$, the variance of the projection of the error score onto the subspace, and $Q_{11} - (\mathbf{J}^{-1})_{11}$, the variance of the error in estimating the error score from the measurement scores

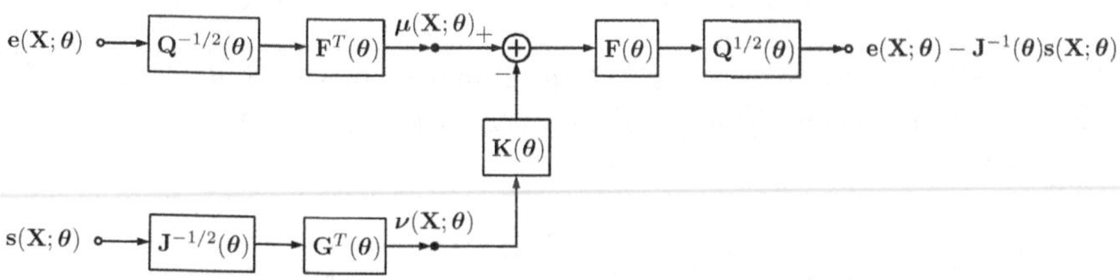

Fig. 10.3 A redrawing of Fig. 10.1 in canonical coordinates. The elements of diagonal $\mathbf{K}$ are the r principal angles between the subspaces $\langle e_1(\mathbf{X},\boldsymbol{\theta}), \ldots, e_r(\mathbf{X},\boldsymbol{\theta})\rangle$ and $\langle s_1(\mathbf{X};\boldsymbol{\theta}), \ldots, s_n(\mathbf{X};\boldsymbol{\theta})\rangle$

$E[\boldsymbol{\mu}(\mathbf{X};\boldsymbol{\theta})\boldsymbol{\nu}^T(\mathbf{X};\boldsymbol{\theta})] = \mathbf{F}^T(\boldsymbol{\theta})\mathbf{Q}^{-1/2}(\boldsymbol{\theta})\mathbf{J}^{-1/2}(\boldsymbol{\theta})\mathbf{G}(\boldsymbol{\theta}) = \mathbf{K}(\boldsymbol{\theta})$, where the $\mathbf{K}(\boldsymbol{\theta})$ are canonical correlations. The LMMSE estimator of $\boldsymbol{\mu}(\mathbf{X};\boldsymbol{\theta})$ from $\boldsymbol{\nu}(\mathbf{X};\boldsymbol{\theta})$ is $\mathbf{K}(\boldsymbol{\theta})\boldsymbol{\nu}(\mathbf{X};\boldsymbol{\theta})$, as illustrated. This factorization gives the factorization of the estimator $\mathbf{J}^{-1}(\boldsymbol{\theta})\mathbf{s}(\mathbf{X};\boldsymbol{\theta})$ as $\mathbf{Q}^{1/2}(\boldsymbol{\theta})\mathbf{F}(\boldsymbol{\theta})\mathbf{K}(\boldsymbol{\theta})\mathbf{G}^T(\boldsymbol{\theta})\mathbf{J}^{-1/2}(\boldsymbol{\theta})\mathbf{s}(\mathbf{X};\boldsymbol{\theta})$. But more importantly, it shows that the eigenvalues of the matrix $\mathbf{Q}^{-1/2}(\boldsymbol{\theta})\mathbf{J}^{-1}(\boldsymbol{\theta})\mathbf{Q}^{-1/2}(\boldsymbol{\theta})$ are cosine-squareds, or squared coherences, between the subspaces $\langle\mathbf{U}\rangle$ and $\langle\mathbf{V}\rangle$. But these principal angles are invariant to coordinate transformations within the subspaces. So these eigenvalues are the cosine-squareds of the principal angles between the subspaces $\langle e_1(\mathbf{X},\boldsymbol{\theta}), \ldots, e_r(\mathbf{X},\boldsymbol{\theta})\rangle$ and $\langle s_1(\mathbf{X};\boldsymbol{\theta}), \ldots, s_r(\mathbf{X};\boldsymbol{\theta})\rangle$.

Projection of Measurement Score $s_1(\mathbf{X};\boldsymbol{\theta})$ onto the Span of Measurement Scores $s_2(\mathbf{X};\boldsymbol{\theta}), \ldots, s_r(\mathbf{X};\boldsymbol{\theta})$. What do we expect of the cosine-squared $(\mathbf{J}^{-1})_{11}/Q_{11}$? To answer this question, parse the Fisher scores as $[s_1(\mathbf{X};\boldsymbol{\theta})\ \mathbf{s}_2^T(\mathbf{X};\boldsymbol{\theta})]$, where $\mathbf{s}_2^T(\mathbf{X};\boldsymbol{\theta}) = [s_2(\mathbf{X};\boldsymbol{\theta})\ \cdots\ s_r(\mathbf{X};\boldsymbol{\theta})]$. The Fisher matrix parses as follows:

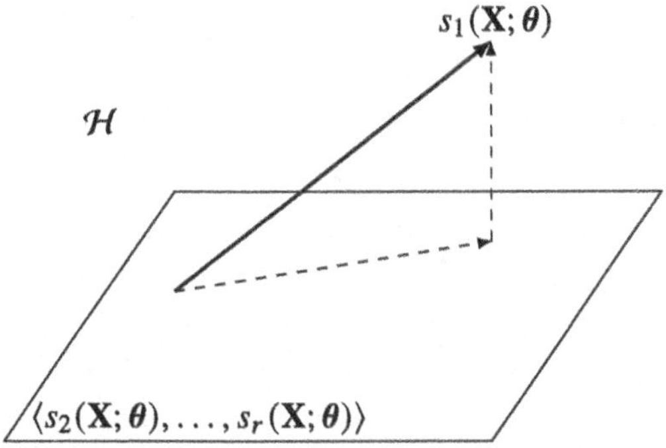

Fig. 10.4 Estimating the measurement score $s_1(\mathbf{X};\boldsymbol{\theta})$ from the measurement scores $s_2(\mathbf{X};\boldsymbol{\theta}),\ldots,s_r(\mathbf{X};\boldsymbol{\theta})$ by projecting $s_1(\mathbf{X};\boldsymbol{\theta})$ onto the subspace $\langle s_2(\mathbf{X};\boldsymbol{\theta}),\ldots,s_r(\mathbf{X};\boldsymbol{\theta})\rangle$

$$\mathbf{J}(\boldsymbol{\theta}) = \mathrm{E}\left[\begin{bmatrix} s_1(\mathbf{X},\boldsymbol{\theta}) \\ \mathbf{s}_2(\mathbf{X};\boldsymbol{\theta})\end{bmatrix}\begin{bmatrix} s_1(\mathbf{X},\boldsymbol{\theta}) \ \mathbf{s}_2^T(\mathbf{X};\boldsymbol{\theta})\end{bmatrix}\right] = \begin{bmatrix} J_{11} & \mathbf{J}_{12} \\ \mathbf{J}_{12}^T & \mathbf{J}_{22}\end{bmatrix}.$$

The LMMSE estimator of the score $s_1(\mathbf{X};\boldsymbol{\theta})$ from the scores $s_2(\mathbf{X};\boldsymbol{\theta}),\ldots,s_r(\mathbf{X};\boldsymbol{\theta})$ is $\mathbf{J}_{12}\mathbf{J}_{22}^{-1}\mathbf{s}_2(\mathbf{X};\boldsymbol{\theta})$, and the MSE of this estimator is $J_{11} - \mathbf{J}_{12}\mathbf{J}_{22}^{-1}\mathbf{J}_{12}^T$. The inverse of $\mathbf{J}(\boldsymbol{\theta})$ may be written as

$$\mathbf{J}^{-1}(\boldsymbol{\theta}) = \begin{bmatrix} (J_{11} - \mathbf{J}_{12}\mathbf{J}_{22}^{-1}\mathbf{J}_{12}^T)^{-1} & * \\ * & * \end{bmatrix}.$$

So $(\mathbf{J}^{-1})_{11} = (J_{11} - \mathbf{J}_{12}\mathbf{J}_{22}^{-1}\mathbf{J}_{12}^T)^{-1}$. In the CRB, $Q_{11} \geq (\mathbf{J}^{-1})_{11}$ is large when the MSE in estimating $s_1(\mathbf{X};\boldsymbol{\theta})$ from $\mathbf{s}_2(\mathbf{X};\boldsymbol{\theta})$ is small, which means the score $s_1(\mathbf{X};\boldsymbol{\theta})$ is nearly linearly dependent upon the remaining scores. That is, the score $s_1(\mathbf{X};\boldsymbol{\theta})$ lies nearly in the span of the scores $\mathbf{s}_2(\mathbf{X};\boldsymbol{\theta})$ as illustrated in Fig. 10.4. A variation on these geometric arguments may be found in [304].

10.3 MVN Model

In the multivariate normal measurement model, measurements are distributed as $\mathbf{X} \sim \mathcal{N}_n(\mathbf{m}(\boldsymbol{\theta}), \mathbf{R}(\boldsymbol{\theta}))$, where the mean vector and the covariance matrix are parameterized by $\boldsymbol{\theta} \in \boldsymbol{\Theta} \subseteq \mathbb{R}^r$. To simplify the points to be made, we shall consider two cases: $\mathbf{X} \sim \mathcal{N}_n(\mathbf{m}(\boldsymbol{\theta}), \sigma^2\mathbf{I}_n)$ and $\mathbf{X} \sim \mathcal{N}_n(\mathbf{0}, \mathbf{R}(\boldsymbol{\theta}))$. The first case is the parameterization of the mean only, with the covariance matrix assumed to be $\sigma^2\mathbf{I}_n$. The second is the parameterization of the covariance matrix only, with the mean assumed to be $\mathbf{0}$.

Parameterization of the Mean. Consider the MVN model $\mathbf{x} \sim \mathcal{N}_n(\mathbf{m}(\boldsymbol{\theta}), \sigma^2\mathbf{I}_n)$. The Fisher score $s_i(\mathbf{X}; \boldsymbol{\theta})$ is

$$s_i = \frac{\partial}{\partial\theta_i}\left(-\frac{1}{2\sigma^2}(\mathbf{X} - \mathbf{m}(\boldsymbol{\theta}))^T(\mathbf{X} - \mathbf{m}(\boldsymbol{\theta}))\right)$$
$$= \frac{1}{\sigma^2}(\mathbf{X} - \mathbf{m}(\boldsymbol{\theta}))^T\frac{\partial}{\partial\theta_i}\mathbf{m}(\boldsymbol{\theta})$$

These may be organized into the Fisher score $\mathbf{s}^T(\mathbf{X}; \boldsymbol{\theta}) = \frac{1}{\sigma^2}(\mathbf{X} - \mathbf{m}(\boldsymbol{\theta})^T\mathbf{G}$, where

$$\mathbf{G} = \begin{bmatrix} \frac{\partial}{\partial\theta_1}m_1(\boldsymbol{\theta}) & \dots & \frac{\partial}{\partial\theta_r}m_1(\boldsymbol{\theta}) \\ \vdots & \ddots & \vdots \\ \frac{\partial}{\partial\theta_1}m_n(\boldsymbol{\theta}) & \dots & \frac{\partial}{\partial\theta_r}m_n(\boldsymbol{\theta}) \end{bmatrix} = \begin{bmatrix}\mathbf{g}_1(\boldsymbol{\theta}) \cdots \mathbf{g}_r(\boldsymbol{\theta})\end{bmatrix} = \begin{bmatrix}\mathbf{g}_1(\boldsymbol{\theta})\ \mathbf{G}_2(\boldsymbol{\theta})\end{bmatrix},$$

with $\mathbf{G}_2(\boldsymbol{\theta}) = [\mathbf{g}_2(\boldsymbol{\theta}) \cdots \mathbf{g}_r(\boldsymbol{\theta})]$. The Fisher matrix is the Gramian

$$\mathbf{J}(\boldsymbol{\theta}) = \mathrm{E}\left[\frac{1}{\sigma^2}\mathbf{G}^T(\boldsymbol{\theta})(\mathbf{X} - \mathbf{m}(\boldsymbol{\theta}))(\mathbf{X} - \mathbf{m}(\boldsymbol{\theta}))^T\mathbf{G}(\boldsymbol{\theta})\frac{1}{\sigma^2}\right] = \frac{1}{\sigma^2}\mathbf{G}^T(\boldsymbol{\theta})\mathbf{G}(\boldsymbol{\theta}),$$

which may be written as

$$\mathbf{J}(\boldsymbol{\theta}) = \frac{1}{\sigma^2}\begin{bmatrix}\mathbf{g}_1^T\mathbf{g}_1 & \mathbf{g}_1^T\mathbf{G}_2 \\ \mathbf{G}_2^T\mathbf{g}_1 & \mathbf{G}_2^T\mathbf{G}_2\end{bmatrix} = \begin{bmatrix}J_{11} & \mathbf{J}_{12} \\ \mathbf{J}_{12}^T & \mathbf{J}_{22}\end{bmatrix}.$$

The inverse of this matrix is

$$\mathbf{J}^{-1}(\boldsymbol{\theta}) = \sigma^2\begin{bmatrix}(\mathbf{g}_1^T(\mathbf{I} - \mathbf{P}_{\mathbf{G}_2})\mathbf{g}_1)^{-1} & * \\ * & *\end{bmatrix}.$$

That is, $Q_{11} \geq (\mathbf{J}^{-1})_{11} = \sigma^2/\mathbf{g}_1^T\mathbf{P}_{\mathbf{G}_2}^{\perp}\mathbf{g}_1$. The LMMSE estimator of $s_1(\mathbf{X}; \boldsymbol{\theta})$ from $\mathbf{s}_2(\mathbf{X}; \boldsymbol{\theta})$ is $\mathbf{J}_{12}\mathbf{J}_{22}^{-1}\mathbf{s}_2(\mathbf{X}; \boldsymbol{\theta})$, and its error covariance is $\frac{1}{\sigma^2}\mathbf{g}_1^T\mathbf{P}_{\mathbf{G}_2}^{\perp}\mathbf{g}_1$. The error covariance for estimating $e_1(\mathbf{X}; \boldsymbol{\theta})$ from $\mathbf{s}_2(\mathbf{X}; \boldsymbol{\theta})$ is the inverse of this.

Had only the parameter θ_1 been unknown, the CRB would have been $(J_{11})^{-1} = \sigma^2/\mathbf{g}_1^T\mathbf{g}_1$. The sine-squared of the angle between $\mathbf{g}_1(\boldsymbol{\theta})$ and the subspace $\langle\mathbf{G}_2(\boldsymbol{\theta})\rangle$ is $\mathbf{g}_1^T\mathbf{P}_{\mathbf{G}_2}^{\perp}\mathbf{g}_1/\mathbf{g}_1^T\mathbf{g}_1$. So the ratio of the CRBs, given by $(\mathbf{J}^{-1})_{11}/(J_{11})^{-1}$, is the inverse of this sine-squared. In this case the Hilbert space geometry of Fig. 10.4 is the Euclidean geometry of Fig. 10.5. When the variation of the mean vector $\mathbf{m}(\boldsymbol{\theta})$ with respect to θ_1 lies near the variations with respect to the remaining parameters, then the sine-squared is small, the dependence of the mean value vector on θ_1 is hard to distinguish from dependence on the other parameters, and the CRB is large accordingly [304].

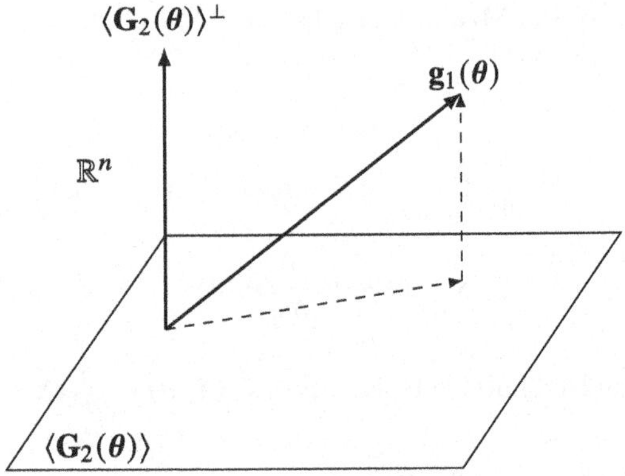

Fig. 10.5 The Euclidean space geometry of estimating measurement score $s_1(\mathbf{X};\boldsymbol{\theta})$ from measurement scores $s_2(\mathbf{X};\boldsymbol{\theta}),\ldots,s_r(\mathbf{X};\boldsymbol{\theta})$ when the Fisher matrix is the Gramian $\mathbf{J}(\boldsymbol{\theta}) = \mathbf{G}^T(\boldsymbol{\theta})\mathbf{G}(\boldsymbol{\theta})/\sigma^2$, as in the MVN model $\mathbf{X} \sim \mathcal{N}_n(\mathbf{m}(\boldsymbol{\theta}), \sigma^2\mathbf{I}_n)$

When $\mathbf{m}(\boldsymbol{\theta}) = \mathbf{H}\boldsymbol{\theta}$, with $\mathbf{H} = [\mathbf{h}_1 \ \cdots \ \mathbf{h}_r]$, then $\mathbf{g}_i(\boldsymbol{\theta}) = \mathbf{h}_i$, and it is the cosine-squareds of the angles between the columns of the model matrix $\mathbf{H}$ that determine performance.

Parameterization of the Covariance. In this case, the Fisher scores are [318]

$$s_i(\mathbf{X};\boldsymbol{\theta}) = -\operatorname{tr}\left(\mathbf{R}^{-1}(\boldsymbol{\theta})\frac{\partial}{\partial\theta_i}\mathbf{R}(\boldsymbol{\theta})\right) + \operatorname{tr}\left(\mathbf{R}^{-1}(\boldsymbol{\theta})\left(\frac{\partial}{\partial\theta_i}\mathbf{R}(\boldsymbol{\theta})\right)\mathbf{R}^{-1}(\boldsymbol{\theta})\mathbf{X}\mathbf{X}^T\right).$$

The entries in the Fisher information matrix are

$$\begin{aligned}
J_{il}(\boldsymbol{\theta}) &= \operatorname{tr}\left(\mathbf{R}^{-1}(\boldsymbol{\theta})\left(\frac{\partial}{\partial\theta_i}\mathbf{R}(\boldsymbol{\theta})\right)\mathbf{R}^{-1}(\boldsymbol{\theta})\left(\frac{\partial}{\partial\theta_l}\mathbf{R}(\boldsymbol{\theta})\right)\right)\\
&= \operatorname{tr}\left(\mathbf{R}^{-1/2}(\boldsymbol{\theta})\left(\frac{\partial}{\partial\theta_i}\mathbf{R}(\boldsymbol{\theta})\right)\mathbf{R}^{-1/2}(\boldsymbol{\theta})\mathbf{R}^{-1/2}(\boldsymbol{\theta})\left(\frac{\partial}{\partial\theta_l}\mathbf{R}(\boldsymbol{\theta})\right)\mathbf{R}^{-1/2}(\boldsymbol{\theta})\right).
\end{aligned}$$

These may be written as the inner products $J_{il}(\boldsymbol{\theta}) = \operatorname{tr}(\mathbf{D}_i(\boldsymbol{\theta})\mathbf{D}_l^T(\boldsymbol{\theta}))$ in the inner product space of Hermitian matrices, where $\mathbf{D}_i(\boldsymbol{\theta})$ are the Hermitian matrices $\mathbf{D}_i(\boldsymbol{\theta}) = \mathbf{R}^{-1/2}(\boldsymbol{\theta})\left(\frac{\partial}{\partial\theta_i}\mathbf{R}(\boldsymbol{\theta})\right)\mathbf{R}^{-1/2}(\boldsymbol{\theta})$. The Fisher matrix is again a Gramian. It may be written

$$\mathbf{J}(\boldsymbol{\theta}) = \begin{bmatrix} J_{11} & \mathbf{J}_{12} \\ \mathbf{J}_{12}^T & \mathbf{J}_{22} \end{bmatrix},$$

where $J_{11} = \text{tr}(\mathbf{D}_1(\boldsymbol{\theta})\mathbf{D}_1^T(\boldsymbol{\theta}))$, $\mathbf{J}_{21}^T = [\text{tr}(\mathbf{D}_1(\boldsymbol{\theta})\mathbf{D}_2^T(\boldsymbol{\theta})) \cdots \text{tr}(\mathbf{D}_1(\boldsymbol{\theta})\mathbf{D}_r^T(\boldsymbol{\theta}))]$, and

$$\mathbf{J}_{22} = \begin{bmatrix} \text{tr}(\mathbf{D}_2(\boldsymbol{\theta})\mathbf{D}_2^T(\boldsymbol{\theta})) & \cdots & \text{tr}(\mathbf{D}_2(\boldsymbol{\theta})\mathbf{D}_r^T(\boldsymbol{\theta})) \\ \vdots & \ddots & \vdots \\ \text{tr}(\mathbf{D}_r(\boldsymbol{\theta})\mathbf{D}_2^T(\boldsymbol{\theta})) & \cdots & \text{tr}(\mathbf{D}_r(\boldsymbol{\theta})\mathbf{D}_r^T(\boldsymbol{\theta})) \end{bmatrix}.$$

The estimator of the score $s_1(\mathbf{X};\boldsymbol{\theta})$ from the scores $s_2(\mathbf{X};\boldsymbol{\theta}),\ldots,s_r(\mathbf{X};\boldsymbol{\theta})$ is $\mathbf{J}_{12}\mathbf{J}_{22}^{-1}\mathbf{s}_2(\mathbf{X};\boldsymbol{\theta})$, and the error covariance matrix of this estimator is $J_{11}-\mathbf{J}_{12}\mathbf{J}_{22}^{-1}\mathbf{J}_{12}^T$. The estimator of $e_1(\mathbf{X};\boldsymbol{\theta})$ from the scores $\mathbf{s}(\mathbf{X};\boldsymbol{\theta})$ is $\mathbf{J}_{12}\mathbf{J}_{22}^{-1}\mathbf{s}(\mathbf{X};\boldsymbol{\theta})$, and the error covariance for this estimator is $(J_{11} - \mathbf{J}_{12}\mathbf{J}_{22}^{-1}\mathbf{J}_{12}^T)^{-1}$. This may be written as $(||\mathbf{P}_{\mathcal{D}_2(\boldsymbol{\theta})}^{\perp}\mathbf{D}_1(\boldsymbol{\theta})||)^{-1}$, where $\mathbf{P}_{\mathcal{D}_2(\boldsymbol{\theta})}\mathbf{D}_1(\boldsymbol{\theta}) = \mathbf{J}_{12}\mathbf{J}_{22}^{-1}\mathbf{D}_1(\boldsymbol{\theta})$ is the projection of $\mathbf{D}_1(\boldsymbol{\theta})$ onto the span of $\mathcal{D}_2(\boldsymbol{\theta}) = (\mathbf{D}_2(\boldsymbol{\theta}),\ldots,\mathbf{D}_r(\boldsymbol{\theta}))$. As before, Hilbert space inner products are replaced by Euclidean space inner products. Treating the $\mathbf{D}_i(\boldsymbol{\theta})$ as vectors in a vector space, the Euclidean geometry is unchanged from the geometry of Fig. 10.5. This insight is due to S. Howard in [174], where a more general account is given of the Euclidean space geometry in this MVN case.

10.4 Accounting for Bias

When the bias $\mathbf{b}(\boldsymbol{\theta}) = E[\mathbf{t}(\mathbf{X})] - \boldsymbol{\theta}$ is not zero, then the derivative of this bias with respect to parameters $\boldsymbol{\theta}$ is

$$\frac{\partial}{\partial\boldsymbol{\theta}}\mathbf{b}(\boldsymbol{\theta}) = \int \mathbf{t}(\mathbf{x})\frac{\frac{\partial}{\partial\boldsymbol{\theta}}p(\mathbf{x};\boldsymbol{\theta})}{p(\mathbf{x};\boldsymbol{\theta})}p(\mathbf{x};\boldsymbol{\theta})d\mathbf{x} - \mathbf{I}_r.$$

That is, $E[\mathbf{t}(\mathbf{X})\mathbf{s}^T(\mathbf{X};\boldsymbol{\theta})] = \mathbf{I}_r + \frac{\partial}{\partial\boldsymbol{\theta}}\mathbf{b}(\boldsymbol{\theta})$. So, in fact, $E[(\mathbf{t}(\mathbf{X}) - \boldsymbol{\mu}(\boldsymbol{\theta}))\mathbf{s}^T(\mathbf{X};\boldsymbol{\theta})] = \mathbf{I}_r+\frac{\partial}{\partial\boldsymbol{\theta}}\mathbf{b}(\boldsymbol{\theta})$, where $\boldsymbol{\mu}(\boldsymbol{\theta}) = E[\mathbf{t}(\mathbf{X})]$ is the mean of $\mathbf{t}(\mathbf{X})$. The bias is $\mathbf{b}(\boldsymbol{\theta}) = \boldsymbol{\mu}(\boldsymbol{\theta})-\boldsymbol{\theta}$.

The composite covariance matrix for the zero-mean error score $\mathbf{t}(\mathbf{X}) - \boldsymbol{\mu}(\boldsymbol{\theta})$ and the zero-mean measurement score is now

$$\begin{aligned}\mathbf{C}(\boldsymbol{\theta}) &= E\left[\begin{bmatrix}\mathbf{t}(\mathbf{X}) - \boldsymbol{\mu}(\boldsymbol{\theta}) \\ \mathbf{s}(\mathbf{X};\boldsymbol{\theta})\end{bmatrix}\begin{bmatrix}(\mathbf{t}(\mathbf{X}) - \boldsymbol{\mu}(\boldsymbol{\theta}))^T & \mathbf{s}^T(\mathbf{X};\boldsymbol{\theta})\end{bmatrix}\right] \\ &= \begin{bmatrix}\mathbf{Q}(\boldsymbol{\theta}) & \mathbf{I}_r + \frac{\partial}{\partial\boldsymbol{\theta}}\mathbf{b}(\boldsymbol{\theta}) \\ (\mathbf{I}_r + \frac{\partial}{\partial\boldsymbol{\theta}}\mathbf{b}(\boldsymbol{\theta}))^T & \mathbf{J}(\boldsymbol{\theta})\end{bmatrix},\end{aligned}$$

where $\mathbf{Q}(\boldsymbol{\theta}) = E[(\mathbf{t}(\mathbf{X}) - \boldsymbol{\mu}(\boldsymbol{\theta}))(\mathbf{t}(\mathbf{X}) - \boldsymbol{\mu}(\boldsymbol{\theta})^T)]$ is the covariance matrix of the zero-mean estimator $\mathbf{t}(\mathbf{X}) - \boldsymbol{\mu}(\boldsymbol{\theta})$. The Fisher information matrix $\mathbf{J}(\boldsymbol{\theta})$ is assumed to be positive definite. Therefore, this covariance matrix is non-negative definite if the Schur complement $\mathbf{Q}(\boldsymbol{\theta}) - (\mathbf{I}_r + \frac{\partial}{\partial\boldsymbol{\theta}}\mathbf{b}(\boldsymbol{\theta}))\mathbf{J}^{-1}(\boldsymbol{\theta})(\mathbf{I}_r + \frac{\partial}{\partial\boldsymbol{\theta}}\mathbf{b}(\boldsymbol{\theta}))^T \succeq \mathbf{0}$. That is,

$$\mathbf{Q}(\boldsymbol{\theta}) \succeq \left(\mathbf{I}_r + \frac{\partial}{\partial \boldsymbol{\theta}}\mathbf{b}(\boldsymbol{\theta})\right)\mathbf{J}^{-1}(\boldsymbol{\theta})\left(\mathbf{I}_r + \frac{\partial}{\partial \boldsymbol{\theta}}\mathbf{b}(\boldsymbol{\theta})\right)^T,$$

with equality iff $\mathbf{t}(\mathbf{X}) - \boldsymbol{\mu}(\boldsymbol{\theta}) = (\mathbf{I}_r + \frac{\partial}{\partial \boldsymbol{\theta}}\mathbf{b}(\boldsymbol{\theta}))\mathbf{J}^{-1}(\boldsymbol{\theta})\mathbf{s}(\mathbf{X}, \boldsymbol{\theta})$. It follows that the covariance matrix of the actual estimator errors $\mathbf{t}(\mathbf{X}) - \boldsymbol{\theta}$ is

$$\begin{aligned}
\mathrm{E}[(\mathbf{t}(\mathbf{X}) - \boldsymbol{\theta})(\mathbf{t}(\mathbf{X}) - \boldsymbol{\theta})^T] &= \mathbf{Q}(\boldsymbol{\theta}) + \mathbf{b}(\boldsymbol{\theta})\mathbf{b}^T(\boldsymbol{\theta}) \\
&\succeq \left(\mathbf{I}_r + \frac{\partial}{\partial \boldsymbol{\theta}}\mathbf{b}(\boldsymbol{\theta})\right)\mathbf{J}^{-1}(\boldsymbol{\theta})\left(\mathbf{I}_r + \frac{\partial}{\partial \boldsymbol{\theta}}\mathbf{b}(\boldsymbol{\theta})\right)^T + \mathbf{b}(\boldsymbol{\theta})\mathbf{b}^T(\boldsymbol{\theta}),
\end{aligned}$$

where $\mathbf{Q}(\boldsymbol{\theta})$ is the covariance of zero-mean $\mathbf{t}(\mathbf{X}) - \boldsymbol{\mu}(\boldsymbol{\theta})$ and $\mathbf{Q}(\boldsymbol{\theta}) + \mathbf{b}(\boldsymbol{\theta})\mathbf{b}^T(\boldsymbol{\theta})$ is a mean squared-error matrix for $\mathbf{t}(\mathbf{X}) - \boldsymbol{\theta}$. This is the CRB on the covariance matrix of the error $\mathbf{t}(\mathbf{X}) - \boldsymbol{\theta}$ when the bias of the estimator $\mathbf{t}(\mathbf{X})$ is $\mathbf{b}(\boldsymbol{\theta}) = \mathrm{E}[\mathbf{t}(\mathbf{X})] - \boldsymbol{\theta} \neq \mathbf{0}$. No assumption has been made about the estimator $\mathbf{t}(\mathbf{X})$, except that its mean is $\boldsymbol{\mu}(\boldsymbol{\theta})$.

All of the previous accounts of efficiency, invariances, and nuisance parameters are easily reworked with these modifications of the covariance between the zero-mean score $\mathbf{t}(\mathbf{X}) - \boldsymbol{\mu}(\boldsymbol{\theta})$ and the zero-mean score $\mathbf{s}(\mathbf{X}, \boldsymbol{\theta})$.

10.5 More General Quadratic Performance Bounds

There is no reason Fisher score may not be replaced by some other function of the pair $\mathbf{x}$ and $\boldsymbol{\theta}$, but of course any such replacement would have to be defended, a point to which we shall turn in due course. In the same vein, we may consider the estimator $\mathbf{t}(\mathbf{X})$ to be an estimator of the function $\mathbf{g}(\boldsymbol{\theta})$ with mean $\mathrm{E}[\mathbf{t}(\mathbf{X})] = \boldsymbol{\mu}(\boldsymbol{\theta}) \neq \mathbf{g}(\boldsymbol{\theta})$. Once choices for the measurement score $\mathbf{s}(\mathbf{X}; \boldsymbol{\theta})$ and the error score $\mathbf{t}(\mathbf{X}) - \boldsymbol{\mu}(\boldsymbol{\theta})$ have been made, we may appeal to the two-channel experiment of Fig. 10.1 and construct the composite covariance matrix

$$\mathrm{E}\left[\begin{bmatrix}\mathbf{t}(\mathbf{X}) - \boldsymbol{\mu}(\boldsymbol{\theta}) \\ \mathbf{s}(\mathbf{X}; \boldsymbol{\theta})\end{bmatrix}\left[(\mathbf{t}(\mathbf{X}) - \boldsymbol{\mu}(\boldsymbol{\theta}))^T \;\; \mathbf{s}^T(\mathbf{X}; \boldsymbol{\theta})\right]\right] = \begin{bmatrix}\mathbf{Q}(\boldsymbol{\theta}) & \mathbf{T}(\boldsymbol{\theta}) \\ \mathbf{T}^T(\boldsymbol{\theta}) & \mathbf{J}(\boldsymbol{\theta})\end{bmatrix}.$$

This equation defines the error covariance matrix $\mathbf{Q}(\boldsymbol{\theta})$, the *sensitivity matrix* $\mathbf{T}(\boldsymbol{\theta})$, and the information matrix $\mathbf{J}(\boldsymbol{\theta})$. The composite covariance matrix is non-negative definite, and the information matrix is assumed to be positive definite. It follows that the Schur complement $\mathbf{Q}(\boldsymbol{\theta}) - \mathbf{T}(\boldsymbol{\theta})\mathbf{J}^{-1}(\boldsymbol{\theta})\mathbf{T}^T(\boldsymbol{\theta})$ is non-negative definite, from which the quadratic covariance bound $\mathbf{Q}(\boldsymbol{\theta}) \succeq \mathbf{T}(\boldsymbol{\theta})\mathbf{J}^{-1}(\boldsymbol{\theta})\mathbf{T}^T(\boldsymbol{\theta})$ follows.

As noted by Weiss and Weinstein [375], the CRB and the bounds of Bhattacharyya [36], Barankin [23], and Bobrovsky and Zakai [40] fit this quadratic structure with appropriate choice of score.

10.5.1 Good Scores and Bad Scores

Let's conjecture that a good score should be zero mean. Add to it a non-zero perturbation $\boldsymbol{\epsilon}$ that is independent of the measurement $\mathbf{x}$. It is straightforward to show that the sensitivity matrix $\mathbf{T}$ remains unchanged by this change in score. However the information matrix is now $\mathbf{J}(\boldsymbol{\theta}) + \boldsymbol{\epsilon}\boldsymbol{\epsilon}^T$. It follows that the quadratic covariance bound $\mathbf{T}(\mathbf{J}(\boldsymbol{\theta}) + \boldsymbol{\epsilon}\boldsymbol{\epsilon}^T)^{-1}\mathbf{T}^T \preceq \mathbf{T}\mathbf{J}(\boldsymbol{\theta})^{-1}\mathbf{T}^T$, resulting in a looser bound. Any proposed score should be mean centered to improve its quadratic covariance bound [239].

As shown by Todd McWhorter in [239], a good score must be a function of a sufficient statistic $\mathbf{Z}$ for the unknown parameters. Otherwise, it may be *Rao-Blackwellized* as $\mathrm{E}[\mathbf{s}(\mathbf{X}; \boldsymbol{\theta})|\mathbf{Z}]$, where the expectation is with respect to the distribution of $\mathbf{Z}$. This Rao-Blackwellized score produces a larger quadratic covariance bound than does the original score $\mathbf{s}(\mathbf{X}; \boldsymbol{\theta})$.

It is also shown in [239] that the addition of more scores to a given score never decreases a quadratic covariance bound. In summary, a good score must be a zero mean score that is a function of a sufficient statistic for the parameters, and the more the better.

The Fisher, Barankin, and Bobrovsky Scores. The Fisher score is zero mean and a function of $p(\mathbf{X}; \boldsymbol{\theta})$, which is always a sufficient statistic. The Barankin score has components $s_i(\mathbf{X}; \boldsymbol{\theta}) = p(\mathbf{X}; \boldsymbol{\theta}_i)/p(\mathbf{X}; \boldsymbol{\theta})$, where $\boldsymbol{\theta}_i \in \boldsymbol{\Theta}$ are selected test points in $\mathbb{R}^r$. Each of these components has mean 1. Bobrovsky and Zakai center the Barankin score to obtain the score $s_i(\mathbf{X}; \boldsymbol{\theta}) - 1 = p(\mathbf{X}; \boldsymbol{\theta}_i)/p(\mathbf{X}; \boldsymbol{\theta}) - 1 = \frac{p(\mathbf{X};\boldsymbol{\theta}_i) - p(\mathbf{X};\boldsymbol{\theta})}{p(\mathbf{X};\boldsymbol{\theta})}$. So the Barankin score is a function of a sufficient statistic, but it is not zero mean. The Bobrovsky and Zakai score is a function of a sufficient statistic, and it is zero mean.

10.5.2 Properties and Interpretations

Quadratic covariance bounds are invariant to non-singular transformation of their scores. An estimator is efficient with respect to a defined score if the quadratic covariance bound is met with equality. The effect of nuisance parameters is the same as it is for Fisher score, with a different definition of the information matrix. The geometric arguments remain essentially unchanged, with one small variation: the projection of $\mathbf{t}(\mathbf{X}) - \boldsymbol{\mu}(\boldsymbol{\theta})$ onto the subspace spanned by the scores is defined as $\mathbf{T}(\boldsymbol{\theta})\mathbf{J}^{-1}(\boldsymbol{\theta})\mathbf{s}(\mathbf{X}; \boldsymbol{\theta})$, with the sensitivity matrix $\mathbf{T}(\boldsymbol{\theta}) = \mathrm{E}[(\mathbf{t}(\mathbf{X}) - \boldsymbol{\mu}(\boldsymbol{\theta}))\mathbf{s}^T(\mathbf{X}; \boldsymbol{\theta})]$ and the information matrix $\mathbf{J}(\boldsymbol{\theta}) = \mathrm{E}[\mathbf{s}(\mathbf{X}; \boldsymbol{\theta})\mathbf{s}^T(\mathbf{X}; \boldsymbol{\theta})]$ determined by the choice of score. In [239], these geometries are further refined by an integral operator representation that gives even more insight into the geometry of quadratic covariance bounds.

10.6 Information Geometry

So far, we have analyzed covariance bounds on parameter estimation, without ever addressing directly the resolvability of the underlying pdfs, or equivalently the resolvability of log-likelihoods, $\log p(\mathbf{X}; \boldsymbol{\theta})$. Therefore, we conclude this chapter with an account of the *information geometry* of log-likelihood. This requires us to consider the manifold of log-likelihoods $\{\log p(\mathbf{X}; \boldsymbol{\theta}) \mid \boldsymbol{\theta} \in \boldsymbol{\Theta} \subseteq \mathbb{R}^r$. This is a manifold, where each point on the manifold is a log-likelihood random variable. These random variables are assumed to be vectors in a Hilbert space of second-order random variables. To scan through the parameter space is to scan the manifold of log-likelihood random variables.

Begin with the manifold of parameters $\boldsymbol{\Theta}$, illustrated as the plane at the bottom of Fig. 10.6. The tangent space at point $\boldsymbol{\theta}$ is a copy of $\mathbb{R}^r$ translated to $\boldsymbol{\theta}$. More general manifolds of parameters and their tangent spaces may be considered as in [327, 328].

The function $\log p(\mathbf{X}; \boldsymbol{\theta})$ is assumed to be an injective map from the parameter manifold $\boldsymbol{\Theta}$ to the log-likelihood manifold $\mathcal{M}$. The manifold $\mathcal{M}$ is depicted in Fig. 10.6 as a curved surface. To each point on the manifold $\mathcal{M}$, attach the *tangent space* $T_{\boldsymbol{\theta}}\mathcal{M} = \left\langle \frac{\partial}{\partial \theta_1} \log p(\mathbf{X}; \boldsymbol{\theta}), \ldots, \frac{\partial}{\partial \theta_r} \log p(\mathbf{X}; \boldsymbol{\theta}) \right\rangle$. This is the linear space of

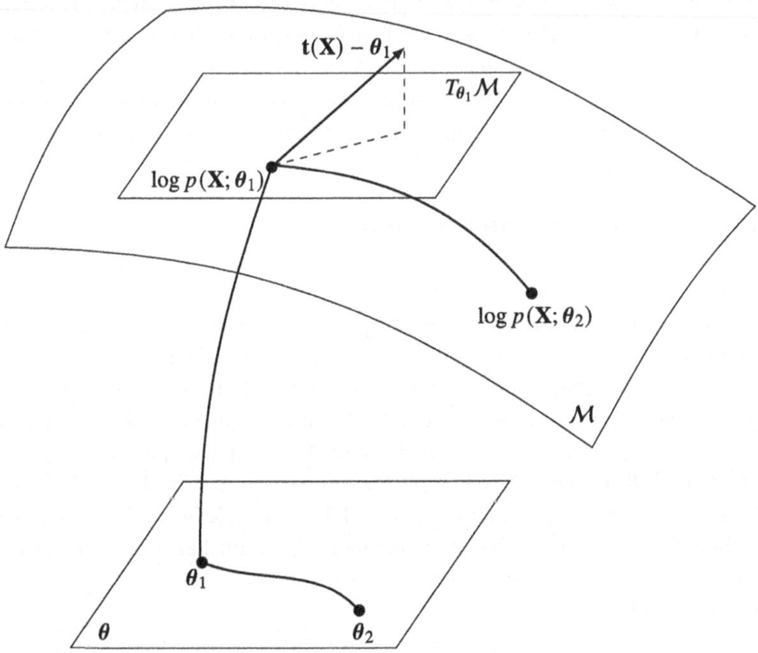

Fig. 10.6 Illustrating the interplay between the parameter space, the log-likelihood manifold, and its tangent space $T_{\boldsymbol{\theta}}\mathcal{M}$, which is the span of the Fisher scores at $\boldsymbol{\theta}$

dimension r spanned by the r Fisher scores. The tangent space is generated by passing the *derivative operator* $\left(\frac{\partial}{\partial\theta_1}, \ldots, \frac{\partial}{\partial\theta_r}\right)$ over the manifold to generate the *tangent bundle*. The tangent space $T_{\boldsymbol{\theta}}\mathcal{M}$ is then a fiber of this bundle obtained by reading the bundle at the manifold point $\log p(\mathbf{X}; \boldsymbol{\theta})$. The tangent space $T_{\boldsymbol{\theta}}\mathcal{M}$ is the set of all second-order random variables of the form $\sum_{i=1}^{r} a_i \frac{\partial}{\partial\theta_i} \log p(\mathbf{X}; \boldsymbol{\theta})$. These are tangent vectors. If a favored tangent vector is identified in each tangent plane, then the result is a *vector field* over the manifold. This vector field is called a *section* of the tangent bundle. Corresponding to the map from the manifold of parameters to the manifold of log-likelihoods, denoted $\boldsymbol{\Theta} \longrightarrow \mathcal{M}$, is the map from tangent space to tangent space, denoted $T_{\boldsymbol{\theta}}\, \boldsymbol{\Theta} \longrightarrow T_{\boldsymbol{\theta}}\mathcal{M}$. This latter map is called the *push forward* at $\boldsymbol{\theta}$ by $\log p$. It generalizes the notion of a Jacobian.

The inner product between any two tangent vectors in the subspace $T_{\boldsymbol{\theta}}\mathcal{M}$ is taken to be the inner product between second-order random variables:

$$\mathrm{E}\left[\sum_{i=1}^{r} a_i \frac{\partial}{\partial\theta_i} \log p(\mathbf{X}; \boldsymbol{\theta}) \sum_{l=1}^{r} b_l \frac{\partial}{\partial\theta_l} \log p(\mathbf{X}; \boldsymbol{\theta})\right] = \sum_{i,l=1}^{r} a_i J_{il} b_l.$$

This expectation is computed with respect to the pdf $p(\mathbf{x}; \boldsymbol{\theta})$, which is to say each tangent space $T_{\boldsymbol{\theta}}\mathcal{M}$ carries along its own definition of inner product determined by $p(\mathbf{x}; \boldsymbol{\theta})$. The norm induced by this inner product is $\sqrt{\sum_{i,l=1}^{r} a_i J_{il} a_l}$. This makes $T_{\boldsymbol{\theta}}\mathcal{M}$ an inner product space. The Fisher information matrix $\mathbf{J}(\boldsymbol{\theta})$ determines a Riemannian metric on the manifold $\mathcal{M}$ by assigning to each point $\log p(\mathbf{X}; \boldsymbol{\theta})$ on the manifold an inner product between any two vectors in the tangent space $T_{\boldsymbol{\theta}}\mathcal{M}$. The set $\{\mathbf{J}(\boldsymbol{\theta}) \mid \boldsymbol{\theta} \in \boldsymbol{\Theta}\}$ is a matrix-valued function on $\boldsymbol{\Theta}$ that induces a Riemannian metric tensor on $\mathcal{M}$. It generalizes the Hessian.

The incremental distance between two values of log-likelihood $\log p(\mathbf{X}; \boldsymbol{\theta} + d\boldsymbol{\theta})$ and $\log p(\mathbf{X}; \boldsymbol{\theta})$ may be modeled to first order as $\sum_{i=1}^{r} \frac{\partial}{\partial\theta_i} \log p(\mathbf{X}; \boldsymbol{\theta}) d\theta_i$. The square of this distance is the expectation $d\boldsymbol{\theta}^T \mathbf{J}(\boldsymbol{\theta}) d\boldsymbol{\theta}$. This is the norm-squared induced on the parameter manifold $\boldsymbol{\Theta}$ by the map $\log p$. As illustrated in Fig. 10.6, pick two points on the manifold $\mathcal{M}$, $\log p(\mathbf{X}; \boldsymbol{\theta}_1)$ and $\log p(\mathbf{X}; \boldsymbol{\theta}_2)$. Define a route between them along the trajectory $\log p(\mathbf{X}; \boldsymbol{\theta}(t))$, with $t \in [0, 1]$, $\boldsymbol{\theta}(0) = \boldsymbol{\theta}_1$, and $\boldsymbol{\theta}(1) = \boldsymbol{\theta}_2$. The distance traveled on the manifold is

$$d(\log p(\mathbf{X}; \boldsymbol{\theta}_1), \log p(\mathbf{X}; \boldsymbol{\theta}_2)) = \int_{\boldsymbol{\theta}(t), t \in [0,1]} \sqrt{d\boldsymbol{\theta}^T(t) \mathbf{J}(\boldsymbol{\theta}(t)) d\boldsymbol{\theta}(t)}.$$

This is an integral along a path in parameter space of the metric $\sqrt{d\boldsymbol{\theta}^T \mathbf{J}(\boldsymbol{\theta}) d\boldsymbol{\theta}}$ induced by the transformation $\log p(\mathbf{X}; \boldsymbol{\theta})$. A fanciful path in $\boldsymbol{\Theta}$ is illustrated at the bottom of Fig. 10.6. If there is a minimum distance over all paths $\boldsymbol{\theta}(t)$, with $t \in [0, 1]$, it is called the geodesic distance between the two log-likelihoods. It is *not* generally the KL divergence between the likelihoods $p(\mathbf{X}; \boldsymbol{\theta}_1)$ and $p(\mathbf{X}; \boldsymbol{\theta}_2)$, and it is not generally determined by a straight-line path from $\boldsymbol{\theta}_1$ to $\boldsymbol{\theta}_2$ in $\boldsymbol{\Theta}$.

Summary. So, what has information geometry brought to our understanding of parameter estimation? It has enabled us to interpret the Fisher information matrix as a metric on the manifold of log-likelihood random variables. This metric then determines the intrinsic distance between two log-likelihoods. This gives a global picture of the significance of the FIM $\mathbf{J}(\boldsymbol{\theta})$, $\boldsymbol{\theta} \in \boldsymbol{\Theta}$, a picture that is not otherwise apparent. But, perhaps, there is a little more intuition to be had.

To the point $\log p(\mathbf{X}; \boldsymbol{\theta}_1)$, we may attach the estimator error $\mathbf{t}(\mathbf{X}) - \boldsymbol{\theta}_1$, as illustrated in Fig. 10.6. This vector of second-order random variables lies off the tangent plane. The LMMSE estimator of $\mathbf{t}(\mathbf{X}) - \boldsymbol{\theta}_1$ from the Fisher scores $s_1(\mathbf{X}; \boldsymbol{\theta}_1), \ldots, s_r(\mathbf{X}; \boldsymbol{\theta}_1)$, is the projection onto the tangent plane $T_{\boldsymbol{\theta}_1}\mathcal{M}$, namely, $\mathbf{J}^{-1}(\boldsymbol{\theta}_1)\mathbf{s}(\mathbf{X}; \boldsymbol{\theta}_1)$. The error covariance matrix is bounded as $\mathbf{Q}(\boldsymbol{\theta}_1) \succeq \mathbf{J}^{-1}(\boldsymbol{\theta}_1)$. The tangent space $T_{\boldsymbol{\theta}_1}\mathcal{M}$ is invariant to transformation of coordinates, so the projection of $\mathbf{t}(\mathbf{X}) - \boldsymbol{\theta}_1$ onto this subspace is invariant to a transformation of coordinates in the tangent space.

Upstairs, in the tangent space, one reasons as one reasons in the two-channel representation of error score and measurement score. Downstairs on the manifold, the Fisher information matrix determines intrinsic distance between any two log-likelihoods. This intrinsic distance is a path integral in the parameter space, with a metric induced by the map $\log p(\mathbf{X}; \boldsymbol{\theta})$. This metric, the Fisher information matrix, is the Hessian of the transformation $\log p(\mathbf{X}; \boldsymbol{\theta})$ from $\boldsymbol{\Theta}$ to $\mathcal{M}$.

So, we have come full circle: the second-order reasoning and LMMSE estimation in the Hilbert space of second-order random variables produced the CRB. There is a two-channel representation. When this second-order picture is attached to the tangent space at a point on the manifold of log-likelihood random variables, then Fisher scores are seen to be a basis for the tangent space. The Fisher information matrix $\mathbf{J}(\boldsymbol{\theta})$ determines inner products between tangent vectors in $T_{\boldsymbol{\theta}}\mathcal{M}$, it determines the Riemannian metric on the manifold, and it induces a metric on the parameter space. This is the metric that determines the intrinsic distance between two log-likelihood random variables.

The MVN Model for Intuition. Suppose $\mathbf{X} \sim \mathcal{N}_n(\mathbf{H}\boldsymbol{\theta}, \mathbf{R})$. Then the measurement score is $\mathbf{s}(\mathbf{X}; \boldsymbol{\theta}) = \mathbf{H}^T\mathbf{R}^{-1}(\mathbf{X} - \mathbf{H}\boldsymbol{\theta})$, and the covariance matrix of this score is the Fisher matrix $\mathbf{J} = \mathbf{H}^T\mathbf{R}^{-1}\mathbf{H}$, which is independent of $\boldsymbol{\theta}$. The ML estimator of $\boldsymbol{\theta}$ is $\mathbf{t}(\mathbf{X}) = (\mathbf{H}^T\mathbf{R}^{-1}\mathbf{H})^{-1}\mathbf{H}^T\mathbf{R}^{-1}\mathbf{X}$, and its expected value is $\boldsymbol{\theta}$. Thus $\mathbf{t}(\mathbf{X}) - \boldsymbol{\theta} = (\mathbf{H}^T\mathbf{R}^{-1}\mathbf{H})^{-1}\mathbf{H}^T\mathbf{R}^{-1}(\mathbf{X} - \mathbf{H}\boldsymbol{\theta}) = \mathbf{J}^{-1}\mathbf{s}(\mathbf{X}; \boldsymbol{\theta})$. This makes $\mathbf{t}(\mathbf{X})$ efficient, with error covariance matrix $\mathbf{Q}(\boldsymbol{\theta}) = \mathbf{J}^{-1}$. The induced metric on $\boldsymbol{\Theta}$ is $d\boldsymbol{\theta}^T\mathbf{J}d\boldsymbol{\theta}$, and the distance between the distributions $\mathcal{N}_n(\mathbf{H}\boldsymbol{\theta}_1, \mathbf{R})$ and $\mathcal{N}_n(\mathbf{H}\boldsymbol{\theta}_2, \mathbf{R})$ is

$$d(\log p(\mathbf{X}; \boldsymbol{\theta}_1), \log p(\mathbf{X}; \boldsymbol{\theta}_2)) = \int_{\boldsymbol{\theta}(t), t\in[0,1]} \sqrt{d\boldsymbol{\theta}^T(t)(\mathbf{H}^T\mathbf{R}^{-1}\mathbf{H})d\boldsymbol{\theta}(t)}.$$

The minimizing path is $\boldsymbol{\theta}(t) = \boldsymbol{\theta}_1 + t(\boldsymbol{\theta}_2 - \boldsymbol{\theta}_1)$, in which case $d\boldsymbol{\theta}(t) = dt(\boldsymbol{\theta}_2 - \boldsymbol{\theta}_1)$. The distance between distributions is then

$$d(\log p(\mathbf{X};\boldsymbol{\theta}_1), \log p(\mathbf{X};\boldsymbol{\theta}_2)) = \int_{t=0}^{1} \sqrt{(\boldsymbol{\theta}_2 - \boldsymbol{\theta}_1)^T(\mathbf{H}^T\mathbf{R}^{-1}\mathbf{H})(\boldsymbol{\theta}_2 - \boldsymbol{\theta}_1)dt^2}$$
$$= \sqrt{(\boldsymbol{\theta}_2 - \boldsymbol{\theta}_1)^T(\mathbf{H}^T\mathbf{R}^{-1}\mathbf{H})(\boldsymbol{\theta}_2 - \boldsymbol{\theta}_1)}.$$

It is not hard to show that this is also the KL divergence between two distributions $\mathcal{N}_n(\mathbf{H}\boldsymbol{\theta}_1, \mathbf{R})$ and $\mathcal{N}_n(\mathbf{H}\boldsymbol{\theta}_2, \mathbf{R})$. This is a special case.

10.7 Chapter Notes

The aim of this chapter has been to bring geometric insight to the topic of Fisher information and the Cramér-Rao bound and to extend this insight to a more general class of quadratic performance bounds. The chapter has left uncovered a vast number of related topics in performance bounding. Among them we identify and annotate the following:

1. **Bayesian bounds.** When a prior distribution is assigned to the unknown parameters $\boldsymbol{\theta}$, then the FIM is replaced by the so-called Fisher-Bayes information matrix, and the CRB is replaced by the so-called Fisher-Bayes bound. These variations are treated comprehensively in the edited volume [359], which contains original results by the editors. A comparison of Fisher-Bayes bounds and CRBs may be found in [318].
2. **Constraints and more general parameter spaces.** When there are constraints on the parameters to be identified, then these constraints may be accounted for in a modified CRB. Representative papers are [143, 233, 335].
3. **CRBs on manifolds.** CRBs for nonlinear parameter estimation on manifolds have been derived and applied in [327]. These bounds have been generalized in [328] to a broad class of quadratic performance bounds on manifolds. These are intrinsic bounds within the Weiss-Weinstein class of quadratic bounds, and they extend the Bhattacharyya, Barankin, and Bobrovsky-Zakai bounds to manifolds.
4. **Efficiency.** The CRB and its related quadratic performance bounds make no claim to tightness, except in those cases where the ML estimator is efficient. In many problems of signal processing and machine learning, the parameters of interest are *mode parameters*, such as frequency, wavenumber, direction of arrival, etc. For these problems, it is typical to find that the CRB is tight at high SNR, but far from tight at SNR below a threshold value where the CRB is said to breakdown. In an effort to study this threshold and to predict performance below threshold, Richmond [286] used a method of intervals due to Van Trees [357] to predict performance. The net effect is that the CRB, augmented with the method of intervals, is a useful way to predict performance in those problems for which the CRB does not accurately predict performance below a threshold SNR.
5. **Model mismatch.** CRBs, and their generalization to the general class of quadratic performance bounds, assume a known distribution for measurements. This raises the question of performance when the measurements are drawn

from a distribution that is not matched to the assumed distribution. Richmond and Horowitz [287] extend Huber's so-called sandwich inequality [181] for the performance of ML to sandwich inequalities for the CRB and other quadratic performance bounds under conditions of model mismatch.

6. **Compression and its consequences.** How much information is retained (or lost) when n measurements $\mathbf{x}$ are linearly compressed as $\mathbf{\Phi x}$, where $\mathbf{\Phi}$ is an $m \times n$ matrix and $m < n$. One way to address this question is to analyze the effect of random compression on the Fisher information matrix and the CRB when the measurements are drawn from the MVN distribution $\mathbf{x} \sim CN_n(\boldsymbol{\mu}(\boldsymbol{\theta}), \mathbf{I}_n)$ and the random matrix is drawn from the class of distributions that is invariant to right-unitary transformations. That is, the distribution of $\mathbf{\Phi U}$ is the distribution of $\mathbf{\Phi}$ for $\mathbf{U}$ an $n \times n$ unitary matrix. These include i.i.d. draws of spherically invariant matrix rows, including, for example, i.i.d. draws of standard complex normal matrix elements. In [257], it is shown that the resulting random Fisher matrix and CRB, suitably normalized, are distributed as matrix Beta random matrices. Concentration ellipses quantify loss in performance.

11 Variations on Coherence

In this chapter, we illustrate the use of coherence and its generalizations to compressed sensing, multiset CCA, kernel methods, and time-frequency analysis. The concept of coherence in compressed sensing and matrix completion quantifies the idea that signals or matrices having a sparse representation in one domain must be spread out in the domain in which they are acquired. Intuitively, this means that the basis used for sensing and the basis used for representation should have low coherence. For example, the Kronecker delta pulse is maximally incoherent with discrete-time sinusoids. The intuition is that the energy in a Kronecker pulse is uniformly spread over sinusoids of all frequencies. Random matrices used for sensing are essentially incoherent with any other basis used for signal representation. These intuitive ideas are made clear in compressed sensing problems by the restricted isometry property and the concept of coherence index, which are discussed in this chapter.

We also consider in this chapter multiview learning, in which the aim is to extract a low-dimensional latent subspace from a series of views of a common information source. The basic tool for fusing data from different sources is multiset canonical correlation analysis (MCCA). In the two-channel case, we have seen in Sect. 3.9 that squared canonical correlations are coherences between unit-variance uncorrelated canonical variates. In the multiset case, several extensions and generalizations of coherence are possible. Popular MCCA formulations include the sum of correlations (SUMCOR) and the maximum variance (MAXVAR) approaches.

Coherence is a measure that can be extended to any reproducing kernel Hilbert space (RKHS). The notion of kernel-induced vector spaces is the cornerstone of kernel-based machine learning techniques. We present in this chapter two kernel methods in which coherence between pairs of nonlinearly transformed vectors plays a prominent role: kernelized CCA (KCCA) and kernel LMS adaptive filtering (KLMS).

In the last section of the chapter, it is argued that complex coherence between values of a time series and values of its Fourier transform defines a useful *complex* time-frequency distribution with *real* non-negative marginals.

D. Ramírez et al., *Coherence*, https://doi.org/10.1007/978-3-031-13331-2_11

11.1 Coherence in Compressed Sensing

Compressed sensing is a mature signal processing technique for the efficient recovery of sparse signals from a reduced set of measurements. For the existence of a unique reconstruction of a sparse signal, the measurement matrix has to satisfy certain incoherence conditions. These incoherence conditions are typically based on the restricted isometry property (RIP) of the measurement matrix, or on the *coherence index* obtained from its Gramian. As we shall see, the concept of coherence therefore plays a fundamental role in determining uniqueness in compressed sensing and sparse recovery problems. The objective of this section is to illustrate this point.

Compressed Sensing Problem. A vector $\mathbf{x} \in \mathbb{C}^N$ is said to be k-sparse if $\|\mathbf{x}\|_0 \leq k$, with $k < N$. In a noiseless situation, the compressed sensing scheme produces $M \ll N$ measurements as

$$\mathbf{y} = \mathbf{A}\mathbf{x},$$

where $\mathbf{A} = [\mathbf{a}_1 \ \cdots \ \mathbf{a}_N]$ is an $M \times N$ measurement or sensing matrix. Without loss of generality, we assume that the measurement matrix is normalized to have unit-norm columns, $\|\mathbf{a}_n\|_2^2 = 1, \forall n$. We also assume that the N columns of $\mathbf{A}$ form a frame in $\mathbb{C}^M$, so there are positive real numbers $0 < A \leq B < \infty$ such that

$$A \leq \frac{\|\mathbf{A}\mathbf{x}\|_2^2}{\|\mathbf{x}\|_2^2} \leq B.$$

Let $\mathsf{k} = \{n_1, \ldots, n_k\}$ denote the non-zero positions of $\mathbf{x}$, and let $\mathbf{A}_\mathsf{k}$ be an $M \times k$ submatrix of the full measurement matrix $\mathbf{A}$ formed by the columns that correspond to the positions of the non-zero elements of $\mathbf{x}$ that are selected by the set k. Call k a support set.

Uniqueness of the Solution. If the set k is known, for the recovery of the non-zero elements $\mathbf{x}_\mathsf{k}$, the system of equations $\mathbf{y} = \mathbf{A}_\mathsf{k}\mathbf{x}_\mathsf{k}$ needs to be solved. For $M \geq k$, the LS solution is

$$\mathbf{x}_\mathsf{k} = \left(\mathbf{A}_\mathsf{k}^H \mathbf{A}_\mathsf{k}\right)^{-1} \mathbf{A}_\mathsf{k}^H \mathbf{y}, \tag{11.1}$$

where $\mathbf{A}_\mathsf{k}^H \mathbf{A}_\mathsf{k}$ is the $k \times k$ Gramian matrix whose elements are inner products between the columns of $\mathbf{A}_\mathsf{k}$. It is then clear that the existence of a solution to this problem requires $\text{rank}(\mathbf{A}_\mathsf{k}^H \mathbf{A}_\mathsf{k}) = k$.[1]

[1] For noisy measurements, the robustness of the rank condition is achieved if the condition number of $\mathbf{A}_\mathsf{k}^H \mathbf{A}_\mathsf{k}$, denoted as $\text{cond}(\mathbf{A}_\mathsf{k}^H \mathbf{A}_\mathsf{k})$, is close to unity.

If the set K is unknown, a direct approach to check the uniqueness of the solution would be to consider all possible $\binom{N}{k}$ combinations of the k possible non-zero positions out of N and find the LS solution (11.1). The solution of the problem is unique if there is only one support set K that produces zero error $\mathbf{y}_{\mathsf{K}} - \mathbf{A}_{\mathsf{K}}\mathbf{x}_{\mathsf{K}}$. This direct approach is infeasible in practice for obvious computational reasons.

An alternative approach to the study of uniqueness of the solution is the following. Consider a k-sparse vector $\mathbf{x}$ such that its k-dimensional reduced form, $\mathbf{x}_{\mathsf{K}}$, is a solution to $\mathbf{y} = \mathbf{A}_{\mathsf{K}}\mathbf{x}_{\mathsf{K}}$. Assume that the solution is not unique. Then, there exists a different k-dimensional vector, $\mathbf{x}_{\mathsf{K}'}$, with non-zero elements at different positions $\mathsf{K}' = \{n'_1, \ldots, n'_k\}$, with $\mathsf{K}' \cap \mathsf{K} = \emptyset$, such that $\mathbf{y} = \mathbf{A}_{\mathsf{K}'}\mathbf{x}_{\mathsf{K}'}$. Then

$$\mathbf{A}_{\mathsf{K}}\mathbf{x}_{\mathsf{K}} - \mathbf{A}_{\mathsf{K}'}\mathbf{x}_{\mathsf{K}'} = \mathbf{A}_{2\mathsf{K}}\mathbf{x}_{2\mathsf{K}} = \mathbf{0}, \tag{11.2}$$

where $\mathbf{A}_{2\mathsf{K}}$ is now an $M \times 2k$ submatrix formed by $2k$ columns of $\mathbf{A}$ and $\mathbf{x}_{2\mathsf{K}}$ is a $2k$-dimensional vector. A nontrivial solution of (11.2) indicates that the k-sparse solution of the system is not unique. If $\operatorname{rank}(\mathbf{A}_{2\mathsf{K}}^H\mathbf{A}_{2\mathsf{K}}) = 2k$ for all $\binom{N}{2k}$ combinations of the $2k$ non-zero positions out of N, the solution is unique. The RIP is tested in a similar combinatorial way by checking the norm $\|\mathbf{A}_{2\mathsf{K}}\mathbf{x}_{2\mathsf{K}}\|_2^2/\|\mathbf{x}_{2\mathsf{K}}\|_2^2$ instead of $\operatorname{rank}(\mathbf{A}_{2\mathsf{K}}^H\mathbf{A}_{2\mathsf{K}})$.

The Restricted Isometry Property (RIP). Begin with the usual statement of RIP(k, ϵ) [56]: for $\mathbf{A} \in \mathbb{C}^{M\times N}$, $M < N$ and for all k-sparse $\mathbf{x}$,

$$(1-\epsilon)\|\mathbf{x}\|_2^2 \le \|\mathbf{A}\mathbf{x}\|_2^2 \le (1+\epsilon)\|\mathbf{x}\|_2^2. \tag{11.3}$$

The RIP constant is the smallest ϵ such that (11.3) holds for all k-sparse vectors. For ϵ not too close to one, the measurement matrix $\mathbf{A}$ approximately preserves the Euclidean norm of k-sparse signals, which in turn implies that k-sparse vectors cannot be in the null space of $\mathbf{A}$, since otherwise there would be no hope of uniquely reconstructing these vectors.

This intuitive idea can be formalized by saying that a k-sparse solution is unique if the measurement matrix satisfies the RIP$(2k, \epsilon)$ condition for a value of the constant ϵ sufficiently smaller than one. When this property holds, all pairwise distances between k-sparse signals are well preserved in the measurement space. That is,

$$(1-\epsilon)\|\mathbf{x}_1 - \mathbf{x}_2\|_2^2 \le \|\mathbf{A}\mathbf{x}_1 - \mathbf{A}\mathbf{x}_2\|_2^2 \le (1+\epsilon)\|\mathbf{x}_1 - \mathbf{x}_2\|_2^2,$$

holds for all k-sparse vectors $\mathbf{x}_1$ and $\mathbf{x}_2$. The RIP$(2k, \epsilon)$ condition can be rewritten this way: for each of the $\binom{N}{2k}$ $2k$-column subsets $\mathbf{A}_{2\mathsf{K}}$ of $\mathbf{A}$, and for all $2k$-vectors $\mathbf{x}_{2\mathsf{K}}$,

$$(1-\epsilon)\|\mathbf{x}_{2\mathsf{K}}\|_2^2 \le \|\mathbf{A}_{2\mathsf{K}}\mathbf{x}_{2\mathsf{K}}\|_2^2 \le (1+\epsilon)\|\mathbf{x}_{2\mathsf{K}}\|_2^2.$$

The constant ϵ can be related to the principal angles between columns of $\mathbf{A}_{2k}$ this way: Isolate $\mathbf{a}$, an arbitrary column of $\mathbf{A}_{2k}$, and let $\mathbf{B}$ denote the remaining columns of $\mathbf{A}_{2k}$. Reorder $\mathbf{A}_{2k}$ as $\mathbf{A}_{2k} = [\mathbf{a}\ \mathbf{B}]$. Now choose $\mathbf{x}_{2k}$ to be the $2k$-vector $\mathbf{x}_{2k} = (\mathbf{A}_{2k}^H\mathbf{A}_{2k})^{-1/2}\mathbf{e}$. The resulting RIP$(2k, \epsilon)$ condition is, for all $2k$-dimensional vectors $\mathbf{e}$,

$$(1-\epsilon)\mathbf{e}^H(\mathbf{A}_{2k}^H\mathbf{A}_{2k})^{-1}\mathbf{e} \le \mathbf{e}^H\mathbf{e} \le (1+\epsilon)\mathbf{e}^H(\mathbf{A}_{2k}^H\mathbf{A}_{2k})^{-1}\mathbf{e}.$$

The Gramian $\mathbf{A}_{2k}^H\mathbf{A}_{2k}$ is structured. So its inverse may be written as

$$(\mathbf{A}_{2k}^H\mathbf{A}_{2k})^{-1} = \begin{bmatrix} \frac{1}{\mathbf{a}^H\mathbf{P}_{\mathbf{B}}^{\perp}\mathbf{a}} & * \\ * & * \end{bmatrix},$$

where $\mathbf{P}_{\mathbf{B}}^{\perp} = \mathbf{I} - \mathbf{P}_{\mathbf{B}}$, with $\mathbf{P}_{\mathbf{B}} = \mathbf{B}(\mathbf{B}^H\mathbf{B})^{-1}\mathbf{B}^H$. Choose $\mathbf{e}$ to be the first standard basis vector and write, for all $\mathbf{A}_{2k} \in \mathbb{C}^{M\times 2k}$ and for every decomposition of $\mathbf{A}_{2k}$,

$$(1-\epsilon)\frac{1}{\mathbf{a}^H\mathbf{P}_{\mathbf{B}}^{\perp}\mathbf{a}} \le 1 \le (1+\epsilon)\frac{1}{\mathbf{a}^H\mathbf{P}_{\mathbf{B}}^{\perp}\mathbf{a}}.$$

According to our definition of coherence in Chap. 3, (see Eq. (3.2)), $\mathbf{a}^H\mathbf{P}_{\mathbf{B}}^{\perp}\mathbf{a}/\mathbf{a}^H\mathbf{a}$ (or simply $\mathbf{a}^H\mathbf{P}_{\mathbf{B}}^{\perp}\mathbf{a}$ since we are considering a measurement matrix with unit-norm columns, i.e., $\|\mathbf{a}\|_2^2 = 1$) is the coherence or cosine-squared of the principal angle between the subspaces $\langle\mathbf{a}\rangle$ and $\langle\mathbf{B}^{\perp}\rangle$, and hence it is the sine-squared of the principal angle between the subspaces $\langle\mathbf{a}\rangle$ and $\langle\mathbf{B}\rangle$. So we write

$$(1-\epsilon) \le \sin^2\theta \le (1+\epsilon). \tag{11.4}$$

The upper bound is trivial, but the lower bound is not. A small value of ϵ in the RIP condition RIP$(2k, \epsilon)$ ensures the angle between $\langle\mathbf{a}\rangle$ and $\langle\mathbf{B}\rangle$ is close to $\pi/2$ for all submatrices $\mathbf{A}_{2k}$. In practice, to verify the RIP constant ϵ is small would require checking $\binom{N}{2k}$ combinations of the submatrices $\mathbf{A}_{2k}$ and $2k$ principal angles for each submatrix. For moderate size problems, this is computationally prohibitive, which motivates the use of a computationally feasible criterion such as the *coherence index*.

The Coherence Index. The coherence index is defined as [105]

$$\rho = \max_{\substack{n,l \\ n\neq l}} |\langle\mathbf{a}_n, \mathbf{a}_l\rangle| = \max_{\substack{n,l \\ n\neq l}} |\cos\theta_{nl}|\,.$$

So, for normalized matrices, the coherence index is the maximum absolute off-diagonal element of the Gramian $\mathbf{A}^H\mathbf{A}$, or the maximum absolute value of the cosine of the angle between the columns of $\mathbf{A}$. If the sensing matrix does not have unit norm columns, the coherence index is

$$\rho = \max_{\substack{n,l \\ n \neq l}} \frac{|\langle \mathbf{a}_n, \mathbf{a}_l \rangle|}{\|\mathbf{a}_n\|_2 \|\mathbf{a}_l\|_2}.$$

The reconstruction of a k-sparse signal is unique if the coherence index ρ satisfies [105]

$$k < \frac{1}{2}\left(1 + \frac{1}{\rho}\right). \tag{11.5}$$

The computation of the coherence index requires only $\binom{N}{2}$ evaluations of angle cosines between pairs of columns of the sensing matrix, which makes it the preferred way to assess the quality of a measurement matrix. Condition (11.5) is usually proved based on the Gershgorin disk theorem. A more intuitive derivation is provided in [333]. Let us begin by considering the initial estimate

$$\mathbf{x}_0 = \mathbf{A}^H \mathbf{y} = \mathbf{A}^H \mathbf{A} \mathbf{x},$$

which can be used to estimate the non-zero positions of the k-sparse signal $\mathbf{x}$. The off-diagonal terms of the Gramian $\mathbf{A}^H\mathbf{A}$ should be as small as possible compared to the unit diagonal elements to ensure that the largest k elements of $\mathbf{x}_0$ coincide with the non-zero elements in $\mathbf{x}$. Consider the case where $k = 1$ and the only non-zero element of $\mathbf{x}$ is at position n_1. To correctly detect this position from the largest element of $\mathbf{x}_0$, the coherence index must satisfy $\rho < 1$. Assume now that the signal $\mathbf{x}$ is 2-sparse. In this case, the correct non-zero positions in $\mathbf{x}$ will always be detected if the original unit amplitude reduced by ρ is greater than the maximum possible disturbance 2ρ. That is, $1 - \rho > 2\rho$. Following the same argument, for a general k-sparse signal, the position of the largest element of $\mathbf{x}$ will be correctly detected in $\mathbf{x}_0$ if

$$1 - (k-1)\rho > k\rho,$$

which yields the uniqueness condition (11.5).

Welch Bounds. Begin with a frame or sensing matrix $\mathbf{A} \in \mathbb{C}^{M\times N}$ with unit-norm columns (signals of unit energy), $||\mathbf{a}_n||_2^2 = 1$, $n = 1, \ldots, N$. Assume $N \geq M$. Construct the rank-deficient Gramian $\mathbf{G} = \mathbf{A}^H\mathbf{A}$ and note that $\mathrm{tr}(\mathbf{G}) = N$. The fundamental Welch bound is [377]

$$N^2 = \mathrm{tr}^2(\mathbf{G}) = \left(\sum_{m=1}^{M} ev_m(\mathbf{G})\right)^2 \overset{(a)}{\leq} M \sum_{m=1}^{M} ev_m^2(\mathbf{G}) = M\,\mathrm{tr}(\mathbf{G}^H\mathbf{G}),$$

where (a) is the Cauchy-Schwarz inequality. This lower bounds the sum of the squares of the magnitude of inner products:

$$\operatorname{tr}(\mathbf{G}^H\mathbf{G}) = \sum_{n,l=1}^{N} |\langle \mathbf{a}_n, \mathbf{a}_l \rangle|^2 \geq \frac{N^2}{M}. \tag{11.6}$$

Equivalently, a lower bound on the sum of the off-diagonal terms of the Gramian $\mathbf{G}$ is

$$N + \sum_{n \neq l} |\langle \mathbf{a}_n, \mathbf{a}_l \rangle|^2 \geq \frac{N^2}{M} \quad \Rightarrow \quad \sum_{n \neq l} |\langle \mathbf{a}_n, \mathbf{a}_l \rangle|^2 \geq \frac{N(N-M)}{M}.$$

Since the mean of a set of non-negative numbers is smaller than their maximum, i.e.,

$$\frac{1}{N(N-1)} \sum_{n \neq l} |\langle \mathbf{a}_n, \mathbf{a}_l \rangle|^2 \leq \max_{\substack{n,l \\ n \neq l}} |\langle \mathbf{a}_n, \mathbf{a}_l \rangle|^2 = \rho^2,$$

it follows that the Welch bound is also a lower bound for the coherence index of any frame or, equivalently, for how small the cross correlation of a set of signals of unit energy can be. That is,

$$\rho^2 \geq \frac{N-M}{M(N-1)}.$$

This bound can be generalized to

$$\rho^{2\alpha} \geq \frac{1}{N-1} \left[\frac{N}{\binom{M+\alpha-1}{\alpha}} - 1 \right],$$

as shown in [377].

From the cyclic property of the trace, it follows that

$$\operatorname{tr}(\mathbf{G}^H\mathbf{G}) = \operatorname{tr}(\mathbf{A}^H\mathbf{A}\mathbf{A}^H\mathbf{A}) = \operatorname{tr}(\mathbf{A}\mathbf{A}^H\mathbf{A}\mathbf{A}^H) = \operatorname{tr}(\mathbf{F}\mathbf{F}^H)$$

where $\mathbf{F} = \mathbf{A}\mathbf{A}^H$ can be interpreted as an $M \times M$ row Gramian since its elements are inner products between rows of $\mathbf{A}$. This simple idea can be used to show that the Welch bound in (11.6) holds with equality if $\mathbf{F} = \frac{N}{M}\mathbf{I}_M$ [234]. A frame of unit vectors $\mathbf{A}$ for which $\mathbf{F} = \mathbf{A}\mathbf{A}^H = \frac{N}{M}\mathbf{I}_M$ is said to be *tight* or a Welch Bound Equality (WBE) frame. A frame is WBE if it is tight. This is a tautology, as WBE and tightness are identical definitions. In this case,

$$\operatorname{tr}(\mathbf{G}^H\mathbf{G}) = \operatorname{tr}(\mathbf{F}\mathbf{F}^H) = \frac{N^2}{M^2}\operatorname{tr}(\mathbf{I}_M) = \frac{N^2}{M}.$$

The vectors $\mathbf{a}_1, \ldots, \mathbf{a}_N$ form a tight frame, in which case for $K = M/N$,

$$\|\mathbf{z}\|_2^2 = K \sum_{n=1}^{N} |\langle \mathbf{z}, \mathbf{a}_n \rangle|^2, \quad \forall \mathbf{z} \in \mathbb{C}^M.$$

Then, a good choice for $\mathbf{A}$ will be a WBE frame [95]. Additionally, if $|\langle \mathbf{a}_n, \mathbf{a}_l \rangle|^2$ are equal for all (n, l), with $n \neq l$, then the frame is said to be equiangular. From a geometrical perspective, an equiangular tight frame is a set of unit norm vectors forming equal angles, therefore, having identical correlation, which is also the smallest possible. This special structure makes equiangular tight frames (ETFs) particularly important in signal processing, communications, and quantum information theory. However, ETFs do not exist for arbitrary frame dimensions, and even if they exist their construction may be difficult. Tight frames that are not equiangular are not necessarily good.

Coherence in Matrix Completion. Closely related to the compressed sensing setting is the problem of recovering a low-rank matrix given very few linear functionals of the matrix (e.g., a subset of the entries of the matrix). In general, there is no hope to recover *all* low-rank matrices from a subset of sampled entries. Think, for instance, of a matrix with one entry equal to 1 and all other entries equal to zero. Clearly, this rank-one matrix cannot be recovered unless we observe all its entries. As we have seen in the compressed sensing problem, to be able to recover a low-rank matrix, this matrix cannot be in the null space of the "sampling operator" that records a subset of the matrix entries. Candès and Recht studied this problem in [57] and showed that the number of observations needed to recover a low-rank matrix $\mathbf{X}$ is small when its singular vectors are spread, that is, uncorrelated or incoherent with the standard basis. This intuitive idea is made concrete in [57] with the definition of the *coherence* for matrix completion problems.

Definition 11.1 (Candès and Recht) Let $\mathbf{X} \in \mathbb{R}^{n \times n}$ be a rank-r matrix and let $\langle \mathbf{X} \rangle$ be its column space, which is a subspace of dimension r with orthogonal projection matrix $\mathbf{P}_r$. Then, the coherence between $\langle \mathbf{X} \rangle$ and the standard Euclidean basis is defined to be

$$\rho^2(\langle \mathbf{X} \rangle) = \frac{n}{r} \max_{1 \leq i \leq n} \|\mathbf{P}_r \mathbf{e}_i\|^2. \tag{11.7}$$

Note that the coherence, as defined in (11.7), takes values in $1 \leq \rho^2(\langle \mathbf{X} \rangle) \leq \frac{n}{r}$. The smallest coherence is achieved, for example, when the column space of $\mathbf{X}$ is spanned by vectors whose entries all have magnitude $1/\sqrt{n}$. The largest possible coherence value is n/r, which would correspond to any subspace that contains a standard basis element. In matrix completion problems, the interest is in matrices whose row and column spaces have low coherence, since these matrices cannot be in the null space of the sampling operator.

11.2 Multiset CCA

Two-channel CCA can be extended to allow for the simultaneous consideration of multiple datasets offering distinct views of the common information sources or factors. Work on this topic dates back to Horst [171] and to the classic work by J. R. Kettering in the early 1970s [195], who studied the most popular multiset or multiview CCA (MCCA) generalizations, namely, the maximum correlation method, or MAXVAR, and the sum of pairwise correlations method, or SUMCOR. Before describing these methods, let us briefly review two-channel CCA. The reader can find a more detailed description of CCA in Sect. 3.9.

11.2.1 Review of Two-Channel CCA

In two-channel CCA, we consider a pair of zero-mean random vectors $\mathbf{x}_1 \in \mathbb{C}^{d_1}$ and $\mathbf{x}_2 \in \mathbb{C}^{d_2}$. The goal is to find linear transformations of the observations to d-dimensional canonical coordinates or canonical variates $\mathbf{z}_1 = \mathbf{U}_1^H \mathbf{x}_1$ and $\mathbf{z}_2 = \mathbf{U}_2^H \mathbf{x}_2$, such that $E[\mathbf{z}_1 \mathbf{z}_2^H] = \text{diag}(k_1, \ldots, k_d)$ and $E[\mathbf{z}_1 \mathbf{z}_1^H] = E[\mathbf{z}_2 \mathbf{z}_2^H] = \mathbf{I}_d$. The dimension d is $d = \min(d_1, d_2)$.

The columns of $\mathbf{U}_1 = [\mathbf{u}_{11} \; \cdots \; \mathbf{u}_{1d}] \in \mathbb{C}^{d_1 \times d}$ and $\mathbf{U}_2 = [\mathbf{u}_{21} \; \cdots \; \mathbf{u}_{2d}] \in \mathbb{C}^{d_2 \times d}$ are the canonical vectors. For example, the first canonical vectors of $\mathbf{U}_1$ and $\mathbf{U}_2$ maximize the correlation between the random variables $z_{11} = \mathbf{u}_{11}^H \mathbf{x}_1$ and $z_{21} = \mathbf{u}_{21}^H \mathbf{x}_2$, subject to $E[|z_{11}|^2] = E[|z_{21}|^2] = 1$. Clearly, this is equivalent to maximizing the squared coherence or squared correlation coefficient

$$\rho^2 = \frac{|\mathbf{u}_{11}^H \mathbf{R}_{12} \mathbf{u}_{21}|^2}{\left(\mathbf{u}_{11}^H \mathbf{R}_{11} \mathbf{u}_{11}\right)\left(\mathbf{u}_{21}^H \mathbf{R}_{22} \mathbf{u}_{21}\right)},$$

where $\mathbf{R}_{il} = E[\mathbf{x}_i \mathbf{x}_l^H]$. This is k_1^2.

In Sect. 3.9, we saw that the canonical vectors and the canonical correlations are given, respectively, by the singular vectors and singular values of the coherence matrix

$$\mathbf{C}_{12} = E\left[(\mathbf{R}_{11}^{-1/2} \mathbf{x}_1)(\mathbf{R}_{22}^{-1/2} \mathbf{x}_2)^H\right] = \mathbf{R}_{11}^{-1/2} \mathbf{R}_{12} \mathbf{R}_{22}^{-1/2} = \mathbf{F} \mathbf{K} \mathbf{G}^H.$$

More concretely, $\mathbf{U}_1 = \mathbf{R}_{11}^{-1/2} \mathbf{F}$ and $\mathbf{U}_2 = \mathbf{R}_{22}^{-1/2} \mathbf{G}$, so that the canonical variates are $\mathbf{z}_1 = \mathbf{U}_1^H \mathbf{x}_1 = \mathbf{F}^H \mathbf{R}_{11}^{-1/2} \mathbf{x}_1$ and $\mathbf{z}_2 = \mathbf{U}_2^H \mathbf{x}_2 = \mathbf{G}^H \mathbf{R}_{22}^{-1/2} \mathbf{x}_2$.

In practice, only samples of the random vectors $\mathbf{x}_1$ and $\mathbf{x}_2$ are observed. Let $\mathbf{X}_1 \in \mathbb{C}^{d_1 \times N}$ and $\mathbf{X}_2 \in \mathbb{C}^{d_2 \times N}$ be matrices containing as columns the samples of $\mathbf{x}_1$ and $\mathbf{x}_2$, respectively. The canonical vectors and canonical correlations[2] are obtained through

[2] These are *sample* canonical vectors and *sample* canonical correlations, but the qualifier sample is dropped when there is no risk of confusion between population (true) canonical correlations and sample canonical correlations.

the SVD of the sample coherence matrix $\hat{\mathbf{C}}_{12} = (\mathbf{X}_1\mathbf{X}_1^H)^{-1/2}\mathbf{X}_1\mathbf{X}_2^H(\mathbf{X}_2\mathbf{X}_2^H)^{-1/2}$. With some abuse of notation, we will also denote the SVD of the sample coherence matrix as $\hat{\mathbf{C}}_{12} = \mathbf{FKG}^H$.

Two-Channel CCA as a Generalized Eigenvalue Problem. Let $\mathbf{U}_1 = (\mathbf{X}_1\mathbf{X}_1^H)^{-1/2}\mathbf{F}$ and $\mathbf{U}_2 = (\mathbf{X}_2\mathbf{X}_2^H)^{-1/2}\mathbf{G}$ be loading matrices with the canonical vectors as columns, where $\mathbf{F}$ and $\mathbf{G}$ are the left and right singular vectors of the sample coherence matrix. Let us take the ith column of $\mathbf{U}_1$ and of $\mathbf{U}_2$ and form the $2d \times 1$ column vector $\mathbf{v}_i = [\mathbf{u}_{1i}^T\ \mathbf{u}_{2i}^T]^T$. The CCA solution satisfies

$$\begin{aligned}(\mathbf{X}_1\mathbf{X}_2^H)\mathbf{U}_2 &= \mathbf{X}_1\mathbf{X}_2^H(\mathbf{X}_2\mathbf{X}_2^H)^{-1/2}\mathbf{G} = (\mathbf{X}_1\mathbf{X}_1^H)^{1/2}\underbrace{(\mathbf{X}_1\mathbf{X}_1^H)^{-1/2}\mathbf{X}_1\mathbf{X}_2^H(\mathbf{X}_2\mathbf{X}_2^H)^{-1/2}}_{\hat{\mathbf{C}}_{12}=\mathbf{FKG}^H}\mathbf{G} \\ &= (\mathbf{X}_1\mathbf{X}_1^H)^{1/2}\mathbf{FK} = (\mathbf{X}_1\mathbf{X}_1^H)(\mathbf{X}_1\mathbf{X}_1^H)^{-1/2}\mathbf{FK} = (\mathbf{X}_1\mathbf{X}_1^H)\mathbf{U}_1\mathbf{K},\end{aligned}$$

and similarly $(\mathbf{X}_2\mathbf{X}_1^H)\mathbf{U}_1 = (\mathbf{X}_2\mathbf{X}_2^H)\mathbf{U}_2\mathbf{K}$. This means that $(\mathbf{v}_i, k_i)$ is a generalized eigenvector-eigenvalue pair of the GEV problem

$$\begin{bmatrix} \mathbf{0} & \mathbf{X}_1\mathbf{X}_2^H \\ \mathbf{X}_2\mathbf{X}_1^H & \mathbf{0} \end{bmatrix}\mathbf{v} = \lambda \begin{bmatrix} \mathbf{X}_1\mathbf{X}_1^H & \mathbf{0} \\ \mathbf{0} & \mathbf{X}_2\mathbf{X}_2^H \end{bmatrix}\mathbf{v}. \tag{11.8}$$

This can be formulated in terms of the matrices

$$\mathbf{S} = \begin{bmatrix} \mathbf{X}_1\mathbf{X}_1^H & \mathbf{X}_1\mathbf{X}_2^H \\ \mathbf{X}_2\mathbf{X}_1^H & \mathbf{X}_2\mathbf{X}_2^H \end{bmatrix} \quad \text{and} \quad \mathbf{D} = \begin{bmatrix} \mathbf{X}_1\mathbf{X}_1^H & \mathbf{0} \\ \mathbf{0} & \mathbf{X}_2\mathbf{X}_2^H \end{bmatrix},$$

as

$$(\mathbf{S} - \mathbf{D})\mathbf{v} = \lambda\mathbf{D}\mathbf{v}. \tag{11.9}$$

The generalized eigenvalues of (11.8) or (11.9) are $\lambda_i = \pm k_i$. We assume they are ordered as $\lambda_1 \geq \cdots \geq \lambda_d \geq \lambda_{d+1} \geq \cdots \geq \lambda_{2d}$, with $k_i = \lambda_i = -\lambda_{d+i}$. A scaled version of the canonical vectors is extracted from the generalized eigenvectors corresponding to positive eigenvalues in the eigenvector matrix $\mathbf{V} = [\mathbf{v}_1\ \cdots\ \mathbf{v}_d]$. This scaling is irrelevant since the canonical correlations are not affected by scaling, either together or independently, the canonical vectors $\mathbf{u}_{1i}$ and $\mathbf{u}_{2i}$, $i = 1, \ldots, d$. The eigenvector matrix $\mathbf{V} = [\mathbf{v}_1\ \cdots\ \mathbf{v}_d]$ obtained by solving (11.8) or (11.9) satisfies $\mathbf{V}^H\mathbf{D}\mathbf{V} = \mathbf{I}_d$. So the canonical vectors extracted from $\mathbf{V} = [\mathbf{U}_1^T\ \mathbf{U}_2^T]^T$ satisfy in turn

$$\mathbf{V}^H\mathbf{D}\mathbf{V} = \mathbf{U}_1^H(\mathbf{X}_1\mathbf{X}_1^H)\mathbf{U}_1 + \mathbf{U}_2^H(\mathbf{X}_2\mathbf{X}_2^H)\mathbf{U}_2 = \mathbf{I}_d,$$

and the canonical vectors obtained through the SVD of the coherence matrix satisfy $\mathbf{U}_1^H(\mathbf{X}_1\mathbf{X}_1^H)\mathbf{U}_1 = \mathbf{I}_d$, and $\mathbf{U}_2^H(\mathbf{X}_2\mathbf{X}_2^H)\mathbf{U}_2 = \mathbf{I}_d$.

Optimization Problems for Two-Channel CCA. Two-channel CCA solves the following optimization problem [157]

$$\begin{aligned} \text{P1:} \quad & \underset{\mathbf{U}_1,\mathbf{U}_2}{\text{minimize}} && \left\| \mathbf{U}_1^H \mathbf{X}_1 - \mathbf{U}_2^H \mathbf{X}_2 \right\|^2, \\ & \text{subject to} && \mathbf{U}_i^H \mathbf{X}_i \mathbf{X}_i^H \mathbf{U}_i = \mathbf{I}_d, \quad i = 1, 2. \end{aligned} \tag{11.10}$$

So CCA minimizes the distance between linear transformations $\mathbf{U}_1^H\mathbf{X}_1$ and $\mathbf{U}_2^H\mathbf{X}_2$ subject to norm constraints of the form $\mathbf{U}_i^H\mathbf{X}_i\mathbf{X}_i^H\mathbf{U}_i = \mathbf{I}_d$. It is easy to check that problem (11.10) is equivalent to

$$\begin{aligned} \text{P1:} \quad & \underset{\mathbf{U}_1,\mathbf{U}_2}{\text{maximize}} && \operatorname{tr}\left(\mathbf{U}_1^H \mathbf{X}_1 \mathbf{X}_2^H \mathbf{U}_2 \right), \\ & \text{subject to} && \mathbf{U}_i^H \mathbf{X}_i \mathbf{X}_i^H \mathbf{U}_i = \mathbf{I}_d, \quad i = 1, 2. \end{aligned} \tag{11.11}$$

The solution of P1 is obtained for canonical vectors $\mathbf{U}_1 = (\mathbf{X}_1\mathbf{X}_1^H)^{-1/2}\mathbf{F}$ and $\mathbf{U}_2 = (\mathbf{X}_2\mathbf{X}_2^H)^{-1/2}\mathbf{G}$, in which case the trace function in (11.11) attains its maximum value $\operatorname{tr}\left(\mathbf{U}_1^H\mathbf{X}_1\mathbf{X}_2^H\mathbf{U}_2\right) = \sum_{i=1}^d k_i$ and the distance between linear transformations in (11.10) attains its minimum value $\left\|\mathbf{U}_1^H\mathbf{X}_1 - \mathbf{U}_2^H\mathbf{X}_2\right\|^2 = 2(d - \sum_{i=1}^d k_i)$. This formulation points to the SUMCOR-CCA generalization to multiple datasets, as we shall see.

Instead of minimizing the distance between $\mathbf{U}_1^H\mathbf{X}_1$ and $\mathbf{U}_2^H\mathbf{X}_2$, one can look for a d-dimensional common subspace that approximates in some optimal manner the row spaces of the transformations $\mathbf{U}_1^H\mathbf{X}_1$ and $\mathbf{U}_2^H\mathbf{X}_2$. Let $\mathbf{V}_d \in \mathbb{S}t(d, \mathbb{C}^N)$ be a unitary basis for such a central subspace. The two-channel CCA solution then solves the problem

$$\begin{aligned} \text{P2:} \quad & \underset{\mathbf{U}_1,\mathbf{U}_2,\mathbf{V}_d}{\text{minimize}} && \sum_{i=1}^{2} \left\| \mathbf{U}_i^H \mathbf{X}_i - \mathbf{V}_d^H \right\|^2, \\ & \text{subject to} && \mathbf{V}_d^H \mathbf{V}_d = \mathbf{I}_d. \end{aligned} \tag{11.12}$$

For a fixed central subspace $\mathbf{V}_d$, the $\mathbf{U}_i^H$ minimizers are $\mathbf{U}_i = (\mathbf{X}_i\mathbf{X}_i^H)^{-1}\mathbf{X}_i\mathbf{V}_d$, $i = 1, 2$. Substituting these values in (11.12), the best d-dimensional subspace $\mathbf{V}_d$ that explains the canonical variates subspace is obtained by solving

$$\underset{\mathbf{V}_d \in \mathbb{S}t(d,\mathbb{C}^N)}{\text{minimize}} \quad \operatorname{tr}\left(\mathbf{V}_d^H \overline{\mathbf{P}} \mathbf{V}_d\right), \tag{11.13}$$

where $\overline{\mathbf{P}}$ is an average of orthogonal projection matrices onto the columns spaces of $\mathbf{X}_1^H$ and $\mathbf{X}_2^H$, namely,

$$\overline{\mathbf{P}} = \frac{1}{2}(\mathbf{P}_1 + \mathbf{P}_2) = \frac{1}{2}\left[\mathbf{X}_1^H(\mathbf{X}_1\mathbf{X}_1^H)^{-1}\mathbf{X}_1 + \mathbf{X}_2^H(\mathbf{X}_2\mathbf{X}_2^H)^{-1}\mathbf{X}_2\right],$$

with eigendecomposition $\overline{\mathbf{P}} = \mathbf{W}\boldsymbol{\Lambda}\mathbf{W}^H$. The problem of finding the central subspace according to an extrinsic distance measure was discussed in detail in Chap. 9. There, we saw that the solution to (11.13) is given by the first d eigenvectors of $\mathbf{W}$, that is, $\mathbf{V}_d^* = \mathbf{W}_d = [\mathbf{w}_1 \cdots \mathbf{w}_d]$. When solving the CCA problem this way, we obtain scaled versions of the canonical vectors. In particular, if the canonical vectors are taken as $\mathbf{U}_i = (\mathbf{X}_i\mathbf{X}_i^H)^{-1}\mathbf{X}_i\mathbf{W}_d$, where $\mathbf{W}_d$ is the central subspace that minimizes (11.13), then

$$\mathbf{U}_1^H(\mathbf{X}_1\mathbf{X}_1^H)\mathbf{U}_1 = \mathbf{W}_d^H\mathbf{X}_1^H(\mathbf{X}_1\mathbf{X}_1^H)^{-1}\mathbf{X}_1\mathbf{W}_d = \mathbf{W}_d^H\mathbf{P}_1\mathbf{W}_d = \mathbf{W}_d^H(2\overline{\mathbf{P}} - \mathbf{P}_2)\mathbf{W}_d,$$

and, therefore,

$$\mathbf{U}_1^H(\mathbf{X}_1\mathbf{X}_1^H)\mathbf{U}_1 + \mathbf{U}_2^H(\mathbf{X}_2\mathbf{X}_2^H)\mathbf{U}_2 = 2\boldsymbol{\Lambda}_d,$$

where $\boldsymbol{\Lambda}_d$ is the $d \times d$ Northwest block of $\boldsymbol{\Lambda}$ containing along its diagonal the d largest eigenvalues of $\overline{\mathbf{P}}$. With only two sets, the central subspace is equidistant from the two row spaces of the canonical variates and hence $\mathbf{U}_i^H(\mathbf{X}_i\mathbf{X}_i^H)\mathbf{U}_i = \boldsymbol{\Lambda}_d$. Appropriately rescaling the canonical vectors would yield the same solution provided by the SVD of the coherence matrix. Clearly, the canonical correlations are invariant, and they are not affected by this rescaling. The extension of this formulation to multiple datasets yields the MAXVAR-CCA generalization.

11.2.2 Multiset CCA (MCCA)

In the two-channel case, we have seen that CCA may be formulated as several different optimization problems, each of which leads to the unique solution for the canonical vectors that maximize the pairwise correlation between canonical variates, subject to orthogonality conditions between the canonical variates. We could well say that CCA is essentially two-channel PCA.

The situation is drastically different when there are more than two datasets, and we wish to find maximally correlated transformations of these datasets. First of all, there are obviously multiple pairwise correlations, and it is therefore possible to optimize different functions of them, imposing also different orthogonality conditions between the canonical variates of the different sets. In the literature, these multiset extensions to CCA are called generalized CCA (GCCA) or multiset CCA (MCCA).

In this section, we present two of these generalizations, probably the most popular, which are natural extensions of the cost functions presented for two-channel CCA in the previous subsection. The first one maximizes the sum of pairwise correlations and is called SUMCOR. The second one seeks a shared low-dimensional representation, or shared central subspace, for the multiple data views, and is called the maximum variance or MAXVAR formulation. Each is a story of coherence between datasets. Both, but especially MAXVAR, have been successfully applied to image processing [247], machine learning [157], and communications problems [364], to name just a few applications.

SUMCOR-MCCA. Consider M datasets $\mathbf{X}_m \in \mathbb{R}^{d_m \times N}$, $m = 1, \ldots, M$, with $M > 2$. The ith column of $\mathbf{X}_m$ corresponds to the ith datum of the mth view or mth dataset. We assume that all datasets are centered. GCCA or MCCA looks for matrices $\mathbf{U}_m \in \mathbb{R}^{d_m \times d}$, $m = 1, \ldots, M$, with $d \leq \min(d_1, \ldots, d_M)$, such that some function of the pairwise correlation matrices between linear transformations, $\mathbf{U}_m^H \mathbf{X}_m \mathbf{X}_n^H \mathbf{U}_n$, is optimized. In particular, the sum-of-correlations (SUMCOR) MCCA problem is

$$\text{SUMCOR-MCCA:} \quad \begin{array}{ll} \underset{\mathbf{U}_1,\ldots,\mathbf{U}_M}{\text{maximize}} & \displaystyle\sum_{1 \leq m < n \leq M} \operatorname{tr}\left(\mathbf{U}_m^H \mathbf{X}_m \mathbf{X}_n^H \mathbf{U}_n\right), \\ \text{subject to} & \mathbf{U}_m^H \mathbf{X}_m \mathbf{X}_m^H \mathbf{U}_m = \mathbf{I}_d, \quad m = 1, \ldots, M. \end{array} \tag{11.14}$$

Suppose, for example, that $d = 1$, so that we seek one-dimensional transformations $\mathbf{z}_m = \mathbf{u}_m^H \mathbf{X}_m$. The SUMCOR-MCCA problem is equivalent to

$$\underset{\mathbf{u}_1,\ldots,\mathbf{u}_M}{\text{maximize}} \quad \sum_{1 \leq m < n \leq M} \frac{\mathbf{u}_m^H \mathbf{X}_m \mathbf{X}_n^H \mathbf{u}_n}{\sqrt{\mathbf{u}_m^H \mathbf{X}_m \mathbf{X}_m^H \mathbf{u}_m}\sqrt{\mathbf{u}_n^H \mathbf{X}_n \mathbf{X}_n^H \mathbf{u}_n}} = \sum_{1 \leq m < n \leq M} \rho_{nm}, \tag{11.15}$$

which is a sum of pairwise correlation coefficients. Observe that the solution of (11.15) is invariant to independent scaling of the canonical vectors. This means that we have the freedom to impose the constraints $\mathbf{u}_m^H \mathbf{X}_m \mathbf{X}_m^H \mathbf{u}_m = 1$, which only affects the norm of the solution. Problem (11.14) is a simple extension of this idea to d sequentially uncorrelated projections. Equivalently, the SUMCOR-MCCA can be formulated as a pairwise distance matching problem [63]

$$\text{SUMCOR-MCCA:} \quad \begin{array}{ll} \underset{\mathbf{U}_1,\ldots,\mathbf{U}_M}{\text{minimize}} & \displaystyle\sum_{1 \leq m < n \leq M} \|\mathbf{U}_m^H \mathbf{X}_m - \mathbf{U}_n^H \mathbf{X}_n\|^2, \\ \text{subject to} & \mathbf{U}_m^H \mathbf{X}_m \mathbf{X}_m^H \mathbf{U}_m = \mathbf{I}_d, \quad m = 1, \ldots, M. \end{array}$$

The SUMCOR-MCCA cannot be solved analytically, and, in fact, it was shown in [294] that it is NP-hard in general. A block coordinate ascent algorithm to solve (11.14), which is both scalable and amenable to distributed implementations, was proposed in [126].

MAXVAR-MCCA. The MAXVAR-MCCA was originally proposed by J. D. Carrol in 1968 as a way to find a shared latent subspace among M datasets [63], so we will begin with this approach to the problem. We have seen in (11.12) that two-channel CCA admits such an interpretation, so the generalization to M datasets is straightforward. The idea is to search for a common or shared subspace among the views that solves the problem [71]

$$\text{MAXVAR-MCCA:} \quad \underset{\mathbf{U}_1,\ldots,\mathbf{U}_M,\mathbf{V}_d}{\text{minimize}} \quad \sum_{m=1}^{M} \left\| \mathbf{U}_m^H \mathbf{X}_m - \mathbf{V}_d^H \right\|^2, \qquad \text{subject to} \quad \mathbf{V}_d^H \mathbf{V}_d = \mathbf{I}_d, \tag{11.16}$$

where $\mathbf{V}_d \in \mathbb{S}t(d, \mathbb{C}^N)$ is a unitary basis for the latent subspace. The MAXVAR-MCCA problem can be solved analytically repeating the steps for the two-channel case. Fixing $\mathbf{V}_d$, the minimizers are $\mathbf{U}_m = (\mathbf{X}_m\mathbf{X}_m^H)^{-1}\mathbf{X}_m\mathbf{V}_d$. Substituting these values in (11.16), a unitary basis for the latent subspace, solves the problem in (11.13), which is repeated here

$$\underset{\mathbf{V}_d \in \mathbb{S}t(d,\mathbb{C}^N)}{\text{minimize}} \quad \text{tr}\left(\mathbf{V}_d^H \overline{\mathbf{P}} \mathbf{V}_d\right).$$

Here $\overline{\mathbf{P}} = \sum_{m=1}^{M} \mathbf{P}_m / M = \mathbf{W}\boldsymbol{\Lambda}\mathbf{W}^H$ is an average of orthogonal projection matrices $\mathbf{P}_m = \mathbf{X}_m^H(\mathbf{X}_m\mathbf{X}_m^H)^{-1}\mathbf{X}_m$. Therefore, a unitary basis for the latent subspace is given by the dominant d eigenvectors of $\overline{\mathbf{P}}$, namely, $\mathbf{W}_d$. A comment worth noting about the normalization of the canonical variates made by this MAXVAR-CCA formulation is the following. Instead of individual constraints of the form $\mathbf{U}_m^H\mathbf{X}_m\mathbf{X}_m^H\mathbf{U}_m = \mathbf{I}_d$, the solution of (11.16) puts constraints on the averaged (or aggregated) canonical variates for the M datasets. Define the average canonical variates as

$$\begin{aligned}\overline{\mathbf{Z}} &= \frac{1}{M}\sum_{m=1}^{M} \mathbf{U}_m^H \mathbf{X}_m = \frac{1}{M}\sum_{m=1}^{M} \mathbf{W}_d^H \mathbf{X}_m^H (\mathbf{X}_m\mathbf{X}_m^H)^{-1}\mathbf{X}_m = \mathbf{W}_d^H \left(\frac{1}{M}\sum_{m=1}^{M} \mathbf{P}_m\right) \\ &= \mathbf{W}_d^H \overline{\mathbf{P}} = \mathbf{W}_d^H(\mathbf{W}\boldsymbol{\Lambda}\mathbf{W}^H) = \boldsymbol{\Lambda}_d \mathbf{W}_d^H.\end{aligned}$$

So the average of the canonical variates is the dominant eigenvector of $\overline{\mathbf{P}}$ (i.e., the directions of the latent subspace) scaled by its eigenvalues. Moreover, the averaged canonical variates satisfy

$$\overline{\mathbf{Z}}\overline{\mathbf{Z}}^H = \mathbf{W}_d^H \overline{\mathbf{P}}^2 \mathbf{W}_d = \boldsymbol{\Lambda}_d^2,$$

so they are uncorrelated but not unit-norm. It is easy to scale them to satisfy $\overline{\mathbf{Z}}\overline{\mathbf{Z}}^H = \mathbf{I}_d$. This is the constraint for the MAXVAR-CCA solution.

The MAXVAR-MCCA approach based on the search for a latent subspace, which is common to all data views, provides interesting insights and puts it in connection with the subspace averaging methods of Chap. 9. However, there are other approaches to MAXVAR-CCA that lead to the same solution and are also of interest in their own right. For instance, the MAXVAR-CCA problem can be formulated as a generalized eigenvalue problem like the one in (11.9) for CCA. Stack the canonical vectors for the kth linear transformation in the vector $\mathbf{v}_k = [\mathbf{u}_{1k}^T \cdots \mathbf{u}_{kM}^T]^T$, $k = 1, \ldots, d$. Then

$$\frac{1}{M-1}(\mathbf{S} - \mathbf{D})\,\mathbf{v}_k = \rho_k \mathbf{D}\mathbf{v}_k, \tag{11.17}$$

where

$$\mathbf{S} = \begin{bmatrix} \mathbf{X}_1\mathbf{X}_1^H & \ldots & \mathbf{X}_1\mathbf{X}_M^H \\ \vdots & \ddots & \vdots \\ \mathbf{X}_M\mathbf{X}_1^H & \ldots & \mathbf{X}_M\mathbf{X}_M^H \end{bmatrix} \quad \text{and} \quad \mathbf{D} = \text{blkdiag}\left(\mathbf{X}_1\mathbf{X}_1^H, \ldots, \mathbf{X}_M\mathbf{X}_M^H\right).$$

The factor $1/(M-1)$ normalizes the eigenvalues $-1 \le \rho_k \le 1$, so that they can be interpreted as correlation coefficients. Note that $\mathbf{v}_k$ is also an eigenvector of the composite coherence matrix

$$\mathbf{D}^{-1/2}(\mathbf{S} - \mathbf{D})\,\mathbf{D}^{-1/2} = \begin{bmatrix} \mathbf{0} & \hat{\mathbf{C}}_{12} & \cdots & \hat{\mathbf{C}}_{1M} \\ \hat{\mathbf{C}}_{21} & \mathbf{0} & \cdots & \hat{\mathbf{C}}_{2M} \\ \vdots & \vdots & \ddots & \vdots \\ \hat{\mathbf{C}}_{M1} & \cdots & \hat{\mathbf{C}}_{MM-1} & \mathbf{0} \end{bmatrix},$$

where each block $\hat{\mathbf{C}}_{nm} = (\mathbf{X}_n\mathbf{X}_n^H)^{-1/2}\mathbf{X}_n\mathbf{X}_m^H(\mathbf{X}_m\mathbf{X}_m^H)^{-1/2}$ is a sample coherence matrix. The MAXVAR-CCA solution obtained by solving (11.17) coincides with the solution of (11.16) up to a scaling of the average canonical variates. That is, the canonical vectors obtained by solving (11.17) satisfy $\sum_{m=1}^{M} \mathbf{U}_m^H\mathbf{X}_m\mathbf{X}_m\mathbf{U}_m = \mathbf{I}_d$, and therefore $\overline{\mathbf{Z}\mathbf{Z}}^H = \mathbf{I}_d/M^2$.

There is yet another equivalent formulation for the MAXVAR-MCCA problem based on Euclidean distances between linear transformations:

$$\text{MAXVAR-MCCA:} \quad \underset{\mathbf{U}_1,\ldots,\mathbf{U}_M}{\text{minimize}} \quad \sum_{m,n=1}^{M} \left\| \mathbf{U}_m^H\mathbf{X}_m - \mathbf{U}_n^H\mathbf{X}_n \right\|^2,$$

$$\text{subject to} \quad \sum_{m=1}^{M} \mathbf{U}_m^H\mathbf{X}_m\mathbf{X}_m\mathbf{U}_m = \mathbf{I}_d.$$

As shown in [365], this interpretation enables the derivation of RLS-like adaptive algorithms that extract the canonical vectors with multiple datasets in an online fashion, that is, for sequentially observed data.

11.3 Coherence in Kernel Methods

It is often desirable to generalize the traditional Euclidean inner product to more flexible inner products characterized by properly chosen kernel functions. This notion of kernel-induced vector spaces is the cornerstone of kernel-based machine learning techniques, also known simply as kernel methods, which have been successfully applied over the last decades to many nonlinear classification and regression problems [210, 316]. It is not surprising, therefore, that the concept of coherence, measured in the transformed space through the kernel function, is relevant to kernel methods.

In this section, we first review the basics of kernel methods and then present two examples of the application of coherence in this field: (i) the kernelized version of CCA (KCCA) and (ii) the kernelized version of the least mean squares (LMS) adaptive filtering algorithm, known as KLMS, in which a coherence-based criterion is commonly applied to limit the complexity of the resulting kernel expansion.

11.3.1 Kernel Functions, Reproducing Kernel Hilbert Spaces (RKHS), and Mercer's Theorem

Begin with complex vectors $\mathbf{x}, \mathbf{y} \in \mathcal{X}$, where the input space $\mathcal{X}$ can be assumed to be a subset of $\mathbb{C}^L$. The traditional Euclidean inner product is $\mathbf{x}^H\mathbf{y}$, which is a mapping from $\mathcal{X} \times \mathcal{X}$ into $\mathbb{C}$. This inner product may be replaced by $k(\mathbf{x}, \mathbf{y}) : \mathcal{X} \times \mathcal{X} \to \mathbb{C}$ for a suitably defined function k. If k is a non-negative definite operator, which is to say $\sum_{i,l=1}^{n} c_i^* k(\mathbf{x}_i, \mathbf{x}_l) c_l \geq 0$ for all n and complex c_i, then it may be expanded as a uniformly convergent series on $\mathcal{X} \times \mathcal{X}$

$$k(\mathbf{x}, \mathbf{y}) = \sum_{m=1}^{\infty} \psi_m(\mathbf{x}) \lambda_m \psi_m^*(\mathbf{y}). \tag{11.18}$$

The $\psi(\mathbf{x})$ are orthonormal eigenfunctions that satisfy the first-order Fredholm integral equation

$$\int_{\mathcal{X}} k(\mathbf{x}, \mathbf{y}) \psi_m(\mathbf{y}) d\mathbf{y} = \psi_m^*(\mathbf{x}) \lambda_m.$$

This is Mercer's theorem [241]. The interpretation as an infinite-dimensional EVD is obvious. With the Mercer expansion, the kernelized inner product $k(\mathbf{x}, \mathbf{y})$ between two Euclidean vectors of dimension L is actually an inner product in an infinite-dimensional Hilbert space $\mathcal{H}$ of functions. What can be said about this space?

Define the Hilbert space $\mathcal{H}$ to be the space of functions $f : \mathcal{X} \to \mathbb{C}$, endowed with inner product $\langle \cdot, \cdot \rangle$. This space is called a reproducing kernel Hilbert space (RKHS) if there exists a kernel k with the properties

1. k has the *reproducing property*

$$\langle f, k(\mathbf{x}, \cdot)\rangle = f(\mathbf{x}), \quad \forall f \in \mathcal{H}.$$

2. k spans $\mathcal{H}$, i.e., $\mathcal{H} = \text{span}\{k(\mathbf{x}, \cdot), \mathbf{x} \in \mathcal{X}\}$.

The term *kernel* stems from its use in integral operators in functional analysis as studied by Hilbert. The Riesz representation theorem and the Moore-Aronszajn theorem [15] establish that the RKHS uniquely determines the kernel function (Riesz) and vice versa (Moore-Aronszajn).

From the Mercer expansion (11.18), it follows that there exists a RKHS $\mathcal{H}$ and a mapping

$$\begin{aligned} \boldsymbol{\phi} &: \mathbf{x} \rightarrow \boldsymbol{\phi}(\mathbf{x}) \in \mathcal{H}, \\ \mathbf{x} &\rightarrow \boldsymbol{\phi}(\mathbf{x}) = \left[\sqrt{\lambda_1}\psi_1(\mathbf{x})\ \sqrt{\lambda_2}\psi_2(\mathbf{x})\ \cdots\right]^T, \end{aligned}$$

such that $k(\mathbf{x}, \mathbf{y})$ corresponds to an inner product in $\mathcal{H}$: $k(\mathbf{x}, \mathbf{y}) = \langle \boldsymbol{\phi}(\mathbf{x}), \boldsymbol{\phi}(\mathbf{y})\rangle$. In kernel methods for machine learning, the RKHS $\mathcal{H}$ is usually called the *feature space* or intrinsic space (since its eigenfunctions ψ_m are independent of the training data), and $\boldsymbol{\phi}(\cdot)$ is the *feature map* that maps an input vector $\mathbf{x}$ to a possibly infinite dimensional feature vector $\boldsymbol{\phi}(\mathbf{x})$.

Example 11.1 (Bandlimited square integrable functions) Let $\mathcal{H}$ be the Hilbert space of real square integrable functions $f(t)$ such that the support of the Fourier transform $F(\omega)$ is included in $[-B, B]$. The inner product is $\langle f(t), g(t)\rangle = \int_{-\infty}^{\infty} f(t)g(t)dt$. This is a RKHS with reproducing kernel

$$k(s, t) = \frac{\sin(B(s-t))}{B(s-t)}.$$

Example 11.2 (Gaussian kernel functions) A typical kernel function used in machine learning is the Gaussian kernel function

$$k(\mathbf{x}, \mathbf{y}) = \exp\left(-\frac{\|\mathbf{x}-\mathbf{y}\|_2^2}{2\sigma^2}\right), \tag{11.19}$$

which can be expanded into the following power series

$$k(\mathbf{x}, \mathbf{y}) = \exp\left(-\frac{\|\mathbf{x}\|_2^2}{2\sigma^2}\right)\left(\sum_{m=1}^{\infty} \frac{1}{(m-1)!}\left(\frac{\langle \mathbf{x}, \mathbf{y}\rangle}{\sigma^2}\right)^{m-1}\right)\exp\left(-\frac{\|\mathbf{y}\|_2^2}{2\sigma^2}\right),$$

where all coefficients in the expansion are positive. Therefore k is a Mercer kernel. The mapping $\boldsymbol{\phi}(\cdot)$ takes the form

$$\boldsymbol{\phi}(\mathbf{x}) = \exp\left(-\frac{\|\mathbf{x}\|_2^2}{2\sigma^2}\right)[\phi_1(\mathbf{x})\ \phi_2(\mathbf{x})\ \cdots]^T ,$$

where each entry $\phi_m(\mathbf{x})$, $m = 1, 2, \ldots, \infty$, is a multivariate polynomial scaled by some positive coefficient.

The Kernel Trick. As the above example has shown, the mapping $\boldsymbol{\phi}(\mathbf{x})$ may be difficult or even impossible to obtain in explicit form for some kernels. Fortunately, one rarely needs to know the mapping $\boldsymbol{\phi}(\mathbf{x})$, as in most cases scalar products, distances, and projections in the induced RKHS can be obtained through the kernel function between input patterns, $k(\mathbf{x}, \mathbf{y})$. In fact, any algorithm that only depends on inner or dot products, i.e., any algorithm that is rotationally invariant, can be *kernelized*. This is the so-called *kernel trick*, which amounts to replacing the original kernel function, typically the linear inner product in the input space $\langle \mathbf{x}, \mathbf{y} \rangle$, by a nonlinear kernel $k(\mathbf{x}, \mathbf{y}) = \langle \boldsymbol{\phi}(\mathbf{x}), \boldsymbol{\phi}(\mathbf{y}) \rangle$, thus adding more flexibility to the solution.

11.3.2 Kernel CCA

Let $\mathbf{X}_1 \in \mathbb{R}^{d_1 \times N}$ and $\mathbf{X}_2 \in \mathbb{R}^{d_2 \times N}$ be two input or training datasets representing two different views of the same underlying latent function or object. They could be, for instance, two sets of N documents paired in terms of a common semantic concept, each document of the paired dataset in a different language. The dimensions of the input vectors are d_1 and d_2, respectively. By seeking transformations that maximize correlation between the two datasets, we may hope to extract the underlying common semantic content or, in general, the underlying latent factors. This is achieved by two-channel canonical correlation analysis (CCA). We have seen in Sect. 11.2.1 that the CCA solution for the dominant canonical correlation k_1 finds the transformation that maximizes correlation by solving

$$\begin{aligned}
&\underset{\mathbf{w}_1, \mathbf{w}_2}{\text{maximize}} && \mathbf{w}_1^T \mathbf{X}_1 \mathbf{X}_2^T \mathbf{w}_2, \\
&\text{subject to} && \mathbf{w}_i^T \mathbf{X}_i \mathbf{X}_i^T \mathbf{w}_i = 1, \quad i = 1, 2.
\end{aligned}$$

Let us express the canonical vectors $\mathbf{w}_1$ and $\mathbf{w}_2$ in terms of their respective input samples as $\mathbf{w}_1 = \mathbf{X}_1 \boldsymbol{\alpha}_1$ and $\mathbf{w}_2 = \mathbf{X}_2 \boldsymbol{\alpha}_2$, where $\boldsymbol{\alpha}_1 \in \mathbb{R}^N$ and $\boldsymbol{\alpha}_2 \in \mathbb{R}^N$. Using these variables, the dual CCA problem formulation is

$$\begin{aligned} & \underset{\boldsymbol{\alpha}_1,\boldsymbol{\alpha}_2}{\text{maximize}} && \boldsymbol{\alpha}_1^T \mathbf{X}_1^T \mathbf{X}_1 \mathbf{X}_2^T \mathbf{X}_2 \boldsymbol{\alpha}_2, \\ & \text{subject to} && \boldsymbol{\alpha}_i^T \mathbf{X}_i^T \mathbf{X}_i \mathbf{X}_i^T \mathbf{X}_i \boldsymbol{\alpha}_i = 1, \quad i = 1, 2. \end{aligned}$$

Clearly, CCA admits a formulation in terms of inner products through the $N \times N$ Gram matrices $\mathbf{X}_1^T \mathbf{X}_1$ and $\mathbf{X}_2^T \mathbf{X}_2$. We can now apply the *kernel trick* and use any nonlinear kernel function, such as the Gaussian kernel in (11.19), and solve the kernel CCA problem [157]

$$\begin{aligned} & \underset{\boldsymbol{\alpha}_1,\boldsymbol{\alpha}_2}{\text{maximize}} && \boldsymbol{\alpha}_1^T \mathbf{K}_1 \mathbf{K}_2 \boldsymbol{\alpha}_2, \\ & \text{subject to} && \boldsymbol{\alpha}_i^T \mathbf{K}_i^2 \boldsymbol{\alpha}_i = 1, \quad i = 1, 2, \end{aligned}$$

where $\mathbf{K}_i$ is the kernel matrix with entries given by all kernel inner products between the columns of $\mathbf{X}_i$. The coherence in the feature space is

$$\rho = \frac{\boldsymbol{\alpha}_1^T \mathbf{K}_1 \mathbf{K}_2 \boldsymbol{\alpha}_2}{\sqrt{\boldsymbol{\alpha}_1^T \mathbf{K}_1^2 \boldsymbol{\alpha}_1}\sqrt{\boldsymbol{\alpha}_2^T \mathbf{K}_2^2 \boldsymbol{\alpha}_2}}.$$

Applying Lagrangian techniques, the solution for the dual coefficients is

$$\boldsymbol{\alpha}_1 = \frac{1}{\lambda} \mathbf{K}_1^{-1} \mathbf{K}_2 \boldsymbol{\alpha}_2,$$

and so $\mathbf{K}_2^2 \boldsymbol{\alpha}_2 - \lambda^2 \mathbf{K}_2^2 \boldsymbol{\alpha}_2 = \mathbf{0}$, which holds for all vectors $\boldsymbol{\alpha}_2$ with $\lambda = 1$ when the kernel matrices are full rank and invertible. For example, this is always true for a Gaussian kernel. In other words, when the dimension of the feature space, $\dim(\mathcal{H})$, is much larger than the number of training data, $\dim(\mathcal{H}) \gg N$, the feature vectors will be linearly independent with high probability. Hence, it is always possible to find perfect correlations between arbitrary transformations of one dataset and an appropriate choice of transformations of the other dataset. This is a problem of *overfitting*. In fact, it is also known that in the low sample support case, when the sample covariance matrices are not full rank, some of the canonical correlations become one [198, 264, 278, 321]. The solution of this overfitting problem is to regularize the problem by adding a penalty on the norms of the canonical vectors and solving

$$\begin{aligned} & \underset{\boldsymbol{\alpha}_1,\boldsymbol{\alpha}_2}{\text{maximize}} && \boldsymbol{\alpha}_1^T \mathbf{K}_1 \mathbf{K}_2 \boldsymbol{\alpha}_2, \\ & \text{subject to} && \boldsymbol{\alpha}_i^T \mathbf{K}_i^2 \boldsymbol{\alpha}_i + c\, \boldsymbol{\alpha}_i^T \mathbf{K}_i \boldsymbol{\alpha}_i = 1, \quad i = 1, 2, \end{aligned}$$

where $c > 0$ is the regularization parameter that limits the flexibility of the projection mappings. Hence, coherence between canonical variates in the feature space is given by

$$\rho = \frac{\boldsymbol{\alpha}_1^T \mathbf{K}_1 \mathbf{K}_2 \boldsymbol{\alpha}_2}{\sqrt{\boldsymbol{\alpha}_1^T (\mathbf{K}_1^2 + c\mathbf{K}_1)\boldsymbol{\alpha}_1}\sqrt{\boldsymbol{\alpha}_2^T (\mathbf{K}_2^2 + c\mathbf{K}_2)\boldsymbol{\alpha}_2}}.$$

The regularized kernel CCA problem can alternatively be formulated as the following generalized eigenvalue problem

$$\begin{bmatrix} \mathbf{0} & \mathbf{K}_1\mathbf{K}_2 \\ \mathbf{K}_2\mathbf{K}_1 & \mathbf{0} \end{bmatrix} \begin{bmatrix} \boldsymbol{\alpha}_1 \\ \boldsymbol{\alpha}_2 \end{bmatrix} = \rho \begin{bmatrix} \mathbf{K}_1(\mathbf{K}_1 + c\mathbf{I}_N) & \mathbf{0} \\ \mathbf{0} & \mathbf{K}_2(\mathbf{K}_2 + c\mathbf{I}_N) \end{bmatrix} \begin{bmatrix} \boldsymbol{\alpha}_1 \\ \boldsymbol{\alpha}_2 \end{bmatrix}.$$

11.3.3 Coherence Criterion in KLMS

By applying linear adaptive filtering principles in the kernel feature space, powerful kernel adaptive filtering (KAF) algorithms can be obtained [221,360]. The simplest among the family of KAF algorithms is the *kernelized* version of the least mean square (LMS) algorithm [381], which is known as the kernel least mean square or KLMS algorithm [220, 283]. The approach employs the traditional kernel trick. Essentially, a nonlinear function $\boldsymbol{\phi}(\cdot)$ maps the time-embedded input time series $\mathbf{x}_n = [x_n \; x_{n-1} \; \cdots \; x_{n-L+1}]^T$ from the input space to the feature RKHS space $\mathcal{H}$ with kernel function $k(\cdot,\cdot)$. Let $\mathbf{w}_{\mathcal{H}}$ be the weight vector in the RKHS and define the filter output at time n as $y_n = \mathbf{w}_{\mathcal{H},n-1}^T \boldsymbol{\phi}(\mathbf{x}_n)$, where $\mathbf{w}_{\mathcal{H},n-1}$ is the estimate of $\mathbf{w}_{\mathcal{H}}$ at the previous time instant $n-1$. Given a desired response d_n, we wish to minimize the squared loss with respect to $\mathbf{w}_{\mathcal{H}}$. The stochastic gradient descent update rule is the well-known LMS rule

$$\mathbf{w}_{\mathcal{H},n} = \mathbf{w}_{\mathcal{H},n-1} + \mu e_n \boldsymbol{\phi}(\mathbf{x}_n),$$

where $\mu > 0$ is the step size or learning rate. By initializing the solution as $\mathbf{w}_{\mathcal{H},0} = \mathbf{0}$ (and hence $e_0 = d_0 = 0$), the solution after $n-1$ iterations is

$$\mathbf{w}_{\mathcal{H},n-1} = \mu \sum_{i=1}^{n-1} e_i \boldsymbol{\phi}(\mathbf{x}_i) = \sum_{i=1}^{n-1} \alpha_i \boldsymbol{\phi}(\mathbf{x}_i), \tag{11.20}$$

where we have introduced the dual variables $\alpha_i = \mu e_i$. Equation (11.20) shows that the filter solution in the feature space can be expressed as a linear combination of the transformed data, which is the statement of the representer theorem [200, 368]. In words, the representer theorem tells us that the solution to some regularization problems in high or infinite dimensional vector spaces lie in finite dimensional subspaces spanned by the representers of the data [369].

The filter output is

$$y_n = \sum_{i=1}^{n-1} \alpha_i \langle \boldsymbol{\phi}(\mathbf{x}_i), \boldsymbol{\phi}(\mathbf{x}_n) \rangle = \sum_{i=1}^{n-1} \alpha_i k(\mathbf{x}_i, \mathbf{x}_n), \tag{11.21}$$

where in the second equality we have used the *kernel trick*. That is, the output of the filter in the RKHS to a new input can be solely expressed in terms of inner products between transformed inputs. Then, it can readily be computed in the input space. Defining $\boldsymbol{\alpha}_{n-1} = [\alpha_1 \cdots \alpha_{n-1}]^T$ and $\mathbf{k}_{n-1} = [k(\mathbf{x}_1, \mathbf{x}_n)\, k(\mathbf{x}_2, \mathbf{x}_n) \cdots k(\mathbf{x}_{n-1}, \mathbf{x}_n)]^T$, the filter output can be expressed in vector form as $y_n = \boldsymbol{\alpha}_{n-1}^T \mathbf{k}_{n-1}$, and the vector of dual variables is updated after each iteration as

$$\boldsymbol{\alpha}_n = \begin{bmatrix} \boldsymbol{\alpha}_{n-1} \\ \mu e_n \end{bmatrix}. \tag{11.22}$$

Update (11.22) emphasizes the growing nature of the KLMS filter, which precludes its direct implementation in practice. In order to design a practical KLMS algorithm, the number of terms in the kernel expansion (11.21) should stop growing over time. This can be achieved by implementing online sparsification techniques, whose aim is to identify kernel functions whose removal is expected to have negligible effect on the quality of the model. One of these sparsification techniques, originally proposed in [283], is based on the *coherence* criterion. In a kernel-based context, the coherence between a new datum $\mathbf{x}_n$ and a dictionary of already stored data $\mathcal{D}_{n-1} = \{\mathbf{x}_1, \ldots, \mathbf{x}_{n-1}\}$ is defined as

$$\rho = \max_{i \in \mathcal{D}_{n-1}} |k(\mathbf{x}_i, \mathbf{x}_n)| \tag{11.23}$$

where we have assumed a unit-norm kernel, such as the Gaussian kernel, that satisfies $k(\mathbf{x}, \mathbf{x}) = 1$. Otherwise, (11.23) should be normalized as $\frac{|k(\mathbf{x}_i, \mathbf{x}_n)|}{\sqrt{k(\mathbf{x}_i, \mathbf{x}_i)}\sqrt{k(\mathbf{x}_n, \mathbf{x}_n)}}$. The coherence measures the maximum cosine of the angle between the new datum and the dictionary data in the RKHS. Alternatively, it is the largest absolute value of the off-diagonal entries in the Gramian or kernel matrix formed by the new datum and the dictionary. It reflects the largest cross-correlation in the updated dictionary.

When the coherence between the new datum $\mathbf{x}_n$ and the dictionary elements at time $n-1$, $\mathcal{D}_{n-1}$, is below a given threshold

$$\max_{i \in \mathcal{D}_{n-1}} |k(\mathbf{x}_i, \mathbf{x}_n)| \leq \epsilon,$$

then the coherence-based KLMS includes $\mathbf{x}_n$ into the dictionary, and the filter coefficients are updated as in (11.22), with $e_n = d_n - y_n = d_n - \boldsymbol{\alpha}_{n-1}^T \mathbf{k}_{n-1}$. When the coherence is above the threshold, the new datum is not included in the dictionary, and the coefficients of the expansion are updated as

$$\boldsymbol{\alpha}_n = \boldsymbol{\alpha}_{n-1} + \mu e_n \mathbf{k}_{n-1}.$$

As proved in [283], the dimension of the dictionary with a coherence-based sparsification rule remains finite. The KLMS algorithm with coherence-based sparsification criterion is summarized in Algorithm 9.

Algorithm 9: KLMS with coherence criterion

Input: Training data $\{\mathbf{x}_n, d_n\}_{n=1}^N$, Gaussian kernel width σ, step size μ, threshold for coherence ϵ
Initialize $\mathcal{D}_1 = \{\mathbf{x}_1\}$, $\boldsymbol{\alpha}_1 = \mu d_1$
for $n = 2, 3, \ldots$ **do**
  Select input-output training pattern $(\mathbf{x}_n, d_n)$
  Compute kernels vector $\mathbf{k}_{n-1} = [k(\mathbf{x}_1, \mathbf{x}_n) \cdots k(\mathbf{x}_{n-1}, \mathbf{x}_n)]^T$
  Compute prediction $y_n = \boldsymbol{\alpha}_{n-1}^T \mathbf{k}_{n-1}$
  Compute error $e_n = d_n - y_n$
  Compute coherence $\rho = \max_{i \in \mathcal{D}_{n-1}} |k(\mathbf{x}_i, \mathbf{x}_n)|$
  if $\rho < \epsilon$ **then**
    New datum is included in the dictionary $\mathcal{D}_n = \mathcal{D}_{n-1} \cup \{\mathbf{x}_n\}$
    Update coefficients $\boldsymbol{\alpha}_n = \begin{bmatrix} \boldsymbol{\alpha}_{n-1} \\ \mu e_n \end{bmatrix}$
  else
    Dictionary is maintained $\mathcal{D}_n = \mathcal{D}_{n-1}$
    Update coefficients $\boldsymbol{\alpha}_n = \boldsymbol{\alpha}_{n-1} + \mu e_n \mathbf{k}_{n-1}$

11.4 Mutual Information as Coherence

Throughout this book, *coherence* between random variables has been treated as a normalized inner product in the Hilbert space of second-order random variables. Perhaps, the basic idea extends to an *information-theoretic coherence* based on mutual information.

Let us consider two zero-mean continuous real random variables, x and y, and recall the error variance of the LMMSE estimator of x from y and the error variance of the LMMSE estimator of y from x. These are

$$\sigma_{x|y}^2 = \sigma_x^2(1 - \rho_{xy}^2), \qquad \sigma_{y|x}^2 = \sigma_y^2(1 - \rho_{xy}^2),$$

where

$$\rho_{xy} = \frac{\mathrm{E}[xy]}{\sigma_x \sigma_y}$$

is the coherence between x and y. Here, $\sigma_x^2 = \mathrm{E}[x^2]$ is the variance of x, and $\mathrm{E}[xy]$ is the covariance between x and y. From these formulas it is easy to see that

$$\log \sigma_x^2 - \log \sigma_{x|y}^2 = -\log(1 - \rho_{xy}^2), \quad \log \sigma_y^2 - \log \sigma_{y|x}^2 = -\log(1 - \rho_{xy}^2).$$

Perhaps there is a connection with entropy and mutual information. Define the following (differential) entropies for the random variables x and y [87]

$$h_x = \mathrm{E}\left[\log \frac{1}{p(x)}\right], \qquad h_y = \mathrm{E}\left[\log \frac{1}{p(y)}\right],$$

and conditional (differential) entropies [87]

$$h_{x|y} = \mathrm{E}\left[\log \frac{1}{p(x|y)}\right], \qquad h_{y|x} = \mathrm{E}\left[\log \frac{1}{p(y|x)}\right].$$

The entropy h_x can be interpreted as a measure of uncertainty about x, and the entropy $h_{x|y}$ is a measure of uncertainty about x, given y. The mutual information between the random variables x and y is [87]

$$I_{xy} = \mathrm{E}\left[\log \frac{p(x,y)}{p(x)p(y)}\right] = h_x - h_{x|y} = h_y - h_{y|x}.$$

The base of the logarithm in these formulas determines the units of entropy and mutual information. If the base is 2, the units are bits, and if the base is e, the units are nats.

A comparison of these entropy formulas with the variance formulas for LMMSE estimation suggests that information-theoretic squared coherence $(\rho_{xy}^I)^2$ may be defined as

$$-\log\left(1 - \left(\rho_{xy}^I\right)^2\right) = I_{xy} \quad \text{or} \quad \left(\rho_{xy}^I\right)^2 = 1 - 2^{-I_{xy}},$$

where it is assumed that the mutual information is measured in bits. Note that $0 \leq (\rho_{xy}^I)^2 \leq 1$, so the transformation of mutual information into a squared coherence makes the latter a more interpretable quantity. For instance, if $y = g(x)$, where $g(\cdot)$ is a deterministic function, we know that $I_{xy} = \infty$, which implies $(\rho_{xy}^I)^2 = 1$. It is also clear that for independent random variables $(\rho_{xy}^I)^2 = 0$. Then, $(\rho_{xy}^I)^2$ can be interpreted as a measure of independence bounded between 0 and 1.

In the bivariate normal case, information-theoretic coherence is standard Hilbert space coherence, but this is not the case if the random variables are non-Gaussian, as the following example shows.

Example 11.3 Let us consider two independent uniform random variables $x \sim U[-1, 1]$ and $y \sim U[-1, 1]$, and apply a clockwise rotation by an angle $0 \leq \theta \leq \pi/4$ to generate z and w as

$$\begin{bmatrix} z \\ w \end{bmatrix} = \begin{bmatrix} \cos\theta & \sin\theta \\ -\sin\theta & \cos\theta \end{bmatrix} \begin{bmatrix} x \\ y \end{bmatrix}.$$

The random variables z and w are dependent for $\theta \neq 0$. However, the squared coherence between z and w is $\rho_{zw}^2 = 0$, regardless of θ. The differential entropies and the mutual information (in bits) are

$$h_z = h_w = (1 + \cos\theta) + \frac{\log(e)\tan\theta}{2},$$

$$I_{zw} = 2\log(\cos\theta) + \log(e)\tan\theta.$$

Therefore, the information-theoretic squared coherence as a function of θ is

$$\left(\rho_{zw}^I\right)^2 = 1 - \frac{1}{e^{\tan\theta}\cos^2\theta}.$$

The definition of the information-theoretic squared coherence generalizes to random vectors $\mathbf{x}$ and $\mathbf{y}$. In the vector-valued case, the LMMSE error covariance matrices and the coherence matrix are

$$\mathbf{Q}_{\mathbf{xx}|\mathbf{y}} = \mathbf{R}_{\mathbf{xx}} - \mathbf{R}_{\mathbf{xy}}\mathbf{R}_{\mathbf{yy}}^{-1}\mathbf{R}_{\mathbf{yx}} = \mathbf{R}_{\mathbf{xx}}^{1/2}(\mathbf{I} - \mathbf{C}_{\mathbf{x}|\mathbf{y}}\mathbf{C}_{\mathbf{x}|\mathbf{y}}^H)\mathbf{R}_{\mathbf{xx}}^{1/2},$$
$$\mathbf{Q}_{\mathbf{yy}|\mathbf{x}} = \mathbf{R}_{\mathbf{yy}} - \mathbf{R}_{\mathbf{yx}}\mathbf{R}_{\mathbf{xx}}^{-1}\mathbf{R}_{\mathbf{xy}} = \mathbf{R}_{\mathbf{yy}}^{1/2}(\mathbf{I} - \mathbf{C}_{\mathbf{y}|\mathbf{x}}\mathbf{C}_{\mathbf{y}|\mathbf{x}}^H)\mathbf{R}_{\mathbf{yy}}^{1/2},$$

where $\mathbf{C}_{\mathbf{x}|\mathbf{y}} = \mathbf{R}_{\mathbf{xx}}^{-1/2}\mathbf{R}_{\mathbf{xy}}\mathbf{R}_{\mathbf{yy}}^{-1/2}$ and $\mathbf{C}_{\mathbf{y}|\mathbf{x}} = \mathbf{R}_{\mathbf{yy}}^{-1/2}\mathbf{R}_{\mathbf{yx}}\mathbf{R}_{\mathbf{xx}}^{-1/2}$. It follows that

$$\log\det(\mathbf{R}_{\mathbf{xx}}) - \log\det(\mathbf{Q}_{\mathbf{xx}|\mathbf{y}}) = -\log\det(\mathbf{I} - \mathbf{C}_{\mathbf{x}|\mathbf{y}}\mathbf{C}_{\mathbf{x}|\mathbf{y}}^H).$$

The mutual information between the random vectors $\mathbf{x}$ and $\mathbf{y}$ is

$$I_{\mathbf{xy}} = \mathrm{E}\left[\log\frac{p(\mathbf{x},\mathbf{y})}{p(\mathbf{x})p(\mathbf{y})}\right] = h_{\mathbf{x}} - h_{\mathbf{x}|\mathbf{y}} = h_{\mathbf{y}} - h_{\mathbf{y}|\mathbf{x}},$$

which suggests a definition of information-theoretic coherence $(\rho_{\mathbf{xy}}^I)^2$:

$$-\log\left(1 - \left(\rho_{\mathbf{xy}}^I\right)^2\right) = I_{\mathbf{xy}} \longleftrightarrow (\rho_{\mathbf{xy}}^I)^2 = 1 - 2^{-I_{\mathbf{xy}}}$$

In the multivariate normal case, information-theoretic squared coherence is Hilbert space squared coherence.

A generalization of this concept to partial coherence is also possible. Recall the LMMSE estimator of random vector $\mathbf{x}$ from $\mathbf{y}$ and $\mathbf{z}$

$$\hat{\mathbf{x}}(\mathbf{y},\mathbf{z}) = \hat{\mathbf{x}}(\mathbf{z}) + \mathbf{R}_{\mathbf{xy}|\mathbf{z}}\mathbf{R}_{\mathbf{yy}|\mathbf{z}}^{-1}\mathbf{R}_{\mathbf{xy}|\mathbf{z}}^H(\mathbf{y} - \hat{\mathbf{y}}(\mathbf{z})),$$

and its corresponding error covariance matrix, given by

$$\mathbf{Q}_{\mathbf{xx}|\mathbf{yz}} = \mathbf{Q}_{\mathbf{xx}|\mathbf{z}} - \mathbf{R}_{\mathbf{xy}|\mathbf{z}}\mathbf{Q}_{\mathbf{yy}|\mathbf{z}}^{-1}\mathbf{R}_{\mathbf{xy}|\mathbf{z}}^H = \mathbf{Q}_{\mathbf{xx}|\mathbf{z}}^{1/2}\left(\mathbf{I} - \mathbf{C}_{\mathbf{xy}|\mathbf{z}}\mathbf{C}_{\mathbf{xy}|\mathbf{z}}^H\right)\mathbf{Q}_{\mathbf{xx}|\mathbf{z}}^{1/2}.$$

The matrix $\mathbf{C}_{\mathbf{xy}|\mathbf{z}} = \mathbf{Q}_{\mathbf{xx}|\mathbf{z}}^{-1/2}\mathbf{R}_{\mathbf{xy}|\mathbf{z}}\mathbf{Q}_{\mathbf{yy}|\mathbf{z}}^{-1/2}$ is the partial coherence matrix, $\mathbf{Q}_{\mathbf{xx}|\mathbf{z}}$ is the error covariance matrix for estimating $\mathbf{x}$ from $\mathbf{z}$, and $\mathbf{R}_{\mathbf{xy}|\mathbf{z}}$ is the cross-covariance between $\mathbf{x} - \hat{\mathbf{x}}(\mathbf{z})$ and $\mathbf{y} - \hat{\mathbf{y}}(\mathbf{z})$. It follows that

$$\log\det(\mathbf{Q}_{\mathbf{xx}|\mathbf{z}}) - \log\det(\mathbf{Q}_{\mathbf{xx}|\mathbf{yz}}) = -\log\det\left(\mathbf{I} - \mathbf{C}_{\mathbf{xy}|\mathbf{z}}\mathbf{C}_{\mathbf{xy}|\mathbf{z}}^H\right).$$

Define the conditional mutual information

$$I_{\mathbf{xy}|\mathbf{z}} = \mathrm{E}\left[\log \frac{p(\mathbf{x},\mathbf{y}|\mathbf{z})}{p(\mathbf{x}|\mathbf{z})p(\mathbf{y}|\mathbf{z})}\right] = h_{\mathbf{x}|\mathbf{z}} - h_{\mathbf{x}|\mathbf{y},\mathbf{z}}.$$

Again, this suggests the following definition of information-theoretic partial coherence

$$-\log\left(1-\left(\rho^I_{\mathbf{xy}|\mathbf{z}}\right)^2\right) = I_{\mathbf{xy}|\mathbf{z}} \longleftrightarrow \left(\rho^I_{\mathbf{xy}|\mathbf{z}}\right)^2 = 1 - 2^{-I_{\mathbf{xy}|\mathbf{z}}}.$$

Note that $0 \leq (\rho^I_{\mathbf{xy}|\mathbf{z}})^2 \leq 1$.

In summary, the theory of LMMSE estimation leads to natural definitions of coherence and partial coherence as functions of variance or covariance. In the MVN case, entropy is a logarithmic function of the determinant of error covariance. By comparing formulas for error covariance in LMMSE estimation with formulas for entropy, it is found that there is a natural definition of information-theoretic squared coherence. This definition extends to information-theoretic partial coherence. Information-theoretic squared coherence is bounded between 0 and 1, making it perhaps more interpretable than mutual information. This does not mean it is any easier to compute than mutual information. It is not.

11.5 Coherence in Time-Frequency Modeling of a Nonstationary Time Series

Consider a nonstationary, zero-mean, second-order random sequence $\{x[n]\}$ with two-index correlation sequence $\{r[n_1,n_2]\}$, re-parameterized as $\{r[n,k]\}$, where $r[n,k] = \mathrm{E}[x[n]x^*[n-k]]$. The question is whether one can define what might be called a time-frequency distribution.

Define the discrete-time Fourier transforms

$$V[n, e^{j\theta}) = \sum_{k=0}^{N-1} r[n,k]e^{-jk\theta} = \mathrm{E}\left[x[n]\left(X(e^{j\theta})e^{jn\theta}\right)^*\right],$$

$$A(e^{j\nu}, k] = \sum_{n=0}^{N-1} r[n,k]e^{-jn\nu} = \mathrm{E}\left[\int_{-\pi}^{\pi} X(e^{j(\theta+\nu)})X^*(e^{j\theta})e^{jk\theta}\frac{d\theta}{2\pi}\right],$$

and

$$S(e^{j(\theta+\nu)}, e^{j\theta}) = \sum_{n=0}^{N-1}\sum_{k=0}^{N-1} r[n,k]e^{-j(k\theta+n\nu)} = \mathrm{E}[X(e^{j(\nu+\theta)})X^*(e^{j\theta})].$$

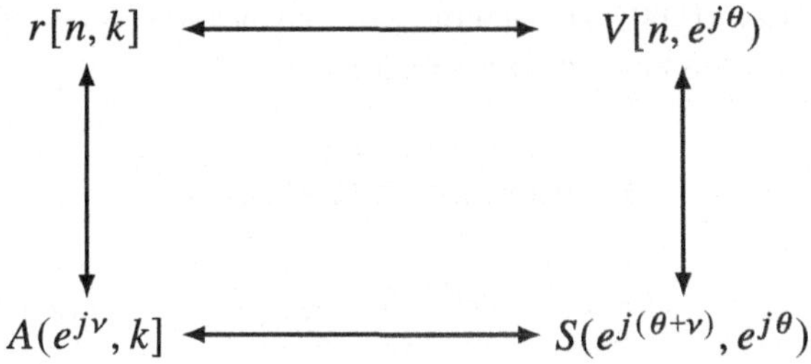

Fig. 11.1 Relation between time-frequency functions

In these Fourier transforms, $X(e^{j\theta}) = \sum_{n=0}^{N-1} x[n]e^{-jn\theta}$ is the discrete-time Fourier transform (DTFT) of $x[n], n = 0, \ldots, N-1$. These Fourier transforms are placed on the four corners diagram of Fig. 11.1. As indicated in the figure, each of A and V is a Fourier transform of r, and S is a Fourier transform of each of A and V. Each Fourier transform is invertible.

It is easily established that the marginals of $V[n, e^{j\theta})$ are these

$$\sum_{n=0}^{N-1} V[n, e^{j\theta}) = S(e^{j\theta}, e^{j\theta}) = \mathrm{E}[|X(e^{j\theta})|^2] \geq 0, \tag{11.24}$$

$$\int_{-\pi}^{\pi} V[n, e^{j\theta}) \frac{d\theta}{2\pi} = r[n, 0] = \mathrm{E}[|x[n]|^2] \geq 0. \tag{11.25}$$

The function $V[n, e^{j\theta})$ is a time-frequency correlation at global time n and global frequency θ; the function $A(e^{j\nu}, k]$ is a Doppler-delay ambiguity function at local frequency ν and local time k; the function $S(e^{j(\theta+\nu)}, e^{j\theta})$ is a two-dimensional power spectrum at global frequency θ and local frequency ν; and it is sometimes called a modulation transfer function. The notation $[\cdot, \cdot)$ is used to indicate that the leftmost variable is discrete and the rightmost is continuous. It may be noted that in Fourier transforms, a term like $n\nu$ is a term in global time n and local frequency ν, as is usual in Fourier analysis, where variation of a function over global time n determines a Fourier transform at local frequency ν. Similarly, a term like $k\theta$ is a term in local time k and global frequency θ, as is usual in (inverse) Fourier transforms, where variation in a Fourier transform over global frequency θ determines function value at local time k.

The sequence value $x[n]$ and Fourier transform term $X(e^{j\theta})e^{jn\theta}$ appearing in the NE corner of the four-corner diagram may be organized into the two-dimensional vector $[x[n] \quad X(e^{j\theta})e^{j\theta n}]^T$, with correlation matrix

$$\mathrm{E}\left[\begin{bmatrix} x[n] \\ X(e^{j\theta})e^{j\theta n} \end{bmatrix} \begin{bmatrix} x^*[n] & (X(e^{j\theta})e^{j\theta n})^* \end{bmatrix}\right] = \begin{bmatrix} r[n, 0] & V[n, e^{j\theta}) \\ V^*[n, e^{j\theta}) & S(e^{j\theta}, e^{j\theta}) \end{bmatrix}.$$

The LMMSE estimator of $x[n]$ from the one-term Fourier component $X(\theta)e^{jn\theta}$ and its corresponding error variance are

$$\hat{x}[n] = \frac{V[n, e^{j\theta})}{S(e^{j\theta}, e^{j\theta})} X(e^{j\theta})e^{jn\theta},$$

$$Q[n, e^{j\theta}) = r[n, 0] - \frac{|V[n, e^{j\theta})|^2}{S(e^{j\theta}, e^{j\theta})} = r[n, 0](1 - |\rho[n, e^{j\theta})|^2),$$

where

$$|\rho[n, e^{j\theta})|^2 = \frac{|V[n, e^{j\theta})|^2}{r[n, 0]S(e^{j\theta}, e^{j\theta})} = \frac{|V[n, e^{j\theta})|^2}{(\int_{-\pi}^{\pi} V[n, e^{j\theta})\frac{d\theta}{2\pi})(\sum_{n=0}^{N-1} V[n, e^{j\theta}))}$$

$$= \frac{|\,\mathrm{E}[x[n](X(e^{j\theta})e^{jn\theta})^*]|^2}{\mathrm{E}[|x[n]|^2]\,\mathrm{E}[|X(e^{j\theta})|^2]} \le 1.$$

These results have several interpretations:

1. The time-frequency correlation $V[n, e^{j\theta})$ measures the correlation between the time series value $x[n]$ and the single frequency phasor $X(e^{j\theta})e^{jn\theta}$ in the Hilbert space of second-order random variables; it may be called a stochastic Rihaczek distribution.
2. The time-frequency correlation $V[n, e^{j\theta})$ is complex, but its marginals in (11.24) and (11.25) are real and non-negative.
3. Suitably normalized, the Rihaczek distribution is the coherence $|\rho[n, e^{j\theta})|^2$, or cosine-squared of the angle between the time series random variable $x[n]$ and the phasor $X(e^{j\theta})e^{jn\theta}$ in the Hilbert space of second-order random variables.
4. The complex Rihaczek spectrum determines a complex time-varying LMMSE filter for estimating a value of a time series from a value of its one-term Fourier expansion.
5. The magnitude squared of the stochastic Rihaczek distribution determines the mean-squared error of the estimator and provides a basis for determining the local spectrum of the time series.

As long in global time n as the magnitude-squared coherence $|\rho[n, e^{j\theta})|^2$ remains near to one, then the time series $\{x[n]\}$ is estimable by the rotating phasor $X(e^{j\theta})e^{jn\theta}$. As long in global frequency θ as $|\rho[n, e^{j\theta})|^2$ is near to one, the Fourier transform $X(e^{j\theta})$ is estimable from the rotating phasor $x[n]e^{-jn\theta}$. These effects are illustrated in Fig. 11.2, where it is suggested that a time series that hugs the phasor $X(e^{j\theta})e^{jn\theta}$ over a short time interval has a squared coherence near to one and may be said to have a spectrum concentrated at the value $X(e^{j\theta})$ over that time interval. This seems a reasonable way to think about a time-frequency distribution: the complex coherence $\rho[n, e^{j\theta})$, and its magnitude squared, measures the similarity between a time series value and the value that would be predicted from a single frequency phasor.

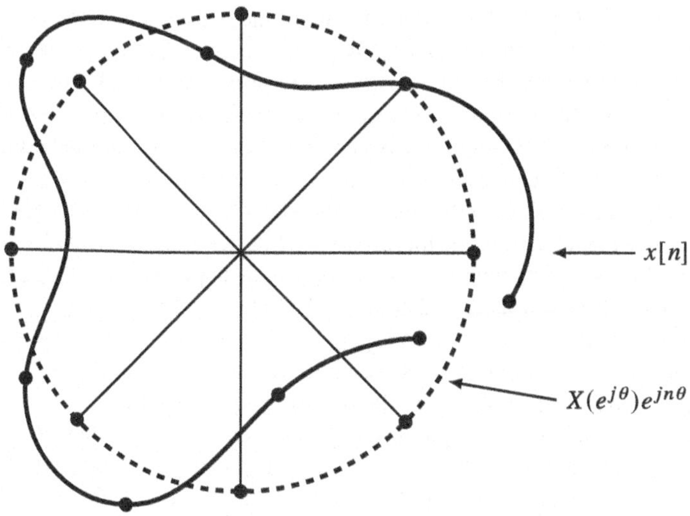

Fig. 11.2 Approximation of $x[n]$ (solid line) by the rotating phasor $X(\theta)e^{jn\theta}$ (dashed line), for $\theta = 2\pi/8$. Observations are denoted by bullets

This account may be called a finite-dimensional account of the stochastic Rihaczek distribution [288]. A more comprehensive account of its geometry and its connection to harmonizable random sequences may be found in [310]. The Rihaczek distribution figures prominently in the work of Martin, Flandrin, and Hlawatsch [121, 122, 166, 231, 232].

11.6 Chapter Notes

1. Coherence was used by Mallat and Zhang in the early 1990s as a heuristic quantity for matching pursuit [225]. Its prominence in compressed sensing was made clear by Donoho and colleagues in [106] and [105]. Coherence was also used by Tropp in [344] to characterize a dictionary in linear sparse approximation problems. An excellent review by Candés and Wakin may be found in [60]. The definition of restricted isometries first appeared in [58]. The derivation of the coherence index in Sect. 11.1 is based on [333].
2. In 1961, Horst [171] first proposed multiset canonical correlation analysis (MCCA) to estimate the pairwise correlations of multiple datasets. He provided two formulations: the sum of correlations (SUMCOR) and the maximum variance (MAXVAR). Carroll [63] in 1968 proposed to find a shared latent correlated space, which was shown to be identical to MAXVAR-MCCA. More recent papers considering this subspace-based approach to MCCA are [71] and [329]. In 1971, Kettenring [195] added three new formulations to MCCA that maximize the sum of the squared correlations (SSQCOR), minimize the smallest eigenvalue

(MINVAR), and minimize the determinant of the correlation matrix (GENVAR), respectively. The formulation of CCA as optimization problems was discussed in [157], where kernel-CCA extensions were proposed as well. The observation that MAXVAR-MCCA can be reformulated, with the appropriate constraint, as an Euclidean distance function between linear transformations, and equivalently as a set of coupled regression problems, was put forward in [365].

3. The theory of positive definite functions was developed by Mercer [241] and gave rise to many applications in the theory of Fourier transforms and topological groups. Aronszajn's seminal paper [15], which appeared in 1950, provides a nice introduction and historical background. The last decades have witnessed a real explosion in the use of reproducing kernels and kernel methods in machine learning, with the support vector machine (SVM) developed by V. Vapnik and colleagues [86, 362] identified as one of its most important uses.
4. Kernel CCA was originally proposed by Hardoon, Szedmak, and Shawe-Taylor in [157]. The coherence-based criterion was originally proposed for kernel adaptive filtering algorithms by Richard, Bermudez, and Honeine in [283].

Epilogue

12

Many of the results in this book have been derived from maximum likelihood reasoning in the multivariate normal model. With interesting parametric constraints on means and covariances, this reasoning produces detectors and estimators that are quite unexpected functions of measured data, often involving complicated functions of eigenvalues. It is quite common for a subspace geometry in a Euclidean or Hilbert space to emerge, and this geometry brings evocative insights. The corresponding distribution theory for estimators and detectors is the theory of distributions for functions of normal random variables. Many coherence statistics are distributed as beta random variables or products of beta-distributed random variables.

But perhaps it is the geometry that is fundamental, and not the distribution theory that produced it. This suggests that geometric reasoning, detached from distribution theory, may provide a way to address vexing problems in signal processing and machine learning, especially when there is no theoretical basis for assigning a distribution to data. With this point of view, geometric reasoning is followed by performance analysis based on a hypothetical distribution for data. This is certainly the way conventional least squares and minimum mean-squared error theory proceed. Much of measurement theory in science proceeds this way. Measurements are made, and the search for structure begins. If it is found, then the search for a physical, chemical, or biological basis for this structure ensues. The suggestion that coherence is an organizing principle is a suggestion that a way to look for structure is to look for coherence in one of its many guises. And in many cases, it seems that a way to look for coherence is to look for statistics that are invariant to transformation groups. If a coherence statistic is found from this reasoning, then an idealized model for the distribution of the data may be used to study the performance of a coherence statistic as a detector of effect or mechanism.

In the processing of signals from many sensors, classical measures of space-time coherence between signals seem an obvious way to determine whether there is a common source for the measured signals. In this case the definition of coherence may seem obvious. But as the results of this book show, coherence statistics are not always so obvious. Perhaps reasoning from a hypothesized distribution reveals other

D. Ramírez et al., *Coherence*, https://doi.org/10.1007/978-3-031-13331-2_12

measures of coherence, such as complicated functions of eigenvalues, trace, and determinant. This suggests a synergy between geometric reasoning and reasoning from distribution theory.

Notation

A

Sets

$\mathbb{Z}$	Set of integers
$\mathbb{R}$	Set of real numbers
$\mathbb{C}$	Set of complex numbers
$\mathbb{R}^M$	Euclidean space of real M-dimensional vectors
$\mathbb{C}^M$	Euclidean space of complex M-dimensional vectors
$\mathcal{S}_{+}^{p}$	Set of Hermitian (symmetric in the real case) $p \times p$ positive semidefinite matrices
$\mathcal{S}_{++}^{p}$	Set of Hermitian (symmetric in the real case) $p \times p$ positive definite matrices
S^{M-1}	Unit sphere in $\mathbb{R}^M$
$U(M)$	Set of $M \times M$ unitary matrices (or unitary group)
$O(M)$	Set of $M \times M$ orthogonal matrices (or orthogonal group)
$\mathbb{S}t(p, \mathbb{F}^n)$	Stiefel manifold of p frames in $\mathbb{F}^n$ ($\mathbb{F} = \mathbb{R}$ or $\mathbb{F} = \mathbb{C}$)
$\mathbb{G}r(p, \mathbb{F}^n)$	Grassmann manifold of p-dimensional subspaces of $\mathbb{F}^n$ ($\mathbb{F} = \mathbb{R}$ or $\mathbb{F} = \mathbb{C}$)
$GL(\mathbb{F}^n)$	General linear group of nonsingular $n \times n$ matrices in $\mathbb{F}^n$ ($\mathbb{F} = \mathbb{R}$ or $\mathbb{F} = \mathbb{C}$)
$\ell_2(\mathbb{Z})$	Hilbert space of the square summable signals with inner product $\langle x, y \rangle = \sum_{n=-\infty}^{\infty} x[n]y^*[n]$ and norm $\|x\| = \sqrt{\langle x, x \rangle}$
$L_2(T)$	Hilbert space of square integrable signals in $[-T/2, T/2]$, with inner product $\langle x, y \rangle = \frac{1}{T}\int_{-T/2}^{T/2} x(t)y^*(t)\mathrm{d}t$ and norm $\|x\| = \sqrt{\langle x, x \rangle}$
$L_2(\mathbb{R})$	Hilbert space of square integrable signals in $\mathbb{R}$, with inner product $\langle x, y \rangle = \int_{-\infty}^{\infty} x(t)y^*(t)\mathrm{d}t$ and norm $\|x\| = \sqrt{\langle x, x \rangle}$

D. Ramírez et al., *Coherence*, https://doi.org/10.1007/978-3-031-13331-2

Scalars, Vectors, Matrices, and Functions

a	Real or complex number
$\|a\|$	Magnitude (modulus) of a
a^*	Conjugate of a
$\mathrm{Re}\{a\}$	Real part of a
$\mathrm{Im}\{a\}$	Imaginary part of a
j	Imaginary unit $j = \sqrt{-1}$; or sometimes $i = \sqrt{-1}$
$(a_1, \ldots, a_M)$	a list of numbers
$\mathbf{a} = [a_1 \cdots a_M]^T$	M-dimensional complex (or real) vector; a column vector
$\langle \mathbf{a}, \mathbf{b} \rangle_{\mathbb{C}^M} = \mathbf{a}^H \mathbf{b} = \sum_{m=1}^{M} a_m^* b_m$	Inner product between the M-dimensional complex (or real) vectors $\mathbf{a}$ and $\mathbf{b}$
$\\|\mathbf{a}\\|_2 = \sqrt{\langle \mathbf{a}, \mathbf{a} \rangle_{\mathbb{C}^M}}$	Euclidean norm of the M-dimensional complex (or real) vector $\mathbf{a}$; or sometimes $\\|\mathbf{a}\\|$
$\\|\mathbf{a}\\|_0 = \dim(\{k \mid a_k \neq 0\})$	ℓ_0 norm of the M-dimensional complex (or real) vector $\mathbf{a}$
$\mathbf{A}$	Complex (or real) matrix
A_{il}, $(\mathbf{A})_{il}$, or a_{il}	ilth element of matrix $\mathbf{A}$
$\mathbf{A}^H$	Hermitian transpose (or conjugate transpose) of $\mathbf{A}$
$\mathbf{A}^{-1}$	Inverse of $\mathbf{A}$
$\mathbf{A}^\#$	Moore-Penrose pseudo-inverse of $\mathbf{A}$
$[\mathbf{A}, \mathbf{B}]$	Commutator of linear operators (or matrices) $\mathbf{A}$ and $\mathbf{B}$
$\mathrm{tr}(\mathbf{A})$	Trace of the complex (or real) matrix $\mathbf{A}$
$\det(\mathbf{A})$	Determinant of the complex (or real) matrix $\mathbf{A}$
$\mathrm{etr}(\mathbf{A})$	Exponential trace of the complex (or real) matrix $\mathbf{A}$
$\\|\mathbf{A}\\|$	Frobenius norm of the complex (or real) matrix $\mathbf{A}$
$\\|\mathbf{A}\\|_2$	Spectral (or ℓ_2) norm of the complex (or real) matrix $\mathbf{A}$
$\\|\mathbf{A}\\|_*$	Nuclear norm of the complex (or real) matrix $\mathbf{A}$
$ev_i(\mathbf{A})$	ith largest eigenvalue of the Hermitian complex (or symmetric real) matrix $\mathbf{A}$; when clear from the context, also denoted λ_i with $\lambda_i = ev_i(\mathbf{A})$
$\mathbf{A} \odot \mathbf{B}$	Hadamard (elementwise) product between $\mathbf{A}$ and $\mathbf{B}$
$\mathbf{A} \otimes \mathbf{B}$	Kronecker product between $\mathbf{A}$ and $\mathbf{B}$
$\mathbf{A} \succeq \mathbf{B}$	The matrix $\mathbf{A} - \mathbf{B}$ is positive semidefinite
$\mathrm{diag}(a_1, a_2, \ldots, a_M)$	Diagonal matrix built from the scalars a_1, $a_2, \ldots, a_M$
$\mathrm{diag}(\mathbf{A})$	Diagonal matrix built from the diagonal elements of $\mathbf{A}$

$\text{blkdiag}(\mathbf{A}_1, \mathbf{A}_2, \ldots, \mathbf{A}_M)$	Block-diagonal matrix built from the matrices $\mathbf{A}_1, \mathbf{A}_2, \ldots, \mathbf{A}_M$
$\text{blkdiag}_L(\mathbf{A})$	Block-diagonal matrix built from the $L \times L$ blocks on the diagonal of $\mathbf{A}$
$\text{vec}(\mathbf{A})$	Vectorization operator by stacking the columns of $\mathbf{A}$
$\mathbf{F}_N$	Fourier matrix of dimension n, i.e., $(\mathbf{F}_N)_{kn} = e^{-j2\pi(k-1)(n-1)/N}$
$\mathbf{F}_N\mathbf{x}$	Discrete Fourier transform (DFT) of the vector $\mathbf{x} = [x[0] \cdots x[N-1]]^T$
$\langle \mathbf{V} \rangle$	Subspace spanned by the columns of $\mathbf{V}$
$\langle \mathbf{V} \rangle^\perp$	Subspace orthogonal to $\langle \mathbf{V} \rangle$
$\mathbf{P_V} = \mathbf{V}(\mathbf{V}^H\mathbf{V})^{-1}\mathbf{V}^H$	Orthogonal projection matrix onto $\langle \mathbf{V} \rangle$
$\mathbf{P}_\mathbf{V}^\perp = \mathbf{I} - \mathbf{P_V}$	Orthogonal projection matrix onto the subspace orthogonal to $\langle \mathbf{V} \rangle$
$(x)^+$, $(\mathbf{x})^+$, or $(\mathbf{X})^+$	Elementwise operation $\max(\cdot, 0)$
$\Gamma(x)$	Gamma function
$\Gamma_p(x)$	Multivariate gamma function
$\tilde{\Gamma}_p(x)$	Complex multivariate gamma function
p_{fa}	Probability of false alarm; sometimes P_{FA}
p_d	Probability of detection; sometimes P_D

Probability, Random Variables, and Distributions

$\overset{d}{=}$	Equality in distribution
$Pr[S]$	Probability of the event S
$\mathrm{E}[\cdot]$	Mathematical expectation
$\langle x, y \rangle = \mathrm{E}[xy^*]$	Inner product between the random variables x and y (defined in the Hilbert space $\mathcal{H}$ of second-order complex random variables)
$D_{KL}(P\|\|Q)$	Kullback-Leibler divergence between the distributions P and Q
$r_{xy} = \langle x, y \rangle = \mathrm{E}[xy^*]$	Covariance between the (zero-mean) complex random variables x and y
$\tilde{r}_{xy} = \langle x, y^* \rangle = \mathrm{E}[xy]$	Complementary covariance between the (zero-mean) complex random variables x and y
$\mathbf{r}_{x\mathbf{y}} = \langle x, \mathbf{y} \rangle = \mathrm{E}[x\mathbf{y}^H]$	Covariance vector between the (zero-mean) complex random variable x and the (zero-mean) complex random vector $\mathbf{y}$; a row vector
$\tilde{\mathbf{r}}_{x\mathbf{y}} = \langle x, \mathbf{y}^* \rangle = \mathrm{E}[x\mathbf{y}^T]$	Complementary covariance vector between the complex (zero-mean) random variable x and the complex (zero-mean) random vector $\mathbf{y}$

$\mathbf{R}_{\mathbf{xy}} = \langle \mathbf{x}, \mathbf{y} \rangle = \mathrm{E}[\mathbf{x}\mathbf{y}^H]$	Covariance matrix between the complex (zero-mean) random vectors $\mathbf{x}$ and $\mathbf{y}$
$\tilde{\mathbf{R}}_{\mathbf{xy}} = \langle \mathbf{x}, \mathbf{y}^* \rangle = \mathrm{E}[\mathbf{x}\mathbf{y}^T]$	Complementary covariance matrix between the complex (zero-mean) random vectors $\mathbf{x}$ and $\mathbf{y}^*$
$\mathbf{C}_{\mathbf{xy}} = \mathbf{R}_{\mathbf{xx}}^{-1/2}\mathbf{R}_{\mathbf{xy}}\mathbf{R}_{\mathbf{yy}}^{-1/2}$	Coherence matrix between the (zero-mean) random vectors $\mathbf{x}$ and $\mathbf{y}$; often $\mathbf{C}$ without subscripts
$\mathrm{Exp}(\lambda)$	Exponential distribution of parameter λ (mean $1/\lambda$)
$\mathcal{U}[a, b]$	Uniform distribution on the interval $[a, b]$
$\Gamma(p, q)$	Gamma distribution with parameters p and q
$GIG(a, b, r)$	Generalized inverse Gaussian distribution with parameters a, b, and r
$\mathrm{Beta}(p, q)$	Beta distribution with parameters p and q
$\mathrm{Beta}(p, q, \lambda)$	Noncentral beta distribution with shape parameters p and q and noncentrality parameter λ
$\mathrm{Beta}_n(p, q)$	$n \times n$ matrix-variate beta distribution with shape parameters p and q
$C\mathrm{Beta}_n(p, q)$	Complex $n \times n$ central matrix-variate beta distribution with shape parameters p and q
$C\mathrm{Beta}_n(p, q, \mathbf{M}\mathbf{M}^H)$	Complex noncentral matrix-variate beta distribution with shape parameters p and q and noncentrality parameter $\mathbf{M}\mathbf{M}^H$
$\mathcal{N}_n(\boldsymbol{\mu}, \boldsymbol{\Sigma})$	n-dimensional Gaussian distribution with mean $\boldsymbol{\mu}$ and covariance matrix $\boldsymbol{\Sigma}$
$C\mathcal{N}_n(\boldsymbol{\mu}, \boldsymbol{\Sigma})$	n-dimensional proper complex Gaussian distribution with mean $\boldsymbol{\mu}$ and covariance matrix $\boldsymbol{\Sigma}$
$\mathcal{N}_{n\times p}(\mathbf{M}, \boldsymbol{\Sigma}_r \times \boldsymbol{\Sigma}_c)$	$n \times p$ matrix-variate Gaussian distribution whose vectorization has mean $\mathrm{vec}(\mathbf{M})$ and covariance matrix $\boldsymbol{\Sigma}_r \otimes \boldsymbol{\Sigma}_c$
$C\mathcal{N}_{n\times p}(\mathbf{M}, \boldsymbol{\Sigma}_r \times \boldsymbol{\Sigma}_c)$	$n \times p$ proper complex matrix-variate Gaussian distribution whose vectorization has mean $\mathbf{M}$ and covariance matrix $\boldsymbol{\Sigma}_r \times \boldsymbol{\Sigma}_c$
χ^2_ν	χ^2 distribution with ν degrees of freedom
$\chi^2_\nu(\lambda)$	Noncentral χ^2 distribution with ν degrees of freedom and noncentrality parameter λ
$\mathcal{W}_n(\boldsymbol{\Sigma}, m)$	n-dimensional Wishart distribution with m degrees of freedom and parameter $\boldsymbol{\Sigma}$
$C\mathcal{W}_n(\boldsymbol{\Sigma}, m)$	n-dimensional complex Wishart distribution with m degrees of freedom and parameter $\boldsymbol{\Sigma}$
t_p	Student's t-distribution with p degrees of freedom
$\mathcal{T}_{n\times p}(\mathbf{M}, \boldsymbol{\Sigma}_r \times \boldsymbol{\Sigma}_c)$	$n \times p$ matrix-variate t-distribution whose vectorization has mean $\mathrm{vec}(\mathbf{M})$ and covariance matrix $\boldsymbol{\Sigma}_r \times \boldsymbol{\Sigma}_c$
$\mathcal{F}(p, q)$	F (or Fisher-Snedecor) distribution with degrees of freedom p and q

$\mathcal{F}_n(p, q)$	$n \times n$ matrix-variate F-distribution with degrees of freedom p and q
$\mathcal{L}_{n\times p}(\mathbf{H})$	Matrix Langevin distribution on the Stiefel $\mathbb{S}_t(p, \mathbb{R}^n)$ with parameter $\mathbf{H}$
$\mathcal{B}_{n\times p}(\mathbf{H})$	Matrix Bingham distribution on the Stiefel $\mathbb{S}_t(p, \mathbb{R}^n)$ with parameter $\mathbf{H}$
$ACG(\boldsymbol{\Sigma})$	Angular central Gaussian on the Stiefel $\mathbb{S}_t(1, \mathbb{R}^n)$ with parameter $\boldsymbol{\Sigma}$
$MACG(\boldsymbol{\Sigma})$	Matrix angular central Gaussian on the Stiefel $\mathbb{S}_t(p, \mathbb{R}^n)$ with parameter $\boldsymbol{\Sigma}$

Basic Results in Matrix Algebra

B

In this appendix, we summarize some basic results in matrix theory. Some are general and others are specialized results that are used to derive results in the book. Excellent references for further reading are [142] and [170].

B.1 Matrices and their Diagonalization

An $n \times m$ matrix $\mathbf{A}$ is a rectangular array of real or complex numbers written as

$$\mathbf{A} = \begin{bmatrix} a_{11} & a_{12} & \cdots & a_{1m} \\ a_{21} & a_{22} & \cdots & a_{2m} \\ \vdots & \vdots & \ddots & \vdots \\ a_{n1} & a_{n2} & \cdots & a_{nm} \end{bmatrix}.$$

Sometimes, A_{il} or $(\mathbf{A})_{il}$ is used to denote a_{il}. All elements of a matrix are taken to be complex unless otherwise stated, that is, $\mathbf{A} \in \mathbb{C}^{n\times m}$. If the columns and rows of a matrix $\mathbf{A}$ are interchanged, the resulting matrix is called the *transpose* of $\mathbf{A}$ and is denoted by $\mathbf{A}^T$. The conjugate transpose (or Hermitian transpose) of an $n \times m$ matrix $\mathbf{A}$ is the $m \times n$ matrix obtained by taking the transpose and then taking the complex conjugate of each entry and is denoted by $\mathbf{A}^H$.

If $m = n$, $\mathbf{A}$ is called square of order or dimension n. A square matrix is said to be a diagonal matrix if all off-diagonal elements are zero, and it will be denoted as $\mathbf{A} = \operatorname{diag}(a_1, \dots, a_n)$. If all the a_i are equal to 1, it is called an identity matrix, denoted $\mathbf{I}_n$ (or simply $\mathbf{I}$ if the dimension is understood). If all elements of $\mathbf{A}$ are zero, then it is a zero matrix or null matrix of dimension n, denoted $\mathbf{0}_n$ (or simply $\mathbf{0}$ if the dimension is understood). For $m \neq n$, the null matrix is a matrix of all zero elements. A square matrix is said to be *Hermitian* if $\mathbf{A} = \mathbf{A}^H$. When all elements

D. Ramírez et al., *Coherence*, https://doi.org/10.1007/978-3-031-13331-2

of $\mathbf{A}$ are real, the matrix is said to be symmetric if $\mathbf{A} = \mathbf{A}^T$. A square matrix is said to be *skew-Hermitian* (*skew-symmetric* in the real case) if $\mathbf{A} = -\mathbf{A}^H$ ($\mathbf{A} = -\mathbf{A}^T$). It is clear that if $\mathbf{A}$ is skew-Hermitian, then the elements on the main diagonal are pure imaginary $a_{ii} = -a_{ii}^*$, $i = 1, \ldots, n$ ($a_{ii} = 0$ if the matrix is skew-symmetric). An $n \times n$ square matrix is said to be *unitary* if $\mathbf{A}^H\mathbf{A} = \mathbf{A}\mathbf{A}^H = \mathbf{I}_n$.

Diagonalizable Matrices. An $n \times n$ matrix $\mathbf{A}$ is said to be diagonalizable if there exists an $n \times n$ nonsingular (invertible) matrix $\mathbf{P}$ such that $\mathbf{P}^{-1}\mathbf{A}\mathbf{P} = \mathbf{\Lambda}$, where $\mathbf{\Lambda}$ is diagonal. The columns of $\mathbf{P}$ are a non-orthogonal basis for $\mathbb{C}^n$ and the rows of $\mathbf{P}^{-1}$ are a non-orthogonal dual basis. It follows that $\mathbf{AP} = \mathbf{P\Lambda}$, which column-by-column may be written $\mathbf{A}\mathbf{p}_k = \mathbf{p}_k\lambda_k$. The matrix $\mathbf{A}$ may be written as $\mathbf{A} = \sum_{k=1}^n \lambda_k \mathbf{p}_k \mathbf{q}_k^H$, where $\mathbf{p}_k$ is the kth column of $\mathbf{P}$ and $\mathbf{q}_k^H$ is the kth row of $\mathbf{P}^{-1}$.

When is a matrix diagonalizable, and how is its diagonal representation computed? Begin with the matrix $\mathbf{A}$ and construct the *characteristic polynomial*

$$\det(\lambda\mathbf{I}_n - \mathbf{A}) = p(\lambda) = \lambda^n + c_{n-1}\lambda^{n-1} + \cdots + c_1\lambda + c_0.$$

By the fundamental theorem of algebra, this polynomial has n roots $\lambda_1, \ldots, \lambda_n$. For each of them, the matrix $\lambda_k\mathbf{I}_n - \mathbf{A}$ is rank deficient, and therefore there exists at least one vector $\mathbf{p}_k$ such that $(\lambda_k\mathbf{I}_n - \mathbf{A})\mathbf{p}_k = \mathbf{0}$. If the vectors $\mathbf{p}_1, \ldots, \mathbf{p}_n$ are a basis for $\mathbb{C}^n$, which is to say $\mathbf{P}$ is nonsingular, then $\mathbf{P}$ diagonalizes $\mathbf{A}$. The vector $\mathbf{p}_k$ is called an eigenvector of $\mathbf{A}$ and λ_k is called an eigenvalue. That is, $\mathbf{A}\mathbf{p}_k = \mathbf{p}_k\lambda_k$, $k = 1, \ldots, n$.

The Cayley-Hamilton theorem states that any square diagonalizable matrix satisfies its own characteristic equation, so

$$p(\mathbf{A}) = \mathbf{A}^n + c_{n-1}\mathbf{A}^{n-1} + \cdots + c_1\mathbf{A} + c_0\mathbf{I}_n = \mathbf{0}_n.$$

That is, $\mathbf{A}^n = -c_{n-1}\mathbf{A}^{n-1} - \cdots - c_1\mathbf{A} - c_0\mathbf{I}_n$. This result may be iterated to expand $\mathbf{A}^p$, for any $p \geq n$, as a linear combination of $\mathbf{A}^{n-1}, \ldots, \mathbf{A}, \mathbf{I}$, with coefficients in the linear combination determined entirely by $c_{n-1}, \ldots, c_1, c_0$.

Note that the trace and the determinant of the matrix always appear as coefficients of the characteristic polynomial; the constant coefficient is $c_0 = (-1)^n \det(\mathbf{A})$ and the coefficient of λ^{n-1} is $c_{n-1} = -\operatorname{tr}(\mathbf{A})$.

Example B.1 For a 2×2 matrix $\mathbf{A}$, the characteristic polynomial is

$$\lambda^2 - \operatorname{tr}(\mathbf{A})\lambda + \det(\mathbf{A}),$$

and the Cayley-Hamilton theorem yields

$$\mathbf{A}^2 - \operatorname{tr}(\mathbf{A})\mathbf{A} + \det(\mathbf{A})\mathbf{I}_2 = \mathbf{0}_2.$$

If $\mathbf{A}$ is invertible, its inverse can be found as

$$\mathbf{A}^{-1} = -\frac{1}{\det(\mathbf{A})}(\mathbf{A} - \text{tr}(\mathbf{A})\mathbf{I}_2).$$

Normal Matrices. An $n \times n$ complex square matrix $\mathbf{A}$ is *normal* if it commutes with its Hermitian, i.e., $\mathbf{A}\mathbf{A}^H = \mathbf{A}^H\mathbf{A}$. A matrix is normal iff it is unitarily similar to a diagonal matrix; that is, there exists a unitary matrix $\mathbf{U}$ and a diagonal matrix $\mathbf{\Lambda}$ of eigenvalues λ_k ordered as $|\lambda_1| \geq |\lambda_2| \geq \cdots \geq |\lambda_n|$ such that $\mathbf{A} = \mathbf{U}\mathbf{\Lambda}\mathbf{U}^H = \sum_{k=1}^n \lambda_k \mathbf{u}_k \mathbf{u}_k^H$. This is called the spectral theorem for normal matrices.

Eigenvalues of special normal matrices are special.

- Unitary: $\mathbf{A}\mathbf{A}^H = \mathbf{I}_n \Rightarrow |\lambda_k| = 1, k = 1, \ldots, n$
- Hermitian: $\mathbf{A} = \mathbf{A}^H \Rightarrow \lambda_k \in \mathbb{R}, k = 1, \ldots, n$
- Hermitian positive definite: $\mathbf{v}^H\mathbf{A}\mathbf{v} > 0 \Rightarrow \lambda_k > 0, k = 1, \ldots, n$
- Hermitian positive semidefinite: $\mathbf{v}^H\mathbf{A}\mathbf{v} \geq 0 \Rightarrow \lambda_k \geq 0, k = 1, \ldots, n$

A more detailed characterization of Hermitian matrices and their eigenvalues is presented in Sect. B.2.

Simultaneous Diagonalization. Two diagonalizable matrices $\mathbf{A}$ and $\mathbf{B}$ are said to be simultaneously diagonalizable, if there exists a single invertible matrix $\mathbf{P}$ such that $\mathbf{P}\mathbf{A}\mathbf{P}^{-1}$ is diagonal and $\mathbf{P}\mathbf{B}\mathbf{P}^{-1}$ is diagonal. The diagonalizable matrices $\mathbf{A}$ and $\mathbf{B}$ are simultaneously diagonalizable, iff they commute: $\mathbf{AB} = \mathbf{BA}$. If each of the matrices $\mathbf{A}$ and $\mathbf{B}$ is normal, which is to say $\mathbf{A}^H\mathbf{A} = \mathbf{A}\mathbf{A}^H$ and $\mathbf{B}^H\mathbf{B} = \mathbf{B}\mathbf{B}^H$, then they are simultaneously diagonalized by a unitary matrix iff they commute. These results are proved in standard textbooks on matrix analysis.

A General Characterization of Eigenvalues of Complex Matrices. Another characterization of the eigenvalues of a square $n \times n$ diagonalizable matrix $\mathbf{A}$ is given by the Gershgorin disks or circle theorem.

Theorem (Gershgorin Disks.) *Let* $\mathbf{A}$ *be an* $n \times n$ *complex matrix and let* λ *be an eigenvalue of* $\mathbf{A}$. *Define the Gershgorin disks of* $\mathbf{A}$ *as*

$$\mathcal{D}_i = \left\{ \mu \mid |\mu - a_{ii}| \leq \sum_{l \neq i} |a_{il}| \right\}.$$

Then, $\lambda \in \cup_{i=1}^n \mathcal{D}_i$. *Moreover, if the union of* k *of the sets* $\mathcal{D}_i$ *is disjoint from the others, then that union contains exactly* k *eigenvalues of* $\mathbf{A}$.

Proof Let **u** and λ be an eigenvector-eigenvalue pair of **A**. Let u_i be the component of **u** with maximum modulus. From the ith row of equation $\mathbf{Au} = \lambda\mathbf{u}$, we have

$$a_{ii}u_i + \sum_{l \neq i} a_{il}u_l = \lambda u_i,$$

or

$$a_{ii} + \sum_{l \neq i} a_{il}\frac{u_l}{u_i} = \lambda.$$

Taking absolute values and noticing that $\left|\frac{u_l}{u_i}\right| \leq 1$, we find that

$$|\lambda - a_{ii}| \leq \sum_{l \neq i} |a_{il}|,$$

which means that $\lambda \in \mathcal{D}_i$. The statement about the disjoint union can be established by a continuity argument (see [334] for a proof). □

B.2 Hermitian Matrices and their Eigenvalues

An $n \times n$ complex matrix is said to be Hermitian if $\mathbf{A} = \mathbf{A}^H$. Clearly, it is normal, which is to say $\mathbf{AA}^H = \mathbf{A}^H\mathbf{A}$. Therefore, by the spectral theorem for normal matrices, it may be diagonalized with eigenvalues $\lambda_i, i = 1, \ldots, n$. All eigenvalues are real. If all of these eigenvalues are greater than zero, then **A** is said to be Hermitian positive definite, which is to say $\mathbf{x}^H\mathbf{Ax} > 0$, for all $\mathbf{x} \in \mathbb{C}^n$ (see Sect. B.2.2). If all of these eigenvalues are greater than or equal to zero, then **A** is said to be Hermitian positive semidefinite, which is to say $\mathbf{x}^H\mathbf{Ax} \geq 0$, for all $\mathbf{x} \in \mathbb{C}^n$.

Write the $n \times n$ Hermitian matrix as $\mathbf{A} = \mathbf{C} + j\mathbf{S}$. The real matrices **C** and **S** may be determined as $\mathbf{C} = (\mathbf{A} + \mathbf{A}^*)/2$ and $\mathbf{S} = (\mathbf{A} - \mathbf{A}^*)/2j$. This is the matrix generalization of the Euler identity $e^{j\theta} = (e^{j\theta} + e^{-j\theta})/2 + j(e^{j\theta} - e^{-j\theta})/2j = \cos\theta + j\sin\theta$. Since **A** is Hermitian, then $\mathbf{C}^T = \mathbf{C}$ and $\mathbf{S}^T = -\mathbf{S}$. That is, the real matrix **C** is symmetric and the real matrix **S** is skew-symmetric. Moreover, the quadratic form $\mathbf{x}^H\mathbf{Ax}$ is real, which shows $\mathbf{x}^H\mathbf{Sx} = 0$. Conversely, if **A** is skew-Hermitian, denoted $\mathbf{A} = -\mathbf{A}^H$, then $\mathbf{C}^T = -\mathbf{C}$ and $\mathbf{S}^T = \mathbf{S}$. Additionally, the quadratic form $\mathbf{x}^H\mathbf{Ax}$ is imaginary, which shows $\mathbf{x}^H\mathbf{Cx} = 0$.

Matrix Exponentials of Hermitian Matrices Are Unitary. The matrix exponential of an $n \times n$ diagonalizable matrix $\mathbf{A} = \mathbf{P\Lambda P}^{-1}$, with eigenvalues λ_i, is $\exp(\mathbf{A}) = \mathbf{P}\exp(\mathbf{\Lambda})\mathbf{P}^{-1}$, with eigenvalues e^{λ_i}. The matrix exponential $\mathbf{U} = \exp(-j\mathbf{A})$ is unitary iff **A** is Hermitian. For the if part, assume **A** is Hermitian. Then $\mathbf{U}^H\mathbf{U} = \exp(j\mathbf{A}^H)\exp(-j\mathbf{A}) = \mathbf{I}_n$. For the only if part, assume $\mathbf{U} = \exp(-j\mathbf{A})$ is unitary. Then $\mathbf{I}_n = \exp(\mathbf{0}_n) = \exp(j\mathbf{A}^H)\exp(-j\mathbf{A})$. Expand these exponentials in a power series to show that $\mathbf{A}^H = \mathbf{A}$.

This result has important consequences for the representation of real orthogonal matrices. Begin with a skew-symmetric matrix $\mathbf{S} = -\mathbf{S}^T$. Construct the special Hermitian matrix $\mathbf{A} = j\mathbf{S}$. Then $\mathbf{U} = \exp(-j\mathbf{A})$ is unitary, which is to say $\mathbf{U} = \exp(\mathbf{S})$ is a real orthogonal matrix.

B.2.1 Characterization of Eigenvalues of Hermitian Matrices

The eigenvalues of Hermitian matrices, besides being real, may be characterized as the solutions of a series of optimization problems. Let us denote the real eigenvalues of an $n \times n$ Hermitian matrix $\mathbf{A}$ as

$$\lambda_{max} = \lambda_1 \geq \lambda_2 \geq \cdots \geq \lambda_{n-1} \geq \lambda_n = \lambda_{min}. \tag{B.1}$$

The largest and the smallest eigenvalues are characterized as the solutions to constrained maximization and minimization problems, respectively, as shown in the following theorem. Proofs of the results presented in this section may be found in [170].

Theorem (Rayleigh-Ritz) *Let* $\mathbf{A}$ *be an* $n \times n$ *Hermitian matrix whose eigenvalues are ordered as in* (B.1). *Then,*

$$\lambda_n \mathbf{x}^H\mathbf{x} \leq \mathbf{x}^H\mathbf{A}\mathbf{x} \leq \lambda_1 \mathbf{x}^H\mathbf{x}, \quad \forall \mathbf{x} \in \mathbb{C}^n,$$

where

$$\lambda_{max} = \lambda_1 = \max_{\mathbf{x}\neq\mathbf{0}} \frac{\mathbf{x}^H\mathbf{A}\mathbf{x}}{\mathbf{x}^H\mathbf{x}} = \max_{\mathbf{x}^H\mathbf{x}=1} \mathbf{x}^H\mathbf{A}\mathbf{x},$$

$$\lambda_{min} = \lambda_n = \min_{\mathbf{x}\neq\mathbf{0}} \frac{\mathbf{x}^H\mathbf{A}\mathbf{x}}{\mathbf{x}^H\mathbf{x}} = \min_{\mathbf{x}^H\mathbf{x}=1} \mathbf{x}^H\mathbf{A}\mathbf{x}.$$

The Rayleigh-Ritz theorem shows that λ_1 (λ_n) is the largest (smallest) value of the quadratic function $\mathbf{x}^H\mathbf{A}\mathbf{x}$ as $\mathbf{x}$ takes values over the unit sphere in $\mathbb{C}^n$, which is a compact set. The Courant-Fischer theorem, or "min-max theorem," provides a characterization of the rest of the eigenvalues of the Hermitian matrix $\mathbf{A}$.

Theorem (Courant-Fischer) *Let* $\mathbf{A}$ *be an* $n \times n$ *Hermitian matrix with eigenvalues ordered as in* (B.1), *and let* k *be an integer in* $1 \leq k \leq n$. *Then,*

$$\lambda_k = \min_{\mathbf{w}_1,\ldots,\mathbf{w}_{n-k}\in\mathbb{C}^n} \quad \max_{\substack{\mathbf{x}\neq\mathbf{0} \\ \mathbf{x}\perp\mathbf{w}_1,\ldots,\mathbf{w}_{n-k}}} \quad \frac{\mathbf{x}^H\mathbf{A}\mathbf{x}}{\mathbf{x}^H\mathbf{x}},$$

or, alternatively,

$$\lambda_k = \max_{\mathbf{w}_1,\ldots,\mathbf{w}_{k-1}\in\mathbb{C}^n} \min_{\substack{\mathbf{x}\neq\mathbf{0}\\ \mathbf{x}\perp\mathbf{w}_1,\ldots,\mathbf{w}_{k-1}}} \frac{\mathbf{x}^H\mathbf{A}\mathbf{x}}{\mathbf{x}^H\mathbf{x}}.$$

If $k = n$ or $k = 1$, the outer optimization is omitted and the Courant-Fischer theorem reduces to the Rayleigh-Ritz theorem. The Courant-Fischer theorem has many applications. Among them, one of the most important is the problem of comparing the eigenvalues of a Hermitian matrix $\mathbf{A}$ with those of an additively perturbed version $\mathbf{A} + \mathbf{E}$. The Weyl theorem gives two-sided bounds for the eigenvalues of $\mathbf{A} + \mathbf{E}$.

Theorem (Weyl) *Let* $\mathbf{A}$ *and* $\mathbf{E}$ *be* $n \times n$ *Hermitian matrices with eigenvalues arranged in decreasing order. For each* $k = 1, \ldots, n$,

$$ev_k(\mathbf{A}) + ev_n(\mathbf{E}) \leq ev_k(\mathbf{A}+\mathbf{E}) \leq ev_k(\mathbf{A}) + ev_1(\mathbf{E}).$$

A consequence of this theorem is that

$$|ev_k(\mathbf{A}+\mathbf{E}) - ev_k(\mathbf{A})| \leq ev_1(\mathbf{E}) = \|\mathbf{E}\|_2,$$

so a perturbation in eigenvalues due to a perturbation in a Hermitian matrix is bounded by $ev_1(\mathbf{E})$. As another corollary of Weyl theorem, the following result states that the eigenvalues of a Hermitian matrix increase if a positive semidefinite matrix is added to it. It is called the *monotonicity theorem.*

Corollary *Let* $\mathbf{A}$ *and* $\mathbf{E}$ *be* $n \times n$ *Hermitian matrices with eigenvalues arranged in decreasing order. Assume that* $\mathbf{E}$ *is positive semidefinite, then*

$$ev_k(\mathbf{A}+\mathbf{E}) \geq ev_k(\mathbf{A}).$$

The following theorem gives bounds for the eigenvalues of the perturbed matrix $\mathbf{A} + \mathbf{E}$ when the perturbation $\mathbf{E}$ has rank at most r.

Theorem B.1 (Theorem 4.3.6 in [170]) *Let* $\mathbf{A}$ *and* $\mathbf{E}$ *be* $n \times n$ *Hermitian matrices with eigenvalues arranged in decreasing order, and suppose that* $\mathbf{E}$ *has rank at most* r. *Then,*

$$ev_{k+2r}(\mathbf{A}+\mathbf{E}) \leq ev_{k+r}(\mathbf{A}) \leq ev_k(\mathbf{A}+\mathbf{E}), \quad k = 1, \ldots, n-2r,$$

and

$$ev_{k+2r}(\mathbf{A}) \leq ev_{k+r}(\mathbf{A}+\mathbf{E}) \leq ev_k(\mathbf{A}), \quad k = 1, \ldots, n-2r.$$

Theorem (Poincare Separation or Interlacing) *Let* $\mathbf{A}$ *be an* $n \times n$ *Hermitian matrix and* $\mathbf{B}$ *an* $n \times r$ *unitary slice, that is,* $\mathbf{B}^H\mathbf{B} = \mathbf{I}_r$. *Denote the real eigenvalues of* $\mathbf{A}$ *as* $\lambda_1 \geq \lambda_2 \geq \cdots \geq \lambda_n$ *and the eigenvalues of* $\mathbf{B}^H\mathbf{A}\mathbf{B}$ *as* $\mu_1 \geq \mu_2 \geq \cdots \geq \mu_r$. *The eigenvalues of* $\mathbf{B}^H\mathbf{A}\mathbf{B}$ *satisfy*

$$\lambda_{n-r+k} \leq \mu_k \leq \lambda_k, \quad k = 1, \ldots, r.$$

The theorem is called the *interlacing* theorem because when $r = n - 1$, the eigenvalues λ_k and μ_k interlace:

$$\lambda_n \leq \mu_{n-1} \leq \lambda_{n-1} \leq \cdots \leq \mu_2 \leq \lambda_2 \leq \mu_1 \leq \lambda_1.$$

B.2.2 Hermitian Positive Definite Matrices

An $n \times n$ Hermitian matrix $\mathbf{A}$ is said to be positive (negative) definite if $\mathbf{x}^H\mathbf{A}\mathbf{x} > 0$ (< 0) for all n-dimensional vectors $\mathbf{x} \neq \mathbf{0}$. The notation $\mathbf{A} \succ \mathbf{0}$ ($\prec \mathbf{0}$) is commonly used. The matrix $\mathbf{A}$ is said to be positive (negative) semidefinite if $\mathbf{x}^H\mathbf{A}\mathbf{x} \geq 0$ (≤ 0) for all $\mathbf{x} \neq \mathbf{0}$, denoted $\mathbf{A} \succeq \mathbf{0}$ ($\preceq \mathbf{0}$). Covariance matrices $\mathbf{R}_{\mathbf{xx}} = E[\mathbf{x}\mathbf{x}^H]$ are positive semidefinite. The set of Hermitian (symmetric in the real case) $n \times n$ positive semidefinite matrices is a convex set (a cone) and it is often denoted as $\mathcal{S}^n_+$. The set of positive definite matrices is denoted as $\mathcal{S}^n_{++}$. To say $\mathbf{A} \succeq \mathbf{B}$ is to say the matrix $\mathbf{A} - \mathbf{B}$ is Hermitian positive semidefinite. To say $\mathbf{A} \succ \mathbf{B}$ is to say the matrix $\mathbf{A} - \mathbf{B}$ is Hermitian positive definite.

We now summarize some well-known properties.

(i) If $\mathbf{A} \succ \mathbf{0}$, then $\mathbf{A}^{-1} \succ \mathbf{0}$.
(ii) If $\mathbf{A} \succeq \mathbf{0}$, its eigenvalues are non-negative.
(iii) For any matrix $\mathbf{B}$, $\mathbf{B}\mathbf{B}^H \succeq \mathbf{0}$.
(iv) If $\mathbf{A} \succ \mathbf{0}$, $\mathbf{B} \succ \mathbf{0}$, and $\mathbf{A} - \mathbf{B} \succ \mathbf{0}$, then $\mathbf{B}^{-1} - \mathbf{A}^{-1} = \mathbf{B}^{-1}(\mathbf{A} - \mathbf{B})\mathbf{A}^{-1}$ is the product of three positive definite matrices and hence $\mathbf{B}^{-1} - \mathbf{A}^{-1} \succ \mathbf{0}$.
(v) If $\mathbf{A} \succ \mathbf{0}$, $\mathbf{B} \succ \mathbf{0}$, and $\mathbf{A} - \mathbf{B} \succ \mathbf{0}$, then $\det(\mathbf{A}) > \det(\mathbf{B})$.
(vi) If $\mathbf{A} \succ \mathbf{0}$ and $\mathbf{B} \succ \mathbf{0}$, then $\det(\mathbf{A} + \mathbf{B}) \geq \det(\mathbf{A}) + \det(\mathbf{B})$.

Matrix Square Root for Positive Semidefinite Hermitian Matrix. A positive semidefinite Hermitian matrix may be factored as the Gramian $\mathbf{A} = (\mathbf{L}\mathbf{D}^{1/2})(\mathbf{D}^{1/2}\mathbf{L}^H)$, where $\mathbf{L}^H$ is upper triangular and the diagonal elements of $\mathbf{D}^{1/2}$ are non-negative. This is commonly called a Cholesky or LDU factorization. The matrix $\mathbf{L}\mathbf{D}^{1/2}$ qualifies as a square root of $\mathbf{A}$, denoted $\mathbf{A}^{1/2}$, with the property $\mathbf{A} = \mathbf{A}^{1/2}\mathbf{A}^{H/2}$. But this square root is neither Hermitian nor positive semidefinite. The *unique positive semidefinite square root* of the Hermitian matrix $\mathbf{A}$ is $\mathbf{A}^{1/2} = \mathbf{U}\mathbf{\Lambda}^{1/2}\mathbf{U}^H$, where $\mathbf{U}\mathbf{\Lambda}\mathbf{U}^H$ is the unique EVD of $\mathbf{A}$.

B.3 Traces

The trace of a square $n \times n$ matrix is defined as

$$\mathrm{tr}(\mathbf{A}) = \sum_{i=1}^{n} a_{ii}.$$

The most elementary identities for trace are these:

(i) $\mathrm{tr}(\mathbf{A}^H) = (\mathrm{tr}(\mathbf{A}))^*$; for real matrices $\mathrm{tr}(\mathbf{A}^T) = \mathrm{tr}(\mathbf{A})$
(ii) $\mathrm{tr}(c\mathbf{A}) = c\,\mathrm{tr}(\mathbf{A})$
(iii) $\mathrm{tr}(\mathbf{A}+\mathbf{B}) = \mathrm{tr}(\mathbf{A}) + \mathrm{tr}(\mathbf{B})$
(iv) $\mathrm{tr}(\mathbf{ABC}) = \mathrm{tr}(\mathbf{CAB}) = \mathrm{tr}(\mathbf{BCA})$; this is the *cyclic property of trace*. As a result of this property, we have that the trace is similarity-invariant. That is, for any $\mathbf{A}$ and any invertible $\mathbf{B}$ of the same dimensions, $\mathrm{tr}(\mathbf{B}^{-1}\mathbf{AB}) = \mathrm{tr}(\mathbf{ABB}^{-1}) = \mathrm{tr}(\mathbf{A})$.
(v) For diagonalizable $\mathbf{A} = \mathbf{P\Lambda P}^{-1}$, $\mathrm{tr}(\mathbf{A}) = \sum_{i=1}^{n} \lambda_i$.
(vi) If $\mathbf{A}$ and $\mathbf{B}$ are two $n \times m$ matrices, then

$$\mathrm{tr}(\mathbf{A}^H\mathbf{B}) = \mathrm{tr}(\mathbf{BA}^H) = (\mathrm{tr}(\mathbf{AB}^H))^*.$$

For complex $n \times 1$ vectors $\mathbf{a}$ and $\mathbf{b}$, this property means that the trace of the outer product is equivalent to the inner product: $\mathrm{tr}(\mathbf{ab}^H) = \mathrm{tr}(\mathbf{b}^H\mathbf{a}) = \mathbf{b}^H\mathbf{a}$.
(vii) $\lambda_n(\mathbf{A})\,\mathrm{tr}(\mathbf{B}) \leq \mathrm{tr}(\mathbf{AB}) \leq \lambda_1(\mathbf{A})\,\mathrm{tr}(\mathbf{B})$, for $n \times n$ Hermitian positive semidefinite matrices $\mathbf{A}$ and $\mathbf{B}$.
(viii) For an $n \times n$ Hermitian positive definite $\mathbf{A}$, $\mathrm{tr}(\mathbf{A})\,\mathrm{tr}(\mathbf{A}^{-1}) \geq n^2$.

B.4 Inverses

If $\mathbf{A}$ is an $n \times n$ matrix with $\det(\mathbf{A}) \neq 0$, $\mathbf{A}$ is called a nonsingular matrix. In this case, there exists a unique matrix $\mathbf{B}$ such that $\mathbf{AB} = \mathbf{BA} = \mathbf{I}_n$. The matrix $\mathbf{B}$ is the inverse of $\mathbf{A}$ and is denoted by $\mathbf{A}^{-1}$.

The following are basic properties:

(i) $\mathbf{AA}^{-1} = \mathbf{A}^{-1}\mathbf{A} = \mathbf{I}_n$.
(ii) $(\mathbf{A}^{-1})^H = (\mathbf{A}^H)^{-1}$.
(iii) If $\mathbf{A}$ and $\mathbf{B}$ are nonsingular $n \times n$ matrices, then $(\mathbf{AB})^{-1} = \mathbf{B}^{-1}\mathbf{A}^{-1}$.
(iv) $\det(\mathbf{A}^{-1}) = 1/\det(\mathbf{A})$.
(v) If $\mathbf{A}$ is a unitary matrix, then $\mathbf{A}^{-1} = \mathbf{A}^H$.
(vi) If $\mathbf{A}$ is an $n \times n$ upper triangular matrix, then $\mathbf{A}^{-1}$ is also upper triangular and its diagonal elements are a_{ii}^{-1}, $i = 1, \ldots, n$.
(vii) For $\mathbf{M} = \mathrm{blkdiag}(\mathbf{A}, \mathbf{B})$, with nonsingular $\mathbf{A}$ and $\mathbf{B}$ matrices, then $\mathbf{M}^{-1} = \mathrm{blkdiag}(\mathbf{A}^{-1}, \mathbf{B}^{-1})$.

(viii) For complex matrix $\mathbf{A} = \mathbf{C} + j\mathbf{S}$, where $\mathbf{C}$ and $\mathbf{S}$ are $n \times n$ real matrices, the inverse of $\mathbf{A}$ is

$$\mathbf{A}^{-1} = \left(\mathbf{C} + \mathbf{S}\mathbf{C}^{-1}\mathbf{S}\right)^{-1} - j\left(\mathbf{S} + \mathbf{C}\mathbf{S}^{-1}\mathbf{C}\right)^{-1}.$$

Moore-Penrose Pseudo-Inverse. The notion of matrix inverse may be generalized with the *Moore-Penrose pseudo-inverse*. For $\mathbf{A} \in \mathbb{C}^{n\times m}$ of rank $r \leq \min(n, m)$, the pseudo-inverse $\mathbf{A}^{\#} \in \mathbb{C}^{m\times n}$ exists and is unique. Its properties are these:

(i) $\mathbf{A}^{\#}\mathbf{A} = (\mathbf{A}^{\#}\mathbf{A})^H$
(ii) $\mathbf{A}\mathbf{A}^{\#} = (\mathbf{A}\mathbf{A}^{\#})^H$
(iii) $\mathbf{A}^{\#}\mathbf{A}\mathbf{A}^{\#} = \mathbf{A}^{\#}$
(iv) $\mathbf{A}\mathbf{A}^{\#}\mathbf{A} = \mathbf{A}$

The pseudo-inverse may be computed from a Gram-Schmidt factorization of $\mathbf{A}$, or from the singular value decomposition (SVD) of $\mathbf{A}$. The SVD and the pseudo-inverse are discussed in more detail in Appendix C.

B.4.1 Patterned Matrices and their Inverses

Consider a $(p+q) \times (p+q)$ matrix $\mathbf{M}$, partitioned as

$$\mathbf{M} = \begin{bmatrix} \mathbf{A} & \mathbf{B} \\ \mathbf{C} & \mathbf{D} \end{bmatrix}, \tag{B.2}$$

where $\mathbf{A}$ is $p \times p$, $\mathbf{B}$ is $p \times q$, $\mathbf{C}$ is $q \times p$, and $\mathbf{D}$ is $q \times q$. $\mathbf{A}$ and $\mathbf{D}$ are assumed to be nonsingular. This matrix has block-Cholesky factorizations:

$$\begin{aligned}\begin{bmatrix} \mathbf{A} & \mathbf{B} \\ \mathbf{C} & \mathbf{D} \end{bmatrix} &= \begin{bmatrix} \mathbf{I}_p & \mathbf{B}\mathbf{D}^{-1} \\ \mathbf{0} & \mathbf{I}_q \end{bmatrix} \begin{bmatrix} \mathbf{A} - \mathbf{B}\mathbf{D}^{-1}\mathbf{C} & \mathbf{0} \\ \mathbf{0} & \mathbf{D} \end{bmatrix} \begin{bmatrix} \mathbf{I}_p & \mathbf{0} \\ \mathbf{D}^{-1}\mathbf{C} & \mathbf{I}_q \end{bmatrix} \\ &= \begin{bmatrix} \mathbf{I}_p & \mathbf{0} \\ \mathbf{C}\mathbf{A}^{-1} & \mathbf{I}_q \end{bmatrix} \begin{bmatrix} \mathbf{A} & \mathbf{0} \\ \mathbf{0} & \mathbf{D} - \mathbf{C}\mathbf{A}^{-1}\mathbf{B} \end{bmatrix} \begin{bmatrix} \mathbf{I}_p & \mathbf{A}^{-1}\mathbf{B} \\ \mathbf{0} & \mathbf{I}_q \end{bmatrix}.\end{aligned} \tag{B.3}$$

These may also be written as the Cholesky factorizations that take the partitioned matrix to block-diagonal form, i.e.,

$$\begin{bmatrix} \mathbf{I}_p & -\mathbf{B}\mathbf{D}^{-1} \\ \mathbf{0} & \mathbf{I}_q \end{bmatrix} \begin{bmatrix} \mathbf{A} & \mathbf{B} \\ \mathbf{C} & \mathbf{D} \end{bmatrix} \begin{bmatrix} \mathbf{I}_p & \mathbf{0} \\ -\mathbf{D}^{-1}\mathbf{C} & \mathbf{I}_q \end{bmatrix} = \begin{bmatrix} \mathbf{A} - \mathbf{B}\mathbf{D}^{-1}\mathbf{C} & \mathbf{0} \\ \mathbf{0} & \mathbf{D} \end{bmatrix},$$

and

$$\begin{bmatrix} \mathbf{I}_p & \mathbf{0} \\ -\mathbf{C}\mathbf{A}^{-1} & \mathbf{I}_q \end{bmatrix} \begin{bmatrix} \mathbf{A} & \mathbf{B} \\ \mathbf{C} & \mathbf{D} \end{bmatrix} \begin{bmatrix} \mathbf{I}_p & -\mathbf{A}^{-1}\mathbf{B} \\ \mathbf{0} & \mathbf{I}_q \end{bmatrix} = \begin{bmatrix} \mathbf{A} & \mathbf{0} \\ \mathbf{0} & \mathbf{D} - \mathbf{C}\mathbf{A}^{-1}\mathbf{B} \end{bmatrix}.$$

The $p \times p$ matrix $\mathbf{A} - \mathbf{B}\mathbf{D}^{-1}\mathbf{C}$ is the Schur complement of the $q \times q$ block $\mathbf{D}$ of $\mathbf{M}$, and the $q \times q$ matrix $\mathbf{D} - \mathbf{C}\mathbf{A}^{-1}\mathbf{B}$ is the Schur complement of the $p \times p$ block $\mathbf{A}$ of $\mathbf{M}$. The corresponding Cholesky factorizations of the inverse of the partitioned matrix are

$$\begin{bmatrix} \mathbf{A} & \mathbf{B} \\ \mathbf{C} & \mathbf{D} \end{bmatrix}^{-1} = \begin{bmatrix} \mathbf{I}_p & \mathbf{0} \\ -\mathbf{D}^{-1}\mathbf{C} & \mathbf{I}_q \end{bmatrix} \begin{bmatrix} (\mathbf{A} - \mathbf{B}\mathbf{D}^{-1}\mathbf{C})^{-1} & \mathbf{0} \\ \mathbf{0} & \mathbf{D}^{-1} \end{bmatrix} \begin{bmatrix} \mathbf{I}_p & -\mathbf{B}\mathbf{D}^{-1} \\ \mathbf{0} & \mathbf{I}_q \end{bmatrix} \tag{B.4}$$

$$= \begin{bmatrix} \mathbf{I}_p & -\mathbf{A}^{-1}\mathbf{B} \\ \mathbf{0} & \mathbf{I}_q \end{bmatrix} \begin{bmatrix} \mathbf{A}^{-1} & \mathbf{0} \\ \mathbf{0} & (\mathbf{D} - \mathbf{C}\mathbf{A}^{-1}\mathbf{B})^{-1} \end{bmatrix} \begin{bmatrix} \mathbf{I}_p & \mathbf{0} \\ -\mathbf{C}\mathbf{A}^{-1} & \mathbf{I}_q \end{bmatrix}. \tag{B.5}$$

Then, the inverse of the partitioned matrix may be written as

$$\begin{bmatrix} \mathbf{A} & \mathbf{B} \\ \mathbf{C} & \mathbf{D} \end{bmatrix}^{-1} = \begin{bmatrix} (\mathbf{A} - \mathbf{B}\mathbf{D}^{-1}\mathbf{C})^{-1} & -\mathbf{A}^{-1}\mathbf{B}(\mathbf{D} - \mathbf{C}\mathbf{A}^{-1}\mathbf{B})^{-1} \\ -\mathbf{D}^{-1}\mathbf{C}(\mathbf{A} - \mathbf{B}\mathbf{D}^{-1}\mathbf{C})^{-1} & (\mathbf{D} - \mathbf{C}\mathbf{A}^{-1}\mathbf{B})^{-1} \end{bmatrix}.$$

The Schur complement $\mathbf{A} - \mathbf{B}\mathbf{D}^{-1}\mathbf{C}$ may be read out as the inverse of the Northwest block of the patterned inverse, and the Schur complement $\mathbf{D} - \mathbf{C}\mathbf{A}^{-1}\mathbf{B}$ may be read out as the inverse of the Southeast block of the patterned inverse.

The following are particular cases of this result when $\mathbf{B} = \mathbf{0}$ or $\mathbf{C} = \mathbf{0}$:

$$\begin{bmatrix} \mathbf{A} & \mathbf{0} \\ \mathbf{C} & \mathbf{D} \end{bmatrix}^{-1} = \begin{bmatrix} \mathbf{A}^{-1} & \mathbf{0} \\ -\mathbf{D}^{-1}\mathbf{C}\mathbf{A}^{-1} & \mathbf{D}^{-1} \end{bmatrix},$$

$$\begin{bmatrix} \mathbf{A} & \mathbf{B} \\ \mathbf{0} & \mathbf{D} \end{bmatrix}^{-1} = \begin{bmatrix} \mathbf{A}^{-1} & -\mathbf{A}^{-1}\mathbf{B}\mathbf{D}^{-1} \\ \mathbf{0} & \mathbf{D}^{-1} \end{bmatrix}.$$

If $\mathbf{A} = \mathbf{I}_p$ and $\mathbf{D} = \mathbf{I}_p$, the expressions further simplify to

$$\begin{bmatrix} \mathbf{I}_p & \mathbf{0} \\ \mathbf{C} & \mathbf{I}_p \end{bmatrix}^{-1} = \begin{bmatrix} \mathbf{I}_p & \mathbf{0} \\ -\mathbf{C} & \mathbf{I}_p \end{bmatrix},$$

$$\begin{bmatrix} \mathbf{I}_p & \mathbf{B} \\ \mathbf{0} & \mathbf{I}_p \end{bmatrix}^{-1} = \begin{bmatrix} \mathbf{I}_p & -\mathbf{B} \\ \mathbf{0} & \mathbf{I}_p \end{bmatrix}.$$

B.4.2 Matrix Inversion Lemma or Woodbury Identity

Consider the matrix $\mathbf{A}+\mathbf{BDC}$, with $\mathbf{A}$ a $p \times p$ invertible matrix, $\mathbf{D}$ a $q \times q$ invertible matrix, and $\mathbf{B}$, $\mathbf{C}$ $p \times q$ and $q \times p$ matrices, respectively. The matrix inversion lemma reads

$$(\mathbf{A}+\mathbf{BDC})^{-1} = \mathbf{A}^{-1} - \mathbf{A}^{-1}\mathbf{B}(\mathbf{D}^{-1} + \mathbf{CA}^{-1}\mathbf{B})^{-1}\mathbf{CA}^{-1}.$$

There are special cases.

- For $\mathbf{A} = \mathbf{I}_p$ and $\mathbf{D} = \mathbf{I}_q$,

$$(\mathbf{I}_p + \mathbf{BC})^{-1} = \mathbf{I}_p - \mathbf{B}(\mathbf{I}_q + \mathbf{CB})^{-1}\mathbf{C}.$$

- When $\mathbf{D} = d$ is scalar, then

$$(\mathbf{A} + \mathbf{b}d\mathbf{c}^H)^{-1} = \mathbf{A}^{-1} - \mathbf{A}^{-1}\mathbf{b}(d^{-1} + \mathbf{c}^H\mathbf{A}^{-1}\mathbf{b})^{-1}\mathbf{c}^H\mathbf{A}^{-1},$$

  which is commonly called the Sherman-Morrison identity. In this case, the term $\mathbf{b}d\mathbf{c}^H$ is a rank-one adjustment to $\mathbf{A}$.

By equating the Northwest blocks of the two Cholesky factorizations of the partitioned inverse in (B.4) and (B.5), the matrix inversion lemma is derived and written alternatively as

$$(\mathbf{A} - \mathbf{BD}^{-1}\mathbf{C})^{-1} = \mathbf{A}^{-1} + \mathbf{A}^{-1}\mathbf{B}(\mathbf{D} - \mathbf{CA}^{-1}\mathbf{B})^{-1}\mathbf{CA}^{-1}. \tag{B.6}$$

For $\mathbf{A} = \mathbf{I}_p$ and $\mathbf{D} = \mathbf{I}_q$, (B.6) becomes

$$(\mathbf{I}_p - \mathbf{BC})^{-1} = \mathbf{I}_p + \mathbf{B}(\mathbf{I}_q - \mathbf{CB})^{-1}\mathbf{C}.$$

Substituting $\mathbf{D} = d$ in (B.6), the Sherman-Morrison identity is written as

$$(\mathbf{A} - \mathbf{b}d^{-1}\mathbf{c}^H)^{-1} = \mathbf{A}^{-1} + \mathbf{A}^{-1}\mathbf{b}(d - \mathbf{c}^H\mathbf{A}^{-1}\mathbf{b})^{-1}\mathbf{c}^H\mathbf{A}^{-1}.$$

Push-Through Identity. Let $\mathbf{B}$ and $\mathbf{C}$ be $p \times q$ and $q \times p$ matrices, respectively. Then,

$$\mathbf{B}(\mathbf{I}_q + \mathbf{CB}) = (\mathbf{I}_p + \mathbf{BC})\mathbf{B}.$$

If $(\mathbf{I}_q + \mathbf{CB})$ is invertible so is $(\mathbf{I}_p + \mathbf{BC})$, in which case the previous equation gives the "push-through" identity:

$$(\mathbf{I}_p + \mathbf{BC})^{-1}\mathbf{B} = \mathbf{B}(\mathbf{I}_q + \mathbf{CB})^{-1}.$$

The name "push-through" identity comes from the observation that in the matrix $\mathbf{B} + \mathbf{BCB}$, $\mathbf{B}$ may be pushed through from the left as $\mathbf{B}(\mathbf{I}_q + \mathbf{CB})$ or from the right as $(\mathbf{I}_p + \mathbf{BC})\mathbf{B}$.

B.5 Determinants

The determinant of a square $n \times n$ matrix $\mathbf{A}$ is defined by

$$\det(\mathbf{A}) = \sum_{\sigma} (-1)^{N(\boldsymbol{\sigma})} \prod_{l=1}^{n} a_{l\sigma_l},$$

where $\sum_{\sigma}$ denotes the sum over the $n!$ permutations of the numbers $1, \ldots, n$, and $N(\boldsymbol{\sigma})$ is the total number of inversions of a permutation. An inversion of a permutation is a pair of numbers with the property that $\sigma_i > \sigma_l$ when $i > l$ in the permutation $\sigma_1, \sigma_2, \ldots, \sigma_n$ of $1, 2, \ldots, n$. For example, the inversions of $(2, 1, 4, 3)$ are $(2, 1)$ and $(4, 3)$; $N(2, 1, 4, 3) = 2$. The value $\text{sgn}(\boldsymbol{\sigma}) = (-1)^{N(\sigma)}$ is called the signature of the permutation $\boldsymbol{\sigma}$, which takes value $\text{sgn}(\boldsymbol{\sigma}) = 1$ whenever the reordering given by $\boldsymbol{\sigma}$ is achieved by successively interchanging two entries an even number of times, and $\text{sgn}(\boldsymbol{\sigma}) = -1$ whenever it is achieved by an odd number of such interchanges.

The absolute value of the determinant, $|\det(\mathbf{A})|$, is the volume of the parallelepiped formed by the columns of $\mathbf{A}$. The following are properties that follow from the definition of $\det(\mathbf{A})$:

(i) If the ith row or column of an $n \times n$ matrix $\mathbf{A}$ is multiplied by a constant c, the determinant is multiplied by c. As a consequence, $\det(c\mathbf{A}) = c^n \det(\mathbf{A})$.
(ii) If any rows (or columns) of a matrix are interchanged, the sign of the determinant is changed.
(iii) If two rows (or columns) of $\mathbf{A}$ are identical, then $\det(\mathbf{A}) = 0$.
(iv) $\det(\mathbf{AB}) = \det(\mathbf{A}) \det(\mathbf{B}) = \det(\mathbf{BA})$, for $n \times n$ matrices $\mathbf{A}$ and $\mathbf{B}$.
(v) For diagonalizable $\mathbf{A} = \mathbf{PAP}^{-1}$, $\det(\mathbf{A}) = \prod_{i=1}^{n} \lambda_i$.
(vi) The determinant is invariant under transposition $\det(\mathbf{A}) = \det(\mathbf{A}^T)$, but $\det(\mathbf{A}^H) = \det(\mathbf{A})^*$.
(vii) $\det(\mathbf{AA}^H) \geq 0$, where $\mathbf{A}$ is any $n \times m$ matrix.
(viii) $\det(\mathbf{I}) = 1$. Note that if $\mathbf{U}$ is a unitary matrix, then $\mathbf{UU}^H = \mathbf{I}$ and hence $|\det(\mathbf{U})| = 1$.
(ix) If $\mathbf{A}$ is invertible, then $\det(\mathbf{A}^{-1}) = 1/\det(\mathbf{A})$.
(x) Let $\mathbf{A}$ be a square $n \times n$ matrix with QR decomposition $\mathbf{A} = \mathbf{QR}$, then $\det(\mathbf{A}) = \prod_{i=1}^{n} r_{ii}$, where r_{ii} are the diagonal elements of the upper triangular matrix $\mathbf{R}$.
(xi) The following useful relation for diagonalizable matrices connects the trace to the determinant of the associated matrix exponential:

$$\det(\exp(\mathbf{A})) = \exp(\mathrm{tr}(\mathbf{A})).$$

This identity is not as mysterious as it appears: $\prod_{i=1}^{n} e^{\lambda_i} = e^{\sum_{i=1}^{n} \lambda_i}$. There is no need for $\mathbf{A}$ to be invertible. Only diagonalizable.

B.5.1 Some Useful Determinantal Identities and Inequalities

Schur's Determinant Identity. Let us consider the $(p+q) \times (p+q)$ partitioned matrix $\mathbf{M}$ defined in (B.2), and repeated here for convenience,

$$\mathbf{M} = \begin{bmatrix} \mathbf{A} & \mathbf{B} \\ \mathbf{C} & \mathbf{D} \end{bmatrix},$$

where $\mathbf{A}$ is $p \times p$, $\mathbf{B}$ is $p \times q$, $\mathbf{C}$ is $q \times p$, and $\mathbf{D}$ is $q \times q$. $\mathbf{A}$ and $\mathbf{D}$ are assumed to be nonsingular. The matrix $\mathbf{M}$ can be factored as in (B.3):

$$\mathbf{M} = \begin{bmatrix} \mathbf{A} & \mathbf{B} \\ \mathbf{C} & \mathbf{D} \end{bmatrix} = \begin{bmatrix} \mathbf{I}_p & \mathbf{B}\mathbf{D}^{-1} \\ \mathbf{0} & \mathbf{I}_q \end{bmatrix} \begin{bmatrix} \mathbf{A} - \mathbf{B}\mathbf{D}^{-1}\mathbf{C} & \mathbf{0} \\ \mathbf{0} & \mathbf{D} \end{bmatrix} \begin{bmatrix} \mathbf{I}_p & \mathbf{0} \\ \mathbf{D}^{-1}\mathbf{C} & \mathbf{I}_q \end{bmatrix}.$$

The determinants of the first and third matrices in the right-hand side are 1, which yields the Schur's determinant identity:

$$\det\left(\begin{bmatrix} \mathbf{A} & \mathbf{B} \\ \mathbf{C} & \mathbf{D} \end{bmatrix}\right) = \det(\mathbf{D}) \det(\mathbf{A} - \mathbf{B}\mathbf{D}^{-1}\mathbf{C}).$$

Clearly, using similar arguments,

$$\det\left(\begin{bmatrix} \mathbf{A} & \mathbf{B} \\ \mathbf{C} & \mathbf{D} \end{bmatrix}\right) = \det(\mathbf{A}) \det(\mathbf{D} - \mathbf{C}\mathbf{A}^{-1}\mathbf{B}).$$

When either the Southwest block or the Northeast block of $\mathbf{M}$ is the zero matrix, Schur's determinant identity is

$$\det\left(\begin{bmatrix} \mathbf{A} & \mathbf{B} \\ \mathbf{0} & \mathbf{D} \end{bmatrix}\right) = \det\left(\begin{bmatrix} \mathbf{A} & \mathbf{0} \\ \mathbf{C} & \mathbf{D} \end{bmatrix}\right) = \det(\mathbf{A}) \det(\mathbf{D}).$$

If $q = 1$, the matrix $\mathbf{M}$ is

$$\mathbf{M} = \begin{bmatrix} \mathbf{A} & \mathbf{b} \\ \mathbf{c}^T & d \end{bmatrix},$$

where d is a scalar, $\mathbf{b}$ is a $p \times 1$ column vector, and $\mathbf{c}^T$ is a $1 \times p$ row vector. Then, a direct application of Schur's determinant identity reads

$$\det\left(\begin{bmatrix} \mathbf{A} & \mathbf{b} \\ \mathbf{c}^T & d \end{bmatrix}\right) = \det(\mathbf{A})\left(d - \mathbf{c}^T\mathbf{A}^{-1}\mathbf{b}\right).$$

If $\mathbf{A} = \alpha\mathbf{I}_p$, the determinant is

$$\det\left(\begin{bmatrix} \alpha\mathbf{I}_p & \mathbf{b} \\ \mathbf{c}^T & d \end{bmatrix}\right) = \alpha^p\left(d - \frac{\mathbf{c}^T\mathbf{b}}{\alpha}\right).$$

Schur's Determinant Identity for Positive Definite Matrices. Suppose now that $\mathbf{M}$ is a Hermitian matrix partitioned as

$$\mathbf{M} = \begin{bmatrix} \mathbf{A} & \mathbf{B} \\ \mathbf{B}^H & \mathbf{D} \end{bmatrix}.$$

Schur's determinant identity is

$$\det(\mathbf{M}) = \det(\mathbf{A})\det(\mathbf{D} - \mathbf{B}^H\mathbf{A}^{-1}\mathbf{B}) = \det(\mathbf{D})\det(\mathbf{A} - \mathbf{B}\mathbf{D}^{-1}\mathbf{B}^H).$$

Then, as a corollary of Schur's determinant identity, we have that $\mathbf{M}$ is positive definite iff $\mathbf{A} \succ \mathbf{0}$ and $\mathbf{D} \succ \mathbf{B}^H\mathbf{A}^{-1}\mathbf{B}$ or, equivalently, iff $\mathbf{D} \succ \mathbf{0}$ and $\mathbf{A} \succ \mathbf{B}\mathbf{D}^{-1}\mathbf{B}^H$. Thus, if the partitioned matrix

$$\mathbf{M} = \begin{bmatrix} \mathbf{A} & \mathbf{B} \\ \mathbf{B}^H & \mathbf{D} \end{bmatrix},$$

is positive definite, we have

$$\mathbf{A} \succ \mathbf{0}, \quad \mathbf{D} \succ \mathbf{0}, \quad \mathbf{A} \succ \mathbf{B}\mathbf{D}^{-1}\mathbf{B}^H, \quad \text{and} \quad \mathbf{D} \succ \mathbf{B}^H\mathbf{A}^{-1}\mathbf{B}.$$

Matrix Determinant Lemma. Let $\mathbf{u}$ and $\mathbf{v}$ be column vectors in $\mathbb{C}^n$. Then, the matrix determinant lemma is

$$\det\left(\mathbf{I}_p + \mathbf{u}\mathbf{v}^H\right) = 1 + \mathbf{v}^H\mathbf{u}.$$

The proof follows from the equality

$$\begin{bmatrix} \mathbf{I}_n & \mathbf{0} \\ \mathbf{v}^H & 1 \end{bmatrix}\begin{bmatrix} \mathbf{I}_n + \mathbf{u}\mathbf{v}^H & \mathbf{u} \\ \mathbf{0}^T & 1 \end{bmatrix}\begin{bmatrix} \mathbf{I}_n & \mathbf{0} \\ -\mathbf{v}^H & 1 \end{bmatrix} = \begin{bmatrix} \mathbf{I}_n & \mathbf{u} \\ -\mathbf{0}^T & 1 + \mathbf{v}^H\mathbf{u} \end{bmatrix}.$$

The determinant of the right-hand side is $1+\mathbf{v}^H\mathbf{u}$. The determinants of the first and third matrix of the left-hand side are 1, and the determinant of the middle matrix is $\det\left(\mathbf{I}_n+\mathbf{u}\mathbf{v}^H\right)$, which proves the lemma.

Note now that if $\mathbf{A}$ is an $n\times n$ invertible matrix, we have

$$\begin{aligned}\det\left(\mathbf{A}+\mathbf{u}\mathbf{v}^H\right) &= \det(\mathbf{A})\det\left(\mathbf{I}_n+(\mathbf{A}^{-1}\mathbf{u})\mathbf{v}^H\right)\\ &= \det(\mathbf{A})(1+\mathbf{v}^H\mathbf{A}^{-1}\mathbf{u}).\end{aligned} \tag{B.7}$$

This equation provides an efficient way to compute the determinant of a matrix updated or corrected by a rank-one matrix $\mathbf{u}\mathbf{v}^H$. If $\mathbf{U}$ and $\mathbf{V}$ are $n\times m$ matrices, (B.7) generalizes to

$$\det\left(\mathbf{A}+\mathbf{U}\mathbf{V}^H\right)=\det(\mathbf{A})\det\left(\mathbf{I}_m+\mathbf{V}^H\mathbf{A}^{-1}\mathbf{U}\right).$$

A special case is this: let $\mathbf{A}=\mathbf{I}_n$, $\mathbf{U}=\mathbf{H}$, and $\mathbf{V}^H=\mathbf{H}^H$; then

$$\det(\mathbf{I}_n+\mathbf{H}\mathbf{H}^H)=\det(\mathbf{I}_m+\mathbf{H}^H\mathbf{H})\geq 1.$$

This identity often appears as

$$\det(\mathbf{I}_n+\mathbf{R}_{\mathbf{xx}}^{-1/2}\mathbf{H}\mathbf{H}^H\mathbf{R}_{\mathbf{xx}}^{-1/2})=\det(\mathbf{I}_m+\mathbf{H}^H\mathbf{R}_{\mathbf{xx}}^{-1}\mathbf{H}).$$

Cauchy-Binet Identity. Let $\mathbf{A}$ be a $p\times N$ matrix and let $\mathbf{B}$ be an $N\times p$ matrix. Assume $p<N$. The matrix $\mathbf{AB}$ is a $p\times p$ matrix. Given any subset $S\subset\{1,\ldots,N\}$ having p-elements, form the two $p\times p$ matrices $\mathbf{A}_S$ and $\mathbf{B}_S$, obtained by just using the rows of $\mathbf{A}$ and columns of $\mathbf{B}$ indexed by the set S. Note that the number of different size-p subsets is $|S|=\binom{N}{p}$. Then, the Cauchy-Binet identity is

$$\det(\mathbf{AB})=\sum_S\det(\mathbf{A}_S\mathbf{B}_S)=\sum_S\det(\mathbf{A}_S)\det(\mathbf{B}_S),$$

where the sum ranges over all choices of S. The Cauchy-Binet identity implies that the determinant of the Gram matrix $\mathbf{A}\mathbf{A}^H$ is the sum of the determinants of smaller Gramians computed from all subsets of p columns of $\mathbf{A}$, that is,

$$\det(\mathbf{A}\mathbf{A}^H)=\sum_S\det\left(\mathbf{A}_S\mathbf{A}_S^H\right).$$

Hadamard's Inequality. Suppose that $\mathbf{A}$ is a Hermitian positive definite matrix partitioned as

$$\mathbf{A} = \begin{bmatrix} \mathbf{B} & \mathbf{C} \\ \mathbf{C}^H & \mathbf{D} \end{bmatrix}.$$

Then,

$$\det(\mathbf{A}) = \det(\mathbf{B})\det(\mathbf{D} - \mathbf{C}^H\mathbf{B}^{-1}\mathbf{C}) \leq \det(\mathbf{B})\det(\mathbf{D}), \tag{B.8}$$

with equality iff $\mathbf{C} = \mathbf{0}$. The last inequality in (B.8) follows because $\mathbf{D} \succ \mathbf{D} - \mathbf{C}^H\mathbf{B}^{-1}\mathbf{C} \succ \mathbf{0}$. This result is known as Fisher's inequality [170] and allows us to prove Hadamard's inequality as follows. Partition an $n \times n$ positive definite matrix $\mathbf{A}$ as

$$\mathbf{A} = \begin{bmatrix} \mathbf{F} & \mathbf{g} \\ \mathbf{g}^H & a_{nn} \end{bmatrix}$$

where a_{nn} is a positive scalar. Fisher's inequality then states that

$$\det(\mathbf{A}) \leq a_{nn}\det(\mathbf{F}), \tag{B.9}$$

with equality iff $\mathbf{g} = \mathbf{0}$. By repeated application of (B.9), we obtain Hadamard's inequality:

$$\det(\mathbf{A}) \leq \prod_{i=1}^{n} a_{ii},$$

with equality iff $\mathbf{A}$ is diagonal.

Other useful determinant inequalities are these:

(i) $\det(\mathbf{A})^{1/n} \leq \frac{1}{n}\operatorname{tr}(\mathbf{A})$, known as the *arithmetic mean-geometric mean inequality*.
(ii) For $\mathbf{A}$ and $\mathbf{B}$ positive semidefinite Hermitian $n \times n$ matrices with respective eigenvalues $\lambda_1 \geq \cdots \geq \lambda_n$ and $\mu_1 \geq \cdots \geq \mu_n$,

$$1 \leq \det(\mathbf{I}_n + \mathbf{AB}) = \det(\mathbf{I}_n + \mathbf{BA}) \leq \prod_{i=1}^{n}(1 + \lambda_i\mu_i),$$

with equality iff $\mathbf{A}$ and $\mathbf{B}$ commute.

(iii) For $\mathbf{A}$ and $\mathbf{B}$ Hermitian positive definite $n \times n$ matrices, $\det(\mathbf{A}+\mathbf{B})^{1/n} \geq \det(\mathbf{A})^{1/n} + \det(\mathbf{B})^{1/n}$, with equality iff $\mathbf{A} = c\mathbf{B}$, for some positive constant c. This is the *Minkowski determinant inequality*.

B.6 Kronecker Products

Many of the results in this book can be expressed more concisely in terms of the Kronecker product of matrices. The definition and some basic properties of this product are reviewed in this section. A more in-depth analysis of the Kronecker product and its properties can be found in [246].

Let $\mathbf{A}$ be a $p \times q$ matrix and $\mathbf{B}$ be an $r \times s$ matrix. Then, the Kronecker product of $\mathbf{A}$ and $\mathbf{B}$, denoted by $\mathbf{A} \otimes \mathbf{B}$, is the $pr \times qs$ matrix:

$$\mathbf{A} \otimes \mathbf{B} = \begin{bmatrix} a_{11}\mathbf{B} & a_{12}\mathbf{B} & \cdots & a_{1q}\mathbf{B} \\ a_{21}\mathbf{B} & a_{22}\mathbf{B} & \cdots & a_{2q}\mathbf{B} \\ \vdots & \vdots & \ddots & \vdots \\ a_{p1}\mathbf{B} & a_{p2}\mathbf{B} & \cdots & a_{pq}\mathbf{B} \end{bmatrix}.$$

A special Kronecker product that arises often in this book is the following:

$$\mathbf{I} \otimes \boldsymbol{\Sigma} = \begin{bmatrix} \boldsymbol{\Sigma} & \mathbf{0} & \cdots & \mathbf{0} \\ \mathbf{0} & \boldsymbol{\Sigma} & \cdots & \mathbf{0} \\ \vdots & \vdots & \ddots & \vdots \\ \mathbf{0} & \mathbf{0} & \cdots & \boldsymbol{\Sigma} \end{bmatrix} = \operatorname{blkdiag}(\boldsymbol{\Sigma}, \ldots, \boldsymbol{\Sigma}).$$

Several properties of the Kronecker product may be summarized:

(i) $(\alpha\mathbf{A}) \otimes (\beta\mathbf{B}) = \alpha\beta(\mathbf{A} \otimes \mathbf{B})$, for any scalars $\alpha, \beta \in \mathbb{C}$,
(ii) $(\mathbf{A}+\mathbf{B}) \otimes \mathbf{C} = \mathbf{A} \otimes \mathbf{C} + \mathbf{B} \otimes \mathbf{C}$,
(iii) $(\mathbf{A} \otimes \mathbf{B})^H = \mathbf{A}^H \otimes \mathbf{B}^H$,
(iv) If $\mathbf{A}$ and $\mathbf{B}$ are square nonsingular matrices, then $(\mathbf{A} \otimes \mathbf{B})^{-1} = \mathbf{A}^{-1} \otimes \mathbf{B}^{-1}$,
(v) If $\mathbf{A}$ is $p \times p$ with eigenvalues $\alpha_1, \ldots, \alpha_p$, and $\mathbf{B}$ is $q \times q$ with eigenvalues $\beta_1, \ldots, \beta_q$, then the eigenvalues of $\mathbf{A} \otimes \mathbf{B}$ are

$$\alpha_i \beta_l, \qquad i = 1, \ldots, p, \quad l = 1, \ldots, q.$$

It follows from this property that the trace, the determinant, and the rank of the Kronecker product are

$$\begin{aligned} \operatorname{tr}(\mathbf{A} \otimes \mathbf{B}) &= \operatorname{tr}(\mathbf{A}) \operatorname{tr}(\mathbf{B}), \\ \det(\mathbf{A} \otimes \mathbf{B}) &= \det(\mathbf{A})^q \det(\mathbf{B})^p, \\ \operatorname{rank}(\mathbf{A} \otimes \mathbf{B}) &= \operatorname{rank}(\mathbf{A}) \operatorname{rank}(\mathbf{B}). \end{aligned}$$

Note that the exponent in det(**A**) is the order of **B** and the exponent in det(**B**) is the order of **A**.

(vi) If **A**, **B**, **C**, and **D** are matrices of appropriate sizes $(\mathbf{A} \otimes \mathbf{B})(\mathbf{C} \otimes \mathbf{D}) = (\mathbf{AC}) \otimes (\mathbf{BD})$. This is called the *mixed-product property* of the Kronecker product because it mixes the Kronecker and the ordinary matrix product.

(vii) Consider the equation $\mathbf{AXB} = \mathbf{C}$, where **X** is the unknown matrix, then

$$(\mathbf{B}^T \otimes \mathbf{A})\mathrm{vec}(\mathbf{X}) = \mathrm{vec}(\mathbf{C}),$$

where vec(**X**) denotes the "vectorization" operator that stacks the columns of **X** into a single column vector.

B.7 Projection Matrices

Projection matrices play a prominent role in this book and they have many useful properties. Let **V** be an $n \times p$ matrix, $p < n$, whose columns form a unitary basis for the subspace $\langle \mathbf{V} \rangle$ such that $\mathbf{V}^H\mathbf{V} = \mathbf{I}_p$. Then, $\mathbf{P_V} = \mathbf{VV}^H$ is the $n \times n$ complex projection matrix that projects vectors $\mathbf{x} \in \mathbb{C}^n$ onto the subspace $\langle \mathbf{V} \rangle$. The projection matrix $\mathbf{P}_\mathbf{V}^\perp = \mathbf{I}_n - \mathbf{P_V}$ projects onto the orthogonal subspace. If the columns of **V** do not form an orthogonal basis for $\langle \mathbf{V} \rangle$, then $\mathbf{P_V} = \mathbf{V}\left(\mathbf{V}^H\mathbf{V}\right)^{-1}\mathbf{V}^H$.

Projection matrices enjoy a number of important properties:

(i) $\mathbf{P_V}$ is Hermitian: $\mathbf{P_V} = \mathbf{P}_\mathbf{V}^H$,
(ii) $\mathbf{P_V}$ is idempotent: $\mathbf{P}_\mathbf{V}^2 = \mathbf{P_V}$,
(iii) the eigenvalues of $\mathbf{P_V}$ consist of p ones and $n - p$ zeros; the eigenvalues of $\mathbf{P}_\mathbf{V}^\perp$ consist of $n - p$ ones and p zeros,
(iv) $\mathbf{P_V}\mathbf{P}_\mathbf{V}^\perp = \mathbf{P}_\mathbf{V}^\perp\mathbf{P_V} = \mathbf{0}$,
(v) let **H** be an $n \times p$ matrix with SVD $\mathbf{H} = \mathbf{FKG}^H$ and pseudo-inverse $\mathbf{H}^\#$; then, $\mathbf{HH}^\#$ is a rank-p projection matrix onto the p-dimensional subspace spanned by the first p columns of **F**, and $\mathbf{H}^\#\mathbf{H}$ is a rank-p projection matrix onto the p-dimensional subspace spanned by the first p columns of **G**.

Write $\mathbf{V} = \mathbf{A} + j\mathbf{B}$ and note

$$\mathbf{V}^H\mathbf{V} = (\mathbf{A}^T - j\mathbf{B}^T)(\mathbf{A} + j\mathbf{B}) = (\mathbf{A}^T\mathbf{A} + \mathbf{B}^T\mathbf{B}) + j(\mathbf{A}^T\mathbf{B} - (\mathbf{A}^T\mathbf{B})^T) = \mathbf{I}_p.$$

It follows that $\mathbf{A}^T\mathbf{A} + \mathbf{B}^T\mathbf{B} = \mathbf{I}_n$ and $\mathbf{A}^T\mathbf{B}$ is symmetric. These are constraints on the construction of complex **V** from real **A** and **B**. The projection $\mathbf{P_V}$ may be written as

$$\begin{aligned}\mathbf{P_V} = \mathbf{VV}^H &= (\mathbf{A} + j\mathbf{B})(\mathbf{A}^T - j\mathbf{B}^T) = (\mathbf{AA}^T + \mathbf{BB}^T) + j(\mathbf{BA}^T - (\mathbf{BA}^T)^T)\\ &= \mathbf{C} + j\mathbf{D}.\end{aligned}$$

The real part of $\mathbf{P_V}$ is symmetric and the imaginary part is skew-symmetric. Moreover, these matrices satisfy $\mathbf{CC}^T + \mathbf{DD}^T = \mathbf{C}$ and $\mathbf{DC}^T - \mathbf{CD}^T = \mathbf{D}$.

Example B.2 (Rank-One Projection in $\mathbb{C}^2$) Define the unit column vector:

$$\mathbf{v} = \frac{1}{\sqrt{2}}\begin{bmatrix} 1 \\ e^{j\theta} \end{bmatrix} = \frac{1}{\sqrt{2}}\begin{bmatrix} 1 \\ \cos\theta \end{bmatrix} + j\frac{1}{\sqrt{2}}\begin{bmatrix} 1 \\ \sin\theta \end{bmatrix} = \mathbf{a} + j\mathbf{b}.$$

Clearly, $\mathbf{a}^T\mathbf{a} + \mathbf{b}^T\mathbf{b} = 1$, and $\mathbf{a}^T\mathbf{b} = \mathbf{b}^T\mathbf{a}$. The rank-1 matrix $\mathbf{P_v} = \mathbf{vv}^H$ may be written as

$$\mathbf{P_v} = \frac{1}{2}\begin{bmatrix} 1 & e^{-j\theta} \\ e^{j\theta} & 1 \end{bmatrix} = \frac{1}{2}\begin{bmatrix} 1 & \cos\theta \\ \cos\theta & 1 \end{bmatrix} + j\frac{1}{2}\begin{bmatrix} 0 & -\sin\theta \\ \sin\theta & 0 \end{bmatrix} = \mathbf{C} + j\mathbf{D}.$$

The matrix $\mathbf{P_v}$ is Hermitian, $\mathbf{P}^H = \mathbf{P}$, and idempotent, $\mathbf{P}^2 = \mathbf{P}$, making $\mathbf{P}$ a projection matrix. Its real part is symmetric, and its imaginary part is skew-symmetric. It is algebra to show that $\mathbf{CC}^T + \mathbf{DD}^T = \mathbf{C}$ and $\mathbf{DC}^T - \mathbf{CD}^T = \mathbf{D}$.

Example B.3 (Centering matrix)] The $n \times n$ matrix

$$\mathbf{P}_{\mathbf{1}_n} = \mathbf{1}_n(\mathbf{1}_n^T\mathbf{1}_n)^{-1}\mathbf{1}_n^T = \frac{1}{n}\mathbf{1}_n\mathbf{1}_n^T,$$

is a rank-one projection matrix onto the subspace spanned by $\mathbf{1}_n = [1 \; \cdots \; 1]^T$. When there is no possible confusion with the dimensions, we may use $\mathbf{1}$ and $\mathbf{P_1}$. Let $\mathbf{x} \in \mathbb{C}^n$, then

$$\mathbf{P_1x} = \mathbf{1}\overline{x} = \begin{bmatrix} \overline{x} \\ \vdots \\ \overline{x} \end{bmatrix},$$

where $\overline{x} = \frac{1}{n}\sum_{i=1}^n x_i$ is the vector mean. The $n \times n$ matrix $\mathbf{P}_\mathbf{1}^\perp = \mathbf{I}_n - \mathbf{P_1}$ is called a centering matrix since $\mathbf{P}_\mathbf{1}^\perp\mathbf{x}$ is a zero-mean vector. For an $n \times m$ matrix $\mathbf{X}$, the multiplication $\mathbf{P}_{\mathbf{1}_n}^\perp\mathbf{X}$ removes the means from each of the m columns, whereas $\mathbf{XP}_{\mathbf{1}_m}^\perp$ removes the means from each of the n rows. For an $n \times n$ matrix, $\mathbf{X}$, the multiplication

$$\tilde{\mathbf{X}} = \mathbf{P}_\mathbf{1}^\perp\mathbf{XP}_\mathbf{1}^\perp,$$

yields a doubly centered matrix where both row and column means are equal to zero.

The centering operation can be applied over $n \times n$ Gramians $\mathbf{G} = \mathbf{X}^H\mathbf{X}$ as

$$\tilde{\mathbf{G}} = \mathbf{P}_{\mathbf{1}}^{\perp}\mathbf{G}\mathbf{P}_{\mathbf{1}}^{\perp},$$

or kernel matrices in reproducing kernel Hilbert spaces (RKHS)

$$\tilde{\mathbf{K}} = \mathbf{P}_{\mathbf{1}}^{\perp}\mathbf{K}\mathbf{P}_{\mathbf{1}}^{\perp},$$

where $k_{il} = \langle \boldsymbol{\phi}(\mathbf{x}_i), \boldsymbol{\phi}(\mathbf{x}_l)\rangle$ and the centering is performed in the feature space.

B.7.1 Gramian, Pseudo-Inverse, and Projection

From the $n \times p$ matrix $\mathbf{H}$ and the $n \times q$ matrix $\mathbf{S}$, construct the $n \times m$ matrix $\mathbf{X} = [\mathbf{H}\ \mathbf{S}]$, with $m = p + q \leq n$. Assume the matrix $\mathbf{X}$ has rank m, which ensures the $m \times m$ Gramian $\mathbf{G} = \mathbf{X}^H\mathbf{X}$ is full rank. The Gramian of $\mathbf{X}$ is the patterned matrix:

$$\mathbf{G} = \mathbf{X}^H\mathbf{X} = \begin{bmatrix} \mathbf{H}^H\mathbf{H} \mathbf{H}^H\mathbf{S} \\ \mathbf{S}^H\mathbf{H} \mathbf{S}^H\mathbf{S} \end{bmatrix}.$$

This matrix, and its inverse, may be block Cholesky factored as

$$\mathbf{G} = \begin{bmatrix} \mathbf{I}_p & \mathbf{H}^H\mathbf{S}(\mathbf{S}^H\mathbf{S})^{-1} \\ \mathbf{0} & \mathbf{I}_q \end{bmatrix} \begin{bmatrix} \mathbf{H}^H(\mathbf{I}_n - \mathbf{P}_\mathbf{S})\mathbf{H} & \mathbf{0} \\ \mathbf{0} & \mathbf{S}^H\mathbf{S} \end{bmatrix} \begin{bmatrix} \mathbf{I}_p & \mathbf{0} \\ (\mathbf{S}^H\mathbf{S})^{-1}\mathbf{S}^H\mathbf{H} & \mathbf{I}_q \end{bmatrix},$$

and

$$\mathbf{G}^{-1} = \begin{bmatrix} \mathbf{I}_p & \mathbf{0} \\ -(\mathbf{S}^H\mathbf{S})^{-1}\mathbf{S}^H\mathbf{H} & \mathbf{I}_q \end{bmatrix} \begin{bmatrix} (\mathbf{H}^H(\mathbf{I}_n - \mathbf{P}_\mathbf{S})\mathbf{H})^{-1} & \mathbf{0} \\ \mathbf{0} & (\mathbf{S}^H\mathbf{S})^{-1} \end{bmatrix} \begin{bmatrix} \mathbf{I}_p & -\mathbf{H}^H\mathbf{S}(\mathbf{S}^H\mathbf{S})^{-1} \\ \mathbf{0} & \mathbf{I}_q \end{bmatrix},$$

respectively. The determinant of $\mathbf{G}$ is $\det(\mathbf{G}) = \det(\mathbf{H}^H(\mathbf{I}_n - \mathbf{P}_\mathbf{S})\mathbf{H})\det(\mathbf{S}^H\mathbf{S})$. These factorizations produce the following results for the pseudo-inverse $\mathbf{X}^\#$ and projection $\mathbf{P}_\mathbf{X} = \mathbf{X}\mathbf{X}^\#$:

$$\mathbf{X}^\# = \mathbf{G}^{-1}\mathbf{X}^H = \begin{bmatrix} (\mathbf{H}^H(\mathbf{I}_n - \mathbf{P}_\mathbf{S})\mathbf{H})^{-1}\mathbf{H}^H(\mathbf{I}_n - \mathbf{P}_\mathbf{S}) \\ (\mathbf{S}^H(\mathbf{I}_n - \mathbf{P}_\mathbf{H})\mathbf{S})^{-1}\mathbf{S}^H(\mathbf{I}_n - \mathbf{P}_\mathbf{H}) \end{bmatrix} = \begin{bmatrix} \mathbf{H}_\mathbf{S}^\# \\ \mathbf{S}_\mathbf{H}^\# \end{bmatrix},$$

$$\mathbf{P}_{\mathbf{HS}} = \mathbf{X}\mathbf{X}^\# = \mathbf{H}\mathbf{H}_\mathbf{S}^\# + \mathbf{S}\mathbf{S}_\mathbf{H}^\# = \mathbf{E}_{\mathbf{HS}} + \mathbf{E}_{\mathbf{SH}}$$

In these formulas, $\mathbf{H}_\mathbf{S}^\# = (\mathbf{H}^H(\mathbf{I}_n - \mathbf{P}_\mathbf{S})\mathbf{H})^{-1}\mathbf{H}^H(\mathbf{I}_n - \mathbf{P}_\mathbf{S})$ and $\mathbf{S}_\mathbf{H}^\# = (\mathbf{S}^H(\mathbf{I}_n - \mathbf{P}_\mathbf{H})\mathbf{S})^{-1}\mathbf{S}^H(\mathbf{I}_n - \mathbf{P}_\mathbf{H})$ are *oblique pseudo-inverses*, with properties

$\mathbf{H}_{\mathbf{S}}^{\#}\mathbf{S} = \mathbf{0}, \mathbf{H}_{\mathbf{S}}^{\#}\mathbf{H} = \mathbf{I}_p$, and $\mathbf{S}_{\mathbf{H}}^{\#}\mathbf{H} = \mathbf{0}, \mathbf{S}_{\mathbf{H}}^{\#}\mathbf{S} = \mathbf{I}_q$; $\mathbf{E}_{\mathbf{HS}} = \mathbf{H}\mathbf{H}_{\mathbf{S}}^{\#}$ and $\mathbf{E}_{\mathbf{SH}} = \mathbf{S}\mathbf{S}_{\mathbf{H}}^{\#}$ are *oblique projections*, with properties $\mathbf{E}_{\mathbf{HS}}\mathbf{S} = \mathbf{0}, \mathbf{E}_{\mathbf{HS}}\mathbf{H} = \mathbf{H}$, and $\mathbf{E}_{\mathbf{SH}}\mathbf{H} = \mathbf{0}, \mathbf{E}_{\mathbf{SH}}\mathbf{S} = \mathbf{S}$. The reader may check that $\mathbf{E}_{\mathbf{HS}}$ and $\mathbf{E}_{\mathbf{SH}}$ are idempotent, but not Hermitian.

B.8 Toeplitz, Circulant, and Hankel Matrices

In problems of signal processing and machine learning, one often finds matrices with special structure. In this section, we review some of the most common structured matrices.

A *Toeplitz matrix* $\mathbf{T}$ is a matrix in which each descending diagonal from left to right is constant ($t_{il} = t_{i+1,l+1} = t_{i-l}$). A real $n \times n$ Toeplitz matrix $\mathbf{T}$ is determined by $2n - 1$ elements, $t_{-n+1}, \ldots, t_0, \ldots, t_{n-1}$, and is given by

$$\mathbf{T} = \begin{bmatrix} t_0 & t_1 & \cdots & t_{n-1} \\ t_{-1} & \ddots & \ddots & \vdots \\ \vdots & \ddots & \ddots & t_1 \\ t_{-n+1} & \cdots & t_{-1} & t_0 \end{bmatrix}.$$

Example B.4 (Wide-Sense Stationary (WSS) process) Let $\{x[k]\}$ be a complex zero-mean wide-sense stationary time series. Then, the covariance matrix of $\mathbf{x}[k] = [x[k] \ \cdots \ x[k-p+1]]^T$ has the following (Hermitian) Toeplitz structure:

$$\mathbf{R}_{\mathbf{xx}} = \mathrm{E}[\mathbf{x}[k]\mathbf{x}^H[k]] = \begin{bmatrix} r_{xx}[0] & r_{xx}[1] & \cdots & r_{xx}[p-1] \\ r_{xx}^*[1] & \ddots & \ddots & \vdots \\ \vdots & \ddots & \ddots & r_{xx}[1] \\ r_{xx}^*[p-1] & \cdots & r_{xx}^*[1] & r_{xx}[0] \end{bmatrix},$$

where $r_{xx}[s] = \mathrm{E}[x[k]x^*[k-s]]$. WSS multichannel and multidimensional processes have covariance matrices of block-Toeplitz form.

Example B.5 (Linear Time-Invariant (LTI) filter) A convolution with a causal linear time-invariant (LTI) filter, $y[k] = (x * h)[k]$, can be described by a Toeplitz matrix product. For example,

$$\begin{bmatrix} y[0] \\ y[1] \\ y[2] \\ y[3] \\ y[4] \end{bmatrix} = \begin{bmatrix} h[0] & 0 & 0 \\ h[1] & h[0] & 0 \\ h[2] & h[1] & h[0] \\ 0 & h[2] & h[1] \\ 0 & 0 & h[2] \end{bmatrix} \begin{bmatrix} x[0] \\ x[1] \\ x[2] \end{bmatrix} = \begin{bmatrix} x[0] & 0 & 0 \\ x[1] & x[0] & 0 \\ x[2] & x[1] & x[0] \\ 0 & x[2] & x[1] \\ 0 & 0 & x[2] \end{bmatrix} \begin{bmatrix} h[0] \\ h[1] \\ h[2] \end{bmatrix}.$$

This example can clearly be extended to noncausal LTI filters. In fact, every LTI filter corresponds to a Toeplitz linear operator.

Example B.6 (Uniform Linear Array) Let us consider a linear array of L equally spaced sensors (uniform linear array or ULA). The array receives signals from K narrowband sources distant enough to be regarded as planar waves when they arrive at the array. For this array geometry, the array response (also called steering vector) for a planar wave has the form

$$\mathbf{a}(\theta) = \left[1 \; e^{-j2\pi \sin(\theta)(d/\lambda)} \; \cdots \; e^{-j2\pi(L-1)\sin(\theta)(d/\lambda)}\right]^T,$$

where λ is the signal wavelength, d is the sensor spacing, and θ is the angle of arrival of the signal with respect to broadside. If the signals are uncorrelated with powers $\sigma_1^2, \ldots, \sigma_K^2$, and the additive noise is spatially white with noise variance σ^2, then the covariance matrix is

$$\mathbf{R} = \sum_{k=1}^{K} \sigma_k^2 \mathbf{a}(\theta_k)\mathbf{a}^H(\theta_k) + \sigma^2 \mathbf{I}_L$$

which is Hermitian and Toeplitz. This structure results from the uniform linear array geometry, the incoherent signal model, and the white noise assumption.

Example B.7 (Slepian matrix) Consider the $n \times n$ positive semidefinite Hermitian matrix:

$$\mathbf{R} = \frac{1}{2\pi} \int_{-\beta\pi}^{\beta\pi} \boldsymbol{\psi}(\theta)\boldsymbol{\psi}^H(\theta)d\theta,$$

where $\boldsymbol{\psi}(\theta) = [1 \; e^{j\theta} \; \cdots \; e^{j(n-1)\theta}]^T$, $0 < \beta \leq 1$. Its elements are

$$r_{kl} = 2\beta\pi \frac{\sin(k-l)\beta\pi}{(k-l)\beta\pi}.$$

This is the Toeplitz correlation matrix for an n-sample version of a wide-sense stationary time series with bandlimited spectrum $S(\theta) = 1, -\beta\pi < \theta \leq \beta\pi$, and zero elsewhere on the Nyquist band $-\pi < \theta \leq \pi$. In the spectral representation $\mathbf{R} = \mathbf{U}\boldsymbol{\Lambda}\mathbf{U}^H$, the columns of $\mathbf{U}$ are the discrete prolate spheroidal sequences, or *Slepian sequences* [326]. The trace of $\mathbf{R}$ is $\text{tr}(\mathbf{R}) = \sum_{i=1}^{n} \lambda_i = n\beta$. The rank of $\mathbf{R}$ is approximately the integer part of $n\beta$, the first $n\beta$ eigenvalues are near to one, and the remaining are near to zero. The time-bandwidth product of this n-sample version is $2\beta\pi n$.

A *circulant matrix* is a square matrix in which each row is a circularly shifted version of the preceding row. A circulant matrix is a particular kind of Toeplitz matrix. An $n \times n$ circulant matrix $\mathbf{C}$ takes the form

$$\mathbf{C} = \begin{bmatrix} c_0 & c_1 & \cdots & \cdots & c_{n-1} \\ c_{n-1} & c_0 & c_1 & \cdots & c_{n-2} \\ c_{n-2} & c_{n-1} & c_0 & \cdots & c_{n-3} \\ \vdots & \ddots & \ddots & \ddots & \vdots \\ c_1 & c_2 & \cdots & c_{n-1} & c_0 \end{bmatrix}. \tag{B.10}$$

An important property of circulant matrices is that they are diagonalized by the Fourier (or DFT) matrix. More precisely, an $n \times n$ circulant matrix $\mathbf{C}$ can be factored as

$$\mathbf{C} = \mathbf{U}_n \, \mathrm{diag}(\mathbf{F}_n \mathbf{c}_1) \mathbf{U}_n^H, \tag{B.11}$$

where $\mathbf{c}_1^T$ denotes the first row of $\mathbf{C}$ and $\mathbf{U}_n = \frac{1}{\sqrt{n}} \mathbf{F}_n$. The matrix $\mathbf{F}_n$ is the discrete Fourier transform (DFT) matrix with Vandermonde structure:

$$\mathbf{F}_n = \begin{bmatrix} 1 & 1 & 1 & 1 & \cdots & 1 \\ 1 & \omega & \omega^2 & \omega^3 & \cdots & \omega^{n-1} \\ 1 & \omega^2 & \omega^4 & \omega^6 & \cdots & \omega^{2(n-1)} \\ \vdots & \vdots & \vdots & \vdots & \ddots & \vdots \\ 1 & \omega^{n-1} & \omega^{2(n-1)} & \omega^{3(n-1)} & \cdots & \omega^{(n-1)(n-1)} \end{bmatrix},$$

where $\omega = e^{-j2\pi/n}$. Note that the eigenvalues of $\mathbf{C}$ are given by the DFT of the first row of $\mathbf{C}$.

This result follows from writing the $n \times n$ circulant matrix $\mathbf{C}$ in (B.10) as

$$\mathbf{C} = c_0 \mathbf{I}_n + c_1 \mathbf{S} + \cdots + c_{n-1} \mathbf{S}^{n-1},$$

where $\mathbf{S}$ is the $n \times n$ circular shift matrix:

$$\mathbf{S} = \begin{bmatrix} 0 & 1 & 0 & \cdots & 0 \\ 0 & 0 & 1 & \cdots & 0 \\ \vdots & \vdots & \vdots & \ddots & \vdots \\ 0 & 0 & 0 & \cdots & 1 \\ 1 & 0 & 0 & \cdots & 0 \end{bmatrix}.$$

The circular shift matrix has DFT factorization $\mathbf{S} = \frac{1}{n} \mathbf{F}_n \mathbf{D} \mathbf{F}_n^H = \mathbf{U}_n \mathbf{D} \mathbf{U}_n^H$, where $\mathbf{F}_n$ is the DFT matrix and $\mathbf{D} = \mathrm{diag}\left(1, e^{-j2\pi/n}, \ldots, e^{-j2\pi(n-1)/n}\right)$. Moreover, $\mathbf{S}^k = \frac{1}{n} \mathbf{F}_n \mathbf{D}^k \mathbf{F}_n^H$. It follows that

$$\begin{aligned}\mathbf{C} &= c_0\mathbf{U}_n\mathbf{U}_n^H + c_1\mathbf{U}_n\mathbf{D}\mathbf{U}_n^H + \cdots + c_{n-1}\mathbf{U}_n\mathbf{D}^{n-1}\mathbf{U}_n^H \\ &= \mathbf{U}_n(c_0\mathbf{I}_n + c_1\mathbf{D} + \cdots + c_{n-1}\mathbf{D}^{n-1})\mathbf{U}_n^H.\end{aligned}$$

The matrix $c_0\mathbf{I}_p + c_1\mathbf{D} + \cdots + c_{n-1}\mathbf{D}^{n-1}$ is diagonal, with the DFT coefficient $C_k = c_0 + c_1 e^{-j2\pi k/n} + \cdots + c_{n-1}e^{-j2\pi k(n-1)/n}$ at position k on the diagonal. That is, $\mathbf{C} = \mathbf{U}_n \operatorname{diag}(\mathbf{F}_n\mathbf{c}_1)\mathbf{U}_n^H$, where $\mathbf{F}_n\mathbf{c}_1$ is the n-point DFT for the first row of the circulant matrix $\mathbf{C}$, as in (B.11).

It is known that Toeplitz matrices are asymptotically equivalent to the simpler, more structured, circulant matrices, meaning that their eigenvalues behave similarly. For a more detailed review of Toeplitz and circulant matrices, as well as their asymptotic equivalence, the reader is directed to the monograph by R. M. Gray [148].

A class of structured matrices that encompasses Toeplitz and circulant matrices is the class of *persymmetric* matrices. An $n \times n$ matrix $\mathbf{P}$ is said to be persymmetric if it is symmetric with respect to the antidiagonal or northeast-southwest diagonal. For example, a 4×4 persymmetric matrices has the form

$$\begin{bmatrix} p_{11} & p_{12} & p_{13} & p_{14} \\ p_{21} & p_{22} & p_{23} & p_{13} \\ p_{31} & p_{32} & p_{22} & p_{12} \\ p_{41} & p_{31} & p_{21} & p_{11} \end{bmatrix}.$$

Clearly, Toeplitz matrices are persymmetric. An $n \times n$ persymmetric matrix $\mathbf{P}$ satisfies

$$\mathbf{P} = \mathbf{J}_n\mathbf{P}^T\mathbf{J}_n,$$

where $\mathbf{J}_n$ is the exchange matrix ($n \times n$ antidiagonal matrix with 1s on its antidiagonal). Persymmetric covariance matrices model clutter covariance matrices in radar systems using a symmetrically spaced linear array [53, 79].

A *Hankel matrix* is a matrix in which each ascending diagonal from left to right (southwest to northeast) is constant. A real $n \times n$ Hankel matrix $\mathbf{H}$ has the form

$$\mathbf{H} = \begin{bmatrix} h_0 & h_1 & h_2 & \cdots & h_{n-1} \\ h_1 & h_2 & \iddots & \iddots & h_n \\ h_2 & \iddots & \iddots & \iddots & \vdots \\ \vdots & \iddots & \iddots & \iddots & h_{2n-3} \\ h_{n-1} & h_n & \cdots & h_{2n-3} & h_{2n-2} \end{bmatrix}.$$

If $\mathbf{H}$ is an $n \times n$ Hankel matrix, then $\mathbf{H} = \mathbf{T}\mathbf{J}_n$, where $\mathbf{T}$ is an $n \times n$ Toeplitz matrix and $\mathbf{J}_n$ is the exchange matrix.

Example B.8 When elements of a time series $\{x[k]\}$ are organized into L-dimensional vectors as

$$\mathbf{x}[k] = \left[x[k]\ x[k+1]\ \cdots\ x[k+L-1]\right]^T,$$

and the resulting vectors are stored as columns of matrix $\mathbf{X}$, then $\mathbf{X}$ has a Hankel structure, i.e.,

$$\mathbf{X} = \left[\cdots\ \mathbf{x}[0]\ \mathbf{x}[1]\ \mathbf{x}[2]\ \cdots\right] = \begin{bmatrix} \cdots & x[0] & x[1] & x[2] & \cdots \\ \iddots & x[1] & x[2] & x[3] & \iddots \\ \iddots & x[2] & x[3] & \iddots & \iddots \\ \vdots & \vdots & \vdots & \vdots & \vdots \\ \cdots & x[L-1] & x[L] & x[L+1] & \cdots \end{bmatrix}.$$

B.9 Important Matrix Optimization Problems

B.9.1 Trace Optimization

Let $\mathbf{A} = \mathbf{U}\boldsymbol{\Lambda}\mathbf{U}^H$ and $\mathbf{B} = \mathbf{V}\boldsymbol{\Sigma}\mathbf{V}^H$ be $n \times n$ Hermitian matrices with respective eigenvalues $\boldsymbol{\lambda} = [\lambda_1\ \cdots\ \lambda_n]^T$ and $\boldsymbol{\sigma} = [\sigma_1\ \cdots\ \sigma_n]^T$ in decreasing order. Then

$$\boldsymbol{\lambda}^T \mathbf{J}_n \boldsymbol{\sigma} \leq \operatorname{tr}(\mathbf{AB}) \leq \boldsymbol{\lambda}^T \boldsymbol{\sigma},$$

with equality at the lower (upper) bound if and only if $\mathbf{V} = \mathbf{U}\mathbf{J}_n$ ($\mathbf{V} = \mathbf{U}$), where $\mathbf{J}_n$ is the exchange matrix of order n. This result is proved in [341].

The upper bound can be stated as the solution of the following trace maximization problem (cf. [230]):

$$\underset{\mathbf{F}^H\mathbf{F}=\mathbf{I}_n}{\text{maximize}} \quad \operatorname{tr}\left(\mathbf{F}^H \mathbf{A} \mathbf{F} \mathbf{B}\right),$$

where the maximum value $\boldsymbol{\lambda}^T\boldsymbol{\sigma} = \sum_{i=1}^n \lambda_i \sigma_i$ is attained at $\mathbf{F} = \mathbf{U}\mathbf{V}^H$. Analogously, the lower bound can be stated as the solution of the following trace minimization problem:

$$\underset{\mathbf{F}^H\mathbf{F}=\mathbf{I}_n}{\text{minimize}} \quad \operatorname{tr}\left(\mathbf{F}^H \mathbf{A} \mathbf{F} \mathbf{B}\right),$$

where the minimum value $\boldsymbol{\lambda}^T \mathbf{J}_n \boldsymbol{\sigma} = \sum_{i=1}^n \lambda_i \sigma_{n-i+1}$ is attained at $\mathbf{F} = \mathbf{U}\mathbf{J}_n\mathbf{V}^H$.

B.9.2 Determinant Optimization

Maximize a Determinant Under a Trace Constraint. The maximization of a determinant subject to a trace constraint is a problem that frequently arises in information theory and communications. An instance of this problem is as follows. Let $\mathbf{Q}$ be an $m \times m$ Hermitian positive semidefinite matrix, and let $\mathbf{H}$ be a complex $n \times m$ matrix, with compact SVD $\mathbf{H} = \mathbf{U}\boldsymbol{\Lambda}^{1/2}\mathbf{V}^H$, where $\boldsymbol{\Lambda}^{1/2} = \operatorname{diag}\left(\lambda_1^{1/2}, \ldots, \lambda_m^{1/2}\right)$, with $\lambda_1 \geq \lambda_2 \geq \cdots \geq \lambda_m \geq 0$. We assume that $n \geq m$. The problem is

$$\underset{\mathbf{Q}\in\mathcal{S}_+^m}{\text{maximize}} \det\left(\mathbf{I}_n + \mathbf{H}\mathbf{Q}\mathbf{H}^H\right), \quad \text{s.t.} \quad \operatorname{tr}(\mathbf{Q}) \leq 1.$$

The identity $\det\left(\mathbf{I}_n + \mathbf{H}\mathbf{Q}\mathbf{H}^H\right) = \det\left(\mathbf{I}_m + \mathbf{H}^H\mathbf{H}\mathbf{Q}\right)$ is valid for congruent matrices. The matrix $\mathbf{G} = \mathbf{H}^H\mathbf{H}$ is Hermitian positive semidefinite with EVD $\mathbf{G} = \mathbf{V}\boldsymbol{\Lambda}\mathbf{V}^H$. The determinant is maximized when $\mathbf{G}$ and $\mathbf{Q}$ commute, so the eigenspaces of $\mathbf{Q}$ and $\mathbf{G}$ must be identical. Therefore, the eigendecomposition of $\mathbf{Q}$ can be written as $\mathbf{Q} = \mathbf{V}\boldsymbol{\Sigma}\mathbf{V}^H$. The eigenvalues of $\mathbf{Q}$, $\sigma_1^2 \geq \sigma_2^2 \geq \cdots \geq \sigma_m^2 \geq 0$, are the solution to

$$\underset{\{\sigma_i^2\geq 0\}_{i=1}^m}{\text{maximize}} \sum_{i=1}^m \log\left(1 + \lambda_i\sigma_i^2\right), \quad \text{s.t.} \quad \sum_{i=1}^m \sigma_i^2 \leq 1.$$

Taking derivatives and equating to zero, we obtain the well-known *waterfilling* solution [87]:

$$\sigma_i^2 = \left(\frac{1}{\nu} - \frac{1}{\lambda_i}\right)^+, \quad i = 1, \ldots, m, \tag{B.12}$$

where ν is a Lagrange multiplier (a.k.a. waterlevel) determined so that the solution satisfies the trace constraint with equality $\sum_{i=1}^m \sigma_i^2 = 1$.

In information theory and communications, this is the solution for the transmit covariance matrix $\mathbf{Q}$ that maximizes the capacity of a multiple-input multiple-output (MIMO) channel when the channel $\mathbf{H}$ is known at the transmitter side (channel state information at the transmitter or CSIT). The capacity achieving distribution is $\mathbf{x} \sim C\mathcal{N}_m(\mathbf{0}, \mathbf{Q})$ [340]. The problem can trivially be extended to maximize $\det\left(\mathbf{R}_n + \mathbf{H}\mathbf{Q}\mathbf{H}^H\right)$, with $\mathbf{R}_n$ a known Hermitian positive semidefinite matrix by defining $\tilde{\mathbf{H}} = \mathbf{R}_n^{-1/2}\mathbf{H}$, or to a trace constraint of the form $\operatorname{tr}(\mathbf{G}\mathbf{Q}\mathbf{G}^H) \leq P$, with $\mathbf{G}$ a known $n \times m$ complex matrix.

H. S. Witsenhausen found in 1975 the solution to this problem under a slightly different formulation [386]. The formulation in [386] interchanges the roles of $\mathbf{Q}$ and $\mathbf{H}$ and solves the equivalent problem:

$$\underset{\mathbf{H}\in\mathbb{C}^{n\times m}}{\text{maximize}} \det\left(\mathbf{I}_n + \mathbf{HQH}^H\right), \quad \text{s.t.} \quad \text{tr}(\mathbf{HH}^H) \leq 1.$$

The solution is $\mathbf{H} = \mathbf{U}\,\text{diag}(\sigma_1, \ldots, \sigma_m)\mathbf{V}^H$, where $\mathbf{V}$ is the eigenspace of $\mathbf{Q}$, $\mathbf{U}$ is an arbitrary unitary matrix, and σ_i^2, $i = 1, \ldots, m$, are given by the waterfilling solution (B.12), with the λ_i being now the eigenvalues of the known matrix $\mathbf{Q}$. Let r be the largest integer for which the returned waterfilling solution for σ_i^2 is positive; then it is easy to check that the waterlevel satisfies

$$\nu^{-1} = r^{-1}\left(1 + \sum_{i=1}^{r} \lambda_i^{-1}\right),$$

and the maximum of the determinant is therefore attained at

$$\begin{aligned}\det\left(\mathbf{I}_n + \mathbf{HQH}^H\right) &= \prod_{i=1}^{r}\left(1 + \lambda_i\sigma_i^2\right) = \prod_{i=1}^{r}\frac{\lambda_i}{\nu} \\ &= r^{-r}\left(1 + \sum_{i=1}^{r}\lambda_i^{-1}\right)^r \prod_{i=1}^{r}\lambda_i.\end{aligned}$$

This is the solution of the max-det problem provided in [386].

Minimizing a Determinant in a First-Order MVN Model. Begin with the problem of minimizing

$$\det\left[\mathbf{I}_L + (\mathbf{Z} - \mathbf{UX})(\mathbf{Z} - \mathbf{UX})^H\right],$$

with respect to $\mathbf{X} \in \mathbb{C}^{p\times N}$, where here $\mathbf{Z}, \mathbf{U}$, and $\mathbf{X}$ are complex matrices of respective dimensions $L \times N$, $L \times p$, and $p \times N$, with $N > L > p$. Construct the following matrix, consisting of $N \times N$ and $p \times p$ blocks on the diagonals, and off-diagonal blocks that match

$$\mathbf{A} = \begin{bmatrix}\mathbf{I}_N & -\mathbf{X}^H \\ \mathbf{0} & \mathbf{I}_p\end{bmatrix}\begin{bmatrix}\mathbf{I}_N + \mathbf{ZZ}^H & \mathbf{Z}^H\mathbf{U} \\ \mathbf{U}^H\mathbf{Z} & \mathbf{U}^H\mathbf{U}\end{bmatrix}\begin{bmatrix}\mathbf{I}_N & \mathbf{0} \\ -\mathbf{X} & \mathbf{I}_p\end{bmatrix} \tag{B.13}$$

$$= \begin{bmatrix}\mathbf{I}_N + (\mathbf{Z} - \mathbf{UX})^H(\mathbf{Z} - \mathbf{UX}) & \mathbf{Z}^H\mathbf{U} - \mathbf{X}^H\mathbf{U}^H\mathbf{U} \\ \mathbf{U}^H\mathbf{Z} - \mathbf{U}^H\mathbf{UX} & \mathbf{U}^H\mathbf{U}\end{bmatrix}. \tag{B.14}$$

Now $\det(\mathbf{A})$ is invariant to $\mathbf{X}$ (see (B.13)), and it may be written as the Schur formula:

$$\begin{aligned}\det(\mathbf{A}) &= \det\left(\mathbf{UU}^H\right)\det\left(\mathbf{I}_N + \mathbf{Z}^H\mathbf{Z} - \mathbf{Z}^H\mathbf{P}_\mathbf{U}\mathbf{Z}\right) \\ &= \det\left(\mathbf{UU}^H\right)\det\left(\mathbf{I}_N + \mathbf{Z}^H(\mathbf{I}_L - \mathbf{P}_\mathbf{U})\mathbf{Z}\right)\end{aligned}$$

$$= \det\left(\mathbf{U}\mathbf{U}^H\right)\det\left(\mathbf{I}_L + (\mathbf{I}_L - \mathbf{P}_\mathbf{U})\mathbf{Z}\mathbf{Z}^H(\mathbf{I}_L - \mathbf{P}_\mathbf{U})\right)$$
$$= \det\left(\mathbf{U}\mathbf{U}^H\right)\det\left(\mathbf{I}_L + \mathbf{P}_\mathbf{U}^\perp\mathbf{Z}\mathbf{Z}^H\mathbf{P}_\mathbf{U}^\perp\right).$$

Moreover, it is bounded as (see (B.14))

$$\det(\mathbf{A}) \leq \det\left(\mathbf{U}\mathbf{U}^H\right)\det\left(\mathbf{I}_N + (\mathbf{Z}-\mathbf{U}\mathbf{X})^H(\mathbf{Z}-\mathbf{U}\mathbf{X})\right)$$
$$= \det(\mathbf{U}\mathbf{U}^H)\det\left(\mathbf{I}_L + (\mathbf{Z}-\mathbf{U}\mathbf{X})(\mathbf{Z}-\mathbf{U}\mathbf{X})^H\right),$$

with equality iff $\mathbf{U}^H\mathbf{Z} = \mathbf{U}^H\mathbf{U}\mathbf{X}$, or $\mathbf{U}\mathbf{X} = \mathbf{P}_\mathbf{U}\mathbf{Z}$. The minimizing solution for $\mathbf{X}$ is $\mathbf{X} = (\mathbf{U}^H\mathbf{U})^{-1}\mathbf{U}^H\mathbf{Z}$, and

$$\min\det\left(\mathbf{I}_L + (\mathbf{Z}-\mathbf{U}\mathbf{X})(\mathbf{Z}-\mathbf{U}\mathbf{X})^H\right) = \det(\mathbf{U}\mathbf{U}^H)\det\left(\mathbf{I}_L + (\mathbf{I}_L - \mathbf{P}_\mathbf{U})\mathbf{Z}\mathbf{Z}^H(\mathbf{I}_L - \mathbf{P}_\mathbf{U})\right)$$
$$= \det(\mathbf{U}\mathbf{U}^H)\det\left(\mathbf{I}_L + \mathbf{P}_\mathbf{U}^\perp\mathbf{Z}\mathbf{Z}^H\mathbf{P}_\mathbf{U}^\perp\right). \tag{B.15}$$

B.9.3 Minimize Trace or Determinant of Error Covariance in Reduced-Rank Least Squares

Begin with random vectors $\mathbf{x} \in \mathbb{C}^m$ and $\mathbf{y} \in \mathbb{C}^n$, with composite covariance matrix:

$$E\left[\begin{bmatrix}\mathbf{x}\\ \mathbf{y}\end{bmatrix}\begin{bmatrix}\mathbf{x}^H & \mathbf{y}^H\end{bmatrix}\right] = \begin{bmatrix}\mathbf{R}_{\mathbf{xx}} & \mathbf{R}_{\mathbf{xy}}\\ \mathbf{R}_{\mathbf{xy}}^H & \mathbf{R}_{\mathbf{yy}}\end{bmatrix}.$$

Error Covariance. Let $\mathbf{A}\mathbf{y}$ be a linear estimator of $\mathbf{x} \in \mathbb{C}^m$ from $\mathbf{y} \in \mathbb{C}^n$. Define $\mathbf{Q}_{\mathbf{xx}}(\mathbf{A})$ to be the error covariance matrix for this estimator, given by

$$\mathbf{Q}_{\mathbf{xx}}(\mathbf{A}) = E\left[(\mathbf{x}-\mathbf{A}\mathbf{y})(\mathbf{x}-\mathbf{A}\mathbf{y})^H\right] = \mathbf{R}_{\mathbf{xx}} - \mathbf{A}\mathbf{R}_{\mathbf{xy}}^H - \mathbf{R}_{\mathbf{xy}}\mathbf{A}^H + \mathbf{A}\mathbf{R}_{\mathbf{yy}}\mathbf{A}^H,$$
$$= (\mathbf{A}\mathbf{R}_{\mathbf{yy}}^{1/2} - \mathbf{R}_{\mathbf{xy}}\mathbf{R}_{\mathbf{yy}}^{-1/2})(\mathbf{A}\mathbf{R}_{\mathbf{yy}}^{1/2} - \mathbf{R}_{\mathbf{xy}}\mathbf{R}_{\mathbf{yy}}^{-1/2})^H + \mathbf{Q}_*,$$

where $\mathbf{Q}_* = \mathbf{R}_{\mathbf{xx}} - \mathbf{R}_{\mathbf{xy}}\mathbf{R}_{\mathbf{yy}}^{-1}\mathbf{R}_{\mathbf{xy}}^H = \mathbf{R}_{\mathbf{xx}} - \mathbf{A}_*\mathbf{R}_{\mathbf{yy}}\mathbf{A}_*^H$, and $\mathbf{A}_* = \mathbf{R}_{\mathbf{xy}}\mathbf{R}_{\mathbf{yy}}^{-1}$. The second line in the identity for $\mathbf{Q}_{\mathbf{xx}}(\mathbf{A})$ is obtained by completing the square (see Appendix 2.A).

The Trace Problem. The problem is to minimize $\operatorname{tr}(\mathbf{Q}_{\mathbf{xx}}(\mathbf{A}))$ under the constraint that the rank of $\mathbf{A}$ be no greater than r, that is,

$$\underset{\mathbf{A}\in\mathbb{C}^{m\times n},\ \operatorname{rank}(\mathbf{A})\leq r}{\text{minimize}} \quad \operatorname{tr}\left[(\mathbf{A}\mathbf{R}_{\mathbf{yy}}^{1/2} - \mathbf{R}_{\mathbf{xy}}\mathbf{R}_{\mathbf{yy}}^{-1/2})(\mathbf{A}\mathbf{R}_{\mathbf{yy}}^{1/2} - \mathbf{R}_{\mathbf{xy}}\mathbf{R}_{\mathbf{yy}}^{-1/2})^H + \mathbf{Q}_*\right].$$

The solution is for $\mathbf{A}\mathbf{R}_{\mathbf{yy}}^{1/2}$ to be the best rank-r approximation to the *half-canonical correlation matrix* $\mathbf{R}_{\mathbf{xy}}\mathbf{R}_{\mathbf{yy}}^{-1/2}$ [302, 308]. That is, $\mathbf{A}\mathbf{R}_{\mathbf{yy}}^{1/2} = \mathbf{U}\boldsymbol{\Sigma}_r\mathbf{V}^H$, where $\mathbf{U}\boldsymbol{\Sigma}\mathbf{V}^H$ is the SVD of $\mathbf{R}_{\mathbf{xy}}\mathbf{R}_{\mathbf{yy}}^{-1/2}$, and $\boldsymbol{\Sigma}_r$ is obtained from $\boldsymbol{\Sigma}$ by zeroing the trailing $m-r$ singular values of $\boldsymbol{\Sigma}$ if $m \le n$, or its trailing $n-r$ singular values if $n \le m$. The resulting solution for $\mathbf{A}$ is $\mathbf{A} = \mathbf{U}\boldsymbol{\Sigma}_r\mathbf{V}^H\mathbf{R}_{\mathbf{yy}}^{-1/2}$. Then, $\mathbf{Q}_{\mathbf{xx}}(\mathbf{A}) = \mathbf{Q}_* + \mathbf{U}(\boldsymbol{\Sigma} - \boldsymbol{\Sigma}_r)^2\mathbf{U}^H$ and

$$\operatorname{tr}(\mathbf{Q}_{\mathbf{xx}}(\mathbf{A})) = \operatorname{tr}(\mathbf{Q}_*) + \operatorname{tr}((\boldsymbol{\Sigma} - \boldsymbol{\Sigma}_r)^2) = \operatorname{tr}(\mathbf{Q}_*) + \sum_{i=r+1}^{\min(m,n)} \sigma_i^2.$$

The extra term in $\operatorname{tr}(\mathbf{Q}_{\mathbf{xx}}(\mathbf{A}))$ is the performance cost due to rank reduction [302].

The Determinant Problem. The problem is to minimize $\det(\mathbf{Q}_{\mathbf{xx}}(\mathbf{A}))$, under the constraint that the rank of $\mathbf{A}$ be no greater than r. Let $\mathbf{C} = \mathbf{R}_{\mathbf{xx}}^{-1/2}\mathbf{R}_{\mathbf{xy}}\mathbf{R}_{\mathbf{yy}}^{-1/2}$ be the coherence matrix with SVD $\mathbf{C} = \mathbf{F}\mathbf{K}\mathbf{G}^H$ and replace $\mathbf{Q}_{\mathbf{xx}}(\mathbf{A})$ by $\mathbf{F}^H\mathbf{R}_{\mathbf{xx}}^{-1/2}\mathbf{Q}_{\mathbf{xx}}(\mathbf{A})\mathbf{R}_{\mathbf{xx}}^{-1/2}\mathbf{F}$. This only scales the determinant by $\det(\mathbf{R}_{\mathbf{xx}}^{-1})$. Write this as

$$\begin{aligned}
&\mathbf{F}^H\mathbf{R}_{\mathbf{xx}}^{-1/2}\mathbf{Q}_{\mathbf{xx}}(\mathbf{A})\mathbf{R}_{\mathbf{xx}}^{-1/2}\mathbf{F} = \\
&\quad =(\mathbf{I} - \mathbf{K}^2) + \mathbf{F}^H(\mathbf{R}_{\mathbf{xx}}^{-1/2}\mathbf{A}\mathbf{R}_{\mathbf{yy}}^{1/2} - \mathbf{C})\mathbf{G}\mathbf{G}^H(\mathbf{R}_{\mathbf{xx}}^{-1/2}\mathbf{A}\mathbf{R}_{\mathbf{yy}}^{1/2} - \mathbf{C})^H\mathbf{F} \\
&\quad =(\mathbf{I} - \mathbf{K}^2) + (\mathbf{F}^H\mathbf{R}_{\mathbf{xx}}^{-1/2}\mathbf{A}\mathbf{R}_{\mathbf{yy}}^{1/2}\mathbf{G} - \mathbf{K})(\mathbf{F}^H\mathbf{R}_{\mathbf{xx}}^{-1/2}\mathbf{A}\mathbf{R}_{\mathbf{yy}}^{1/2}\mathbf{G} - \mathbf{K})^H.
\end{aligned}$$

This is the sum of two positive semidefinite matrices. The determinant of this matrix is minimized by the rank-r matrix $\mathbf{A} = \mathbf{R}_{\mathbf{xx}}^{1/2}\mathbf{F}\mathbf{K}_r\mathbf{G}^H\mathbf{R}_{\mathbf{yy}}^{-1/2}$, in which case

$$\det(\mathbf{Q}_{\mathbf{xx}}(\mathbf{A})) = \det(\mathbf{I} - \mathbf{K}_r)^2\det(\mathbf{R}_{\mathbf{xx}}) = \det(\mathbf{Q}_*)\frac{1}{\prod_{i=r+1}^{\min(m,n)}(1 - k_i^2)}.$$

As in the trace problem, the cost is inflated by a factor that depends on the smallest canonical correlations. See [178] for the original derivation of this result by different methods.

B.9.4 Maximum Likelihood Estimation in a Factor Model

Suppose measurements $\mathbf{X} = [\mathbf{x}_1 \ \cdots \ \mathbf{x}_N]$, with $\mathbf{x}_i \in \mathbb{C}^n$ and $n \le N$, are drawn independently from a multivariate normal distribution $CN_n(\mathbf{0}, \mathbf{R})$, with $\mathbf{R} = \mathbf{G}\mathbf{G}^H + \boldsymbol{\Psi}$. This is said to be a factor model for a normal random vector $\mathbf{x}$, which is to say $\mathbf{x} = \mathbf{G}\mathbf{u} + \mathbf{e}$, where $\mathbf{G} \in \mathbb{C}^{n\times r}$ contains *factor loadings* in its columns, and $\mathbf{u} = [u_1 \ \cdots \ u_r]^T$ contains common factors u_i. On the other hand, $\mathbf{e} = [e_1 \ \cdots \ e_n]^T$

contains specific or unique factors. That is, $\mathbf{x} = \sum_{i=1}^{r} \mathbf{g}_i u_i + \mathbf{e}$. The specific factors are assumed to be uncorrelated so the covariance matrix $\boldsymbol{\Psi}$ is a diagonal matrix with diagonal elements $\psi_{11} > \psi_{22} > \cdots > \psi_{nn} > 0$. The common factors are also assumed to be uncorrelated and are each standardized to have variance 1 so that $\mathrm{E}[\mathbf{u}\mathbf{u}^H] = \mathbf{I}_r$.

If $\boldsymbol{\Psi}$ is known, the ML estimate of $\mathbf{G}$ can be obtained in closed-form. It is as if the measurement $\boldsymbol{\Psi}^{-1/2}\mathbf{x}_i$ is drawn from the distribution $CN_n(\mathbf{0}, \boldsymbol{\Sigma})$, with $\boldsymbol{\Sigma} = \mathbf{F}\mathbf{F}^H + \mathbf{I}_n$ and $\mathbf{G} = \boldsymbol{\Psi}^{1/2}\mathbf{F}$. Define the Hermitian matrix $\mathbf{S} = \boldsymbol{\Psi}^{-1/2}\mathbf{X}\mathbf{X}^H\boldsymbol{\Psi}^{-1/2}/N$ and give it the EVD $\mathbf{S} = \mathbf{U}\boldsymbol{\Lambda}\mathbf{U}^H$, with eigenvalues $\lambda_1 > \lambda_2 > \cdots > \lambda_n$. The matrix $\mathbf{F}$ is to be determined as the solution of

$$\underset{\mathbf{F}}{\text{minimize}}\ \mathrm{tr}(\boldsymbol{\Sigma}^{-1}\mathbf{S}) - \log\det(\boldsymbol{\Sigma}^{-1}\mathbf{S}), \quad \text{s.t.} \quad \mathbf{F}^H\mathbf{F} \quad \text{is diagonal},$$

where this cost function is a monotone function of Gaussian likelihood and the diagonal constraint is included to obtain a unique solution. The solution is [13]

$$\mathbf{F} = \mathbf{U}\mathbf{D}^{1/2},$$

where $\mathbf{D} = \mathrm{diag}[(\lambda_1 - 1)^+, \ldots, (\lambda_r - 1)^+, 0, \ldots, 0]$. It is easy to check that $\boldsymbol{\Sigma} = \mathbf{F}\mathbf{F}^H + \mathbf{I}_n = \mathbf{U}(\mathbf{D} + \mathbf{I}_n)\mathbf{U}^H$, where $\mathbf{D} + \mathbf{I}_n = \mathrm{diag}(\lambda_1, \lambda_2, \ldots, \lambda_{\tilde{r}}, 1, \ldots, 1)$, with $\tilde{r} = \min(\arg\max_i(\lambda_i \geq 1), r)$, i.e., the index of smallest eigenvalue that is larger than one, or r if this index is below r. This amounts to eigenvalue shaping in the EVD $\mathbf{S} = \mathbf{U}\boldsymbol{\Lambda}\mathbf{U}^H$ for the model $\mathbf{F}\mathbf{F}^H + \mathbf{I}_n$. This solution holds for all $r \leq n$, so the solution for $\mathbf{F}$ automatically returns a rank-$\tilde{r}$ matrix of dimension $n \times \tilde{r}$.

If $\boldsymbol{\Psi}$ is unknown, $\mathbf{G}$ and $\boldsymbol{\Psi}$ have to be determined as the minimizers of

$$V(\mathbf{G}, \boldsymbol{\Psi}) = \mathrm{tr}(\mathbf{R}^{-1}\mathbf{S}) - \log\det(\mathbf{R}^{-1}\mathbf{S}),$$

where $\mathbf{R} = \mathbf{G}\mathbf{G}^H + \boldsymbol{\Psi}$ and $\mathbf{S} = \mathbf{X}\mathbf{X}^H/N$. Note that $V(\mathbf{G}, \boldsymbol{\Psi})$ can be expressed as

$$V(\mathbf{G}, \boldsymbol{\Psi}) = r(a - \log g - 1),$$

where a and g are the arithmetic and geometric means of the eigenvalues of $\mathbf{R}^{-1}\mathbf{S}$ [227].

B.10 Matrix Derivatives

B.10.1 Differentiation with Respect to a Real Matrix

This subsection covers differentiation of a number of expressions with respect to $\mathbf{X} \in \mathbb{R}^{n\times m}$. They have been taken from a number of sources, mainly [246, 302], and [263]. It is assumed here that the matrix has no special structure (Toeplitz, positive definite, etc).

Let $a(\mathbf{X})$ denote a scalar function of the matrix $\mathbf{X}$; the gradient of $a(\mathbf{X})$ with respect to $\mathbf{X}$ is the $n \times m$ matrix:

$$\frac{\partial a(\mathbf{X})}{\partial \mathbf{X}} = \begin{bmatrix} \frac{\partial a(\mathbf{X})}{\partial x_{11}} & \frac{\partial a(\mathbf{X})}{\partial x_{12}} & \cdots & \frac{\partial a(\mathbf{X})}{\partial x_{1m}} \\ \frac{\partial a(\mathbf{X})}{\partial x_{21}} & \cdots & \cdots & \frac{\partial a(\mathbf{X})}{\partial x_{2m}} \\ \vdots & \vdots & \ddots & \vdots \\ \frac{\partial a(\mathbf{X})}{\partial x_{n1}} & \cdots & \cdots & \frac{\partial a(\mathbf{X})}{\partial x_{nm}} \end{bmatrix}.$$

A few important special cases follow.

Derivatives of a Determinant. Unless stated otherwise, it is assumed that $\mathbf{X}$ is square and invertible:

$$\frac{\partial \det(\mathbf{X})}{\partial \mathbf{X}} = \det(\mathbf{X})\mathbf{X}^{-T}$$

$$\frac{\partial \det(\mathbf{X})^k}{\partial \mathbf{X}} = k \det(\mathbf{X})^k \mathbf{X}^{-T}$$

$$\frac{\partial \det(\mathbf{AXB})}{\partial \mathbf{X}} = \det(\mathbf{AXB})\mathbf{X}^{-T}$$

$$\frac{\partial \det(\mathbf{X}^T\mathbf{AX})}{\partial \mathbf{X}} = 2 \det(\mathbf{X}^T\mathbf{AX})\mathbf{X}^{-T}$$

$$\frac{\partial \log |\det(\mathbf{X})|}{\partial \mathbf{X}} = \mathbf{X}^{-T}$$

If $\mathbf{X}$ is not square, then

$$\frac{\partial \det(\mathbf{X}^T\mathbf{AX})}{\partial \mathbf{X}} = \det(\mathbf{X}^T\mathbf{AX}) \left(\mathbf{AX}\left(\mathbf{X}^T\mathbf{AX}\right)^{-1} + \mathbf{A}^T\mathbf{X}\left(\mathbf{X}^T\mathbf{A}^T\mathbf{X}\right)^{-1}\right)$$

$$\frac{\partial \log \det(\mathbf{X}^T\mathbf{X})}{\partial \mathbf{X}} = 2(\mathbf{X}^{\#})^T$$

A useful first-order approximation is

$$\det(\mathbf{I}_n + t\mathbf{X}) = 1 + t \operatorname{tr}(\mathbf{X}) + o(t^2),$$

and, therefore, the derivative of $\det(\mathbf{I}_n + t\mathbf{X})$ at $t = 0$ is $\operatorname{tr}(\mathbf{X})$.

Derivatives of an Inverse. It is assumed that $\mathbf{X}$ is square and invertible:

$$\frac{\partial \mathbf{a}^T\mathbf{X}^{-1}\mathbf{b}}{\partial \mathbf{X}} = -\mathbf{X}^{-T}\mathbf{a}\mathbf{b}^T\mathbf{X}^{-T}$$

$$\frac{\partial \det(\mathbf{X}^{-1})}{\partial \mathbf{X}} = -\det(\mathbf{X}^{-1})\mathbf{X}^{-T}$$

$$\frac{\partial \operatorname{tr}(\mathbf{A}\mathbf{X}^{-1}\mathbf{B})}{\partial \mathbf{X}} = -\left(\mathbf{X}^{-1}\mathbf{B}\mathbf{A}\mathbf{X}^{-1}\right)^T$$

Derivatives of Traces. For congruent matrices, then

$$\frac{\partial \operatorname{tr}(\mathbf{X})}{\partial \mathbf{X}} = \mathbf{I}_n$$

$$\frac{\partial \operatorname{tr}(\mathbf{X}\mathbf{A})}{\partial \mathbf{X}} = \mathbf{A}^T$$

$$\frac{\partial \operatorname{tr}(\mathbf{A}\mathbf{X}\mathbf{B})}{\partial \mathbf{X}} = \mathbf{A}^T\mathbf{B}^T$$

$$\frac{\partial \operatorname{tr}(\mathbf{A}\mathbf{X}^T\mathbf{B})}{\partial \mathbf{X}} = \mathbf{B}\mathbf{A}.$$

If $\mathbf{A}$ is $m \times m$ and $\mathbf{X}$ is $n \times n$, then

$$\frac{\partial \operatorname{tr}(\mathbf{A} \otimes \mathbf{X})}{\partial \mathbf{X}} = \operatorname{tr}(\mathbf{A})\mathbf{I}_n.$$

B.10.2 Differentiation with Respect to a Complex Matrix

To extend the results of the previous section to the case of complex matrices, we need to introduce the concept of *generalized complex derivatives*. The material in this section is based on [318] and [165].

Analytic and Nonanalytic Functions. It is known from classical complex analysis that a complex function $f(z)$ is differentiable iff it is *analytic*. If $f(z)$ satisfies the Cauchy-Riemann conditions, then it is analytic. For the case of scalar functions, the Cauchy-Riemann equations can be formulated as

$$\frac{\partial f(z)}{\partial z^*} = 0,$$

meaning that $f(z)$ *cannot* depend on z^*. Unfortunately, in signal processing and machine learning, we often deal with *real* cost functions, such as likelihoods or regularized loss functions, which are not analytic. For instance, consider the squared distance $f(z) = |z|^2 = zz^*$. This function is not differentiable because the value of the limit

$$\lim_{\Delta z \to 0} \frac{f(z_0 + \Delta z) - f(z_0)}{\Delta z} = \lim_{\Delta z \to 0} \frac{|z_0 + \Delta z|^2 - |z_0|^2}{\Delta z}$$
$$= \lim_{\Delta z \to 0} \frac{(\Delta z) z_0^* + z_0(\Delta z)^* + |\Delta z|^2}{\Delta z}, \tag{B.16}$$

depends on how $\Delta z = \Delta x + j\Delta y$ approaches 0. Let Δz approach 0 such that first $\Delta x \to 0$, then the value of the last limit in (B.16) is

$$\lim_{\Delta y \to 0} \frac{j\Delta y z_0^* - jz_0\Delta y + (\Delta y)^2}{j\Delta y} = \lim_{\Delta y \to 0} z_0^* - z_0 - j\Delta y = z_0^* - z_0.$$

However, if we let Δz approach 0 such that first $\Delta y \to 0$, then the value of the limit is $z_0^* + z_0$ and, therefore, this limit does not exits.

There are two alternatives for finding the derivative of a scalar real-valued function $f(z)$ with respect to z. The first one is to rewrite $f(z)$ as a function of the real and imaginary parts of z and then find the derivatives of the real-valued bidimensional function $f(x, y)$ with respect to the real variables x and y. A more elegant way to solve the problem was developed by the Austrian mathematician Wilhelm Wirtinger, using what is now called *Wirtinger calculus*. The main idea of Wirtinger calculus is to treat $f(z)$ as a function of two independent variables z and z^*. Notice that any real-valued function that depends on z must also be explicitly or implicitly dependent on z^*. The squared distance function $f(z) = |z|^2 = zz^*$ is a clear example. A generalized complex derivative is then formally defined as the derivative with respect to z while treating z^* as constant. Another generalized derivative is defined as the derivative with respect to z^* while treating z as constant.

Definition Let $z = x + jy$, where $x, y \in \mathbb{R}$. Then, the generalized complex differential operators, or Wirtinger derivatives, are defined as

$$\frac{\partial f(z)}{\partial z} = \frac{1}{2}\left(\frac{\partial f(z)}{\partial x} - j\frac{\partial f(z)}{\partial y}\right), \qquad \frac{\partial f(z)}{\partial z^*} = \frac{1}{2}\left(\frac{\partial f(z)}{\partial x} + j\frac{\partial f(z)}{\partial y}\right).$$

For example, when $f(z) = |z|^2$, it is easy to find that

$$\frac{\partial f(z)}{\partial z} = z^*, \quad \text{and} \quad \frac{\partial f(z)}{\partial z^*} = z.$$

The generalized complex derivative equals the normal derivative, when $f(z)$ is an analytic function. In this case, the conjugate generalized complex derivative equals zero.

These ideas are carried over to derivatives of real-valued functions that depend on complex vectors or matrices. In fact, nothing prevents us from applying the Wirtinger derivatives to complex-valued functions as well. Therefore, in the following, we no longer assume that $f(z)$ is a real-valued function.

Let us start by presenting some common derivatives of scalar functions with respect to a complex vector $\mathbf{x}$. It is assumed that $\mathbf{a}$ and $\mathbf{A}$ are independent of $\mathbf{x}$:

$$\frac{\partial \mathbf{a}^H \mathbf{x}}{\partial \mathbf{x}} = \mathbf{a}^H \quad \text{and} \quad \frac{\partial \mathbf{a}^H \mathbf{x}}{\partial \mathbf{x}^*} = \mathbf{0}$$

$$\frac{\partial \mathbf{x}^H \mathbf{a}}{\partial \mathbf{x}} = \mathbf{0} \quad \text{and} \quad \frac{\partial \mathbf{x}^H \mathbf{a}}{\partial \mathbf{x}^*} = \mathbf{a}^T$$

$$\frac{\partial \mathbf{x}^H \mathbf{A}\mathbf{x}}{\partial \mathbf{x}} = \mathbf{x}^H \mathbf{A} \quad \text{and} \quad \frac{\partial \mathbf{x}^H \mathbf{A}\mathbf{x}}{\partial \mathbf{x}^*} = \mathbf{x}^T \mathbf{A}^T$$

$$\frac{\partial \mathbf{x}^T \mathbf{A}\mathbf{x}}{\partial \mathbf{x}} = \mathbf{x}^T \left(\mathbf{A} + \mathbf{A}^T\right) \quad \text{and} \quad \frac{\partial \mathbf{x}^T \mathbf{A}\mathbf{x}}{\partial \mathbf{x}^*} = \mathbf{0}$$

$$\frac{\partial \exp\left(-\mathbf{x}^H \mathbf{A}\mathbf{x}\right)}{\partial \mathbf{x}} = -\exp\left(-\mathbf{x}^H \mathbf{A}\mathbf{x}\right) \mathbf{x}^H \mathbf{A}$$

$$\frac{\partial \ln\left(\mathbf{x}^H \mathbf{A}\mathbf{x}\right)}{\partial \mathbf{x}} = \left(\mathbf{x}^H \mathbf{A}\mathbf{x}\right)^{-1} \mathbf{x}^H \mathbf{A}$$

For complex matrices, the following are important special cases:

$$\frac{\partial \operatorname{tr}(\mathbf{X}^H \mathbf{A}\mathbf{X})}{\partial \mathbf{X}} = \mathbf{A}^T \mathbf{X}^* \quad \text{and} \quad \frac{\partial \operatorname{tr}(\mathbf{X}^H \mathbf{A}\mathbf{X})}{\partial \mathbf{X}^*} = \mathbf{A}\mathbf{X}$$

$$\frac{\partial \det(\mathbf{X}^H \mathbf{A}\mathbf{X})}{\partial \mathbf{X}} = \det(\mathbf{X}^H \mathbf{A}\mathbf{X}) \left((\mathbf{X}^H \mathbf{A}\mathbf{X})^{-1} \mathbf{X}^H \mathbf{A}\right)^T$$

and

$$\frac{\partial \det(\mathbf{X}^H \mathbf{A}\mathbf{X})}{\partial \mathbf{X}^*} = \det(\mathbf{X}^H \mathbf{A}\mathbf{X}) \mathbf{A}\mathbf{X} (\mathbf{X}^H \mathbf{A}\mathbf{X})^{-1}$$

The SVD

C

C.1 The Singular Value Decomposition

Begin with a matrix $\mathbf{H} \in \mathbb{C}^{L \times p}$. The singular value decomposition (SVD)[1] factors this matrix as $\mathbf{H} = \mathbf{FKG}^H$, where $\mathbf{F} \in U(L), \mathbf{G} \in U(p)$, and $\mathbf{K} \in \mathbb{R}^{L \times p}$ is a matrix of non-negative real values along its main diagonal, and zeros elsewhere. The columns of $\mathbf{F}$ are termed left singular vectors, the columns of $\mathbf{G}$ are termed right singular vectors, and the entries in the diagonal of $\mathbf{K}$ are termed singular values. They are denoted by $k_1, \ldots, k_{\min(L,p)}$. In this book, we assume the singular values are ordered, $k_1 \geq \cdots \geq k_{\min(L,p)}$. More on the matrix $\mathbf{K}$ will follow.

Think of $\mathbf{H}$ as a linear map from the space $\mathbb{C}^p$ into the space $\mathbb{C}^L$. The SVD factors this map as a resolution of a vector in $\mathbb{C}^p$ onto a special basis $\mathbf{G}$ for that space, a scaling of the resulting coefficients by the singular values, and a reconstruction in the basis $\mathbf{F}$. This is an analysis-synthesis interpretation: analyze a signal $\mathbf{s}$ onto the columns of $\mathbf{G}$ to produce coordinate values $\mathbf{G}^H\mathbf{s}$, scale these coordinate-by-coordinate to produce the coefficients $\mathbf{KG}^H\mathbf{s}$, and resynthesize onto the basis $\mathbf{F}$ to produce $\mathbf{y} = \mathbf{Hs} = \mathbf{FKG}^H\mathbf{s}$.

Rank and Nullity. The rank of $\mathbf{H}$ is r, with $r \leq \min(L, p)$, which is the dimension of the range space of $\mathbf{H}$ acting as a linear map from $\mathbb{C}^p$ to $\mathbb{C}^L$. The nullity is $L - r$, which is the dimension of the null space of $\mathbf{H}$. The SVD shows that the leftmost r-slice of $\mathbf{F}$, corresponding to the nonzero singular values, is a unitary basis for the range space, and the rightmost $(L-r)$-slice of $\mathbf{G}$ is a unitary basis for the null space.

[1] A version of the SVD was discovered by E. Beltrami and C. Jordan, but the SVD for rectangular and complex matrices was evidently discovered by C. Eckart and G. J. Young. There have been many contributions to its efficient and stable numerical computation.

D. Ramírez et al., *Coherence*, https://doi.org/10.1007/978-3-031-13331-2

Singular Value Matrix. The matrix $\mathbf{K}$ is an $L \times p$ matrix of singular values, structured as

$$\mathbf{K} = \begin{bmatrix} k_1 & 0 & \cdots & 0 \\ 0 & k_2 & \cdots & 0 \\ \vdots & \vdots & \ddots & \vdots \\ 0 & 0 & \cdots & k_p \\ 0 & 0 & \cdots & 0 \\ \vdots & \vdots & \cdots & \vdots \\ 0 & 0 & \cdots & 0 \end{bmatrix},$$

when $L \geq p$; and for $L \leq p$, it is structured as

$$\mathbf{K} = \begin{bmatrix} k_1 & 0 & \cdots & 0 & 0 & \cdots & 0 \\ 0 & k_2 & \cdots & 0 & 0 & \cdots & 0 \\ \vdots & \vdots & \ddots & \vdots & \vdots & \cdots & \vdots \\ 0 & 0 & \cdots & k_L & 0 & \cdots & 0 \end{bmatrix}.$$

Along the diagonals of these matrices, $k_i = 0$ for $i > r$, where r is the rank of the matrix $\mathbf{H}$.

Gramians. The corresponding factorizations of the Gramians $\mathbf{H}\mathbf{H}^H$ and $\mathbf{H}^H\mathbf{H}$ are $\mathbf{H}\mathbf{H}^H = \mathbf{F}\mathbf{K}\mathbf{K}^T\mathbf{F}^H$ and $\mathbf{H}^H\mathbf{H} = \mathbf{G}\mathbf{K}^T\mathbf{K}\mathbf{G}^H$, a result that shows the nonzero eigenvalues of these two Gramians to be identical. Each of these matrices is Hermitian and positive semidefinite, so the left and right singular vectors of the SVD serve as eigenvectors of these Gramians, and the squares of the singular values serve as their non-negative eigenvalues.

Pseudo-Inverse. In our study of least squares approximation, we shall need the matrix $\mathbf{K}^\#$, defined to be the transpose of the matrices $\mathbf{K}$ defined above, with each nonzero k_i replaced by $1/k_i$ and each 0 remaining unchanged. It is easy to see that $\mathbf{K}^\#$ satisfies all conditions of a Moore-Penrose pseudo-inverse. However, it is perhaps more insightful to note that

$$\mathbf{K}\mathbf{K}^\# = \text{blkdiag}(\mathbf{I}_p, \mathbf{0}) \quad \text{and} \quad \mathbf{K}^\#\mathbf{K} = \mathbf{I}_p,$$

for $L \geq p$, and

$$\mathbf{K}\mathbf{K}^\# = \mathbf{I}_L \quad \text{and} \quad \mathbf{K}^\#\mathbf{K} = \text{blkdiag}(\mathbf{I}_L, \mathbf{0}),$$

for $L \leq p$. The matrix $\text{blkdiag}(\mathbf{I}_p, \mathbf{0})$ is a rank-p projection onto the dimension-p subspace spanned by the first p standard basis vectors.

The pseudo-inverse of the matrix $\mathbf{H}$ is $\mathbf{H}^{\#} = \mathbf{G}\mathbf{K}^{\#}\mathbf{F}^H$, and from this definition, it follows that

$$\mathbf{H}\mathbf{H}^{\#} = \mathbf{F}\,\text{blkdiag}(\mathbf{I}_p, \mathbf{0})\,\mathbf{F}^H \quad \text{and} \quad \mathbf{H}^{\#}\mathbf{H} = \mathbf{I}_p,$$

for $L \geq p$, and

$$\mathbf{H}\mathbf{H}^{\#} = \mathbf{I}_L \quad \text{and} \quad \mathbf{H}^{\#}\mathbf{H} = \mathbf{G}\,\text{blkdiag}(\mathbf{I}_L, \mathbf{0})\,\mathbf{G}^H,$$

for $L \leq p$. The matrix $\mathbf{H}\mathbf{H}^{\#} = \mathbf{F}\,\text{blkdiag}(\mathbf{I}_p, \mathbf{0})\,\mathbf{F}^H$ is a rank-p projection onto the p-dimensional subspace spanned by the first p columns of $\mathbf{F}$, and the matrix $\mathbf{H}^{\#}\mathbf{H} = \mathbf{G}\,\text{diag}(\mathbf{I}_L, \mathbf{0})\,\mathbf{G}^H$ is a rank-L projection onto the L-dimensional subspace spanned by the first L columns of $\mathbf{G}$.

Thin SVD. The *thin* SVD factors the matrix $\mathbf{H}$ as $\mathbf{H} = \mathbf{F}_r\mathbf{K}_r\mathbf{G}_r^H$, where $\mathbf{F}_r$ is the leftmost $L \times r$ slice of $\mathbf{F}$, $\mathbf{G}_r$ is the leftmost $p \times r$ slice of $\mathbf{G}$, and $\mathbf{K}_r$ is either the topmost $r \times r$ block of $\mathbf{K}$ or the leftmost $r \times r$ block of $\mathbf{K}$. Then, $\mathbf{H}$ may be written as $\mathbf{H} = \sum_{i=1}^{r} \mathbf{f}_i\, k_i\, \mathbf{g}_i^H$, and the pseudo-inverse may be written as $\mathbf{H}^{\#} = \sum_{i=1}^{r} \mathbf{g}_i\, k_i^{-1}\, \mathbf{f}_i^H$.

The thin SVD simplifies some notation, whereas the fat SVD illuminates some geometry. So both are useful.

Polar Decomposition. Let $\mathbf{H}$ be an $L \times p$ matrix of rank p. The thin SVD of $\mathbf{H}$ may be written as

$$\mathbf{H} = \mathbf{F}\mathbf{K}\mathbf{G}^H = (\mathbf{F}\mathbf{G}^H)(\mathbf{G}\mathbf{K}\mathbf{G}^H) = \mathbf{Q}\mathbf{P},$$

where $\mathbf{F}$ is an $L \times p$ slice of a unitary matrix, $\mathbf{K}$ is $p \times p$, and $\mathbf{G} \in U(p)$. With this rewriting, $\mathbf{P} = \mathbf{G}\mathbf{K}\mathbf{G}^H$ is non-negative definite, a generalization of a non-negative magnitude, and the matrix $\mathbf{Q} = \mathbf{F}\mathbf{G}^H$ is a $L \times p$ slice of a unitary matrix, a generalization of a unimodular complex number. So the decomposition of $\mathbf{H}$ as $\mathbf{H} = \mathbf{Q}\mathbf{P}$ is called a (left) polar decomposition, and the SVD gives it. If we had started with an $L \times p$ matrix $\mathbf{H}$ of rank L, we would have ended up with the (right) polar decomposition $\mathbf{H} = \mathbf{P}\mathbf{Q}^H$, with $\mathbf{P} = \mathbf{F}\mathbf{K}\mathbf{F}^H$ non-negative definite, and $\mathbf{Q}^H = \mathbf{F}\mathbf{G}^H$.

Connection to Eigenvalue Decomposition. Consider a (square) normal matrix, $\mathbf{N}$. That is, $\mathbf{N}\mathbf{N}^H = \mathbf{N}^H\mathbf{N}$. The spectral theorem for normal matrices says that $\mathbf{N}$ may be unitarily diagonalized as $\mathbf{N} = \mathbf{G}\boldsymbol{\Lambda}\mathbf{G}^H$, with $\mathbf{G}$ unitary and $\boldsymbol{\Lambda}$ a diagonal matrix of complex eigenvalues. Factor the complex eigenvalue matrix $\mathbf{\Lambda}$ as $\mathbf{\Lambda} = \mathbf{\Phi K}$, where the diagonal matrix $\mathbf{\Phi}$ is a phasing matrix of unimodular coefficients and $\mathbf{K}$ is a diagonal matrix of non-negative magnitudes. Then, the factorization of $\mathbf{N}$ may be written as the SVD $\mathbf{N} = \mathbf{FKG}^H$, where $\mathbf{F} = \mathbf{G\Phi}$ remains unitary. Conversely, begin with an SVD of $\mathbf{N}$ as $\mathbf{N} = \mathbf{FKG}^H$. Rewrite this as $\mathbf{G\Phi KG}^H = \mathbf{G\Lambda G}^H$, where $\mathbf{G\Phi} = \mathbf{F}$ and $\mathbf{\Lambda} = \mathbf{\Phi K}$.

Consider the power series $f(x) = a_0 + a_1 x + a_2 x^2 + \cdots$. The matrix-valued power series in the normal matrix $\mathbf{N}$ is $f(\mathbf{N}) = a_0\mathbf{I} + a_1\mathbf{G\Lambda G}^H + a_2\mathbf{G\Lambda}^2\mathbf{G}^H + \cdots = \mathbf{G}\, f(\mathbf{\Lambda})\, \mathbf{G}^H$, where $f(\mathbf{\Lambda}) = a_0\mathbf{I} + a_1\mathbf{\Lambda} + \cdots$. An important special case is $f(x) = \exp(x) = 1 + x + x^2/2! + \cdots$, in which case $f(\mathbf{N}) = \exp(\mathbf{N}) = \mathbf{G}\exp(\mathbf{\Lambda})\,\mathbf{G}^H$.

Among commonly encountered normal matrices are the Hermitian, Hermitian positive semidefinite, and unitary matrices. Their eigenvalues and some of their properties have been extensively discussed in Appendix B.

C.2 Low-Rank Matrix Approximation

From the SVD $\mathbf{H} = \mathbf{FKG}^H$, construct the rank-$r$ approximation $\mathbf{H}_r = \mathbf{FK}_r\mathbf{G}^H$, where $\mathbf{K}_r$ is the matrix $\mathbf{K}$ with all singular values $k_i, i > r$, set to zero. This rank-r approximation forms the basis for rank reduction, noise-cleaning, or de-noising. It is the minimum Frobenius norm approximation and the minimum ℓ_2 approximation to $\mathbf{H}$, as the following arguments, based on a theorem of Weyl, demonstrate.[2]

Theorem *Let* $\mathbf{H} = \mathbf{FKG}^H$ *be an* $L \times p$ *matrix, whose rank-r approximation is* $\mathbf{H}_r = \mathbf{FK}_r\mathbf{G}^H$. *Let* $\mathbf{M}$ *be any* $L \times p$ *matrix of rank no more than r, then*

$$\|\mathbf{H} - \mathbf{H}_r\| \leq \|\mathbf{H} - \mathbf{M}\|,$$

and

$$\|\mathbf{H} - \mathbf{H}_r\|_2 \leq \|\mathbf{H} - \mathbf{M}\|_2.$$

Proof The proof of the theorem is based on the following lemma, which is essentially a restatement of one of Weyl's inequalities (cf. Theorem B.1 in Appendix B).

Lemma *Let* $\mathbf{H}$ *denote an* $L \times p$ *matrix of rank p, and* $\mathbf{M}$ *an* $L \times p$ *matrix of rank no more than r. Denote a singular value of* $\mathbf{H}$ *by* $sv(\mathbf{H})$. *Then, for* $r \leq p$, $sv_{i+r}(\mathbf{H}) \leq sv_i(\mathbf{H} - \mathbf{M})$.

[2] This demonstration follows the web posting of Victor Chen (http://www.victorchen.org).

From this lemma, it follows that no other rank-r approximation to $\mathbf{H}$ has smaller Frobenius norm than $\mathbf{H}_r$, i.e.,

$$\begin{aligned}\mathrm{tr}[(\mathbf{H}-\mathbf{H}_r)^H(\mathbf{H}-\mathbf{H}_r)] &= \sum_{i=1}^{p-r} sv_{i+r}^2(\mathbf{H}) \leq \sum_{i=1}^{p-r} sv_i^2(\mathbf{H}-\mathbf{M}) \\ &\leq \mathrm{tr}[(\mathbf{H}-\mathbf{M})^H(\mathbf{H}-\mathbf{M})],\end{aligned}$$

where we have assumed $L \geq p$. Similar arguments can be used for $L \leq p$.

The lemma also establishes that $\mathbf{H}_r$ also minimizes the ℓ_2 norm of a rank-r approximation since $sv_1^2(\mathbf{H}-\mathbf{H}_r) = sv_{r+1}^2(\mathbf{H}) \leq sv_1^2(\mathbf{H}-\mathbf{M})$. □

C.3 The CS Decomposition and the GSVD

The abbreviations CS and GSVD stand for *cosine-sine* decomposition and *generalized singular value decomposition*. In this section, the CS and GSVD decompositions are defined in the context of two-channel models for data matrices.

A unitary slice $\mathbf{Q}$ may be taken to be a unitary representation of a nonsingular channel matrix in a QR factorization $\mathbf{H} = \mathbf{QR}$. The CS decomposition may be taken to be a *coupled* QR-like factorization of channel matrices $\mathbf{H}_X$ and $\mathbf{H}_Y$. In a similar vein, the generalized singular value decomposition (GSVD) may be said to be a *coupled* two-channel version of the singular value decomposition (SVD).[3]

C.3.1 CS Decomposition

Every $L \times p$ unitary slice $\mathbf{Q} \in \mathbb{C}^{L\times p}$, with $L \geq p$, has a trivial SVD, $\mathbf{Q} = \mathbf{U\Sigma V}^H$, where $\mathbf{U} = \mathbf{Q}$ and $\mathbf{\Sigma V}$ is a $p \times p$ identity matrix. But in two-channel problems, the matrix $\mathbf{Q}$ parses naturally into two submatrices $\mathbf{A} \in \mathbb{C}^{L_X\times p}$ and $\mathbf{B} \in \mathbb{C}^{L_Y\times p}$, with $L_X \geq p$ and $L_Y \geq p$, as

$$\mathbf{Q} = \begin{bmatrix}\mathbf{A}\\ \mathbf{B}\end{bmatrix} \in \mathbb{C}^{(L_X+L_Y)\times p},$$

where $\mathbf{Q}^H\mathbf{Q} = \mathbf{A}^H\mathbf{A} + \mathbf{B}^H\mathbf{B} = \mathbf{I}_p$. It follows that the singular values of $\mathbf{A}$ are less than or equal to 1. Give $\mathbf{A}$ the SVD $\mathbf{A} = \mathbf{U}_A\mathbf{C}\mathbf{V}_A^H$, and represent $\mathbf{Q}$ as

[3] These extensions are reminiscent of canonical correlation analysis (CCA) as a two-channel version of principal component analysis (PCA).

$$\mathbf{Q} = \begin{bmatrix} \mathbf{U}_A\mathbf{C} \\ \mathbf{B}\mathbf{V}_A \end{bmatrix} \mathbf{V}_A^H.$$

In this representation, $(\mathbf{U}_A\mathbf{C})^H(\mathbf{U}_A\mathbf{C}) = \mathbf{C}^H\mathbf{C}$ is diagonal with diagonal elements less than or equal to 1. The quadratic form $(\mathbf{B}\mathbf{V}_A)^H(\mathbf{B}\mathbf{V}_A)$ may be resolved as

$$\begin{aligned}(\mathbf{B}\mathbf{V}_A)^H(\mathbf{B}\mathbf{V}_A) &= \mathbf{V}_A^H(\mathbf{B}^H\mathbf{B})\mathbf{V}_A = \mathbf{V}_A^H(\mathbf{I}_p - \mathbf{A}^H\mathbf{A})\mathbf{V}_A \\ &= \mathbf{V}_A^H(\mathbf{I}_p - \mathbf{V}_A\mathbf{C}^2\mathbf{V}_A^H)\mathbf{V}_A = \mathbf{I}_p - \mathbf{C}^2,\end{aligned}$$

where we can define $\mathbf{S}^2 = \mathbf{I}_p - \mathbf{C}^2$. Then, $\mathbf{B}\mathbf{V}_A\mathbf{S}^{-1} = \mathbf{U}_B$ is a unitary slice, and $\mathbf{Q}$ may be written as

$$\mathbf{Q} = \begin{bmatrix} \mathbf{U}_A\mathbf{C} \\ \mathbf{U}_B\mathbf{S} \end{bmatrix} \mathbf{V}_A^H.$$

This is the CS decomposition of the unitary slice $\mathbf{Q}$. The matrices $\mathbf{U}_A$ and $\mathbf{U}_B$ are unitary slices, $\mathbf{C}$ and $\mathbf{S}$ are diagonal, with diagonal elements less than or equal to 1, and $\mathbf{C}^2 + \mathbf{S}^2 = \mathbf{I}_p$.

As an operator, the matrix $\mathbf{Q}$ operates on vectors $\mathbf{u} \in \mathbb{C}^p$ as

$$\mathbf{Q}\mathbf{u} = \begin{bmatrix} \mathbf{U}_A\mathbf{C}\mathbf{V}_A^H\mathbf{u} \\ \mathbf{U}_B\mathbf{S}\mathbf{V}_A^H\mathbf{u} \end{bmatrix} = \begin{bmatrix} \mathbf{U}_A \\ & \mathbf{U}_B \end{bmatrix} \begin{bmatrix} \mathbf{C} & \mathbf{0} \\ \mathbf{0} & \mathbf{S} \end{bmatrix} \begin{bmatrix} \mathbf{v} \\ \mathbf{v} \end{bmatrix},$$

where $\mathbf{v} = \mathbf{V}_A^H\mathbf{u}$. That is, the vector $[\mathbf{v}^T\ \mathbf{v}^T]^T \in \mathbb{C}^{2p}$ is rotated by the $2p \times 2p$ rotation matrix blkdiag$(\mathbf{C}, \mathbf{S})$ and reconstructed on the unitary basis $[\mathbf{U}_A^T\ \mathbf{U}_B^T]^T$. The net is to represent the respective channel maps $\mathbf{A}\mathbf{u}$ and $\mathbf{B}\mathbf{u}$ as $\mathbf{A}\mathbf{u} = \mathbf{U}_A\mathbf{C}\mathbf{V}_A^H\mathbf{u}$ and $\mathbf{B}\mathbf{u} = \mathbf{U}_B\mathbf{S}\mathbf{V}_A^H\mathbf{u}$. The key assumption is that $\mathbf{Q}$ is unitary.

C.3.2 The GSVD

Begin with two channels, X and Y. Their respective channel matrices are $\mathbf{H}_X$ and $\mathbf{H}_Y$, and the composite channel matrix $\mathbf{H}$ is

$$\mathbf{H} = \begin{bmatrix} \mathbf{H}_X \\ \mathbf{H}_Y \end{bmatrix}.$$

The composite Gramian $\mathbf{H}^H\mathbf{H} = \mathbf{H}_X^H\mathbf{H}_X + \mathbf{H}_Y^H\mathbf{H}_Y$ is taken to be positive definite. Write

$$\mathbf{H}_X = \mathbf{H}_X(\mathbf{H}_X^H\mathbf{H}_X)^{-1/2}(\mathbf{H}_X^H\mathbf{H}_X)^{1/2}(\mathbf{H}_X^H\mathbf{H}_X + \mathbf{H}_Y^H\mathbf{H}_Y)^{-1/2}(\mathbf{H}_X^H\mathbf{H}_X + \mathbf{H}_Y^H\mathbf{H}_Y)^{1/2},$$

and group terms to write this as $\mathbf{H}_X = \mathbf{U}_X \mathbf{A}_X \mathbf{G}^H$, where $\mathbf{U}_X = \mathbf{H}_X(\mathbf{H}_X^H \mathbf{H}_X)^{-1/2}$ is unitary, $\mathbf{A}_X = (\mathbf{H}_X^H \mathbf{H}_X)^{1/2}(\mathbf{H}_X^H \mathbf{H}_X + \mathbf{H}_Y^H \mathbf{H}_Y)^{-1/2}$, and $\mathbf{G}^H = (\mathbf{H}_X^H \mathbf{H}_X + \mathbf{H}_Y^H \mathbf{H}_Y)^{1/2}$. Use the same procedure to write $\mathbf{H}_Y = \mathbf{U}_Y \mathbf{A}_Y \mathbf{G}^H$, where $\mathbf{U}_Y = \mathbf{H}_Y(\mathbf{H}_Y^H \mathbf{H}_Y)^{-1/2}$, $\mathbf{A}_Y = (\mathbf{H}_Y^H \mathbf{H}_Y)^{1/2}(\mathbf{H}_X^H \mathbf{H}_X + \mathbf{H}_Y^H \mathbf{H}_Y)^{-1/2}$, which yields

$$\mathbf{H} = \begin{bmatrix} \mathbf{H}_X \\ \mathbf{H}_Y \end{bmatrix} = \begin{bmatrix} \mathbf{U}_X \mathbf{A}_X \\ \mathbf{U}_Y \mathbf{A}_Y \end{bmatrix} \mathbf{G}^H.$$

It is easy to verify that $\mathbf{A}_X^H \mathbf{A}_X + \mathbf{A}_Y^H \mathbf{A}_Y = \mathbf{I}_p$, and $\mathbf{A}_X^H \mathbf{U}_X^H \mathbf{U}_X \mathbf{A}_X + \mathbf{A}_Y^H \mathbf{U}_Y^H \mathbf{U}_Y \mathbf{A}_Y = \mathbf{I}_p$. Then, using the CS decomposition, this may be written as

$$\mathbf{H} = \begin{bmatrix} \mathbf{H}_X \\ \mathbf{H}_Y \end{bmatrix} = \begin{bmatrix} \mathbf{U}_A \mathbf{C} \\ \mathbf{U}_B \mathbf{S} \end{bmatrix} \mathbf{V}_A^H \mathbf{G}^H,$$

where $\mathbf{U}_A \mathbf{C} \mathbf{V}_A^H$ is the SVD of $\mathbf{U}_X \mathbf{A}_X$, $\mathbf{U}_B \mathbf{S} = \mathbf{U}_Y \mathbf{A}_Y \mathbf{V}_A$, and $\mathbf{S} = (\mathbf{I} - \mathbf{C}^2)^{1/2}$.

The interpretations of the GSVD are the interpretations of the CS. The essential difference between the GSVD and the CS decomposition is that in the GSVD, the Gramian $\mathbf{H}^H \mathbf{H} = \mathbf{H}_X^H \mathbf{H}_X + \mathbf{H}_Y^H \mathbf{H}_Y$ is assumed only positive semidefinite, whereas in the CS decomposition, the Gramian $\mathbf{Q}^H \mathbf{Q} = \mathbf{A}^H \mathbf{A} + \mathbf{B}^H \mathbf{B}$ is assumed to be identity. The geometrical picture of this decomposition is developed comprehensively in [113].

Normal Distribution Theory and Related

D

D.1 Introduction

Our experience is that multivariate statistical theory is not central to a modern education in signal processing and machine learning. So, in this appendix, we have assembled a breezy introduction to those of its elements that are essential for the reading of this book. The theory itself has reached its apotheosis in the writings of some of its most important innovators (in chronological order):

- R. A. Fisher, *Contributions to Mathematical Statistics,* John Wiley and Sons, New York, 1950.
- A. T. James, "Normal multivariate analysis and the orthogonal group," *Annals of Mathematical Statistics,* vol. 25, no. 1, pp. 40–75, 1954.
- T. W. Anderson, *Introduction to Multivariate Statistical Analysis,* John Wiley and Sons, New York, 1958.
- S. S. Wilks, *Mathematical Statistics,* John Wiley and Sons, New York, 1964.
- C. R. Rao, *Linear Statistical Inference and its Applications,* Wiley, New York, 1965.
- M. G. Kendall and A. Stuart, *The Advanced Theory of Statistics,* Hafner Publishing Co., New York, 1968.
- A. M. Kshirsagar, *Multivariate Analysis,* Marcel-Dekker, New York, 1972.
- M. S. Srivastava and C. G. Khatri, *An Introduction to Multivariate Statistics,* North Holland, New York, 1979.
- K. V. Mardia, J. T. Kent, and J. M. Bibby, *Multivariate Analysis,* Academic Press, London, 1979.
- R. J. Muirhead, *Aspects of Multivariate Statistical Theory,* John Wiley and Sons, New York, 1982.
- M. L. Eaton, *Multivariate Statistics: A Vector Space Approach,* John Wiley and Sons, New York, 1983.

D. Ramírez et al., *Coherence*, https://doi.org/10.1007/978-3-031-13331-2

- C. G. Khatri and C. R. Rao, "Effects of estimated noise covariance matrix in optimal signal detection," *IEEE Transactions on Acoustics, Speech, and Signal Processing,* vol. 35, no. 5, pp. 671–679, May 1987.

In the theory of probability, one begins with a measure space $(\Omega, \mathcal{F}, \mu)$ and defines a random variable (rv) X to be a measurable mapping from Ω to $\mathbb{R}$. This defines a new measure space $(\mathbb{R}, \mathcal{B}, F_X)$, where the cumulative distribution function (cdf) $F_X(x)$ is defined to be $F_X(x) = \Pr[X \leq x]$. To say that the rv X is measurable is to say that $\Pr[X \leq x] = \mu(A)$, where A, the inverse image of the set $\{X \leq x\}$, is a set in the sigma field $\mathcal{F}$. All questions about the probability that X lies in an open or closed set, or a finite union or intersection of such sets, may be answered with $F_X(x)$. Such sets are the sets of the Borel field $\mathcal{B}$. This model building extends to random vectors $\mathbf{x} \in \mathbb{R}^L$ and random matrices $\mathbf{X} \in \mathbb{R}^{L \times N}$. Then, probability statements are statements about cylinder sets. In fact, it extends to other fields than the real field $\mathbb{R}$. For our purposes, the field of interest is often the complex field $\mathbb{C}$.

In much of signal processing and machine learning, these important technicalities may be dispensed with and replaced by a study of the cdf F_X. When the rv X is continuous, meaning its cdf is absolutely continuous with respect to Lebesgue measure, then the cdf F_X may be written as $F_X(x) = \int_{-\infty}^{x} f_X(z)dz$, where $dF(x) = f_X(x)dx$; $f_X(x)$ is called the probability density function (pdf). The interpretation is that $f_X(x)dx$ is the probability that *a draw, or realization,* of the random variable X from the distribution $F_X(x)$ will take a value in the interval $(x, x + dx)$. The pdf $f_X(x)$ may be viewed as the inverse Fourier transform of the characteristic function (chf) $\phi_X(\omega) = \mathrm{E}[e^{j\omega X}] = \int_{-\infty}^{\infty} f_X(x)e^{j\omega x}dx$, which we denote

$$f_X(x) = \int_{-\infty}^{\infty} \phi_X(\omega)e^{-j\omega x}\frac{d\omega}{2\pi} \longleftrightarrow \phi_X(\omega) = \int_{-\infty}^{\infty} f_X(x)e^{j\omega x}dx.$$

The double arrow notation means the pdf and chf are a Fourier transform pair. This definition of the characteristic function generalizes to the study of pdfs and cdfs for vector- and matrix-valued random variables.

As in much of applied science and engineering, including signal processing and machine learning, it is often easier to derive the characteristic function for a random variable than it is to derive its pdf. In fact, it is not uncommon to encounter problems where the characteristic function may be derived, but it may not be inverted for its pdf, except by numerical means. Nonetheless, the *distribution* of a rv X is said to be known if either its pdf or chf is known.

In this discussion, we have been meticulous about distinguishing a random variable X from its realization x. So a distribution statement about a random variable X is a device for evaluating which realizations are likely and which are not. A function of a random variable, $\Lambda(X)$, may be called a *statistic*. But its realization $\Lambda(x)$ is a value of the statistic. For example, multiple realizations $x_n, n = 1, \ldots, N$, of a rv X may be used to *estimate* the mean of the rv X as $\overline{x} = N^{-1}(x_1 + x_2 + \cdots + x_N)$. The corresponding *statistic or estimator* is

$\overline{X} = N^{-1}(X_1 + X_2 + \cdots + X_N)$, which is to say the sample mean random variable is the average of the random variables in a random experiment.

A strict adherence to notational convention would use $X \sim F_X(x)$ to indicate the random variable X has distribution F_X with cdf $F_X(x)$ and pdf $f_X(x)$; $\mathbf{X} \sim F_\mathbf{X}(\mathbf{x})$ to indicate the random matrix $\mathbf{X}$ has distribution $F_\mathbf{X}$ with cdf $F_\mathbf{X}(\mathbf{x})$, etc. But these notational conventions can become cumbersome, so when the context is clear, we relax them. For example, we might say $u \sim \mathcal{N}_1(0, 1)$ with pdf $f(u)$, when strictly speaking we should say $U \sim \mathcal{N}_1(0, 1)$ with pdf $f_U(u)$.

In this appendix, we study the distribution of normal, or Gaussian, random variables, vectors, and matrices. Then, we use the elegant machinery of spherically invariant normal experiments to derive many of the most important distributions of multivariate analysis.

The Basic Data Structure. The basic data structure will be $\mathbf{X} \in \mathbb{R}^{L \times N}$, where L is the number of channels and N is the number of temporal measurements. The matrix might be called a space-time matrix, and the matrix $\mathbf{X}^T$ a time-space matrix. When $L = 2$ and $N = 1$, the associated experiment is one of making one measurement in two channels or sensors. This is a bivariate experiment. When $L > 2$ and $N = 1$, the associated experiment is one of making a single measurement or snapshot in L channels. This is a multivariate experiment. When $L > 1$ and $N > L$, the associated experiment is one of making N measurements in L channels. If we associate N with temporal measurements and L with channels, then this is a space-time experiment. For all cases, the real field $\mathbb{R}$ may be replaced by the complex field $\mathbb{C}$, and for proper complex random variables, formulas for pdfs remain essentially unchanged except for a doubling in parameters of the pdf. This will be clarified in Appendix E.

D.2 The Normal Random Variable

Begin with the definition of a univariate normal rv u with mean 0 and variance 1, denoted $u \sim \mathcal{N}(0, 1)$. The pdf and chf for this random variable are

$$f(u) = \frac{1}{\sqrt{2\pi}} \exp\left(-\frac{u^2}{2}\right) \longleftrightarrow \phi(\omega) = \exp(-\omega^2/2),$$

where $-\infty < u, \omega < \infty$. The transformed rv $x = \sigma u + \mu$ is then a univariate normal rv, denoted $x \sim \mathcal{N}(\mu, \sigma^2)$, with pdf and chf given by

$$f(x) = \frac{1}{\sqrt{2\pi\sigma^2}} \exp\left\{-\frac{(x-\mu)^2}{2\sigma^2}\right\} \longleftrightarrow \phi(\omega) = \exp\left(j\omega\mu - \frac{\omega^2\sigma^2}{2}\right),$$

where $-\infty < x, \omega < \infty$, and $\sigma^2 > 0$. The mean and variance of this rv are, respectively, μ and σ^2.

D.3 The Multivariate Normal Random Vector

A random vector $\mathbf{x} \in \mathbb{R}^L$ is said to be normal if for every $\mathbf{a} \in \mathbb{R}^L$, the random variable $\mathbf{a}^T\mathbf{x}$ is univariate normal, denoted $\mathcal{N}(m, v)$, where $m = \mathrm{E}[\mathbf{a}^T\mathbf{x}]$ is the mean of $\mathbf{a}^T\mathbf{x}$ and $v = \mathrm{E}\left[(\mathbf{a}^T\mathbf{x} - m)^2\right]$ is its variance. These may be written as $m = \mathbf{a}^T\boldsymbol{\mu}$ and $v = \mathbf{a}^T\boldsymbol{\Sigma}\mathbf{a}$, where $\boldsymbol{\mu} = \mathrm{E}[\mathbf{x}]$ and $\boldsymbol{\Sigma} = \mathrm{E}\left[(\mathbf{x} - \boldsymbol{\mu})(\mathbf{x} - \boldsymbol{\mu})^T\right]$ are, respectively, the mean and covariance matrix of the random vector $\mathbf{x}$. We say the random vector $\mathbf{x}$ is distributed as $\mathcal{N}_L(\boldsymbol{\mu}, \boldsymbol{\Sigma})$ with pdf

$$
\begin{aligned}
f(\mathbf{x}) &= \frac{1}{(2\pi)^{L/2}\det(\boldsymbol{\Sigma})^{1/2}} \exp\left\{-\frac{1}{2}(\mathbf{x} - \boldsymbol{\mu})^T\boldsymbol{\Sigma}^{-1}(\mathbf{x} - \boldsymbol{\mu})\right\} \\
&= \frac{1}{(2\pi)^{L/2}\det(\boldsymbol{\Sigma})^{1/2}} \exp\left\{\mathrm{tr}\left(-\frac{1}{2}\boldsymbol{\Sigma}^{-1}(\mathbf{x} - \boldsymbol{\mu})(\mathbf{x} - \boldsymbol{\mu})^T\right)\right\}.
\end{aligned}
$$

The expression $\exp\{\mathrm{tr}(\cdot)\}$ is often written $\mathrm{etr}(\cdot)$. This is an elliptically contoured pdf, constant on the *level sets* $\{\mathbf{x} \in \mathbb{R}^L \mid (\mathbf{x} - \boldsymbol{\mu})^T\boldsymbol{\Sigma}^{-1}(\mathbf{x} - \boldsymbol{\mu}) = \text{const.}\}$. The quadratic form $(\mathbf{x} - \boldsymbol{\mu})^T\boldsymbol{\Sigma}^{-1}(\mathbf{x} - \boldsymbol{\mu})$ is the squared Mahalanobis distance between $\mathbf{x}$ and its mean $\boldsymbol{\mu}$. All points in a level set are equidistant from the mean, as measured by the Mahalanobis distance. The characteristic function of this random vector is

$$
\phi(\boldsymbol{\omega}) = \exp\left(j\boldsymbol{\omega}^T\boldsymbol{\mu} - \frac{1}{2}\boldsymbol{\omega}^T\boldsymbol{\Sigma}\,\boldsymbol{\omega}\right).
$$

The chf $\phi(\boldsymbol{\omega})$ is well-defined for all positive semidefinite covariance matrices $\boldsymbol{\Sigma}$, whereas the pdf is well-defined only for positive definite $\boldsymbol{\Sigma}$.

When $\boldsymbol{\mu} = \mathbf{0}$ and $\boldsymbol{\Sigma} = \mathbf{I}_L$, then

$$
\phi(\boldsymbol{\omega}) = \exp\left(-\sum_{k=1}^{L}\omega_k^2/2\right) = \prod_{k=1}^{L}\exp\left(-\omega_k^2/2\right),
$$

which is a product of characteristic functions for independent univariate $\mathcal{N}(0, 1)$ random variables. That is, the random variables $x_1, x_2, \ldots, x_L$, are independent and identically distributed (i.i.d.) normal random variables, each with mean 0 and variance 1. The random vector $\mathbf{x}$ is said to be a *white* normal random vector, or a white Gaussian vector. Its pdf is *spherically contoured*, and its distribution is said to be *spherically invariant*, as the distribution of $\mathbf{Qx}$ is the distribution of $\mathbf{x}$ for $\mathbf{Q} \in O(L)$, any orthogonal $L \times L$ matrix. Importantly, to say jointly normal random variables are *uncorrelated* is to say they are *independent*, and vice versa. This equivalence is not generally true for other multivariate distributions.

D.3.1 Linear Transformation of a Normal Random Vector

A linearly transformed version of $\mathbf{x}$, namely, $\mathbf{y} = \mathbf{A}\mathbf{x}$, with $\mathbf{A} \in \mathbb{R}^{p \times L}$, has characteristic function

$$\phi(\omega) = \exp\left\{ j\omega^T \mathbf{A}\boldsymbol{\mu} - \frac{1}{2}\omega^T \mathbf{A}\boldsymbol{\Sigma}\mathbf{A}^T \omega \right\}.$$

This makes $\mathbf{y}$ multivariate normal (MVN), with mean $\mathbf{A}\boldsymbol{\mu}$ and covariance matrix $\mathbf{A}\boldsymbol{\Sigma}\mathbf{A}^T$, which need not be positive definite. We denote this $\mathbf{y} \sim \mathcal{N}_p(\mathbf{A}\boldsymbol{\mu}, \mathbf{A}\boldsymbol{\Sigma}\mathbf{A}^T)$. When $\mathbf{A}\boldsymbol{\Sigma}\mathbf{A}^T$ is positive definite, then the characteristic function may be inverted, yielding

$$f(\mathbf{y}) = \frac{1}{(2\pi)^{p/2} \det(\mathbf{A}\boldsymbol{\Sigma}\,\mathbf{A}^T)^{1/2}} \exp\left\{ -\frac{1}{2}(\mathbf{y} - \mathbf{A}\boldsymbol{\mu})^T (\mathbf{A}\boldsymbol{\Sigma}\mathbf{A}^T)^{-1} (\mathbf{y} - \mathbf{A}\boldsymbol{\mu}) \right\}.$$

D.3.2 The Bivariate Normal Random Vector

Denote $\mathbf{x} = [x_1\ x_2]^T$ with $\boldsymbol{\mu} = [\mu_1\ \mu_2]^T$. Partition the covariance matrix $\boldsymbol{\Sigma}$ as

$$\boldsymbol{\Sigma} = \begin{bmatrix} \mathrm{E}[(x_1 - \mu_1)^2] & \mathrm{E}[(x_1 - \mu_1)(x_2 - \mu_2)] \\ \mathrm{E}[(x_1 - \mu_1)(x_2 - \mu_2)] & \mathrm{E}[(x_2 - \mu_2)^2] \end{bmatrix} = \begin{bmatrix} \sigma_1^2 & \sigma_1\sigma_2\rho \\ \sigma_1\sigma_2\rho & \sigma_2^2 \end{bmatrix}$$

where

$$\rho = \frac{\mathrm{E}[(x_1 - \mu_1)(x_2 - \mu_2)]}{\sqrt{\mathrm{E}[(x_1 - \mu_1)^2]\,\mathrm{E}[(x_2 - \mu_2)^2]}}.$$

In this partition, the diagonal terms are variances and the cross-terms are cross-covariances. The scalar ρ is the *correlation coefficient*. The determinant of $\boldsymbol{\Sigma}$ is $\det(\boldsymbol{\Sigma}) = \sigma_1^2\sigma_2^2(1 - \rho^2) > 0$, for $-1 < \rho < 1$.

The covariance matrix $\boldsymbol{\Sigma}$ may be Cholesky factored two ways, one LDU and the other UDL, as

$$\begin{aligned} \boldsymbol{\Sigma} &= \begin{bmatrix} 1 & 0 \\ \rho\sigma_2/\sigma_1 & 1 \end{bmatrix} \begin{bmatrix} \sigma_1^2 & 0 \\ 0 & \sigma_2^2(1 - \rho^2) \end{bmatrix} \begin{bmatrix} 1 & \rho\sigma_2/\sigma_1 \\ 0 & 1 \end{bmatrix} \\ &= \begin{bmatrix} 1 & \rho\sigma_1/\sigma_2 \\ 0 & 1 \end{bmatrix} \begin{bmatrix} \sigma_1^2(1 - \rho^2) & 0 \\ 0 & \sigma_2^2 \end{bmatrix} \begin{bmatrix} 1 & 0 \\ \rho\sigma_1/\sigma_2 & 1 \end{bmatrix}. \end{aligned}$$

The corresponding factorizations of $\boldsymbol{\Sigma}^{-1}$ are

$$\boldsymbol{\Sigma}^{-1} = \begin{bmatrix} 1 & -\rho\sigma_2/\sigma_1 \\ 0 & 1 \end{bmatrix} \begin{bmatrix} 1/\sigma_1^2 & 0 \\ 0 & 1/\sigma_2^2(1 - \rho^2) \end{bmatrix} \begin{bmatrix} 1 & 0 \\ -\rho\sigma_2/\sigma_1 & 1 \end{bmatrix}$$

$$= \begin{bmatrix} 1 & 0 \\ -\rho\sigma_1/\sigma_2 & 1 \end{bmatrix} \begin{bmatrix} 1/\sigma_1^2(1-\rho^2) & 0 \\ 0 & 1/\sigma_2^2 \end{bmatrix} \begin{bmatrix} 1 & -\rho\sigma_1/\sigma_2 \\ 0 & 1 \end{bmatrix}.$$

These factorizations of $\mathbf{\Sigma}^{-1}$ may be used to write the bivariate normal pdf in two ways as

$$f(\mathbf{x}) = \frac{1}{\sqrt{2\pi\sigma_1^2}} \exp\left\{-\frac{(x_1-\mu_1)^2}{2\sigma_1^2}\right\}$$

$$\times \frac{1}{\sqrt{2\pi\sigma_2^2(1-\rho^2)}} \exp\left\{-\frac{[(x_2-\mu_2)-\rho(\sigma_2/\sigma_1)(x_1-\mu_1)]^2}{2\sigma_2^2(1-\rho^2)}\right\},$$

and

$$f(\mathbf{x}) = \frac{1}{\sqrt{2\pi\sigma_2^2}} \exp\left\{-\frac{(x_2-\mu_2)^2}{2\sigma_2^2}\right\}$$

$$\times \frac{1}{\sqrt{2\pi\sigma_1^2(1-\rho^2)}} \exp\left\{-\frac{[(x_1-\mu_1)-\rho(\sigma_1/\sigma_2)(x_2-\mu_2)]^2}{2\sigma_1^2(1-\rho^2)}\right\}.$$

These pdfs show that the conditional distribution of x_2, given x_1, is univariate normal with mean $\mu_2 + \rho(\sigma_2/\sigma_1)(x_1 - \mu_1)$, and variance $\sigma_2^2(1-\rho^2)$. This makes $\rho(\sigma_2/\sigma_1)(x_1 - \mu_1)$ the minimum variance estimator of $x_2 - \mu_2$ from x_1. The conditional and unconditional variance are $\sigma_2^2(1-\rho^2)$ and σ_2^2, respectively. This is a reduction in the variance of x_2 by a factor of $1-\rho^2$, which shows the value of regression or filtering. It is easy to reverse the roles of x_2 and x_1 to make similar statements about the minimum variance estimator of x_1 from x_2.

The elliptical level set $\{\mathbf{x} \in \mathbb{R}^L \mid (\mathbf{x}-\boldsymbol{\mu})^T\mathbf{\Sigma}^{-1}(\mathbf{x}-\boldsymbol{\mu}) = \text{const}\}$ may now be written:

$$\frac{(x_1-\mu_1)^2}{2\sigma_1^2} + \frac{[(x_2-\mu_2)-\rho(\sigma_2/\sigma_1)(x_1-\mu_1)]^2}{2\sigma_2^2(1-\rho^2)} = \text{const},$$

or

$$\frac{(x_2-\mu_2)^2}{2\sigma_2^2} + \frac{[(x_1-\mu_1)-\rho(\sigma_1/\sigma_2)(x_2-\mu_2)]^2}{2\sigma_1^2(1-\rho^2)} = \text{const}.$$

For insight, choose $\mu_1 = \mu_2 = 0$, and const $= 1$. Then, the estimating line $x_2 = \rho(\sigma_2/\sigma_1)x_1$ intersects the rectangle $[(-\sigma_1, 0), (\sigma_1, 0)] \times [(0, -\sigma_2), (0, \sigma_2)]$ at the points $(\sigma_1, \rho\sigma_2)$ and $(-\sigma_1, -\rho\sigma_2)$, and the estimating line $x_1 = \rho(\sigma_1/\sigma_2)x_2$ intersects the square at $(\rho\sigma_1, \sigma_2)$ and $(-\rho\sigma_1, -\sigma_2)$. Figure D.1 shows how these

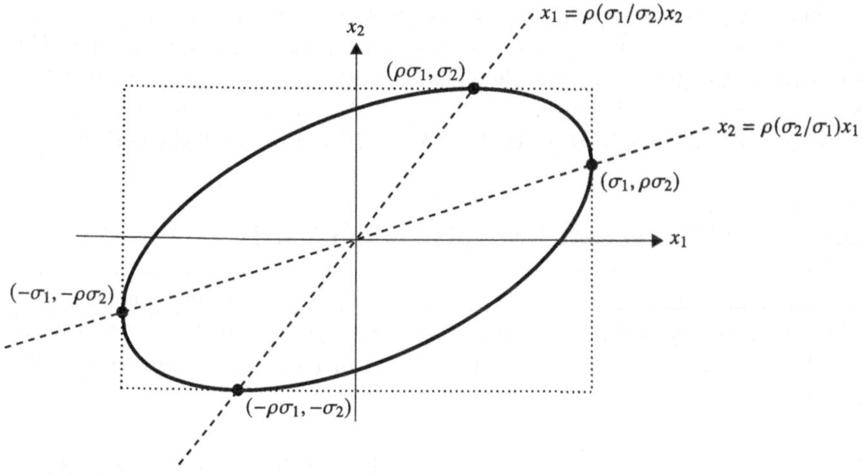

Fig. D.1 Elliptical level set for a bivariate normal vector

estimating lines determine the elliptical level set. The area of the ellipse enclosed by this level set is $\pi\sigma_1\sigma_2(1 - \rho^2)$, and it scales quadratically with const. It is called a *concentration ellipse.* The probability that a draw of $\mathbf{x}$ lies within this concentration ellipse is the probability that the random vector $\mathbf{u} = \boldsymbol{\Sigma}^{-1/2}\mathbf{x}$ lies within a circle of radius const. It is a simple calculation to show that this probability is $1 - \exp(-\text{const}^2)$. For small values of const, this probability is approximately $\text{const}^2/2$, and this is the value of $f(\mathbf{x})$, evaluated at $\mathbf{x} = \mathbf{0}$, times the area of the concentration ellipse. This is analytical justification for an intuitive idea. Always a good thing.

D.3.3 Analysis and Synthesis

Call $\mathbf{u} \sim \mathcal{N}_L(\mathbf{0}, \mathbf{I}_L)$ a vector of i.i.d. $\mathcal{N}(0, 1)$ random variables. Its level sets are circular. Give a positive definite matrix $\boldsymbol{\Sigma}$ the EVD $\boldsymbol{\Sigma} = \mathbf{FKF}^T$, where $\mathbf{F}$ is an $L \times L$ orthogonal matrix, and $\mathbf{K} = \text{diag}(k_1, k_2, \ldots, k_L)$, with $k_1 \geq \cdots \geq k_L > 0$. The random vector $\mathbf{x} = \mathbf{FK}^{1/2}\mathbf{u}$ is MVN with mean zero and covariance matrix $E[\mathbf{xx}^T] = \mathbf{FKF}^T = \boldsymbol{\Sigma}$. This is one way to synthesize a zero-mean MVN random vector of covariance $\boldsymbol{\Sigma}$ from a set of i.i.d. scalar normal random variables of zero mean and variance 1. Another is to Cholesky factor (or LDU decompose) the covariance matrix as $\boldsymbol{\Sigma} = \mathbf{HDH}^T$, and synthesize $\mathbf{x}$ as $\mathbf{x} = \mathbf{HD}^{1/2}\mathbf{u}$. With these factorizations, we may call $\mathbf{FK}^{1/2}$ and $\mathbf{HD}^{1/2}$ square roots of $\boldsymbol{\Sigma}$, and denote them $\boldsymbol{\Sigma}^{1/2}$. The matrix $\mathbf{H}$ is lower triangular, giving the synthesis $\mathbf{HD}^{1/2}\mathbf{u}$ an interpretation as a sequence of order-increasing moving average filterings of $\mathbf{D}^{1/2}\mathbf{u}$. The matrix $\mathbf{F}$ is orthogonal, giving the synthesis $\mathbf{FK}^{1/2}\mathbf{u}$ the interpretation of rotations of $\mathbf{K}^{1/2}\mathbf{u}$. This latter interpretation illuminates the transformation of circular level sets $\{\mathbf{u} \mid \mathbf{u}^T \mathbf{I}_L \mathbf{u} = \text{const}\}$ for the white random vector $\mathbf{u}$ into elliptical level sets $\{\mathbf{x} \mid \mathbf{x}^T \boldsymbol{\Sigma}^{-1}\mathbf{x} = \mathbf{x}^T \mathbf{F}\mathbf{K}^{-1}\mathbf{F}^T\mathbf{x} = \text{const}\}$ for the random vector $\mathbf{x}$. This synthesis is called *coloring* of a white random vector $\mathbf{u}$ for a colored random vector $\mathbf{x}$. The argument is easily reversed for the *analysis* of a colored random vector $\mathbf{x}$ for a white vector $\mathbf{u}$, i.e., $\mathbf{u} = \mathbf{D}^{-1/2}\mathbf{H}^{-1}\mathbf{x} = \mathbf{K}^{-1/2}\mathbf{F}^T\mathbf{x}$. This is called *whitening*.

D.4 The Multivariate Normal Random Matrix

A real random matrix $\mathbf{X} \in \mathbb{R}^{L\times N}$ is said to be normally distributed with mean value $E[\mathbf{X}] = \mathbf{M}$ and *separable covariance matrix* $\boldsymbol{\Sigma}_r \otimes \boldsymbol{\Sigma}_c$, where $\mathbf{M} \in \mathbb{R}^{L\times N}$ and $\boldsymbol{\Sigma}_r$ and $\boldsymbol{\Sigma}_c$ are, respectively, $N \times N$ and $L \times L$ positive definite matrices, if its pdf is [244]:

$$f(\mathbf{X}) = \frac{1}{(2\pi)^{LN/2}\det(\boldsymbol{\Sigma}_r)^{L/2}\det(\boldsymbol{\Sigma}_c)^{N/2}} \operatorname{etr}\left\{-\frac{1}{2}\boldsymbol{\Sigma}_c^{-1}(\mathbf{X}-\mathbf{M})\boldsymbol{\Sigma}_r^{-1}(\mathbf{X}-\mathbf{M})^T\right\}. \tag{D.1}$$

The notation for the normally distributed matrix $\mathbf{X}$ is $\mathbf{X} \sim \mathcal{N}_{L\times N}(\mathbf{M}, \boldsymbol{\Sigma}_r \otimes \boldsymbol{\Sigma}_c)$. This formula is more intuitive if $\mathbf{M} = \mathbf{0}$, and the quadratic form in the etr term is written as $\boldsymbol{\Sigma}_c^{-1}\mathbf{X}\boldsymbol{\Sigma}_r^{-1}\mathbf{X}^T = \mathbf{N}\mathbf{N}^T$, with $\mathbf{N} = \boldsymbol{\Sigma}_c^{-1/2}\mathbf{X}\boldsymbol{\Sigma}_r^{-1/2}$ an $L \times N$ random matrix of independent $\mathcal{N}(0, 1)$ random variables.

Define the $LN \times 1$ vector $\mathbf{x} = \text{vec}(\mathbf{X})$ to be a vectorization of the matrix $\mathbf{X}$ by columns. Then, the density of $\mathbf{x}$ is

$$f(\mathbf{x}) = \frac{1}{(2\pi)^{LN/2}\det(\boldsymbol{\Sigma}_r \otimes \boldsymbol{\Sigma}_c)^{1/2}} \exp\left\{-\frac{1}{2}(\mathbf{x}-\boldsymbol{\mu})^T(\boldsymbol{\Sigma}_r \otimes \boldsymbol{\Sigma}_c)^{-1}(\mathbf{x}-\boldsymbol{\mu})\right\}. \tag{D.2}$$

where $E[\mathbf{x}] = \boldsymbol{\mu} = \text{vec}(\mathbf{M})$ is the $LN \times 1$-dimensional mean of $\mathbf{x}$ and $E[(\mathbf{x}-\boldsymbol{\mu})(\mathbf{x}-\boldsymbol{\mu})^T] = \boldsymbol{\Sigma}_r \otimes \boldsymbol{\Sigma}_c$ is the $LN \times LN$-dimensional covariance of $\mathbf{x}$. To establish that (D.2) is the same as (D.1), define $\mathbf{A} = (\mathbf{X} - \mathbf{M})$ and use the Kronecker identities:[1]

$$\begin{aligned}
\operatorname{tr}\left(\boldsymbol{\Sigma}_c^{-1}\mathbf{A}\boldsymbol{\Sigma}_r^{-1}\mathbf{A}^T\right) &= \text{vec}(\mathbf{A})^T\left(\boldsymbol{\Sigma}_r^{-1} \otimes \boldsymbol{\Sigma}_c^{-1}\right)\text{vec}(\mathbf{A}),\\
(\boldsymbol{\Sigma}_r \otimes \boldsymbol{\Sigma}_c)^{-1} &= \boldsymbol{\Sigma}_r^{-1} \otimes \boldsymbol{\Sigma}_c^{-1},\\
\det(\boldsymbol{\Sigma}_r \otimes \boldsymbol{\Sigma}_c) &= (\det(\boldsymbol{\Sigma}_r))^L(\det(\boldsymbol{\Sigma}_c))^N.
\end{aligned}$$

Denote the random matrix $\mathbf{X}$ as $\mathbf{X} = [\mathbf{x}_1\ \mathbf{x}_2\ \cdots\ \mathbf{x}_N]$. One interpretation, consistent with the applications in this book, is that each column represents an array snapshot and the temporal sequence of N such array snapshots compose the matrix. In some cases, the snapshots may be independent samples of a random vector $\mathbf{x}$, but they need not be.

[1] A full account of the Kronecker product and its properties can be found in Sect. B.6.

D.4.1 Analysis and Synthesis

The synthesis of normal random matrices from a white random matrix is an extension of the synthesis of a normal random vector from a white random vector. Consider a matrix $\mathbf{U} \in \mathbb{R}^{L\times N}$, consisting of i.i.d. normal $\mathcal{N}(0, 1)$ random variables, and assume $N \geq L$. The pdf of $\mathbf{U}$ is proportional to $\text{etr}(-\mathbf{U}\mathbf{I}_N\mathbf{U}^T/2)$. Organize $\mathbf{U}$ as $\mathbf{U} = [\mathbf{u}_1\ \mathbf{u}_2\ \cdots\ \mathbf{u}_N]$, where each of the $\mathbf{u}_n$, $n = 1, \ldots, N$, is an $L \times 1$ column vector. Vectorize this matrix by stacking columns to obtain an $LN \times 1$ vector. The covariance matrix of this vector is $\mathbf{I}_N \otimes \mathbf{I}_L$, which is a block-structured matrix consisting of $\mathbf{I}_L$ repeated along the N-dimensional identity $\mathbf{I}_N$. So the matrix $\mathbf{U}$ is said to be distributed as $\mathcal{N}_{L\times N}(\mathbf{0}, \mathbf{I}_N \otimes \mathbf{I}_L)$.

If we color each of the columns $\mathbf{u}_n$ by the common $L \times L$ filter $\boldsymbol{\Sigma}_c^{1/2}$ to produce $\mathbf{V} = \boldsymbol{\Sigma}_c^{1/2}\mathbf{U}$, with inversion $\mathbf{U} = \boldsymbol{\Sigma}_c^{-1/2}\mathbf{V}$, then the pdf of $\mathbf{V}$ is proportional to $\text{etr}(-\boldsymbol{\Sigma}_c^{-1/2}\mathbf{V}\mathbf{I}_N\mathbf{V}^T\boldsymbol{\Sigma}_c^{-1/2}/2)$. The covariance of $\text{vec}(\mathbf{V})$ is $\mathbf{I}_N \otimes \boldsymbol{\Sigma}_c$, so $\mathbf{V}$ is said to be distributed as $\mathcal{N}_{L\times N}(\mathbf{0}, \mathbf{I}_N \otimes \boldsymbol{\Sigma}_c)$. Now, color the rows of $\mathbf{V}$ with the common filter $\boldsymbol{\Sigma}_r^{1/2}$ to obtain the matrix $\mathbf{X} = \mathbf{V}\boldsymbol{\Sigma}_r^{1/2}$, with inverse $\mathbf{V} = \mathbf{X}\boldsymbol{\Sigma}_r^{-1/2}$. The pdf of $\mathbf{X}$ is proportional to $\text{etr}(-\boldsymbol{\Sigma}_c^{-1/2}\mathbf{X}\boldsymbol{\Sigma}_r^{-1}\mathbf{X}^T\boldsymbol{\Sigma}_c^{-1/2}/2)$ or $\text{etr}(-\boldsymbol{\Sigma}_c^{-1}\mathbf{X}\boldsymbol{\Sigma}_r^{-1}\mathbf{X}^T/2)$. So, beginning with i.i.d. $\mathbf{U}$, the matrix $\mathbf{X} = \boldsymbol{\Sigma}_c^{1/2}\mathbf{U}\boldsymbol{\Sigma}_r^{1/2}$ has pdf proportional to $\text{etr}(-\boldsymbol{\Sigma}_c^{-1}\mathbf{X}\boldsymbol{\Sigma}_r^{-1}\mathbf{X}^T/2)$. We shall call this a $\mathcal{N}_{L\times N}(\mathbf{0}, \boldsymbol{\Sigma}_r \otimes \boldsymbol{\Sigma}_c)$ random matrix, for $\boldsymbol{\Sigma}_r \otimes \boldsymbol{\Sigma}_c$ is the covariance of the column vector $\mathbf{x} = \text{vec}(\mathbf{X})$.

In summary, the action of the transformation $\boldsymbol{\Sigma}_c^{1/2}\mathbf{U}\boldsymbol{\Sigma}_r^{1/2}$ is to transform a $\mathcal{N}_{L\times N}(\mathbf{0}, \mathbf{I}_N \otimes \mathbf{I}_L)$ distribution to a $\mathcal{N}_{L\times N}(\mathbf{0}, \boldsymbol{\Sigma}_r \otimes \boldsymbol{\Sigma}_c)$ distribution. Throughout this discussion, the Jacobians of the transformations and the Kronecker identity, $\det(\boldsymbol{\Sigma}_r \otimes \boldsymbol{\Sigma}_c) = (\det(\boldsymbol{\Sigma}_r))^L(\det(\boldsymbol{\Sigma}_c))^N$ account for the determinants in the resulting multivariate normal densities. From the matrix identity $(\boldsymbol{\Sigma}_r \otimes \boldsymbol{\Sigma}_c)^{-1} = \boldsymbol{\Sigma}_r^{-1} \otimes \boldsymbol{\Sigma}_c^{-1}$ and the mixed-product property $(\mathbf{A} \otimes \mathbf{B})(\mathbf{C} \otimes \mathbf{D}) = (\mathbf{AC}) \otimes (\mathbf{BD})$, it follows that the whitening transformation $\mathbf{U} = \boldsymbol{\Sigma}_c^{-1/2}\mathbf{X}\boldsymbol{\Sigma}_r^{-1/2}$ produces a random matrix $\mathbf{U}$ such that $\text{vec}(\mathbf{U})$ has covariance $(\boldsymbol{\Sigma}_r^{-1} \otimes \boldsymbol{\Sigma}_c^{-1})(\boldsymbol{\Sigma}_r \otimes \boldsymbol{\Sigma}_c) = \mathbf{I}_N \otimes \mathbf{I}_L$.

When normal distributions are spherically invariant, then very special insights are gained by making a coordinate transformation from Euclidean coordinates to polar coordinates, or more accurately to polar decompositions. In the next three sections, normal experiments are conducted with spherically invariant normal random variables, vectors, and matrices.

D.5 The Spherically Invariant Bivariate Normal Experiment

Begin with the bivariate normal random vector $\mathbf{u} = [u_1\ u_2]^T$, with u_1 and u_2 independently distributed as $\mathcal{N}(0, 1)$ rvs. Their joint pdf and chf are

$$f(\mathbf{u}) = \frac{1}{2\pi}\exp(-\mathbf{u}^T\mathbf{u}/2) \longleftrightarrow \phi(\boldsymbol{\omega}) = \exp(-\boldsymbol{\omega}^T\boldsymbol{\omega}/2),$$

where $\boldsymbol{\omega} = [\omega_1\ \omega_2]^T$ and $-\infty < u_1, u_2, \omega_1, \omega_2 < \infty$. These are products of the individual univariate pdfs and chfs.

The distribution of $\mathbf{u}$ is invariant to rotation by a 2×2 orthogonal matrix $\mathbf{Q}$, which is to say $\mathbf{Qu}$ is distributed as $\mathbf{u}$ is distributed. It is common to say the bivariate random vector $\mathbf{u}$ is spherically invariant, which is to say its distribution is spherically invariant, which is to say the level curves of its pdf are spherically contoured. Most of the results to follow hold for any spherically invariant bivariate random vector.

D.5.1 Coordinate Transformation: The Rayleigh and Uniform Distributions

Change coordinates from Cartesian to polar to write the joint density of the rvs $r = \sqrt{u_1^2 + u_2^2}$ and $\theta = \arctan(u_2/u_1)$, with $0 \leq \theta < 2\pi$, as

$$f(r, \theta) = \frac{r}{2\pi} e^{-r^2/2},$$

for $r \geq 0$ and $0 \leq \theta < 2\pi$. The appearance of r in the pdf accounts for the Jacobian of the transformations from (r, θ) to (u_1, u_2).

The marginal density for the radius r is found by marginalizing over the joint distribution for r and θ and is given by

$$f(r) = re^{-r^2/2}, \quad r \geq 0.$$

The marginal density for the angle θ is

$$f(\theta) = \frac{1}{2\pi}, \quad 0 \leq \theta < 2\pi.$$

So the radius is *Rayleigh distributed* and the angle is *uniformly distributed*. The joint density for the radius and angle is the product of the marginals, making them statistically independent. This transformation from Cartesian coordinates to polar coordinates will be extended to vector-valued MVN distributions and to matrix-valued MVN distributions in the sections to follow.

D.5.2 Geometry and Spherical Invariance

The geometry of this coordinate transformation is illuminated by defining the unit-norm vector $\mathbf{t} = [\cos\theta\ \ \sin\theta]^T$. Then, the polar-to-Cartesian coordinate transformation is $\mathbf{u} = \mathbf{t}r$, with $\mathbf{t}^T\mathbf{t} = 1$ and $r^2 = \mathbf{u}^T\mathbf{u} \geq 0$. This is a stochastic representation for $\mathbf{u}$, wherein the rv r is drawn from a distribution with pdf $f(r) = re^{-r^2/2}$, $r \geq 0$, and this draw is spun through an angle θ drawn uniformly from the interval $[0, 2\pi)$ to produce the components of $\mathbf{u} = [u_1\ u_2]^T$. To draw θ uniformly from the interval $[0, 2\pi)$ is to draw $\mathbf{t}$ uniformly with respect to Haar measure on the unit circle S^1. In this case, Haar measure on S^1 is the measure induced by the uniform measure of θ on the interval $[0, 2\pi)$. The interval $[0, 2\pi)$ is a group under group action of cyclic translation, and the unit circle S^1 is a group under the group action of rotation through angle θ.

The factorization $\mathbf{u} = \mathbf{t}r$ encodes for stochastic generation of $\mathbf{t}$: generate $\mathbf{u}$ and compute $\mathbf{t} = \mathbf{u}/(\mathbf{u}^T\mathbf{u})^{1/2}$. Hence, $\mathbf{t}$ is uniformly distributed on S^1, with pdf $f(\mathbf{t}) = 1/2\pi$. The random vector $\mathbf{u}$ may have any spherically invariant distribution, provided $Pr[\mathbf{u} = \mathbf{0}] = 0$. That is, $\mathbf{u}$ need not be spherically invariant normal for $\mathbf{t}$ to be uniformly distributed on the unit circle S^1.

To say $\mathbf{u}$ is spherically invariant is to say the pdf of $\mathbf{u}$ is invariant to orthogonal transformation $\mathbf{Qu}$, where $\mathbf{Q}$ is a 2×2 rotation matrix. But $\mathbf{Qu} = \mathbf{Qt}r$, so to say $\mathbf{u}$ is spherically invariant is to say $\mathbf{t}$ is spherically invariant. That is, for $\mathbf{Q}$ the orthogonal matrix determined by rotation angle ϕ, the transformation

$$\mathbf{Qt} = \begin{bmatrix} \cos\phi & -\sin\phi \\ \sin\phi & \cos\phi \end{bmatrix} \begin{bmatrix} \cos\theta \\ \sin\theta \end{bmatrix} = \begin{bmatrix} \cos(\theta+\phi) \\ \sin(\theta+\phi) \end{bmatrix},$$

is uniformly distributed on S^1 for θ uniformly distributed on $[0, 2\pi)$. This statement holds for any ϕ, whether randomly drawn or not, provided it is independent of θ. This language will be clarified and generalized when we develop similar stochastic representations for L-dimensional random vectors $\mathbf{u}$ and for $L \times N$ random matrices $\mathbf{U}$.

In the paragraphs to follow, many of the derived random variables are a function only of the uniformly distributed random vector $\mathbf{t}$. Therefore, the distributions of these random variables hold for any bivariate random experiment for which $\mathbf{t}$ is uniformly distributed on the circle S^1.

D.5.3 Chi-Squared Distribution of $\mathbf{u}^T\mathbf{u}$

Write $\mathbf{u}^T\mathbf{u} = u_1^2 + u_2^2 = s$. This makes $s \sim \chi_2^2$, with pdf

$$f(s) = \frac{1}{2}e^{-s/2}, \quad s > 0,$$

which is a chi-squared density, χ_k^2, with $k = 2$ degrees of freedom, an exponential density $\text{Exp}(\lambda)$ with parameter $\lambda = 1/2$, or a gamma density $\Gamma(\alpha, \beta)$ with parameters $\alpha = 1$ and $\beta = 1/2$. Let us recall that if a random variable x is distributed as $x \sim \chi_k^2$, its pdf is

$$f(x) = \frac{x^{k/2-1}}{2^{k/2}\Gamma\left(\frac{k}{2}\right)}e^{-x/2}, \quad x > 0. \tag{D.3}$$

If $x \sim \Gamma(\alpha, \beta)$, its pdf is

$$f(x) = \frac{\beta^\alpha x^{\alpha-1}}{\Gamma(\alpha)} e^{-\beta x}, \quad x > 0.$$

The sum of two independent $\Gamma(1/2, \beta)$ random variables is a $\Gamma(1, \beta)$ random variable, and the sum of two χ_1^2 random variables is a χ_2^2 random variable. Note that $\Gamma(1/2) = \sqrt{\pi}$, $\Gamma(1) = 1$, and for n integer $\Gamma(n) = (n-1)!$

D.5.4 Beta Distribution of $\rho^2 = \frac{\mathbf{u}^T \mathbf{P}_1 \mathbf{u}}{\mathbf{u}^T \mathbf{u}}$

Consider an arbitrary rank-1 projection matrix $\mathbf{P}_1$, and construct the random variable

$$\rho^2 = \frac{\mathbf{u}^T \mathbf{P}_1 \mathbf{u}}{\mathbf{u}^T \mathbf{u}}.$$

Write this as $\rho^2 = \mathbf{t}^T \mathbf{P}_1 \mathbf{t}$, where $\mathbf{t} = [\cos\theta \;\; \sin\theta]^T$. The projection $\mathbf{P}_1$ may be written as $\mathbf{Q}\mathbf{e}_1\mathbf{e}_1^T\mathbf{Q}^T$, where $\mathbf{e}_1 = [1 \; 0]^T$; the random vector $\mathbf{Q}^T\mathbf{t}$ is distributed as $\mathbf{t}$. So ρ^2 is the random variable $\rho^2 = \cos^2\theta$. This is the coherence between the random variables u_1 and u_2, distributed as $\rho^2 \sim \mathrm{Beta}(1/2, 1/2)$. Its pdf is

$$f(\rho^2) = \frac{1}{\pi\sqrt{\rho^2(1-\rho^2)}}, \qquad 0 \le \rho^2 \le 1,$$

and its cdf is

$$F(\rho^2) = \frac{2}{\pi} \arcsin(\rho), \qquad 0 \le \rho^2 \le 1.$$

For this reason, the Beta(1/2, 1/2) distribution is sometimes called the arcsine distribution. This is also the distribution for $\sin^2\theta$.

The beta distribution plays a prominent role in the study of coherence in this book. Its definition and several of its properties are summarized in the following paragraphs.

A random variable x is said to be beta-distributed with parameters α and β if its probability density function is

$$f(x) = \frac{\Gamma(\alpha+\beta)}{\Gamma(\alpha)\Gamma(\beta)} x^{\alpha-1}(1-x)^{\beta-1}, \qquad 0 \le x \le 1;$$

its distribution is denoted $x \sim \mathrm{Beta}(\alpha, \beta)$. Its mean and variance are

$$\mathrm{E}[x] = \frac{\alpha}{\alpha+\beta},$$

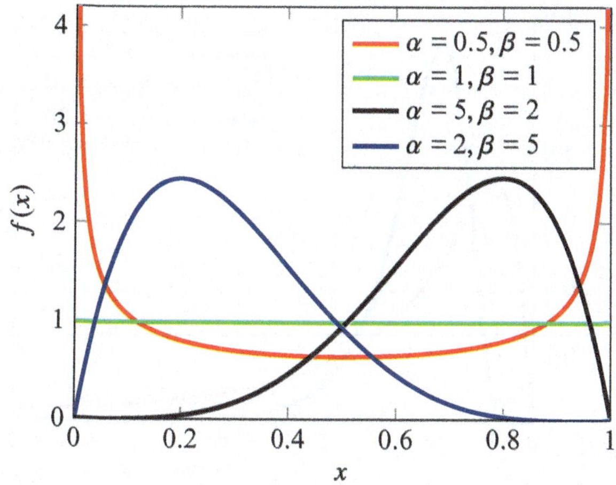

Fig. D.2 Probability density function of $x \sim \text{Beta}(\alpha, \beta)$ for different values of the shape parameters α and β

and

$$\text{E}\left[(x - \text{E}[x])^2\right] = \frac{\alpha\beta}{(\alpha+\beta)^2(\alpha+\beta+1)}.$$

Figure D.2 shows the density $\text{Beta}(\alpha, \beta)$ for different values of α and β, including $\text{Beta}(1/2, 1/2)$, which is the arcsine distribution or Jeffrey's prior, and $\text{Beta}(1, 1)$, which is uniformly distributed in $[0, 1]$. It is clear that if $x \sim \text{Beta}(\alpha, \beta)$, then $1 - x \sim \text{Beta}(\beta, \alpha)$.

If $\rho^2 \sim \text{Beta}(\alpha, \beta)$, then the density of ρ is

$$f(\rho) = 2\frac{\Gamma(\alpha+\beta)}{\Gamma(\alpha)\Gamma(\beta)}\rho^{2\alpha-1}(1-\rho^2)^{\beta-1}, \quad -1 < \rho \leq 1.$$

D.5.5 F-Distribution of $\mathtt{f} = \frac{\mathbf{u}^T(\mathbf{I}_2 - \mathbf{P}_1)\mathbf{u}}{\mathbf{u}^T\mathbf{P}_1\mathbf{u}}$

Write

$$\mathtt{f} = \frac{\mathbf{u}^T(\mathbf{I}_2 - \mathbf{P}_1)\mathbf{u}}{\mathbf{u}^T\mathbf{P}_1\mathbf{u}} = \frac{1-\rho^2}{\rho^2} = \frac{1-\cos^2\theta}{\cos^2\theta},$$

which makes $\mathtt{f} = \tan^2\theta$. This is an F-distributed random variable with parameters $r_1 = 1$ and $r_2 = 1$; that is, $\mathtt{f} \sim \mathcal{F}(1, 1)$, with pdf

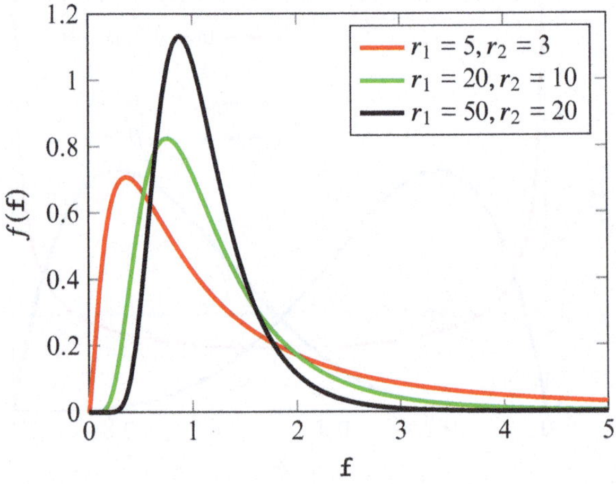

Fig. D.3 Probability density function of $\mathtt{f} \sim \mathcal{F}(r_1, r_2)$ for different values of the numerator and denominator degrees of freedom r_1 and r_2

$$f(\mathtt{f}) = \frac{1}{\pi \mathtt{f}^{1/2}(1+\mathtt{f})}, \qquad \mathtt{f} > 0.$$

If $\mathtt{f} \sim \mathcal{F}(r_1, r_2)$, its density is

$$f(\mathtt{f}) = \frac{\Gamma\left(\frac{r_1+r_2}{2}\right)}{\Gamma\left(\frac{r_1}{2}\right)\Gamma\left(\frac{r_2}{2}\right)} \left(\frac{r_1}{r_2}\right)^{r_1/2} \frac{\mathtt{f}^{\frac{r_1}{2}-1}}{\left(1+\frac{r_1}{r_2}\mathtt{f}\right)^{\frac{r_1+r_2}{2}}}.$$

Figure D.3 shows the density $\mathtt{f} \sim \mathcal{F}(r_1, r_2)$ for different values of (r_1, r_2).

D.5.6 Distributions for Other Derived Random Variables

We may summarize these results and several others derived from them. Distribution results for functions of θ depend only on the uniform distribution of θ.

- $u_1^2 \sim \chi_1^2$
- $u_2^2 \sim \chi_1^2$
- $u_1^2 + u_2^2 \sim \chi_2^2$
- $\cos^2\theta \sim \text{Beta}(1/2, 1/2)$
- $\sin^2\theta \sim \text{Beta}(1/2, 1/2)$
- $\tan^2\theta \sim \mathcal{F}(1, 1)$
- The distribution for the random variable $x = \cos\theta = u_1/r$ is

$$f(x) = \frac{1}{\pi\sqrt{(1-x^2)}}, \qquad -1 \le x \le 1.$$

This is also the density for $y = \sin\theta = u_2/r$.

- The ratio $t = \tan\theta = y/x$ has Cauchy density, given by

$$f(t) = \frac{1}{\pi(1+t^2)}, \quad -\infty < t < \infty,$$

which is invariant under the transformation $z = 1/t$. So this is also the density for the ratio $z = x/y = 1/\tan\theta$. This is a special case, with $r = 1$, of the univariate t-distribution with r degrees of freedom, given by

$$f(t) = \frac{\Gamma\left(\frac{r+1}{2}\right)}{(\pi r)^{1/2}\Gamma(r/2)}\left(1+\frac{t^2}{r}\right)^{-\frac{(r+1)}{2}}. \tag{D.4}$$

D.5.7 Generation of Standard Normal Random Variables

Recall that for a continuous rv x, with distribution function F_x, the transformation $z = F_x(x)$ is uniformly distributed as $z \sim \mathcal{U}[0, 1]$. Moreover, $x = F_x^{-1}(z) \sim F_x(x)$. For example, if $x \sim \text{Exp}(1)$, with cdf $F_x(x) = 1 - e^{-x}, x > 0$, and pdf $f_x(x) = e^{-x}, x > 0$, the rv $z = 1 - e^{-x} \sim \mathcal{U}[0, 1]$, and $x = -\log(1 - z) \sim \text{Exp}(1)$. Slavish application of the principle to the generation of a standard normal rv $x \sim \mathcal{N}(0, 1)$ would have us generate x as the solution to the equation:

$$F_x(x) = \int_{-\infty}^{x} \frac{1}{\sqrt{2\pi}} \exp\left(-y^2/2\right) dy = z.$$

This amounts to finding a realization of the rv x for which the normal integral equals the realization of a uniformly distributed rv z. This is not practical. Perhaps there is an insight from the bivariate normal experiment that provides an alternative.

In the bivariate normal experiment, with u_1 and u_2 independent $\mathcal{N}(0, 1)$ rvs, the radius-squared, $r^2 = u_1^2 + u_2^2$, is distributed as $r^2 \sim \text{Exp}(1/2)$, and the angle θ is distributed as $\mathcal{U}[0, 2\pi]$, which in turn is distributed as the rv $\theta = 2\pi v$, where $v \sim \mathcal{U}[0, 1]$. Moreover, the rvs $u_1 = r\cos(\theta)$ and $u_2 = r\sin(\theta)$ are i.i.d. $\mathcal{N}(0, 1)$ rvs. The exponentially distributed rv r^2 may be synthesized from uniformly distributed $w \sim \mathcal{U}[0, 1]$ as $r^2 = -2\log(1 - w)$ and the rv r may be synthesized as the rv $r = \sqrt{-2\log(1 - w)}$. This gives the Box-Muller method for generating independent $\mathcal{N}(0, 1)$ rvs u_1 and u_2 from a pair of independent $\mathcal{U}[0, 1]$ rvs v and w; that is,

$$u_1 = \sqrt{-2\log(1-w)}\cos(2\pi v), \qquad u_2 = \sqrt{-2\log(1-w)}\sin(2\pi v).$$

The rv $1 - w$ is distributed as w, so $1 - w$ may be replaced by w in this formula.

This is the synthesis story. The analysis story for generating a uniformly distributed rv from u_1 and u_2 is this. Taking into account that $-(u_1^2+u_2^2)/2 = \log w$, it follows that $w = \exp\{-(u_1^2 + u_2^2)/2\}$ is uniformly distributed on $[0, 1]$.

D.6 The Spherically Invariant Multivariate Normal Experiment

The spherically invariant bivariate normal experiment generalizes to the spherically invariant multivariate normal experiment when the independent normal random variables u_1 and u_2 are replaced by the normal random vector $\mathbf{u} \sim \mathcal{N}_L(\mathbf{0}, \mathbf{I}_L)$.

D.6.1 Coordinate Transformation: The Generalized Rayleigh and Uniform Distributions

Begin with the random vector $\mathbf{u} \sim \mathcal{N}_L(\mathbf{0}, \mathbf{I}_L)$. The transformation from polar coordinates to Euclidean coordinates is [244]

$$
\begin{aligned}
u_1 &= r \sin\theta_1 \sin\theta_2 \cdots \sin\theta_{L-2} \sin\theta_{L-1}, \\
u_2 &= r \sin\theta_1 \sin\theta_2 \cdots \sin\theta_{L-2} \cos\theta_{L-1}, \\
u_3 &= r \sin\theta_1 \sin\theta_2 \cdots \cos\theta_{L-2}, \\
&\vdots \\
u_{L-1} &= r \sin\theta_1 \cos\theta_2, \\
u_L &= r \cos\theta_1.
\end{aligned}
$$

That is, $\mathbf{u} = \mathbf{t}r$. By construction, $\mathbf{t}^T\mathbf{t} = 1$ and $r^2 = \mathbf{u}^T\mathbf{u} \geq 0$. The determinant of the Jacobian of the transformation is

$$\det(J(u_1, \ldots, u_L \to r, \theta_1, \ldots, \theta_{L-1})) = r^{L-1}(\sin\theta_1)^{L-2}(\sin\theta_2)^{L-3} \cdots (\sin\theta_{L-2}).$$

For $\mathbf{u} \sim \mathcal{N}_L(\mathbf{0}, \mathbf{I}_L)$, the joint density of the polar random variables $r^2, \theta_1, \theta_2, \ldots, \theta_{L-1}$, is then

$$f(r^2, \theta_1, \ldots, \theta_{L-1}) = \frac{1}{2(2\pi)^{L/2}}(r^2)^{L/2-1}(\sin\theta_1)^{L-2}(\sin\theta_2)^{L-3} \cdots (\sin\theta_{L-2})\, e^{-\frac{r^2}{2}}, \tag{D.5}$$

from which it is evident that $r^2, \theta_1, \theta_2, \ldots, \theta_{L-1}$, are mutually independent, and each angle θ_k has a density proportional to $(\sin\theta_k)^{L-1-k}$. Hence, θ_{L-1} is uniformly distributed on $[0, 2\pi)$. Integrating (D.5) with respect to $\theta_1, \theta_2, \ldots, \theta_{L-1}$, yields the surface area of the unit radius sphere S^{L-1} in $\mathbb{R}^L$, which is $2\pi^{L/2}/\Gamma(L/2)$. Then, it follows that $s = r^2$ has the χ_L^2 density

$$f(s) = \frac{1}{2^{L/2}\Gamma\left(\frac{L}{2}\right)} s^{L/2-1} e^{-s/2}, \quad s \geq 0,$$

and the distribution of r is the generalized Rayleigh distribution with density

$$f(r) = \frac{2}{2^{L/2}\Gamma\left(\frac{L}{2}\right)} r^{L-1} e^{-r^2/2}, \quad r \geq 0.$$

The random vector $\mathbf{t}$ is uniformly distributed on the surface of the sphere S^{L-1} in $\mathbb{R}^L$, meaning its pdf is

$$f(\mathbf{t}) = \frac{\Gamma\left(\frac{L}{2}\right)}{2\pi^{L/2}}, \quad \mathbf{t}^T\mathbf{t} = 1.$$

For $L = 2$, these are the results for the distribution of r and $\mathbf{t}$ in the spherically invariant bivariate normal experiment, described in Sect. D.5.

In [110], it is shown that the marginal pdf for $\mathbf{t}_k$, a k-dimensional subset of $\mathbf{t}$ constructed from k coordinates, is

$$f(\mathbf{t}_k) = \frac{\Gamma\left(\frac{L}{2}\right)}{\Gamma\left(\frac{1}{2}\right)^k \Gamma\left(\frac{L-k}{2}\right)} (1 - \mathbf{t}_k^T \mathbf{t}_k)^{\frac{L-k}{2}-1}, \quad 0 < \mathbf{t}_k^T \mathbf{t}_k \leq 1.$$

This result holds for $1 \leq k < L - 1$. Moreover, the pdf for $\mathbf{t}_k^T \mathbf{t}_k$ is Beta $\left(\frac{k}{2}, \frac{L-k}{2}\right)$. For $k = 1$ and $L = 2$, this is the result for the distribution of u_1^2/r^2 in the spherically invariant bivariate normal experiment. For $k = L - 1$, the pdf for $\mathbf{t}_{L-1}$ is

$$f(\mathbf{t}_{L-1}) = \frac{\Gamma\left(\frac{L}{2}\right)}{\Gamma\left(\frac{1}{2}\right)^L} (1 - \mathbf{t}_{L-1}^T \mathbf{t}_{L-1})^{-1/2}, \quad 0 < \mathbf{t}_{L-1}^T \mathbf{t}_{L-1} \leq 1, \tag{D.6}$$

and $\mathbf{t}_{L-1}^T \mathbf{t}_{L-1} \sim$ Beta $\left(\frac{L-1}{2}, \frac{1}{2}\right)$.

The stochastic representation $\mathbf{u} = \mathbf{t}r$ encodes for the synthesis of $\mathbf{t}$ uniformly distributed on S^{L-1} and $r^2 \sim \chi_L^2$, independent of $\mathbf{t}$: generate $\mathbf{u}$, and compute $r = (\mathbf{u}^T\mathbf{u})^{1/2}$ and $\mathbf{t} = \mathbf{u}/r$. The random vector $\mathbf{u}$ may have any spherically invariant distribution, provided $Pr[r = 0] = 0$. To derive the uniform distribution of $\mathbf{t}$, let $\mathbf{t}_{L-1}$ be a vector with the first $L - 1$ components of $\mathbf{t}$, and proceed as follows. The nonsingular transformation from $(\mathbf{t}_{L-1}, r)$ to $\mathbf{u}$ is

$$\mathbf{u} = \begin{bmatrix} \mathbf{t}_{L-1} r \\ (1 - \mathbf{t}_{L-1}^T \mathbf{t}_{L-1})^{1/2} r \end{bmatrix},$$

with Jacobian

$$J((\mathbf{t}_{L-1}, r) \to \mathbf{u}) = \begin{bmatrix} r\mathbf{I}_{L-1} & -\frac{r}{(1-\mathbf{t}_{L-1}^T\mathbf{t}_{L-1})^{1/2}}\mathbf{t}_{L-1} \\ \mathbf{t}_{L-1}^T & (1-\mathbf{t}_{L-1}^T\mathbf{t}_{L-1})^{1/2} \end{bmatrix}.$$

Using the determinant identity (B.5.1) in Appendix B, the Jacobian determinant is $r^{L-1}(1 - \mathbf{t}_{L-1}^T\mathbf{t}_{L-1})^{-1/2}$. Then, assuming $\mathbf{u} \sim \mathcal{N}_L(\mathbf{0}, \mathbf{I}_L)$,

$$f(\mathbf{t}_{L-1}, r) = \frac{1}{(2\pi)^{L/2}} e^{-r^2/2} r^{L-1} \left(1 - \mathbf{t}_{L-1}^T\mathbf{t}_{L-1}\right)^{-1/2} = f(\mathbf{t}_{L-1}) f_r(r).$$

Integrate over r to marginalize this joint pdf for $f(\mathbf{t}_{L-1})$ in (D.6).

D.6.2 Geometry and Spherical Invariance

If $\mathbf{u}$ is spherically invariant, then the distribution of $\mathbf{u}$ is invariant to left orthogonal transformation $\mathbf{Qu}$, which is to say $\mathbf{Qt}r$ is distributed as $\mathbf{t}r$, which is to say $\mathbf{Qt}$ is distributed as $\mathbf{t}$. That is, $\mathbf{t}$ is spherically invariant, which is to say it is *uniformly distributed with respect to Haar measure* on the surface of the L-dimensional sphere, $S^{L-1} = \{\mathbf{t} \in \mathbb{R}^L \mid \mathbf{t}^T\mathbf{t} = 1\}$. Therefore, for spherically invariant $\mathbf{u} = \mathbf{t}r$, the distribution of $\mathbf{t}$ is the distribution derived for the spherically invariant MVN experiment.

In the paragraphs to follow, many of the derived random variables are a function only of the uniformly distributed random vector $\mathbf{t}$. Therefore, the distributions of these random variables hold for any multivariate random experiment for which $\mathbf{t}$ is uniformly distributed on the sphere S^{L-1}.

D.6.3 Chi-Squared Distribution of $\mathbf{u}^T\mathbf{u}$

The rv $\mathbf{u}^T\mathbf{u}$ is $r^2 = u_1^2 + \cdots + u_L^2$, which is chi-squared distributed as $r^2 \sim \chi_L^2$.

D.6.4 Beta Distribution of $\rho_p^2 = \frac{\mathbf{u}^T\mathbf{P}_p\mathbf{u}}{\mathbf{u}^T\mathbf{u}}$

Let $\mathbf{P}_p$ be an arbitrary rank-p projection matrix. It has representation $\mathbf{P}_p = \mathbf{V}_p\mathbf{V}_p^T$, where $\mathbf{V}_p$ is a column slice from an $L \times L$ orthogonal matrix $\mathbf{Q} = [\mathbf{V}_p \ \mathbf{V}_{L-p}]$. The ratio

$$\rho_p^2 = \frac{\mathbf{u}^T\mathbf{P}_p\mathbf{u}}{\mathbf{u}^T\mathbf{u}}$$

is the cosine-squared of the angle that the random vector $\mathbf{u}$ makes with the subspace $\langle \mathbf{V}_p \rangle$. This ratio may be written as

$$\rho_p^2 = \mathbf{t}^T \mathbf{P}_p \mathbf{t},$$

where $\mathbf{u} = \mathbf{t}r$, and the spherically invariant random vector $\mathbf{t}$ is uniformly distributed on the unit sphere S^{L-1}. The distribution of $\mathbf{t}$ is the distribution of $\mathbf{Qt}$, so the cosine-squared may be written as

$$\rho_p^2 = \mathbf{t}^T \mathbf{Q}^T \mathbf{V}_p \mathbf{V}_p^T \mathbf{Q} \mathbf{t} = \sum_{k=1}^{p} t_k^2.$$

Then, [110] shows that this is distributed as a Beta$(p/2, (L-p)/2)$ random variable and is independent of $\mathbf{u}^T\mathbf{u}$. This result holds for arbitrary rank-p projection $\mathbf{P}_p$. That is, for any choice of $\mathbf{P}_p$, the cosine-squared statistic is distributed as the sum of squares of the first p components, or of any p components, of $\mathbf{t}$. The sine-squared statistic $1 - \rho_p^2 = \frac{\mathbf{u}^T(\mathbf{I}_L - \mathbf{P}_p)\mathbf{u}}{\mathbf{u}^T\mathbf{u}}$ is distributed as a Beta$((L-p)/2, p/2)$. When the identity $\mathbf{I}_L$ is resolved as $\mathbf{I}_L = \mathbf{P}_p + \mathbf{P}_p^{\perp}$, then the cosine-squared statistic may be written as

$$\rho_p^2 = \frac{\mathbf{u}^T \mathbf{P}_p \mathbf{u}}{\mathbf{u}^T \mathbf{P}_p \mathbf{u} + \mathbf{u}^T \mathbf{P}_p^{\perp} \mathbf{u}} \sim \text{Beta}\left(\frac{p}{2}, \frac{L-p}{2}\right).$$

When $\mathbf{P}_p$ is the rank-1 projection $\mathbf{P}_1 = \mathbf{a}(\mathbf{a}^T\mathbf{a})^{-1}\mathbf{a}^T$, then $\rho_1^2 = (\mathbf{a}^T\mathbf{u})^2/[(\mathbf{a}^T\mathbf{a})(\mathbf{u}^T\mathbf{u})] \sim \text{Beta}(1/2, (L-1)/2)$. When

$$\mathbf{a} = [\underbrace{1 \cdots 1}_{p}\ \underbrace{0 \cdots 0}_{L-p}]^T,$$

then this is the distribution for the sum of squares of the first p components of $\mathbf{t}$, and this is the distribution of the ratio of the sum of the first p squares of $\mathbf{u}$ to the sum of all squares of $\mathbf{u}$. An equivalent way of stating this result is as follows. Let $\mathbf{x} \sim \mathcal{N}_p(\mathbf{0}, \mathbf{I}_p)$ and $\mathbf{y} \sim \mathcal{N}_{L-p}(\mathbf{0}, \mathbf{I}_{L-p})$ be two independent real normal vectors, then

$$z = \frac{\mathbf{x}^T\mathbf{x}}{\mathbf{x}^T\mathbf{x} + \mathbf{y}^T\mathbf{y}},$$

has a Beta$(p/2, (L-p)/2)$ distribution and is independent of $\mathbf{x}^T\mathbf{x} + \mathbf{y}^T\mathbf{y}$. All of these results are dependent only on the spherically invariant distribution for $\mathbf{t}$ and the dimension of the projection $\mathbf{P}_r$.

These results have been derived as if the random vector $\mathbf{u}$ is spherically invariant MVN and $\mathbf{P}_p$ is an arbitrary deterministic projection. But, of course, the results hold for any multivariate distribution for which the random vector $\mathbf{t}$ in the factorization $\mathbf{u} = \mathbf{t}r$ is uniformly distributed on the unit sphere S^{L-1}. They hold for $\mathbf{P}_p$ drawn independently of $\mathbf{u}$, from any distribution. Moreover, $\mathbf{t}$ may be drawn uniformly from S^{L-1} by starting with deterministic $\mathbf{t}_0$, and spinning it to $\mathbf{W}\mathbf{t}_0$, with orthogonal $\mathbf{W}$ drawn uniformly from the orthogonal group (this will be clarified in the section on the spherically symmetric matrix distribution). Then, the cosine-squared statistic may be written as

$$\rho_p^2 = \mathbf{t}_0^T\mathbf{W}^T\mathbf{P}_p\mathbf{W}\mathbf{t}_0 = \mathbf{t}_0^T\mathbf{W}^T\mathbf{V}_p\mathbf{V}_p^T\mathbf{W}\mathbf{t}_0.$$

The $L \times p$ matrix $\mathbf{W}_p = \mathbf{W}^T\mathbf{V}_p$ is a uniform draw of an orthogonal frame from the Stiefel manifold of rank-p frames, $\mathbb{S}t(p, \mathbb{R}^L)$, or equivalently the matrix $\mathbf{W}^T\mathbf{P}_p\mathbf{W}$ is the projection onto a subspace uniformly drawn from the Grassmann manifold of dimension-p subspaces, $\mathbb{G}r(p, \mathbb{R}^L)$. So the cosine-squared statistic

$$\rho_p^2 = \mathbf{t}_0^T\mathbf{W}^T\mathbf{P}_p\mathbf{W}\mathbf{t}_0 = \mathbf{t}_0^T\mathbf{W}_p\mathbf{W}_p^T\mathbf{t}_0 \sim \text{Beta}\left(\frac{p}{2}, \frac{L-p}{2}\right)$$

for deterministic $\mathbf{t}_0$ and $\mathbf{W}_p$ is drawn uniformly from the Stiefel manifold.

In summary, the statistic $\rho_p^2 = \frac{\mathbf{u}^T\mathbf{P}_p\mathbf{u}}{\mathbf{u}^T\mathbf{u}} \sim \text{Beta}(p/2, (L-p)/2)$ for (i) $\mathbf{u}$ spherically invariant and $\mathbf{P}_p$ a rank-p projection, deterministic or drawn independently of $\mathbf{u}$ from any distribution, or (ii) $\mathbf{u} = \mathbf{u}_0$ deterministic and $\mathbf{P}_p$ constructed as $\mathbf{P}_p = \mathbf{W}_p\mathbf{W}_p^T$ with $\mathbf{W}_p$ drawn uniformly from the Stiefel manifold of rank-p frames.

D.6.5 F-Distribution of $\mathtt{f}_p = \frac{p}{L-p}\frac{\mathbf{u}^T(\mathbf{I}_L-\mathbf{P}_p)\mathbf{u}}{\mathbf{u}^T\mathbf{P}_p\mathbf{u}}$

Write

$$\mathtt{f}_p = \frac{p}{L-p}\frac{\mathbf{u}^T(\mathbf{I}_L - \mathbf{P}_p)\mathbf{u}}{\mathbf{u}^T\mathbf{P}_p\mathbf{u}},$$

as

$$\mathtt{f}_p = \frac{p}{L-p}\frac{1-\rho_p^2}{\rho_p^2}.$$

This makes $\mathtt{f}_p$ an F-distributed random variable, $\mathtt{f}_p \sim \mathcal{F}(L-p, p)$. For MVN vectors, this result can be rephrased as follows: Let $\mathbf{x} \sim \mathcal{N}_p(\mathbf{0}, \mathbf{I}_p)$ and $\mathbf{y} \sim \mathcal{N}_{L-p}(\mathbf{0}, \mathbf{I}_{L-p})$ be two independent real normal vectors, then $\frac{p}{L-p}\frac{\mathbf{y}^T\mathbf{y}}{\mathbf{x}^T\mathbf{x}} \sim \mathcal{F}(L-p, p)$ and is independent of $\mathbf{x}^T\mathbf{x} + \mathbf{y}^T\mathbf{y}$. The scaling $((L-p)/p)\mathtt{f}_p$ is the tangent-squared of the angle that $\mathbf{u}$ makes with any arbitrary subspace $\langle\mathbf{V}_p\rangle$ of dimension p.

D.6.6 Distributions for Other Derived Random Variables

We may summarize these results and several others derived from them. All but the first two depend only on the uniform distribution of **t**.

- $u_l^2 \sim \chi_1^2, l = 1, \ldots, L$
- $\mathbf{u}^T\mathbf{u} \sim \chi_L^2$
- $\rho_p^2 = \frac{\mathbf{u}^T\mathbf{P}_p\mathbf{u}}{\mathbf{u}^T\mathbf{u}} \sim \text{Beta}(p/2, (L-p)/2)$
- $1 - \rho_p^2 = \frac{\mathbf{u}^T(\mathbf{I}_L - \mathbf{P}_p)\mathbf{u}}{\mathbf{u}^T\mathbf{u}} \sim \text{Beta}((L-p)/2, p/2)$
- $\mathtt{f}_p = \frac{p}{L-p}\frac{\mathbf{u}^T(\mathbf{I}_L - \mathbf{P}_p)\mathbf{u}}{\mathbf{u}^T\mathbf{P}_p\mathbf{u}} \sim \mathcal{F}(L-p, p)$
- Define $w = \cos(\mathbf{u}, \mathbf{a}) = \frac{\mathbf{a}^T\mathbf{u}}{\|\mathbf{u}\| \|\mathbf{a}\|}$, with $\mathbf{a} \in \mathbb{R}^N$, then

$$w \sim \frac{\Gamma\left(\frac{L}{2}\right)}{\Gamma\left(\frac{1}{2}\right)\Gamma\left(\frac{L-1}{2}\right)}(1-w^2)^{(L-1)/2-1}, \quad -1 \leq w \leq 1.$$

  This is also the density for $v = \sin(\mathbf{u}, \mathbf{a}) = \sqrt{1-w^2}$.
- The ratio $t = \tan(\mathbf{u}, \mathbf{a}) = v/w = v/\sqrt{1-v^2}$ has a t-density with $r = L-1$ degrees of freedom $t \sim \text{t}(L-1)$ (see Eq. (D.4)).

D.7 The Spherically Invariant Matrix-Valued Normal Experiment

The spherically invariant bivariate and MVN experiments generalize to the spherically invariant matrix-valued experiment when the normal random vector $\mathbf{u} \sim \mathcal{N}_L(\mathbf{0}, \mathbf{I}_L)$ is replaced by the $L \times N$ normal random matrix $\mathbf{U} \sim \mathcal{N}_{L\times N}(\mathbf{0}, \mathbf{I}_N \otimes \mathbf{I}_L)$, $N \geq L$. It is as if the multichannel random vector $\mathbf{u} \in \mathbb{R}^L$ has been replaced by N independent measurements or snapshots of such a random vector.

D.7.1 Coordinate Transformation: Bartlett's Factorization

To be consistent with the treatment of spherical invariance in the bivariate and multivariate normal cases, we shall let $\mathbf{U}$ denote an $L \times N$ random matrix, populated by independent and identically distributed $\mathcal{N}(0, 1)$ random variables. This matrix may be "QR" factored as $\mathbf{U}^T = \mathbf{TR}$, where $\mathbf{T}$ is an $N \times L$ slice of a random orthogonal matrix and $\mathbf{R}$ is an $L \times L$ upper triangular matrix (the inverse of $\mathbf{R}$ is also upper triangular). This produces an LDU Cholesky factorization of the $L \times L$ Gramian $\mathbf{U}\mathbf{U}^T$ as $\mathbf{U}\mathbf{U}^T = \mathbf{R}^T\mathbf{R}$, and the matrix $\mathbf{R}^T$ may be taken as a uniquely defined lower triangular square root of $\mathbf{U}\mathbf{U}^T$. That is, $\mathbf{R}^T$ may be taken as a uniquely defined definition of $(\mathbf{U}\mathbf{U}^T)^{1/2}$ in the developments to follow. The orthogonal slice $\mathbf{T}$ is uniformly distributed with respect to Haar measure on the Stiefel manifold $\mathbb{St}(L, \mathbb{R}^N)$, which is to say its distribution is invariant to left orthogonal transformation by an $N \times N$ orthogonal matrix $\mathbf{Q} \in O(N)$. Thus, in this experiment, the distribution of $\mathbf{U}^T$ is invariant to left multiplication by an orthogonal matrix, and, therefore, so is the resulting $\mathbf{T} = \mathbf{U}^T\mathbf{R}^{-1}$. The factorization of the tall matrix $\mathbf{U}^T = \mathbf{TR}$ is called a *polar decomposition.*

According to Bartlett's factorization theorem (see Appendix G), the random matrices $\mathbf{T}$ and $\mathbf{R}$ are statistically independent. Moreover, the Gramian $\mathbf{R}^T\mathbf{R}$ is Wishart distributed as $\mathcal{W}_L(\mathbf{I}_L, N)$, and the elements of the Cholesky factor $\mathbf{R}$ are all independent, with r_{ll} distributed as χ^2_{L-l+1}, $l = 1, 2, \ldots, L$, and r_{lm} distributed as $\mathcal{N}(0, 1)$, $1 \leq l < m \leq L$. This is the statistical characterization of the QR factorization of $\mathbf{U}^T$. The factorization $\mathbf{U}^T = \mathbf{TR}$ encodes for a stochastic synthesis of $\mathbf{T}$: generate $\mathbf{U}$ and QR factor it as $\mathbf{U}^T = \mathbf{TR}$, and solve for $\mathbf{T} = \mathbf{U}^T\mathbf{R}^{-1}$. The resulting $\mathbf{T}$ is uniformly distributed with respect to Haar measure on $\mathbb{St}(L, \mathbb{R}^N)$. The uniform pdf for $\mathbf{T}$ is

$$f(\mathbf{T}) = \frac{1}{\mathrm{Vol}(\mathbb{St}(L, \mathbb{R}^N))},$$

where $\mathrm{Vol}(\mathbb{St}(L, \mathbb{R}^N))$ is the volume of the Stiefel manifold [112, 244]:

$$\mathrm{Vol}(\mathbb{St}(L, \mathbb{R}^N)) = \frac{2^L \pi^{LN/2}}{\Gamma_L(N/2)}.$$

The multivariate gamma function $\Gamma_L(x)$ can be expressed as a product of univariate gamma functions (cf. Theorem 2.1.12 of [244]):

$$\Gamma_L(x) = \pi^{L(L-1)/4} \prod_{l=1}^{L} \Gamma\left(x - \frac{(l-1)}{2}\right). \tag{D.7}$$

This volume is the product of areas for unit spheres S^{l-1} in $\mathbb{R}^l$ for $l = N - L + 1, \ldots, N$:[2]

$$\mathrm{Vol}(\mathbb{St}(L, \mathbb{R}^L)) = \prod_{l=N-L+1}^{N} \frac{2\pi^{l/2}}{\Gamma(l/2)}.$$

[2] This interpretation of Alan Edelman's result for volume is evidently due to John Barnett, as reported in a short Internet posting by Jason D. M. Rennie.

D.7.2 Geometry and Spherical Invariance

In the spherically invariant MVN experiment, the $L \times 1$ vector $\mathbf{t}$ was shown to be uniformly distributed on the sphere S^{L-1}. The $N \times L$ matrix $\mathbf{T}$ is uniformly distributed on the Stiefel manifold, which is the set of all orthogonal L-frames in $\mathbb{R}^N$. This set may be thought of as the set of orthogonal slices, $\mathbf{Q} \in \mathbb{R}^{N\times L}, \mathbf{Q}^T\mathbf{Q} = \mathbf{I}_L$. The notation is this: $\mathbb{S}t(L, \mathbb{R}^N) = \{\mathbf{Q} \in \mathbb{R}^{N\times L} \mid \mathbf{Q}^T\mathbf{Q} = \mathbf{I}_L\}$. The topology of $\mathbb{S}t(L, \mathbb{R}^N)$ is the subspace topology inherited from $\mathbb{R}^{N\times L}$, and with this topology, $\mathbb{S}t(L, \mathbb{R}^N)$ is a compact manifold of dimension $NL - L(L+1)/2$. The orthogonal group $O(N)$ acts transitively on the elements of the Stiefel manifold, which is to say for $\mathbf{Q} \in O(N)$, the left orthogonal transformation $\mathbf{QT}$ is another L-frame in $\mathbb{S}t(L, \mathbb{R}^N)$ and starting from any frame $\mathbf{T}$, $\{\mathbf{QT}, \mathbf{Q} \in O(N)\}$ generates $\mathbb{S}t(L, \mathbb{R}^N)$. For $\mathbf{Q} \in O(L)$, the right transformation $\mathbf{TQ}$ leaves $\mathbf{T}$ invariant as a subspace. A representation of the Stiefel manifold is $\mathbb{S}t(L, \mathbb{R}^N) = O(N)/O(N-L)$, which is to say, "begin with an $N \times N$ orthogonal matrix $\mathbf{Q}_N$ drawn from the orthogonal group $O(N)$, extract the first L columns, and you have an L frame from $\mathbb{S}t(L, \mathbb{R}^N)$. The extraction is invariant to rotation of the last $N - L$ columns of $\mathbf{Q}_N$, and this accounts for the mod notation $/O(N-L)$. If this representation is further "mod-ed" as $Q(N)/(Q(N-L) \times Q(L))$, then the subspace $\langle \mathbf{T} \rangle$ is an element of the Grassmannian manifold $\mathbb{G}r(L, \mathbb{R}^N)$. In other words, while the L-frame $\mathbf{T}$ is distinguished from the L-frame $\mathbf{TQ}_L$ on $\mathbb{S}t(L, \mathbb{R}^N)$, the range of $\mathbf{T}$, or equivalently the projection $\mathbf{TT}^T$, is indistinguishable from the range of $\mathbf{TQ}_L$ and the projection $\mathbf{TQ}_L\mathbf{Q}_L^T\mathbf{T}^T = \mathbf{TT}^T$.

D.7.3 Wishart Distribution of $\mathbf{UU}^T$

Begin with $\mathbf{U} \sim N_{L\times N}(\mathbf{0}, \mathbf{I}_N \otimes \mathbf{I}_L)$. The $L \times L$ matrix $\mathbf{S} = \mathbf{UU}^T$ is distributed as a Wishart matrix, denoted $\mathbf{S} \sim \mathcal{W}_L(\mathbf{I}_L, N)$, with pdf

$$f(\mathbf{S}) = \frac{1}{2^{LN/2}\Gamma_L(N/2)} \det(\mathbf{S})^{(N-L-1)/2} \operatorname{etr}\left(-\frac{1}{2}\mathbf{S}\right), \quad \mathbf{S} \succ \mathbf{0}.$$

In the matrix case, the Wishart distribution plays the role of a χ^2 distribution. For a more complete account on the Wishart distribution, see Appendix G.

D.7.4 The Matrix Beta Distribution

Consider the $L \times L$ matrix

$$\mathbf{B} = (\mathbf{UU}^T)^{-1/2}\mathbf{UP}_p\mathbf{U}^T(\mathbf{UU}^T)^{-T/2},$$

where the $N \times N$ matrix $\mathbf{P}_p$ is a rank-p projection. In this experiment, $p \geq L$ and $N - p \geq L$. With $\mathbf{U}^T$ factored as $\mathbf{U}^T = \mathbf{TR}$, then $(\mathbf{UU}^T)^{-1} = \mathbf{R}^{-1}\mathbf{R}^{-T}$, and $(\mathbf{UU}^T)^{-1/2}$ may be taken to be $\mathbf{Q}_L\mathbf{R}^{-T}$ with $\mathbf{Q}_L \in O(L)$ an arbitrary orthogonal matrix. Then it is straightforward to show that $\mathbf{B} = \mathbf{Q}_L\mathbf{T}^T\mathbf{P}_p\mathbf{T}\mathbf{Q}_L^T$. The frame $\mathbf{T}\mathbf{Q}_L^T$ is distributed as $\mathbf{T}$ is distributed, so $\mathbf{B}$ is distributed as $\mathbf{T}^T\mathbf{P}_p\mathbf{T}$ is distributed, with $\mathbf{T}$ uniformly distributed on the Stiefel manifold. This is a stochastic representation of $\mathbf{B}$, which is distributed as a matrix-valued beta random matrix, denoted $\mathbf{B} \sim \text{Beta}_L(p/2, (N-p)/2)$. Its pdf is

$$f(\mathbf{B}) = \frac{\Gamma_L\left(\frac{N}{2}\right)}{\Gamma_L\left(\frac{p}{2}\right)\Gamma_L\left(\frac{N-p}{2}\right)} \det(\mathbf{B})^{(p-L-1)/2} \det(\mathbf{I}_L - \mathbf{B})^{(N-p-L-1)/2}, \quad \mathbf{0} \prec \mathbf{B} \prec \mathbf{I}_L, \tag{D.8}$$

where $\Gamma_L(\cdot)$ is the multivariate gamma function defined in (D.7). A matrix $\mathbf{B}$ with density function (D.8) is said to have the matrix-variate beta of type I distribution with parameters $p/2$ and $(N-p)/2$. When $L = 1$, this is the pdf of the univariate beta random variable $b = \mathbf{t}^T\mathbf{P}_p\mathbf{t} \sim \text{Beta}(p/2, (N-p)/2)$.

These results hold when the matrix $\mathbf{U} \sim N_{L\times N}(\mathbf{0}, \mathbf{I}_N \otimes \mathbf{I}_L)$ is colored to produce

$$\mathbf{X} = \mathbf{\Sigma}^{1/2}\mathbf{U} \sim N_{LN}(\mathbf{0}, \mathbf{I}_N \otimes \mathbf{\Sigma}).$$

Then, the matrix $(\mathbf{XX}^T)^{-1/2}\mathbf{X}\mathbf{P}_p\mathbf{X}^T(\mathbf{XX}^T)^{-T/2}$ may be written as

$$(\mathbf{\Sigma}^{1/2}\mathbf{R}^T\mathbf{R}\mathbf{\Sigma}^{T/2})^{-1/2}\mathbf{\Sigma}^{1/2}\mathbf{R}^T\mathbf{T}^T\mathbf{P}_p\mathbf{T}\mathbf{R}\mathbf{\Sigma}^{T/2}(\mathbf{\Sigma}^{1/2}\mathbf{R}^T\mathbf{R}\mathbf{\Sigma}^{T/2})^{-T/2}.$$

Without loss of generality, we may call $\mathbf{\Sigma}^{1/2}\mathbf{R}^T\mathbf{Q}_L$ the square root of $\mathbf{\Sigma}^{1/2}\mathbf{R}^T\mathbf{R}\mathbf{\Sigma}^{T/2}$, and $\mathbf{Q}_L^T\mathbf{R}^{-T}\mathbf{\Sigma}^{-1/2}$ the square root of $(\mathbf{\Sigma}^{1/2}\mathbf{R}^T\mathbf{R}\mathbf{\Sigma}^{T/2})^{-1}$. Then,

$$(\mathbf{XX}^T)^{-1/2}\mathbf{X}\mathbf{P}_p\mathbf{X}^T(\mathbf{XX})^{-T/2} = \mathbf{Q}_L^T\mathbf{T}^T\mathbf{P}_p\mathbf{T}\mathbf{Q}_L \sim \text{Beta}_L\left(\frac{p}{2}, \frac{N-p}{2}\right).$$

The fact that the matrix-variate beta density is a function solely of its eigenvalues allows us to apply a standard result, originally proved by P. L. Hsu (cf. Theorem 2 in [176]; see also [13]) to obtain the joint density of the eigenvalues of $\mathbf{B}$.

Lemma *If* $\mathbf{B} \sim \text{Beta}_L\left(\frac{p}{2}, \frac{N-p}{2}\right)$, *then the joint density of the eigenvalues* $1 \geq \lambda_1 \geq \cdots \geq \lambda_L \geq 0$ *of* $\mathbf{B}$ *is*

$$f(\lambda_1, \ldots, \lambda_L) =$$

$$\frac{\pi^{L^2/2}\Gamma_L\left(\frac{N}{2}\right)}{\Gamma_L\left(\frac{p}{2}\right)\Gamma_L\left(\frac{N-p}{2}\right)\Gamma_L\left(\frac{L}{2}\right)} \prod_{l=1}^{L}\left[\lambda_l^{(p-L-1)/2}(1-\lambda_l)^{(N-p-L-1)/2}\right]\prod_{l<m}^{L}(\lambda_l - \lambda_m).$$

Some properties of the Wishart distribution have a similar counterpart for the matrix-valued beta distribution. For example, Bartlett's factorization for Wishart matrices shows that if $\mathbf{S} = \mathbf{R}^T\mathbf{R} \sim \mathcal{W}_L(\mathbf{I}, N)$, where $\mathbf{R}$ is upper triangular with positive elements along its diagonal, then r_{ll}^2 is χ_{N-l+1}^2 for $l = 1, \ldots, L$. The following theorem, due to Kshirsagar [207] (see also Theorem 3.3.3 in [244]), provides a similar result for the matrix-valued Beta.

Theorem D.1 *If* $\mathbf{B} \sim \text{Beta}_L\left(\frac{p}{2}, \frac{N-p}{2}\right)$ *is factored as* $\mathbf{B} = \mathbf{R}^T\mathbf{R}$, *where* $\mathbf{R}$ *is upper triangular with positive diagonal elements, then* $r_{11}, \ldots, r_{LL}$, *are all independent and* $r_{ll}^2 \sim \text{Beta}\left(\frac{p-l+1}{2}, \frac{N-p}{2}\right)$, *for* $l = 1, \ldots, L$.

Note that the distribution of the elements r_{lm}, $l \neq m$, of $\mathbf{R}$ in the factorization $\mathbf{B} = \mathbf{R}^T\mathbf{R}$ is, in general, unknown. This is in contrast to Bartlett's factorization result for Wishart matrices.

From this result, the determinant of the beta-distributed random matrix

$$\det(\mathbf{B}) = \frac{\det(\mathbf{U}\mathbf{P}_r\mathbf{U}^T)}{\det(\mathbf{U}\mathbf{U}^T)} \stackrel{d}{=} \det(\mathbf{T}^T\mathbf{P}_r\mathbf{T})$$

is equal in distribution to a product of independent univariate betas. That is,

$$\det(\mathbf{B}) \stackrel{d}{=} \prod_{l=1}^{L} b_l,$$

where

$$b_l \sim \text{Beta}\left(\frac{p-l+1}{2}, \frac{N-p}{2}\right).$$

The statistic $\det(\mathbf{B})$ is a Wilks lambda statistic, and this result is often called the null distribution of the Wilks lambda statistic, as the projection $\mathbf{P}_p$ is assumed deterministic, or if stochastic, independent of $\mathbf{U}$. This null distribution is fundamental in the analysis of the coherence statistic under the null hypothesis that two normal random matrices are independent.

D.7.5 The Matrix F-Distribution

Define the matrix $\mathbf{F} = \mathbf{B}^{-1} - \mathbf{I}_L$ with inverse $\mathbf{B} = (\mathbf{I}_L + \mathbf{F})^{-1}$. The determinant of the Jacobian of the transformation is $\det(J(\mathbf{B} \to \mathbf{F})) = \det(\mathbf{I}_L + \mathbf{F})^{-L-1}$. It is then a change of variable formula to show that $\mathbf{F} \sim \mathcal{F}_L\left(\frac{N-p}{2}, \frac{p}{2}\right)$, meaning $\mathbf{F}$ is a matrix F-distributed random matrix with pdf [98]

$$f(\mathbf{F}) = \frac{\Gamma_L\left(\frac{N}{2}\right)}{\Gamma_L\left(\frac{p}{2}\right)\Gamma_L\left(\frac{N-p}{2}\right)} \frac{\det(\mathbf{F})^{(p-L-1)/2}}{\det(\mathbf{I}_L + \mathbf{F})^{N/2}}, \quad \mathbf{F} \succ \mathbf{0}.$$

The matrix-variate F-distribution is also known as the matrix-variate beta distribution of type II [152].

From the stochastic representation of $\mathbf{B}$ as $\mathbf{B} = \mathbf{T}^T\mathbf{P}_p\mathbf{T}$ and $\mathbf{F} = \mathbf{B}^{-T/2}(\mathbf{I}_L - \mathbf{B})\mathbf{B}^{-1/2}$, the stochastic representation of $\mathbf{F}$ is

$$\mathbf{F} = (\mathbf{T}^T\mathbf{P}_p\mathbf{T})^{-T/2}\mathbf{T}^T\mathbf{P}_p^{\perp}\mathbf{T}(\mathbf{T}^T\mathbf{P}_p\mathbf{T})^{-1/2}.$$

So the the matrix F-distribution is the distribution for this statistic.

D.7.6 Special Cases

It appears that these distribution results require multivariate normality. But, in fact, they hold for any matrix-variate distribution for which the factorization $\mathbf{X}^T = \mathbf{TR}$ returns an orthogonal slice $\mathbf{T}$ whose distribution is invariant to left orthogonal transformation by an orthogonal matrix. There is nothing special about the projection $\mathbf{P}_p$. In fact, it may be deterministically specified, or it may be randomly drawn, independently of $\mathbf{X}$. The argument is this. Conditioned on $\mathbf{P}_p$, the distributions of $\mathbf{B}$ and $\mathbf{F}$ depend only on p, the rank of $\mathbf{P}_p$, leaving the unconditional distributions as given. The following paragraphs demonstrate the power of these results by working out several special cases for special constructions of $\mathbf{X}$ and $\mathbf{P}_p$.

Subset Selection. Let $\mathbf{P}_p = \mathbf{E}_p\mathbf{E}_p^T$, with $\mathbf{E}_p$ the $N \times p$ selection matrix defined as

$$\mathbf{E}_p = \begin{bmatrix} \mathbf{I}_p \\ \mathbf{0}_{(N-p)\times p} \end{bmatrix}.$$

Parse $\mathbf{U}$ as $\mathbf{U} = [\mathbf{U}_p \ \mathbf{U}_{N-p}]$, so that the $L \times p$ matrix $\mathbf{U}_p$ consists of the first p columns of $\mathbf{U}$ (the head), and the $L \times (N-p)$ matrix $\mathbf{U}_{N-p}$ consists of the last $N-p$ columns of $\mathbf{U}$ (the tail). The Gramian $\mathbf{U}\mathbf{U}^T$ may be written $\mathbf{U}\mathbf{U}^T = \mathbf{U}_p\mathbf{U}_p^T + \mathbf{U}_{N-p}\mathbf{U}_{N-p}^T$, where these two terms are independent, and $\mathbf{U}_p\mathbf{U}_p^T = \mathbf{U}\mathbf{P}_p\mathbf{U}^T$. It follows that

$$(\mathbf{U}_p\mathbf{U}_p^T + \mathbf{U}_{N-p}\mathbf{U}_{N-p}^T)^{-1/2}\mathbf{U}_p\mathbf{U}_p^T(\mathbf{U}_p\mathbf{U}_p^T + \mathbf{U}_{N-p}\mathbf{U}_{N-p}^T)^{-T/2} \sim \text{Beta}_L\left(\frac{p}{2}, \frac{N-p}{2}\right).$$

Again, we could have started with $\mathbf{\Sigma}^{1/2}\mathbf{U}$ and derived the same result. That is, for $\mathbf{X} \sim \mathcal{N}_{L\times p}(\mathbf{0}, \mathbf{I}_p \otimes \mathbf{\Sigma})$ and $\mathbf{Y} \sim \mathcal{N}_{L\times(N-p)}(\mathbf{0}, \mathbf{I}_{N-p} \otimes \mathbf{\Sigma})$ independent random matrices, we have

$$(\mathbf{S_{xx}}+\mathbf{S_{yy}})^{-1/2}\mathbf{S_{xx}}(\mathbf{S_{xx}}+\mathbf{S_{yy}})^{-T/2} \sim \text{Beta}_L\left(\frac{p}{2},\frac{N-p}{2}\right),$$

where $\mathbf{S_{xx}} = \mathbf{XX}^T$ is the $L \times L$ (scaled) sample covariance matrix of $\mathbf{X}$ and $\mathbf{S_{yy}} = \mathbf{YY}^T$ is the $L \times L$ (scaled) sample covariance matrix of $\mathbf{Y}$. There is nothing fundamental about the parameterizations p and $N-p$. If they are replaced by N_x and N_y, then $\mathbf{S_{xx}} = \mathbf{XX}^T$ and $\mathbf{S_{yy}} = \mathbf{YY}^T$ are distributed as $\mathcal{W}_L(\mathbf{\Sigma}, N_x)$ and $\mathcal{W}_L(\mathbf{\Sigma}, N_y)$, respectively, which yields

$$(\mathbf{S_{xx}}+\mathbf{S_{yy}})^{-1/2}\mathbf{S_{xx}}(\mathbf{S_{xx}}+\mathbf{S_{yy}})^{-T/2} \sim \text{Beta}_L\left(\frac{N_x}{2},\frac{N_y}{2}\right).$$

This result was proved in [253] (see also Theorem 3.3.1 in [244]).

In [253], it is also shown that if $\mathbf{S_{xx}} = \mathbf{XX}^T \sim \mathcal{W}_L(\sigma^2\mathbf{I}_L, N_x)$ and $\mathbf{S_{yy}} = \mathbf{YY}^T \sim \mathcal{W}_L(\sigma^2\mathbf{I}_L, N_y)$ for some positive σ^2 and $\mathbf{S}_{\mathbf{yy}}^{1/2}$ is taken to be the unique symmetric positive definite square root of $\mathbf{S_{yy}}$, then

$$\mathbf{F} = \mathbf{S}_{\mathbf{yy}}^{-1/2}\mathbf{S_{xx}}\mathbf{S}_{\mathbf{yy}}^{-1/2} \sim \mathcal{F}_L\left(\frac{N_x}{2},\frac{N_y}{2}\right). \tag{D.9}$$

However, this result only holds for $\mathbf{\Sigma} = \sigma^2\mathbf{I}_L$ and further $\mathbf{F}$ is not independent of $\mathbf{S_{xx}}+\mathbf{S_{yy}}$. For arbitrary $\mathbf{\Sigma} \neq \sigma^2\mathbf{I}_L$, the density of $\mathbf{F}$ as defined in (D.9) is not $\mathcal{F}_L\left(\frac{N_x}{2},\frac{N_y}{2}\right)$ (the density is given by Olkin and Rubin [253, Equation 3.2]).

Squared Canonical Correlation Matrix. Begin with the $L_x \times N$ random matrix $\mathbf{X} \sim \mathcal{N}_{L_x\times N}(\mathbf{0}, \mathbf{I}_N \otimes \mathbf{\Sigma_{xx}})$ and the $L_y \times N$ random matrix $\mathbf{Y} \sim \mathcal{N}_{L_y\times N}(\mathbf{0}, \mathbf{I}_N \otimes \mathbf{\Sigma_{yy}})$, where $\mathbf{\Sigma_{xx}}$ is the $L_x \times L_x$ covariance matrix of each of the independent columns of $\mathbf{X}$ and $\mathbf{\Sigma_{yy}}$ is the $L_y \times L_y$ covariance matrix for each of the columns of $\mathbf{Y}$. It is as if $\mathbf{X}^T = \mathbf{T}_x\mathbf{R}_x\mathbf{\Sigma}_{\mathbf{xx}}^{T/2}$, with $\mathbf{R}_x$ $L_x \times L_x$ upper triangular and $\mathbf{T}_x$ an $N \times L_x$ slice of an orthogonal matrix, and $\mathbf{Y}^T = \mathbf{T}_y\mathbf{R}_y\mathbf{\Sigma}_{\mathbf{yy}}^{T/2}$, with $\mathbf{R}_y$ $L_y \times L_y$ upper triangular and $\mathbf{T}_y$ an $N \times L_y$ orthogonal slice. The Gramians are $\mathbf{XX}^T = \mathbf{\Sigma}_{\mathbf{xx}}^{1/2}\mathbf{R}_x^T\mathbf{R}_x\mathbf{\Sigma}_{\mathbf{xx}}^{T/2}$ and $\mathbf{YY}^T = \mathbf{\Sigma}_{\mathbf{yy}}^{1/2}\mathbf{R}_y^T\mathbf{R}_y\mathbf{\Sigma}_{\mathbf{yy}}^{T/2}$. Without loss of generality, we may model the square roots of these Gramians as $(\mathbf{XX}^T)^{1/2} = \mathbf{\Sigma}_{\mathbf{xx}}^{1/2}\mathbf{R}_x^T\mathbf{Q}_x$ and $(\mathbf{Y}^T\mathbf{Y})^{1/2} = \mathbf{\Sigma}_{\mathbf{yy}}^{1/2}\mathbf{R}_y^T\mathbf{Q}_y$, where the $\mathbf{Q}_x$ and $\mathbf{Q}_y$ are respectively $L_x \times L_x$ and $L_y \times L_y$ orthogonal matrices. Then, the sample coherence matrix $\hat{\mathbf{C}} = (\mathbf{XX}^T)^{-1/2}\mathbf{XY}^T(\mathbf{YY}^T)^{-T/2}$ resolves as $\mathbf{Q}_x^T\mathbf{T}_x^T\mathbf{T}_y\mathbf{Q}_y$. The square of this matrix is

$$\hat{\mathbf{C}}\hat{\mathbf{C}}^T = \mathbf{Q}_x^T\mathbf{T}_x^T\mathbf{T}_y\mathbf{T}_y^T\mathbf{T}_x\mathbf{Q}_x = \mathbf{Q}_x^T\mathbf{T}_x^T\mathbf{P}_{L_y}\mathbf{T}_x\mathbf{Q}_x,$$

where $\mathbf{P}_{L_y}$ is the rank-L_y projection $\mathbf{P}_{L_y} = \mathbf{T}_y\mathbf{T}_y^T$. Under the null hypothesis that the random matrices $\mathbf{X}$ and $\mathbf{Y}$ are uncorrelated, and therefore independent in the MVN model, the projection $\mathbf{P}_{L_y}$ is independent of $\mathbf{X}$, and this squared coherence matrix is distributed as

$$\hat{\mathbf{C}}\hat{\mathbf{C}}^T \sim \text{Beta}_{L_x}\left(\frac{L_y}{2}, \frac{N-L_y}{2}\right).$$

This is said to be the null distribution of the squared coherence matrix.

Let us assume $L_x \leq L_y$ and define the squared canonical correlations μ_l, $l = 1, \ldots, L_x$. When $\mathbf{X}$ and $\mathbf{Y}$ are independent normal random matrices, the joint distribution of the squared sample canonical correlations was derived by P. L. Hsu in 1939 [176] and is [244, pp. 559]

$$f(\mu_1, \ldots, \mu_{L_x}) = \frac{\pi^{L_x^2/2}\Gamma_{L_x}\left(\frac{N}{2}\right)}{\Gamma_{L_x}\left(\frac{L_x}{2}\right)\Gamma_{L_x}\left(\frac{N-L_y}{2}\right)\Gamma_{L_x}\left(\frac{L_y}{2}\right)} \times \prod_{l=1}^{L_x}\left[\mu_l^{(L_y-L_x-1)/2}(1-\mu_l)^{(N-L_y-L_x-1)/2}\right]\prod_{l<m}^{L_x}(\mu_l-\mu_m), \tag{D.10}$$

where $\Gamma_{L_x}(x)$ denotes the multivariate gamma function defined in (D.7).

In theory, the marginal distribution for any single squared canonical correlation, or for any ordered or unordered subset of them, can be obtained by integrating the joint pdf in (D.10). However, the integrals involved cannot be solved in closed-form, in general. The marginal for the largest squared canonical correlation μ_1, which is of particular interest in some statistical tests, was derived by T. Sugiyama in terms of the hypergeometric function of a matrix argument [338]. This result is presented in the following proposition.

Proposition D.1 *The largest eigenvalue μ_1 of $\hat{\mathbf{C}}\hat{\mathbf{C}}^T$ has the following density:*

$$f(\mu_1) = K\,\mu_1^{L_x(L_x/2-1)}(1-\mu_1)^{(N-L_x-L_y-1)/2} \times {}_2F_1\left(\frac{L_x+L_y-N+1}{2}, \frac{L_x-1}{2}; \frac{L_x+L_y+1}{2}; \mu_1\mathbf{I}_{L_x-1}\right), \tag{D.11}$$

where ${}_2F_1$ is the hypergeometric function of matrix argument and K is a normalizing constant.

Efficient algorithms for computing these functions have been proposed by Koev and Edelman [203]. Using the MATLAB code provided in [203] to compute the hypergeometric function, this analytic expression perfectly matches the simulation results as shown in Fig. D.4 for an example with two independent normal matrices of dimensions 4×10 ($L_x = L_y = 4$, $N = 10$).

When $L_x = L_y = 2$, the following result fully characterizes the distribution of the squared coherence matrix.

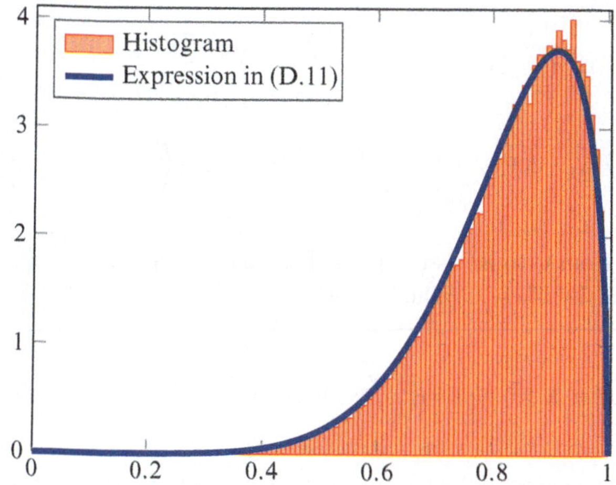

Fig. D.4 Density of the largest squared canonical correlation when $p = 4$ and $N = 10$

Example D.1 Let $\mathbf{X} \sim \mathcal{N}_{2\times N}(\mathbf{0}, \mathbf{I}_N \otimes \mathbf{I}_2)$ and $\mathbf{Y} \sim \mathcal{N}_{2\times N}(\mathbf{0}, \mathbf{I}_N \otimes \mathbf{I}_2)$. The columns of $\mathbf{X}$ and $\mathbf{Y}$ are i.i.d. bivariate random vectors, and the coherence matrix $\hat{\mathbf{C}} = \left(\mathbf{X}\mathbf{X}^T\right)^{-1/2}\left(\mathbf{X}\mathbf{Y}^T\right)\left(\mathbf{Y}\mathbf{Y}^T\right)^{-1/2}$ is a 2×2 matrix. Factor the squared coherence matrix as $\hat{\mathbf{C}}\hat{\mathbf{C}}^T = \mathbf{R}^T\mathbf{R}$, where

$$\mathbf{R} = \begin{bmatrix} r_{11} & r_{12} \\ 0 & r_{22} \end{bmatrix}.$$

Then the distributions of r_{11}, r_{22}, and r_{12} are

$$r_{11}^2 \sim \text{Beta}\left(1, \frac{N-2}{2}\right), \quad r_{22}^2 \sim \text{Beta}\left(\frac{1}{2}, \frac{N-2}{2}\right), \quad r_{12}^2 \stackrel{d}{=} v^2(1 - r_{11}^2)(1 - r_{22}^2),$$

where $v^2 \sim \text{Beta}\left(\frac{1}{2}, \frac{N-3}{2}\right)$. Further, r_{11}^2, r_{22}^2, and v^2 are independent.

The proof of this result proceeds as follows. The 2×2 squared coherence matrix $\hat{\mathbf{C}} = \mathbf{R}^T\mathbf{R}$ is

$$\hat{\mathbf{C}} = \begin{bmatrix} r_{11}^2 & r_{11}r_{12} \\ r_{11}r_{12} & r_{12}^2 + r_{22}^2 \end{bmatrix} \sim \text{Beta}_2\left(1, \frac{N-2}{2}\right),$$

with density

$$f(\hat{\mathbf{C}}) = K \det(\hat{\mathbf{C}})^{-1/2} \det(\mathbf{I}_2 - \hat{\mathbf{C}})^{(N-5)/2},$$

where $K = \frac{\Gamma_2\left(\frac{N}{2}\right)}{\Gamma_2(1)\Gamma_2\left(\frac{N}{2}-1\right)}$ is a normalizing constant. Then,

$$\det(\hat{\mathbf{C}}) = r_{11}^2 r_{22}^2, \quad \det(\mathbf{I}_2 - \hat{\mathbf{C}}) = (1 - r_{11}^2)(1 - r_{22}^2)\left(1 - \frac{r_{12}^2}{(1 - r_{11}^2)(1 - r_{22}^2)}\right).$$

The Jacobian from $\hat{\mathbf{C}}$ to $\mathbf{R}^T$ is (cf. [244, Theorem 2.1.9]) $J(\hat{\mathbf{C}} \to \mathbf{R}^T) = 2^2 r_{11}^2 r_{22}$, so the joint density of r_{11}, r_{22}, and r_{12} is

$$f(r_{11}, r_{22}, r_{12}) = K\, 2^2 r_{11}(1-r_{11}^2)^{\frac{N-5}{2}}(1-r_{22}^2)^{\frac{N-5}{2}}\left(1 - \frac{r_{12}^2}{(1 - r_{11}^2)(1 - r_{22}^2)}\right)^{\frac{N-5}{2}}.$$

Now, make a change of variables from (r_{11}, r_{22}, r_{12}) to (r_{11}, r_{22}, v), where v is

$$v^2 = \frac{r_{12}^2}{(1 - r_{11}^2)(1 - r_{22}^2)}.$$

Then $dr_{12} = (1 - r_{11}^2)^{1/2}(1 - r_{22}^2)^{1/2} dv$. Hence, the joint density of r_{11}, r_{22}, and v is

$$f(r_{11}, r_{22}, v) = K\, 2^2 r_{11}(1 - r_{11}^2)^{\frac{N-4}{2}}(1 - r_{22}^2)^{\frac{N-4}{2}}\left(1 - v^2\right)^{\frac{N-5}{2}}.$$

This shows that r_{11}, r_{22}, and v are all independent. Further, it is immediate to show $r_{11}^2 \sim \text{Beta}\left(1, \frac{N-2}{2}\right)$, $r_{22}^2 \sim \text{Beta}\left(\frac{1}{2}, \frac{N-2}{2}\right)$, and $v^2 \sim \text{Beta}\left(\frac{1}{2}, \frac{N-3}{2}\right)$, thus proving the result.[3] The distribution of r_{11}^2 and r_{22}^2 is obviously in agreement with the result given by Appendix D.1 when $L = 2$.

D.7.7 Summary

The random matrix $\mathbf{X} = \boldsymbol{\Sigma}^{1/2}\mathbf{U} \sim \mathcal{N}_{L\times N}(\mathbf{0}, \mathbf{I}_N \otimes \boldsymbol{\Sigma})$ may be parsed in two ways. In one of the parsings, N is decomposed as $N = N_f + N_g$, so that

$$\mathbf{X} = \begin{bmatrix}\mathbf{F} & \mathbf{G}\end{bmatrix}$$

[3] It is useful to note that if $x^2 \sim \text{Beta}(\alpha, \beta)$, then the density of x is

$$f(x) = 2\frac{\Gamma(\alpha+\beta)}{\Gamma(\alpha)\Gamma(\beta)}x^{2\alpha-1}\left(1 - x^2\right)^{\beta-1}.$$

where $\mathbf{F} \sim \mathcal{N}_{L\times N_f}(\mathbf{0}, \mathbf{I}_{N_f} \otimes \boldsymbol{\Sigma})$ and $\mathbf{G} \sim \mathcal{N}_{L\times N_g}(\mathbf{0}, \mathbf{I}_{N_g} \otimes \boldsymbol{\Sigma})$, with $N_f, N_g \geq L$. In the other parsing, L is decomposed as $L = L_y + L_z$, so that

$$\mathbf{X} = \begin{bmatrix} \mathbf{Y} \\ \mathbf{Z} \end{bmatrix}$$

with $\mathbf{Y} \sim \mathcal{N}_{L_y\times N}(\mathbf{0}, \boldsymbol{\Sigma}_{\mathbf{yy}})$, $\mathbf{Z} \sim \mathcal{N}_{L_z\times N}(\mathbf{0}, \boldsymbol{\Sigma}_{\mathbf{zz}})$, and $\boldsymbol{\Sigma} = \text{blkdiag}(\boldsymbol{\Sigma}_{\mathbf{yy}}, \boldsymbol{\Sigma}_{\mathbf{zz}})$.

We may summarize the most important consequences of the matrix-valued spherically invariant experiment as follows:

(a) For $\mathbf{X} \sim \mathcal{N}_{L\times N}(\mathbf{0}, \mathbf{I}_N \otimes \boldsymbol{\Sigma})$ and $\mathbf{P}_p$ deterministic or stochastically drawn independent of $\mathbf{X}$,
  (a1) $\boldsymbol{\Sigma}^{-1/2}\mathbf{X}\mathbf{X}^T\boldsymbol{\Sigma}^{-T/2} \sim \mathcal{W}_L(\mathbf{I}_L, N)$ and $\mathbf{X}\mathbf{X}^T \sim \mathcal{W}_L(\boldsymbol{\Sigma}, N)$
  (a2) $(\mathbf{X}\mathbf{X}^T)^{-1/2}\mathbf{X}\mathbf{P}_p\mathbf{X}^T(\mathbf{X}\mathbf{X}^T)^{-T/2} \sim \text{Beta}_L\left(\frac{p}{2}, \frac{N-p}{2}\right)$
  (a3) $(\mathbf{X}(\mathbf{I}_N - \mathbf{P}_p)\mathbf{X}^T)^{-1/2}\mathbf{X}\mathbf{P}_p\mathbf{X}^T(\mathbf{X}(\mathbf{I}_N - \mathbf{P}_p)\mathbf{X}^T)^{-T/2} \sim \mathcal{F}_L\left(\frac{p}{2}, \frac{N-p}{2}\right)$

(b) For $\mathbf{X} = [\mathbf{F}\ \mathbf{G}]$, where $\mathbf{F} \sim \mathcal{N}_{L\times N_f}(\mathbf{0}, \mathbf{I}_{N_f} \otimes \boldsymbol{\Sigma})$ and $\mathbf{G} \sim \mathcal{N}_{L\times N_g}(\mathbf{0}, \mathbf{I}_{N_g} \otimes \boldsymbol{\Sigma})$,
  (b1) $(\boldsymbol{\Sigma}^{-1/2}\mathbf{F}\mathbf{F}^T\boldsymbol{\Sigma}^{-T/2} + \boldsymbol{\Sigma}^{-1/2}\mathbf{G}\mathbf{G}^T\boldsymbol{\Sigma}^{-T/2}) \sim \mathcal{W}_L(\mathbf{I}_L, N)$, which is the sum of independent $\mathcal{W}_L(\mathbf{I}_L, N_f)$ and $\mathcal{W}_L(\mathbf{I}_L, N_g)$ random matrices
  (b2) $(\mathbf{F}\mathbf{F}^T + \mathbf{G}\mathbf{G}^T)^{-1/2}\mathbf{F}\mathbf{F}^T(\mathbf{F}\mathbf{F}^T + \mathbf{G}\mathbf{G}^T)^{-1/2} \sim \text{Beta}_L\left(\frac{N_f}{2}, \frac{N_g}{2}\right)$
  (b3) When $\boldsymbol{\Sigma} = \sigma^2\mathbf{I}_L$, $(\mathbf{G}\mathbf{G}^T)^{-1/2}\mathbf{F}\mathbf{F}^T(\mathbf{G}\mathbf{G}^T)^{-T/2} \sim \mathcal{F}_L\left(\frac{N_f}{2}, \frac{N_g}{2}\right)$

(c) For $\mathbf{X} = \begin{bmatrix} \mathbf{Y} \\ \mathbf{Z} \end{bmatrix}$, where $\mathbf{Y} \sim \mathcal{N}_{L_y\times N}(\mathbf{0}, \mathbf{I}_N \otimes \boldsymbol{\Sigma}_{yy})$ and $\mathbf{Z} \sim \mathcal{N}_{N_z\times N}(\mathbf{0}, \mathbf{I}_N \otimes \boldsymbol{\Sigma}_{zz})$,
  (c1) $\hat{\mathbf{C}}\hat{\mathbf{C}}^T = (\mathbf{Y}\mathbf{Y}^T)^{-1/2}\mathbf{Y}\mathbf{Z}^T(\mathbf{Z}\mathbf{Z}^T)^{-1}\mathbf{Z}\mathbf{Y}^T(\mathbf{Y}\mathbf{Y}^T)^{-T/2} \sim \text{Beta}_{L_y}\left(\frac{L_z}{2}, \frac{N-L_z}{2}\right)$.

These specialize to the results of the bivariate experiment in Sect. D.5.6 by letting $L = 2$ and $N = 1$ and to the MVN experiment in Sect. D.6.6 by letting $L > 2$ and $N = 1$.

D.8 Spherical, Elliptical, and Compound Distributions

D.8.1 Spherical Distributions

The spherically invariant MVN random vector is special because so many important functions of it remain spherically invariant. This point is evident in Sects. D.5 through D.7, where the spherically invariant χ^2, Beta, t-, and F-distributions are derived in the context of spherically invariant normal experiments. Moreover, for physical modeling, it is sometimes desired to retain the spherically invariant contours (or level sets) of the MVN distribution while allowing for decays in these contours more general than the decay $\exp(-\mathbf{x}^T\mathbf{x})$. One particularly flexible model is the multivariate t-distribution with r degrees of freedom and pdf given by

$$f(\mathbf{x}) = \frac{\Gamma\left(\frac{N+r}{2}\right)}{(\pi r)^{N/2}\,\Gamma\left(\frac{r}{2}\right)\left(1+\frac{1}{r}\mathbf{x}^T\mathbf{x}\right)^{(N+r)/2}}. \tag{D.12}$$

When (D.12) is specialized to $r = 1$, this is the multivariate Cauchy distribution; when the dimension of $\mathbf{x}$ is specialized to $L = 1$, this is the univariate t-distribution with r degrees of freedom in (D.4).

D.8.2 Elliptical Distributions

The class of elliptical distributions generalizes the class of spherical distributions. The equal density contours of elliptical distributions have the same elliptical shape as the multivariate normal, but their heavier tails make them useful for robust statistics. The density of an elliptical symmetric (ES) random vector $\mathbf{x} \in \mathbb{R}^L$ with mean vector $\boldsymbol{\mu}$ and scatter matrix $\boldsymbol{\Sigma}$ may be written as [250]

$$f(\mathbf{x}) = K\,\det(\boldsymbol{\Sigma})^{-1} g_{\mathbf{x}}\left((\mathbf{x}-\boldsymbol{\mu})^T\boldsymbol{\Sigma}^{-1}(\mathbf{x}-\boldsymbol{\mu})\right), \tag{D.13}$$

where K is a normalizing constant and $g_{\mathbf{x}}(\cdot) : [0,\infty) \to [0,\infty)$ is any density generator function ensuring that (D.13) integrates to one. The standard Gaussian $\mathbf{x} \sim \mathcal{N}_L(\boldsymbol{\mu}, \boldsymbol{\Sigma})$ is a particular case of an ES distribution in which $g_{\mathbf{x}}(a) = e^{-a/2}$ and $K = \pi^{-L/2}$. The function $g_{\mathbf{x}}(\cdot)$ may be used to describe distributions with heavier or lighter tails than the Gaussian distribution. Examples are provided in [250]. A particularly interesting ES distribution is the angular central Gaussian.

The Angular Central Gaussian Distribution. Let $\mathbf{x} \sim \mathcal{N}_L(\mathbf{0}, \boldsymbol{\Sigma})$ be a random L-dimensional normal vector of zero mean and covariance $\boldsymbol{\Sigma}$. The aim is to write this as $\mathbf{x} = \mathbf{t}r$ with $\mathbf{t}$ non-uniformly distributed on the sphere S^{L-1} and r non-negative. The distribution of $\mathbf{x}$ is

$$p(\mathbf{x}) = p(\mathbf{t}r) = \frac{1}{(2\pi)^{L/2}\det(\boldsymbol{\Sigma})^{1/2}} \exp\left(-r^2\frac{\mathbf{t}^T\boldsymbol{\Sigma}^{-1}\mathbf{t}}{2}\right). \tag{D.14}$$

The marginal distribution of $\mathbf{t}$ is obtained by integrating (D.14) along the radial direction using the volume form, which is given by $dV = r^{L-1}dr\,d\Omega$, where $d\Omega = (\sin\theta_1)^{L-2}(\sin\theta_2)^{L-3}\cdots(\sin\theta_{L-2})d\theta_1\cdots d\theta_{L-1}$.

The distribution of $\mathbf{t}$ can be computed as

$$p(\mathbf{t}) = \frac{1}{(2\pi)^{L/2}\det(\boldsymbol{\Sigma})^{1/2}} \int_0^\infty \exp\left\{-r^2\frac{\mathbf{x}^T\boldsymbol{\Sigma}^{-1}\mathbf{x}}{2}\right\} r^{L-1}dr.$$

Taking into account that

$$\int_0^\infty x^{L-1}e^{-ax^2}dx = \frac{\Gamma(L/2)}{2a^{L/2}},$$

the pdf of the unit-norm random vector $\mathbf{t}$ is

$$p(\mathbf{t}) = \frac{\Gamma(L/2)}{2\pi^{L/2}}\det(\boldsymbol{\Sigma})^{-1/2}\left(\mathbf{t}^T\boldsymbol{\Sigma}^{-1}\mathbf{t}\right)^{-L/2}, \quad \mathbf{t}^T\mathbf{t} = 1.$$

This is the pdf for the angular central Gaussian distribution, denoted $ACG(\boldsymbol{\Sigma})$. Note that the angular central Gaussian is an elliptical distribution with density generator function $g_{\mathbf{x}}(a) = a^{-L/2}$ (cf. Eq. (D.13)). When $\boldsymbol{\Sigma} = \mathbf{I}_L$, then this is the uniform distribution on S^{L-1}, with pdf

$$f(\mathbf{t}) = \frac{\Gamma(L/2)}{2\pi^{L/2}}, \quad \mathbf{t}^T\mathbf{t} = 1.$$

The family of angular central Gaussian distributions is an alternative to the family of Bingham distributions for directional statistics [226, 373]. An angular central Gaussian $\mathbf{t}$ with parameter $\boldsymbol{\Sigma}$ can be transformed to a uniform distribution on the sphere by the transformation $\tilde{\mathbf{t}} = \boldsymbol{\Sigma}^{-1/2}\mathbf{t}/\left(\mathbf{t}^T\boldsymbol{\Sigma}^{-1}\mathbf{t}\right)^{1/2}$. A complete statistical analysis of the angular central Gaussian can be found in [352]. In particular, it is shown in [352] that the maximum likelihood estimate of $\boldsymbol{\Sigma}$ based on i.i.d. samples $\mathbf{t}_n \sim ACG(\boldsymbol{\Sigma}), n = 1, \ldots, N$, is the solution to the equation

$$\hat{\boldsymbol{\Sigma}} = \frac{L}{N}\sum_{n=1}^{N}\frac{1}{\mathbf{t}_n^T\hat{\boldsymbol{\Sigma}}^{-1}\mathbf{t}_n}\mathbf{t}_n\mathbf{t}_n^T,$$

which is known as Tyler's estimator, widely used in robust statistics.

The following theorem proved by Y. Chikuse in [73] derives a matrix-variate version of the angular central Gaussian distribution.

Theorem (Theorem 2.2 in [73]) *Suppose* $\mathbf{X} \sim N_{L\times N}(\mathbf{0}, \mathbf{I}_N \otimes \boldsymbol{\Sigma})$, *is a matrix-variate zero-mean normal distribution, where* $\boldsymbol{\Sigma}$ *is an* $L \times L$ *covariance matrix. Assume that* $N \geq L$ (rank($\mathbf{X}$) $= L$). *Write the unique polar decomposition of* $\mathbf{X}^T$ *as*

$$\mathbf{X}^T = \mathbf{T}\mathbf{R},$$

where $\mathbf{T} = \mathbf{X}^T(\mathbf{X}\mathbf{X}^T)^{-1/2}$, *and* $\mathbf{R} = (\mathbf{X}\mathbf{X}^T)^{1/2}$, *so that* $\mathbf{T} \in \mathbb{S}t(L, \mathbb{R}^N)$ *and* $\mathbf{R} = (\mathbf{X}\mathbf{X}^T)^{1/2}$ *denotes the unique positive definite square root of the* $L \times L$ *positive definite matrix* $\mathbf{X}\mathbf{X}^T$. *Then, the marginal of the orientation matrix* $\mathbf{T}$ *has a matrix angular central Gaussian distribution whose density is*

$$f(\mathbf{T}) = K \det(\mathbf{\Sigma})^{-N/2} \det\left(\mathbf{T}^T\mathbf{\Sigma}^{-1}\mathbf{T}\right)^{-L/2},$$

where $K = \text{Vol}\left(\mathbb{S}t(L, \mathbb{R}^N)\right)^{-1}$. *This distribution is denoted in this book as* $MACG(\mathbf{\Sigma})$.

A special case of this result, for $\mathbf{\Sigma} = \mathbf{I}_L$, gives the uniform pdf for orthogonal frames on the Stiefel manifold. The marginal of $\mathbf{R}$ can also be expressed in closed-form and involves the hypergeometric function [73].

D.8.3 Compound Distributions

Among the class of spherically contoured pdfs are random vectors generated as

$$\mathbf{z} = \sqrt{\tau}\mathbf{x}, \tag{D.15}$$

where τ and $\mathbf{x}$ are independent and distributed, respectively, as $\tau \sim f(\tau)$ and $\mathbf{x} \sim \mathcal{N}_L(\mathbf{0}, \mathbf{I}_L)$. If the normal vector is first colored by deterministic $\mathbf{\Sigma}^{1/2}$ to produce $\mathbf{x} \sim \mathcal{N}_L(\mathbf{0}, \mathbf{\Sigma})$, then $\sqrt{\tau}\mathbf{x}$ has an elliptically contoured pdf. The class of elliptically contoured distributions in Sect. D.8.2 admits the following compound representation:

$$\mathbf{z} = \sqrt{\tau}\mathbf{A}\mathbf{t} + \boldsymbol{\mu},$$

where $\mathbf{\Sigma} = \mathbf{A}\mathbf{A}^T$, $\mathbf{t}$ is uniformly distributed on S^{L-1}, τ is a non-negative real random variable, and $\boldsymbol{\mu}$ is the mean of the distribution. As an example of a compound Gaussian distribution, if $\mathbf{x} \sim \mathcal{N}_L(\mathbf{0}, \mathbf{I}_L)$ and τ in (D.15) takes the values σ_1^2 and σ_2^2, with $\sigma_2^2 \gg \sigma_1^2$, with probabilities $1-\epsilon$ and ϵ, then we have a Gaussian mixture model or ϵ-contaminated normal distribution, often used in sparse modeling or sparse deconvolution problems.

Typically, these compound distributions arise in problems with random effects, such as a random scaling of a random vector $\mathbf{x}$ by gain τ with assumed prior distribution $f(\tau)$. For instance, clutter is modeled in radar problems as a compound Gaussian vector [131], where $\mathbf{x}$ in (D.15) is called the *speckle* and the scale random variable τ, independent of $\mathbf{x}$, is called the *texture*. Priors for the texture of particular interest to model radar clutter are the gamma distribution [371], the inverse gamma distribution [19], and the inverse Gaussian [251].

When $\mathbf{x} \sim \mathcal{N}_L(\mathbf{0}, \mathbf{\Sigma})$, the conditional distribution of $\mathbf{z}$ given τ is $\mathbf{z}|\tau \sim \mathcal{N}_L(\mathbf{0}, \tau\mathbf{\Sigma})$, and its marginal is the elliptically contoured pdf given by

$$f(\mathbf{z}) = \frac{1}{(2\pi)^{L/2}\det(\mathbf{\Sigma})^{1/2}} \int \tau^{-L/2} \exp\left(-\frac{\mathbf{x}^T\mathbf{\Sigma}^{-1}\mathbf{x}}{2\tau}\right) f(\tau)d\tau,$$

where $f(\tau)$ is the prior of the texture.

Compound Gaussian with Gamma Density. When the scale, or texture parameter, follows a gamma density $\tau \sim \Gamma(\alpha, \beta)$, with pdf $f(\tau) = \frac{\beta^\alpha}{\Gamma(\alpha)}\tau^{\alpha-1}e^{-\beta\tau}$, the marginal of the elliptically contoured compound Gaussian is

$$f(\mathbf{z}) = \frac{\beta^\alpha}{\Gamma(\alpha)(2\pi)^{L/2}\det(\boldsymbol{\Sigma})^{1/2}} \int \tau^{\alpha-1-L/2} \exp\left\{-\left(\beta\tau + \frac{q}{\tau}\right)\right\} d\tau,$$

where $q = \mathbf{z}^T\boldsymbol{\Sigma}^{-1}\mathbf{z}/2$. This yields a K-distribution [1,371,372]:

$$f(\mathbf{z}) = Cq^{\frac{1}{2}\left(\alpha-\frac{L}{2}\right)} K_{\alpha-\frac{L}{2}}\left(2\sqrt{q\beta}\right), \tag{D.16}$$

where $K_\nu(z)$ is the modified Bessel function of order ν (recall that $K_{-\nu}(z) = K_\nu(z)$), and the normalizing constant is

$$C = \frac{2\beta^{\frac{1}{2}\left(\alpha+\frac{L}{2}\right)}}{(2\pi)^{L/2}\det(\boldsymbol{\Sigma})^{1/2}\Gamma(\alpha)}.$$

To derive this result, the following integral representation for the modified Bessel function is useful ([254, Eq. 10.32.10]):

$$K_\nu(z) = \left(\frac{1}{2}\right)\left(\frac{z}{2}\right)^\nu \int_0^\infty e^{-\left(t-\frac{z^2}{4t}\right)} t^{-\nu-1} dt.$$

If $\alpha = \beta = \nu$, so that the texture has unit mean and variance $1/\nu$, then (D.16) reduces to

$$f(\mathbf{z}) = \frac{2\nu^{\frac{1}{2}\left(\nu+\frac{L}{2}\right)}}{(2\pi)^{L/2}\det(\boldsymbol{\Sigma})^{1/2}\Gamma(\nu)} q^{\frac{1}{2}\left(\nu-\frac{L}{2}\right)} K_{\nu-\frac{L}{2}}\left(2\sqrt{q\nu}\right),$$

and the smaller the value of ν (with $\nu > 0$) is, the heavier-tailed (or spikier) is the K-distribution. On the contrary, when $\nu \to \infty$, the distribution converges to the normal distribution.

Compound Gaussian with Inverse Gamma Density. Another common prior for the scale variable is the inverse gamma density, which is the conjugate prior for the variance when the likelihood is normal [19]. We say the texture τ follows an inverse gamma distribution with parameters α and β, denoted as $\tau \sim \text{Inv}\Gamma(\alpha, \beta)$, when its density is

$$f(\tau) = \frac{\beta^\alpha}{\Gamma(\alpha)}\tau^{-(\alpha+1)}e^{-\frac{\beta}{\tau}}.$$

Clearly, if $\tau \sim \text{Inv}\Gamma(\alpha, \beta)$, then $1/\tau \sim \Gamma(\alpha, \beta)$.

The marginal of a compound Gaussian random vector with inverse gamma prior is

$$f(\mathbf{z}) = \frac{\beta^{\alpha}}{\Gamma(\alpha)(2\pi)^{L/2}\det(\boldsymbol{\Sigma})^{1/2}} \int \tau^{-(\alpha+1+L/2)} e^{-\left(\frac{\beta+q}{\tau}\right)} d\tau,$$

where $q = \mathbf{z}^T \boldsymbol{\Sigma}^{-1}\mathbf{z}/2$. The integral can be computed in closed-form, yielding the marginal

$$f(\mathbf{z}) = \frac{\Gamma\left(\alpha + \frac{L}{2}\right)}{\Gamma(\alpha)(2\pi)^{L/2}\det(\boldsymbol{\Sigma})^{1/2}\beta^{L/2}} \left(1 + \frac{\mathbf{z}^T\boldsymbol{\Sigma}^{-1}\mathbf{z}}{2\beta}\right)^{-(\alpha+L/2)}.$$

Specializing the previous expression to the case $\alpha = \beta = \nu/2$ (note that $\nu > 2$ is required for the inverse gamma to have a finite mean), then the density of the compound Gaussian with inverse gamma prior is

$$f(\mathbf{z}) = \frac{\Gamma\left(\frac{\nu+L}{2}\right)}{\Gamma(\nu)(\pi)^{L/2}\det(\boldsymbol{\Sigma})^{1/2}\nu^{L/2}} \left(1 + \frac{\mathbf{z}^T\boldsymbol{\Sigma}^{-1}\mathbf{z}}{\nu}\right)^{-\left(\frac{\nu+L}{2}\right)},$$

which is a multivariate t-density with ν degrees of freedom. The smaller the number of degrees of freedom, ν, the heavier-tailed is the distribution. When $\nu \to \infty$, the multivariate t-distribution reduces to the multivariate normal distribution.

The vector-valued t-density can be generalized to the matrix-valued t-density (which also belongs to the family of compound Gaussian distributions) as follows. Let us begin with a Gaussian matrix $\mathbf{X} \sim \mathcal{N}_{L\times N}(\mathbf{0}, \boldsymbol{\Sigma}_r \otimes \mathbf{I}_L)$, so their columns are arbitrarily correlated but their rows are uncorrelated. Now, color each of the rows of $\mathbf{X}$ with a covariance matrix drawn from an inverse Wishart $\mathbf{W} \sim \mathcal{W}_L^{-1}(\boldsymbol{\Sigma}_c, \nu + N - 1)$ to produce $\mathbf{Z} = \mathbf{W}^{1/2}\mathbf{X}$. Then $\mathbf{X}$ has a matrix-variate t-density denoted as $\mathcal{T}_{L\times N}(\mathbf{0}, \boldsymbol{\Sigma}_r \otimes \boldsymbol{\Sigma}_c)$, with pdf

$$f(\mathbf{Z}) = \frac{K}{\det(\boldsymbol{\Sigma}_c)^{N/2}\det(\boldsymbol{\Sigma}_r)^{L/2}} \det\left(\mathbf{I}_L + \boldsymbol{\Sigma}_c^{-1}\mathbf{Z}\boldsymbol{\Sigma}_r^{-1}\mathbf{Z}^T\right)^{-\frac{(\nu+L+N-1)}{2}},$$

where

$$K = \frac{\Gamma_L\left(\frac{\nu+L+N-1}{2}\right)}{\pi^{NL/2}\Gamma_L\left(\frac{\nu+N-1}{2}\right)}.$$

MMSE Estimation with Compound Gaussian Models. The covariance matrix for the random vector $\mathbf{z} = \sqrt{\tau}\mathbf{x}$ may be written as $\mathbf{R}_{\mathbf{zz}} = \mathrm{E}[\mathbf{z}\mathbf{z}^T] = \mathrm{E}_\tau[\mathrm{E}[\mathbf{z}\mathbf{z}^T|\tau]] = \mathrm{E}_\tau[\tau]\boldsymbol{\Sigma} = \gamma^2\boldsymbol{\Sigma}$. This means that the linear minimum mean-squared error estimators work the same in the compound Gaussian model as in the Gaussian model. In fact, if the random vector $\mathbf{z}$ is organized into its two-channel components, $\mathbf{z} = [\mathbf{z}_1^T \ \mathbf{z}_2^T]^T$, then all components of $\boldsymbol{\Sigma}$ scale with γ^2, so that $\hat{\mathbf{z}}_1 = \boldsymbol{\Sigma}_{12}\boldsymbol{\Sigma}_{22}^{-1}\mathbf{z}_2$ and the error covariance matrix is $\gamma^2\mathbf{Q}$, where $\mathbf{Q} = \boldsymbol{\Sigma}_{11} - \boldsymbol{\Sigma}_{12}\boldsymbol{\Sigma}_{22}^{-1}\boldsymbol{\Sigma}_{12}^T$. But is this the conditional mean estimator in this compound Gaussian model?

A general result in conditioning is that $\mathrm{E}[\mathbf{z}_1|\mathbf{z}_2] = \mathrm{E}_\tau[\mathrm{E}[\mathbf{z}_1|\mathbf{z}_2, \tau]|\mathbf{z}_2]$, which is read as a conditional expectation over the distribution of τ, given $\mathbf{z}_2$, of the conditional expectation over the distribution of $\mathbf{z}_1$, given $\mathbf{z}_2$ and τ. This may also be written as $\mathrm{E}[\mathbf{z}_1|\mathbf{z}_2] = \mathrm{E}_{\tau|\mathbf{z}_2}[\mathrm{E}_{\mathbf{z}_1|\mathbf{z}_2,\tau}[\mathbf{z}_1]]$. The inner expectation, conditioned on τ, returns $\hat{\mathbf{z}}_1 = \boldsymbol{\Sigma}_{12}\boldsymbol{\Sigma}_{22}^{-1}\mathbf{z}_2$, as the distribution of $[\mathbf{z}_1^T \ \mathbf{z}_2^T]^T$, given τ, is MVN with covariance $\gamma^2\boldsymbol{\Sigma}$. This is independent of τ, so it remains unchanged under the outer expectation. The error covariance matrix is $\gamma^2\mathbf{Q}$. So the conditional mean estimator is the linear MMSE estimator. These results generalize to the entire class of elliptically contoured distributions described in Sect. D.8.2.

The Complex Normal Distribution

E

E.1 The Complex MVN Distribution

Consider the real MVN random vector $\mathbf{x} \sim \mathcal{N}_{2L}(\mathbf{0}, \boldsymbol{\Sigma})$, channelized into $\mathbf{x}_1$ and $\mathbf{x}_2$, each component of which is L-dimensional. That is, $\mathbf{x}^T = [\mathbf{x}_1^T \ \mathbf{x}_2^T]$. From these two real components, construct the complex vector $\mathbf{z} = \mathbf{x}_1 + j\mathbf{x}_2$, and its complex conjugate $\mathbf{z}^* = \mathbf{x}_1 - j\mathbf{x}_2$. These may be organized into the $2L$-dimensional vector $\mathbf{w}^T = [\mathbf{z}^T \ \mathbf{z}^H]$. There is a one-to-one correspondence between $\mathbf{w}$ and $\mathbf{x}$, given by

$$\begin{bmatrix} \mathbf{z} \\ \mathbf{z}^* \end{bmatrix} = \begin{bmatrix} \mathbf{I}_L & j\mathbf{I}_L \\ \mathbf{I}_L & -j\mathbf{I}_L \end{bmatrix} \begin{bmatrix} \mathbf{x}_1 \\ \mathbf{x}_2 \end{bmatrix},$$

$$\begin{bmatrix} \mathbf{x}_1 \\ \mathbf{x}_2 \end{bmatrix} = \frac{1}{2} \begin{bmatrix} \mathbf{I}_L & \mathbf{I}_L \\ -j\mathbf{I}_L & j\mathbf{I}_L \end{bmatrix} \begin{bmatrix} \mathbf{z} \\ \mathbf{z}^* \end{bmatrix}.$$

These transformations may be written as $\mathbf{w} = \mathbf{T}\mathbf{x}$ and $\mathbf{x} = \mathbf{T}^{-1}\mathbf{w}$, with $\mathbf{T}$ and $\mathbf{T}^{-1}$ defined as

$$\mathbf{T} = \begin{bmatrix} \mathbf{I}_L & j\mathbf{I}_L \\ \mathbf{I}_L & -j\mathbf{I}_L \end{bmatrix},$$

$$\mathbf{T}^{-1} = \frac{1}{2} \begin{bmatrix} \mathbf{I}_L & \mathbf{I}_L \\ -j\mathbf{I}_L & j\mathbf{I}_L \end{bmatrix}.$$

The determinant of $\mathbf{T}$ is $\det(\mathbf{T}) = (-2j)^L$.

The connections between the symmetric covariance $\boldsymbol{\Sigma}$ of the real $2L$-dimensional vector $\mathbf{x}$ and the Hermitian covariance $\mathbf{R}_{\mathbf{ww}}$ of the complex $2L$-dimensional vector $\mathbf{w}$ are

D. Ramírez et al., *Coherence*, https://doi.org/10.1007/978-3-031-13331-2

$$\begin{aligned}\mathbf{R}_{\mathbf{ww}} &= E[\mathbf{w}\mathbf{w}^H] = E[\mathbf{T}\mathbf{x}\mathbf{x}^T\mathbf{T}^H] = \mathbf{T}\boldsymbol{\Sigma}\mathbf{T}^H \\ \boldsymbol{\Sigma} &= \mathbf{T}^{-1}\mathbf{R}_{\mathbf{ww}}\mathbf{T}^{-H},\end{aligned}$$

with $\det(\mathbf{R}_{\mathbf{ww}}) = \det(\boldsymbol{\Sigma}) \mid \det(\mathbf{T})|^2 = 2^{2L}\det(\boldsymbol{\Sigma})$.

The Complex Normal pdf. With the correspondence between $\mathbf{z}$, $\mathbf{x}$, and $\mathbf{w}$, and between $\boldsymbol{\Sigma}$ and $\mathbf{R}_{\mathbf{ww}}$, the pdf for the complex random vector $\mathbf{z}$ may be written as

$$\begin{aligned}f(\mathbf{z}) &= \frac{1}{(2\pi)^L \det(\boldsymbol{\Sigma})^{1/2}} \exp\left(-\frac{1}{2}\mathbf{x}^T\boldsymbol{\Sigma}^{-1}\mathbf{x}\right), \quad \mathbf{x} = \begin{bmatrix}\operatorname{Re}\{\mathbf{z}\} \\ \operatorname{Im}\{\mathbf{z}\}\end{bmatrix}, \\ &= \frac{1}{\pi^L \det(\mathbf{R}_{\mathbf{ww}})^{1/2}} \exp\left(-\frac{1}{2}\mathbf{w}^H\mathbf{R}_{\mathbf{ww}}^{-1}\mathbf{w}\right), \quad \mathbf{w} = \begin{bmatrix}\mathbf{z} \\ \mathbf{z}^*\end{bmatrix}.\end{aligned} \tag{E.1}$$

The function $f(\mathbf{z})$ in the second line of (E.1) is defined to be the pdf for the general complex normal distribution, and $\mathbf{z}$ is said to be a complex normal random vector. What does this mean? Begin with the complex vector $\mathbf{z} = \mathbf{x}_1 + j\mathbf{x}_2$; $\mathbf{x}_1$ is the real part of $\mathbf{z}$ and $\mathbf{x}_2$ is the imaginary part of $\mathbf{z}$. The pdf $f(\mathbf{z})$ may be expressed as in the first line of (E.1). Or, begin with $\mathbf{z}$ and define $\mathbf{w} = [\mathbf{z}^T\ \mathbf{z}^H]^T$. The pdf $f(\mathbf{z})$ may be expressed as in the second line of (E.1).

Hermitian and Complementary Covariance Matrices. Let us explore the covariance matrix $\mathbf{R}_{\mathbf{ww}}$ that appears in the pdf $f(\mathbf{z})$. The covariance matrix $\boldsymbol{\Sigma} = E[\mathbf{x}\mathbf{x}^T]$ is patterned as

$$\boldsymbol{\Sigma} = \begin{bmatrix}\boldsymbol{\Sigma}_{11} & \boldsymbol{\Sigma}_{12} \\ \boldsymbol{\Sigma}_{12}^T & \boldsymbol{\Sigma}_{22}\end{bmatrix},$$

where $\boldsymbol{\Sigma}_{12} = E[\mathbf{x}_1\mathbf{x}_2^T]$. The covariance matrix for $\mathbf{w}$ is patterned as

$$\mathbf{R}_{\mathbf{ww}} = \begin{bmatrix}\mathbf{R}_{\mathbf{zz}} & \widetilde{\mathbf{R}}_{\mathbf{zz}} \\ \widetilde{\mathbf{R}}_{\mathbf{zz}}^* & \mathbf{R}_{\mathbf{zz}}^*\end{bmatrix}.$$

In this pattern, the covariance matrix $\mathbf{R}_{\mathbf{zz}}$ is the usual Hermitian covariance matrix for the complex vector $\mathbf{z}$, and $\widetilde{\mathbf{R}}_{\mathbf{zz}}$ is the *complementary* covariance [5, 318]:

$$\begin{aligned}\mathbf{R}_{\mathbf{zz}} &= E[\mathbf{z}\mathbf{z}^H] = E\left[(\mathbf{x}_1 + j\mathbf{x}_2)(\mathbf{x}_1^T - j\mathbf{x}_2^T)\right] = (\boldsymbol{\Sigma}_{11} + \boldsymbol{\Sigma}_{22}) + j(\boldsymbol{\Sigma}_{12}^T - \boldsymbol{\Sigma}_{12}), \\ \widetilde{\mathbf{R}}_{\mathbf{zz}} &= E[\mathbf{z}\mathbf{z}^T] = E\left[(\mathbf{x}_1 + j\mathbf{x}_2)(\mathbf{x}_1^T + j\mathbf{x}_2^T)\right] = (\boldsymbol{\Sigma}_{11} - \boldsymbol{\Sigma}_{22}) + j(\boldsymbol{\Sigma}_{12}^T + \boldsymbol{\Sigma}_{12}).\end{aligned}$$

Importantly, the complementary covariance encodes for the difference in covariances in the two channels in its real part and for the cross-covariance between the two channels in its imaginary part. These formulas are easily inverted for $\mathbf{\Sigma}_{11} = (1/2)\mathrm{Re}\{\mathbf{R_{zz}}+\widetilde{\mathbf{R}}_{\mathbf{zz}}\}$, $\mathbf{\Sigma}_{22} = (1/2)\mathrm{Re}\{\mathbf{R_{zz}}-\widetilde{\mathbf{R}}_{\mathbf{zz}}\}$, and $\mathbf{\Sigma}_{12}^T = (1/2)\mathrm{Im}\{\mathbf{R_{zz}}+\widetilde{\mathbf{R}}_{\mathbf{zz}}\}$. In much of science and engineering, it is assumed that this complementary covariance $\widetilde{\mathbf{R}}$ is zero, although there are many problems in optics, signal processing, and communications where it is now realized that the complementary covariance is not zero (i.e., $\mathbf{\Sigma}_{11} \neq \mathbf{\Sigma}_{22}$ and/or $\mathbf{\Sigma}_{12}^T \neq -\mathbf{\Sigma}_{12}$). Then the general pdf for the complex normal vector $\mathbf{z}$ is the pdf of (E.1).

Bivariate $\mathbf{x} = [x_1\ x_2]^T$ and Univariate $z = x_1 + jx_2$. The covariance matrix for $\mathbf{x}$ is

$$\mathbf{\Sigma} = \begin{bmatrix} \sigma_{11}^2 & \rho\sigma_{11}\sigma_{22} \\ \rho\sigma_{11}\sigma_{22} & \sigma_{22}^2 \end{bmatrix},$$

where $\mathrm{E}[x_1x_2] = \rho\sigma_{11}\sigma_{22}$, and ρ is the correlation coefficient. The Hermitian and complementary variances of z are

$$R_{zz} = \sigma_{zz}^2 = \sigma_{11}^2 + \sigma_{22}^2,$$
$$\widetilde{R}_{zz} = \tilde{\sigma}_{zz}^2 = \sigma_{11}^2 - \sigma_{22}^2 + j2\rho\sigma_{11}\sigma_{22} = \sigma_{zz}^2\kappa e^{j\theta}.$$

In this parameterization, $\kappa = \frac{|\tilde{\sigma}_{zz}^2|}{\sigma_{zz}^2}$ is a circularity coefficient; $0 \leq \kappa \leq 1$. This circularity coefficient is the modulus of the correlation coefficient between z and z^*. With this parameterization,

$$\mathbf{R_{ww}} = \sigma_{zz}^2 \begin{bmatrix} 1 & \kappa e^{j\theta} \\ \kappa e^{-j\theta} & 1 \end{bmatrix},$$

so $\det(\mathbf{R_{ww}}) = \sigma_{zz}^2(1-\kappa^2)$. The inverse of this covariance matrix is

$$\mathbf{R}_{\mathbf{ww}}^{-1} = \frac{1}{\sigma_{zz}^2(1-\kappa^2)} \begin{bmatrix} 1 & -\kappa e^{j\theta} \\ -\kappa e^{-j\theta} & 1 \end{bmatrix}.$$

The general result for the pdf $f(\mathbf{z})$ in (E.1) specializes to

$$f(z) = \frac{1}{\pi\sigma_{zz}^2\sqrt{1-\kappa^2}} \exp\left\{-\frac{|z|^2 - \kappa\mathrm{Re}(z^2e^{-j\theta})}{\sigma_{zz}^2(1-\kappa^2)}\right\}.$$

E.2 The Proper Complex MVN Distribution

If the complementary covariance is zero, then $\boldsymbol{\Sigma}_{11} = \boldsymbol{\Sigma}_{22}$, and $\boldsymbol{\Sigma}_{12}^T = -\boldsymbol{\Sigma}_{12}$. The corresponding covariance matrix $\boldsymbol{\Sigma}$ is therefore skew-symmetric, and the corresponding Hermitian covariance $\mathbf{R_{zz}}$ is $\mathbf{R_{zz}} = 2\boldsymbol{\Sigma}_{11} + j2\boldsymbol{\Sigma}_{12}^T$. We shall parameterize the covariance matrices $\boldsymbol{\Sigma}$, $\mathbf{R_{zz}}$, and $\mathbf{R_{ww}}$ as

$$\boldsymbol{\Sigma} = \begin{bmatrix} \frac{1}{2}\mathbf{A} & -\frac{1}{2}\mathbf{B} \\ \frac{1}{2}\mathbf{B} & \frac{1}{2}\mathbf{A} \end{bmatrix} = \boldsymbol{\Sigma}^T,$$

where $\mathbf{A}^T = \mathbf{A}$ and $\mathbf{B}^T = -\mathbf{B}$. Then $\mathbf{R_{zz}} = \mathbf{A} + j\mathbf{B} = \mathbf{R}_{\mathbf{zz}}^H$, and

$$\mathbf{R_{ww}} = \begin{bmatrix} \mathbf{R_{zz}} & \mathbf{0} \\ \mathbf{0} & \mathbf{R}_{\mathbf{zz}}^* \end{bmatrix}.$$

It follows that the quadratic form $\mathbf{w}^H \mathbf{R}_{\mathbf{ww}}^{-1} \mathbf{w} = 2\mathbf{z}^H \mathbf{R}_{\mathbf{zz}}^{-1} \mathbf{z}$. The consequence is that the pdf in (E.1) simplifies to

$$f(\mathbf{z}) = \frac{1}{\pi^L \det(\mathbf{R_{zz}})} \exp\left(-\mathbf{z}^H \mathbf{R}_{\mathbf{zz}}^{-1} \mathbf{z}\right).$$

This is the pdf of a *proper* complex MVN random vector. It is important to emphasize that this is the MVN pdf for the very special case where the covariances of $\mathbf{x}_1$ and $\mathbf{x}_2$ in the MVN random vector $\mathbf{x} = [\mathbf{x}_1^T\ \mathbf{x}_2^T]^T$ have identical symmetric covariance $\boldsymbol{\Sigma}_{11} = (1/2)\mathbf{A} = \boldsymbol{\Sigma}_{22}$ and cross-covariance $\boldsymbol{\Sigma}_{12} = -(1/2)\mathbf{B}$, where $\mathbf{B}^T = -\mathbf{B}$.

In summary, when we write $\mathbf{z} \sim CN_L(\mathbf{0}, \mathbf{R_{zz}})$, we mean $\mathbf{z} = \mathbf{x}_1 + j\mathbf{x}_2$ has zero mean and Hermitian covariance matrix $\mathbf{R_{zz}} = E[\mathbf{z}\mathbf{z}^H] = \mathbf{A} + j\mathbf{B}$, where $\mathbf{A}^T = \mathbf{A}$ and $\mathbf{B}^T = -\mathbf{B}$. These, in turn, are determined by the covariances of $\mathbf{x}_1$ and $\mathbf{x}_2$, given by $E[\mathbf{x}_1\mathbf{x}_1^T] = E[\mathbf{x}_2\mathbf{x}_2^T] = (1/2)\mathbf{A}$ and $E[\mathbf{x}_1\mathbf{x}_2^T] = (-1/2)\mathbf{B}$.

It is straightforward to generalize the proper complex MVN densities to the case of nonzero mean random vectors and matrices. A proper normal random vector $\mathbf{z} \in \mathbb{C}^L$ with mean $\boldsymbol{\mu}$ and covariance matrix $\mathbf{R_{zz}}$ is denoted $\mathbf{z} \sim CN_L(\boldsymbol{\mu}, \mathbf{R_{zz}})$ with density

$$f(\mathbf{z}) = \frac{1}{\pi^L \det(\mathbf{R_{zz}})} \exp\left[-(\mathbf{z} - \boldsymbol{\mu})^H \mathbf{R}_{\mathbf{zz}}^{-1} (\mathbf{z} - \boldsymbol{\mu})\right].$$

The density function of $\mathbf{Z} \sim CN_{L\times N}(\mathbf{M}, \boldsymbol{\Sigma}_r \otimes \boldsymbol{\Sigma}_c)$ is

$$f(\mathbf{Z}) = \frac{1}{\pi^{LN} \det(\boldsymbol{\Sigma}_r)^L \det(\boldsymbol{\Sigma}_c)^N} \operatorname{etr}\left[\boldsymbol{\Sigma}_c^{-1}(\mathbf{Z} - \mathbf{M})\boldsymbol{\Sigma}_r^{-1}(\mathbf{Z} - \mathbf{M})^H\right].$$

White Noise. To say the complex random vector $\mathbf{z} = \mathbf{x}_1 + j\mathbf{x}_2$ is *white* is to say it is distributed as $\mathbf{z} \sim C\mathcal{N}_L(\mathbf{0}, \mathbf{I}_L)$. That is, $\mathbf{R_{zz}} = \mathbf{I}_L$, which is to say, $\mathbf{A} = \mathbf{I}_L$, and $\mathbf{B} = \mathbf{0}$. This, in turn, says the random variables $\mathbf{x}_1$ and $\mathbf{x}_2$ are independent normal random vectors with common covariances $(1/2)\mathbf{I}_L$. It follows that $\mathbf{z}$ has the representation $\mathbf{z} = \frac{1}{\sqrt{2}}\mathbf{x}_1 + \frac{j}{\sqrt{2}}\mathbf{x}_2$, where $\mathbf{x}_1 \sim \mathcal{N}_L(\mathbf{0}, \mathbf{I}_L)$ and $\mathbf{x}_2 \sim \mathcal{N}_L(\mathbf{0}, \mathbf{I}_L)$. Equivalently, it is the random vector $\sqrt{2}\mathbf{z}$ that is the complex sum of two white real random vectors of covariance $\mathbf{I}_L$. Then, it is the quadratic form $2\mathbf{z}^H\mathbf{z} = \mathbf{x}_1^T\mathbf{x}_1 + \mathbf{x}_2^T\mathbf{x}_2$ that is the sum of two quadratic forms in real white random variables, so it is distributed as $2\mathbf{z}^H\mathbf{z} \sim \chi^2_{2L}$. This caution will become important when we analyze quadratic forms in normal random variables in Sect. F.4. The more general quadratic form $2\mathbf{z}^H\mathbf{P_H}\mathbf{z}$ is distributed as $2\mathbf{z}^H\mathbf{P_H}\mathbf{z} \sim \chi^2_{2p}$, when $\mathbf{P_H}$ is an orthogonal projection matrix onto the p-dimensional subspace $\langle\mathbf{H}\rangle$.

Bivariate $\mathbf{x} = [x_1\ x_2]^T$ and Proper Univariate $z = x_1 + jx_2$. In this case, $R_{zz} = 2\sigma_{11}^2$, and the complex correlation coefficient $\kappa e^{j\theta} = 0$. The pdf for z is

$$f(z) = \frac{1}{\pi\sigma_{zz}^2}\exp\left(-\frac{|z|^2}{\sigma_{zz}^2}\right).$$

The connection between σ_{zz}^2 and κ, and the usual parameterization of the bivariate $\mathbf{x} = [x_1\ x_2]^T$, is

$$\sigma_{zz}^2 = 2\sigma_{11}^2, \qquad \tilde{\sigma}_{zz} = 0.$$

That is, $\sigma_{11}^2 = \sigma_{22}^2$ and $\kappa = 0$. When $\sigma_{zz}^2 = 1$, which is to say z is a complex normal random variable with mean 0 and variance 1, then the variances of x_1 and x_2 are $\sigma_{11}^2 = \sigma_{22}^2 = 1/2$. The density for z may be written as

$$f(z) = \frac{1}{\pi}\exp(-|z|^2) = \frac{1}{\sqrt{2\pi(1/2)}}\exp\left(-\frac{x_1^2}{2(1/2)}\right)\frac{1}{\sqrt{2\pi(1/2)}}\exp\left(-\frac{x_2^2}{2(1/2)}\right),$$

which shows the complex random variable $z = x_1 + jx_2 \sim C\mathcal{N}(0, 1)$ to be composed of real independent random variables $x_1 \sim \mathcal{N}(0, 1/2)$ and $x_2 \sim \mathcal{N}(0, 1/2)$. That is, z is complex normal with mean 0 and variance 1, with independent real and imaginary components, each of which is real normal with mean 0 and variance 1/2. The complex random variable z is said to be circular. The scaled magnitude-squared $2|z|^2$ may be written as $2|z|^2 = (\sqrt{2}x_1)^2 + (\sqrt{2}x_2)^2$, where each of $\sqrt{2}x_1$ and $\sqrt{2}x_2$ is distributed as $\mathcal{N}(0, 1)$. As a consequence, it is the random variable $2|z|^2$ that is distributed as a chi-squared random variable with two degrees of freedom, i.e., $2|z|^2 \sim \chi_2^2$. This accounts for the factor of 2 in many quadratic forms in complex variables.

E.3 An Example from Signal Theory

The question of propriety arises naturally in signal processing, communication theory, and machine learning, where complex signals are composed from two channels of real signals. That is, from the real signals $u(t)$ and $v(t), t \in \mathbb{R}$, the complex signal $z(t)$ is constructed as $z(t) = u(t) + jv(t)$. A particularly interesting choice for $v(t)$ is the *Hilbert transform* of $u(t)$, given by

$$v(t) = \int_{-\infty}^{\infty} \frac{1}{\pi \tau} u(t-\tau) d\tau,$$

where $-\infty < t < \infty$. This convolution is a filtering of $u(t)$. The function $1/\pi t, -\infty < t < \infty$, is the impulse response of a linear time-invariant filter, whose complex frequency response is $-j\,\mathrm{sgn}(\omega), -\infty < \omega < \infty$. As a consequence, the complex signal may be written as

$$z(t) = \int_{-\infty}^{\infty} h(\tau) u(t-\tau) d\tau,$$

where

$$h(t) = \delta(t) + j\frac{1}{\pi t} \longleftrightarrow H(\omega) = 1 + \mathrm{sgn}(\omega) = \begin{cases} 2, & \omega > 0, \\ 0, & \omega \leq 0. \end{cases}$$

As usual, $\delta(t)$ is the Dirac delta function, and the double arrow denotes a Fourier transform pair.

Now, suppose the real signal $u(t)$ is *wide-sense stationary*, with correlation function $r_{uu}(\tau) = \mathrm{E}[u(t)u^*(t-\tau)] \longleftrightarrow S_{uu}(\omega)$. The function $S_{uu}(\omega)$ is the *power spectral density* of the random signal $u(t)$. It is not hard to show that the complex signal $z(t)$ is wide-sense stationary. That is, its Hermitian and complementary correlation functions are

$$r_{zz}(\tau) = \mathrm{E}[z(t)z^*(t-\tau)] \longleftrightarrow S_{zz}(\omega),$$

and

$$\tilde{r}_{zz}(\tau) = \mathrm{E}[z(t)z(t-\tau)] \longleftrightarrow \tilde{S}_{zz}(\omega).$$

The functions $S_{zz}(\omega)$ and $\tilde{S}_{zz}(\omega)$ are called, respectively, the Hermitian and complementary power spectra. These may be written as

$$S_{zz}(\omega) = H(\omega) S_{uu}(\omega) H^*(\omega) = \begin{cases} 4S_{uu}(\omega), & \omega > 0, \\ 0, & \omega \leq 0, \end{cases}$$

and

$$\tilde{S}_{zz}(\omega) = H(\omega)S_{uu}(\omega)H(-\omega) = 0.$$

It follows that the complementary correlation function is zero, meaning the *complex analytic signal* $z(t)$ is *wide-sense stationary and proper* whenever the real signal from which it is constructed is wide-sense stationary.

The power spectrum of the real signal $u(t)$ is real and an even function of ω. The power spectrum of the complex signal $z(t)$ is real, but zero for negative frequencies. If the real signal $u(t)$ is a wide-sense stationary Gaussian signal, a well-defined concept, then the complex signal $z(t)$ is a proper, wide-sense stationary complex Gaussian signal. The complex analytic signal $z(t)$, with one-sided power spectrum $S_{zz}(\omega)$, is a spectrally efficient representation of the real signal $u(t) = \text{Re}\{z(t)\}$.

E.4 Complex Distributions

In this section, we include the complex versions of some of the distributions in Appendix D.

Functions of Complex Multivariate Normal Distributions. Let $\mathbf{x} \sim \mathcal{CN}_{L_x}(\mathbf{0}, \mathbf{I}_{L_x})$ and $\mathbf{y} \sim \mathcal{CN}_{L_y}(\mathbf{0}, \mathbf{I}_{L_y})$ be two independent complex normal random vectors. Then,

(a) $\sqrt{2}\mathbf{x}^H\mathbf{x} \sim \chi^2_{2L_x}$,
(b) $\frac{\mathbf{x}^H\mathbf{P}_p\mathbf{x}}{\mathbf{x}^H\mathbf{x}} \sim \text{Beta}(p, L_x - p)$, where $\mathbf{P}_p$ is a rank-p orthogonal projection matrix,
(c) $\frac{\mathbf{x}^H\mathbf{x}}{\mathbf{x}^H\mathbf{x}+\mathbf{y}^H\mathbf{y}} \sim \text{Beta}(L_x, L_y)$, and it is independent of $\mathbf{x}^H\mathbf{x} + \mathbf{y}^H\mathbf{y}$,
(d) $\frac{L_x}{L_y}\frac{\mathbf{y}^H\mathbf{y}}{\mathbf{x}^H\mathbf{x}} \sim \mathcal{F}(2L_y, 2L_x)$, and it is independent of $\mathbf{x}^H\mathbf{x} + \mathbf{y}^H\mathbf{y}$.

Functions of Complex Matrix-Valued Normal Distributions. Let $\mathbf{X} \sim \mathcal{CN}_{L\times N_x}(\mathbf{0}, \mathbf{I}_{N_x} \otimes \mathbf{I}_L)$ and $\mathbf{Y} \sim \mathcal{CN}_{L\times N_y}(\mathbf{0}, \mathbf{I}_{N_y} \otimes \mathbf{I}_L)$ be two independent complex normal random matrices with $N_x, N_y > L$. Denote their (scaled) sample covariance matrices as $\mathbf{S_{xx}} = \mathbf{XX}^H$ and $\mathbf{S_{yy}} = \mathbf{YY}^H$. Then,

(a) $\mathbf{S_{xx}} \sim \mathcal{CW}_L(\mathbf{I}_L, N_x)$ with density

$$f(\mathbf{S_{xx}}) = \frac{\det(\mathbf{S_{xx}})^{N_x-L}}{\tilde{\Gamma}(N_x)} \text{etr}\left(-\mathbf{S_{xx}}\right), \quad \mathbf{S_{xx}} \succ \mathbf{0},$$

(b) $\mathbf{U} = \left(\mathbf{S_{xx}} + \mathbf{S_{yy}}\right)^{-1/2} \mathbf{S_{xx}} \left(\mathbf{S_{xx}} + \mathbf{S_{yy}}\right)^{-1/2} \sim \mathcal{CB}_L(N_x, N_y)$ with density

$$f(\mathbf{U}) = \frac{\tilde{\Gamma}(N_x + N_y)}{\tilde{\Gamma}(N_x)\tilde{\Gamma}(N_y)} \det(\mathbf{U})^{N_x - L} \det(\mathbf{I}_L - \mathbf{U})^{N_y - L}, \quad \mathbf{I}_L \succ \mathbf{U} \succ \mathbf{0},$$

(c) $\mathbf{U} = \mathbf{S_{yy}}^{-1/2} \mathbf{S_{xx}} \mathbf{S_{yy}}^{-1/2} \sim \mathcal{CF}_L(N_x, N_y)$ with density

$$f(\mathbf{U}) = \frac{\tilde{\Gamma}(N_x + N_y)}{\tilde{\Gamma}(N_x)\tilde{\Gamma}(N_y)} \frac{\det(\mathbf{U})^{N_x - L}}{\det(\mathbf{I}_L + \mathbf{U})^{N_x + N_y}}, \quad \mathbf{U} \succ \mathbf{0}.$$

Complex Compound Gaussian Distributions. Let $\mathbf{z} = \sqrt{\tau}\mathbf{x}$ be a complex compound Gaussian vector with speckle component modeled as $\mathbf{x} \sim \mathcal{CN}_L(\mathbf{0}, \boldsymbol{\Sigma})$ and texture (or scale) $\tau > 0$, with prior distribution $f(\tau)$. When τ follows a gamma density with unit mean and variance $1/\nu$, $\tau \sim \Gamma(\nu, \nu)$, then $\mathbf{z}$ follows a multivariate complex K-distribution given by Abramovich and Besson [1] and Olilla et al. [250]

$$f(\mathbf{z}) = \frac{2\,\nu^{\frac{\nu+L}{2}}}{\pi^L \det(\boldsymbol{\Sigma})\Gamma(\nu)} (\mathbf{z}^H \boldsymbol{\Sigma}^{-1}\mathbf{z})^{\frac{\nu-L}{2}} K_{\nu-L}\left(2\sqrt{\nu(\mathbf{z}^H \boldsymbol{\Sigma}^{-1}\mathbf{z})}\right),$$

where $K_{\nu-L}$ is the modified Bessel function of order $\nu - L$. When the prior for the texture τ follows an inverse gamma distribution with parameters $\alpha = \beta = \nu$, $\tau \sim \mathrm{Inv}\Gamma\,(\nu, \nu)$, then the compound-Gaussian distribution is a complex multivariate t-density with ν degrees of freedom and density

$$f(\mathbf{z}) = \frac{\Gamma\,(\nu + L)}{\pi^L \det(\boldsymbol{\Sigma})\Gamma(\nu)\nu^L} \left(1 + \frac{\mathbf{z}^H \boldsymbol{\Sigma}^{-1}\mathbf{z}}{\nu}\right)^{-(\nu+L)}.$$

Quadratic Forms, Cochran's Theorem, and Related

F

F.1 Quadratic Forms and Cochran's Theorem

Consider the quadratic form $z = \mathbf{x}^T \boldsymbol{\Sigma}^{-1}\mathbf{x}$, where the L-dimensional MVN random vector $\mathbf{x}$ is distributed as $\mathbf{x} \sim \mathcal{N}_L(\mathbf{0}, \boldsymbol{\Sigma})$, with positive definite covariance matrix $\boldsymbol{\Sigma}$. The random vector $\mathbf{x}$ may be synthesized as $\mathbf{x} = \boldsymbol{\Sigma}^{1/2}\mathbf{u}$, where $\mathbf{u} \sim \mathcal{N}_L(\mathbf{0}, \mathbf{I}_L)$, so this quadratic form is $z = \mathbf{u}^T \boldsymbol{\Sigma}^{1/2}\boldsymbol{\Sigma}^{-1}\boldsymbol{\Sigma}^{1/2}\mathbf{u} = \mathbf{u}^T\mathbf{u}$. The quadratic form z is then the sum of L i.i.d. random variables, each distributed as χ_1^2, and thus $z \sim \chi_L^2$. This result generalizes to non-central normal vectors [244], as shown next.

Theorem *If* $\mathbf{x} \sim \mathcal{N}_L(\boldsymbol{\mu}, \boldsymbol{\Sigma})$, *where* $\boldsymbol{\Sigma}$ *is nonsingular, then* $(\mathbf{x}-\boldsymbol{\mu})^T\boldsymbol{\Sigma}^{-1}(\mathbf{x}-\boldsymbol{\mu}) \sim \chi_L^2$, *and* $\mathbf{x}^T\boldsymbol{\Sigma}^{-1}\mathbf{x} \sim \chi_L^2(\delta)$, *a non-central* χ_L^2 *with noncentrality parameter* $\delta = \boldsymbol{\mu}^T\boldsymbol{\Sigma}^{-1}\boldsymbol{\mu}$.

Begin with the vector-valued normal variable $\mathbf{u} \sim \mathcal{N}_L(\mathbf{0}, \mathbf{I}_L)$ and build the quadratic form $\mathbf{u}^T\mathbf{P}\mathbf{u}$, where $\mathbf{P}$ is a positive semidefinite matrix. By diagonalizing $\mathbf{P}$, it can be seen that the quadratic form is statistically equivalent to $\sum_{l=1}^{L}\lambda_l u_l^2$, where λ_l is the lth eigenvalue of $\mathbf{P}$ and $u_l \sim \mathcal{N}(0, 1)$ are independent random variables. Cochran's theorem states that a necessary and sufficient condition for $\mathbf{u}^T\mathbf{P}\mathbf{u}$ to be distributed as χ_p^2 is that $\mathbf{P}$ is rank-p and idempotent, i.e., $\mathbf{P}^2 = \mathbf{P}$. In other words, $\mathbf{P}$ must be a projection matrix of rank p, in which case $\mathbf{P}$ has p unit eigenvalues. The sufficiency is demonstrated by writing $\mathbf{P}$ as $\mathbf{P} = \mathbf{V}_p\mathbf{V}_p^T$, with $\mathbf{V}_p$ a p-column slice of an $L \times L$ orthogonal matrix. The quadratic form $z = \mathbf{u}^T\mathbf{P}\mathbf{u}$ may be written as $z = \mathbf{w}^T\mathbf{w}$, where $\mathbf{w} = \mathbf{V}_p^T\mathbf{x}$, which is distributed as $\mathbf{w} \sim \mathcal{N}_p(\mathbf{0}, \mathbf{I}_p)$, yielding $z \sim \chi_p^2$.

This result generalizes as follows. Decompose the identity as $\mathbf{I}_L = \mathbf{P}_1 + \mathbf{P}_2 + \cdots + \mathbf{P}_k$, where the $\mathbf{P}_i$ are projection matrices of respective ranks p_i, and $p_1 + p_2 + \cdots + p_k = L$. This requires $\mathbf{P}_i\mathbf{P}_l = \mathbf{0}$ for all $i \neq l$. Define $z = \mathbf{u}^T\mathbf{u} = \sum_{i=1}^{k}\mathbf{u}^T\mathbf{P}_i\mathbf{u} = \sum_{i=1}^{k} z_i$. The random variable z is distributed as $z \sim \chi_L^2$ and each of the z_i is distributed as $z_i \sim \chi_{p_i}^2$. Moreover, the random variables z_i are independent because

D. Ramírez et al., *Coherence*, https://doi.org/10.1007/978-3-031-13331-2

$\mathbf{P}_i\mathbf{x}$ and $\mathbf{P}_l\mathbf{x}$ are independent for all $i \neq l$. Therefore, z is the sum of k independent $\chi^2_{p_i}$ random variables, and the sum of such random variables is distributed as χ^2_L. A generalized version of Cochran's theorem can be stated as follows [207].

Theorem *Let $\sum_{i=1}^k z_i = \sum_{i=1}^k \mathbf{x}^T\mathbf{P}_i\mathbf{x}$ be a sum of quadratic forms in $\mathbf{x} \sim \mathcal{N}_L(\mathbf{0}, \mathbf{I}_L)$. Then, for the quadratic forms $z_i = \mathbf{x}^T\mathbf{P}_i\mathbf{x}$ to be independently distributed as $\chi^2_{p_i}$, $p_i = \text{rank}(\mathbf{P}_i)$, any of the following three equivalent conditions is necessary and sufficient:*

1. $\mathbf{P}_i^2 = \mathbf{P}_i, \quad \forall i,$
2. $\mathbf{P}_i\mathbf{P}_l = \mathbf{0}, \quad \forall i \neq l,$
3. $\sum_{i=1}^k \text{rank}(\mathbf{P}_i) = L.$

F.2 Decomposing a Measurement into Signal and Orthogonal Subspaces

Begin with the L-dimensional measurement $\mathbf{u} \sim \mathcal{N}_L(\mathbf{0}, \mathbf{I}_L)$, and the $L \times L$ orthogonal matrix $\mathbf{Q} = [\mathbf{Q}_1\ \mathbf{Q}_2]$. Assume $\mathbf{Q}_1$ is an $L \times p$ slice of this orthogonal matrix and $\mathbf{Q}_2$ is the remaining $L - p$ slice. That is,

$$\mathbf{Q}^T\mathbf{Q} = \text{blkdiag}\left(\mathbf{I}_p, \mathbf{I}_{L-p}\right) = \mathbf{I}_L,$$

$$\mathbf{Q}\mathbf{Q}^T = \mathbf{Q}_1\mathbf{Q}_1^T + \mathbf{Q}_2\mathbf{Q}_2^T = \mathbf{P}_1 + \mathbf{P}_2 = \mathbf{I}_L,$$

where $\mathbf{P}_1 = \mathbf{Q}_1\mathbf{Q}_1^T$ is a rank-p orthogonal projection matrix and $\mathbf{P}_2 = \mathbf{Q}_2\mathbf{Q}_2^T$ is a rank-$(L - p)$ orthogonal projection matrix. Together they resolve the identity. The random vector $\mathbf{Q}^T\mathbf{u}$ is distributed as $\mathbf{Q}^T\mathbf{u} \sim \mathcal{N}_L(\mathbf{0}, \mathbf{I}_L)$, which shows that the random variables $\mathbf{Q}_1^T\mathbf{u}$ and $\mathbf{Q}_2^T\mathbf{u}$ are uncorrelated, and therefore independent in this normal model. As a consequence, $\mathbf{u}^T\mathbf{P}_1\mathbf{u}$ is independent of $\mathbf{u}^T\mathbf{P}_2\mathbf{u}$. The random vector $\mathbf{Q}_1^T\mathbf{u}$ is a resolution of $\mathbf{u}$ for its coordinates in the p-dimensional subspace $\langle\mathbf{Q}_1\rangle$, and $\mathbf{P}_1\mathbf{u}$ is a projection of $\mathbf{u}$ onto the subspace $\langle\mathbf{Q}_1\rangle$. These same interpretations apply to $\mathbf{Q}_2^T\mathbf{u}$ and $\mathbf{P}_2\mathbf{u}$ as resolutions in and projections onto the $(L - p)$-dimensional subspace $\langle\mathbf{Q}_2\rangle$. The projections $\mathbf{P}_1$ and $\mathbf{P}_2$ decompose $\mathbf{u}$ into orthogonal components, one called the component in the *signal subspace* $\langle\mathbf{Q}_1\rangle$, and the other called the component in the *orthogonal subspace* $\langle\mathbf{Q}_2\rangle$. This orthogonal subspace is sometimes called the noise subspace, a misnomer. With this construction, $\mathbf{u}^T\mathbf{u} = \mathbf{u}_1^T\mathbf{P}_1\mathbf{u} + \mathbf{u}^T\mathbf{P}_2\mathbf{u}$. That is, the *power* $\mathbf{u}^T\mathbf{u}$ is decomposed into *signal subspace power* $\mathbf{u}^T\mathbf{P}_1\mathbf{u}$ and its *orthogonal subspace power* $\mathbf{u}^T\mathbf{P}_2\mathbf{u}$. In this Pythagorean decomposition, $\mathbf{u}^T\mathbf{u} \sim \chi^2_L$, $\mathbf{u}^T\mathbf{P}_1\mathbf{u} \sim \chi^2_p$, and $\mathbf{u}^T\mathbf{P}_2\mathbf{u} \sim \chi^2_{L-p}$. Moreover, $\mathbf{Q}_1^T\mathbf{u}$ is independent of $\mathbf{u}^T(\mathbf{I}_L - \mathbf{P}_1)\mathbf{u} = (\mathbf{Q}_2^T\mathbf{u})^T(\mathbf{Q}_2^T\mathbf{u})$, by virtue of the independence of $\mathbf{Q}_1^T\mathbf{u}$ and $\mathbf{Q}_2^T\mathbf{u}$. This last result is often stated as a theorem.

Theorem F.1 (See p. 31 of [207]) *If* $\mathbf{u} \sim \mathcal{N}_L(\mathbf{0}, \mathbf{I}_L)$ *and* $\mathbf{Q}_1 \in \mathbb{R}^{L\times p}$ *is an orthogonal matrix such that* $\mathbf{Q}_1^T\mathbf{Q}_1 = \mathbf{I}_p$*, then* $\mathbf{Q}_1^T\mathbf{u} \sim \mathcal{N}_p(\mathbf{0}, \mathbf{I}_p)$*, and* $(\mathbf{Q}_1^T\mathbf{u})^T(\mathbf{Q}_1^T\mathbf{u}) \sim \chi_p^2$ *is independent of* $\mathbf{u}^T\mathbf{u} - (\mathbf{Q}_1^T\mathbf{u})^T(\mathbf{Q}_1^T\mathbf{u})$*, which has a* χ_{L-p}^2 *distribution.*

F.3 Distribution of Squared Coherence

Begin with two normal random vectors $\mathbf{x}$ and $\mathbf{y}$, marginally distributed as $\mathbf{x} \sim \mathcal{N}_L(\mathbf{1}\mu_x, \sigma_x^2\mathbf{I}_L)$ and $\mathbf{y} \sim \mathcal{N}_L(\mathbf{1}\mu_y, \sigma_y^2\mathbf{I}_L)$. Under the null hypothesis, they are uncorrelated, and therefore independent, with joint distribution:

$$\begin{bmatrix}\mathbf{x}\\ \mathbf{y}\end{bmatrix} \sim \mathcal{N}_{2L}\left(\begin{bmatrix}\mathbf{1}\mu_x\\ \mathbf{1}\mu_y\end{bmatrix}, \begin{bmatrix}\sigma_x^2\mathbf{I}_L & \mathbf{0}\\ \mathbf{0} & \sigma_y^2\mathbf{I}_L\end{bmatrix}\right).$$

A standard statistic for testing this null hypothesis is the *coherence* statistic:

$$\rho^2 = \frac{(\mathbf{x}^T\mathbf{P}_1^\perp\mathbf{y})^2}{(\mathbf{x}^T\mathbf{P}_1^\perp\mathbf{x})(\mathbf{y}^T\mathbf{P}_1^\perp\mathbf{y})}. \tag{F.1}$$

This statistic bears comment. The vector $\mathbf{1}$ is the *ones vector*, $\mathbf{1} = [1 \ \cdots \ 1]^T$, $(\mathbf{1}^T\mathbf{1})^{-1}\mathbf{1}^T$ is its pseudo-inverse, $\mathbf{P}_1 = \mathbf{1}(\mathbf{1}^T\mathbf{1})^{-1}\mathbf{1}^T = \mathbf{1}\mathbf{1}^T/L$ is the orthogonal projector onto the dimension-1 subspace $\langle\mathbf{1}\rangle$, and $\mathbf{P}_1^\perp = \mathbf{I}_L - \mathbf{P}_1$ is the projector onto its orthogonal complement. The vectors $\mathbf{P}_1^\perp\mathbf{x}$ and $\mathbf{P}_1^\perp\mathbf{y}$ are mean-centered versions of $\mathbf{x}$ and $\mathbf{y}$, i.e., $\mathbf{P}_1^\perp\mathbf{x} = \mathbf{x} - \mathbf{1}(\mathbf{1}^T\mathbf{x}/L)$ and $\mathbf{P}_1^\perp\mathbf{y} = \mathbf{y} - \mathbf{1}(\mathbf{1}^T\mathbf{y}/L)$. That is, the coherence statistic ρ^2 in (F.1) is the coherence between mean-centered versions of $\mathbf{x}$ and $\mathbf{y}$. Moreover, as the statistic is invariant to scale of $\mathbf{x}$ and scale of $\mathbf{y}$, we may without loss of generality assume $\sigma_x^2 = \sigma_y^2 = 1$.

Following the lead of Cochran's theorem, let us decompose the identity $\mathbf{I}_L$ into three mutually orthogonal projections of respective ranks 1, 1, and $L-2$. That is, $\mathbf{I}_L = \mathbf{P}_1 + \mathbf{P}_2 + \mathbf{P}_3$, where the projection matrices are defined as

$$\mathbf{P}_1 = \mathbf{1}(\mathbf{1}^T\mathbf{1})^{-1}\mathbf{1}^T, \qquad \mathbf{P}_2 = \mathbf{P}_1^\perp\mathbf{y}(\mathbf{y}^T\mathbf{P}_1^\perp\mathbf{y})^{-1}\mathbf{y}^T\mathbf{P}_1^\perp, \qquad \mathbf{P}_3 = \mathbf{U}_3\mathbf{U}_3^T,$$

with $\mathbf{U}_3^T\mathbf{1} = \mathbf{0}$ and $\mathbf{U}_3^T(\mathbf{P}_1^\perp\mathbf{y}) = \mathbf{0}$. It is clear that $\mathbf{P}_1^\perp = \mathbf{P}_2 + \mathbf{P}_3$. So the squared coherence statistic may be written as

$$\rho^2 = \frac{\mathbf{x}^T\mathbf{P}_2\mathbf{x}}{\mathbf{x}^T\mathbf{P}_2\mathbf{x} + \mathbf{x}^T\mathbf{P}_3\mathbf{x}}.$$

By Cochran's theorem, the quadratic forms in $\mathbf{P}_2$ and $\mathbf{P}_3$ are independently distributed as χ_1^2 and χ_{L-2}^2 random variables, making ρ^2 distributed as Beta $\left(\frac{1}{2}, \frac{L-2}{2}\right)$.

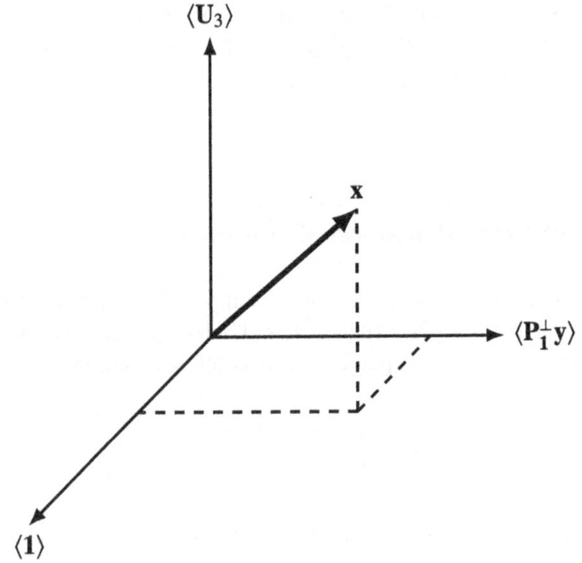

Fig. F.1 Decomposition of $\mathbf{x}$ into the three independent components

This result holds for all $\mathbf{y}$, so that when $\mathbf{y}$ is random, this distribution is a conditional distribution. But this distribution is independent of $\mathbf{y}$, making it the unconditional distribution of ρ^2 as well. This simple derivation shows the power of Cochran's theorem. It is worth noting in this derivation that the quadratic forms $\mathbf{x}^T\mathbf{P}_2\mathbf{x}$ and $\mathbf{x}^T\mathbf{P}_3\mathbf{x}$ are quadratic forms in a zero-mean normal random vector, whereas the quadratic form $\mathbf{x}^T\mathbf{P}_1\mathbf{x}$ is a quadratic form in a mean $\mathbf{1}\mu_x$ random variable. So, in the resolution $\mathbf{x}^T\mathbf{x} = \mathbf{x}^T\mathbf{P}_1\mathbf{x} + \mathbf{x}^T\mathbf{P}_2\mathbf{x} + \mathbf{x}^T\mathbf{P}_3\mathbf{x}$, the non-central distribution of $\mathbf{x}^T\mathbf{x}$ is $\mathbf{x}^T\mathbf{x} \sim \chi_L^2(L\mu_x^2)$, with $\mathbf{x}^T\mathbf{P}_1\mathbf{x} \sim \chi_1^2(L\mu_x^2)$, $\mathbf{x}^T\mathbf{P}_2\mathbf{x} \sim \chi_1^2$, and $\mathbf{x}^T\mathbf{P}_3\mathbf{x} \sim \chi_{L-2}^2$. In other words, the noncentrality parameter is carried in just one of the quadratic forms, and this quadratic form does not enter into the construction of the squared coherence ρ^2. Figure F.1 shows the decomposition of $\mathbf{x}$ into three independent components.

F.4 Cochran's Theorem in the Proper Complex Case

Cochran's theorem goes through essentially unchanged when a real normal random vector is replaced by a proper complex normal vector, and a real projection is replaced by a Hermitian projector (see Sect. B.7). To outline the essential arguments, begin with the proper complex MVN random vector $\mathbf{x} \sim CN_L(\mathbf{0}, \boldsymbol{\Sigma})$. It may be synthesized as $\mathbf{x} = \boldsymbol{\Sigma}^{1/2}\mathbf{u}$ with $\mathbf{u} \sim CN_L(\mathbf{0}, \mathbf{I}_L)$, so $z = \mathbf{x}^H\boldsymbol{\Sigma}^{-1}\mathbf{x}$ may be written as $z = \mathbf{u}^H\mathbf{u}$. The random vector $\mathbf{u}$ is composed as $\mathbf{u} = \mathbf{u}_1 + j\mathbf{u}_2$, where the real and imaginary parts are independent and distributed as $\mathcal{N}_L\left(\mathbf{0}, \frac{1}{2}\mathbf{I}_L\right)$. Therefore, the quadratic form $2z$ is the sum of $2L$ i.i.d. normals $\mathcal{N}(0, 1)$, and hence $2z \sim \chi^2_{2L}$.

Now, construct the $L \times L$, rank-p, Hermitian matrix $\mathbf{P} = \mathbf{V}\mathbf{V}^H$, where $\mathbf{V}$ is an $L \times p$ slice of a unitary matrix. Cochran's theorem establishes that $2\mathbf{u}^H\mathbf{P}\mathbf{u}$ is distributed as $\sim \chi^2_{2p}$, with $p \leq L$, iff the matrix $\mathbf{P}$ is a rank-p projection matrix. More generally, with $\mathbf{I}_L$ decomposed into orthogonal complex projections, $\mathbf{I}_L = \mathbf{P}_1 + \mathbf{P}_2 + \cdots + \mathbf{P}_k$, with $\text{rank}(\mathbf{P}_i) = p_i$, and $\sum_{i=1}^k p_i = L$, the χ^2_{2L} random variable $2\mathbf{u}^H\mathbf{u} = 2\mathbf{u}^H\mathbf{P}_1\mathbf{u} + 2\mathbf{u}^H\mathbf{P}_2\mathbf{u} + \cdots + 2\mathbf{u}^H\mathbf{P}_k\mathbf{u}$ is distributed as the sum of independent $\chi^2_{2p_i}$ random variables. An easy demonstration of $2\mathbf{u}^H\mathbf{P}_i\mathbf{u} \sim \chi^2_{2p_i}$ is this. Factor $\mathbf{P}_i$ as $\mathbf{V}_i\mathbf{V}_i^H$ and write $2\mathbf{u}^H\mathbf{P}_i\mathbf{u}$ as $(\sqrt{2}\mathbf{V}_i^H\mathbf{u})^H(\sqrt{2}\mathbf{V}_i^H\mathbf{u}_i)$, where the random vector $\sqrt{2}\mathbf{V}_i^H\mathbf{u}$ is distributed as $\sqrt{2}\mathbf{V}_i^H\mathbf{u} \sim \mathcal{CN}_{p_i}(\mathbf{0}, 2\mathbf{I}_{p_i})$. This proper complex random vector is composed of real and imaginary parts, each distributed as $\mathcal{N}_{p_i}(\mathbf{0}, \mathbf{I}_{p_i})$. Its squared magnitude is the sum of $2p_i$ real normal random variables, each of variance one, making it distributed as $\chi^2_{2p_i}$.

The Wishart Distribution, Bartlett's Factorization, and Related

G

In this appendix, we follow Kshirsagar's derivation of the Wishart distribution [207] and in the bargain get Bartlett's decomposition of an $L \times N$ real Gaussian data matrix $\mathbf{X}$ into its independently distributed QR factors. These results are then extended to complex Wishart matrices.

G.1 Bartlett's Factorization

The aim is to show that when $\mathbf{X}$ is an $L \times N$ matrix of i.i.d. $\mathcal{N}(0, 1)$ random variables, $N \geq L$, then $\mathbf{X}^T$ may be factored as $\mathbf{X}^T = \mathbf{QR}$, where the scaled sample covariance matrix $\mathbf{S} = \mathbf{XX}^T = \mathbf{R}^T\mathbf{R}$ is Wishart distributed, $\mathcal{W}_L(\mathbf{I}_L, N)$, and the unitary slice $\mathbf{Q}$ is uniformly distributed on the Stiefel manifold $\mathbb{S}t(L, \mathbb{R}^N)$ with respect to Haar measure. That is, $\mathbf{Q}$ is an orthogonal L-frame whose distribution is invariant to $N \times N$ left orthogonal transformations.

We assume N, the sample size, to be greater than L, the dimensionality of the input data. The $L \times L$ matrix $\mathbf{R}$ is upper triangular with positive diagonal elements, and the $N \times L$ matrix $\mathbf{Q}$ is an L-column slice of an orthogonal matrix, i.e., $\mathbf{Q}^T\mathbf{Q} = \mathbf{I}_L$. Hence, the $L \times L$ scaled sample covariance matrix has the LU decomposition $\mathbf{S} = \mathbf{XX}^T = \mathbf{R}^T\mathbf{R}$, where the diagonal elements of $\mathbf{R}$ are positive with probability one. The approach to the distribution of $\mathbf{S}$ will be to find the distribution of the components of $\mathbf{R}$ and then to find the Jacobian of the transformation from elements of $\mathbf{R}$ to elements of $\mathbf{S} = \mathbf{R}^T\mathbf{R}$.

The matrix $\mathbf{S} = \mathbf{XX}^T$ is the $L \times L$ scaled sample covariance matrix, determined by its $L(L + 1)/2$ unique elements, L on its diagonal and $L(L - 1)/2$ on its lower (or upper) triangle. It is the joint distribution of these elements that we seek.

The lth column of upper triangular $\mathbf{R}$ consists of l nonzero terms, denoted by the column vector $\mathbf{r}_l$, followed by $L - l$ zeros, that is,

$$[\underbrace{r_{1l}\; r_{2l}\; \cdots\; r_{ll}}_{\mathbf{r}_l^T}\; \underbrace{0\; \cdots\; 0}_{L-l}]^T. \tag{G.1}$$

D. Ramírez et al., *Coherence*, https://doi.org/10.1007/978-3-031-13331-2

From the construction of the QR factorization, it is clear that the first l columns of $\mathbf{Q}$ depend only on the first l columns of $\mathbf{X}^T$, making them independent of the remaining columns, and the column vector $\mathbf{r}_l$ depends only on the first l columns of $\mathbf{X}^T$, making it independent of the remaining columns.

Denote the lth column of $\mathbf{X}^T$ as $\mathbf{v}_l$, the lth column of $\mathbf{R}$ as in (G.1), the $\mathbf{r}_l$ vector as $\mathbf{r}_l = [\tilde{\mathbf{r}}_{l-1}^T \; r_{ll}]^T$, where $\tilde{\mathbf{r}}_{l-1}$ is a vector with the first $l-1$ elements of $\mathbf{r}_l$, and the leftmost $N \times l$ slice of $\mathbf{Q}$ as $\mathbf{Q}_l = [\mathbf{Q}_{l-1} \; \mathbf{q}_l]$. It follows that $\mathbf{v}_l = \mathbf{Q}_l \mathbf{r}_l$, and $\mathbf{v}_l^T \mathbf{v}_l = \tilde{\mathbf{r}}_{l-1}^T \tilde{\mathbf{r}}_{l-1} + r_{ll}^2$. Moreover, $\tilde{\mathbf{r}}_{l-1} = \mathbf{Q}_{l-1}^T \mathbf{v}_l$, so that $\tilde{\mathbf{r}}_{l-1}^T \tilde{\mathbf{r}}_{l-1} = \mathbf{v}_l^T \mathbf{Q}_{l-1} \mathbf{Q}_{l-1}^T \mathbf{v}_l$ and $r_{ll}^2 = \mathbf{v}_l^T \mathbf{q}_l \mathbf{q}_l^T \mathbf{v}_l$. Each of these is a quadratic form in a projection, and the two projections $\mathbf{Q}_{l-1}\mathbf{Q}_{l-1}^T$ and $\mathbf{q}_l \mathbf{q}_l^T$ are orthogonal of respective ranks $l-1$ and 1. By Cochran's theorem (see Appendix F), the χ_N^2 random variable $\mathbf{v}_l^T \mathbf{v}_l \sim \chi_N^2$ is the sum of two independent random variables $\tilde{\mathbf{r}}_{l-1}^T \tilde{\mathbf{r}}_{l-1} \sim \chi_{l-1}^2$ and $r_{ll}^2 \sim \chi_{N-(l-1)}^2$. The random variables r_{il}, $i = 1, \ldots, l-1$, are independently distributed as $\mathcal{N}(0, 1)$ random variables, and each is independent of r_{ll}. The pdf of r_{ll} is the distribution of the square root of a $\chi_{N-(l-1)}^2$ random variable with density

$$f(r_{ll}) = \frac{r_{ll}^{N-l}}{2^{(N-l-1)/2}\,\Gamma\left(\frac{N-l+1}{2}\right)} e^{-r_{ll}^2/2}.$$

These arguments hold for all $l = 1, \ldots, L$, so the the pdf of $\mathbf{R}$ is

$$f(\mathbf{R}) = \prod_{l=1}^{L} \prod_{i=1}^{l-1} \left(\frac{1}{\sqrt{2\pi}} e^{-r_{il}^2/2} \right) \prod_{k=1}^{L} \frac{r_{kk}^{N-k}}{2^{(N-k-1)/2}\,\Gamma\left(\frac{N-k+1}{2}\right)} e^{-r_{kk}^2/2}.$$

More graphically, the stochastic representation of $\mathbf{R}$ is

$$\mathbf{R} \overset{d}{=} \begin{bmatrix} \sqrt{\chi_N^2} & \mathcal{N}(0,1) & \mathcal{N}(0,1) & \cdots & \mathcal{N}(0,1) \\ 0 & \sqrt{\chi_{N-1}^2} & \mathcal{N}(0,1) & \cdots & \mathcal{N}(0,1) \\ 0 & 0 & \sqrt{\chi_{N-2}^2} & \mathcal{N}(0,1) & \vdots \\ \vdots & \vdots & \ddots & \ddots & \vdots \\ 0 & 0 & \cdots & \sqrt{\chi_{N-L+2}^2} & \mathcal{N}(0,1) \\ 0 & 0 & \cdots & \cdots & \sqrt{\chi_{N-L+1}^2} \end{bmatrix}.$$

Now, transform from the variables r_{il} of $\mathbf{R}$ to the variables s_{il} of $\mathbf{S} = \mathbf{R}^T\mathbf{R}$ by computing the Jacobian determinant:

$$\det(J(\mathbf{R} \to \mathbf{S})) = \frac{1}{\det(J(\mathbf{S} \to \mathbf{R}))} = 2^{-L} \prod_{l=1}^{L} r_{ll}^{l-L-1}.$$

Then, taking into account that the determinant and trace of $\mathbf{S}$ are

$$\det(\mathbf{S}) = \prod_{l=1}^{L} r_{ll}^2 \text{ and } \operatorname{tr}(\mathbf{S}) = \sum_{i \le l}^{L} r_{il}^2 = \sum_{l=1}^{L}\sum_{i=1}^{l} r_{il}^2,$$

the pdf of $\mathbf{S}$ is

$$f(\mathbf{S}) = \frac{1}{K(L,N)} \det(\mathbf{S})^{(N-L-1)/2} \operatorname{etr}(-\mathbf{S}/2), \tag{G.2}$$

for $\mathbf{S} \succ \mathbf{0}$, and zero otherwise. Here, the constant $K(L,N)$ is

$$K(L,N) = 2^{LN/2}\Gamma_L\left(\frac{N}{2}\right),$$

where $\Gamma_L(x)$ is the multivariate gamma function defined in (D.7). The random matrix $\mathbf{S}$ is said to be a Wishart-distributed random matrix, denoted $\mathbf{S} \sim \mathcal{W}_L(\mathbf{I}_L, N)$.

The stochastic representation of $\mathbf{Q}$ is $\mathbf{Q} = \mathbf{X}^T\mathbf{R}^{-1}$, with $\mathbf{Q}^T\mathbf{Q} = \mathbf{I}_L$. The stochastic representation of $\mathbf{Q}$ is invariant to left orthogonal transformation by an $N \times N$ orthogonal matrix, as the distribution of $\mathbf{X}^T$ is invariant to this transformation. This makes $\mathbf{Q}$ uniformly distributed on the Stiefel manifold $\mathbb{St}(L, \mathbb{R}^N)$. This is Bartlett's factorization of $\mathbf{X}^T$ into independently distributed factors $\mathbf{Q}$ and $\mathbf{R}$.

G.2 Real Wishart Distribution and Related

More generally, the matrix $\mathbf{X}$ is a real $L \times N$ random sample from a $\mathcal{N}_{L\times N}(\mathbf{0}, \mathbf{I}_N \otimes \boldsymbol{\Sigma})$ distribution. So, the matrix $\mathbf{X}$ is composed of N independent samples of the L-variate vector $\mathbf{x} \sim \mathcal{N}_L(\mathbf{0}, \boldsymbol{\Sigma})$. We have the following definition.

Definition (Wishart Distribution) Let $\mathbf{X} \sim \mathcal{N}_{L\times N}(\mathbf{0}, \mathbf{I}_N \otimes \boldsymbol{\Sigma})$, $N \ge L$, $\boldsymbol{\Sigma} \succ \mathbf{0}$, and let $\mathbf{S} = \mathbf{X}\mathbf{X}^T$ be the scaled sample covariance matrix. Then, the pdf of $\mathbf{S}$ is given by

$$f(\mathbf{S}) = \frac{1}{2^{LN/2}\Gamma_L\left(\frac{N}{2}\right)\det(\boldsymbol{\Sigma})^{N/2}} \det(\mathbf{S})^{(N-L-1)/2} \operatorname{etr}\left(-\frac{1}{2}\boldsymbol{\Sigma}^{-1}\mathbf{S}\right), \tag{G.3}$$

which is known as the Wishart distribution $\mathcal{W}_L(\boldsymbol{\Sigma}, N)$ with N degrees of freedom.

The argument is this. Begin with $\mathbf{X} \sim \mathcal{N}_{L\times N}(\mathbf{0}, \mathbf{I}_N \otimes \boldsymbol{\Sigma})$ and $\mathbf{Y} = \boldsymbol{\Sigma}^{-1/2}\mathbf{X}$. Then $\mathbf{Y}\mathbf{Y}^T \sim \mathcal{W}_L(\mathbf{I}_L, N)$ with distribution given by (G.2). But $\mathbf{Y}\mathbf{Y}^T = \boldsymbol{\Sigma}^{-1/2}\mathbf{S}\boldsymbol{\Sigma}^{-1/2}$. The Jacobian determinant of the transformation is

$$\det(J(\mathbf{Y}\mathbf{Y}^T \to \mathbf{S})) = \det(\boldsymbol{\Sigma})^{(L+1)/2}. \tag{G.4}$$

The determinant and trace of $\mathbf{YY}^T$ are

$$\det(\mathbf{YY}^T) = \frac{\det(\mathbf{S})}{\det(\mathbf{\Sigma})} \text{ and } \operatorname{tr}(\mathbf{YY}^T) = \operatorname{tr}\left(\mathbf{\Sigma}^{-1}\mathbf{S}\right). \tag{G.5}$$

Using (G.4) and (G.5) to transform the pdf of $\mathbf{YY}^T$ in (G.2), we obtain (G.3). Note that when $\mathbf{\Sigma} = \mathbf{I}_L$ and $L = 1$, we recover the χ_N^2 distribution in (D.3). The Wishart distribution in the particular case $L = 2$ was first derived by Fisher in 1915 [118], and for a general $L \geq 2$ was derived by Wishart in 1928 [385].

Definition (Inverse Wishart) If $\mathbf{S} \sim \mathcal{W}_L(\mathbf{\Sigma}, N)$, $N \geq L$, then $\mathbf{G} = \mathbf{S}^{-1}$ is said to have an inverse Wishart or inverted Wishart distribution $\mathbf{G} \sim \mathcal{W}_L^{-1}(\mathbf{\Sigma}, N)$. Using the Jacobian determinant $\det(J(\mathbf{G} \to \mathbf{S})) = \det(\mathbf{G})^{-L-1}$, the density of $\mathbf{G}$ is

$$f(\mathbf{G}) = \frac{\det(\mathbf{G})^{-(N+L+1)/2} \operatorname{etr}\left(-\frac{1}{2}\mathbf{\Sigma}^{-1}\mathbf{G}^{-1}\right)}{2^{LN/2}\Gamma_L\left(\frac{N}{2}\right)\det(\mathbf{\Sigma})^{N/2}}.$$

The mean value of $\mathbf{G}$ is $\mathrm{E}[\mathbf{G}] = \mathbf{\Sigma}^{-1}/(N - L - 1)$.

The inverse Wishart distribution is used in Bayesian statistics as the conjugate prior for the covariance matrix of a multivariate normal distribution. If $\mathbf{S} = \mathbf{XX}^T \sim \mathcal{W}_L(\mathbf{\Sigma}, N)$ and $\mathbf{\Sigma}$ is given an inverse Wishart distribution, $\mathbf{\Sigma} \sim \mathcal{W}_L^{-1}(\mathbf{\Phi}, \nu)$, then the posterior distribution for the covariance matrix $\mathbf{\Sigma}$, given the data $\mathbf{S}$, follows an inverse Wishart distribution with $N + \nu$ degrees of freedom and parameter $\mathbf{\Phi} + \mathbf{S}$. That is, $\mathbf{\Sigma} \mid \mathbf{S} \sim \mathcal{W}_L^{-1}(\mathbf{\Phi} + \mathbf{S}, \nu + N)$.

The Joint Distribution of the Eigenvalues of Wishart Matrices. If $\mathbf{S} \sim \mathcal{W}_L(\mathbf{I}_L, N)$, the Wishart pdf in (G.3) can be expressed as

$$f(\mathbf{S}) = \frac{1}{2^{LN/2}\Gamma_L\left(\frac{N}{2}\right)} \exp\left(-\frac{1}{2}\sum_{l=1}^{L}\lambda_l\right)\left(\prod_{l=1}^{L}\lambda_l\right)^{(N-L-1)/2},$$

which is a function solely of the eigenvalues of $\mathbf{S}$. Then, the application of [13, Theorem 13.3.1] by Anderson to this particular case gives the following result. If $\mathbf{S} \sim \mathcal{W}_L(\mathbf{I}_L, N)$, $N \geq L$, the joint density of the eigenvalues $\lambda_1 \geq \cdots \geq \lambda_L$ of $\mathbf{S}$ is

$$f(\lambda_1, \ldots, \lambda_L) = \frac{1}{G(L, N)} \exp\left(-\frac{1}{2}\sum_{l=1}^{L}\lambda_l\right)\left(\prod_{l=1}^{L}\lambda_l\right)^{(N-L-1)/2} \prod_{l<i}^{L}(\lambda_l - \lambda_i), \tag{G.6}$$

where

$$G(L, N) = \frac{2^{LN/2}\Gamma_L\left(\frac{N}{2}\right)\Gamma_L\left(\frac{L}{2}\right)}{\pi^{L^2/2}}$$

and the term $\prod_{l<i}^{L}(\lambda_l - \lambda_i)$ comes from the integral of the Jacobian of the transformation from the matrix space to its eigenvalue-eigenvector space.

The conventional method to generate random draws from the joint pdf in (G.6) is to generate a Gaussian random matrix, $\mathbf{X} \sim \mathcal{N}_{L\times N}(\mathbf{0}, \mathbf{I}_N \otimes \mathbf{I}_L)$, calculate the Wishart matrix $\mathbf{S} = \mathbf{X}\mathbf{X}^T$, and then calculate its eigenvalues. This procedure is clearly computationally demanding. When $L = 2$, a much more efficient sampling procedure has been proposed in [295]. The eigenvalues $\lambda_1 \geq \lambda_2 \geq 0$ of a 2×2 Wishart matrix $\mathbf{S}$ satisfy the characteristic polynomial:

$$\det(\mathbf{S} - \lambda \mathbf{I}_2) = \lambda^2 - \lambda\, \mathrm{tr}(\mathbf{S}) + \det(\mathbf{S}),$$

whose roots are

$$\lambda_i = \frac{1}{2}\,\mathrm{tr}(\mathbf{S}) \left(1 \pm \sqrt{1 - \frac{\det(\mathbf{S})}{\left(\frac{1}{2}\,\mathrm{tr}(\mathbf{S})\right)^2}}\right),$$

where λ_1 (λ_2) corresponds to the root with positive (negative) sign. The term inside the square root, $\eta = \frac{\det(\mathbf{S})}{\left(\frac{1}{2}\mathrm{tr}(\mathbf{S})\right)^2}$, is the sphericity statistic introduced in Chap. 4, which is distributed as $\eta \sim \mathrm{Beta}\,((N-1)/2, 1)$, and it is independent of $\mathrm{tr}(\mathbf{S}) = \mathrm{tr}(\mathbf{X}\mathbf{X}^T) \sim \chi^2_{2N}$. Note that if $\eta \sim \mathrm{Beta}\,((N-1)/2, 1)$, then $1 - \eta \sim \mathrm{Beta}\,(1, (N-1)/2)$. Therefore, to sample from the pdf $f(\lambda_1, \lambda_2)$ is to generate $s \sim \mathrm{Beta}\,(1, (N-1)/2)$ and $t \sim \chi^2_{2N}$, and then calculate λ_1 and λ_2 as

$$\lambda_1 = \frac{1}{2}t\left(1 + \sqrt{s}\right), \qquad \lambda_2 = \frac{1}{2}t\left(1 - \sqrt{s}\right).$$

This is a stochastic representation of the eigenvalues of 2×2 Wishart matrices.

Some Useful Properties. The Wishart distribution has an additive property similar to that of the chi-squared distribution. If $\mathbf{S}_1, \ldots, \mathbf{S}_k$, are $L \times L$ matrices having independent Wishart distributions $\mathcal{W}_L(\boldsymbol{\Sigma}, N_i)$, $i = 1, \ldots, k$, then the matrix $\mathbf{S} = \sum_{i=1}^k \mathbf{S}_i \sim \mathcal{W}_L(\boldsymbol{\Sigma}, N)$, with $N = \sum_{i=1}^k N_i$ degrees of freedom.

Theorem (Theorem 3.2.5 in [244]) *If* $\mathbf{S} \sim \mathcal{W}_L(\boldsymbol{\Sigma}, N)$ *and* $\mathbf{M}$ *is a fixed* $k \times L$ *matrix of rank* k, *then* $\mathbf{M}\mathbf{S}\mathbf{M}^T \sim \mathcal{W}_k(\mathbf{M}\boldsymbol{\Sigma}\mathbf{M}^T, N)$.

In particular, if $\mathbf{M} = [\mathbf{I}_m\ \mathbf{0}]$, then $\mathbf{M}\mathbf{S}\mathbf{M}^T$ is the sample covariance matrix of $\mathbf{M}\mathbf{X}$, namely, the Northwest $m \times m$ submatrix of $\mathbf{S}$. The distribution of $\mathbf{M}\mathbf{S}\mathbf{M}^T$ is $\mathcal{W}_m(\boldsymbol{\Sigma}_m, N)$, where $\boldsymbol{\Sigma}_m$ is the Northwest $m \times m$ sub-block of $\boldsymbol{\Sigma}$. This result shows that the marginal distribution of any submatrix of $\mathbf{S}$ located on the diagonal of $\mathbf{S}$ is Wishart. In particular, $s_{ll} \sim \mathcal{W}_1(\sigma_{ll}^2, N)$, that is, $s_{ll}/\sigma_{ll}^2 \sim \chi^2_N$, for $l = 1, \ldots, L$. These random variables, however, are not independent unless $\boldsymbol{\Sigma}$ is diagonal. Noticing that $\mathbf{a}^T\mathbf{S}\mathbf{a} \sim \mathcal{W}_1(\mathbf{a}^T\boldsymbol{\Sigma}\mathbf{a}, N)$, we have the following results.

Theorem (Theorem 3.4.2 in [244]) *If* $\mathbf{S} \sim \mathcal{W}_L(\boldsymbol{\Sigma}, N)$, $N \geq L$, *and* $\mathbf{a}$ *is a fixed* $L \times 1$ *vector such that* $\mathbf{a}^T \boldsymbol{\Sigma} \mathbf{a} \neq 0$, *then*

$$\frac{\mathbf{a}^T \mathbf{S} \mathbf{a}}{\mathbf{a}^T \boldsymbol{\Sigma} \mathbf{a}} \sim \chi_N^2.$$

Theorem (Theorem 3.4.7 in [227]) *If* $\mathbf{S} \sim \mathcal{W}_L(\boldsymbol{\Sigma}, N)$, $N \geq L$, *and* $\mathbf{a}$ *is a fixed* $L \times 1$ *vector such that* $\mathbf{a}^T \boldsymbol{\Sigma}^{-1} \mathbf{a} \neq 0$, *then*

$$\frac{\mathbf{a}^T \boldsymbol{\Sigma}^{-1} \mathbf{a}}{\mathbf{a}^T \mathbf{S}^{-1} \mathbf{a}} \sim \chi_{N-L+1}^2.$$

As a result of Bartlett's factorization result, we have the following important theorem.

Theorem *If* $\mathbf{S} \sim \mathcal{W}_L(\boldsymbol{\Sigma}, N)$, $N \geq L$, *then* $\det(\boldsymbol{\Sigma}^{-1/2}\mathbf{S}\boldsymbol{\Sigma}^{-1/2}) = \det(\mathbf{S})/\det(\boldsymbol{\Sigma})$ *is distributed as the product of L independent* χ^2 *random variables with degrees of freedom* $N, N-1, \ldots, N-L+1$; *that is,*

$$\frac{\det(\mathbf{S})}{\det(\boldsymbol{\Sigma})} \stackrel{d}{=} \prod_{l=1}^{L} \chi_{N-l+1}^2.$$

Proof The matrix $\mathbf{A} = \boldsymbol{\Sigma}^{-1/2}\mathbf{S}\boldsymbol{\Sigma}^{-1/2} \sim \mathcal{W}_L(\mathbf{I}_L, N)$. Applying Bartlett's factorization result, $\det(\mathbf{A}) = \prod_{l=i}^{L} r_{ll}^2$, where the $r_{ll}^2 \sim \chi_{N-l+1}^2$. □

Asymptotically in N, $\log(\det(\mathbf{S})/\det(\boldsymbol{\Sigma}))$ is normally distributed [244].

A few useful moments for the trace and the determinant of central real Wishart matrices are provided in the following proposition, which specializes the results for complex Wishart matrices in [350] to the real case.

Proposition *Let* $\mathbf{S} \sim \mathcal{W}_L(\mathbf{I}_L, N)$. *If* $N \geq L$ *then*

(i) $\mathrm{E}[\mathrm{tr}(\mathbf{S})] = LN$,
(ii) $\mathrm{E}[\mathrm{tr}(\mathbf{S}^2)] = LN(L + N + 1)$,
(iii) $\mathrm{E}[\mathrm{tr}(\mathbf{S})^2] = LN(LN + 2)$,
(iv) $\mathrm{E}[\det(\mathbf{S})^k] = 2^{kL} \prod_{l=0}^{L-1} \frac{\Gamma\left(\frac{1}{2}(N-l)+k\right)}{\Gamma\left(\frac{1}{2}(N-l)\right)}$. *For* $k = 1$, *the expression reduces to* $\mathrm{E}[\det(\mathbf{S})] = \frac{N!}{(N-L)!}$, *which can be checked by noticing that*

$$\det(\mathbf{S}) \stackrel{d}{=} \chi_N^2\, \chi_{N-1}^2 \cdots \chi_{N-L+1}^2.$$

(v) If $N > L + 1$ *then* $\mathrm{E}[\mathrm{tr}(\mathbf{S}^{-1})] = \frac{L}{N-L-1}$.

G.3 Complex Wishart Distribution and Related

Distribution results for real Wishart matrices can easily be generalized to complex Wishart matrices.

Bartlett's Factorization for Complex Wishart Matrices. Let $\mathbf{X} \in \mathbb{C}^{L\times N}$ be a random sample from the complex normal distribution $\mathbf{X} \sim \mathcal{CN}_{L\times N}(\mathbf{0}, \mathbf{I}_N \otimes \mathbf{I}_L)$. The $L \times L$ matrix $\mathbf{S} = \mathbf{X}\mathbf{X}^H$ can be Cholesky factored as $\mathbf{S} = \mathbf{R}^H\mathbf{R}$, where $\mathbf{R}$ is upper triangular with positive diagonal elements. By repeating the steps for the real case in Sect. G.1, we find that $\mathbf{R}$ has independent entries with the following stochastic representation:

$$\mathbf{R} \stackrel{d}{=} \begin{bmatrix} \sqrt{\frac{1}{2}\chi^2_{2N}} & \mathcal{CN}(0,1) & \mathcal{CN}(0,1) & \cdots & \mathcal{CN}(0,1) \\ 0 & \sqrt{\frac{1}{2}\chi^2_{2(N-1)}} & \mathcal{CN}(0,1) & \cdots & \mathcal{CN}(0,1) \\ 0 & 0 & \sqrt{\frac{1}{2}\chi^2_{2(N-2)}} & \mathcal{CN}(0,1) & \vdots \\ \vdots & \vdots & \ddots & \ddots & \vdots \\ 0 & 0 & \cdots & \sqrt{\frac{1}{2}\chi^2_{2(N-L+2)}} & \mathcal{CN}(0,1) \\ 0 & 0 & \cdots & \cdots & \sqrt{\frac{1}{2}\chi^2_{2(N-L+1)}} \end{bmatrix}.$$

Definition (Complex Wishart Distribution) Let $\mathbf{X} \sim \mathcal{CN}_{L\times N}(\mathbf{0}, \mathbf{I}_N \otimes \boldsymbol{\Sigma})$ be a proper complex $L \times N$ matrix, where $\boldsymbol{\Sigma}$ is now the Hermitian covariance matrix. Then, the density of $\mathbf{S} = \mathbf{X}\mathbf{X}^H$ is given by

$$f(\mathbf{S}) = \frac{1}{\tilde{\Gamma}_L(N)\det(\boldsymbol{\Sigma})^N} \det(\mathbf{S})^{(N-L)} \operatorname{etr}\left(-\boldsymbol{\Sigma}^{-1}\mathbf{S}\right), \tag{G.7}$$

where the complex multivariate gamma function is

$$\tilde{\Gamma}_L(x) = \pi^{L(L-1)/2} \prod_{l=1}^{L} \Gamma(x-l+1). \tag{G.8}$$

The pdf (G.7) is the pdf for a complex random matrix $\mathbf{S}$ that is said to be distributed as a Wishart matrix, denoted $\mathbf{S} \sim \mathcal{CW}_L(\boldsymbol{\Sigma}, N)$, with N degrees of freedom.

Definition (Complex Inverse Wishart) If $\mathbf{S} \sim \mathcal{CW}_L(\boldsymbol{\Sigma}, N)$, $N \geq L$, then $\mathbf{G} = \mathbf{S}^{-1}$ is said to have a complex inverse Wishart distribution $\mathbf{G} \sim \mathcal{CW}_L^{-1}(\boldsymbol{\Sigma}, N)$. Using the Jacobian determinant in the complex case, $\det(J(\mathbf{G} \to \mathbf{S})) = \det(\mathbf{G})^{-2L}$, the density of $\mathbf{G}$ is

$$f(\mathbf{G}) = \frac{1}{\tilde{\Gamma}_L(N)\det(\boldsymbol{\Sigma})^N}\det(\mathbf{G})^{(N+L)}\,\mathrm{etr}\left(-\boldsymbol{\Sigma}^{-1}\mathbf{G}\right).$$

where $\tilde{\Gamma}_L(x)$ is the complex multivariate gamma function defined in (G.8). The mean value of $\mathbf{G}$ is $\mathrm{E}[\mathbf{G}] = \boldsymbol{\Sigma}^{-1}/(N-L)$.

The Joint Distribution of the Eigenvalues of Complex Wishart Matrices. Let $\mathbf{S} \sim CW_L(\mathbf{I}_L, N)$ be a complex Wishart matrix. Then, the joint density of the eigenvalues $\lambda_1 \geq \cdots \geq \lambda_L$ of $\mathbf{S}$ is [184]

$$f(\lambda_1,\ldots,\lambda_L) = \frac{\pi^{L(L-1)}}{\tilde{\Gamma}_L(N)\,\tilde{\Gamma}_L(L)}\exp\left(-\sum_{l=1}^{L}\lambda_l\right)\left(\prod_{l=1}^{L}\lambda_l\right)^{N-L}\prod_{l<i}^{L}(\lambda_l-\lambda_i)^2.$$

A few useful trace and determinant moments of complex Wishart matrices are given in the following proposition [350].

Proposition *Let* $\mathbf{S} \sim CW_L(\mathbf{I}_L, N)$, *with* $N \geq L$. *Then*

(i) $\mathrm{E}[\mathrm{tr}(\mathbf{S})] = LN$,
(ii) $\mathrm{E}[\mathrm{tr}(\mathbf{S}^2)] = LN(L+N)$,
(iii) $\mathrm{E}[\mathrm{tr}(\mathbf{S})^2] = LN(LN+1)$,
(iv) $\mathrm{E}[\det(\mathbf{S})^k] = \prod_{l=0}^{L-1}\frac{\Gamma(N-l+k)}{\Gamma(N-l)}$. *For* $k=1$, $\mathrm{E}[\det(\mathbf{S})] = \frac{N!}{(N-L)!}$, *i.e., the same result as in the real case. This is readily seen by noticing that in the complex case, we have*

$$\det(\mathbf{S}) \stackrel{d}{=} \frac{1}{2}\chi^2_{2N}\,\frac{1}{2}\chi^2_{2(N-1)}\cdots\frac{1}{2}\chi^2_{2(N-L+1)}.$$

(v) If $N > L$ *then* $\mathrm{E}[\mathrm{tr}(\mathbf{S}^{-1})] = \frac{L}{N-L}$.

G.4 Distribution of Sample Mean and Sample Covariance

The following classic result gives the distribution of the sample mean and the sample covariance matrix in a multivariate normal model. It is an easy consequence of Cochran's theorem.

Theorem *Let* $\mathbf{x}_1,\ldots,\mathbf{x}_N$ *be independent and identically distributed as* $\mathbf{x}_n \sim N_L(\boldsymbol{\mu},\boldsymbol{\Sigma})$ *random vectors, with* $N \geq L$. *Define the sample mean* $\bar{\mathbf{x}}$ *and sample covariance matrix* $\mathbf{S}$:

$$\bar{\mathbf{x}} = \frac{1}{N}\sum_{n=1}^{N}\mathbf{x}_n \quad \text{and} \quad \mathbf{S} = \frac{1}{N}\sum_{n=1}^{N}(\mathbf{x}_n - \bar{\mathbf{x}})(\mathbf{x}_n - \bar{\mathbf{x}})^T .$$

Then, the distributions of the sample mean and the sample covariance matrix are

$$\bar{\mathbf{x}} \sim \mathcal{N}_L\left(\boldsymbol{\mu}, \frac{1}{N}\boldsymbol{\Sigma},\right) \quad \textit{and} \quad N\mathbf{S} \sim \mathcal{W}_L(\boldsymbol{\Sigma}, N-1).$$

Further, $\bar{\mathbf{x}}$ *and* $\mathbf{S}$ *are independent.*

Example G.1 (Real univariate case) In the univariate case, the sample mean and variance are independent and distributed as

$$\overline{x} = \frac{1}{N}\sum_{n=1}^{N}x_n \sim \mathcal{N}(\mu, \sigma^2/N), \quad \text{and} \quad Ns^2 = \sum_{n=1}^{N}(x_n - \bar{x})^2 \sim \mathcal{W}_1(\sigma^2, N-1),$$

or, equivalently, $Ns^2/\sigma^2 \sim \chi^2_{N-1}$. The demonstration of this result is an exercise in the application of Cochran's theorem. Begin with i.i.d. rvs $x_n \sim \mathcal{N}(\mu, \sigma^2)$, $n = 1, \ldots, N$, organized into the column vector $\mathbf{x} = [x_1 \, \cdots \, x_N]^T$. Then, define $u_n = (x_n - \mu)/\sigma \sim \mathcal{N}(0, 1)$, and organize these into the column vector $\mathbf{u} = (\mathbf{x} - \mathbf{1}\mu)/\sigma$, where $\mathbf{1} = [1 \, \cdots \, 1]^T$ is the ones-vector. Define the rank-1 projection matrix $\mathbf{P_1} = \mathbf{1}(\mathbf{1}^T\mathbf{1})^{-1}\mathbf{1}^T$ and the rank-$(N-1)$ projection matrix $\mathbf{I}_N - \mathbf{P_1}$. Note $\mathbf{P_1}(\mathbf{I}_N - \mathbf{P_1}) = \mathbf{0}$ and $\mathbf{P_1} + (\mathbf{I}_N - \mathbf{P_1}) = \mathbf{I}_N$.

By Cochran's theorem, it follows that $\mathbf{u}^T\mathbf{u} = \mathbf{u}^T\mathbf{P_1}\mathbf{u} + \mathbf{u}^T(\mathbf{I}_N - \mathbf{P_1})\mathbf{u}$ decomposes $\mathbf{u}^T\mathbf{u} \sim \chi^2_N$ into the sum of independent random variables $\mathbf{u}^T\mathbf{P_1}\mathbf{u} \sim \chi^2_1$ and $\mathbf{u}^T(\mathbf{I}_N - \mathbf{P_1})\mathbf{u} \sim \chi^2_{N-1}$. The projected random vector $\mathbf{P_1}\mathbf{u}$ may be written as $\mathbf{1}(\overline{x} - \mu)/\sigma$, where $\overline{x} = (\mathbf{1}^T\mathbf{1})^{-1}\mathbf{1}^T\mathbf{x} = N^{-1}\sum_{n=1}^{N} x_n$ is the sample mean of the rvs $x_n, n = 1, \ldots, N$. The projected random vector $(\mathbf{I}_N - \mathbf{P_1})\mathbf{u}$ may be written as $(\mathbf{x} - \mathbf{1}\overline{x})/\sigma$. That is, $\mathbf{u}^T\mathbf{u} = \mathbf{u}^T\mathbf{P_1}\mathbf{u} + \mathbf{u}^T(\mathbf{I}_N - \mathbf{P_1})\mathbf{u}$ may be written as

$$\frac{1}{\sigma^2}\sum_{n=1}^{N}(x_n - \mu)^2 = \frac{N}{\sigma^2}(\overline{x} - \mu)^2 + \frac{1}{\sigma^2}\sum_{n=1}^{N}(x_n - \overline{x})^2.$$

It follows that the sample mean $\overline{x} = N^{-1}\sum_{n=1}^{N} x_n$ and the sample variance $s^2 = N^{-1}\sum_{n=1}^{N}(x_n - \overline{x})^2$ are independent random variables. The distribution of $\overline{x}$ is $\overline{x} \sim \mathcal{N}(\mu, \sigma^2/N)$ and the distribution of Ns^2/σ^2 is $Ns^2/\sigma^2 \sim \chi^2_{N-1}$.

Null Distribution of Coherence Statistics

H

In this appendix, the null distributions for various coherence statistics are derived. These null distributions are stated as the distributions of products of independent beta-distributed random variables, which makes the sampling from any of the distributions a problem of sampling from independent beta-distributed random variables and then taking their products. To say the distribution is the distribution of a product of independent beta-distributed random variables is to say the statistic itself has a stochastic representation as the product of independent beta-distributed random variables and vice versa.

Besides their use for sampling from null distributions for coherence statistics, stochastic representations may be used to compute moments. Hence, the asymptotic distribution of Wilks [384] may be modified to obtain a more accurate approximation, as proposed in [43]. Additionally, the moments of a coherence statistic may be used to derive saddlepoint approximations of its null distribution [202].

H.1 Null Distribution of the Tests for Independence

This section derives stochastic representations for the GLRs in Sect. 4.8, under the null hypothesis. We start with the case of random variables; see Sect. 4.8.1, where the GLR is given by (4.9). Then, the case of random vectors is considered; see Sect. 4.8.2 and the GLR in (4.10). In the case of random variables, the GLR tests whether the covariance matrix is diagonal, whereas in the case of random vectors, the GLR tests whether the covariance matrix is *block-diagonal*.[1]

The derivations of this appendix are based on distribution results described in previous appendices and on the linear prediction theory of Cholesky factors and Gram determinants [302].

[1] Using the ideas in this section, these distribution results may be generalized to the case where the blocks are themselves block-diagonal matrices, an idea that may be iterated indefinitely.

D. Ramírez et al., *Coherence*, https://doi.org/10.1007/978-3-031-13331-2

H.1.1 Testing Independence of Random Variables

The setup is this: The random matrix $\mathbf{X}$ is an $L \times N$ matrix:

$$\mathbf{X} = \begin{bmatrix} \mathbf{x}_1 & \cdots & \mathbf{x}_N \end{bmatrix} = \begin{bmatrix} \boldsymbol{\rho}_1^H \\ \vdots \\ \boldsymbol{\rho}_L^H \end{bmatrix},$$

where $\boldsymbol{\rho}_l^H$, the lth row of $\mathbf{X}$, is a $1 \times N$ random sample of the lth random variable x_l.[2]

The hypothesis test is $\mathcal{H}_0 : \mathbf{X} \sim \mathcal{CN}_{L\times N}(\mathbf{0}, \mathbf{I}_N \otimes \text{diag}(r_{11}, \ldots, r_{LL}))$ vs. $\mathcal{H}_1 : \mathbf{X} \sim \mathcal{CN}_{L\times N}(\mathbf{0}, \mathbf{I}_N \otimes \mathbf{R})$, where $r_{ll} > 0$ and $\mathbf{R} \succ \mathbf{0}$. The GLR for this test is the Hadamard ratio, which is given in (4.9) and repeated here for convenience:

$$\lambda_I = \frac{\det(\mathbf{S})}{\prod_{l=1}^{L} s_{ll}}.$$

The sample variance s_{ll} is the lth element on the diagonal of the sample covariance matrix $\mathbf{S} = \mathbf{X}\mathbf{X}^H/N$. By virtue of the invariance of λ_I to arbitrary scaling of the components of $\mathbf{x}_n$, we assume in the following that, under $\mathcal{H}_0$, $\mathbf{X} \sim \mathcal{CN}_{L\times N}(\mathbf{0}, \mathbf{I}_N \otimes \mathbf{I}_L)$. That is, the elements of $\mathbf{X}$ are i.i.d. complex Gaussians with zero mean and unit variance.

Write the Hadamard ratio as

$$\lambda_I = \frac{\det(\mathbf{X}\mathbf{X}^H)}{\prod_{l=1}^{L} \boldsymbol{\rho}_l^H \boldsymbol{\rho}_l}.$$

and parse the data matrix $\mathbf{X}$ as

$$\mathbf{X} = \begin{bmatrix} \mathbf{W}_{l-1}^H \\ \boldsymbol{\rho}_l^H \\ \vdots \end{bmatrix}.$$

[2] The reader may be puzzled by the use of a row vector like $\boldsymbol{\rho}_l^H$ when its entries are realizations of the random variable x_l, and not x_l^*. The first justification is that there is freedom in the choice of variable names, and this choice leads to expressions that are easy to read. The second justification is a reassurance: in all of the work of this appendix, the distribution theory of the random variable x_l^* is the same as the distribution theory of the random variable x_l, and hypothesis tests regarding the distribution of the random variables $x_l^*, l = 1, \ldots, L$ are the same as hypothesis tests regarding the distribution of the random variables $x_l, l = 1, \ldots, L$, so the matrix $\mathbf{X}$ could as well be a matrix of random samples of the random variables $x_l^*, l = 1, \ldots, L$, and the notation $\boldsymbol{\rho}_l^H$ would cause no concern.

The matrix $\mathbf{W}_{l-1}^H$ is an $(l-1) \times N$ block of rows $\boldsymbol{\rho}_1^H$ through $\boldsymbol{\rho}_{l-1}^H$. The numerator of λ_I can be written in terms of error variances as [302]

$$\det(\mathbf{X}\mathbf{X}^H) = \prod_{l=1}^{L} \sigma_l^2 = \prod_{l=1}^{L} \boldsymbol{\rho}_l^H (\mathbf{I}_N - \mathbf{P}_{l-1})\boldsymbol{\rho}_l.$$

In this expression, $\mathbf{P}_{l-1} = \mathbf{W}_{l-1}(\mathbf{W}_{l-1}^H\mathbf{W}_{l-1})^{-1}\mathbf{W}_{l-1}^H$ is the rank-$(l-1)$ projection onto the subspace of $\mathbb{C}^N$ spanned by the $l-1$ columns of $\mathbf{W}_{l-1}$. To simplify notation, we have used $\mathbf{P}_{l-1}$ to denote this projection, rather than $\mathbf{P}_{\mathbf{W}_{l-1}}$.

Using the decomposition of the determinant in terms of error variances, the GLR may be written as

$$\lambda_I = \frac{\prod_{l=1}^{L} \boldsymbol{\rho}_l^H (\mathbf{I}_N - \mathbf{P}_{l-1})\boldsymbol{\rho}_l}{\prod_{l=1}^{L} \boldsymbol{\rho}_l^H \boldsymbol{\rho}_l} = \prod_{l=2}^{L} \frac{\boldsymbol{\rho}_l^H (\mathbf{I}_N - \mathbf{P}_{l-1})\boldsymbol{\rho}_l}{\boldsymbol{\rho}_l^H \boldsymbol{\rho}_l},$$

where the term corresponding to $l = 1$ in this product is 1. This finding is sometimes paraphrased as "the distribution of λ_I is invariant to the distribution of the first channel measurements $\boldsymbol{\rho}_1^H$."

Under the null hypothesis $\mathcal{H}_0$, every normal random variable in $\mathbf{X}$ is independent of every other, so that the random vectors $\{\boldsymbol{\rho}_l, l = 1, \ldots, L\}$ are mutually uncorrelated. Importantly, the random vector $\boldsymbol{\rho}_l$ is independent of the random projection matrix $\mathbf{P}_{l-1}$. Then, recall that for a proper complex normal random vector $\boldsymbol{\rho} \sim CN_N(\mathbf{0}, \mathbf{I}_N)$, and rank-$r$ projection matrix $\mathbf{P}_r$, the random variable $2\boldsymbol{\rho}^H\boldsymbol{\rho}$ is distributed as χ_{2N}^2, and it decomposes as $2\boldsymbol{\rho}^H\boldsymbol{\rho} = 2\boldsymbol{\rho}^H\mathbf{P}_r\boldsymbol{\rho} + 2\boldsymbol{\rho}^H\mathbf{P}_r^{\perp}\boldsymbol{\rho}$, where $2\boldsymbol{\rho}^H\mathbf{P}_r\boldsymbol{\rho}$ is distributed as χ_{2r}^2, and the random variable $2\boldsymbol{\rho}^H\mathbf{P}_r^{\perp}\boldsymbol{\rho}$ is distributed as $\chi_{2(N-r)}^2$. It follows that $\boldsymbol{\rho}^H\mathbf{P}_r\boldsymbol{\rho}/\boldsymbol{\rho}^H\mathbf{P}_r^{\perp}\boldsymbol{\rho} \sim F_{2r,2(N-r)}$ and $\boldsymbol{\rho}^H\mathbf{P}_r^{\perp}\boldsymbol{\rho}/\boldsymbol{\rho}^H\boldsymbol{\rho} \sim \mathrm{Beta}(N-r, r)$. This result holds for arbitrary projections, provided they are chosen independently of $\boldsymbol{\rho}$. So the distribution of the random variable $\boldsymbol{\rho}_l^H(\mathbf{I}_N - \mathbf{P}_{l-1})\boldsymbol{\rho}_l$ is only dependent on the distribution of $\boldsymbol{\rho}_l$, and the sequence of $\boldsymbol{\rho}_l$ is a sequence of independent random vectors. Thus, under the null, it follows that λ_I is distributed as the product of independent beta-distributed random variables:

$$\lambda_I = \prod_{l=2}^{L} U_l,$$

where $U_l \sim \mathrm{Beta}(N-(l-1), l-1)$. This is the complex version of the Anderson result [13], derived by somewhat different means in [77]. Alternatively, the stochastic representation of the Hadamard ratio may be rewritten as $\lambda_I = \prod_{l=1}^{L-1} U_l$, where $U_l \sim \mathrm{Beta}(N-l, l)$.

H.1.2 Testing Independence of Random Vectors

This section generalizes the result above to the case of random vectors. The stochastic representation for the real case was derived in [13] and generalized to complex vectors in [201]. The setup, which is described in more detail in Sect. 4.8.2, is this. The observation matrix is a $PL \times N$ matrix:

$$\mathbf{X} = \begin{bmatrix} \mathbf{U}_1 \\ \vdots \\ \mathbf{U}_P \end{bmatrix},$$

where each of the $\mathbf{U}_p$ is an $L \times N$ random sample of the pth random vector $\mathbf{u}_p \in \mathbb{C}^L$. The hypothesis test is $\mathcal{H}_0 : \mathbf{X} \sim C\mathcal{N}_{PL\times N}(\mathbf{0}, \mathbf{I}_N \otimes \text{blkdiag}(\mathbf{R}_{11}, \ldots, \mathbf{R}_{PP}))$ vs. $\mathcal{H}_1 : \mathbf{X} \sim C\mathcal{N}_{PL\times N}(\mathbf{0}, \mathbf{I}_N \otimes \mathbf{R})$, where $\mathbf{R}_{pp} \succ \mathbf{0}$ and $\mathbf{R} \succ \mathbf{0}$. The GLR for this test is

$$\lambda_I = \frac{\det(\mathbf{S})}{\prod_{p=1}^{P} \det(\mathbf{S}_{pp})} = \frac{\det(\mathbf{X}\mathbf{X}^H)}{\prod_{p=1}^{P} \det(\mathbf{U}_p\mathbf{U}_p^H)}.$$

Again, due to the problem invariances, we can assume that, under $\mathcal{H}_0$, $\mathbf{X} \sim C\mathcal{N}_{PL\times N}(\mathbf{0}, \mathbf{I}_N \otimes \mathbf{I}_{PL})$, the elements of $\mathbf{X}$ are i.i.d. complex Gaussians with zero mean and unit variance.

Each of the determinants in the denominator can be written as in the previous section:

$$\det\left(\mathbf{U}_p\mathbf{U}_p^H\right) = \prod_{l=1}^{L} \sigma_{p,l}^2 = \prod_{l=1}^{L} \boldsymbol{\rho}_{p,l}^H (\mathbf{I}_N - \mathbf{P}_{p,l-1}) \boldsymbol{\rho}_{p,l},$$

where $\boldsymbol{\rho}_{p,l}^H$ is the lth row of $\mathbf{U}_p$ and $\mathbf{P}_{p,l-1} = \mathbf{W}_{p,l-1}(\mathbf{W}_{p,l-1}^H \mathbf{W}_{p,l-1})^{-1} \mathbf{W}_{p,l-1}^H$, with $\mathbf{W}_{p,l-1}^H$ an $(l-1) \times N$ block of rows $\boldsymbol{\rho}_{p,1}^H$ through $\boldsymbol{\rho}_{p,l-1}^H$. Similarly, we can compute the numerator as

$$\det(\mathbf{X}\mathbf{X}^H) = \prod_{p=1}^{P} \prod_{l=1}^{L} \boldsymbol{\rho}_{(p-1)L+l}^H (\mathbf{I}_N - \mathbf{P}_{(p-1)L+l-1}) \boldsymbol{\rho}_{(p-1)L+l},$$

where $\boldsymbol{\rho}_k^H$ is the kth row of $\mathbf{X}$ and $\mathbf{P}_{k-1} = \mathbf{V}_{k-1}(\mathbf{V}_{k-1}^H \mathbf{V}_{k-1})^{-1} \mathbf{V}_{k-1}^H$, with $\mathbf{V}_{k-1}^H$ being a $(k-1) \times N$ block of rows $\boldsymbol{\rho}_1^H$ through $\boldsymbol{\rho}_{k-1}^H$. Moreover, it is clear that $\boldsymbol{\rho}_{(p-1)L+l} = \boldsymbol{\rho}_{p,l}$ and that the subspace spanned by the columns of $\mathbf{W}_{p,l-1}$ is contained in the subspace spanned by the columns of $\mathbf{V}_{(p-1)L+l-1}$, which yields $\mathbf{P}_{p,l-1}\mathbf{P}_{(p-1)L+l-1} = \mathbf{P}_{p,l-1}$.

Using the previous determinant decompositions, we can write the generalized Hadamard ratio as

$$\lambda_I = \frac{\prod_{p=1}^{P}\prod_{l=1}^{L} \boldsymbol{\rho}_{(p-1)L+l}^H (\mathbf{I}_N - \mathbf{P}_{(p-1)L+l-1})\boldsymbol{\rho}_{(p-1)L+l}}{\prod_{p=1}^{P}\prod_{l=1}^{L} \boldsymbol{\rho}_{p,l}^H (\mathbf{I}_N - \mathbf{P}_{p,l-1})\boldsymbol{\rho}_{p,l}}$$

$$= \prod_{p=1}^{P}\prod_{l=1}^{L} \frac{\boldsymbol{\rho}_{(p-1)L+l}^H (\mathbf{I}_N - \mathbf{P}_{(p-1)L+l-1})\boldsymbol{\rho}_{(p-1)L+l}}{\boldsymbol{\rho}_{(p-1)L+l}^H (\mathbf{I}_N - \mathbf{P}_{p,l-1})\boldsymbol{\rho}_{(p-1)L+l}}.$$

For $p = 1$, $\mathbf{P}_{p,l-1} = \mathbf{P}_{(p-1)L+l-1}$, which allows us to remove the corresponding terms:

$$\lambda_I = \prod_{p=2}^{P}\prod_{l=1}^{L} \frac{\boldsymbol{\rho}_{(p-1)L+l}^H (\mathbf{I}_N - \mathbf{P}_{(p-1)L+l-1})\boldsymbol{\rho}_{(p-1)L+l}}{\boldsymbol{\rho}_{(p-1)L+l}^H (\mathbf{I}_N - \mathbf{P}_{p,l-1})\boldsymbol{\rho}_{(p-1)L+l}}.$$

It is easily shown that $(\mathbf{I}_N - \mathbf{P}_{(p-1)L+l-1}) = (\mathbf{I}_N - \mathbf{P}_{p,l-1})(\mathbf{I}_N - \mathbf{P}_{(p-1)L+l-1})(\mathbf{I}_N - \mathbf{P}_{p,l-1})$. Therefore, each term in this double product may be written as

$$\frac{\boldsymbol{\rho}_{(p-1)L+l}^H (\mathbf{I}_N - \mathbf{P}_{(p-1)L+l-1})\boldsymbol{\rho}_{(p-1)L+l}}{\boldsymbol{\rho}_{(p-1)L+l}^H (\mathbf{I}_N - \mathbf{P}_{p,l-1})\boldsymbol{\rho}_{(p-1)L+l}} = \frac{\boldsymbol{\xi}_{(p-1)L+l}^H (\mathbf{I}_N - \mathbf{P}_{(p-1)L+l-1})\boldsymbol{\xi}_{(p-1)L+l}}{\boldsymbol{\xi}_{(p-1)L+l}^H \boldsymbol{\xi}_{(p-1)L+l}},$$

where $\boldsymbol{\xi}_{(p-1)L+l}^H = \boldsymbol{\rho}_{(p-1)L+l}^H (\mathbf{I}_N - \mathbf{P}_{p,l-1})$. The GLR is

$$\lambda_I = \prod_{p=2}^{P}\prod_{l=1}^{L} \frac{\boldsymbol{\xi}_{(p-1)L+l}^H (\mathbf{I}_N - \mathbf{P}_{(p-1)L+l-1})\boldsymbol{\xi}_{(p-1)L+l}}{\boldsymbol{\xi}_{(p-1)L+l}^H \boldsymbol{\xi}_{(p-1)L+l}}.$$

For each pair (p, l), the random vector $\boldsymbol{\xi}_{(p-1)L+l}$ is independent of the projection $\mathbf{P}_{(p-1)L+l-1}$, and therefore each ratio in λ_I is beta-distributed. It follows that λ_I is distributed as the product of independent beta-distributed random variables:

$$\lambda_I = \prod_{p=2}^{P}\prod_{l=1}^{L} U_{p,l},$$

where $U_{p,l} \sim \text{Beta}(N - (p-1)L - (l-1), (p-1)L)$. This proof is a variation on the proofs of Anderson [13] and Klausner [201]. As before, the distribution can be simplified as

$$\lambda_I = \prod_{p=1}^{P-1}\prod_{l=1}^{L} U_{p,l},$$

where $U_{p,l} \sim \text{Beta}(N - pL - l + 1, pL)$. The distribution of λ_I for different dimensions, i.e., $\mathbf{u}_p \in \mathbb{C}^{L_p}$, can be obtained by similar means and is given by

$$\lambda_I \stackrel{d}{=} \prod_{p=1}^{P-1} \prod_{l=1}^{L_{p+1}} U_{p,l},$$

where

$$U_{p,l} \sim \text{Beta}\left(N - l + 1 - \sum_{i=1}^{p} L_i, \sum_{i=1}^{p} L_i\right).$$

H.2 Testing for Block-Diagonal Matrices of Different Block Sizes

This section addresses a variation on the problems of the previous section that is required for the analysis of the GLR for the detection of cyclostationarity in Chap. 8.

The setup here is that the covariance matrix, under both hypotheses, is

$$\mathbf{R} = \text{blkdiag}(\mathbf{R}^{(1)}, \ldots, \mathbf{R}^{(M)}),$$

where the mth block, $\mathbf{R}^{(m)}$, is a $Q_m \times Q_m$ positive definite matrix. Under the alternative, each of these blocks has no further structure, but under the null, each is also block-diagonal:

$$\mathbf{R}^{(m)} = \text{blkdiag}\left(\mathbf{R}_{11}^{(m)}, \ldots, \mathbf{R}_{P_m P_m}^{(m)}\right).$$

Each of the $\mathbf{R}_{pp}^{(m)}$ is an $L_p^{(m)} \times L_p^{(m)}$ positive definite matrix without further structure. Obviously, $Q_m = \sum_{p=1}^{P_m} L_p^{(m)}$. This scenario is more general than that in Chap. 8 since the block sizes can be different.

The GLR for this test follows along the lines presented in previous chapters. Parsing the sample covariance matrix as the covariance matrix $\mathbf{R}$ is parsed, the GLR is given by

$$\lambda_{BD} = \prod_{m=1}^{M} \frac{\det\left(\mathbf{S}^{(m)}\right)}{\prod_{p=1}^{P_m} \det\left(\mathbf{S}_{pp}^{(m)}\right)},$$

which is the product of M independent, but not identically distributed, GLRs for testing independence of random vectors. Hence, the stochastic representation is

$$\lambda_{BD} \stackrel{d}{=} \prod_{m=1}^{M} \prod_{p=1}^{P_m - 1} \prod_{l=1}^{L_{p+1}^{(m)}} U_{p,l}^{(m)},$$

where

$$U_{p,l}^{(m)} \sim \text{Beta}\left(N - l + 1 - \sum_{i=1}^{p} L_i^{(m)}, \sum_{i=1}^{p} L_i^{(m)}\right).$$

H.3 Testing for Block-Sphericity

This section addresses the stochastic representation of the GLR for the block-sphericity test (4.7), following the lines of [85]. By virtue of the problem invariances, we shall assume that the observations are distributed as $\mathbf{X} \sim \mathcal{CN}_{PL\times N}(\mathbf{0}, \mathbf{I}_N \otimes \mathbf{I}_{PL})$.

The first step is to rewrite (4.7) as

$$\lambda_S = \overbrace{\frac{\det(\mathbf{S})}{\prod_{p=1}^{P}\det(\mathbf{S}_{pp})}}^{\lambda_I} \times \overbrace{\frac{\prod_{p=1}^{P}\det(\mathbf{S}_{pp})}{\det\left(\frac{1}{P}\sum_{p=1}^{P}\mathbf{S}_{pp}\right)^{P}}}^{\lambda_E},$$

where λ_I is the GLR for testing the independence of random vectors, presented in Sect. 4.8.2, and λ_E is the GLR for testing equality of covariance matrices for independent random vectors of Sect. 4.7. We have therefore decomposed the GLR into the statistic for the independence test and that of the test for equality of covariance matrices, conditioned on the random vectors being independent. Following along the lines in [319, Appendix A], which is based on Basu's theorem [27], we shall show that λ_I and λ_E are independent under the null. Hence, the stochastic representation of λ_S is the product of the stochastic representation of λ_I and the stochastic representation of λ_E. Let us start by introducing Basu's theorem.

Theorem (Basu) *If* $\mathbf{W}$ *is a complete and sufficient statistic for the family of distributions* $\{f(\mathbf{x};\boldsymbol{\theta}) \mid \boldsymbol{\theta} \in \boldsymbol{\Theta}\}$, *then it is independent of any ancillary statistic.*[3]

In the test for block-sphericity, the family of distributions under the null is

$$f(\mathbf{x};\mathbf{R}_{\mathbf{uu}}) = \frac{1}{\pi^{PLM}\det(\mathbf{R}_{\mathbf{uu}})^{PM}} \exp\left\{-M\,\mathrm{tr}\left(\mathbf{R}_{\mathbf{uu}}^{-1}\sum_{p=1}^{P}\mathbf{S}_{pp}\right)\right\},$$

where $\mathbf{R}_{\mathbf{uu}}$ is any positive definite matrix. Thus, the set of sample covariance matrices $\mathbf{S}_{11}, \ldots, \mathbf{S}_{PP}$, is a complete and sufficient statistic. To show that λ_I is ancillary, rewrite λ_I as

[3] An ancillary statistic is a function of the sampled data whose distribution does not depend on the parameters of the model, $\boldsymbol{\theta}$.

$$\lambda_I = \frac{\det(\mathbf{S})}{\prod_{p=1}^{P} \det(\mathbf{S}_{pp})} = \frac{\det\left(\tilde{\mathbf{S}}\right)}{\prod_{p=1}^{P} \det\left(\tilde{\mathbf{S}}_{pp}\right)},$$

where $\tilde{\mathbf{S}} = \left(\mathbf{I}_P \otimes \mathbf{R}_{\mathbf{uu}}^{-1/2}\right) \mathbf{S} \left(\mathbf{I}_P \otimes \mathbf{R}_{\mathbf{uu}}^{-1/2}\right)$. Taking into account that $\mathbf{S} \sim \mathcal{CW}_{PL}(\mathbf{I}_P \otimes \mathbf{R}_{\mathbf{uu}}, N)$, it is easy to show that $\tilde{\mathbf{S}} \sim \mathcal{CW}_{PL}(\mathbf{I}_P \otimes \mathbf{I}_L, N)$, making λ_I an ancillary statistic because its distribution does not depend on $\mathbf{R}_{\mathbf{uu}}$ under the null. Then, according to Basu's theorem, λ_I is independent of $\mathbf{S}_{11}, \ldots, \mathbf{S}_{PP}$, and, as a consequence, λ_I and λ_E are independent under the null.

The stochastic representation of λ_I has been obtained in Sect. H.1. The stochastic representation of λ_E is an extension to complex variables of the distribution for the real case derived in [13]. The proof is omitted here as it is rather technical and does not provide any additional insight with respect to that in [13] for the real case. The stochastic representation is

$$\lambda_E \stackrel{d}{=} P^{LP} \prod_{p=1}^{P-1} \prod_{l=1}^{L} A_{p,l}^{p} \left(1 - A_{p,l}\right) B_{p,l}^{p+1},$$

where

$$A_{p,l} \sim \text{Beta}(Np - l + 1, N - l + 1), \quad B_{p,l} \sim \text{Beta}(N(p+1) - 2l + 2, l - 1),$$

are independent. Hence, the stochastic representation of the block-sphericity GLR is

$$\lambda_S \stackrel{d}{=} P^{LP} \prod_{p=1}^{P-1} \prod_{l=1}^{L} U_{p,l} A_{p,l}^{p} \left(1 - A_{p,l}\right) B_{p,l}^{p+1}, \tag{H.1}$$

where $A_p^{(l)}$ and $B_p^{(l)}$ are defined above. As a reminder, all of the beta-distributed random variables are independent.

The stochastic representation in (H.1) can be specialized to the sphericity test by simply considering $L = 1$. Hence, the stochastic representation of the sphericity test is

$$\lambda_S \stackrel{d}{=} P^P \prod_{p=1}^{P-1} U_p A_p^p \left(1 - A_p\right), \tag{H.2}$$

where $U_p \sim \text{Beta}(N - p, p)$ and $A_p \sim \text{Beta}(Np, N)$. Whenever the second parameter of a beta distribution is zero, it is deterministic at 1. This result is much more complicated than the result in Sect. 4.5, but both are stochastic representations of the same statistic. In the following, we show that the moments of both stochastic representations are identical.

The rth moment of (H.2) is given by

$$\mathrm{E}[\lambda_S^r] = P^{Pr} \prod_{p=1}^{P-1} \mathrm{E}\left[U_p^r\right] \mathrm{E}\left[A_p^{pr} (1 - A_p)^r\right],$$

which requires the computation of two expectations. The first is

$$\mathrm{E}\left[U_p^r\right] = \frac{\Gamma(N-p+r)\Gamma(N)}{\Gamma(N-p)\Gamma(N+r)},$$

and the second is

$$\mathrm{E}\left[A_p^{pr} (1 - A_p)^r\right] = \frac{\Gamma((N+r)p)\Gamma(N+r)\Gamma(N(p+1))}{\Gamma(Np)\Gamma(N)\Gamma((N+r)(p+1))}.$$

Then, $\mathrm{E}[\lambda_S^r]$ is

$$\begin{aligned}
\mathrm{E}[\lambda_S^r] &= P^{Pr} \prod_{p=1}^{P-1} \frac{\Gamma(N-p+r)}{\Gamma(N-p)} \frac{\Gamma((N+r)p)}{\Gamma((N+r)(p+1))} \frac{\Gamma(N(p+1))}{\Gamma(Np)} \\
&= P^{Pr} \frac{\Gamma(N+r)}{\Gamma((N+r)P)} \frac{\Gamma(NP)}{\Gamma(N)} \prod_{p=1}^{P-1} \frac{\Gamma(N-p+r)}{\Gamma(N-p)} \\
&= P^{Pr} \frac{\Gamma(NP)}{\Gamma((N+r)P)} \prod_{p=0}^{P-1} \frac{\Gamma(N-p+r)}{\Gamma(N-p)}.
\end{aligned}$$

After a change of variable (in p) and a substitution of L for P, this is the rth moment of the stochastic representation in Sect. 4.5.

References

1. Y.I. Abramovich, O. Besson, Fluctuating target detection in fluctuating K-distributed clutter. IEEE Signal Process. Lett. **22**(10), 1791–1795 (2015)
2. Y.I. Abramovich, A. Gorokhov, Expected-likelihood versus maximum-likelihood estimation for adaptive detection with an unconditional (stochastic) Gaussian interference model, in *Asilomar Conference on Signals, Systems, and Computers* (2005)
3. Y.I. Abramovich, N. Spencer, A. Gorokhov, Bounds on maximum likelihood ratio — Part I: Application to antenna array detection-estimation with perfect wavefront coherence. IEEE Trans. Signal Process. **52**(6), 1524–1536 (2014)
4. P.A. Absil, R. Mahony, R. Sepulchre, *Optimization Algorithms on Matrix Manifolds* (Princeton University Press, Princeton, 2008)
5. T. Adali, P.J. Schreier, L.L. Scharf, Complex-valued signal processing: The proper way to deal with impropriety. IEEE Trans. Signal Process. **59**(11), 5101–5125 (2011)
6. P. Abbaddo, D. Orlando, G. Ricci, L.L. Scharf. A unified theory of adaptive subspace detection. Part II: Numerical results. IEEE Trans. Signal Process. **70**(10), 4939–4950 (2022)
7. S. Ali, D. Ramírez, M. Jansson, G. Seco-Granados, J.A. López-Salcedo, Multi-antenna spectrum sensing by exploiting spatio-temporal correlation. EURASIP J. Adv. Signal Process. 160 (2014)
8. J. Alvarez-Vizoso, R. Arn, M. Kirby, C. Peterson, B. Draper, Geometry of curves in $\mathbb{R}^n$ from the local singular value decomposition. Linear Algebr. Appl. **571**, 180–202 (2019)
9. J. Alvarez-Vizoso, M. Kirby, C. Peterson, Local eigenvalue decomposition for embedded Riemannian manifolds. Linear Algebra Appl. **604**, 21–51 (2020)
10. J. Alvarez-Vizoso, M. Kirby, C. Peterson, Manifold curvature learning from hypersurface integral invariants. Linear Algebra Appl. **602**, 179–205 (2020)
11. S. Amari, *Differential-Geometrical Methods in Statistics.* Lecture Notes in Statistics (Springer, Berlin, 2000)
12. S. Amari, H. Nagaoka, *Methods of Information Geometry* Translations of Mathematical Monographs, vol. 191 (Oxford University Press, Oxford, 2000)
13. T.W. Anderson, *An Introduction to Multivariate Statistical Analysis* (Wiley, New York, 1958)
14. T.W. Anderson, Asymptotic theory for principal component analysis. Ann. Math. Statist. **34**(1), 122–148 (1963)
15. N. Aronszajn, Theory of reproducing kernels. Trans. Amer. Math. Soc. **68**, 307–404 (1950)
16. S. Ashrafulla, J.P. Haldar, A.A. Joshi, R.M. Leahy, Canonical Granger causality between regions of interest. NeuroImage **83**, 189–199 (2013)
17. E. Axell, E.G. Larsson, Multiantenna spectrum sensing of a second-order cyclostationary signal, in *IEEE International Workshop on Computational Advances in Multi-Sensor Adaptive Processing* (2011), pp. 329–332
18. E. Axell, G. Leus, E.G. Larsson, H.V. Poor, Spectrum sensing for cognitive radio: state-of-the-art and recent advances. IEEE Signal Process. Mag. **29**(3), 101–116 (2012)
19. A. Balleri, A. Nehorai, J. Wang, Maximum likelihood estimation for compound Gaussian clutter with inverse gamma texture. IEEE Trans. Aero. Electr. Syst. **43**(2), 775–779 (2007)

D. Ramírez et al., *Coherence*, https://doi.org/10.1007/978-3-031-13331-2

20. F. Bandiera, O. Besson, D. Orlando, G. Ricci, L.L. Scharf, GLRT-based direction detectors in homogeneous noise and subspace interference. IEEE Trans. Signal Process. **55**(6), 2386–2394 (2007)
21. F. Bandiera, A. De Maio, A.S. Greco, G. Ricci, Adaptive radar detection of distributed targets in homogeneous and partially homogeneous noise plus subspace interference. IEEE Trans. Signal Process. **55**(4), 1223–1237 (2007)
22. O. Baneerje, L. El Gahoui, A. d'Aspremont, Model selection through sparse maximum likelihood estimation for multivariate Gaussian or binary data. J. Mach. Learn. Res. **9**, 485–516 (2008)
23. E.W. Barankin. Locally best unbiased estimates. Ann. Math. Statist. **20**, 477–501 (1949)
24. L. Barnett, A.B. Barrett, A.K. Seth, Granger causality and transfer entropy are equivalent for Gaussian variables. Phys. Rev. Lett. **103**, 238701 (2009)
25. L. Barnett, A.K. Seth, Granger causality for state-space models. Phys. Rev. E **91**, 040101 (2015)
26. R. Basri, D.W. Jacobs, Lambertian reflectance and linear subspaces. IEEE Trans. Pattern Anal. Mach. Intell. **2**(25), 218–233 (2003)
27. D. Basu, On statistics independent of a complete sufficient statistic. Sankhyā Indian J. Statist. (1933–1960) **15**(4), 377–380 (1955)
28. R.T. Behrens L.L. Scharf, Signal processing applications of oblique projection operators. IEEE Trans. Signal Process. **42**(6), 1413–1424 (1994)
29. M. Beko, J. Xavier, V.A.N. Barroso, Noncoherent communications in multiple-antenna systems: receiver design and codebook construction. IEEE Trans. Signal Process. **55**(12), 5703–5715 (2007)
30. J. Benesty, J. Chen, Y. Huang, Estimation of the coherence function with the MVDR approach, in *IEEE International Conference on Acoustics, Speech, and Signal Processing* (2006), pp. 500–503
31. O. Besson, L.L. Scharf, CFAR matched direction detector. IEEE Trans. Signal Process. **54**(7), 2840–2844 (2006)
32. O. Besson, L.L. Scharf, S. Kraut, Adaptive detection of a signal known only to lie on a line in a known subspace, when primary and secondary data are partially homogenous. IEEE Trans. Signal Process. **54**(12), 4698–4705 (2005)
33. O. Besson, L.L. Scharf, F. Vincent, Matched direction detectors and estimators for array processing with subspace steering vector uncertainties. IEEE Trans. Signal Process. **53**(12), 4453–4463 (2005)
34. O. Besson, S. Kraut, L.L. Scharf, Detection of an unknown rank-one component in white noise. IEEE Trans. Signal Process. **54**(7), 2835–2839 (2006)
35. O. Besson, A. Coluccia, E. Chaumette, G. Ricci, F. Vincent, Generalized likelihood ratio test for detection of Gaussian rank-one signals in Gaussian noise with unknown statistics. IEEE Trans. Signal Process. **65**(4), 1082–1092 (2016)
36. A. Bhattacharyya, On some analogues of the amount of information and their use in statistical estimation. Sankhyā Indian J. Statist. (1933-1960) **8**, 1–14 (1946)
37. C. Bingham, An antipodally symmetric distribution on the sphere. Ann. Statist. **2**, 1201–1225 (1974)
38. A. Björk, G.H. Golub, Numerical methods for computing angles between linear subspaces. Math. Comput. **37**(123), 579–594 (1973)
39. D.W. Bliss, P.A. Parker, Temporal synchronization of MIMO wireless communication in the presence of interference. IEEE Trans. Signal Process. **58**(3), 1794–1806 (2010)
40. B. Bobrovsky, M. Zakai, A lower bound on the estimation error for certain diffusion processes. IEEE Trans. Inf. Theory **22**(1), 45–52 (1976)
41. S. Bose, A.O. Steinhardt, A maximal invariant framework for adaptive detection with structured and unstructured covariance matrices. IEEE Trans. Signal Process. **43**(9), 2164–2175 (1995)
42. S. Bose, A.O. Steinhardt, Adaptive array detection of uncertain rank-one waveforms. IEEE Trans. Signal Process. **44**(11), 2164–2175 (1996)

43. G.E.P. Box, A general distribution theory for a class of likelihood criteria. Biometrika **36**(3/4), 317–346 (1949)
44. S. Boyd, L. Vandenberghe, *Convex Optimization* (Cambridge University Press, Cambridge, 1972.)
45. P.S. Bradley, O.L. Mangasarian, K-plane clustering. J. Global Opt. **16**(1), 23–32 (2000)
46. Y. Bresler, Maximum likelihood estimation of a linearly structured covariance with application to antenna array processing, in *Annual ASSP Work. Spectrum Estimation and Modeling* (1988), pp. 172–175
47. Y. Bresler, A. Makovski, Exact maximum likelihood parameter estimation of superimposed exponential signals in noise. IEEE Trans. Acoust. Speech Signal Process. **34**(5), 307–310 (1986)
48. E. Broszkiewicz-Suwaj, Methods for determining the presence of periodic correlation based on the bootstrap methodology. Technical Report Research Report HSC/03/2, Wroclaw University of Technology (2003)
49. E. Broszkiewicz-Suwaj, A. Makagon, R. Weron, A. Wylomańska, On detecting and modeling periodic correlation in financial data. Phys. A Statist. Mech. Appl. **336**, 196–205 (2004)
50. K.A. Burgess, B.D. Van Veen, Subspace-based adaptive generalized likelihood ratio detectors. IEEE Trans. Signal Process. **44**(2), 912–927 (1996)
51. R.W. Butler, P. Pakrooh, L.L. Scharf, A MIMO version of the Reed-Yu detector and its connection to the Wilks Lambda and Hotelling t^2 statistics. IEEE Trans. Signal Process. **68**, 2925–2934 (2020)
52. D. Cabric, Addressing the feasibility of cognitive radios. IEEE Signal Process. Mag. **25**(6), 85–93 (2008)
53. L. Cai, H. Wang, A persymmetric multiband GLR algorithm. IEEE Trans. Aero. Electr. Syst. **28**, 3253–3256 (1992)
54. T.T. Cai, L. Wang, Orthogonal matching pursuit for sparse signal recovery with noise. IEEE Trans. Inf. Theory **57**(7), 4680–4688 (2011)
55. R.B. Calinski, J. Harabasz, A dendrite method for cluster analysis. Commun. Statist. **3**, 1–27 (1974)
56. E.J. Candès, The restricted isometry property and its implications for compressed sensing. Comptes Rendus. Mathematique **346**, 589–592 (2008)
57. E.J. Candès, B. Recht, Exact matrix completion via convex optimization. Found. Comput. Math. **9**, 717–772 (2009)
58. E.J. Candès, T. Tao, Decoding by linear programming. IEEE Trans. Inf. Theory **51**, 4203–4215 (2005)
59. E.J. Candès, T. Tao, The Dantzig selector: Statistical estimation when p is much larger than n. Ann. Statist. **35**(6), 2313–2351 (2007)
60. E.J. Candès, M.B. Wakin, An introduction to compressive sampling. IEEE Signal Process. Mag. **25**(2), 21–30 (2008)
61. E.J. Candès, M.B. Wakin, S.P. Boyd, Enhancing sparsity by reweighted l_1 minimization. J. Fourier Anal. App. **14**, 877–905 (2008)
62. L. Cardeño, D.K. Nagar, Testing block sphericity of a covariance matrix. Divulgaciones Matemáticas **9**(1), 25–34 (2001)
63. J.D. Carrol, Generalization of canonical correlation analysis to three or more sets of variables, in *Proceedings of the 76th Annual Convention of the American Psychological Association* (1968), pp. 227–228
64. G.C. Carter, Coherence and time delay estimation. Proc. IEEE **75**, 236–255 (1987)
65. G.C. Carter, A.H. Nuttall, P.G. Cable, The smoothed coherence transform. Proc. IEEE **61**, 1497–1498 (1973)
66. S. Chandna, A.T. Walden, A frequency domain test for propriety of complex-valued vector time series. IEEE Trans. Signal Process. **65**(6), 1425–1436 (2017)
67. D.G. Chapman, H. Robbins, Minimum variance estimation without regularity assumptions. Ann. Math. Statist. **22**(4), 581–586 (1951)
68. K.-C. Chen, R. Prasad, *Cognitive Radio Networks* (Wiley, Hoboken, 2009)

69. J.Y. Chen, I.S. Reed, A detection algorithm for optical targets in clutter. IEEE Trans. Aero. Electr. Syst. **23**(1), 46–59 (1987)
70. W.-S. Chen, I.S. Reed, A new CFAR detection test for radar. Digital Signal Process. **1**(4), 198–214 (1991)
71. J. Chen, G. Wang, G.B. Giannakis, Graph multiview canonical correlation analysis. IEEE Trans. Signal Process. **67**(11), 2826–2838 (2019)
72. Y. Chi, L.L. Scharf, A. Pezeshki, A.R. Calderbank, Sensitivity to basis mismatch in compressed sensing. IEEE Trans. Signal Process. **59**(5), 2182–2195 (2011)
73. Y. Chikuse, The matrix angular central Gaussian distribution. Multivariate Analy. **33**, 265–274 (1990)
74. Y. Chikuse, *Statistics on Special Manifolds* (Springer, Berlin, 2003)
75. D.S. Coates, P.J. Diggle, Tests for comparing two estimated spectral densities. J. Time Ser. Analy. **7**(1), 7–20 (1986)
76. D. Cochran, H. Gish, Multiple-channel detection using generalized coherence, in *IEEE International Conference on Acoustics, Speech and Signal Processing*, vol. 5 (1989), pp. 2883–2886
77. D. Cochran, H. Gish, D. Sinno, A geometric approach to multiple-channel signal detection. IEEE Trans. Signal Process. **43**(9), 2049–2057 (1995)
78. P.C. Consul, The exact distribution of likelihood criteria of different hypotheses, in ed. by P.R. Krishnaian, *Multivariate Analysis* (Academic, New York, 1969), pp. 171–181
79. E. Conte, A. De Maio, Exploiting persymmetry for CFAR detection in compound-Gaussian clutter. IEEE Trans. Aero. Electr. Syst. **39**, 719–724 (2003)
80. E. Conte, M. Lops, G. Ricci, Asymptotically optimum radar detection in compound Gaussian clutter. IEEE Trans. Aero. Electr. Syst. **31**(2), 617–625 (1995)
81. E. Conte, M. Lops, G. Ricci, Adaptive matched filter detection in spherically invariant noise. IEEE Signal Process. Lett. **3**(8), 912–927 (1996)
82. E. Conte, A. De Maio, G. Ricci, GLRT-based adaptive detection algorithms for range-spread targets. IEEE Trans. Signal Process. **49**(7), 1336–1348 (2001)
83. E. Conte, A. De Maio, G. Ricci, CFAR detection of distributed targets in non-Gaussian disturbance. IEEE Trans. Aero. Electr. Syst. **38**(2), 612–621 (2002)
84. J.H. Conway, R.H. Hardin, N.J.A. Sloane, Packing lines, planes, etc.: Packings in Grassmannian spaces. Exper. Math. **5**(2), 139–159 (1996)
85. B.R. Correia, C.A. Coelho, F.J. Marques, Likelihood ratio test for the hyper-block matrix sphericity covariance structure — Characterization of the exact distribution and development of near-exact distributions for the test statistic. REVSTAT - Statist. J. **16**(3), 365–403 (2018)
86. C. Cortes, V. Vapnik, Support-vector networks. Mach. Learn. **20**(3), 273–297 (1995)
87. T.M. Cover, J.A. Thomas, *Elements of Information Theory* (Wiley-Interscience, Hoboken, 2006)
88. H. Cox, Resolving power and sensitivity to mismatch of optimum array processors. Acoust. Soc. Amer. J. **54**(3), 771 (1973)
89. H. Cox, R. Zeskind, M. Owen, Robust adaptive beamforming. IEEE Trans. Acoust. Speech Signal Process. **35**(10), 1365–1376 (1987)
90. H. Cramér, *Mathematical Methods of Statistics* (Princeton University Press, Princeton, 1946)
91. I. Csiszár, J. Körner, *Information Theory: Coding Theorems for Discrete Memoryless Systems* (Cambridge University Press, Cambridge, 2011)
92. G. Cui, J. Liu, H. Li, B. Himed, Signal detection with noisy reference for passive sensing. Signal Process. **108**, 389–399 (2015)
93. A.V. Dandawaté, G.B. Giannakis, Statistical tests for presence of cyclostationarity. IEEE Trans. Signal Process. **42**(9), 2355–2369 (1994)
94. S. Dasgupta, A. Gupta, An elementary proof of a theorem of Johnson and Lindenstrauss. Random Struct. Algor. **22**(1), 60–65 (2002)
95. S. Datta, S. Howard, D. Cochran, Geometry of the Welch bounds. Linear Algebra Appl. **437**(10), 2455–2470 (2012)

96. D.L. Davies, D.W. Bouldin, A cluster separation measure. IEEE Trans. Pattern Anal. Mach. Intell. **1**(2), 224–227 (1979)
97. G. Davis, S. Mallat, Z. Zhang, Adaptive time-frequency decompositions with matching pursuits. Optical Eng. **33**(7), 2183–2191 (1993)
98. A.P. Dawid, Some matrix-variate distribution theory: Notational considerations and a Bayesian application. Biometrika **68**, 265–274 (1981)
99. K. Dedecius, *Partial Forgetting in Bayesian Estimation*. PhD thesis, Czech Technical University, Prague, Czech Republic, 2010
100. I.S. Dhillon, R.W. Heath Jr., T. Strohmer, J.A. Tropp, Constructing packings in Grassmannian manifolds via alternating projection. Exper. Math. **17**, 9–35 (2008)
101. G. Dietl, W. Utschick, On reduced-rank approaches to matrix Wiener filters in MIMO systems, in *IEEE International Symposium on Signal Processing and Information Technology* (2003), pp. 82–85
102. P.J. Diggle, *Time Series* (Oxford University Press, Oxford, 1990)
103. P.J. Diggle, N.I. Fisher, Nonparametric comparison of cumulative periodograms. J. R. Stat. Soc. Ser. C (App. Stat.) **40**(3), 423–434 (1991)
104. R.A. Dobie, Analysis of auditory evoked potentials by magnitude squared coherence. Ear Hear. **10**(1), 2–13 (1989)
105. D.L. Donoho, Compressed sensing. IEEE Trans. Inf. Theory **52**(4), 1289–1306 (2006)
106. D.L. Donoho, X. Huo, Uncertainty principles and ideal atomic decomposition. IEEE Trans. Inf. Theory **47**(7), 2845–2862 (2001)
107. T.D. Downs, Orientation statistics. Biometrika **59**, 665–676 (1972)
108. B. Draper, M. Kirby, J. Marks, T. Marrinan, C. Peterson, A flag representation for finite collections of subspaces of mixed dimensions. Numer. Linear Algebra Appl. **451**, 15–32 (2014)
109. R.O. Duda, P.E. Hart, D.G. Stork, *Pattern Classification* (Wiley, Hoboken, 2001)
110. M.L. Eaton, On the projection of isotropic distributions. Ann. Statist. **9**(2), 391–400 (1981)
111. M.L. Eaton, *Multivariate Statistics* Institute of Mathematical Statistics (1983)
112. A. Edelman, Volumes and integration. Finite random matrix theory (Handout notes) (2005). http://web.mit.edu/18.325/www/handouts.html, Accessed 20 Oct 2021
113. A. Edelman, Y. Wang, The GSVD: Where are the ellipses?, matrix trigonometry, and more. SIAM J. Matrix Anal. Appl. **41**(4), 1826–1856 (2020)
114. A. Edelman, T. Arias, S.T. Smith, The geometry of algorithms with orthogonality constraints. SIAM J. Matrix Anal. Appl. **20**(2), 303–353 (1998)
115. Y.C. Eldar, A. Beck, Hidden convexity based near maximum-likelihood CDMA detection, in *IEEE International Workshop on Signal Processing Advances in Wireless Communications* (2005)
116. S. Enserink, D. Cochran, On detection of cyclostationary signals, in *IEEE International Conference on Acoustics, Speech, and Signal Processing* (1995), pp. 2004–2007
117. R.P. Feynman, *QED: The Strange Theory of Light and Matter* (Princeton University Press, Princeton, 1985)
118. R.A. Fisher, Frequency distribution of the values of the correlation coefficient in samples from an indefinitely large population. Biometrika **10**(4), 507–521 (1915)
119. R.A. Fisher, The general sampling distribution of the multiple correlation coefficient. Proc. R. Soc. Lond. **121**, 654–673 (1928)
120. F.H.P. Fitzek, M.D. Katz (eds.), *Cooperation in Wireless Networks: Principles and Applications* (Springer, Berlin, 2006)
121. P. Flandrin, *Temps-Fréquence*. Hermes, Paris, France (1993)
122. P. Flandrin, *Time-Frequency/Time-Scale Analysis*, vol. 10 (Academic, San Diego, 1998)
123. K. Fokianos, A. Savvides, On comparing several spectral densities. Technometrics **50**(3), 317–331 (2008)
124. G.E. Forsythe, G.H. Golub, On the stationary values of a second-degree polynomial on the unit sphere. J. Soc. Indust. Appl. Math. **13**(4), 1050–1068 (1965)

125. R. Frieden, Restoring with maximum likelihood and maximum entropy. J. Optical Soc. of Amer. **62**(4), 511–518 (1972)
126. X. Fu, K. Huan, E.E. Paplexakis, H.A. Song, P. Talukdar, N.D. Sidiropoulos, C. Faloutsos, T. Mitchel, Efficient and distributed generalized canonical correlation analysis for big multiview data. IEEE Trans. Knowl. Data Eng. **31**(12), 2304–2318 (2019)
127. W.A. Gardner, A unifying view of coherence in signal processing. Signal Process. **29**(2), 113–140 (1992)
128. V. Garg, I. Santamaria, D. Ramírez, L.L. Scharf, Subspace averaging and order determination for source enumeration. IEEE Trans. Signal Process. **67**, 3028–3041 (2019)
129. J. Geweke, Measurement of linear dependence and feedback between multiple time series. J. Am. Stat. Assoc. **77**(378), 304–313 (1982)
130. J. Geweke, Measures of conditional linear dependence and feedback between time series. J. Am. Stat. Assoc. **79**(388), 907–915 (1984)
131. F. Gini, M. Greco, Covariance matrix estimation for CFAR detection in correlated heavy tailed clutter. Signal Process. **82**, 1847–1859 (2002)
132. N. Giri, On the complex analogues of T^2 and R^2 tests. Ann. Math. Statist. **36**, 664–670 (1965)
133. H. Gish, D. Cochran, Generalized coherence, in *IEEE International Conference on Acoustics, Speech, and Signal Processing*, vol. 5 (1987), pp. 2745–2748
134. E.D. Gladyshev, Periodically correlated random sequences. Soviet Math. Dokl. **2**, 385–388 (1961)
135. S. Gogineni, P. Setlur, M. Rangaswamy, R.R. Nadakuditi, Passive radar detection with noisy reference channel using principal subspace similarity. IEEE Trans. Aero. Electr. Syst. **454**(1), 18–36 (2018)
136. R.H. Gohary, T.N. Davidson, Noncoherent MIMO communication: Grassmannian constellations and efficient detection. IEEE Trans. Inf. Theory **55**(3), 1176–1205 (2009)
137. L. Goldfarb, A unified approach to pattern recognition. Pattern Recog. **17**(5), 575–582 (1984)
138. A. Goldsmith, S.A. Jafar, I. Maric, S. Srinivasa, Breaking spectrum gridlock with cognitive radios: an information theoretic perspective. Proc. IEEE **97**(5), 894–914 (2009)
139. J.S. Goldstein, I.S. Reed, Reduced-rank adaptive filtering. IEEE Trans. Signal Process. **45**(2), 492–496 (1997)
140. J.S. Goldstein, I.S. Reed, L.L. Scharf, A multistage representation of the Wiener filter based on orthogonal projections. IEEE Trans. Inf. Theory **44**(7), 2943–2949 (1998)
141. G.H. Golub, C.F. Van Loan, An analysis of the total least squares problem. SIAM J. Num. Analy. **17**, 883–893 (1983)
142. G.H. Golub, C.F. Van Loan, *Matrix Computations* (The Johns Hopkins University Press, Baltimore, 1983)
143. J.D. Gorman, A. Hero, Lower bounds for parametric estimation with constraints. IEEE Trans. Inf. Theory **26**(6), 1285–1301 (1990)
144. I.F. Gorodnitsky, B.D. Rao, Sparse signal reconstruction from limited data using FOCUSS: a re-weighted minimum norm algorithm. IEEE Trans. Signal Process. **45**(3), 600–616 (1997)
145. J.C. Gower, Some distance properties of latent roots and vector methods in multivariate analysis. Biometrika **53**, 315–328 (1966)
146. J.C. Gower, G.B. Dijksterhuis, *Procrustes Problems* (Oxford University Press, Oxford, 2004)
147. C.W.J. Granger, Investigating causal relations by econometric models and cross-spectral methods. Econometrica **37**(3), 424–438 (1969)
148. R.M. Gray, Toeplitz and circulant matrices: a review. Found. Trends Commun. Inf. Theory **2**(3), 155–239 (2006)
149. U. Grenander, G. Szegö, *Toeplitz Forms and Their Applications* (University of California Press, Berkeley, 1958)
150. H.D. Griffiths, C.J. Baker, Passive coherent location radar systems. Part 1: performance prediction. IEE Proc. Radar Sonar Navig. **152**(3), 124–132 (2005)
151. H.D. Griffiths, N.R.W. Long, Television-based bistatic radar. IEE Proc. F (Comm., Radar Signal Process.) **133**, 649–657 (1986)

152. A.K. Gupta, D.K. Nagar, *Matrix Variate Distributions* (Chapman & Hall/CRC, Boca Raton, 2000)
153. D.E. Hack, L.K. Patton, B. Himed, M.A. Saville, Detection in passive MIMO radar networks. IEEE Trans. Signal Process. **62**(11), 2999–3012 (2014)
154. R.R. Hagege, J.M. Francos, Universal manifold embedding for geometric deformed functions. IEEE Trans. Inf. Theory **62**(6), 3676–3684 (2016)
155. J.M. Hammersley, On estimating restricted parameters. J. R. Stat. Soc. Ser. B (Methodological) **12**(2), 192–240 (1950)
156. M.T. Harandi, M. Saltzmann, S. Jayasumana, R. Hartley, H. Li, Expanding the family of Grassmannian kernels: An embedding perspective, in *European Conference Computer Vision* (2014)
157. D.R. Hardoon, S. Szedmak, J. Shawe-Taylor, Canonical correlations analyisis: an overview with application to learning methods. Neural Comput. **16**(12), 2639–2664 (2004)
158. T. Hastie, R. Tibshirani, J. Friedman, *The Elements of Statistical Learning* (Springer Series in Statistics, New York, 2001)
159. L.D. Haugh, Checking the independence of two covariance-stationary time series: a univariate residual cross-correlation approach. Journal Ame. Stat. Assoc. **71**(354), 378–385 (1976)
160. U. Hemke, K. Hüper, J. Trumpf, Newton's method on Grassmann manifolds (2007). arXiv:0709.2205v2
161. M.A. Herman, T. Strohmer, General deviants: an analysis of perturbations in compressed sensing. IEEE J. Sel. Topics Signal Process. **4**(2), 342–349 (2010)
162. M.R. Hestenes, E. Stielfel, Methods of conjugate gradient for solving linear systems. J. Res. Nat. Bureau Standards **49**(6), 409–436 (1952)
163. S. Hiltunen, P. Loubaton, Asymptotic analysis of a GLR test for detection with large sensor arrays: New results, in *IEEE International Conference on Acoustics, Speech, and Signal Processing* (2017)
164. S. Hiltunen, P. Loubaton, P. Chevalier, Large system analysis of a GLRT for detection with large sensor arrays in temporally white noise. IEEE Trans. Signal Process. **63**(20), 5409–5423 (2015)
165. A. Hjørungnes, *Complex-Valued Matrix Derivatives* (Cambridge University Press, Cambridge, 2011)
166. F. Hlawatsch, *Time-Frequency Analysis and Synthesis of Linear Signal Spaces: Time-Frequency Filters, Signal Detection and Estimation, and Range-Doppler Estimation* (Kluwer Academic Publishers, Dordrecht, 1998)
167. F. Hlawatsch, W. Kozek, Time-frequency projection filters and time-frequency signal expansions. IEEE Trans. Signal Process. **42**(12), 3321–3334 (1994)
168. P.D. Hoff, Simulation of the matrix Bingham-von Mises-Fisher distribution, with applications to multivariate data and relational data. J. Comp. Graph. Stats. **18**(2), 438–456 (2009)
169. Y. Hong, Testing for independence between two covariance stationary time series. Biometrika **83**(3), 615–625 (1996)
170. R.A. Horn, C.R. Johnson, *Matrix Analysis* (Cambridge University Press, Cambridge, 1985)
171. P. Horst, Generalized canonical correlation analysis and their applications to experimental data. J. Clin. Psychol. **17**(4), 331–347 (1961)
172. S. Horstmann, D. Ramírez, P.J. Schreier, Two-channel passive detection of cyclostationary signals. IEEE Trans. Signal Process. **68**, 2340–2355 (2020)
173. H. Hotelling, Relations between two sets of variates. Biometrika **28**, 321–377 (1936)
174. S.D. Howard, W. Moran, P. Pakrooh, L.L. Scharf, Hilbert space geometry of quadratic performance bounds, in *Asilomar Conference on Signals, Systems, and Computers* (2017), pp. 1578–158
175. S.D. Howard, S. Sirianunpiboon, D. Cochran, The geometry of coherence and its application to cyclostationary time series, in *IEEE Workshop Statistical Signal Processing* (2018)
176. P.L. Hsu, On the distribution of roots of certain determinantal equations. Ann. Eugenics **9**, 250–258 (1939)

177. L.K. Hua, *Harmonic Analysis of Functions of Several Complex Variables in Classical Domains* (American Mathematical Society, Providence, 1963)
178. Y. Hua, M. Nikpour, P. Stoica, Optimal reduced-rank estimation and filtering. IEEE Trans. Signal Process. **49**(3), 457–469 (2001)
179. L. Huang, H.C. So, Source enumeration via MDL criterion based on linear shrinkage estimation of noise subspace covariance matrix. IEEE Trans. Signal Process. **61**(19), 4806–4821 (2013)
180. L. Huang, Y. Xiao, H.C. So, J.-K. Zhang, Bayesian information criterion for source enumeration in large-scale adaptive antenna array. IEEE Trans. Vehic. Tech. **65**(5), 3018–3032 (2016)
181. P.J. Huber, The behavior of maximum likelihood estimates under nonstandard conditions, in *Berkeley Symposium on Mathematical Statistics and Probability* (1967), pp. 221–233
182. H.L. Hurd, N.L. Gerr, Graphical methods for determining the presence of periodic correlation. J. Time Ser. Analy. **12**(4), 337–350 (1991)
183. S. Huzurbazar, R.W. Butler, Importance sampling for p-value computations in multivariate tests. J. Comp. Graph. Stats. **7**(3), 342–355 (1998)
184. A.T. James, Distributions of matrix variates and latent roots derived from normal samples. Ann. Math. Statist. **35**(2), 475–501 (1964)
185. Y. Jin, B. Friedlander, A CFAR adaptive subspace detector for second-order Gaussian signals. IEEE Trans. Signal Process. **53**(3), 871–884 (2005)
186. S. John, Some optimal multivariate tests. Biometrika **58**(1), 123–127 (1971)
187. W.B. Johnson, J. Lindestrauss, Extensions of Lipzchitz maps into Hilbert space. Contemp. Math. **26**, 189–206 (1984)
188. K.G. Jöreskog, Testing a simple structure hypotheses in factor analysis. Psychometrika **31**, 165–178 (1966)
189. K.G. Jöreskog, Some contributions to maximum likelihood factor analysis. Psychometrika **32**, 443–482 (1967)
190. H. Karcher, Riemannian center of mass and mollifier smoothing. Comm. Pure Appl. Math. **5**(30), 509–541 (1977)
191. S. Kay, Exponentially embedded families: new approaches to model order estimation. IEEE Trans. Aero. Electr. Syst. **41**(1), 333–345 (2005)
192. S.M. Kay, J.R. Gabriel, An invariance property of the generalized likelihood ratio test. IEEE Signal Process. Lett. **10**(12), 352–355 (2003)
193. E.J. Kelly, An adaptive detection algorithm. IEEE Trans. Aero. Electr. Syst. **22**(2), 115–127 (1986)
194. E.J. Kelly, K. Forsythe, Adaptive detection and parameter estimation for multidimensional signal models. Technical Report 848, MIT Lincoln Labs (1989)
195. J.R. Kettenring, Canonical analysis of several sets of variables. Biometrika **58**(3), 433–451 (2019)
196. K. Khamaru, R. Mazumder, Computation of the maximum likelihood estimator in low-rank factor analysis. Math. Program. **176**(1), 279–310 (2019)
197. C.G. Khatri, Distribution of the largest or the smallest characteristic root under null hypothesis concerning complex multivariate normal populations. Ann. Math. Statist. **35**(4), 1807–1810 (1964)
198. C.G. Khatri, Notes on multiple and canonical correlation for a singular covariance matrix. Psychometrika **41**(4), 465–470 (1976)
199. C.G. Khatri, C.R. Rao, Effects of estimated noise covariance matrix in optimal signal detection. IEEE Trans. Acoust. Speech Signal Process. **35**(5), 671–679 (1987)
200. G. Kimeldorf, G. Wahba, A correspondence between Bayesian estimation of stochastic processes and smoothing by splines. Ann. Math. Statist. **41**, 495–502 (1970)
201. N. Klausner, M.R. Azimi-Sadjadi, L.L. Scharf, Detection of spatially-correlated time series from a network of sensor arrays. IEEE Trans. Signal Process. **62**(6), 1396–1407 (2014)

202. N. Klausner, M.R. Azimi-Sadjadi, L.L. Scharf, Saddlepoint approximations for correlation testing among multiple Gaussian random vectors. IEEE Signal Process. Lett. **23**(5), 703–707 (2016)
203. P. Koev, A. Edelman, The efficient evaluation of the hypergeometric function of a matrix argument. Math. Comput. **75**(274), 833–846 (2006)
204. S. Kraut, L.L. Scharf, The CFAR adaptive subspace detector is a scale-invariant GLRT. IEEE Trans. Signal Process. **47**(9), 2538–2541 (1999)
205. S. Kraut, L.L. Scharf, L.T. McWhorter, Adaptive subspace detectors. IEEE Trans. Signal Process. **49**(1), 1–16 (2001)
206. S. Kraut, L.L. Scharf, R.W. Butler, The adaptive coherence estimator: a uniformly most-powerful-invariant adaptive detection statistic. IEEE Trans. Signal Process. **53**(2), 427–438 (2005)
207. A.N. Kshirsagar, *Multivariate Analysis* (Dekker, New York, 1972)
208. R. Kumaresan, D.W. Tufts, Estimating the parameters of exponentially damped sinuoids and pole-zero modeling in noise. IEEE Trans. Acoust. Speech Signal Process. **30**, 833–840 (1982)
209. R. Kumaresan, L.L. Scharf, A.K. Shaw, An algorithm for pole-zero modelling and spectral analysis. IEEE Trans. Acoust. Speech Signal Process. **34**, 637–640 (1986)
210. S.Y. Kung, *Kernel Methods and Machine Learning* (Cambridge University Press, Cambridge, 2014)
211. N. Laneman, D. Tse, G. Wornell, Cooperative diversity in wireless networks: efficient protocols and outage behavior. IEEE Trans. Inf. Theory **50**(12), 3062–3080 (2004)
212. D.N. Lawley, A.E. Maxwell, Factor analysis as a statistical method. J. R. Stat. Soc. Ser. D (The Statistician) **12**(3), 209–229 (1962)
213. D.N. Lawley, A.E. Maxwell, *Factor Analysis as a Statistical Method* (American Elsevier, 1971)
214. E.L. Lehmann, Some principles of the theory of testing hypotheses. Ann. Math. Statist. **21**, 1–26 (1950)
215. E.L. Lehmann, J.P. Romano, *Testing Statistical Hypotheses* (Springer, Berlin, 2005)
216. A. Leshem, A.-J. van der Veen, Multichannel detection of Gaussian signals with uncalibrated receivers. IEEE Signal Process. Lett. **8**(4), 120–122 (2001)
217. J. Li, P. Stoica, *MIMO Radar Signal Processing* (Wiley-IEEE Press, Hoboken, 2008)
218. F. Li, R.J. Vaccaro, Analysis of min-norm and MUSIC with arbitrary array geometries. IEEE Trans. Aero. Electr. Syst. **26**, 976–985 (1990)
219. F. Li, R.J. Vaccaro, Unified analysis for DOA estimation algorithms in array signal processing. Signal Process. **25**, 147–169 (1991)
220. W. Liu, P.P. Pokharel, J.C. Principe, The kernel least mean square algorithm. IEEE Trans. Signal Process. **56**(2), 543–554 (2008)
221. W. Liu, J.C. Principe, S. Haykin, *Kernel Adaptive Filtering* (Wiley, Hoboken, 2010)
222. J. Liu, H. Li, B. Himed, On the performance of the cross-correlation detector for passive radar applications. Signal Process. **113**, 32–37 (2015)
223. M. Loève, *Probability Theory II*, 4th edn. (Springer, New York, 1978)
224. Z. Lu, A.M. Zoubir, Generalized Bayesian information criterion for source enumeration in array processing. IEEE Trans. Signal Process. **61**(6), 1470–1480 (2013)
225. S.G. Mallat, Z. Zhang, Matching pursuits with time-frequency dictionaries. IEEE Trans. Signal Process. **41**(12), 3397–3415 (1993)
226. K.V. Mardia, *Statistics of Directional Data* (Academic, New York, 1972)
227. K.V. Mardia, J.T. Kent, J.M. Bibby, *Multivariate Analysis* (Academic, New York, 1979)
228. F.J. Marques, C.A. Coelho, P. Marques, The block-matrix sphericity test: Exact and near-exact distributions for the test statistic, in ed.by P.E. Oliveira, M.T. da Graca, C. Henriques, M. Vichi, *Recent Developments in Modeling and Applications in Statistics* (Springer, Berlin, 2013), pp. 169–177
229. T. Marrinan, J.R. Beveridge, B. Draper, M. Kirby, C. Peterson, Finding the subspace mean or median to fit your need, in *IEEE Conference on Computer Vision and Pattern Recognition* (2014), pp. 1082–1089

230. A.W. Marshall, I. Olkin, B.C. Arnold, *Inequalities: Theory of Majorization and Its Application* (Springer, Berlin, 2011)
231. W. Martin, Time-frequency analysis of random signals, in *IEEE International Conference on Acoustics, Speech, and Signal Processing*, vol. 7 (1982), pp. 1325–1328
232. W. Martin, P. Flandrin, Wigner-Ville spectral analysis of nonstationary processes. IEEE Trans. Acoust. Speech Signal Process. **33**(6), 1461–1470 (1985)
233. T.L. Marzetta, A simple derivation of the constrained multiple parameter Cramer-Rao bound. IEEE Trans. Signal Process. **41**(6), 2247–2249 (1993)
234. J.L. Massey, T. Mittelholzer, Welch's bound and sequence sets for code-division multiple-access systems, in ed. by R. Capocelli, A. De Santis, U. Vaccaro, *Sequences II* (Springer, Berlin, 1993), pp. 63–78
235. A.M. Mathai, P.N. Rathie, The exact distribution for the sphericity test. J. Statist. Res. **4**, 140–159 (1970)
236. J. Mauchly, Significance test for sphericity of a normal n-variate distribution. Ann. Math. Statist. **11**, 204–209 (1940)
237. J.H. McClellan, D. Lee, Exact equivalence of the Steiglitz-McBride iteration and IQML. IEEE Trans. Signal Process. **39**, 509–512 (1991)
238. L.T. McWhorter, L.L. Scharf, Matched subspace detectors for stochastic signals, in *Annual Adaptive Sensor Array Processing Workshop* (2003)
239. L.T. McWhorter, L.L. Scharf, Properties of quadratic performance bounds, in *Asilomar Conference on Signals, Systems, and Computers* (1993)
240. L.T. McWhorter, L.L. Scharf, L.J. Griffiths, Adaptive coherence estimation for radar signal processing, in *Asilomar Conference on Signals, Systems, and Computers* (1996)
241. J. Mercer, Functions of positive and negative type, and their connection with the theory of integral equations. Philosoph. Trans. Royal Soc. A **209**, 415–446 (1909)
242. H. Mheidat, M. Uysa, N. Al-Dhahir, Equalization techniques for distributed space-time block codes with amplify-and-forward relaying. IEEE Trans. Signal Process. **55**(5), 1839–1852 (2007)
243. J. Mitola, G.Q. Maguire Jr., Cognitive radio: making software radios more personal. IEEE Pers. Comm. **6**, 13–18 (1999)
244. R.J. Muirhead, *Aspects of Multivariate Statistical Theory* (Wiley, Hoboken, 2005)
245. R.R. Nadakuditi, A. Edelman, Sample eigenvalue based detection of high-dimensional signals in white noise with relatively few samples. IEEE Trans. Signal Process. **56**(7), 2625–2638 (2008)
246. H. Neudecker, Some theorems on matrix differentiation with special reference to Kronecker matrix products. Journal Ame. Stat. Assoc. **64**, 953–963 (1969)
247. A.A. Nielsen, Multiset canonical correlations analysis and multispectral, truly multitemporal remote sensing data. IEEE Trans. Image Process. **11**(3), 293–305 (2002)
248. F. Nielsen, An elementary introduction to information geometry. Entropy **22**(10) (2020)
249. A.H. Nuttall, Invariance of distribution of coherence estimate to second-channel statistics. IEEE Trans. Acoust. Speech Signal Process. **29**(1), 120–122 (1981)
250. E. Olilla, D.E. Tyler, V. Koivunen, H.V. Poor, Complex elliptical symmetric distributions: survey, new results and applications. IEEE Trans. Signal Process. **60**(11), 5597–5625 (2012)
251. E. Olilla, D.E. Tyler, V. Koivunen, H.V. Poor, Compound-Gaussian clutter modeling with an inverse-Gaussian texture distribution. IEEE Signal Process. Lett. **19**(12), 876–879 (2012)
252. I. Olkin, Testing and estimation for structures which are circularly symmetric in blocks. ETS Res. Bull. Ser. **1972**(2), i–20 (1972)
253. I. Olkin, H. Rubin, Multivariate beta distributions and independence properties of the Wishart distribution. Ann. Math. Statist. **35**, 261–269 (1964)
254. F.W.J. Olver, D.W. Lozier, R.F. Boisvert, C.W. Clark (eds.), *NIST Handbook of Mathematical Functions* (National Institute of Standards and Technology and Cambridge University Press, Cambridge, 2010)
255. D. Orlando, G. Ricci, L.L. Scharf. A unified theory of adaptive subspace detection. Part I: Detector designs. IEEE Trans. Signal Process. **70**(10), 4925–4938 (2022)

256. P.W. Otter, On Wiener-Granger causality, information and canonical correlation. Econ. Lett. **35**, 187–191 (1991)
257. P. Pakrooh, A. Pezeshki, L.L. Scharf, D. Cochran, S.D. Howard, Analysis of Fisher information and the Cramer-Rao bound for nonlinear parameter estimation after random compression. IEEE Trans. Signal Process. **63**(23), 6423–6428 (2015)
258. P. Pakrooh, L. Scharf, M. Cheney, A. Homan, M. Ferrara, The adaptive coherence estimator for detection in wind turbine clutter, in *IEEE Radar Conference* (2017)
259. P. Pakrooh, L.L. Scharf, R.W. Butler, Distribution results for a multirank version of the Reed-Yu detector, in *Asilomar Conference on Signals, Systems, and Computers* (2017)
260. Y.C. Pati, R. Rezaiifar, P.S. Krishnaprasad, Orthogonal matching pursuit: Recursive function approximation with applications to wavelet decomposition, in *Asilomar Conference on Signals, Systems, and Computers* (1993)
261. A. Paulraj, R. Roy, T. Kailath, A subspace rotation approach to signal parameter estimation. Proc. IEEE **74**, 1044–1045 (1986)
262. E. Pekalska, P. Paclick, R.P.W. Duin, A generalized kernel approach to dissimilarity-based classification. J. Mach. Learn. Res. **2**, 175–211 (2001)
263. K.B. Petersen, M.S. Pedersen, *The Matrix Cookbook* (Technical University of Denmark, Lyngby, 2012)
264. A. Pezeshki, L.L. Scharf, M.R. Azimi-Sadjadi, M. Lundberg, Empirical canonical correlation analysis in subspaces, in *Asilomar Conference on Signals, Systems, and Computers* (2004), pp. 994–997
265. R. Price, Introduction: Welcome to spacetime, in *The Future of Spacetime* (W. W. Norton, London, 2002)
266. A. Pries, D. Ramírez, P.J. Schreier, LMPIT-inspired tests for detecting a cyclostationary signal in noise with spatio-temporal structure. IEEE Trans. Wirel. Comm. **17**(9), 6321–6334 (2018)
267. R. Prony, Essai experimental et analytique: Sur es lois de la delatabile de fluides elastique et sur celles de la force expansive de la vapeur de l'alkool, a differentes temperatures. J. de l'Ecole Polytechnique Floreal et Pairiala **1**, 24–76, (1795)
268. D. Ramírez, J. Vía, I. Santamaria, L.L. Scharf, Detection of spatially-correlated Gaussian time series. IEEE Trans. Signal Process. **58**(10), 5006–5015 (2010)
269. D. Ramírez, J. Vía, I. Santamaria, L.L. Scharf, Multiple-channel detection of a Gaussian time series over frequency-flat channels, in *IEEE International Conference on Acoustics, Speech, and Signal Processing* (2011)
270. D. Ramírez, G. Vazquez-Vilar, R. López-Valcarce, J. Vía, I. Santamaria, Detection of rank-P signals in cognitive radio networks with uncalibrated multiple antennas. IEEE Trans. Signal Process. **59**(1), 3764–3774 (2011)
271. D. Ramírez, J. Vía, I. Santamaria, The locally most powerful test for multiantenna spectrum sensing with uncalibrated receivers, in *IEEE International Conference on Acoustics, Speech, and Signal Processing* (2012)
272. D. Ramírez, J. Iscar, J. Vía, I. Santamaria, L.L. Scharf, The locally most powerful invariant test for detecting a rank-P Gaussian signal in white noise, in *IEEE Sensor Array and Multichannel Signal Processing Workshop* (2012)
273. D. Ramírez, J. Vía, I. Santamaria, L.L. Scharf, Locally most powerful invariant tests for correlation and sphericity of Gaussian vectors. IEEE Trans. Inf. Theory **59**(4), 2128–2141 (2013)
274. D. Ramírez, P.J. Schreier, J. Vía, I. Santamaria, L.L. Scharf, Detection of multivariate cyclostationarity. IEEE Trans. Signal Process. **63**(20), 5395–5408 (2015)
275. D. Ramírez, D. Romero, J. Vía, R. López-Valcarce, I. Santamaria, Testing equality of multiple power spectral density matrices. IEEE Trans. Signal Process. **66**(23), 6268–6280 (2018)
276. D. Ramírez, I. Santamaria, S. Van Vaerenbergh, L.L. Scharf, Multi-channel factor analysis with common and unique factors. IEEE Trans. Signal Process. **68**, 113–126 (2020)
277. C.R. Rao, Information and the accuracy attainable in the estimation of statistical parameters. Bull. Calcutta Math. Soc. **37**, 81–89 (1945)

278. C.R. Rao, A lemma on G-inverse of a matrix and computation of correlation-coefficients in the singular case. Commun. Statist. Part A Theory Methods **10**(1), 1–10 (1981)
279. M. Razaviyayn, M. Hong, Z.Q. Luo, A unified convergence analysis of block successive minimization methods for nonsmooth optimization. SIAM J. Opt. **23**(2), 1126–1153 (2013)
280. I.S. Reed, X. Yu, Adaptive multiple-band CFAR detection of an optical pattern with unknown spectral distribution. IEEE Trans. Acoust. Speech Signal Process. **38**(10), 1760–1770 (1990)
281. I.S. Reed, J.D. Mallet, L.E. Brennan, Rapid convergence rate in adaptive arrays. IEEE Trans. Aero. Electr. Syst. **10**(6), 1760–1770 (1974)
282. G. Ricci, L.L. Scharf, Adaptive radar detection of extended Gaussian targets, in *Annual Adaptive Sensor Array Processing Workshop* (2004)
283. C. Richard, J.C.M. Bermudez, P. Honeine, Online prediction of time series data with kernels. IEEE Trans. Signal Process. **57**(3), 1058–1067 (2009)
284. C.D. Richmond, *Adaptive array signal processing and performance analysis in non-Gaussian environments*. PhD Thesis, MIT, Cambridge, 1996
285. C.D. Richmond, Statistical performance analysis of the adaptive sidelobe blanker detection algorithm, in *Asilomar Conference on Signals, Systems, and Computers* (1997), pp. 872–876
286. C.D. Richmond, Mean-squared error and threshold SNR prediction of maximum likelihood signal parameter estimation with estimated colored noise covariances. IEEE Trans. Inf. Theory **52**(5), 2146–2164 (2006)
287. C.D. Richmond, L.L. Horowitz, Parameter bounds on estimation accuracy under model misspecification. IEEE Trans. Signal Process. **63**(9), 2263–2278 (2015)
288. A. Rihaczek, Signal energy distribution in time and frequency. IEEE Trans. Inf. Theory **14**(3), 369–374 (1968)
289. F.C. Robey, D.R. Fuhrmann, E.J. Kelly, R. Nitzberg, A CFAR adaptive matched filter detector. IEEE Trans. Aero. Electr. Syst. **28**(1), 208–216 (1992)
290. A. Rosuel, P. Vallet, P. Loubaton, X. Mestre, On the detection of low-rank signal in the presence of spatially uncorrelated noise: a frequency domain approach. IEEE Trans. Signal Process. **69**, 4458–4473 (2021)
291. P.J. Rousseeuw, Silhouettes: a graphical aid to the interpretation and validation of cluster analysis. Comp. App. Math. **20**, 53–65 (1987)
292. S.N. Roy, *Some Aspects of Multivariate Analysis* (Wiley, Hoboken, 1957)
293. R. Roy, T. Kailath, ESPRIT–Estimation of signal parameters via rotational invariance techniques. IEEE Trans. Acoust. Speech Signal Process. **37**(7), 984–995 (1989)
294. J. Rupnik, P. Skraba, J. Shawe-Taylor, S. Guettes, A comparison of relaxations of multiset canonical correlation analysis and applications (2013). arXiv:1302.0974v1
295. I. Santamaria, V. Elvira, An efficient sampling scheme for the eigenvalues of dual Wishart matrices. IEEE Signal Process. Lett. **28**, 2177–2181 (2021)
296. I. Santamaria, J. Vía, Estimation of the magnitude squared coherence spectrum based on reduced-rank canonical coordinates, in *IEEE International Conference on Acoustics, Speech, and Signal Processing* (2007)
297. I. Santamaria, L.L. Scharf, D. Cochran, J. Vía, Passive detection of rank-one signals with a multiantenna reference signal, in *European Signal Processing Conference* (2016)
298. I. Santamaria, L.L. Scharf, M. Kirby, C. Peterson, J. Francos, An order fitting rule for optimal subspace averaging, in *IEEE Statistical Signal Processing Workshop* (2016)
299. I. Santamaria, L.L. Scharf, J. Vía, Y. Wang, H. Wang, Passive detection of correlated subspace signals in two MIMO channels. IEEE Trans. Signal Process. **65**(20), 5266–5280 (2017)
300. I. Santamaria, D. Ramírez, L.L. Scharf, Subspace averaging for source enumeration in large arrays, in *IEEE Statistical Signal Processing Workshop* (2018)
301. I. Santamaria, L.L. Scharf, D. Ramírez, Scale-invariant subspace detectors based on first- and second-order statistical models. IEEE Trans. Signal Process. **68**, 6432–6443 (2020)
302. L.L. Scharf, *Statistical Signal Processing: Detection, Estimation and Time Series Analysis* (Addison-Wesley, Boston, 1991)
303. L.L. Scharf, B. Friedlander, Matched subspace detectors. IEEE Trans. Signal Process. **42**(8), 2146–2157 (1994)

304. L.L. Scharf, L.T. McWhorter, Geometry of the Cramér-Rao bound. Signal Process. **31**(3), 301–311 (1993)
305. L.L. Scharf, L.T. McWhorter, Adaptive matched subspace detectors and adaptive coherence estimators, in *Asilomar Conference on Signals, Systems, and Computers* (1996)
306. L.L. Scharf, C.T. Mullis, Canonical coordinates and the geometry of inference, rate, and capacity. IEEE Trans. Signal Process. **48**(3), 824–831 (2000)
307. L.L. Scharf, P. Pakrooh, Multipulse subspace detectors, in *Asilomar Conference on Signals, Systems, and Computers* (2017)
308. L.L. Scharf, J.K. Thomas, Wiener filters in canonical coordinates for transform coding, filtering, and quantizing. IEEE Trans. Signal Process. **46**(3), 647–654 (1998)
309. L.L. Scharf, Y. Wang, Testing for causality using a partial coherence statistic (2021). arXiv:2112.03987v1
310. L.L. Scharf, B. Friedlander, P. Flandrin, A. Hanssen, The Hilbert space geometry of the stochastic Rihaczek distribution, in *Asilomar Conference on Signals, Systems, and Computers*, vol. 1 (2001), pp. 720–725
311. L.L. Scharf, E.K.P. Chong, J.S. Goldstein, M.D. Zoltowski, I.S. Reed, Subspace expansion and the equivalence between conjugate direction and multistage Wiener filters. IEEE Trans. Signal Process. **56**(10), 5013–5019 (2008)
312. L.L. Scharf, E.K.P. Chong, A. Pezeshki, J.R. Luo, Sensitivity considerations in compressed sensing, in *Asilomar Conference on Signals, Systems, and Computers* (2011)
313. L.L. Scharf, T. McWhorter, J. Given, M. Cheney, General first-order framework for passive detection with two sensor arrays, in *Asilomar Conference on Signals, Systems, and Computers* (2019)
314. S.V. Schell, W.A. Gardner, Detection of the number of cyclostationary signals in unknown interference and noise, in *Asilomar Conference on Signals, Systems, and Computers*, vol. 1 (1990), pp. 473–477
315. I.J. Schoenberg, Remarks to Maurice Fréchet's article "Sur la definition axiomatique d'une classe d'espace distanciés vectoriellement applicable sur l'espace de Hilbert. Ann. Math. **36**, 724–732 (1935)
316. B. Schölkopf, A.J. Smola, *Learning with Kernels: Support Vector Machines, Regularization, Optimization, and Beyond* (MIT Press, Cambridge, 2001)
317. P.J. Schreier, A unifying discussion of correlation analysis for complex random vectors. IEEE Trans. Signal Process. **56**(4), 1327–1336 (2006)
318. P.J. Schreier, L.L. Scharf, *Statistical Signal Processing of Complex-Valued Data: The Theory of Improper and Noncircular Signals* (Cambridge University Press, Cambridge, 2010)
319. S. Sedighi, A. Taherpour, J. Sala-Alvarez, T. Khattab, On the performance of Hadamard ratio detector-based spectrum sensing for cognitive radios. IEEE Trans. Signal Process. **63**(14), 3809–3824 (2015)
320. E. Serpedin, F. Panduru, I. Sarı, G.B. Giannakis, Bibliography on cyclostationarity. Signal Process. **85**(12), 2233–2303 (2005)
321. V. Seshadri, G.P.H. Styan, Canonical correlations, rank additivity and characterization of multivariate normality, in *Colloquia Mathematica Societatis János Bolyai, vol. 21: Analytic Function Methods in Probability Theory (Debrecen, Hungary, Aug. 1977), J. Bolyai, Budapest and North-Holland, Amsterdam* (1980), pp. 331–344
322. J.C. Shaw, Correlation and coherence analysis of the EEG: a selective tutorial review. Int. J. Psychophysiol. **1**(3), 255–266 (1984)
323. S. Sirianunpiboon, S.D. Howard, D. Cochran, Multiple-channel detection of signals having known rank, in *IEEE International Conference on Acoustics, Speech, and Signal Processing* (2013), pp. 6536–6540
324. S. Sirianunpiboon, S.D. Howard, D. Cochran, Detection in multiple channels having unequal noise power, in *IEEE Statistical Signal Processing Workshop* (2016)
325. S. Sirianunpiboon, S.D. Howard, D. Cochran, Detection of cyclostationarity using generalized coherence, in *IEEE International Conference on Acoustics, Speech, and Signal Processing* (2018)

326. D. Slepian, Prolate spheroidal wave functions, Fourier analysis, and uncertainty — V: The discrete case. Bell. Syst. Techn. J. **57**(5), 1371–1430 (1978)
327. S.T. Smith, Covariance, subspace, and intrinsic Cramer-Rao bounds. IEEE Trans. Signal Process. **53**(5), 1610–1630 (2005)
328. S.T. Smith, L.L. Scharf, L.T. McWhorter, Intrinsic quadratic performance bounds on manifolds, in *IEEE International Conference on Acoustics, Speech, and Signal Processing*, Toulouse, France (2006), pp. 1013–1016
329. M. Sorensen, C.I. Kanatsoulis, N.D. Sidiropoulos, Generalized canonical correlation analysis: a subspace intersection approach. *IEEE Trans. Signal Process.* **69**, 2452–2467 (2021)
330. C. Spearman, The proof and measurement of association between two things. Amer. J. Psychol. **15**(1), 72–101 (1904)
331. A. Srivastava, E. Klassen, Monte Carlo extrinsic estimators of manifold-valued parameters. IEEE Trans. Signal Process. **50**(2), 299–308 (2002)
332. M.S. Srivastava, C.G. Khatri, *An Introduction to Multivariate Statistics* (North Holland, Amsterdam, 1979)
333. L. Stankovic, D.P. Mandic, M. Dakovic, I. Kisil, Demystifying the coherence index in compressive sensing. IEEE Signal Process. Mag. **37**(1), 152–162 (2020)
334. G.W. Stewart, *Matrix Algorithms, Vol. II: Eigensystems* (Society for Industrial and Applied Mathematics, Philadelphia, 2001)
335. P. Stoica, B.C. Ng, On the Cramer-Rao bound under parametric constraints. IEEE Signal Process. Lett. **5**(7), 177–179 (1998)
336. P. Stoica, M. Viberg, Maximum likelihood parameter and rank estimation in reduced-rank multivariate linear regressions. IEEE Trans. Signal Process. **44**(12), 3069–3078 (1996)
337. P. Stoica, K.M. Wong, Q. Wu, On a nonparametric detection method for array signal processing in correlated noise fields. IEEE Trans. Signal Process. **44**(4), 1030–1032 (1996)
338. T. Sugiyama, Distribution of the largest latent root and the smallest latent root of the generalized B statistic and F statistic in multivariate analysis. Ann. Math. Statist. **38**(4), 1152–1159 (1967)
339. Y. Sun, P. Babu, P. Palomar, Majorization-minimization algorithms in signal processing, communications, and machine learning. IEEE Trans. Signal Process. **65**(3), 794–816 (2017)
340. E. Telatar, Capacity of multi-antenna Gaussian channels. Eur. Trans. Telecommun. **10**(6), 585–595 (1999)
341. C.M. Theobald, An inequality for the trace of the product of two symmetric matrices. Proc. Camb. Philos. Soc. **77**, 265–267 (1975)
342. R. Tibshirani, Regression selection and shrinkage via the Lasso. J. R. Stat. Soc. Ser. B (Methodological) **58**(1), 267–288 (1996)
343. W.S. Torgerson, *Theory and Methods of Scaling* (Wiley, Hoboken, 1958)
344. J. Tropp, Greed is good: algorithmic results for sparse approximation. IEEE Trans. Inf. Theory **50**(10), 2231–2242 (2004)
345. R.D. Trueblood, D.L. Alspach, Multiple coherence as a detection statistic. Technical Report, Naval Ocean Systems Center (1978)
346. P. Tseng, Nearest q-flat to m points. J. Optim. Theory Appl. **105**(1), 249–252 (2000)
347. D.W. Tufts, R. Kumaresan, Estimation of frequencies of multiple sinusoids: Making linear prediction work like maximum likelihood. Proc. IEEE **70**, 975–989 (1982)
348. D.W. Tufts, R. Kumaresan, Singular value decomposition and improved frequency estimation using linear prediction. IEEE Trans. Acoust. Speech Signal Process. **30**, 671–675 (1982)
349. J.K. Tugnait, Comparing multivariate complex random signals: algorithm, performance analysis and application. IEEE Trans. Signal Process. **64**(4), 934–947 (2016)
350. A.M. Tulino, S. Verdú, Random matrix theory and wireless communications. Found. Trends Commun. Inf. Theory **1**(1), 1–182 (2004)
351. P. Turaga, A. Veeraraghavan, A. Srivastava, R. Chellappa, Statistical computations on Grassmann and Stiefel manifolds for image and video-based recognition. IEEE Trans. Pattern Anal. Mach. Intell. **33**(11), 2273–2286 (2011)

352. D.E. Tyler, Statistical analysis for the angular central Gaussian distribution on the sphere. Biometrika **74**(3), 579–589 (1987)
353. P. Urriza, E. Rebeiz, D. Cabric, Multiple antenna cyclostationary spectrum sensing based on the cyclic correlation significance test. IEEE J. Sel. Areas Comm. **31**(11), 2185–2195 (2013)
354. R.J. Vaccaro, Y. Ding, Optimal subspace-based parameter estimation, in *IEEE International Conference on Acoustics, Speech, and Signal Processing* (1993)
355. S. Van Huffel, J. Vandewalle, Analysis and solution of the nongeneric total least squares problem. SIAM J. Matrix Anal. Appl. **9**(3), 360–372 (1988)
356. S. Van Huffel, J. Vandewalle, *The Total Least Squares Problem: Computational Aspects and Analysis* (Society for Industrial and Applied Mathematics, Philadelphia, 1991)
357. H.L. Van Trees, *Detection, Estimation and Modulation Theory: Detection, Estimation, and Filtering Theory (Part I)* (Wiley, Hoboken, 1968)
358. H.L. Van Trees, *Detection, Estimation and Modulation Theory: Optimum Array Processing (Part IV)* (Wiley, Hoboken, 2002)
359. H.L. Van Trees, K.L. Bell (eds.), *Bayesian Bounds for Parameter Estimation and Nonlinear Filtering/Tracking* (IEEE Press and Wiley Interscience, Hoboken, 2007)
360. S. Van Vaerenbergh, I. Santamaria, A comparative study of kernel adaptive filtering algorithms, in *IEEE Digital Signal Processing and Signal Processing Education Meeting* (2013), pp. 181–186
361. B.D. Van Veen, K.M. Buckley, Beamforming: a versatile approach to spatial filtering. IEEE ASSP Mag. **5**(2), 4–24 (1988)
362. V. Vapnik, *The Nature of Statistical Learning Theory* (Springer, Berlin, 1995)
363. G. Vazquez-Vilar, R. López-Valcarce, J. Sala-Alvarez, Multiantenna spectrum sensing exploiting spectral a priori information. IEEE Trans. Wirel. Comm. **10**(12), 4345–4355 (2011)
364. J. Vía, I. Santamaria, J. Pérez, Deterministic CCA-based algorithms for blind equalization of FIR-MIMO channels. IEEE Trans. Signal Process. **55**(7), 3867–3878 (2007)
365. J. Vía, I. Santamaria, J. Pérez, A learning algorithm for adaptive canonical correlation analysis of several data sets. Neural Netw. **20**(1), 139–152 (2007)
366. R. Vidal, Subspace clustering. IEEE Signal Process. Mag. **28**(2), 52–68 (2011)
367. R. Vidal, Y. Ma, S.S. Sastry, *Generalized Principal Component Analysis* (Springer, Berlin, 2016)
368. G. Wahba, *Spline Models for Observational Data* (Society for Industrial and Applied Mathematics, Philadelphia, 1990)
369. G. Wahba, Y. Wang, Representer theorem, in *Wiley StatsRef: Statistics Reference Online* (2019), pp. 1–11
370. Y. Wang, I. Santamaria, L.L. Scharf, H. Wang, Canonical coordinates for target detection in a passive radar network, in *Asilomar Conference on Signals, Systems, and Computers* (2016)
371. K.D. Ward, Compound representation of high resolution sea clutter. Electr. Lett. **17**(16), 561–563 (1981)
372. K.D. Ward, R.J.A. Tough, S. Watts, Sea clutter: Scattering, the k distribution and radar performance. Waves Random Complex Media **17**(2), 233–234 (2006)
373. G.S. Watson, *Statistics on Spheres* (Wiley, Hoboken, 1983)
374. M. Wax, T. Kailath, Detection of signals by information theoretic criteria. IEEE Trans. Acoust. Speech Signal Process. **33**(2), 387 (1985)
375. E. Weinstein, A.J. Weiss, A general class of lower bounds in parameter estimation. IEEE Trans. Inf. Theory **32**(2), 338–342 (1988)
376. M.E. Weippert, J.D. Hiemstra, J.S. Goldstein, M.D. Zoltowski, Insights from the relationship between the multistage Wiener filter and the method of conjugate gradients, in *Sensor Array and Multichannel Signal Processing Workshop* (2002), pp. 388–392
377. L. Welch, Lower bounds on the maximum cross correlation of signals. IEEE Trans. Inf. Theory **20**(3), 397–399 (1974)
378. J. Whittaker, *Graphical Models in Applied Multivariate Statistics* (Wiley, Hoboken, 1946)
379. P. Whittle, Estimation and information in time series analysis. Skand. Aktuar. **35**, 48–60 (1952)

380. P. Whittle, Gaussian estimation in stationary time series. Bull. Inst. Int. Statist. **39**, 105–129 (1962)
381. B. Widrow, M.E. Hoff, Adaptive switching circuits, in *1960 IRE WESCON Convention Record* (1960), pp. 96–104
382. S.S. Wilks, Certain generalizations in the analysis of variance. Biometrika **24**(3/4), 471–494 (1932)
383. S.S. Wilks, On the independence of k sets of normally distributed statistical variables. Econometrica **3**, 309–325 (1935)
384. S.S. Wilks, The large-sample distribution of the likelihood ratio for testing composite hypotheses. Ann. Math. Statist. **9**(1), 60–62 (1938)
385. J. Wishart, The generalised product moment distribution in samples from a normal multivariate population. Biometrika **20**(1/2), 32–52 (1928)
386. H.S. Witsenhausen, A determinant maximization problem occurring in the theory of data communication. SIAM J. Appl. Math. **29**(3), 515–522 (1975)
387. L. Wolf, A. Shashua, Learning over sets using kernel principal angles. J. Mach. Learn. Res. **4**, 913–931 (2003)
388. G.R. Wu, F. Chen, D. Kang, X. Zhang, D. Marinazzo, H. Chen, Multiscale causal connectivity analysis by canonical correlation: theory and application to epileptic brain. IEEE Trans. Biomed. Eng. **58**(11), 3088–3096 (2011)
389. C. Xu, S. Kay, Source enumeration via the EEF criterion. IEEE Signal Process. Lett. **15**, 569–572 (2008)
390. G. Young, A.S. Householder, Discussion of a set of points in terms of their mutual distances. Psychometrika **3**, 19–22 (1938)
391. F. Zernike, The concept of degree of coherence and its application to optical problems. Physica **5**(8), 785–795 (1938)
392. J. Zhang, G. Zhu, R.W. Heath Jr., K. Huang, Grassmannian learning: Embedding geometry awareness in shallow and deep learning (2018). arXiv:180802229v2
393. F. Zhao, L. Guibas, *Wireless Sensor Networks: An Information Processing Approach* (Elsevier, Amsterdam, 2004)
394. Z. Zhu, S. Kay, On Bayesian exponentially embedded family for model order selection. IEEE Trans. Signal Process. **66**(4), 933–943 (2018)

Alphabetical Index

D. Ramírez et al., *Coherence*, https://doi.org/10.1007/978-3-031-13331-2